Lecture Notes in Computer Science 8370

Commenced Publication in 1973
Founding and Former Series Editors:
Gerhard Goos, Juris Hartmanis, and Jan van Leeuwen

T0223463

Adrian-Horia Dediu Carlos Martín-Vide
José-Luis Sierra-Rodríguez Bianca Truthe (Eds.)

Language and Automata Theory and Applications

8th International Conference, LATA 2014
Madrid, Spain, March 10-14, 2014
Proceedings

 Springer

Volume Editors

Adrian-Horia Dediu
Rovira i Virgili University, Research Group on Mathematical Linguistics
Avinguda Catalunya, 35, 43002 Tarragona, Spain
E-mail: adrian.dediu@urv.cat

Carlos Martín-Vide
Rovira i Virgili University, Research Group on Mathematical Linguistics
Avinguda Catalunya, 35, 43002 Tarragona, Spain
E-mail: carlos.martin@urv.cat

José-Luis Sierra-Rodríguez
Complutense University of Madrid, School of Computer Science
Department of Software Engineering and Artificial Intelligence
Profesor José García Santesmases, 9, 28040 Madrid, Spain
E-mail: jlsierra@fdi.ucm.es

Bianca Truthe
Otto-von-Guericke-Universität Magdeburg, Fakultät für Informatik
Institut für Wissens- und Sprachverarbeitung
Universitätsplatz 2, 39106 Magdeburg, Germany
E-mail: truthe@iws.cs.uni-magdeburg.de

ISSN 0302-9743   e-ISSN 1611-3349
ISBN 978-3-319-04920-5   e-ISBN 978-3-319-04921-2
DOI 10.1007/978-3-319-04921-2
Springer Cham Heidelberg New York Dordrecht London

Library of Congress Control Number: 2014930764

LNCS Sublibrary: SL 1 – Theoretical Computer Science and General Issues

Typesetting: Camera-ready by author, data conversion by Scientific Publishing Services, Chennai, India

Printed on acid-free paper

Springer is part of Springer Science+Business Media (www.springer.com)

Preface

These proceedings contain the papers that were presented at the 8th International Conference on Language and Automata Theory and Applications (LATA 2014), held in Madrid, Spain, during March 10–14, 2014.

The scope of LATA is rather broad, including: algebraic language theory; algorithms for semi-structured data mining; algorithms on automata and words; automata and logic; automata for system analysis and program verification; automata, concurrency, and Petri nets; automatic structures; cellular automata; codes; combinatorics on words; compilers; computability; computational complexity; data and image compression; decidability issues on words and languages; descriptional complexity; digital libraries and document engineering; DNA and other models of bio-inspired computing; foundations of finite state technology; foundations of XML; fuzzy and rough languages; grammars (Chomsky hierarchy, contextual, unification, categorial, etc.); grammatical inference and algorithmic learning; graphs and graph transformation; language varieties and semigroups; language-based cryptography; language-theoretic foundations of artificial intelligence and artificial life; natural language and speech automatic processing; parallel and regulated rewriting; parsing; patterns; power series; quantum, chemical and optical computing; semantics; string and combinatorial issues in computational biology and bioinformatics; string processing algorithms; symbolic dynamics; symbolic neural networks; term rewriting; transducers; trees, tree languages and tree automata; weighted automata.

LATA 2014 received 116 submissions. Each one was reviewed by three Program Committee members, many of whom consulted with external referees. After a thorough and vivid discussion phase, the committee decided to accept 45 papers (which represents an acceptance rate of 38.79%). The conference program also included four invited talks and one invited tutorial. Part of the success in the management of such a large number of submissions is due to the excellent facilities provided by the EasyChair conference management system.

We would like to thank all invited speakers and authors for their contributions, the Program Committee and the reviewers for their cooperation, and Springer for its very professional publishing work.

December 2013

Adrian-Horia Dediu
Carlos Martín-Vide
José-Luis Sierra
Bianca Truthe

Organization

LATA 2014 was organized by the Research Group on Implementation of Language-Driven Software and Applications, ILSA, from Complutense University of Madrid and the Research Group on Mathematical Linguistics, GRLMC, from Rovira i Virgili University, Tarragona.

Program Committee

Dana Angluin	Yale University at New Haven, USA
Eugene Asarin	Paris Diderot University, France
Jos Baeten	CWI, Amsterdam, The Netherlands
Christel Baier	Dresden University of Technology, Germany
Jin-Yi Cai	University of Wisconsin at Madison, USA
Marek Chrobak	University of California at Riverside, USA
Andrea Corradini	University of Pisa, Italy
Mariangiola Dezani	University of Turin, Italy
Ding-Zhu Du	University of Texas at Dallas, USA
Michael R. Fellows	Charles Darwin University, Darwin, Australia
Jörg Flum	University of Freiburg, Germany
Nissim Francez	Technion-Israel Institute of Technology, Haifa, Israel
Jürgen Giesl	RWTH Aachen University, Germany
Annegret Habel	University of Oldenburg, Germany
Kazuo Iwama	Kyoto University, Japan
Sampath Kannan	University of Pennsylvania, Philadelphia, USA
Ming-Yang Kao	Northwestern University, Evanston, USA
Deepak Kapur	University of New Mexico, Albuquerque, USA
Joost-Pieter Katoen	RWTH Aachen University, Germany
S. Rao Kosaraju	Johns Hopkins University, Baltimore, USA
Evangelos Kranakis	Carleton University, Ottawa, Canada
Gad M. Landau	University of Haifa, Israel
Andrzej Lingas	Lund University, Sweden
Jack Lutz	Iowa State University, Ames, USA
Ian Mackie	École Polytechnique, Palaiseau, France
Carlos Martín-Vide	Rovira i Virgili University, Tarragona, Spain
Giancarlo Mauri	University of Milano-Bicocca, Milan, Italy
Faron G. Moller	Swansea University, UK
Paliath Narendran	University at Albany, SUNY, USA
Enno Ohlebusch	Ulm University, Germany
Helmut Prodinger	Stellenbosch University, South Africa
Jean-François Raskin	Free University of Brussels, Belgium

Wolfgang Reisig   Humboldt University, Berlin, Germany
Marco Roveri   Bruno Kessler Foundation, Trento, Italy
Michaël Rusinowitch   LORIA, Nancy, France
Yasubumi Sakakibara   Keio University, Japan
Davide Sangiorgi   University of Bologna, Italy
Colin Stirling   University of Edinburgh, UK
Jianwen Su   University of California at Santa Barbara, USA
Jean-Pierre Talpin   IRISA, Rennes, France
Andrzej Tarlecki   University of Warsaw, Poland
Rick Thomas   University of Leicester, UK
Sophie Tison   University of Lille 1, France
Rob van Glabbeek   NICTA, Sydney, Australia
Helmut Veith   Vienna University of Technology, Austria

External Reviewers

Aldinucci, Marco
Amit, Mika
Anantharaman, Siva
Arana, Andrew
Artale, Alessandro
Beccuti, Marco
Bednarczyk, Marek A.
Belkhir, Walid
Beller, Timo
Bernardinello, Luca
Blunsom, Phil
Bollig, Benedikt
Bouchard, Christopher
Brenguier, Romain
Bruni, Roberto
Carle, Benjamin
Chaiken, Seth
Chrząstowski-Wachtel, Piotr
Clemente, Lorenzo
Dang, Zhe
Dehnert, Christian
Delbot, François
de'Liguoro, Ugo
Dennunzio, Alberto
Devillers, Raymond
Didier, Gilles
Dubslaff, Clemens
Fatès, Nazim

Filiot, Emmanuel
Flick, Nils Erik
Floderus, Peter
Frosini, Andrea
Fuhs, Carsten
Furia, Carlo A.
Gadducci, Fabio
Genet, Thomas
Gero, Kimberly
Gierds, Christian
Griggio, Alberto
Haar, Stefan
Hagge Cording, Patrick
Hai, Zhao
Hertrampf, Ulrich
Hibbs, Peter
Ibarra, Oscar H.
Iosif, Radu
Joosten, Joost J.
Joshi, Prachi
Kaminski, Michael
Kari, Jarkko
Kestler, Hans
Khomenko, Victor
Klein, Joachim
Klüppelholz, Sascha
Kowaluk, Mirosław
Křetínský, Jan

Krishnamoorthy, Mukkai
Kutsia, Temur
Lecroq, Thierry
Leporati, Alberto
Levcopoulos, Christos
Linker, Sven
Lundell, Eva-Marta
Mairesse, Jean
Manea, Florin
Martyugin, Pavel
Mayr, Richard
Mazza, Damiano
Minkov, Einat
Moelle, Andre
Monmege, Benjamin
Morcira, Nelma
Mover, Sergio
Mukund, Madhavan
Müller, David
Naldi, Aurélien
Nepomnyachiy, Sergey
Niehren, Joachim
Núñez Queija, Rudesindo
Palano, Beatrice
Paparo, Omer
Persson, Mia
Pin, Jean-Éric
Plandowski, Wojciech
Porreca, Antonio E.
Prüfer, Robert
Radke, Hendrik

Ranise, Silvio
Roos, Yves
Rozenberg, Liat
Sammartino, Matteo
Sankowski, Piotr
Sankur, Ocan
Schneider-Kamp, Peter
Servais, Frédéric
Shavrukov, Volodya
Shoukourian, Hayk
Shukla, Sandeep
Sledneu, Dzmitry
Sokol, Dina
Sun, Yutian
Sürmeli, Jan
Swaminathan, Mani
Szczuka, Marcin
Tarasenko, Sergey
Thomas, Wolfgang
Titov, Ivan
Tonetta, Stefano
Tuosto, Emilio
Turuani, Mathieu
Valiron, Benoît
Vandin, Andrea
Vigneron, Laurent
Vinju, Jurgen
Vouillon, Jérôme
Vuillon, Laurent
Wintner, Shuly

Organizing Committee

Adrian-Horia Dediu, Tarragona
Ana Fernández-Pampillón, Madrid
Carlos Martín-Vide, Tarragona (Co-chair)
Antonio Sarasa, Madrid
José-Luis Sierra, Madrid (Co-chair)
Bianca Truthe, Magdeburg
Lilica Voicu, Tarragona

Table of Contents

A Brief History of Strahler Numbers

Javier Esparza, Michael Luttenberger, and Maximilian Schlund

Fakultät für Informatik, Technische Universität München, Germany

Abstract. The Strahler number or Horton-Strahler number of a tree, originally introduced in geophysics, has a surprisingly rich theory. We sketch some milestones in its history, and its connection to arithmetic expressions, graph traversing, decision problems for context-free languages, Parikh's theorem, and Newton's procedure for approximating zeros of differentiable functions.

1   The Strahler Number

In 1945, the geophysicist Robert Horton found it useful to associate a *stream order* to a system of rivers (geophysicists seem to prefer the term 'stream") [20].

> *Unbranched fingertip tributaries are always designated as of order 1, tributaries or streams of the 2d order receive branches or tributaries of the 1st order, but these only; a 3d order stream must receive one or more tributaries of the 2d order but may also receive 1st order tributaries. A 4th order stream receives branches of the 3d and usually also of lower orders, and so on.*

Several years later, Arthur N. Strahler replaced this ambiguous definition by a simpler one, very easy to compute [26]:

> *The smallest, or "finger-tip", channels constitute the first-order segments. [...]. A second-order segment is formed by the junction of any two first-order streams; a third-order segment is formed by the joining of any two second order streams, etc.*

Streams of lower order joining a higher order stream do not change the order of the higher stream. Thus, if a first-order stream joins a second-order stream, it remains a second-order stream. Figure 1 shows the Strahler number for a fragment of the course of the Elbe river with some of its tributaries. The stream system is of order 4.

From a computer science point of view, stream systems are just trees.

Definition 1. *Let t be a tree with root r. The* Strahler number *of t, denoted by $S(t)$, is inductively defined as follows.*

- *If r has no children (i.e., t has only one node), then $S(t) = 0$.*
- *If r has children $r_1, \ldots, r_n$, then let $t_1, \ldots, t_n$ be the subtrees of t rooted at $r_1, \ldots, r_n$, and let $k = \max\{S(t_1), \ldots, S(t_n)\}$: if exactly one of $t_1, \ldots, t_n$ has Strahler number k, then $S(t) = k$; otherwise, $S(t) = k + 1$.*

A.-H. Dediu et al. (Eds.): LATA 2014, LNCS 8370, pp. 1–13, 2014.

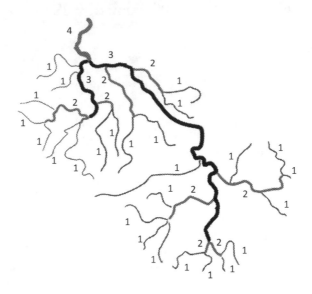

Fig. 1. Strahler numbers for a fragment of the Elbe river

Note that in this formal definition the Strahler number of a simple chain (a "finger-tip") is *zero*, and not *one*. This allows another characterization of the Strahler number of a tree t as the height of the largest minor of t that is a perfect binary tree (i.e., a rooted tree where every inner node has two children and all leaves have the same distance to the root): Roughly speaking, such a binary tree is obtained by, starting at the root, following paths along which the Strahler number never decreases by more than one unit at a time, and then contracting all nodes with only one child. If t itself is a binary tree, then this minor is unique. We leave the details as a small exercise.

Figure 2 shows trees with Strahler number 1, 2, and 3, respectively. Each node is labeled with the Strahler number of the subtree rooted at it.

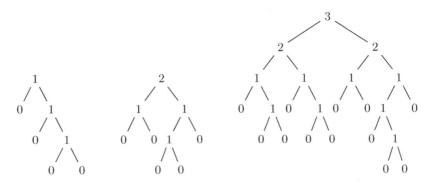

Fig. 2. Trees of Strahler number 1, 2, and 3

Together with other parameters, like bifurcation ratio and mean stream length, Horton and Strahler used stream orders to derive quantitative empirical laws for stream systems. Today, geophysicists speak of the Strahler number (or Horton-Strahler number) of a stream system. According to the excellent Wikipedia article on the Strahler number (mainly due to David Eppstein), the Amazon and the Mississippi have Strahler numbers of 10 and 12, respectively.

2 Strahler Numbers and Tree Traversal

The first appearance of the Strahler number in Computer Science seems to be due to Ershov in 1958 [8], who observed that the number of registers needed to evaluate an arithmetic expression is given by the Strahler number of its syntax tree. For instance, the syntax tree of $(x + y \cdot z) \cdot t$, shown on the left of Figure 3, has Strahler number 2, and indeed can be computed with just two registers R_1, R_2 by means of the code shown on the right.

$$R_1 \leftarrow y$$
$$R_2 \leftarrow z$$
$$R_2 \leftarrow R_1 \times R_2$$
$$R_2 \leftarrow x$$
$$R_1 \leftarrow R_1 + R_2$$
$$R_2 \leftarrow w$$
$$R_1 \leftarrow R_1 \times R_2$$

Fig. 3. An arithmetic expression of Strahler number 2

The strategy for evaluating a expression $e = e_1 \; op \; e_2$ is easy: start with the subexpression whose tree has lowest Strahler number, say e_1; store the result in a register, say R_1; reuse all other registers to evaluate e_2; store the result in R_2; store the result of $R_1 \, op \, R_2$ in R_1.

Ershov's observation is recalled by Flajolet, Raoult and Vuillemin in [16], where they add another observation of their own: the Strahler number of a *binary* tree is the minimal stack size required to traverse it. Let us attach to each node of the tree the Strahler number of the subtree rooted at it. The traversing procedure follows again the "lowest-number-first" policy (notice that arithmetic expressions yield binary trees). If a node with number k has two children, then the traversing procedure moves to the child with lowest number, and pushes the (memory address of the) other child onto the stack. If the node is a leaf, then the procedure pops the top node of the stack and jumps to it. To prove that the stack size never exceeds the Strahler number, we observe that, if a node of number k has two children, then at least one of its children has number smaller than k. So the procedure only pushes a node onto the stack when it moves to a node of strictly smaller number, and we are done.

Notice, however, that the "lowest-number-first" policy requires to know the Strahler number of the nodes. If these are unknown, all we can say is that a nondeterministic traversing procedure always needs a stack of size at least equal to the Strahler number, and that it *may* succeed in traversing the tree with a stack of exactly that size.

2.1 Distribution of Strahler Numbers

The goal of Flajolet, Raoult and Vuillemin's paper is to study the distribution of Strahler numbers in the binary trees with a fixed number n of leaves. Let S_n be the random variable corresponding to the Strahler number of a binary tree (every node has either two or 0 children) with n internal nodes chosen uniformly at random. Since the Strahler number of t is the height of the largest perfect binary tree embeddable in t, we immediately have $S_n \leq \lfloor \log_2(n+1) \rfloor$. The paper shows that

$$Exp[S_n] \approx \log_4 n \qquad \text{and} \qquad Var[S_n] \in \mathcal{O}(1) \ .$$

In other words, when n grows the Strahler number of most trees becomes increasingly closer to $\log_4 n$. Independently of Flajolet, Raoult and Vuillemin, also Kemp derives in [21] the same asymptotic behaviour of the expected Strahler number of a random binary tree. Later, Flajolet and Prodinger extend the analysis to trees with both binary and unary inner nodes [17]. Finally, Devroye and Kruszewski show in [5] that the probability that the Strahler number of a random binary tree with n nodes deviates by at least k from the expected Strahler number of $\log_4 n$ is bounded from above by $\frac{2}{4^k}$, that is, the Strahler number is highly concentrated around its expected value.

2.2 Strahler Numbers in Language Theory: Derivation Indices, Caterpillars, and Dimensions

Derivation indices and caterpillars. The Strahler number has been rediscovered (multiple times!) by the formal language community. In [19], Ginsburg and Spanier introduce the index of a derivation $S \Rightarrow \alpha_1 \Rightarrow \alpha_2 \Rightarrow \cdots \Rightarrow w$ of a given grammar as the maximal number of variables occurring in any of the sentential forms α_i (see also [27]). For instance, consider the grammar $X \to aXX \mid b$. The index of the derivations

$$X \Rightarrow aXX \Rightarrow aXaXX \Rightarrow abaXX \Rightarrow ababX \Rightarrow ababb$$

$$X \Rightarrow aXX \Rightarrow abaXX \Rightarrow abaXX \Rightarrow ababX \Rightarrow ababb$$

is 3 (because of $aXaXX$) and 2, respectively. For context-free grammars, where we have the notion of derivation tree of a word, we define the index of a derivation tree as the minimal index of its derivations. If the grammar is in Chomsky normal form, then a derivation tree has index k if and only if its Strahler number is $(k-1)$.

A first use of the Strahler number of derivation trees can be found in [4], where Chytil and Monien, apparently unaware of the Strahler number, introduce k-*caterpillars* as follows:

> *A caterpillar is an ordered tree in which all vertices of outdegree greater than one occur on a single path from the root to a leaf. A 1-caterpillar is simply a caterpillar and for $k > 1$ a k-caterpillar is a tree obtained from a caterpillar by replacing each hair by a tree which is at most $(k-1)$-caterpillar.*

Clearly, a tree is a k-caterpillar if and only if its Strahler number is equal to k.

Let $L_k(G)$ be the subset of words of $L(G)$ having a derivation tree of Strahler number at most k (or, equivalently, being a k-caterpillar). Chytil and Monien prove that there exists a nondeterministic Turing machine with language $L(G)$ that recognizes $L_k(G)$ in space $O(k \log |G|)$. Assume for simplicity that G is in Chomsky normal form. In order to nondeterministically recognize $w = a_1 a_2 \ldots a_n \in L_k(G)$, we guess on-the-fly (i.e., while traversing it) a derivation tree of w with Strahler number at most k, using a stack of height at most k. The traversing procedure follows the "smaller-number-first" policy. More precisely, the nodes of the tree are triples (X, i, j) with intended meaning "X generates a tree with yield $a_i \ldots a_j$". We start at node $(S, 1, n)$. At a generic node (X, i, j), we proceed as follows. If $i = j$, then we check that $X \to a_i$ is a production, pop a new node, and jump to it. If $i < j$, then we guess a production, say $X \to YZ$, and an index $i \le l \le j$, guess which of $(Y, 1, i)$ and (Z, l, j) generates the subtree of lowest number, say (Y, i, l), and jump to it, pushing (Z, l, j) onto the stack.

The traversing procedure can also be used to check emptiness of $L_k(G)$ in nondeterministic logarithmic space (remember: k is not part of the input) [13]. In this case we do not even need to guess indices: if the current node is labeled by X, then we proceed as follows. If X has no productions, then we stop. If G has a production $X \to a$ for some terminal a, we pop the next node from the stack and jump to it. If G has productions for X, but only of the form $X \to YZ$, then we guess one of them and proceed as above. Notice that checking emptiness of $L(G)$ is a P-complete problem, and so unlikely to be solvable in logarithmic space.

Tree dimension. The authors of this paper are also guilty of rediscovering the Strahler number. In [9] we defined the *dimension* of a tree, which is . . . nothing but its Strahler number.[1] Several papers [9,11,18,13] have used tree dimension

---

[1] The name dimension was chosen to reflect that trees with Strahler number 1 are a chain (with hairs), trees of dimension 2 are chains of chains (with hairs), that can be nicely drawn in the plane, trees of dimension 3 are chains of chains of chains (with hairs), with can be nicely displayed in 3-dimensional space, etc.

(that is, they have used the Strahler number) to show that $L_{n+1}(G)$, where n is the number of variables of a grammar G in Chomsky normal form, has interesting properties[2]:

(1) Every $w \in L(G)$ is a scattered subword of some $w' \in L_{n+1}(G)$ [13].
(2) For every $w_1 \in L(G)$ there exists $w_2 \in L_{n+1}(G)$ such that w_1 and w_2 have the same *Parikh image*, where the Parikh image of a word w is the function $\Sigma \to \mathbb{N}$ that assigns to every terminal the number of times it occurs in w. Equivalently, w and w' have the same Parikh image if w' can be obtained from w by reordering its letters [9].

The first property has already found at least one interesting application in the theory of formal verification (see [13]). The second property has been used in [12] to provide a simple "constructive" proof of Parikh's theorem. Parikh's theorem states that for every context-free language L there is a regular language L' such that L and L' have the same Parikh image (i.e., the set of Parikh images of the words of L and L' coincide). For instance, if $L = \{a^n b^n \mid n \geq 0\}$, then we can take $L' = (ab)^*$.

The proof describes a procedure to construct this automaton. By property (2), it suffices to construct an automaton A such that $L(A)$ and $L_{k+1}(G)$ have the same Parikh image. We construct A so that its runs "simulate" the derivations of G of index at most $k+1$. Consider for instance the context-free grammar with variables A_1, A_2 (and so $k = 2$), terminals a, b, c, axiom A_1, and productions

$$A_1 \ \to \ A_1 A_2 | a \qquad A_2 \ \to \ b A_2 a A_2 | c A_1$$

Figure 4 shows on the left a derivation of index 3, and on the right the run of A simulating it. The states store the current number of occurrences of A_1 and A_2, and the transitions keep track of the terminals generated at each derivation step. The run of A generates $bacaaca$, which has the same Parikh image as $abcaaca$.

$$
\begin{array}{lcl}
A_1 & & (0,1) \\
\Rightarrow A_1 A_2 & \xrightarrow{\epsilon} & (1,1) \\
\Rightarrow A_1 b A_2 a A_2 & \xrightarrow{ba} & (1,2) \\
\Rightarrow A_1 b c A_1 a A_2 & \xrightarrow{c} & (2,1) \\
\Rightarrow abc A_1 a A_2 & \xrightarrow{a} & (1,1) \\
\Rightarrow abcaa A_2 & \xrightarrow{a} & (0,1) \\
\Rightarrow abcaac A_1 & \xrightarrow{c} & (1,0) \\
\Rightarrow abcaaca & \xrightarrow{a} & (0,0)
\end{array}
$$

Fig. 4. A derivation and its "simulation"

---

[2] For an arbitrary grammar G, the same properties hold for $L_{nm+1}(G)$, where m is the maximal number of variables on the right-hand-side of a production, minus 1. If G is in Chomsky normal form, then $m \leq 1$.

The complete automaton is shown in Figure 5.

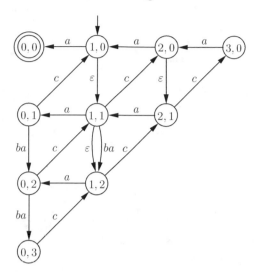

Fig. 5. The Parikh automaton of $A_1 \to A_1A_2|a$, $A_2 \to bA_2aA_2|cA_1$ with axiom A_1

3   Strahler Numbers and Newton's Method

Finally, we present a surprising connection between the Strahler number and Newton's method to numerically approximate a zero of a function. The connection works for multivariate functions, but in this note we just consider the univariate case.

Consider an equation of the form $X = f(X)$, where $f(X)$ is a polynomial with nonnegative real coefficients. Since the right-hand-side is a monotonic function, by Knaster-Tarski's or Kleene's theorem the equation has exactly one smallest solution (possibly equal to ∞). We denote this solution by μf. It is perhaps less known that μf can be given a "language-theoretic" interpretation. We explain this by means of an example (see [14] for more details).

Consider the equation

$$X = \frac{1}{4}X^2 + \frac{1}{4}X + \frac{1}{2} \tag{1}$$

It is equivalent to $(X - 1)(X - 2) = 0$, and so its least solution is $X = 1$. We introduce identifiers a, b, c for the coefficients, yielding the formal equation

$$X = f(X) := aX^2 + bX + c . \tag{2}$$

We "rewrite" this equation as a context-free grammar in Greibach normal form in the way one would expect:

$$G : X \to aXX \mid bX \mid c , \tag{3}$$

Consider now the derivation trees of this grammar. It is convenient to rewrite the derivation trees as shown in Figure 6: We write a terminal not at a leaf, but at its parent node, and so we now write the derivation tree on the left of the figure in the way shown on the right. Notice that, since each production generates a different terminal, both representations contain exactly the same information.[3]

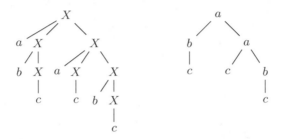

Fig. 6. New convention for writing derivation trees

We assign to each derivation tree t its *value* $V(t)$, defined as the product of the coefficients labeling the nodes. So, for instance, for the tree of Figure 6 we get the value $a^2 \cdot b^2 \cdot c^3 = (1/4)^4(1/2)^3 = 1/128$. Further, we define the value $V(T)$ of a set T of trees as $\sum_{t \in T} V(t)$ (which can be shown to be well defined, even if T is infinite). If we denote by T_G the set of all derivation trees of G, then

$$\mu f = V(T_G). \qquad (4)$$

The earliest reference for the this theorem in all its generality we are aware of is Bozapalidis [2] (Theorem 16) to whom also [6] gives credit.

A well-known technique to approximate μf is *Kleene iteration*, which consists of computing the sequence $\{\kappa_i\}_{i \in \mathbb{N}}$ of *Kleene approximants* given by

$$\kappa_0 = 0$$
$$\kappa_{i+1} = f(\kappa_i) \text{ for every } i \geq 0$$

It is easy to show that this corresponds to evaluating the derivation trees (with our new convention) *by height*. More precisely, if H_i is the set of derivation trees of T_G of height *less than* i, we get

$$\kappa_i = V(H_i) \qquad (5)$$

In other words, the Kleene approximants correspond to evaluating the derivation trees of G by increasing height.

It is well known that convergence of Kleene iteration can be slow: in the worst case, the number of correct digits grows only logarithmically in the number of

---

[3] This little change is necessary, because the tree of the derivation $X \Rightarrow c$ has Strahler number 1 if trees are drawn in the standard way, and 0 according to our new convention.

iterations. *Newton iteration* has much faster convergence (cf. [15,10,25]). Recall that Newton iteration approximates a zero of a differentiable function $g(X)$. For this, given an approximation ν_i of the zero, one geometrically computes the next approximation as follows:

- compute the tangent to $g(X)$ at the point $(\nu_i, g(\nu_i))$;
- take for ν_{i+1} the X-components of the intersection point of the tangent and the x-axis.

For functions of the form $g(X) = f(X) - X$, an elementary calculation yields the sequence $\{\nu_i\}_{i \in \mathbb{N}}$ of *Newton approximants*

$$\nu_0 = 0$$
$$\nu_{i+1} = \nu_i - \frac{f(\nu_i) - \nu_i}{f'(\nu_i) - 1}$$

We remark that in general choosing $\nu_0 = 0$ as the initial approximation may not lead to convergence – only in the special cases of the nonnegative reals or, more generally, ω-continuous semirings, convergence is guaranteed for $\nu_0 = 0$.

A result of [11] (also derived independently in [23]) shows that, if S_i is the set of derivation trees of T_G of Strahler number *less than* i (where trees are drawn according to our new convention), then

$$\nu_i = V(S_i) \tag{6}$$

In other words, the Newton approximants correspond to evaluating the derivation trees of G by increasing Strahler number!

The connection between Newton approximants and Strahler numbers has several interesting consequences. In particular, one can use results on the convergence speed of Newton iteration [3] to derive information on the distribution of the Strahler number in randomly generated trees. Consider for instance random trees generated according to the following rule.

A node has three children with probability 0.1, two children with probability 0.2, one child with probability 0.1, and zero children with probability 0.6.

Let G the context-free grammar

$$X \to aXXX \mid bXX \mid cX \mid d$$

with valuation $V(a) = 0.1, V(b) = 0.2, V(c) = 0.1, V(d) = 0.6$. It is easy to see that the probability of generating a tree t is equal to its value $V(t)$. For instance, the tree t of Figure 7 satisfies $Pr[t] = V(t) = a \cdot b^2 \cdot c \cdot d^5$.

Therefore, the Newton approximants of the equation

$$X = 0.1X^3 + 0.2X^2 + 0.1X + 0.6$$

give the distribution of the random variable S that assigns to each tree its Strahler number. Since $f(X) = 0.1X^3 + 0.2X^2 + 0.1X + 0.6$ and $f'(X) = 0.3X^2 + 0.4X^2 + 0.1$, we get

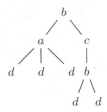

Fig. 7. A tree with probability $a \cdot b^2 \cdot c \cdot d^5$

$$\nu_0 = 0.6$$
$$\nu_{i+1} = \nu_i - \frac{\nu_i^3 + 2\nu_i^2 - 9\nu_i + 6}{3\nu_i^2 + 4\nu_i - 9}$$

and so for the first approximants we easily obtain

$$\nu_0 = Pr[S < 0] = 0$$
$$\nu_1 = Pr[S < 1] = 0.66\overline{7}$$
$$\nu_2 = Pr[S < 2] \approx 0.904$$
$$\nu_3 = Pr[S < 3] \approx 0.985$$
$$\nu_4 = Pr[S < 4] \approx 0.999$$

As we can see, the probability converges very rapidly towards 1. This is not a coincidence. The function $f(X)$ satisfies $\mu f < 1$, and a theorem of [3] shows that for every f satisfying this property, there exist numbers $c > 0$ and $0 < d < 1$ such that

$$Pr[S \geq k] \leq c \cdot d^{2^k} .$$

4   Strahler Numbers and ...

We have exhausted neither the list of properties of the Strahler number, nor the works that have obtained them or used them. To prove the point, we mention some more papers.

In 1978, Ehrenfeucht et al. introduced the same concept for derivation trees w.r.t. ET0L systems in [7] where it was called *tree-rank*.

Meggido et al. introduced in 1981 the *search number* of an undirected tree [22]: the minimal number of police officers required to capture a fugitive when police officers may move along edges from one node to another, and the fugitive can move from an edge to an incident one as long as the common vertex is not blocked by a police officer; the fugitive is captured when he cannot move anymore. For trees, the search number coincides with the better known *pathwidth* (see e.g. [1]), defined for general graphs. In order to relate the pathwidth to the Strahler number, we need to extend the definition of the latter to undirected trees: let the Strahler number $S(t)$ of an *undirected* tree be the minimal Strahler number of all the directed trees obtained by choosing a node as root, and orienting all edges away from it. We can show that for any tree t:

$$pathwidth(t) - 1 \le S(t) \le 2 \cdot pathwidth(t)$$

Currently, we are studying the Strahler number in the context of natural language processing. Recall that the Strahler number measures the minimal height of a stack required to traverse a tree, or, more informally, the minimal amount of memory required to process it. We conjecture that most sentences of a natural language should have a small Strahler number – simply not to overburden the reader or listener. Table 1 contains the results of an examination of several publicly available tree banks (banks of sentences that have been manually parsed by human linguists), which seem to support this conjecture. For each language we have computed the average and maximum Strahler number of the parse trees in the corresponding tree bank. We are currently investigating whether this fact can be used to improve unlexicalized parsing of natural languages.

Table 1. Average and maximum Strahler numbers for several treebanks of natural languages. ‡: SPMRL shared task dataset, ♣: 10% sample from the Penn treebank shipped with python nltk, ♠: TueBa-D/Z treebank.

Language	Source	Average	Maximum
Basque	SPMRL‡	2.12	3
English	Penn♣	2.38	4
French	SPMRL	2.29	4
German	SPMRL	1.94	4
German	TueBa-D/Z♠	2.13	4
Hebrew	SPMRL	2.44	4
Hungarian	SPMRL	2.11	4
Korean	SPMRL	2.18	4
Polish	SPMRL	1.68	3
Swedish	SPMRL	1.83	4

5   Conclusions

We have sketched the history of the Strahler number, which has been rediscovered a surprising number of times, received a surprising number of different names (stream order, stream rank, index, tree rank, tree dimension, k-caterpillar ...), and turns out to have a surprising number of applications and connections (Parikh's theorem, Newton's method, pathwidth ...).

This paper is by no means exhaustive, and we apologize in advance to the many authors we have surely forgotten. We intend to extend this paper with further references. If you know of further work connected to the Strahler number, please contact us.

Acknowledgments.   We thank Carlos Esparza for his help with some calculations.

References

1. Bienstock, D., Robertson, N., Seymour, P., Thomas, R.: Quickly excluding a forest. Journal of Combinatorial Theory, Series B 52(2), 274–283 (1991), http://www.sciencedirect.com/science/article/pii/009589569190068U
2. Bozapalidis, S.: Equational elements in additive algebras. Theory Comput. Syst. 32(1), 1–33 (1999)
3. Brázdil, T., Esparza, J., Kiefer, S., Luttenberger, M.: Space-efficient scheduling of stochastically generated tasks. Inf. Comput. 210, 87–110 (2012)
4. Chytil, M., Monien, B.: Caterpillars and context-free languages. In: Choffrut, C., Lengauer, T. (eds.) STACS 1990. LNCS, vol. 415, pp. 70–81. Springer, Heidelberg (1990)
5. Devroye, L., Kruszewski, P.: A note on the Horton-Strahler number for random trees. Inf. Process. Lett. 56(2), 95–99 (1995)
6. Droste, M., Kuich, W., Vogler, H.: Handbook of Weighted Automata. Springer (2009)
7. Ehrenfeucht, A., Rozenberg, G., Vermeir, D.: On et0l systems with finite tree-rank. SIAM J. Comput. 10(1), 40–58 (1981)
8. Ershov, A.P.: On programming of arithmetic operations. Comm. ACM 1(8), 3–9 (1958)
9. Esparza, J., Kiefer, S., Luttenberger, M.: On fixed point equations over commutative semirings. In: Thomas, W., Weil, P. (eds.) STACS 2007. LNCS, vol. 4393, pp. 296–307. Springer, Heidelberg (2007)
10. Esparza, J., Kiefer, S., Luttenberger, M.: Computing the least fixed point of positive polynomial systems. SIAM J. Comput. 39(6), 2282–2335 (2010)
11. Esparza, J., Kiefer, S., Luttenberger, M.: Newtonian program analysis. J. ACM 57(6), 33 (2010)
12. Esparza, J., Ganty, P., Kiefer, S., Luttenberger, M.: Parikhs theorem: A simple and direct automaton construction. Inf. Process. Lett. 111(12), 614–619 (2011)
13. Esparza, J., Ganty, P., Majumdar, R.: Parameterized verification of asynchronous shared-memory systems. In: Sharygina, Veith (eds.) [24], pp. 124–140
14. Esparza, J., Luttenberger, M.: Solving fixed-point equations by derivation tree analysis. In: Corradini, A., Klin, B., Cîrstea, C. (eds.) CALCO 2011. LNCS, vol. 6859, pp. 19–35. Springer, Heidelberg (2011)
15. Etessami, K., Yannakakis, M.: Recursive markov chains, stochastic grammars, and monotone systems of nonlinear equations. J. ACM 56(1) (2009)
16. Flajolet, P., Raoult, J.-C., Vuillemin, J.: The number of registers required for evaluating arithmetic expressions. Theor. Comput. Sci. 9, 99–125 (1979)
17. Flajolet, P., Prodinger, H.: Register allocation for unary-binary trees. SIAM J. Comput. 15(3), 629–640 (1986)
18. Ganty, P., Majumdar, R., Monmege, B.: Bounded underapproximations. Formal Methods in System Design 40(2), 206–231 (2012)
19. Ginsburg, S., Spanier, E.: Derivation-bounded languages. Journal of Computer and System Sciences 2, 228–250 (1968)
20. Horton, R.E.: Erosional development of streams and their drainage basins: hydrophysical approach to quantitative morphology. Geol. Soc. Am. Bull. 56(3), 275–370 (1945)
21. Kemp, R.: The average number of registers needed to evaluate a binary tree optimally. Acta Informatica 11, 363–372 (1979)

22. Megiddo, N., Hakimi, S.L., Garey, M.R., Johnson, D.S., Papadimitriou, C.H.: The complexity of searching a graph (preliminary version). In: FOCS, pp. 376–385. IEEE Computer Society (1981)
23. Pivoteau, C., Salvy, B., Soria, M.: Algorithms for combinatorial structures: Well-founded systems and newton iterations. J. Comb. Theory, Ser. A 119(8), 1711–1773 (2012)
24. Sharygina, N., Veith, H. (eds.): CAV 2013. LNCS, vol. 8044. Springer, Heidelberg (2013)
25. Stewart, A., Etessami, K., Yannakakis, M.: Upper Bounds for Newton's Method on Monotone Polynomial Systems, and P-Time Model Checking of Probabilistic One-Counter Automata. In: Sharygina, Veith (eds.) [24], pp. 495–510
26. Strahler, A.N.: Hypsometric (area-altitude) analysis of erosional topology. Geol. Soc. Am. Bull. 63(11), 1117–1142 (1952)
27. Yntema, M.K.: Inclusion relations among families of context-free languages. Information and Control 10, 572–597 (1967)

On the Parikh Membership Problem
for FAs, PDAs, and CMs

Oscar H. Ibarra[1,*] and Bala Ravikumar[2]

[1] Department of Computer Science
University of California, Santa Barbara, CA 93106, USA
ibarra@cs.ucsb.edu
[2] Department of Computer & Engineering Science
Sonoma State University, Rohnert Park, CA 94928, USA
ravi@cs.sonoma.edu

Abstract. We consider the problem of determining if a string w belongs to a language L specified by an automaton (NFA, or PDA augmented by reversal-bounded counters, etc.) where the string w is specified by its Parikh vector. If the automaton (PDA augmented with reversal-bounded counters) is fixed and the Parikh vector is encoded in unary (binary), the problem is in $DLOGSPACE$ ($PTIME$). When the automaton is part of the input and the Parikh vector is encoded in binary, we show the following results: if the input is an NFA accepting a letter-bounded language (i.e., $\subseteq a_1^* \cdots a_k^*$ for some distinct symbols $a_1, ..., a_k$), the problem is in $PTIME$, but if the input is an NFA accepting a word-bounded language (i.e., $\subseteq w_1^* \cdots w_m^*$ for some nonnull strings $w_1, ..., w_m$), it is NP-complete. The proofs involve solving systems of linear Diophantine equations with non-negative integer coefficients. As an application of the results, we present efficient algorithms for a generalization of a tiling problem posed recently by Dana Scott. Finally, we give a classification of the complexity of the membership problem for restricted classes of semilinear sets.

Keywords: Parikh vector, NFA, counter machine, reversal-bounded counters, CFG, Chomsky Normal Form, bounded language.

1 Introduction

Membership problems are the most fundamental problems in computation theory. Here we study a variation in which the input string is specified by its Parikh vector - i.e., as a vector $< n_1, n_2, ..., n_k >$ where k is the alphabet size and n_i is the number of occurrences of the i-th letter. A potential application area involves pattern matching in which some symbols are allowed to commute [13]. Our study was also motivated by a tiling problem (posed by Dana Scott) for which a polynomial time algorithm follows from the new version of the membership problem studied here. The problem of membership given by Parikh vector is a natural one and is probably of interest in its own right.

---

* Supported in part by NSF Grants CCF-1143892 and CCF-1117708.

A.-H. Dediu et al. (Eds.): LATA 2014, LNCS 8370, pp. 14–31, 2014.

Membership problem as well as other problems (such as equivalence, containment etc.) have been studied [6] where the input is a Parikh vector. But prior studies have generally assumed that the language is also represented as a semilinear set (in terms of the basis vectors). In this setting, the membership (equivalence) problem has been shown to be NP-complete (Σ_2^p-complete) [6]. [13] and [2] have studied membership problems similar to the ones presented in this paper. The main difference between these papers and ours are as follows: (a) we present an application to a class of tiling problems, (b) we present NP-hardness result even when restricted to NFA's accepting a bounded language and (c) we show the positive result (namely PTIME algorithm) for a wider class of languages (namely those that are accepted by PDA's augmented by reversal-bounded counter machines).

The rest of the paper is organized as follows: In Section 2, we review some basic definitions and concepts used in this paper. In Section 3, we introduce Scott's tiling problem and present an efficient algorithm to solve the unary as well as the binary version. In Section 4, we consider the Parikh vector membership problem when the language is part of the input. We extend the polynomial time membership to NPDA's augmented by reversal-bounded counters. Next we consider the membership problem when the automaton is not fixed (i.e., it is part of the input). In the case of binary encoding of the input (Parikh) vector, the problem is shown to be in $PTIME$ when the input is an NFA that accepts a letter-bounded regular language. It is also in $PTIME$ if the language is k string-bounded for a fixed k. It becomes NP-complete if k is not fixed. Thus, the membership problem exhibits an interesting contrast - when the input NFA is letter-bounded it is solvable in polynomial time, but is NP-complete when it is string-bounded (i.e., a subset of $w_1^*...w_k^*$ for some w_1, ..., w_k). In Section 5, we give a classification of the complexity of the membership problem for some restricted classes of semilinear sets. In Section 6, we conclude with a summary of the main results presented in this work. Our work shows some results on the solvability of systems of linear Diophantine equations with non-negative integer coefficients.

2   Preliminaries

We will assume that the readers are familiar with terms and notation used in formal language, automata theory and complexity theory as presented in standard references such as [5], [3].

Let N be the set of non-negative integers and $n \geq 1$. $Q \subseteq N^n$ is a *linear set* if there is a vector c in N^n (the constant vector) and a set of periodic vectors $V = \{v_1, \ldots, v_r\}$, $r \geq 0$, each v_i in N^n such that $Q = \{c + t_1 v_1 + \cdots + t_r v_r \mid t_1, \ldots, t_r \in N\}$. We denote this set as $Q(c, V)$. A finite union of linear sets is called a *semilinear set*.

Let $\Sigma = \{a_1, \ldots, a_n\}$. For $w \in \Sigma^*$, let $|w|$ be the number of letters (symbols) in w, and $|w|_{a_i}$ denote the number of occurrences of a_i in w.

The *Parikh map* $P(w)$ of w is the vector $(|w|_{a_1}, \ldots, |w|_{a_n})$; similarly, the Parikh image of a language L is defined as $P(L) = \{P(w) \mid w \in L\}$.

3   D. Scott's Problem and Related Problems

Polyominoes are a collection of unit squares forming a connected piece in the sense that each square is reachable from any other by going through adjacent squares. Polyominoes were made popular in the recreational Mathematics literature by [4] and numerous puzzles have been created based on polyominoes. Pentominoes are polyominoes made of five squares. There are 12 distinct pentominoes and they have been labeled using letters P through Z by Conway - see the figure below. Dana Scott [19] pioneered the use of backtracking to solve the problem of placing one copy each of the 12 different pentominoes in the standard 8 by 8 checker-board with a 2 by 2 hole in the center. Of course, such a placement requires leaving no other holes and with no overlap among the tiles. Note that the tiles can be placed on the board in any orientation. There are 63 distinct pentominoes when all possible orientations (rotations and reflections) are counted as distinct. From now on, these will be called oriented pentominoes.

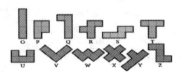

Fig. 1. Twelve pentominoes with Conway's labels

In a lecture at the University of Pennsylvania in April, 2012, Scott introduced a tiling problem: Suppose the 12 pentominoes are labeled 1 through 12. Given a sequence of twelve positive integers $(n_1, n_2, ..., n_{12})$ as input, the problem is to determine if there is a tiling of $5 \times n$ checker-board using exactly n_i copes of tile i (where $n = n_1 + ... + n_{12}$). Scott asked whether this problem is in $PTIME$, NP-complete or possibly a problem of intermediate complexity (with n_i's given in unary). We will show that Scott's problem is actually solvable in $DLOGPSACE$ ($PTIME$) when the input is given in unary (binary).

The main result of this section is that the solutions to the tiling problem can be encoded as a regular language L_{tile}. In the next section, we will use this result to design a polynomial time algorithm for Scott's problem.

We begin by describing the alphabet over which we will define the language L_{tile}. To define this, we first uniquely label all the different 63 pentominoes. For pentomino j, we will assign labels for each of its squares by a label using the following convention: Suppose the pentomino j occupies a total of r columns when placed on the board. (For example, the x-pentomino has three columns - the first one containing one square, the second one containing 3 squares and the third one containing one square.) The label we will assign to column c of pentomino j is the pair (j, c). Thus, the center square of the x-pentomino (assuming its label is j_x) will be labeled $(j_x, 2)$. For a oriented pentomino j, let w_j be its width. For example, the R-pentomino's width is 3 in all its (eight) orientations.

Let $\Sigma = \{[p_1, p_2, p_3, p_4, p_5] \mid p_j = (j, c_j), \ j \in \{1, 2, ..., 63\} \text{ and } c_j \leq w_j\}$. We will now define the language L_{tile} as follows: Consider any valid placement P of the 63 oriented pentominoes to cover a $5 \times m$ checkerboard for some m. The encoding associated with this placement is to replace each of the squares its checkerboard with the corresponding label of the pentomino that occupies that square. For a placement P, let $code(P)$ denote this string. The language L is the collection of all such codes. Formally,

$L_{tile} = \{code(P) \mid P$ is a valid placement of a collection of pentominoes on a $5 \times m$ checkerboard for some $m\}$.

Theorem 1. L_{tile} *is regular.*

Proof. We will describe informally a DFA M to accept L_{tile}. We will start by describing the state set of M. In fact, we will use the set Σ together with a start state q_0 and a dead state q_d as the state set Q of M. We define a *valid opening state* $p \in Q$ as follows: $p = [(r_1, 1), (r_2, 1), (r_3, 1), (r_4, 1), (r_5, 1)]$ is a valid opening state if the following condition is satisfied: the oriented pentominoes whose labels appear in the set $\{r_1, r_2, r_3, r_4, r_5\}$ can be placed on the leftmost column of a checker-board so that each pentomino's label matches the label r_i and the pentonimoes do not overlap each other in any column. Basically, the opening state represents the first column of a valid placement of pentominoes. For example, suppose O_v stand for the vertical orientation of the O-pentomino. The state

$[(O_v, 1), (O_v, 1), (O_v, 1), (O_v, 1), (O_v, 1)]$ represents a valid opening state that corresponding to placing the O-pentomino vertically on the board so that it covers the leftmost column completely. On the other hand, we can also place five copies of O_h, the same O-pentomino in horizontal orientation. In this case, the opening state associated with the leftmost column of this placement would be $[(O_h, 1), (O_h, 1), (O_h, 1), (O_h, 1), (O_h, 1)]$.

Next we define the notion of consistency: We say that a state p is consistent with a state q (note that this relationship is not symmetric) if the following is true: Consider state $p = [(r_1, c_1), (r_2, c_2), (r_3, c_3), (r_4, c_4), (r_5, c_5)]$. Create a 2 by 5 board and place the pentominoes on the squares of the left column so that the pentomino r_j is placed with its column c_j is on square $(1, j)$ for $j = 1$ to 5. Let $q = [(s_1, d_1), (s_2, d_2), (s_3, d_3), (s_4, d_4), (s_5, d_5)]$. The following three conditions should be satisfied for l and q to be consistent:

1. it should be possible to place the pentominoes matching state p as described above.
2. the extensions of pentominoes placed in the above step in column 2 should match the labels associated with state q.
3. it should be possible to fill the 'holes' in the second column by placing pentominoes specified in state q.

We can now describe transitions as follows: a transition from state p on input q (note: here q is the next input as well as the next state) goes to state q if and only if p and q are consistent. On all other inputs, p transitions to the dead state q_d. Finally we describe a test to determine if a state $p = p = [(r_1, c_1), (r_2, c_2), (r_3, c_3), (r_4, c_4), (r_5, c_5)]$ is the accepting state. Apply the first step in checking the consistency between states p and q described above. Recall that this involves placing pentominoes on the left column of a 2 by 5 checker-board so that the column numbers c_1, c_2 etc. match the column numbers of the pentominoes being placed. After this placement, if none of the placed pentominoes extend to the right column, then p is accepting state. It is easy to see that this DFA accepts the set of encodings of a valid placement of a collection of pentominoes on $5 \times m$ checker-board for some m and so the claim is proved. $\square$

We now show that Scott's problem is in $DLOGSPACE$. We begin with the following result:

Theorem 2. *Let M be an NPDA augmented with 1-reversal counters such that $L(M) \subseteq \{a_1, ..., a_k\}^*$. Let $L_M = \{a_1^{n_1} \cdots a_k^{n_k} \mid$ there exists w in $L(M)$ such that for $1 \leq i \leq k$, w has exactly n_i occurrences of a_i $\}$. Then L_M can be accepted by a DFA augmented with 1-reversal counters that runs in linear time.*

Proof. We first construct an NPDA M' augmented with 1-reversal counters that accepts L_M. In addition to the counters used by M, M' has k new 1-reversal counters $C_1, .., C_k$. M' on input $a_1^{n_1} \cdots a_k^{n_k}$ first reads the input and stores $n_1, ..., n_k$ in counters $C_1, C_2, ..., C_k$. Then M' guesses an input w to M symbol-by-symbol and simulates M. It also decrements counter C_i whenever it guesses symbol a_i in w. When all the counters become zero, M' accepts if and only if w is accepted by M.

It is known (see [8]) that any language $B \subseteq w_1^* \cdots w_k^*$, where $k \geq 1$ and $w_1, ..., w_k$ are (not necessarily distinct) nonnull strings, accepted by an NPDA with 1-reversal counters is a semilinear language, i.e., $Q_B = \{(i_1, i_2, \ldots, i_k) \mid w_1^{i_1} w_2^{i_2} \cdots w_k^{i_k} \in B\}$ is a semilinear subset of N^k. Hence, L_M is a semilinear language. It was recently shown in [11] that the following statements are equivalent:

1. B is semilinear language.
2. B is accepted by a DFA augmented with 1-reversal counters.

In [1], it was shown that for every NFA M with 1-reversal counters, there is a constant c such that every string x of length n in L(M) can be accepted by M within cn time (even if x is non-bounded). Hence, a DFA augmented with 1-reversal counters runs in linear time. The theorem follows. $\square$

When the values of the counters are polynomial, they can be stored in log n space; hence:

Corollary 3. $L(M)$ *(where M is as defined in Theorem 2) is in $DLOGSPACE$ and, hence, in $PTIME$.*

We can generalize Theorem 2:

Theorem 4. *Let M be an NPDA augmented with 1-reversal counters, $k \geq 1$, and $w_1, w_2, \ldots, w_k$ be nonnull strings. Let $L_M = \{w_1^{n_1} w_2^{n_2} \cdots w_k^{n_k} \mid$ there exists w in $L(M)$ such that for $1 \leq i \leq k$, w has exactly n_i occurrences of w_i, and all the occurrences of w_1's, $\ldots$, w_k's are not overlapping (hence $|w| = n_1|w_1| + \cdots + n_k|w_k|$) $\}$. Then L_M can be accepted by a DFA augmented with 1-reversal counters that runs in linear time. Hence, L_M is in $DLOGSPACE$ and in $PTIME$.*

Proof. As in the proof of Theorem 2, we first construct an NPDA M' augmented with 1-reversal counters that accepts L_M. M' when given a string x in $w_1^* \cdots w_k^*$, first reads x and stores $n_1, ..., n_k$ in k counters $C_1, ..., C_k$ such that $x = w_1^{n_1} \cdots w_k^{n_k}$. (Note that the n_i's are not unique, since there may be more than one decomposition of x.) Then M' guesses an input w to M segment-by-segment (where a segment is one of $w_1, ..., w_k$) and simulates M. It also decrements counter C_i whenever it guesses segment w_i. When all the counters become zero, M' accepts if and only if w is accepted by M.

Again from [8], the set $L_{M'}$ is a semilinear language. The rest of the proof is the same as in Theorem 2.   ⊓⊔

We use Theorems 1 and 4 to show the main result of this section.

Theorem 5. *There is a $DLOGSPACE$ algorithm that takes as input a vector $(n_1, ..., n_{12})$ (in unary notation) of non-negative integers (where $\sum_j n_j = n$) and answers yes (no) if there is a tiling of $5 \times n$ checker-board using n_j pentominoes of type j.*

Proof. Since the input vector is given in unary, the input vector $(n_1, n_2, ..., n_{12})$ can be viewed as the string $a_1^{n_1}...a_{12}^{n_{12}}$. We describe a nondeterministic 12 counter machine M (which reverses each counter at most twice) that accepts the language $L = \{a_1^{n_1}...a_{12}^{n_{12}} \mid$ there is a tiling of $5 \times n$ checker-board using n_j pentominoes of type $j\}$. We describe M informally. Let M_{tile} be the DFA described in the proof of 1. For simplicity, we can assume that the input is of the form $a_1^*...a_n^*$. (This can be easily checked by M in the finite control.) M guesses (symbol by symbol) a string w that represent a tiling of $5 \times n$ checker-board where n is the length of the input string. After each symbol is guessed it simulates a single step of the DFA M_{tile} and it also moves the input head exactly once after each symbol is guessed. It further increases the counter j if the current input symbol it scans on the input tape is a_j. In addition, in finite control, it keeps track of the last five symbols it guessed, and uses it to decrement j for each tile j that was guessed, and that lies completely to the left of the current input position. This simulation may pose a small problem in that for most of the tile types, the counter values may have to be decremented to a value below 0 since the incrementing may happen later than the decrementing. But this problem can be easily overcome with the standard trick of representing a negative value $-k$ using a counter value k and remembering the sign in the finite control. Note that the incrementing of the counter happens in one block while the input head moves over the block $a_j^{n_j}$ and decrementing happens every time a tile of type j is actually seen on the (guessed) board. However, because of the above trick we use to avoid negative counter values, each counter may be reversed at most twice. This is seen as follows: if a tile of type j is encountered before the input head reads the block a_j, then clearly a symbol will be pushed for each occurrence of tile j in the guessed board, and then when the block of a_j's is reached, a symbol is popped off counter j for each a_j on the input tape. This happens until the counter value reaches 0. From this point, for each a_j on the input tape, the counter will be incremented and each occurrence of tile type j will result in decrementing the counter. When the entire input has been read, if all the counters reach value 0, and the DFA M_{tile} reaches an accepting state, it is clear that the input is a yes instance of the problem and is accepted. It is also clear that no guessing will result in acceptance of the string $a_1^{n_1}...a_{12}^{n_{12}}$ that is a 'no' instance of the problem. Thus it is clear that a NFA N_1 with 12 counters each of which reverse at most twice can accept the language L. It is easy to see that N_1 can be simulated by a 24-counter machine NFA N_2 with counters reversing once. Using Theorem 4, N_2 can be simulated by a DFA augmented by a counter machine and hence we conclude that there is a $DLOGSPACE$ Turing machine for L. □

4   Case of the Input Vector $(n_1, \dots, n_k)$ Encoded in Binary

Throughout this section, we will assume that $n_1, \dots, n_k$ are represented in binary with leading bits 1. So the time complexity involving these numbers will be a function of $log(n_1) + \cdots + log(n_k)$.

We will need the following result of Lenstra in [16]:

Theorem 6. *Let S be a system of linear constraints:*

$$v_{11}x_1 + v_{12}x_2 + \cdots + v_{1m}x_m \leq n_1$$

$$.....$$

$$v_{k1}x_1 + v_{k2}x_2 + \cdots + v_{km}x_m \leq n_k$$

where $k, m \geq 1$ and the n_i's, and the v_{ij}'s are integers (+, -, 0), represented in binary.

When m (the number of variables) or k (the number of equations) is fixed, deciding if the system has an integer solution (+, -, 0) for $x_1, ..., x_m$ is in $PTIME$.

When the n_i's and the v_{ij}'s are non-negative and the inequalities become equalities, we can show:

Theorem 7. *Let S_1 be a system of linear equations:*

$$v_{11}x_1 + v_{12}x_2 + \cdots + v_{1m}x_m = n_1$$

$$.....$$

$$v_{k1}x_1 + v_{k2}x_2 + \cdots + v_{km}x_m = n_k$$

where $k, m \geq 1$ and the n_i's, and the v_{ij}'s are non-negative integers, represented in binary. When m is fixed, deciding if the system S_1 has a non-negative integer solution for $x_1, ..., x_m$ is in $PTIME$.

Proof. We transform system S_1 to a system S of Theorem 6 as follows. For $1 \leq i \leq k$, we convert

$$v_{i1}x_1 + v_{i2}x_2 + \cdots + v_{im}x_m = n_i$$

to two inequalities:

$$v_{i1}x_1 + v_{i2}x_2 + \cdots + v_{im}x_m \leq n_i$$

$$-v_{i1}x_1 - v_{i2}x_2 - \cdots - v_{im}x_m \leq -n_i$$

We also introduce for $1 \leq i \leq m$, the inequality:

$$-x_i \leq 0$$

Then we get a system of inequalities of the form S given in Theorem 6 with m variables and $2k + m$ inequalities. It follows that system S_1 is solvable in polynomial time. $\qquad\square$

When m (number of variables) in the system S_1 in Theorem 7 is not fixed, the problem is NP-hard, even when (the number of equations) $k = 1$ [17]. However, when the coefficients are bounded by a fixed positive integer, we can show:

Theorem 8. *Let S_2 be a system of linear equations:*

$$v_{11}x_1 + v_{12}x_2 + \cdots + v_{1m}x_m = n_1$$

$$.....$$

$$v_{k1}x_1 + v_{k2}x_2 + \cdots + v_{km}x_m = n_k$$

*where $k, m \geq 1$ and the n_i's, and the v_{ij}'s are non-negative integers (represented in binary) such that the v_{ij}'s are bounded by a **fixed** positive integer d. When (the number of equations) k is fixed, deciding if the system S_2 has a non-negative integer solution for $x_1, ..., x_m$ is in $PTIME$.*

Proof. Call the given system of equations as (1). First we will describe a polynomial time algorithm when $k = 1$. Since $v_{1j} \leq d$, we can rewrite the equation

$$v_{11}x_1 + v_{12}x_2 + \cdots + v_{1m}x_m = n_1 \tag{1}$$

as

$$X_1 + 2X_2 + ... + dX_d = n_1 \tag{2}$$

where X_i, $i = 1...d$ are new variables so that the former equation has a solution in non-negative integers if and only if the latter has a solution in non-negative integers. The idea is to add all the variables x_j for which the coefficients in the original equation (1) equal r and set it to X_r. Formally, define

$$X_r = \sum_{j:v_{1j}=r} x_j \qquad (3)$$

and we get equation (2). Suppose (1) has a solution in non-negative integers, then clearly (2) also has a solution by simply setting X_i as in (3). Conversely, if (2) has a solution, then we can find a solution to (1) as follows: Suppose $X_i = d$ in the solution of (2). In equation (3), choose one of the variables x_j that appears on the right-hand side and set it to d and the others to 0. Since each variable x_j appears only once in (3), this assignment is consistent and provides a solution to (1). A polynomial time algorithm to determine if (2) has a solution follows from Theorem 7.

We next show how to extend this idea to k equations. Note that a variable x_j has coefficients $v_{j,i}$ in the set $\{0, 1, ..., d\}$. Define d_i as the column vector associated with x_i:

$$d_i = (v_{1,i}, ..., c_{k,i})$$

Let $D = \{v_1, v_2, ..., v_r\}$ be the set $\{d_i \mid 1 \le i \le m\}$ with duplicates removed. Clearly $r \le (d+1)^k$. Define a map $f : \{1, 2, ..., m\} \to \{1, 2, ..., d\}$ such that v_t is the vector associated with variable x_j where $t = f(j)$.

We introduce new variables $y_1, ... , y_r$ and let

$$y_i = \sum_{j:f(j)=i} x_j$$

We now replace the variables x_j by y_i and rewrite the equations (1). The following illustrates this process. Suppose the original system of equations is:

$$x_1 + x_2 + 2x_3 + x_4 + 2x_5 + 2x_6 + 2x_7 + 2x_8 + 2x_9 = 123$$
$$2x_1 + x_2 + 2x_3 + x_4 + x_5 + 2x_6 + x_7 + 2x_8 + x_9 = 97$$
$$x_1 + 2x_2 + x_3 + 2x_4 + x_5 + 2x_6 + 2x_7 + 2x_8 + x_9 = 104$$

The multi-set $\{d_i \mid 1 \le i \le 9\} = \{(1, 2, 1), (1, 1, 2), (2, 2, 1), (1, 1, 2), (2, 1, 1), (2, 2, 2), (2, 1, 2), (2, 2, 2), (2, 1, 1)\}$ and after removing duplicates, we get:

$D = \{(1, 2, 1), (1, 1, 2), (2, 2, 1), (2, 1, 1), (2, 2, 2), (2, 1, 2)\}$ and the function f defined as:

$f(1) = 1$, $f(2) = f(4) = 2$, $f(3) = 3$, $f(5) = f(9) = 4$, $f(6) = f(8) = 5$ and $f(6) = 7$.

Thus y_i's are related to the original variables x_i's via the equations:

$$y_1 = x_1$$
$$y_2 = x_2 + x_4$$
$$y_3 = x_3 \tag{4}$$
$$y_4 = x_5 + x_9$$
$$y_5 = x_6 + x_8$$
$$y_6 = x_7$$

Using y_i's to eliminate x_j's, we get:

$$y_1 + y_2 + 2(y_3 + y_4 + y_5 + y_6) = 123$$
$$y_2 + y_4 + y_6 + 2(y_1 + y_3 + y_5) = 97 \tag{5}$$
$$y_1 + y_3 + y_4 + 2(y_2 + y_5 + y_6) = 104$$

Note that the number of variables in the new sytem of equations is bounded by a constant r (since both d and k are constants).

In order to conclude the proof, we need to prove the following:

1. Given (1), we can construct (3) in time bounded by a polynomial in the input size.
2. The system (1) has a solution in non-negative integers if and only if the system (3) does.
3. There is a polynomial time algorithm to determine if (3) has a solution in non-negative integers.

We will show each of these claims.

Claim 1. By scanning the equations, a vector v associated with each variable x_j can be created. Then by sorting the vectors and elimiating duplications the set D can be constructed, with label j attached to vector v. Finally, the equations can be rewritten by replacing each variable x_j by y_i where v_i is the vector associated with variable x_j. All these steps can be done in polynomial time.

Claim 2. Note that (3) was arrived at from (1) by replacing sums of the form $x_{i_1} + x_{i_2} + ...$ by new variables y_t. If (1) has a solution, clearly we can find a solution to (3) by taking the assignments for x_j's and using them to compute the corresponding y_t's. Conversely, suppose (3) has a solution in which the variable y_t has assignment v. y_t is a sum of some x's such as $x_{i_1} + x_{i_2} + ...$ We can pick arbitrarily one of the x_i's in this expression and set it to v and the others to 0. The key point is that since the variables that occur in these equations form a partition, such an assignment will not cause any inconsistency. This shows (2).

Claim 3. This readily follows from the fact that in the system (3), the number of variables is bounded by $(d+1)^k$ which is bounded by a constant since k and d are bounded. Thus by Theorem 7, there is a polynomial time algorithm to determine if (3) has a non-negative integer solution. This concludes the proof.   □

Theorem 9. *Let M be an NPDA augmented with 1-reversal counters such that $L(M) \subseteq \{a_1, ..., a_k\}^*$. The problem of deciding, given $n_1, ..., n_k$, whether there exists a string w in $L(M)$ with exactly n_i occurrences of a_i (for $1 \leq i \leq k$) is in PTIME. (Note that the time complexity is a function of $\log(n_1) + \cdots + \log(n_k)$.)*

Proof. As in the proof of Theorem 2, $Q_M = \{(n_1, ..., n_k) \mid a_1^{i_1} \cdots a_k^{i_k} \in L(M)\}$ is a semilinear subset of N^k. Now Q_M is a union of linear sets. Clearly, it sufficient to show the result for the case when Q is a linear set.

So let Q be a linear set specified by a constant vector $c = (c_1, ..., c_k)$ in N^k and periodic vectors $v_1 = (v_{11}, v_{21}, ..., v_{k1}), ..., v_m = (v_{1m}, v_{2m}, ..., v_{km})$ in N^k, That is, $Q = \{(n_1, ..., n_k) \mid (n_1, ..., n_k) = c + v_1 x_1 + \cdots + v_m x_m$ for some nonnegative integers $x_1, ..., x_m\}$. Assume that the c_i's and v_{ij}'s are written in binary.

Thus, given $(n_1, ..., n_k)$, we first check that for $1 \leq i \leq k$, $n_i - c_t \geq 0$, and then determine if a system S_1 of k linear equations with m variables with non-negative integer coefficients of the form given in Theorem 7 has a nonnegative integer solution in $x_1, ..., x_m$, when the right hand side of the equations are $n_1 - c_1, ..., n_k - c_k$, respectively. Since the linear set is fixed, k is fixed. Hence, this problem is in PTIME by Theorem 7.   □

Now we consider the case when an NFA accepting a bounded language is part of the input. In this case, the problem becomes NP-complete. First we need a lemma. The membership problem for linear sets is the following:

> **Given:** A specification of a linear set L as a collection of vectors v_0, v_1, v_2, ..., v_m where each v_i is a k-dimensional vector of integers and a target v, a k-dimensional vector of integers. (The components of the vectors in L as well as v are represented in binary.)
> **Question:** Is v in the linear set spanned by the vectors in L? (i.e, do there exist non-negative integers d_1, ..., d_m such that $v = v_0 + \sum_{j=1}^{m} d_v v_j$?)

The specification of a linear set can be written as a system of expressions:
$$c_1 + v_{11}x_1 + v_{12}x_2 + \cdots + v_{1m}x_m$$

$$....$$

$$c_k + v_{k1}x_1 + v_{k2}x_2 + \cdots + v_{km}x_m$$

where $k, m \geq 1$ and the c_i's and the v_{ij}'s are at most 4. To decide if $(n_1, ..., n_k)$ is in the linear set, we need to determine if there exist non-negative integers $x_1, ..., x_m$ such that the ith expression evaluates to n_i.

Clearly, since the c_i's are at most 4, we can modify the system so that the constant term in each expression is zero, and checking instead if the new system has a solution for $(n_1 - c_1, ..., n_k - c_k)$. So we may assume that $c_1 = \cdots = c_k = 0$. (This is equivalent to the vector v_0 being 0.) It was shown by D. Hyunh in [6] that this problem is NP-complete. Note that the complexity is a function of $|S| + log(n_1) + \cdots + log(n_k)$, where $|S|$ is the size of S. We will now show the following stronger NP-completeness result.

Lemma 10. *The linear set membership is NP-hard even when all the components of the vectors in the specifications of the linear sets and the target vector are bounded by 4.*

Proof. Consider the standard reduction from 3-SAT to subset sum [15], (Theorem 34.15, page 1015). Let the instance C of 3-SAT be $\{C_1, ..., C_m\}$ over the variables $x_1, ..., x_n$. The instance C is turned into a set $S_C = A \cup B$ of integers where each integer (in base 10) with $(n + m)$ digits where n is the number of clauses and m is the number of Boolean variables and a target integer t of $n + m$ digits. The number of integers in S_C is $2(n + m)$ - forming two sets A and B. A contains two integers for each Boolean variable (v_i and v_i' - corresponding to variable x_i and its complement x_i') and hence $|A| = 2n$. B contains two integers for each clause (s_i and s_i' corresponding to clause C_i), so $|B| = 2m$. The digits (in the decimal representation) of the numbers in A and B lie in $\{0, 1, 2\}$.

The reduction is as follows: The integer v_i (v_i') has a 1 in column i and a 1 for column $(n + j)$ for each clause C_j in which the literal x_i (x_i') appears. The integer s_j (s_j') has a 1 (2) in column $n + j$. The target t is the integer whose decimal representation is $1^n 4^m$. It is easy to see that the 3-SAT instance C is satisfiable if and only if there is a subset of S_C with sum $= t$. Shown below is an example (taken from [15]) that shows the reduction for the instance $D = C_1 \wedge C_2 \wedge C_3 \wedge C_4$, where $C_1 = (x_1 \vee \bar{x}_2 \vee \bar{x}_3)$, $C_2 = (\bar{x}_1 \vee \bar{x}_2 \vee \bar{x}_3)$, $C_3 = (\bar{x}_1 \vee \bar{x}_2 \vee x_3)$, and $C_4 = (x_1 \vee x_2 \vee x_3)$.

		x_1	x_2	x_3	C_1	C_2	C_3	C_4
v_1	=	1	0	0	1	0	0	1
v_1'	=	1	0	0	0	1	1	0
v_2	=	0	1	0	0	0	0	1
v_2'	=	0	1	0	1	1	1	0
v_3	=	0	0	1	0	0	1	1
v_3'	=	0	0	1	1	1	0	0
s_1	=	0	0	0	1	0	0	0
s_1'	=	0	0	0	2	0	0	0
s_2	=	0	0	0	0	1	0	0
s_2'	=	0	0	0	0	2	0	0
s_3	=	0	0	0	0	0	1	0
s_3'	=	0	0	0	0	0	2	0
s_4	=	0	0	0	0	0	0	1
s_4'	=	0	0	0	0	0	0	2
t	=	1	1	1	4	4	4	4

Fig. 2. Reduction example

We will adapt this reduction so that it transforms 3-SAT to linear set membership problem. We do this by making v_i, v_i', s_i and s_i' as $(n+m)$ dimensional vectors in which the j-th component will be the j-th digit in the above reduction, and similarly for the target. (Thus, in the figure, each row, except the last, represents a vector in the linear set, and the last row represents the target vector.) However, an additional change is needed to make the reduction work. The reason is that in the subset sum problem, we are allowed to choose each subset at most once which corresponds to taking linear combinations with coefficient 0 or 1. But in the linear set membership problem, we are allowed to multiply each vector by any positive integer (not just 0 or 1). The following modification to the above reduction will overcome this problem: we add m additional columns and extend the vectors as follows: for vectors v_i and v_i', all the components in the additional vectors are 0. For s_j and s_j', there will be a 1 in $(m+n+j)$-th component. The target will now be the vector $(1, 1, ..., 1, 4, 4, ..., 4, 1, 1, ..., 1)$ where the last m components are 1. It is clear that the only linear combinations of vectors from $A \cup B$ that can match t are those with coefficients are 0 or 1 - since multiplication by any integer greater than 1 will create a component greater than 1 in one of the first m- or of the last m-components, and hence can't equal t which has 1 in those positions. Thus the membership problem for linear sets in which each component and target vector is chosen from the set $\{0, 1, 2, 3, 4\}$ is NP-hard. □

Now we show the NP-completeness of Parikh membership problem when the NFA accepting a bounded language is part of the input.

Theorem 11. *The problem of deciding, given an NFA M and $w_1, ..., w_m$ such that $L(M) \subseteq w_1^* ... w_m^*$, where each $w_i \in \{a_1, ..., a_k\}^+$, and $n_1, ..., n_k$, whether there exists a string w in $L(M)$ with exactly n_i occurrences of a_i (for $1 \le i \le k$) is NP-complete.*

Proof. Note that in this version, in addition to the NFA M and the vector $(n_1, n_2, ..., n_k)$, the strings $w_1, ..., w_m$ are given as input. Thus the input size for the problem is $N = |M| + \Sigma_j log_2 n_j + \Sigma_j |w_j|$. For convenience, we will call this problem Parikh membership problem for bounded NFA.

First we show that the problem is in NP. The NP algorithm is as follows:

Step 1: Guess integers $N_1, ..., N_m$ and check that the following equations are satisfied:

$$N_1 a_{11} + N_2 a_{12} + ... + N_m a_{1m} = n_1$$

$$....$$

$$N_1 a_{k1} + N_2 a_{k2} + ... + N_m a_{km} = n_k$$

In the above, a_{ij} is the number of occurrences of symbol a_i in string w_j.) Clearly, the size of a_{ij} are bounded by a polynomial in the input size and it is clear that the equations can be checked in time polynomial in N, the input size.

Step 2: Check that the string $w_1^{N_1}...w_m^{N_m}$ is in $L(M)$ as follows:

For each string w_i, create a matrix M_i of order $q \times q$ (where q is the number of states in M) as follows: $M_i[j,k] = 1$ if $q_k \in \delta(q_j, w_i)$, i.e., q_k is reachable from q_j on string w_i. It is clear that the entries of M_i can be filled in polynomial time by performing a breadth-first search on the graph of M starting from state q_j using the successive symbols of the string w_i.

Next we use repeated squaring technique to perform Boolean matrix exponentiation (using the recursive formula $x^n = x$ (of $n = 1$), $x^{n-1} \times x$ (if n is odd and $n > 1$) and $x^n = (x^{\frac{n}{2}})^2$ (if n is even)) to compute $M_i^{N_i}$ for each i. It is easy to see that the above algorithm can be performed in $O(|M|^3 log\ N_i)$ operations.

Step 3: We create a graph G with $q \times k$ nodes in a grid of $k+1$ columns and q rows and label the nodes by the row and column index pair. Column index starts with 0. We add an edge from $< r-1, s >$ to $< r, t >$ if $M_r^{N_r}[s,t] = 1$. We can create this graph in time $O(qk)$ by reading off the entries from the Boolean matrices computed in Step 2.

Step 4: Let q_0 be the starting state of M. Accept the input $(n_1, n_2, ..., n_k)$ if and only if there is a path from $< 0, 1 >$ in graph G to a vertex $< k, f >$ for some accepting state q_f of M.

It is not hard to see that the above algorithm correctly solves the problem and is an NP algorithm.

Next, we show that the problem is NP-hard. We present a reduction from linear set membership problem (see Lemma 10) to Parikh membership for bounded NFA. Let $a_1, ..., a_k$ be distinct symbols. We construct an NFA M such that $L(M) \subseteq w_1^* w_2^* \cdots w_m^*$ where $w_1 = a_1^{v_{11}} a_2^{v_{21}} \cdots a_k^{v_{k1}}$, $w_2 = a_1^{v_{12}} a_2^{v_{22}} \cdots a_k^{v_{k2}}$, ..., $w_m = a_1^{v_{1m}} a_2^{v_{2m}} \cdots a_k^{v_{km}}$. M operates as follows, given a string w:

Step 1: M executes the following process x_1 times for some nondeterministically chosen x_1 (including $x_1 = 0$): reads $a_1^{v_{11}} a_2^{v_{21}} \cdots a_k^{v_{k1}}$ and then goes to Step 2.

Step 2: M executes the following process x_2 times for some nondeterministically chosen x_2 (including $x_2 = 0$): reads $a_1^{v_{12}} a_2^{v_{22}} \cdots a_k^{v_{k2}}$ and then goes to Step 3.

.....

Step m: M executes the following process x_m times for some nondeterministically chosen x_m (including $x_m = 0$): reads $a_1^{v_{1m}} a_2^{v_{2m}} \cdots a_k^{v_{km}}$ and then enters an accepting state.

Clearly, the size of M is polynomial in the size of S. Moreover, given $n_1, ..., n_k$, there is a string w in $L(M)$ where the number of occurrences of a_i in w is exactly n_i (for $1 \le i \le k$) if and only if the system S generates $(n_1, ..., n_k)$.

It follows that deciding if there is a string w in $L(M)$ such that the number of occurrences of a_i in w is exactly n_i (for $1 \le i \le k$) is NP-hard. This concludes the proof of Theorem 11. $\qquad\square$

Note that in the case of DFA accepting an unbounded language, Parikh membership problem is already NP-hard by a reduction from Hamilton cycle problem [20]. The upper-bound for the general (unbounded) NFA remains NP. One way to see this is as follows: [7] showed that the problem of Parikh membership is in NP when the input is a context-free grammar. Since there is a logspace reduction from NFA to CFG, the claim follows. Thus Parikh membership problem for DFA and NFA are NP-complete. However, when the NFA is augmented by 1-reversal counters, it is not clear if Parikh membership problem is in NP. (Of course, the normal membership problem is in NP since a Turing machine can simulate the given M on the given string x in linear time, but since M is an NFA, the simulation results in an NP algorithm.) The best upper-bound we can obtain for this case is PSPACE. (In this case, an input string x is guessed whose Parikh map is the given vector v, and the counter machine is simulated on x. Since $|x|$ is exponential in the size of v, the simulation can take exponential number of steps, but needs only polynomial space.)

Next consider the class of regular expressions of the form $w_1^* \cdots w_m^*$ over the alphabet $\{a_1, ..., a_k\}$, where $k, m \geq 1$. Moreover, for $1 \leq i \leq m$ and $1 \leq j \leq k$, the number of occurrence of a_j in w_i is at most 3. We denote such a regular expression by R_{km} and the language it denotes by $L(R_{km})$.

Theorem 12

1. *When m or k is fixed, the problem of deciding, given a regular expression R_{km} and $n_1, ..., n_k$, whether there is a string w in $L(R_{km})$ with exactly n_i occurrences of a_i (for $1 \leq i \leq k$) is in PTIME.*
2. *When k and m are not fixed, the problem of deciding, given a regular expression R_{km} and $n_1, ..., n_k$, whether there is a string w in $L(R_{km})$ with exactly n_i occurrences of a_i (for $1 \leq i \leq k$) is NP-complete.*

Proof Given a regular expression $R_{km} = w_1^* \cdots w_m^*$, we first construct another regular expression $R'_{km} = z_1^* \cdots z_m^*$, such that if in w_i, a_j occurs i_j times, $z_i = a_1^{i_1} a_2^{i_2} \cdots a_k^{i_k}$. Clearly, there exists a string w in $L(R_{km})$ with exactly n_i occurrences of a_i if and only if there is string z in $L(R'_{km})$ with exactly n_i occurrences of a_i. From R'_{km}, we then construct in polynomial time a system of k linear equations with m variables with nonnegative integer coefficients of the form given in the proof of Theorem 9. If m (resp., k) is fixed, then the system can be solved in polynomial time by Theorem 7 (resp., Theorem 8).

For part (2), given regular expression R_{km}, we construct in polynomial time an NFA M accepting the language $L(R_{km})$. It follows from Theorem 11 that the problem is in NP. For NP-hardness, we observe that the NFA M constructed in the second part of the proof of Theorem 11 accepts exactly the language denoted by R'_{km} (hence denoted by R_{km}). $\square$

We note that the nondeterminism in the NP proof of Theorem 11 is because we need to "guess" the decomposition of the n_i's. However if the NFA M accepts a letter-bounded language, i.e., subset of $a_1^* \cdots a_k^*$, then there is no nondeterminism involved. Hence, we have:

Corollary 13. *The problem of deciding, given an NFA M such that $L(M) \subseteq a_1^* ... a_k^*$ (where the a_i's are distinct symbols), and $n_1, ..., n_k$, whether there exists a string w in $L(M)$ with exactly n_i occurrences of a_i (for $1 \le i \le k$) is in PTIME.*

The following result concerns the Parikh membership problem for NFA with 1-reversal counters, when the machine and the Parikh vector (in unary notation) are given as input:

Theorem 14. *The problem of deciding, given an NFA M augmented with 1-reversal counters over input alphabet $\Sigma = \{a_1, ..., a_k\}$ and $n_1, ..., n_k$, whether there exists a string in $L(M) \subseteq \Sigma^*$ with exactly n_i occurrences of a_i (for $1 \le i \le k$) is in PTIME.*

Proof. Given M, we constrict an NFA M' augmented with 1-reversal counters (k more counters than M) to accept the the language $L_M = \{a_1^{n_1} \cdots a_k^{n_k} \mid$ there exists w in $L(M)$ such that for $1 \le i \le k$, w has exactly n_i occurrences of a_i $\}$, as in the proof of Theorem 2. Clearly, M' has size polynomial in the size of M, and (from [1]) there is a constant c such that every string in L_M can be accepted by M' within $c(n_1 + \cdots + n_k)$ time (i.e., in linear time). It fiollows that M can be simulated by a nondeterministic TM in $log\ n$ space, which can then be simulated by a deterministic TM in polynomial time.   □

Clearly, if $n_1, ..., n_k$ are given in binary, the Parikh membership problem would be in polynomial space (PSPACE). Whether or not it is in PTIME is an interesting open problem.

5   Complexity of the Semilinear Set Membership Problem

Using the results in Section 4, we can classify the complexity of the membership problem for semilinear sets into four categories.

Clearly, if the membership problem when the semilincar set is a linear set is in PTIME (resp., NP-complete), then the problem for the general case (i.e., the semilinear set is a union of linear sets) is also in PTIME (resp., NP-complete).

So we need only consider the case when the semilinear set is a linear set, specified by $S = (c, V)$, where $c \in N^k$ is the constant vector, and $V = \{v_1, ..., v_m\}$ is the set of periodic vectors, $m \ge 1$, each $v_i \in N^k$.

Theorem 15

1. *For any fixed positive integer m, the membership problem is in PTIME when the number of periodic vectors in S is at most m.*
2. *The membership problem is NP-complete if the number of periodic vectors in S is not bounded, even when (the arity of S) $k = 1$.*
3. *The membership problem is NP-complete if the number of periodic vectors in S is not bounded, even when the components of the periodic vectors have value at most 3 (note that the arity of S is no longer assumed to be bounded in this case).*

4. *For any fixed positive integers k and d, the membership problem is in PTIME when the arity of S is at most k and the components of the periodic vectors of the linear set have value at most d.*

Proof. Part 1 follows from Theorem 7, since the specification S of the linear set can be represented as a system of equations that satisfies Theorem 7 (after moving the constant terms in each equation to the right-hand side.)

Part 2 follows from [17] (see remark before Theorem 8). Part 3 follows from the proof of Theorem 11. Part 4 follows from Theorem 8.   □

6   Conclusion

In this paper, we have studied some problems related to testing membership for regular and other languages. Unlike most of the study on such problems, we assume that the string is specified by its Parikh vector. We studied two versions of this problem in which the Parikh vector is specified in binary (unary). We showed that the problem can be solved in PTIME in the unary case when the fixed language comes from a very broad class (namely, the class of languages accepted by NPDA augmented by reversal-bounded counter machines). When the input vector is specified in binary, the complexity of the problem is already NP-hard in the case of word-bounded regular languages. One of the interesting fact we found is the difference between the letter-bounded regular languages and word-bounded regular languages. The membership problem for the former case is in PTIME while in the latter case it is NP-complete. Our results imply that a classical tiling problem is in DLOGSPACE (PTIME) when the number of tiles of various types are specified in unary (binary) notation.

Acknowledgments. We thank T.-D. Hyunh for pointing out that his proof of the NP-completeness of the linear set membership problem can be modified so that all components of the vectors in the specifications of the linear sets are bounded by a fixed integer. We also thank Maximilian Schlund for bringing the references [2] and [13] to our attention.

References

1. Baker, B.S., Book, R.V.: Reversal-bounded multipushdown machines. J. Comput. Syst. Sci. 8(3), 315–332 (1974)
2. Esparza, J.: Petri nets, commutative context-free grammars and basic parallel processes. Fundamenta Informaticae 30, 23–41 (1997)
3. Ginsburg, S.: The Mathematical Theory of Context-Free Languages. McGraw-Hill, New York (1966)
4. Golomb, S.W.: Polyominoes, 2nd edn. Princeton University Press (1994) ISBN 0-691-02444-8
5. Hopcroft, J.E., Ullman, J.D.: Introduction to Automata, Languages and Computation. Addison-Wesley (1978)

6. Hyunh, T.-D.: The Complexity of semilinear sets. Elektr. Inform.-verarbeitung and Kybern. 6, 291–338 (1982)
7. Hyunh, T.-D.: Commutative Grammars: The Complexity of Uniform Word Problems. Information and Control 57, 21–39 (1983)
8. Ibarra, O.H.: Reversal-bounded multicounter machines and their decision problems. J. Assoc. Comput. Mach. 25, 116–133 (1978)
9. Ibarra, O.H., Ravikumar, B.: On sparseness and ambiguity for acceptors and transducers. In: Monien, B., Vidal-Naquet, G. (eds.) STACS 1986. LNCS, vol. 210, pp. 171–179. Springer, Heidelberg (1985)
10. Ibarra, O.H., Ravikumar, B.: On bounded languages and reversal-bounded automata. In: Dediu, A.-H., Martín-Vide, C., Truthe, B. (eds.) LATA 2013. LNCS, vol. 7810, pp. 359–370. Springer, Heidelberg (2013)
11. Ibarra, O.H., Seki, S.: Characterizations of bounded semilinear languages by one-way and two-way deterministic machines. IJFCS 23, 1291–1306 (2012)
12. Ibarra, O.H., Yen, H.: On the Containment and Equivalence Problems for Two-way Transducers. Theoretical Computer Science 429, 155–163 (2012)
13. Kopczynski, E., To, A.W.: Parikh Images of Grammars: Complexity and Applications. In: Proc. of 25th Annual IEEE Logic in Computer Science, pp. 80–89 (2010)
14. Lavado, G.J., Pighizzini, G., Seki, S.: Converting Nondeterministic Automata and Context-Free Grammars into Parikh Equivalent Deterministic Automata. In: Yen, H.-C., Ibarra, O.H. (eds.) DLT 2012. LNCS, vol. 7410, pp. 284–295. Springer, Heidelberg (2012)
15. Cormen, T., Leiserson, C., Rivest, R., Stein, C.: Introduction to Algorithms, 2nd edn. MIT Press (2001)
16. Lenstra Jr., H.W.: Integer programming with a fixed number of variables. Mathematics of Operations Research 8, 583–548 (1983)
17. Lueker, G.S.: Two NP-Complete Problems in Nonnegative Integer Programming. Report No. 178, Computer Science Laboratory, Princeton University (1975)
18. Parikh, R.J.: On context-free languages. J. Assoc. Comput. Mach. 13, 570–581 (1966)
19. Scott, D.S.: Programming a combinatorial puzzle. Technical Report No. 1, Department of Electrical Engineering. Princeton University (1958)
20. To, A.W.: Parikh Images of Regular Languages: Complexity and Applications (2010) (unpublished manuscript)

Matchings, Random Walks, and Sampling

Sanjeev Khanna

Dept. of Computer & Information Science
University of Pennsylvania, Philadelphia, PA 19104, USA
sanjeev@cis.upenn.edu

The maximum matching problem is among the most well-studied problems in combinatorial optimization with many applications. The matching problem is well-known to be efficiently solvable, that is, there are algorithms that solve the matching problem using polynomial space and time. However, as large data sets become more prevalent, there is a growing interest in sublinear algorithms — these are algorithms whose resource requirements are substantially smaller than the size of the input that they operate on. In this talk, we will describe some results that illustrate surprising effectiveness of randomization in solving exact and approximate matching problems in sublinear space or time.

Specifically, the first part of the talk will focus on the problem of finding a perfect matching in a regular bipartite graph. Regular bipartite graphs are fundamental objects that have been extensively studied in computer science and mathematics. It is well-known that a regular bipartite graph of degree d can be decomposed into exactly d perfect matchings. The problem of finding a perfect matching in a regular graph has applications to routing in switch fabrics, task-assignment, and edge-coloring of general bipartite graphs. The perfect matching problem in regular bipartite graphs is also closely related to the Birkhoff-von Neumann decomposition of doubly stochastic matrices [1, 9]. The first algorithm for finding a perfect matching in a regular bipartite graph dates back to the work of König in 1916 [8] who gave an $O(mn)$ time algorithm; here n denotes the number of vertices and m denotes the number of edges. The well-known bipartite matching algorithm of Hopcroft and Karp [6] can be used to solve this problem in $O(m\sqrt{n})$ time. However, the degree regularity assumption has allowed researchers to develop more efficient algorithms, and a sequence of improvements over two decades have culminated in an $O(n + m)$ time algorithm by Cole, Ost, and Schirra [2]. We will show that sampling and random walks can be used to obtain sublinear-time exact algorithms for this problem [3–5]. In particular, we will describe a simple algorithm that uses random walks to find a perfect matching in $O(n \log n)$ time, independent of the number of edges in the graph. Our algorithm, which is within an $O(\log n)$ factor of output complexity, also yields a fast algorithm for computing Birkhoff-von-Neumann decomposition of doubly stochastic matrices. These ideas have also been used for rounding fractional flow solutions in recently developed fast algorithms for the maximum flow problem in general graphs.

In the second part of the talk, we will consider the problem of estimating the size of a maximum matching using small space in the streaming model where the edges of the input graph arrive one by one. It is easy to approximate the

A.-H. Dediu et al. (Eds.): LATA 2014, LNCS 8370, pp. 32–33, 2014.

matching size to within a constant factor using $O(n)$ space – one can simply maintain a maximal matching of the input graph. It is also known that one can obtain an $\tilde{O}(\sqrt{n})$ approximation to the matching size using poly-logarithmic space. We show that if the edges of the graph are streamed in a random order, then poly-logarithmic space suffices to estimate the maximum matching size to within a poly-logarithmic factor [7]. Our result is based on a new local algorithm for this problem that may be of independent interest.

References

1. Birkhoff, G.: Tres observaciones sobre el algebra lineal. Univ. Nac. Tucumán Rev. Ser. A 5, 147–151 (1946)
2. Cole, R., Ost, K., Schirra, S.: Edge-coloring bipartite multigraphs in $O(E \log D)$ time. Combinatorica 21(1), 5–12 (2001)
3. Goel, A., Kapralov, M., Khanna, S.: Perfect matchings via uniform sampling in regular bipartite graphs. ACM Transactions on Algorithms 6(2) (2010)
4. Goel, A., Kapralov, M., Khanna, S.: Perfect matchings in $O(n \log n)$ time in regular bipartite graphs. SIAM J. Comput. 42(3), 1392–1404 (2013)
5. Goel, A., Kapralov, M., Khanna, S.: Perfect matchings in $\tilde{O}(n^{1.5})$ time in regular bipartite graphs. To appear in Combinatorica (2013)
6. Hopcroft, J., Karp, R.: An $n^{\frac{5}{2}}$ algorithm for maximum matchings in bipartite graphs. SIAM J. Comput. 2(4), 225–231 (1973)
7. Kapralov, M., Khanna, S., Sudan, M.: On the communication and streaming complexity of maximum bipartite matching. In: Proceedings of the Twenty-Fifth Annual ACM-SIAM Symposium on Discrete Algorithms, SODA (2014)
8. König, D.: Über graphen und ihre anwendung auf determinententheorie und mengenlehre. Math. Annalen 77, 453–465 (1916)
9. von Neumann, J.: A certain zero-sum two-person game equivalent to the optimal assignment problem. Contributions to the Optimal Assignment Problem to the Theory of Games 2, 5–12 (1953)

Interprocedural Information Flow Analysis of XML Processors[*]

Helmut Seidl and Máté Kovács

Technische Universität München, Germany
{seidl,kovacsm}@in.tum.de

Abstract. A crucial issue when providing publicly accessible web services is that sensitive data should only be accessible by authorized users. Accessibility of data within an application or information flow can conveniently be formalized as a 2-hyperproperty of a program. Here, we present a technique to interprocedurally analyze information flow in XML processors. Our approach is based on general techniques for program matching, and relational abstract interpretation of the resulting 2-programs. In case of XML processors, the abstract relational semantics then can be practically analyzed by means of finite tree automata.

1   Introduction

Web services accessed by a variety of users such as a conference management system or the information system of a health insurance company maintain sensible data which should not be accessible to everyone. Therefore, rules are provided to formalize which users are authorized to access which pieces of data. *Information flow analysis* allows to verify that the system complies with the rules given by the access policy by inferring which pieces of observable information may depend on secret data. Such dependencies can conveniently be formalized as *2-hyperproperties* of programs, more precisely, 2-hypersafety properties [6]. 2-Hyperproperties do not refer to single executions of a program but to pairs of executions. Information-flow on the other hand can be formalized by comparing the observations of two executions whose starting states only differ in secret data. No information leaks if unauthorized users cannot observe any difference between the two. This observational indistinguishability has also been called *noninterference* [11].

Consider, e.g., the code fragment in Listing 1 written in the Web Services Business Process Execution Language [1] (BPEL). The fragment is meant to update the health status of patients in a database according to the results of their blood tests. The database is stored in the data structure of variable `pList`. The actual patient is identified by the value in the variable `patientId`, while the update depends on the value of `test`. We consider the blood test and the health status of patients as confidential information, while the rest of the database is

---

[*] This work was partially supported by the German Research Foundation (DFG) under the project SpAGAT (grant no. SE 551/14-2) in the priority program "Reliably Secure Software Systems – RS3".

A.-H. Dediu et al. (Eds.): LATA 2014, LNCS 8370, pp. 34–61, 2014.

```
 1 <if name="If">
 2  <condition> <![CDATA[$test < 0.5]]>
 3  </condition>
 4   <assign name="EvalGood">
 5    <copy> <from>"good"</from>
 6     <to> $pList/patientRecord[id=$patientId]
 7                  /health/text()
 8     </to> </copy>
 9   </assign>
10  <else>
11   <assign name="EvalPoor">
12    <copy> <from>"poor"</from>
13     <to> $pList/patientRecord[id=$patientId]
14                  /health/text()
15     </to> </copy>
16   </assign>
17  </else>
18 </if>
```

Listing 1. A BPEL fragment updating the health status of patients

public. Accordingly, the data structure is *heterogeneous*, in that it stores confidential as well as public information. Our goal is to prove that the value of test only interferes with the health status in the description of the patient, but neither with his name nor with the size of the data structure etc. Such an analysis is challenging, because the control states of the program depend on secret data. In the example, the difference still is not observable, since the two branches perform the same manipulations on public data. However, information flow analyses such as [5, 16, 24] which rely on *tainting* of data elements and program points may not be able to verify the given program. Clarkson and Schneider, on the other hand, have observed that the verification of k-hypersafety properties of a program can be reduced to the verification of ordinary safety properties of the *self-composition* of the program [6]. In this paper we follow the approach of [6] and provide techniques for automatically inferring 2-hypersafety properties by means of *decent* self-compositions of programs. Our observation is that precision can be gained by aligning similar parts of the program and then applying *relational* abstract interpretation.

Related techniques based on the analysis of self-compositions of programs have been presented in [2, 3, 15]. Here, we indicate how this idea can be conveniently extended to languages with procedures. Our exposition is based on the approach of [15] from where we borrow key notions and definitions. The self-composition algorithm there is extended to procedures in such a way that the resulting 2-program can be analyzed by means of standard interprocedural analysis. This means that the two executions performed in parallel are synchronized in a way that they could be implemented on a single runtime stack.

The algorithm uses a *tree distance measure* as a heuristic to align syntactically similar pieces of code fragments with each other. As a result, a call to a procedure *proc* is either aligned with another call to *proc*, or a `skip` instruction. Therefore, in the resulting composition, three procedures are introduced corresponding to *proc* in the original pair of programs, namely, the self-composition [*proc*, *proc*] of *proc*, together with [*proc*, `skip`] and [`skip`, *proc*].

Given a 2-program resulting from the self-composition, we may apply any approach to interprocedural analysis, e.g., the functional approach [9, 14, 21] based on function summaries, or the call string approach advocated in [21] based on abstractions of the call-stack.

In this paper we exemplify the summary-based approach for the case where secrecy is analyzed at the level of program variables. However, in the presence of heterogeneous data structures as in Listing 1 secrecy cannot be successfully analyzed in terms of variables. More generally, web services and business workflows implemented, e.g., in BPEL store data in XML documents and support manipulation of XML documents by means of XPath [4] expressions and XSLT [13] transformations. Accordingly, information flow policies may not only refer to values of variables but may specify the secrecy of individual nodes or subtrees of documents explicitly. In order to discuss these issues, we will not refer to complicated high-level web standards, but consider a simple "assembly" language for tree manipulation where procedures are used, e.g., to evaluate XPath queries. In order to argue relationally about differences of subdocuments, we rely on an abstract domain of *public views* of documents. These are represented by sets of trees while the abstract transformers corresponding to operations on document trees are formalized by means of Horn clauses. The abstract domain of sets of trees seems to be too complicated to allow for concise descriptions of procedure summaries. Therefore, we rely on a simple call string 0 approach where procedure calls and returns are essentially modeled by jumps. Given a 2-program (represented as a control-flow graph), the analysis proceeds in two steps. First, implications are generated to specify the relations of abstract values corresponding to the different nodes in the graph. These implications comply with the format of the subclass of $\mathcal{H}_1$ clauses. The least model of a finite set of such clauses is known to consist of regular sets of trees [18, 25]. Therefore in the second step, the normalization procedure from [18] is applied to compute representations of these sets by means of finite tree automata.

To summarize, this paper has the following contributions:

- An algorithm is introduced for the construction of self-compositions of programs having procedure calls.
- The applicability of our method is demonstrated by proving information flow properties of two examples:
  - In the first example summary functions are used to represent the abstract effect of procedures.
  - In the second example global control flow graphs are applied to prove the information flow security of a tree-manipulating program.

Our exposition uses a minimalistic language in order to simplify the presentation and to emphasize the key features of our techniques.

The rest of this paper is organized as follows. In Section 2 the concepts of the programming language and its semantics are formalized. In Section 3 our technique is introduced to compose pairs of programs, and the relation between programs and their compositions is discussed. Section 4 shows how to apply abstract interpretation on self-compositions of programs. As a first interprocedural analysis, an information flow analysis is presented, which interprocedurally tracks on disequalities of variables using function summarization. The application of global control flow graphs is discussed in Section 5. Finally, in Section 6 we relate our work to others and conclude.

2   Preliminaries

We use a structured programming language extended with procedures in order to present our analyses. We assume that all variables are local. All locals are passed to a called procedure where the result computed by a call is passed to the caller via the dedicated variable ret. The syntax of programs f is given by the following grammar:

$$
\begin{array}{lll}
\text{(declarations)} & f ::= \textbf{procedure } \mathit{proc}\{p\} \, f \mid p \\
\text{(program)} & p ::= \varepsilon \mid c ; p \\
\text{(command)} & c ::= \textbf{skip} \mid x := e \mid \textbf{call } \mathit{proc} \mid \textbf{while } b \, \{p\} \mid \\
& \quad \textbf{if } b \, \{p_{\mathtt{tt}}\} \textbf{ else } \{p_{\mathtt{ff}}\}
\end{array} \tag{1}
$$

Thus, a program consists of a sequence of procedure declarations, followed by the main program. The main program and the bodies of procedures are sequences of commands. A command c is either an assignment, a procedure call, a conditional selection of branches if or a while loop. For convenience there is also a skip command that leaves the program state intact. Expressions are used on the right-hand-side of assignments to compute new values based on already existing ones. In later sections we will instantiate the expression language for particular examples.

The semantics of programs is defined by means of a transition relation $cfg_1 \to cfg_2$ between configurations. The transitive closure of the transition relation is denoted by $\to^*$. A configuration is a tuple $\langle p, s \rangle$, where p is a sequence of commands to be executed on the state s. We denote the set of states by S. The execution of a program has terminated, if the configuration is of the form $\langle \varepsilon, s \rangle$, i.e., the remaining sequence of commands to be executed is empty. We abbreviate the configuration $\langle \varepsilon, s \rangle$ by the state s.

The semantics of the programming language is shown in Figure 1. The semantics of a call to a procedure proc, is defined by means of the functions enter : $S \to S$ and combine : $S \times S \to S$. The function enter constructs the initial state for the execution of the procedure, while combine combines the original state at the call site with the return value computed by the procedure. In case the state is a mapping from variables to values, then all variables are passed as

$$\text{A: } \langle x\texttt{:=}e\texttt{;}p, s\rangle \to \langle p, [\![x\texttt{:=}e]\!]s\rangle \qquad \text{SK: } \langle \texttt{skip;}p, s\rangle \to \langle p, s\rangle$$

$$\text{PC: } \frac{\langle p', \mathsf{enter}(s)\rangle \to^* s' \quad \text{where the procedure definition is: } \textbf{procedure } proc\{p'\}}{\langle \texttt{call } proc\texttt{;}p, s\rangle \to \langle p, \mathsf{combine}(s, s')\rangle}$$

$$\text{WT: } \frac{[\![b]\!]s = \texttt{tt}}{\langle \texttt{while } b \ \{p_{\texttt{tt}}\}\texttt{;}p, s\rangle \to \langle p_{\texttt{tt}} \ \texttt{while } b \ \{p_{\texttt{tt}}\}\texttt{;}p, s\rangle}$$

$$\text{WF: } \frac{[\![b]\!]s = \texttt{ff}}{\langle \texttt{while } b \ \{p_{\texttt{tt}}\}\texttt{;}p, s\rangle \to \langle p, s\rangle}$$

$$\text{IT: } \frac{[\![b]\!]s = \texttt{tt}}{\langle \texttt{if } b \ \{p_{\texttt{tt}}\} \ \texttt{else } \{p_{\texttt{ff}}\}\texttt{;}p, s\rangle \to \langle p_{\texttt{tt}} \ \texttt{if } b \ \{p_{\texttt{tt}}\} \ \texttt{else } \{p_{\texttt{ff}}\}\texttt{;}p, s\rangle}$$

$$\text{IF: } \frac{[\![b]\!]s = \texttt{ff}}{\langle \texttt{if } b \ \{p_{\texttt{tt}}\} \ \texttt{else } \{p_{\texttt{ff}}\}\texttt{;}p, s\rangle \to \langle p_{\texttt{ff}} \ \texttt{if } b \ \{p_{\texttt{tt}}\} \ \texttt{else } \{p_{\texttt{ff}}\}\texttt{;}p, s\rangle}$$

Fig. 1. The semantics of the programming language

parameters, and results are communicated back to the caller in the variable `ret`. Therefore, we have:

$$\begin{aligned} \mathsf{enter}(s) \quad &= s \\ \mathsf{combine}(s, s') &= s[\texttt{ret} \mapsto s'(\texttt{ret})] \end{aligned}$$

Above, $s[x \mapsto v]$ denotes the mapping where for all $y \neq x$ it holds that $s[x \mapsto v] = s(y)$ and $s[x \mapsto v](x) = v$.

3   Self-composition of Programs

The information flow properties we want to prove can be formalized as 2-hypersafety properties [6]. Formally, a 2-hypersafety property is given by two relations on program states, the initial ρ_{in} and final ρ_{fi} relations which specify the pairs of states that should be observationally equivalent. A program satisfies the (end-to-end) 2-hypersafety property, if $(s, t) \in \rho_{in}$ entails that $(s', t') \in \rho_{fi}$ whenever $\langle p, s\rangle \to^* s'$ and $\langle p, t\rangle \to^* t'$. The verification of a hypersafety property of a program on the other hand, can be reduced to the verification of an *ordinary* safety property, but now for a self-composition of the program [3,6,23]. Safety properties can be inferred by means of abstract interpretation [8].

Definition 1. *A program pp is a self-composition of the program p if for all pairs of states s and t it holds that whenever $\langle p, s\rangle \to^* s'$ and $\langle p, t\rangle \to^* t'$ then $\langle pp, (s, t)\rangle \to^* (s', t')$.*

A program satisfying the condition in Definition 1 will be called *2-program*. A 2-program manipulates pairs of states. We construct self-compositions for each procedure and the main program separately. In the case of branching constructs it is possible that different branches are executed on the two members of a pair of states. Therefore, now we discuss how to construct the composition of two potentially different sequences of commands p and q in general. We construct the composition of two programs using two mutually recursive functions:

- p2c(p, q) constructs the composition of the two sequences of commands p and q.
- c2c(c, d) computes the composition of the two commands c and d.

3.1   Composing Two Sequences of Commands

The composition of two sequences $p = c_1; ...; c_k$ and $q = d_1; ...; d_l$ is constructed by the function p2c(p, q). An alignment Ω of the two sequences is a sequence of pairs of commands $(c'_1, d'_1); ...; (c'_m, d'_m)$, where each c'_i and d'_j are either a skip operation or a command occurring in the original sequences. The set of all possible alignments $A(p, q)$ is recursively defined by:

$$
\begin{aligned}
A(\varepsilon, \varepsilon) \quad &= \varepsilon \cup \{(\mathrm{skip}, \mathrm{skip}); \Omega \mid \Omega \in A(\varepsilon, \varepsilon)\} \\
A(\varepsilon, d; q) \quad &= \{(\mathrm{skip}, d); \Omega \mid \Omega \in A(\varepsilon, q)\} \cup \\
&\quad\ \{(\mathrm{skip}, \mathrm{skip}); \Omega \mid \Omega \in A(\varepsilon, d; q)\} \\
A(c; p, \varepsilon) \quad &= \{(c, \mathrm{skip}); \Omega \mid \Omega \in A(p, \varepsilon)\} \cup \\
&\quad\ \{(\mathrm{skip}, \mathrm{skip}); \Omega \mid \Omega \in A(c; p, \varepsilon)\} \qquad\qquad (2)\\
A(c; p, d; q) &= \{(c, d); \Omega \mid \Omega \in A(p, q)\} \cup \\
&\quad\ \{(\mathrm{skip}, d); \Omega \mid \Omega \in A(c; p, q)\} \cup \\
&\quad\ \{(c, \mathrm{skip}); \Omega \mid \Omega \in A(p, d; q)\} \cup \\
&\quad\ \{(\mathrm{skip}, \mathrm{skip}); \Omega \mid \Omega \in A(c; p, d; q)\}
\end{aligned}
$$

In order to identify a *decent* alignment, we proceed as in [15] by defining the function p2c. This function chooses a best alignment Ω_{opt} according to some *tree distance measure* td for pairs of commands. Accordingly, we set:

$$
\Omega_{opt} = \arg\min_{\Omega \in A(p,q)} \sum_{i=1}^{|\Omega|} \mathrm{td}(\Omega[i].1, \Omega[i].2)
$$

where $\Omega[i]$ stands for the ith pair in the sequence, and $\Omega[i].1$ and $\Omega[i].2$ stand for the first and second members of the pair, respectively. The best alignment Ω_{opt} according to td is the one, where the sum of the distances between the abstract syntax trees of the members of pairs is minimal. In our implementation, we use for td the *Robust Tree Edit Distance* of [19]. Finally the function c2c$(\Omega_{opt}[i].1, \Omega_{opt}[i].2)$ is called for all pairs of commands in the best alignment in order to construct the corresponding fragment in the composition of the sequences.

3.2   Composing Two Commands

In this section we discuss how the composition of two commands is constructed. The composition of two commands is a fragment of a 2-program that operates on pairs of states in order to meet the requirements of Definition 1.

We regard two commands as *composable*, if both of them are either if constructs, while constructs, skip instructions, invocations of the same procedure, or syntactically equal assignments. In our implementation we furthermore require from branching constructs that their conditions are also syntactically equal.

First, we treat the case when the pair of commands are *not* composable. For this we make use of two functions $\mathsf{skip1}(c)$ and $\mathsf{skip2}(c)$. In case c is an assignment, then $\mathsf{skip1}(c) = $ [skip,c] and $\mathsf{skip2}(c) = $ [c,skip]. In case c is a call to some procedure *proc*, then we set: $\mathsf{skip1}(c) = $ call [skip,*proc*] and $\mathsf{skip2}(c) = $ call [*proc*,skip]. In case c is a branching construct, then skip1 replaces the conditional expression b with [true,b], skip2 replaces b with [b,true] and both of them transform the subtrees of the corresponding ASTs recursively. Accordingly, we have:

$$\mathsf{skip1}(\text{if } b \ \{p_{\mathrm{tt}}\} \text{ else } \{p_{\mathrm{ff}}\}) = \text{if } [\texttt{true},b] \ \{\mathsf{skip1}(p_{\mathrm{tt}})\} \text{ else } \{\mathsf{skip1}(p_{\mathrm{ff}})\}$$
$$\mathsf{skip1}(\text{while } b \ \{p\}) = \text{while } [\texttt{true},b] \ \{\mathsf{skip1}(p)\}$$
$$\mathsf{skip2}(\text{if } b \ \{p_{\mathrm{tt}}\} \text{ else } \{p_{\mathrm{ff}}\}) = \text{if } [b,\texttt{true}] \ \{\mathsf{skip2}(p_{\mathrm{tt}})\} \text{ else } \{\mathsf{skip2}(p_{\mathrm{ff}})\}$$
$$\mathsf{skip2}(\text{while } b \ \{p\}) = \text{while } [b,\texttt{true}] \ \{\mathsf{skip2}(p)\}$$

The application of skip1 and skip2 on sequences $p = c_1; c_2;, \ldots$ is just syntactic sugar for applying them to each command individually, i.e., $\mathsf{skip1}(c_1; c_2;, \ldots) = \mathsf{skip1}(c_1); \mathsf{skip1}(c_2); \ldots$

Now we can define the composition of two commands c and d that are not composable by:

$$\mathsf{c2c}(c, d) = \mathsf{skip1}(c); \mathsf{skip2}(d)$$

In other words, the composition of two commands that are not composable is a sequence of two pairs of commands, where the first manipulates the first member of the pair of states and the second manipulates the second member of the pair of states.

```
 1 if  [b₁,b₂]  {
 2     p2c(p_tt, q_tt)
 3 } else  {
 4     if  [¬b₁,b₂]  {
 5         p2c(p_ff, q_tt)
 6     } else  {
 7         if  [b₁,¬b₂]  {
 8             p2c(p_tt, q_ff)
 9         } else  {
10             p2c(p_ff, q_ff)
11         };
12     };
13 };
```

Listing 2. The composition of two if constructs

Now we consider the case when the two commands are composable. For two equal assignments or procedure calls we have that:

$$c2c(x\!:=\!e, x\!:=\!e) \qquad\qquad = [x\!:=\!e, x\!:=\!e]$$
$$c2c(\texttt{call } proc, \texttt{call } proc) = \texttt{call } [proc, proc]$$

The composition of if b_1 $\{p_{\texttt{tt}}\}$ else $\{p_{\texttt{ff}}\}$ and if b_2 $\{q_{\texttt{tt}}\}$ else $\{q_{\texttt{ff}}\}$ is as follows:

As shown by Listing 2, the composition of two if constructs consists of three if constructs embedded into each other. The idea is that a composition of the bodies of the original branching constructs should be generated for all possible evaluations of the conditional expressions. The composition of while b_1 $\{p\}$ and while b_2 $\{q\}$ is given by Listing 3. By composing two loops we need to take into consideration that it is possible that they are executed a different number of times. Therefore, the first loop at line 1 in Listing 3 handles the case when the original loops execute in synchrony, while the two loops at lines 2 and 3 handle the case when one of the loops has already terminated, but the other has not.

```
1 while  [b₁,b₂]  {  p2c(p,q) };
2 while  [¬b₁,b₂]  {  skip1(p) };
3 while  [b₁,¬b₂]  {  skip2(q) };
```

Listing 3. The composition of two while loops

3.3   Handling Procedures

In the composition of sequences of commands we may find commands of the following forms:

$$\texttt{call } [proc, proc]$$
$$\texttt{call } [\texttt{skip}, proc]$$
$$\texttt{call } [proc, \texttt{skip}]$$

The first command executes the self-composition of the procedure identified by $proc$, the other two model the case when only one of the pair of executions calls the procedure. Therefore, when constructing the self-composition of the main program, self-compositions of the procedures need to be constructed as well where each procedure of the original program gives rise to three procedures in the resulting 2-program. For each procedure declaration procedure $proc\{p\}$ we generate:

$$\texttt{procedure } [proc, proc]\{\texttt{p2c}(p, p)\}$$
$$\texttt{procedure } [\texttt{skip}, proc]\{\texttt{skip1}(p)\}$$
$$\texttt{procedure } [proc, \texttt{skip}]\{\texttt{skip2}(p)\}$$

The body of the procedure $[proc, proc]$ is the self-composition of the body of the original procedure. The other two take care of the case when only one member of a pair of commands is a procedure call and the other is skip.

3.4   Semantics of Self-compositions of Programs

For pairs of Boolean expressions, we define $[\![\,[b_1,b_2]\,]\!](s,t) = [\![b_1]\!]s \wedge [\![b_2]\!]t$. In case of assignments c or skip we set $[\![\,[c,d]\,]\!](s,t) = ([\![c]\!]s, [\![d]\!]t)$. In case of procedure calls the semantics is defined analogously to the original semantics of procedures. Given that a procedure declaration in the composition is procedure r $\{pp\}$ we define $[\![\text{call } r]\!]^{\sharp}(s,t) = \text{combine}_2((s,t),(s',t'))$ where $\langle pp, \text{enter}_2(s,t) \rangle \rightarrow^{*}$ (s',t'). If both states are mappings from variables to values, they are given by:

$$\text{enter}_2(s,t) = (s,t)$$
$$\text{combine}_2((s,t),(s',t')) = (s[\text{ret} \mapsto s'(\text{ret})], t[\text{ret} \mapsto t'(\text{ret})])$$

In general, the functions enter_2, combine_2 should meet the following requirements:

$$\text{enter}_2(s,t) \qquad\qquad = (\text{enter}(s), \text{enter}(t))$$
$$\text{combine}_2((s,t),(s',t')) = (\text{combine}(s,s'), \text{combine}(t,t'))$$

Finally, we define the 2-program $\text{p2c}(f,f)$ for a program f as the collection of procedure declarations for $[proc, proc]$, $[\text{skip}, proc]$, $[proc, \text{skip}]$ ($proc$ declared in f), together with $\text{p2c}(p,p)$ for the main program p of f. Then the following relation between the semantics of a sequence of commands and the self-composition which we have constructed can be proven by induction on the abstract syntax:

Theorem 2. *Any sequence p of commands of a program f and its self-composition $\text{p2c}(p,p)$ satisfy the condition of Definition 1.*   □

By Theorem 2, the information flow analysis of an ordinary program f with procedures can be reduced to inferring interprocedural invariants for the 2-program $\text{p2c}(f,f)$. In the next section we review two examples of such interprocedural analyses.

4   Relational Interprocedural Analysis of Self-compositions of Programs

In order to analyze the self-composition of a program it is convenient to represent the main program as well as the declared procedures of the self-composition by means of control flow graphs (CFG). These CFGs differ from CFGs for ordinary programs only in that edges are not labeled with single assignments, guards or procedures, but with *pairs* of such actions. Thus, one such CFG is a tuple $G = (N, E, n_{in}, n_{fi})$, where N is a finite set of nodes, E is a set of directed and labeled edges, and n_{in} and n_{fi} are the initial and final nodes, respectively. The nodes of the CFG stand for sequences of (pairs of) commands pp that need to be executed on states in a configuration $\langle pp, (s,t) \rangle$, while the edges of the CFG represent the transition relation $\rightarrow$. Whenever $\langle c; pp, (s,t) \rangle \rightarrow \langle pp, (s',t') \rangle$ then the corresponding edge of the CFG is labeled by c if the command is an assignment or a call. Guards bb and $\neg bb$ are used as labels for edges entering the then and else branches of if commands with condition bb, and accordingly, for entering the bodies of while loops with condition bb or exiting those loops, respectively. In the concrete semantics, each label l of an edge corresponds to a state transformer $[\![l]\!] : S \times S \to S \times S$. These state transformers are partial functions, i.e., the guards that hold on a pair of states propagate the pair unmodified, while the result is undefined on those states on which the guard does not hold.

In order to perform abstract interpretation for self-compositions of programs, we choose a complete lattice $(\mathbb{D}, \sqsubseteq)$ whose elements represent the potential invariants at program points. We use a function γ to map an abstract value $D \in \mathbb{D}$ to the set pairs of states for which the invariant D holds. As usual, the function γ is called *concretization*.

Example 1. Assume that we only want to track the set of program variables where the two states possibly differ. For that, we choose $\mathbb{D}$ as the complete lattice $\mathcal{P}(\mathcal{X})_{\perp} = \{\perp\} \cup \mathcal{P}(\mathcal{X})$ consisting of a least element $\perp$ (denoting unreachability) together with all subsets of the set of program variables $\mathcal{X}$. The ordering relation is given by $\perp \sqsubseteq D$ for all $D \in \mathbb{D}$, and for $D \neq \perp \neq D'$, $D \sqsubseteq D'$ iff $D \subseteq D'$.

The concretization $\gamma(V)$ of an abstract value V is then defined by:

$$\gamma(V) = \{(s,t) \mid \forall x \in \mathcal{X} : x \notin V \Rightarrow s(x) = t(x)\}$$
$$\gamma(\perp) = \emptyset \tag{3}$$

Besides a complete lattice of potential invariants, transfer functions $[\![l]\!]^{\sharp}$ are needed for the labels l at control flow edges, which describe how the potential invariants are affected. For a transfer function $[\![l]\!]^{\sharp}$, we require that for all abstract values D and pairs of states (s,t) with $(s,t) \in \gamma(D)$ the following holds:

$$[\![l]\!](s,t) \in \gamma([\![l]\!]^{\sharp} D)$$

Example 2. In the case when only potential disequalities are tracked, all transfer functions $[\![l]\!]^{\sharp}$ preserve the abstract value $\perp$. For $V \neq \perp$, we could have:

$$[\![x\!:=\!e, x\!:=\!e]\!]^{\sharp}(V) = (\mathsf{vars}(e) \cap V \neq \emptyset) ? (V \cup \{x\}) : (V \backslash \{x\})$$
$$[\![x\!:=\!e, \mathsf{skip}]\!]^{\sharp}(V) = V \cup \{x\}$$
$$[\![\mathsf{skip}, x\!:=\!e]\!]^{\sharp}(V) = V \cup \{x\}$$
$$[\![\neg b, b]\!]^{\sharp}(V) \quad = (\mathsf{vars}(b) \cap V \neq \emptyset) ? V : \perp$$
$$[\![b, \neg b]\!]^{\sharp}(V) \quad = (\mathsf{vars}(b) \cap V \neq \emptyset) ? V : \perp$$
$$[\![bb]\!]^{\sharp}(V) \quad = V \qquad \text{for any other guard } bb$$

Here, we use the "? :" operator as in C for representing a conditional choice, i.e.,

$$b ? x : y = \begin{cases} x & \text{if } b \text{ holds} \\ y & \text{otherwise} \end{cases}$$

Note that all listed transformers are indeed *monotonic*. The intuition behind these definitions are as follows. Whenever both assignments are syntactically equal, then the same value will be assigned to the left-hand side variable x, if all variables occurring on the right-hand side agree in their respective values. In this case, x must be removed from the argument set V. Otherwise, it must be added. On the other hand when guards are concerned, we can be sure that the condition is not met only when the guard is of the form $[b,\neg b]$ or $[\neg b,b]$, and all variables occurring in b agree in their respective values. In this case, the guard definitely cannot be taken and therefore $\perp$ can be returned. In all other cases, no information can be extracted.

Besides transformers for assignments and guards, we can also provide abstract enter and combine functions. For the analysis of Example 1, both functions preserve $\perp$ (in each argument); for non-$\perp$ arguments, their results are given by:

$$\begin{aligned}
\mathsf{enter}_2^\sharp(V) &= V \\
\mathsf{combine}_2^\sharp(V_1, V_2) &= (V_1 \setminus \{\texttt{ret}\}) \cup (V_2 \cap \{\texttt{ret}\})
\end{aligned}$$

Since the function enter_2 simply forwards the state to the called procedure, its abstract variant is the identity function as well. Because the values of variable $\texttt{ret}$ in the resulting pair of concrete states only depend on the result of the procedure call, it is first removed from the initial abstract value V_1. In case the results of a pair of procedure calls may differ in the variable $\texttt{ret}$, then this variable may contain different values after evaluating $\mathsf{combine}_2$. Therefore, its abstract variant adds $\texttt{ret}$ to the set of variables possibly affected by the secret.

Although we are thus given abstract versions of enter_2 and $\mathsf{combine}_2$, the abstract transformers for call edges are still not yet known: they additionally depend on the abstract semantics of the called procedures. In order to compute these, we introduce for each program point n of a CFG the unknown $[\![n]\!]^\sharp$, which denotes the transformer that transforms the abstract value at the entry point of the DFG into the abstract value at program point n. These abstract transformers can jointly be approximated by the least solution of the following constraint system:

$$\begin{aligned}
[\![n_{in}]\!]^\sharp &\sqsupseteq \mathsf{Id} && \text{For the initial node } n_{in} \text{ of some CFG.} \\
[\![n]\!]^\sharp &\sqsupseteq H^\sharp([\![n_{fi}^r]\!]^\sharp) \circ [\![m]\!]^\sharp && \text{For an edge: } (m, \texttt{call } r, n) \quad \text{where } n_{fi}^r \text{ is the} \\
& && \text{final node of the CFG of procedure } r. \qquad (4) \\
[\![n]\!]^\sharp &\sqsupseteq [\![l]\!]^\sharp \circ [\![m]\!]^\sharp && \text{For an edge: } (m, l, n) \quad \text{where } l \text{ is a guard or} \\
& && \text{an assignment.}
\end{aligned}$$

Here, Id denotes the identity function, $\circ$ denotes function composition and $H^\sharp$ is an operator which takes the effect of a procedure body and transforms it into the effect of a call. Thus, $H^\sharp$ is defined by:

$$H^\sharp \; g \; D = \mathsf{combine}_2^\sharp(D, g(\mathsf{enter}_2^\sharp(D)))$$

This constraint system does no longer speak about the complete lattice $\mathbb{D}$ itself but about the set of monotonic functions in $\mathbb{D} \to \mathbb{D}$. This set, however, can again be considered as a complete lattice w.r.t. the argumentwise ordering on functions. Likewise, if $\mathbb{D}$ is finite (as in our example) then also the set $\mathbb{D} \to \mathbb{D}$ is finite. Therefore, the least fixpoint solution of the constraint system can be computed by ordinary Kleene iteration.

Now assume that we are given a set V_0 of variables which at program start may contain secrets. Furthermore, assume that h is the abstract transformer computed by the analysis for the exit point of the main program. If $h(V_0) = \bot$, then we can be sure that program exit cannot be simultaneously reached by two executions starting with initial states $(s_0, t_0) \in \gamma(V_0)$. If $h(V_0) \neq \bot$, then all variables that may potentially leak information about the initial secrets are contained in the set $h(V_0)$.

```
1  procedure fib {
2      x := xin;
3      if x = 0 { ret := 0; }
4      else {
5          if x = 1 { ret := 1; }
6          else {
7              one := 1;
8              xin := x-one;
9              call fib;
10             r1 :- ret;
11             xin := xin-one;
12             call fib;
13             r2 := ret;
14             ret := r1+r2;
15         };
16     };
17 }
18
19 xin := secin;
20 call fib;
21 secfib := ret;
22 xin := pubin
23 call fib;
24 pubfib := ret;
```

Listing 4. A program computing Fibonacci numbers

Let us, for example, consider the recursive program from Listing 4 which computes two members of the Fibonacci series from the numbers in variables `secin` and `pubin` of the initial state, respectively. The corresponding two return values then are stored in the variables `secfib` and `pubfib` of the final state, respectively.

```
1  procedure [skip,fib] {
2      [skip,x := xin];
3      if [true,x = 0} { [skip,ret := 0]; }
4      else {
5          if [true,x = 1] { [skip,ret:= 1]; }
6          else {
```

```
 7          [skip,one  := 1];
 8          [skip,xin  := x-one];
 9          call [skip,fib];
10          [skip,r1  := ret];
11          [skip,xin  := x-one];
12          call [skip,fib];
13          [skip,r2  := ret];
14          [skip,ret  := r1+r2];
15       };
16    };
17 }
18
19 procedure [fib,skip] {
20    [x  := xin,skip];
21    if [x = 0,true] { [ret := 0,skip]; }
22    else {
23        if [x = 1,true] { [ret := 1,skip]; }
24        else {
25            [one  := 1,skip];
26            [xin  := x-one,skip];
27            call [fib,skip];
28            [r1  := ret,skip];
29            [xin  := x-one,skip];
30            call [fib,skip];
31            [r2  := ret,skip];
32            [ret  := r1+r2,skip];
33        };
34    };
35 }
36
37 procedure [fib,fib] {
38    [x:=xin,x:=xin];
39    if [x = 0,x = 0] {   [ret := 0,ret := 0];
40    } else {
41        if [¬ x=0, x=0] {
42            [skip,ret:=0];
43            if [x = 1,true] { [ret := 1,skip]; }
44            else {
45                [one  := 1,skip];
46                [xin  := x-one,skip];
47                call [fib,skip];
48                [r1  := ret,skip];
49                [xin  := xin-one,skip];
50                call [fib,skip];
51                [r2  := ret,skip];
52                [ret  := r1+r2,skip];
53            };
54        } else {
55            if [x=0, ¬ x=0] {
56                [ret  := 0,skip]
```

```
57          if [true,x = 1] { [skip,ret:= 1]; }
58          else {
59              [skip,one := 1];
60              [skip,xin := x-one];
61              call [skip,fib];
62              [skip,r1 := ret];
63              [skip,xin := xin-one];
64              call [skip,fib];
65              [skip,r2 := ret];
66              [skip,ret := r1+r2];
67          };
68      } else {
69          if [x = 1,x = 1] { [ret := 1,ret := 1]; }
70          else {
71              if [¬ x=1,x = 1] {
72                  [skip,ret := 1];
73                  [one := 1,skip];
74                  [xin := x-one,skip];
75                  call [fib,skip];
76                  [r1 := ret,skip];
77                  [xin := xin-one,skip];
78                  call [fib,skip];
79                  [r2 := ret,skip];
80                  [ret := r1+r2,skip];
81              } else {
82                  if [x = 1, x=1] {
83                      [ret:= 1,skip];
84                      [skip,one := 1];
85                      [skip,xin := x-one];
86                      call [skip,fib];
87                      [skip,r1 := ret];
88                      [skip,xin := xin-one];
89                      call [skip,fib];
90                      [skip,r2 := ret];
91                      [skip,ret := r1+r2];
92                  } else {
93                      [one := 1,one := 1];
94                      [xin := x-one,xin := x-one];
95                      call [fib,fib];
96                      [r1 := ret,r1 := ret];
97                      [xin := xin-one,xin := xin-one];
98                      call [fib,fib];
99                      [r2 := ret,r2 := ret];
100                     [ret := r1+r2,ret := r1+r2];
101                 };
102             };
103         };
104     };
105 };
106 };
```

```
107 }
108
109 [xin := secin,xin := secin];
110 call [fib,fib];
111 [secfib := ret,secfib := ret];
112 [xin := pubin,xin := pubin];
113 call [fib,fib];
114 [pubfib := ret,pubfib := ret];
```

Listing 5. The self-composition of the program in Listing 4

The self-composition of the program in Listing 4 is shown in Listing 5.

According to the abstract semantics, we can compute the abstract transformers for the end points of the procedures [fib,fib], [skip,fib], and [fib,skip]. Clearly, all of them map the bottom element ⊥ to ⊥. Furthermore, whenever xin is not contained in the argument set V, then ret will not be member of resulting abstract value. On the other hand if xin is member of the argument V, then ret will also be member of the resulting abstract value. As a consequence, the secret value in secin does only influence the variable secfib at program exit, but does not influence the variable pubfib.

5   Analysis Using Global Control Flow Graphs

Summary based interprocedural analysis may be difficult if the complete lattice $\mathbb{D}$ for expressing relational invariants is infinite. An example for this is the analysis of XML processing programs from [15]. Still, this analysis can be extended to programs with procedures by directly abstracting the occurring call-stacks. Here, we only consider the simplest variant of this approach where each call-stack is abstracted by a call site, i.e., the topmost element of the stack. For each node $n \in N$, we then introduce an unknown $D[n]$ representing the abstract value from $\mathbb{D}$ describing an invariant satisfied by all pairs (s, t) of concrete states reaching node n. In order to determine these values, we again put up a constraint system consisting of the following constraints:

$$D[n_{in}] \sqsupseteq D_0$$
$$D[n] \quad \sqsupseteq [\![l]\!]^{\sharp}(D[m]) \qquad \text{For each edge } (m, l, n) \text{ where } l \neq \texttt{call } r.$$

Here, D_0 is the abstract value for pairs of initial states, and l is not a call edge. Moreover for every call edge $(m, \texttt{call } r, n)$, we add the constraints:

$$D[n_{in}^r] \sqsupseteq \mathsf{enter}_2^{\sharp}(D[m])$$
$$D[n] \quad \sqsupseteq \mathsf{combine}_2^{\sharp}(D[m], D[n_{fi}^r])$$

Thus at the price of potential loss in precision, it suffices to compute with values from $\mathbb{D}$ alone — without resorting to abstract lattices of functions.

Let us apply this approach to an interprocedural information flow analysis of programs manipulating tree-structured data. For the sake of simplicity our

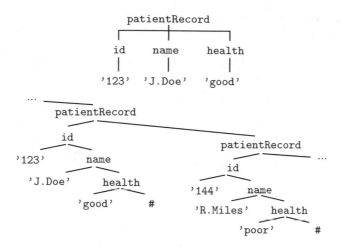

Fig. 2. A record describing a patient in unranked form (top), and its binary representation as a member of a list (bottom)

programming language is designed to manipulate binary trees. There is no loss of generality involved, since XML documents can be encoded as binary trees, e.g., by means of the *first-child-next-sibling* (FCNS) [7] encoding. Figure 2 shows an unranked tree describing a patient in a database, and its binary representation as member of a list. The FCNS encoding maps the first child of a node to its first child, and its sibling to the right to its second child. The symbol # stands for empty unranked forests. It is a requirement that the set of labels of binary and nullary nodes are disjunct. In order to achieve this we put labels of nullary nodes other than # between ' signs and call them *basic values*. We denote the set of labels for binary nodes with Σ_2 and the set of labels for nullary nodes with Σ_0.

$$
\begin{aligned}
&\text{(tree expression)} &e &::= \# \mid x \mid x/1 \mid x/2 \mid \sigma_2(x,y) \mid \lambda_t(x_1, x_2, ...) \\
&\text{(Boolean expression)} &b &::= \texttt{top}(x) = \sigma \mid \lambda_b(x_1, x_2, ...)
\end{aligned}
\tag{5}
$$

The instantiation of the expression language of the programming language shown in (1) can be found in (5). Now there are tree expressions on the right-hand-sides of assignments. A tree expression can be the binary representation of an empty forest #, the value of a variable x, the first or second child of the tree in a variable denoted as $x/1$ or $x/2$ respectively, and a new tree $\sigma_2(x,y)$ composed of two existing trees and a binary alphabet element. λ_t stands for an interpreted tree function which is provided by the execution environment returning a basic value. A Boolean expression can either check whether the root of a tree is labeled with a specific alphabet element, or it can be an interpreted Boolean expression λ_b provided by the execution environment. An example for an interpreted Boolean expression is test < '0.5'. The value of the expression is true if the tree-structured value in the variable test consists of a single leaf labeled with a string that can be interpreted as a floating-point number, and its value is less then 0.5.

The execution state s of the programming language typically is a mapping $s : \mathcal{X} \rightarrow \mathfrak{B}_{\Sigma_2,\Sigma_0} \cup \{\xi\}$ from the set of variables $\mathcal{X}$ to the set of binary trees $\mathfrak{B}_{\Sigma_2,\Sigma_0}$ over the binary alphabet Σ_2 and nullary alphabet Σ_0. It can, however, also equal to ξ — flagging that a runtime error occurred. This happens when the child of a nullary node is queried by an expression of the form $x/1$ or $x/2$.

$$
\text{for all } e \text{ it holds that } [\![x\!:=\!e]\!]\xi = \xi
$$
$$
[\![x\!:=\!y]\!]s = s[x \mapsto s(y)] \qquad [\![x\!:=\!\#]\!]s = s[x \mapsto \#]
$$
$$
[\![x\!:=\!\sigma(x_1,x_2)]\!]s = s[x \mapsto \sigma(s(x_1), s(x_2))]
$$
$$
[\![x\!:=\!y/1]\!]s = \begin{cases} s[x \mapsto \tau_1] & \text{if } s(y) = \sigma(\tau_1,\tau_2) \text{ for some label } \sigma \\ & \text{and trees } \tau_1 \text{ and } \tau_2 \\ \xi & \text{otherwise} \end{cases} \tag{6}
$$
$$
[\![x\!:=\!y/2]\!]s = \begin{cases} s[x \mapsto t_2] & \text{if } s(y) = \sigma(\tau_1,\tau_2) \text{ for some label } \sigma \\ & \text{and trees } \tau_1 \text{ and } \tau_2 \\ \xi & \text{otherwise} \end{cases}
$$
$$
[\![x\!:=\!\lambda_t(x_1,x_2,...)]\!] = s[x \mapsto [\![\lambda_t]\!](s(x_1), s(x_2), ...)]
$$

The formal semantics of assignments is shown in (6). The error state is propagated by assignments without modification.

Now we define the semantics of Boolean expressions. $[\![\mathtt{top}(x)\!=\!\sigma]\!]s = s$ if $s(x) = \sigma(\tau_1,\tau_2)$ for some trees τ_1 and τ_2, or if $\sigma = \#$ and $s(x) = \#$. The basic values of leaves can be examined using interpreted Boolean expressions. The fact whether $[\![\lambda_b(x_1,x_2,...)]\!]s = s$ holds, i.e., if an interpreted Boolean expression holds on a state depends on the predefined semantics in the runtime environment.

Boolean expressions propagate the error state only if they are negated, i.e., they are of the form $\neg b$. The results of positive Boolean expressions on the error state is undefined. This behavior is necessary in order to ensure that loops terminate in case an error occurs.

The only requirement for interpreted functions λ_t and λ_b is that they are deterministic, and do not result in the error state.

We define now the semantics of procedure calls. By entering a procedure the values of variables are unchanged. However, if the state is erroneous, then the procedure is not executed. Therefore we have:

$$
\mathsf{enter}(s) = \begin{cases} s & \text{if } s \neq \xi \\ & \text{otherwise undefined} \end{cases}
$$

By returning from a procedure execution there are two possibilities. If there was no error during the procedure, then the value of variable $\mathtt{ret}$ is updated with that at the end of the procedure execution, the values of other variables are propagated from the call site. On the other hand, if there was an error during the execution of the procedure, then the error state needs to be propagated.

Therefore we have:

$$\text{combine}(s, s') = \begin{cases} s[\text{ret} \mapsto s'(\text{ret})] & \text{if } s \neq \frac{1}{4} \text{ and } s' \neq \frac{1}{4} \\ \frac{1}{4} & \text{otherwise} \end{cases}$$

As an example consider the program from the introduction, now compiled to our assembly language. The program is meant to manipulate the database of an insurance company. The database consists of a list of records as it is shown on the bottom of Figure 2. The program updates the health status of patients depending on the results of blood tests as shown by Listing 6.

```
1  procedure query {
2    empty := #;
3    found := false(empty,empty);
4    while top(found) = false {
5      ret := pList/1;
6      idVal := ret/1;
7      if idVal = patientId {
8        found := true(empty,empty);
9      } else {
10        pList := pList/2;
11      };
12    };
13 }
14
15 procedure update {
16    empty :=#;
17    ret := #;
18    newEntIdVal := newEntry/1;
19    while top(pList) = patientRecord {
20      entry:= pList/1;
21      entIdVal := entry/1;
22      if entIdVal = newEntIdVal {
23        ret := patientRecord(newEntry,ret);
24      } else {
25        ret := patientRecord(entry,ret);
26      };
27      pList := pList/2;
28    };
29 }
30
31 procedure createNewEntry{
32    idVal   := oldEntry/1;
33    name    := oldEntry/2;
34    nameVal := name/1;
35
36    newHealth := health(newHealthVal,empty);
37    newName   := name(nameVal,newHealth);
38    ret       := id(idVal,newName);
39 }
```

```
40
41 empty   :=   #;
42 if  test  <  0.5  {
43     call  query;
44     entry   := ret;
45     newHealthVal  :=   'good';
46     call  createNewEntry;
47     newEntry  := ret;
48     call  update;
49     pList   := ret;
50 } else {
51     call  query;
52     entry   := ret;
53     newHealthVal   :=   'poor';
54     call  createNewEntry;
55     newEntry  := ret;
56     call  update;
57     pList  := ret;
58 };
```

Listing 6. A routine for updating the database of a hospital

The main program starts at line 41. At line 42 a branching decision is made based on the result of the blood test. In both branches, first the database entry with the appropriate identifier is queried using the procedure `query`. The procedure iterates over the list of entries in the database and returns the one with the appropriate identifier. Then, a new entry is constructed using the procedure `createNewEntry`, where the health status is updated according to the result of the test. And finally, the old entry is exchanged with the new one in the database using the procedure `update`.

In our example we consider the result of the blood test and the health status of patients as confidential. Therefore, the goal of our analysis is to prove that these values do not interfere with the names and identifiers of patients, the shape and size of the database etc.

We use *public views* of documents in order to identify subdocuments possibly depending on confidential data. Figure 3 shows a pair of documents and a corresponding public view. Thus, we consider a public view as an abstraction of pairs of trees. A pair of trees (τ_1, τ_2) is a member of the concretization $\gamma(\tau)$ of a public view τ, if τ can be constructed from the pair of concrete trees by replacing

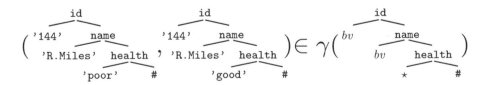

Fig. 3. A pair of documents and a public view of them

potentially different subtrees with a leaf labeled $\star$, and by replacing leaves labeled with equal basic values with leaves labeled bv. Accordingly, a public view is a tree over the binary alphabet Σ_2 and the nullary alphabet $\{\#, bv, \star\}$.

An abstract description of a relation of states is a tuple (T, E), where $T : \mathcal{X} \to P(\mathfrak{B}_{\Sigma_2, \{\#, bv, \star\}})$ is a mapping from variables to sets of public views, and E is an element of the 3-point lattice $\{\checkmark \sqsubset \frac{t}{t} \sqsubset \top\}$. E is necessary in order to represent potential runtime errors. A pair of states (s, t) is a member of the relation given by $(T, \checkmark)$ (denoted[1] $(s, t) \in \gamma(T, \checkmark)$) whenever for each variable $x \in \mathcal{X}$ there is a $\tau \in T(x)$ so that $(s(x), t(x)) \in \gamma(\tau)$. In case $E = \frac{t}{t}$, then $(\frac{t}{t}, \frac{t}{t})$ is also member of the relation. This represents the situation when a runtime error could have potentially occurred independently of the secret. On the other hand, $E = \top$ represents that a runtime error might have occurred *depending* on the secret. Therefore, in this case pairs of states are also members of the relation where only one member is the error state $\frac{t}{t}$.

Given an initial abstract value $(T_0, \checkmark)$ describing the set of public views of pairs of potential initial states, we are interested in computing the sets of potential public views for every reachable node of the CFG of the self-composition of the program. In our analysis, sets of public views for variables x occurring at a node n are described by means of unary predicates $\mathrm{var}_{x,n}$, which are defined by means of Horn clauses. Formally, $\tau \in T(x)$ at node n if $\mathrm{var}_{x,n}(\tau)$ holds. The value of E at node n is represented by means of the unary predicate error_n.

Horn clauses are used for the specification of the analysis in two ways. First, they are used to specify the set of potential initial public views for each variable at the initial node n_{in}. Secondly, they are used to formalize how the views at different program points are related to each other. These clauses are obtained from the CFG of the 2-program.

For our analysis, we assume that information flow policies are defined by *regular sets* of public views, and thus can be described by finite tree automata. We will not define finite tree automata here, but note that their languages can be defined by means of clauses of the form:

$$\mathrm{p}(\sigma_0). \quad \text{or} \quad \mathrm{p}(\sigma_2(X_1, X_2)) \Leftarrow \mathrm{p}_1(X_1), \mathrm{p}_2(X_2).$$

Here, p, p_1 and p_2 are unary predicates corresponding to the states of the automaton, and σ_0 and σ_2 are nullary and binary constructors respectively.

In our running example, the health status of patients in the list of records seen in Figure 2 is confidential. Thus, the information flow policy, the set of public views of all possible databases is given by the predicate qPlist defined by the implications in (7):

---

[1] Here we overload the notation γ and use it for the concretization of public views of document trees, and for the concretization of abstract descriptions of pairs of states.

$$\text{qPlist(patientRecord(L,R))} \Leftarrow \text{qId(L),qPlist(R)}.$$
$$\text{qPlist(\#)}.$$
$$\text{qId(id(L,R))} \Leftarrow \text{qBV(L),qName(R)}. \tag{7}$$
$$\text{qName(name(L,R))} \Leftarrow \text{qBV(L),qHealth(R)}.$$
$$\text{qHealth(health(L,R))} \Leftarrow \text{qStar(L),qEmpty(R)}.$$
$$\text{qBV}(bv). \ \text{qStar}(\star). \ \text{qEmpty(\#)}.$$

qPlist(patientRecord)

qId(id)   qPlist(patientRecord)

qBV(bv) qName(name)   ...   qPlist(#)

qBV(bv) qHealth(health)

qStar($\star$) qEmpty(#)

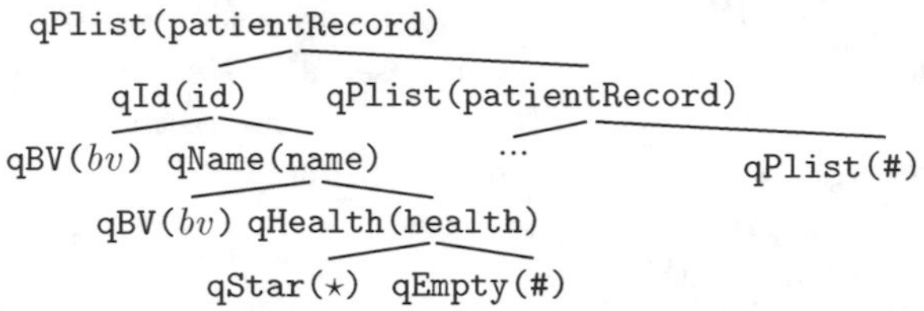

Fig. 4. A run of the automaton specified in (7) accepting the public view of a list of records like that on the bottom of Figure 2 with state qPlist

As Figure 4 illustrates, a list accepted by the predicate qPlist is either a tree with root labeled patientRecord having a first child accepted by qId and a second child accepted by qPlist, or it is a nullary node with label #. The other predicates can be understood similarly. In particular, the predicate qHealth accepts a tree only, if the first child of its root is a leaf labeled $\star$ specifying that the corresponding value is confidential. Supposing that the variable pList contains the database in the initial state of our program, the corresponding information for the initial node n_{in} is defined by: $\text{var}_{\text{pList},n_{in}}(\text{X}) \Leftarrow \text{qPlist(X)}$.

In the following, we describe how the abstract state transformers $[\![l]\!]^\sharp$ of edges (m, l, n) are formalized by means of Horn clauses. In order to do so we need that the set of binary elements Σ_2 potentially occurring in the program is finite and a priori known. This information can be extracted, e.g., from the interface descriptions of web services. Due to the construction of self-compositions of programs edges either refer to assignments or to Boolean expressions, but never to both.

5.1 Horn Clauses for Procedure Calls

Consider a call edge $(m, \text{call } r, n)$ in one of the CFGs of the program. Here we denote the initial and final nodes of the CFG corresponding to procedure r with n_{in}^r and n_{fi}^r, respectively.

The function enter$_2$ propagates values of variables without modification. Therefore, we generate the following implication for each variable x:

$$\text{var}_{x,n_{in}^r}(\text{X}) \Leftarrow \text{var}_{x,m}(\text{X}).$$

For the propagation of the error state we have:

$$\text{error}_{n_{in}^r}(\checkmark) \Leftarrow \text{error}_m(\_).$$

Above, '_' stands for an anonymous variable. The condition of the implication makes sure that the error state $\checkmark$ is only added to the abstract value corresponding to node n^r_{in} if it is reachable.

In order to define $\text{combine}_2^\sharp$ we generate the following implication for all variables x other than ret:

$$\text{var}_{x,n}(\text{X}) \Leftarrow \text{var}_{x,m}(\text{X}).$$

For ret we define:

$$\text{var}_{\text{ret},n}(\text{X}) \Leftarrow \text{var}_{\text{ret},n^r_{fi}}(\text{X}).$$

Thus, the value of variable ret is propagated from the final node of the procedure r to node n, whereas the values of other variables x are propagated from node m. The error state at node n needs to be greater or equal to those at nodes m and n^r_{fi}. Therefore, the error states corresponding to nodes m and n^r_{fi} are joined:

$$\text{error}_n(\text{X}) \Leftarrow \text{error}_m(\text{X}).$$
$$\text{error}_n(\text{X}) \Leftarrow \text{error}_{n^r_{fi}}(\text{X}).$$

The definitions of the clauses for simulating assignments and Boolean expressions are as in |15| which we repeat here for the sake of self-containedness.

5.2   Horn Clauses for Assignments

First, we discuss the case of assignments, i.e., transformers of edges of the following form:

$$(m, [\![x\!:=\!e_1(x_1,\ldots,x_n), y\!:=\!e_2(y_1,\ldots,y_m)]\!]^\sharp, n)$$

If no error occurs, then x and y are updated, the values of other variables remain unchanged. Accordingly, for all variables $z \neq x$ and $z \neq y$ the following clauses are defined, which propagate their values unmodified:

$$\text{var}_{z,n}(\text{X}) \Leftarrow \text{var}_{z,m}(\text{X}).$$

In order to handle errors, the following implication is generated in addition:

$$\text{error}_n(\text{X}) \Leftarrow \text{error}_m(\text{X}). \tag{8}$$

Values of variables on the left hand sides of the assignments are defined by the following clauses.

- For edges with label $(x\!:=\!\#, x\!:=\!\#)$ we have $\text{var}_{x,n}(\#) \Leftarrow \text{var}_{x,m}(\_)$ where '_' denotes an anonymous logic variable. The implication is required to ensure that # is added to the predicate $\text{var}_{x,n}$ only if m may be reachable. At a reachable node n for all variables x there is a tree τ so that $\text{var}_{x,n}(\tau)$ holds.
- For edges with $(x\!:=\!y, x\!:=\!y)$ we have: $\text{var}_{x,n}(\text{X}) \Leftarrow \text{var}_{y,m}(\text{X})$.
- For edges with $(x\!:=\!\sigma_2(y,z), x\!:=\!\sigma_2(y,z))$ we have:

$$\text{var}_{x,n}(\sigma_2(\text{L},\text{R})) \Leftarrow \text{var}_{y,m}(\text{L}), \text{var}_{z,m}(\text{R}).$$

- For edges with $(x\text{:=}y/1, x\text{:=}y/1)$ we have $\mathrm{var}_{x,n}(\mathtt{L}) \Leftarrow \mathrm{var}_{y,m}(\sigma_2(\mathtt{L},\_))$ for all $\sigma_2 \in \Sigma_2$. As an example, let us suppose that the abstract value of variable pList at node m is a model of the predicate qPlist according to the implications in (7). Using the command id := pList/1 we can assign the head of the list into variable id. The implication defining the abstract value of variable id after the assignment is:

$$\mathrm{var}_{\mathrm{id},n}(\mathtt{L}) \Leftarrow \mathrm{var}_{\mathrm{pList},m}(\mathtt{patientRecord}(\mathtt{L},\_)).$$

However, during the analysis the label of the root of the tree in a variable needs to be treated as unknown. Therefore, the implication is repeated for all possible binary alphabet elements $\sigma_2 \in \Sigma_2$.

An error is caused by an expression of the form $x/1$, if the content of x does not have children, i.e., it is a leaf. Therefore, in addition the following is defined:

$$\begin{aligned}
\mathrm{error}_n(\sharp) &\Leftarrow \mathrm{var}_{y,m}(\mathtt{\#}). \\
\mathrm{error}_n(\sharp) &\Leftarrow \mathrm{var}_{y,m}(bv). \\
\mathrm{error}_n(\top) &\Leftarrow \mathrm{var}_{y,m}(\star).
\end{aligned} \qquad (9)$$

- For edges with $(x\text{:=}y/2, x\text{:=}y/2)$ we have $\mathrm{var}_{x,n}(\mathtt{R}) \Leftarrow \mathrm{var}_{y,m}(\sigma_2(\_,\mathtt{R}))$ for all $\sigma_2 \in \Sigma_2$. The implications handling the error state are identical to those in (9).
- By edges with (l,l) where $l = x\text{:=}\lambda_t(x_1, \ldots, x_k)$, it needs to be examined whether the arguments of the function λ_t contain secret. The implications below are used for the purpose, where the second and third lines are defined for all $\sigma_2 \in \Sigma_2$:

$$\begin{aligned}
\mathrm{secret}(\star). & \\
\mathrm{secret}(\sigma_2(\mathtt{L},\_)) &\Leftarrow \mathrm{secret}(\mathtt{L}). \\
\mathrm{secret}(\sigma_2(\_,\mathtt{R})) &\Leftarrow \mathrm{secret}(\mathtt{R}).
\end{aligned}$$

Concerning the resulting value of x we have:

$$\mathrm{var}_{x,n}(bv) \Leftarrow \mathrm{var}_{x_1,m}(\_), \mathrm{var}_{x_2,m}(\_), \ldots, \mathrm{var}_{x_k,m}(\_). \qquad (10)$$

$$\mathrm{var}_{x,n}(\star) \Leftarrow \mathrm{var}_{x_i,m}(\mathtt{X}), \mathrm{secret}(\mathtt{X}), \mathrm{var}_{x_1,m}(\_), \ldots, \mathrm{var}_{x_k,m}(\_). \qquad (11)$$

According to implication (10), the value of x at node n will potentially be bv if all of the arguments $x_1, ..., x_k$ of λ_t are defined. Furthermore, according to implication (11), if any of the input variables depends on the secret, then the resulting abstract value will also contain $\star$. There is an implication of the form (11) defined for all arguments x_i of λ_t.

- By edges with labels $(x\text{:=}e(x_1, \ldots, x_k), \mathtt{skip})$ or $(\mathtt{skip}, x\text{:=}e(x_1, \ldots, x_k))$ we have:

$$\mathrm{var}_{x,n}(\star) \Leftarrow \mathrm{var}_{x_1,m}(\_), \ldots, \mathrm{var}_{x_k,m}(\_).$$

If the effect of an edge consists of an assignment and a skip command, then in the resulting abstract state the value of the variable on the left-hand-side becomes $\star$. This indicates that its value might be different in the corresponding two concrete states. At this point we cannot take the values of variables on the right-hand-side into consideration.

If the expression is of the form $x/1$ or $x/2$, then we have in addition: $\mathtt{error}_n(\top) \Leftarrow \mathtt{var}_{x,m}(\_)$ in order to indicate that an error may occur only in one member of the pair of corresponding concrete states.

5.3   Horn Clauses for Boolean Expressions

Now we discuss abstract transformers with Boolean expressions. In our implementation we treat two branching constructs composable only if their conditional expressions are syntactically equivalent. Therefore, here we only need to treat the corresponding combinations of Boolean expressions.

- By edges labeled (b,b), $(b,\mathtt{skip})$ or $(\mathtt{skip},b)$, where $b = \lambda_b(x_1, x_2, \ldots, x_k)$, the values of all variables y occurring in the program are propagated the following way:

$$\mathtt{var}_{y,n}(\mathtt{X}) \Leftarrow \mathtt{var}_{y,m}(\mathtt{X}), \mathtt{var}_{x_1,m}(\_), \ldots, \mathtt{var}_{x_k,m}(\_).$$

In other words, it is checked whether the input variables of the conditional expression have been defined, in order to ensure that the node m is reachable. The actual values of variables are propagated without modification.
- In case the label of the root of a tree is tested using an edge having label of the form $(\mathtt{top}(x)=\sigma, \mathtt{top}(x)=\sigma)$, then the following clauses are defined to propagate the values of variables $y \neq x$ if $\sigma \in \Sigma_2$:

$$\begin{aligned} \mathtt{var}_{y,n}(\mathtt{X}) &\Leftarrow \mathtt{var}_{y,m}(\mathtt{X}), \mathtt{var}_{x,m}(\sigma(\_,\_)). \\ \mathtt{var}_{y,n}(\mathtt{X}) &\Leftarrow \mathtt{var}_{y,m}(\mathtt{X}), \mathtt{var}_{x,m}(\star). \end{aligned} \tag{12}$$

The value of the variable x is propagated as well:

$$\mathtt{var}_{x,n}(\sigma(\mathtt{L},\mathtt{R})) \Leftarrow \mathtt{var}_{x,m}(\sigma(\mathtt{L},\mathtt{R})). \tag{13}$$

$$\mathtt{var}_{x,n}(\sigma(\star,\star)) \Leftarrow \mathtt{var}_{x,m}(\star). \tag{14}$$

If $\sigma = \#$ then $\sigma(X,Y)$ is exchanged with $\#$ in (12), (13) and (14).
- For edges with $(\mathtt{top}(x)=\sigma, \mathtt{skip})$ or $(\mathtt{skip}, \mathtt{top}(x)=\sigma)$ (12) needs to be repeated for all variables $y \neq x$, and (13) needs to be repeated in order to propagate the value of x as well. In addition $\mathtt{var}_{x,n}(\star) \Leftarrow \mathtt{var}_{x,m}(\star)$ needs to be defined.
- By edges having labels of the form $(\neg\mathtt{top}(x)=\sigma, \neg\mathtt{top}(x)=\sigma)$, $(\neg\mathtt{top}(x)=\sigma, \mathtt{skip})$ or $(\mathtt{skip}, \neg\mathtt{top}(x)=\sigma)$, the values of the variables are propagated only in the case, when the root of the value of x is labeled with some $\delta \neq \sigma$. Therefore, the following implication is defined for all variables y other than x and for all alphabet elements $\delta \in \Sigma_2 \setminus \{\sigma\}$:

$$\mathtt{var}_{y,n}(\mathtt{X}) \Leftarrow \mathtt{var}_{y,m}(\mathtt{X}), \mathtt{var}_{x,m}(\delta(\_,\_)). \tag{15}$$

In order to handle the value of x as well, the following implication is defined for all $\delta \in \Sigma_2 \setminus \{\sigma\}$:

$$\mathtt{var}_{x,n}(\delta(\mathtt{L},\mathtt{R})) \Leftarrow \mathtt{var}_{x,m}(\delta(\mathtt{L},\mathtt{R})). \tag{16}$$

Additionally, we need to define (15) and (16) so that $\delta(X,Y)$ is replaced by $\star$, and if $\sigma \neq \#$ then by $\#$ too.

– In case the two components of the label of an edge are the negations of each other, e.g., $(\neg b(x_1, x_2, ..., x_k), b(x_1, x_2, ..., x_k))$, then the values of variables need to be propagated only in the case, when at least one of the variables in the argument depends on the secret. Assuming that b is a function, the simultaneous execution of the two steps cannot take place otherwise. Accordingly, the following is defined for all variables y occurring in the program and for all variables $x_i \in \{x_1, ..., x_k\}$:

$$\text{var}_{y,n}(\text{X}) \Leftarrow \text{var}_{y,m}(\text{X}), \text{var}_{x_i,m}(\text{Z}), \text{secret}(\text{Z}). \tag{17}$$

The transfer functions of edges labeled with Boolean expressions propagate the error state without modification, therefore we generate implications of the form (8) for each of them.

5.4 Discussion

In [15], we have proven that the abstract transformer $[\![f, g]\!]^\sharp$ for a pair (f, g) as defined by Horn clauses is correct. In other words, if $[\![f]\!](s_0) = s$ and $[\![g]\!]t_0 = t$ where $(s_0, t_0) \in \gamma(D_0)$ and $[\![f, g]\!]^\sharp D_0 = D$, then $(s, t) \in \gamma(D)$ holds. Therefore, noninterference for a particular output variable x holds at program exit n_{fi}, if the predicate $\text{var}_{x,n_{fi}}$ does not accept trees containing $\star$. Algorithmically, therefore, the analysis boils down to computing (or approximating) the least model of the set of Horn clauses defined for the program. Now we observe that the head of each clause possibly generated by our analysis, is of one of the following forms:

$$h \quad ::= \quad \text{p} \mid \text{p}(X) \mid \text{p}(\sigma_0) \mid \text{p}(\sigma_2(X_1, X_2)),$$

where X_1 and X_2 are distinct. Therefore, all of them belong to the class of Horn clauses $\mathcal{H}_1$ [18, 25]. Finite sets of clauses from this class are known to have least models consisting of *regular* sets. Moreover, finite automata characterizing these regular sets can be effectively computed by means of a normalization procedure.

We have carried out the analysis of the self-composition of the program in Listing 6. The initial abstract values of the input variables are defined by the following implications:

$$\begin{aligned} \text{var}_{\text{pList}, n_{in}}(\text{X}) &\Leftarrow \text{qPlist}(\text{X}). \\ \text{var}_{\text{test}, n_{in}}(\star). \ \text{var}_{\text{patientId}, n_{in}}(bv). \end{aligned} \tag{18}$$

The initial values of all other variables x are defined by predicates of the form $\text{var}_{x,n_{in}}(\#)$, the initial error state is defined by $\text{error}_{n_{in}}(\checkmark)$.

The result of the analysis reveals that the final abstract value of the variable pList equals to the initial one defined by the predicate qPlist in (18). In other words, the secret remains in the variable test and in the health status of the records in the list, but does not interfere with other values. Thus, the call string 0 approach is sufficient to verify noninterference. The reason is that the procedure calls of the original program are all within a similar security context. In applications, where this is not the case, more information about calling contexts may be required.

6 Related Work and Conclusion

Information flow analysis can be traced at least as far back as to the fundamental work of Denning [10] who provided a lattice model for information flow, and Volpano et al. who provided a first sound type system to analyze such properties [24]. Since then various extensions of the original ideas have been proposed. For a still valuable overview see, e.g, in [20]. An alternative approach to information flow analysis is through program dependence graphs as, e.g., realized in the system Joana for Java [12].

Both approaches loose information when the control flow depends on secret data. Information flow properties, on the other hand, can be naturally formalized as 2-hypersafety properties. As observed by Clarkson and Schneider, the verification of 2-hypersafety properties can be reduced to the verification of ordinary safety properties for self-compositions of programs [6]. This direct road to a more precise information flow analysis has been further explored by Barthe et al. [2, 3] and Banerjee et al. [17, 22] who provide proof methods based on self-compositions. In our predecessor paper [15], we have introduced syntactic matching of programs to provide appropriate self-compositions and relational abstract interpretation in order to obtain fully automatic analyses. These ideas have been exemplified by means of an intraprocedural analysis of XML transformers. Our present paper has reviewed that approach and extended it with a generic treatment of procedures. In principle, any technique for interprocedural analysis (e.g., [9,14,21]) can be applied together with suitable abstract domains which allow to express potential differences between reaching pairs of program states. For the case of programs manipulating XML trees, we used public views to identify these differences. Analyzing the sets of potentially reaching public views for 2-programs, turned out to be quite analogous to analyzing the sets of potentially reaching XML values for ordinary programs.

References

1. Alves, A., Arkin, A., Askary, S., Barreto, C., Bloch, B., Curbera, F., Ford, M., Goland, Y., Guízar, A., Kartha, N., Liu, C.K., Khalaf, R., Koenig, D., Marin, M., Mehta, V., Thatte, S., Rijn, D., Yendluri, P., Yiu, A.: Web services business process execution language version 2.0 (OASIS standard). WS-BPEL TC OASIS (2007), http://docs.oasis-open.org/wsbpel/2.0/wsbpel-v2.0.html

2. Barthe, G., Crespo, J.M., Kunz, C.: Relational verification using product programs. In: Butler, M., Schulte, W. (eds.) FM 2011. LNCS, vol. 6664, pp. 200–214. Springer, Heidelberg (2011)

3. Barthe, G., Crespo, J.M., Kunz, C.: Beyond 2-safety: Asymmetric product programs for relational program verification. In: Artemov, S., Nerode, A. (eds.) LFCS 2013. LNCS, vol. 7734, pp. 29–43. Springer, Heidelberg (2013)

4. Berglund, A., Boag, S., Chamberlin, D., Fernández, M.F., Kay, M., Robie, J., Siméon, J.: XML path language (XPath) 2.0 (second edition). World Wide Web Consortium, Recommendation REC-xpath20-20101214 (December 14, 2010), http://www.w3.org/TR/2010/REC-xpath20-20101214/

5. Broberg, N., Sands, D.: Paralocks: Role-based information flow control and beyond. In: Proceedings of the 37th ACM SIGPLAN-SIGACT Symposium on Principles of Programming Languages, POPL 2010, pp. 431–444 (2010)

6. Clarkson, M.R., Schneider, F.B.: Hyperproperties. Journal of Computer Security 18(6), 1157–1210 (2010)

7. Comon, H., Dauchet, M., Gilleron, R., Löding, C., Jacquemard, F., Lugiez, D., Tison, S., Tommasi, M.: Tree automata techniques and applications (2007), http://www.grappa.univ-lille3.fr/tata (release October 12, 2007)

8. Cousot, P., Cousot, R.: Abstract interpretation: A unified lattice model for static analysis of programs by construction or approximation of fixpoints. In: Graham, R.M., Harrison, M.A., Sethi, R. (eds.) Conference Record of the Fourth ACM Symposium on Principles of Programming Languages (POPL), pp. 238–252. ACM (1977)

9. Cousot, P., Cousot, R.: Static determination of dynamic properties of recursive procedures. In: Neuhold, E. (ed.) IFIP Conf. on Formal Description of Programming Concepts, pp. 237–277. North- Holland (1977)

10. Denning, D.E.: A lattice model of secure information flow. Commun. ACM 19(5), 236–243 (1976)

11. Goguen, J.A., Meseguer, J.: Security policies and security models. IEEE Symposium on Security and Privacy, 11–20 (1982)

12. Hammer, C., Snelting, G.: Flow-sensitive, context-sensitive, and object-sensitive information flow control based on program dependence graphs. Int. J. Inf. Sec. 8(6), 399–422 (2009)

13. Kay, M.: XSL transformations (XSLT) version 2.0. World Wide Web Consortium, Recommendation REC-xslt20-20070123 (January 23, 2007), http://www.w3.org/TR/xslt20/

14. Knoop, J., Steffen, B.: The interprocedural coincidence theorem. In: Pfahler, P., Kastens, U. (eds.) CC 1992. LNCS, vol. 641, pp. 125–140. Springer, Heidelberg (1992)

15. Kovács, M., Seidl, H., Finkbeiner, B.: Relational abstract interpretation for the verification of 2-hypersafety properties. In: Sadeghi, A.R., Gligor, V.D., Yung, M. (eds.) ACM Conference on Computer and Communications Security (CCS 2013), pp. 211–222. ACM (2013)

16. Myers, A.C.: JFlow: Practical mostly-static information flow control. In: Appel, A.W., Aiken, A. (eds.) Proceedings of the 26th ACM SIGPLAN-SIGACT Symposium on Principles of Programming Languages, POPL 1999, pp. 228–241. ACM (1999)

17. Nanevski, A., Banerjee, A., Garg, D.: Dependent type theory for verification of information flow and access control policies. ACM Trans. Program. Lang. Syst. 35(2), 6 (2013)

18. Nielson, F., Riis Nielson, H., Seidl, H.: Normalizable horn clauses, strongly recognizable relations, and spi. In: Hermenegildo, M.V., Puebla, G. (eds.) SAS 2002. LNCS, vol. 2477, pp. 20–35. Springer, Heidelberg (2002)

19. Pawlik, M., Augsten, N.: RTED: A robust algorithm for the tree edit distance. PVLDB 5(4), 334–345 (2011)

20. Sabelfeld, A., Myers, A.C.: Language-based information-flow security. IEEE Journal on Selected Areas in Communications 21(1), 5–19 (2003)

21. Sharir, M., Pnueli, A.: Two approaches to interprocedural data flow analysis. In: Muchnick, S., Jones, N. (eds.) Program Flow Analysis: Theory and Application, pp. 189–233. Prentice-Hall (1981)

22. Stewart, G., Banerjee, A., Nanevski, A.: Dependent types for enforcement of information flow and erasure policies in heterogeneous data structures. In: Peña, R., Schrijvers, T. (eds.) 15th International Symposium on Principles and Practice of Declarative Programming (PPDP 2013), pp. 145–156. ACM (2013)
23. Terauchi, T., Aiken, A.: Secure information flow as a safety problem. In: Hankin, C., Siveroni, I. (eds.) SAS 2005. LNCS, vol. 3672, pp. 352–367. Springer, Heidelberg (2005)
24. Volpano, D.M., Irvine, C.E., Smith, G.: A sound type system for secure flow analysis. Journal of Computer Security 4(2/3), 167–188 (1996)
25. Weidenbach, C.: Towards an automatic analysis of security protocols in first-order logic. In: Ganzinger, H. (ed.) CADE 1999. LNCS (LNAI), vol. 1632, pp. 314–328. Springer, Heidelberg (1999)

Computing Optimal Reachability Costs in Priced Dense-Timed Pushdown Automata

Parosh Aziz Abdulla, Mohamed Faouzi Atig, and Jari Stenman

Dept. of Information Technology, Uppsala University, Sweden
{parosh.abdulla,mohamed_faouzi.atig,jari.stenman}@it.uu.se

Abstract. We study priced dense-timed pushdown automata that are a generalization of the classic model of pushdown automata, in the sense that they operate on real-valued clocks, and that the stack symbols have real-valued ages. Furthermore, the model allows a cost function that assigns transition costs to transitions and storage costs to stack symbols. We show that the optimal cost, i.e., the infimum of the costs of the set of runs reaching a given control state, is computable.

1 Introduction

Pushdown automata are a widely used model both in language theory and program verification. Recently, several models have been introduced that extend pushdown automata with clocks and real-time constraints [10,12,1]. In the mean time, several works have extended the model of timed automata [5] with *prices* (*weights*) (e.g., [6,7,9]). Weighted timed automata are used in the modeling of embedded systems, where the behavior of the system is usually constrained by the availability of different types of resources.

In this paper, we consider *Priced Dense-Timed Pushdown Automata* (PTPA) that subsume all the above models. PTPA are a generalization of classic pushdown automata with real-valued clocks, timed constraints, and prices for computations. More precisely, a PTPA contains a finite set of global clocks, and each symbol in the stack is equipped with a real number indicating its age. The global clocks admit the same kind of operations as in timed automata, and timed transitions increase the clock values and the ages of stack symbols at the same rate. Pop operations may only be performed if the age of the topmost stack symbol is within a given time interval. Furthermore, the model is priced in the sense that there is a cost function that assigns transition costs to transitions and storage costs to stack symbols.

We study the problem of computing the optimal cost to reach a given control state. In general, a cost-optimal computation may not exist (e.g., even in priced timed automata it can happen that there is no computation of cost 0, but there exist computations of cost $\leq \epsilon$ for every $\epsilon > 0$). However, we show that the infimum of the costs is computable. To do this, we perform a sequence of reductions that ultimately translates the problem to the problem of control state reachability for plain (unpriced and untimed) pushdown automata. The latter problem is known to be decidable [8].

A.-H. Dediu et al. (Eds.): LATA 2014, LNCS 8370, pp. 62–75, 2014.

Related Work. Priced extensions of timed models have been studied in the literature. The paper [4] studies a priced dense-timed extension of Petri nets, where the optimal cost is computed for satisfying coverability objectives (reaching an upward closed set of markings). Proofs for solving the coverability problem in Petri nets are in general quite different from those for the solving control state reachability problem in pushdown systems. This is already the case for the unpriced untimed case, where the former relies on Karp-Miller constructions [11] or backward reachability analysis [3], while the latter uses finite automata constructions [8]. This difference is also reflected in the priced timed case. In particular, [4] (using backward reachability analysis) reduces optimal cost computation to the reachability problem for a more powerful model than plain Petri nets, namely that of Petri nets with one inhibitor arc. In our case, we reduce the problem to the plain pushdown model.

Several timed extensions of pushdown automata have been considered [12,10,1]. Since our model extends these, some of the techniques need to be reused. However, priced timed models are nontrivial extensions of (unpriced) timed models. Here, in a similar manner to priced extensions of timed Petri nets [4] and timed automata [9], we need to reason about special forms of computations, and a nontrivial modification of the (region-based) symbolic encoding is also necessary to represent the infinite state space.

In [2] we study priced *discrete-timed* pushdown automata. In the discrete-time case, time is interpreted as being incremented in discrete steps and thus the clock values and ages of stack symbols are in a countable domain. The method of [2] cannot be extended to the dense time case. It is well-known that, in timed models, using discrete domains represents a substantial simplification compared to using dense time. In particular, the absence of fractional parts in clock values and stack symbol ages leads to a much simpler symbolic representation of the stack. The model of priced discrete-timed pushdown automata is generalized in [13], where the authors consider pushdown systems that can modify the whole stack using transducers.

2   Preliminaries

We use $\mathbb{R}^{\geq 0}$ to denote the non-negative reals. For $r \in \mathbb{R}^{\geq 0}$, where $r = n + r'$ for $n \in \mathbb{N}$, $r' \in \mathbb{R}^{\geq 0}$ and $r' < 1$, we let $\lfloor r \rfloor = n$ denote the *integral part* and $fract\,(r) = r'$ denote the *fractional part* of r. Given a set A, we use 2^A for the powerset of A. For sets A and B, $f : A \to B$ denotes a (possibly partial) function from A to B. If f is undefined at a, we write $f(a) = \bot$. We use $dom\,(f)$ and $range\,(f)$ to denote the domain and range of f. Given a function f and a set A, we use $f(A)$ for the image $\{f(x) \mid x \in A\}$ of A under f. The image $img(f)$ of f is then defined as $f(dom\,(f))$. We write $f[a \leftarrow b]$ to denote the function f' such that $f'(a) = b$ and $f'(x) = f(x)$ for $x \neq a$. Given a function $f : A \times B \to C$, we sometimes write $f(a)(b) = c$ to make explicit that f might be applied partially, i.e. to get a function $f(a) : B \to C$. The set of *intervals* of the form $[a : b]$, $(a : b]$, $[a : b)$, $(a : b)$, $[a : \infty)$ or $(a : \infty)$, where $a, b \in \mathbb{N}$, is denoted by $\mathcal{I}$. Given a set A, we use $\inf(A)$, $\max(A)$ and $\min(A)$ to denote the infimum, the maximum and the minimum of A, respectively.

Let A be an alphabet. We denote by A^*, (resp. A^+) the set of all *words* (resp. non-empty words) over A. The empty word is denoted by ϵ. For a word w, $|w|$ denotes the length of w (we have $|\epsilon| = 0$). For words w_1, w_2, we use $w_1 \cdot w_2$ for the concatenation of w_1 and w_2. We extend $\cdot$ to sets W_1, W_2 of words by defining $W_1 \cdot W_2 = \{w_1 \cdot w_2 \mid w_1 \in W_1, w_2 \in W_2\}$. For a word $w = a_0 \ldots a_n$, and $i \in \{0, \ldots, n\}$, we let $w[i]$ denote a_i. Given a word $w = \langle x_0, y_0 \rangle \ldots \langle x_n, y_n \rangle \in (X \times Y)^*$, we define the *first projection* $proj_1(t) = x_0 \ldots x_n$ and the *second projection* $proj_2(t) = y_0 \ldots y_n$. We define a binary *shuffle operation* $\otimes$ inductively: For $w \in (2^A)^*$, define $w \otimes \epsilon = \epsilon \otimes w = w$. For sets $r_1, r_2 \in 2^A$ and words $w_1, w_2 \in (2^A)^*$, define $(r_1 \cdot w_1) \otimes (r_2 \cdot w_2) = (r_1 \cdot (w_1 \otimes (r_2 \cdot w_2))) \cup (r_2 \cdot ((r_1 \cdot w_1) \otimes w_2)) \cup ((r_1 \cup r_2) \cdot (w_1 \otimes w_2))$.

3 Priced Timed Pushdown Automata

In this section, we introduce PTPA and define cost-optimal reachability.

Model. Formally, a PTPA is a tuple $\mathcal{T} = \langle Q, q_{init}, \Gamma, X, \Delta, Cost \rangle$, where Q is a finite set of states, $q_{init} \in Q$ is the initial state, Γ is a finite stack alphabet, X is a finite set of clocks, and $Cost : (\Gamma \cup \Delta) \to \mathbb{N}$ is a function assigning transition costs to transition rules and storage costs to stack symbols. The set Δ consists of a finite number of transition rules of the form $\langle q, op, q' \rangle$, where $q, q' \in Q$ and op is either (i) *nop*, an operation that does not modify the clocks or the stack, (ii) *push(a)*, where $a \in \Gamma$, which pushes a onto the stack with initial age 0, (iii) *pop(a, I)*, where $a \in \Gamma$ and $I \in \mathcal{I}$, which pops a of the stack if its age is in I, (iv) *test(x, I)*, where $x \in X$ and $I \in \mathcal{I}$, which is only enabled if $x \in I$, or (v) *reset(x)*, where $x \in X$, which sets the value of the clock x to 0.

Semantics. A *clock valuation* is a function $\mathsf{X} : X \to \mathbb{R}^{\geq 0}$ which assigns a concrete value to each clock. A *stack content* is a word $w \in (\Gamma \times \mathbb{R}^{\geq 0})^*$, i.e. a sequence of stack symbols and their corresponding ages. A *configuration* is a tuple $\langle q, \mathsf{X}, w \rangle$, where $q \in Q$ is a state, X is a clock valuation, and w is a stack content. For a configuration $\gamma = \langle q, \mathsf{X}, w \rangle$, define the functions $\mathtt{State}(\gamma) = q$, $\mathtt{ClockVal}(\gamma) = \mathsf{X}$, and $\mathtt{Stack}(\gamma) = w$. For any transition rule $t = \langle q, op, q' \rangle \in \Delta$, define $\mathtt{Op}(t) = op$. Given a PTPA $\mathcal{T}$, we use $Conf(\mathcal{T})$ to denote the set of all configurations of $\mathcal{T}$. Let X_{init} be the clock valuation such that $\mathsf{X}_{init}(x) = 0$ for each $x \in X$. The *initial configuration* γ_{init} is the configuration $\langle q_{init}, \mathsf{X}_{init}, \epsilon \rangle$. The operational semantics of a PTPA $\mathcal{T}$ are defined by a transition relation over the set of configurations $Conf(\mathcal{T})$. It consists of two types of transitions; *timed* transitions, which simulate time passing, and *discrete* transitions, which are applications of the transition rules in Δ.

Timed Transitions. Fix some $r \in \mathbb{R}^{\geq 0}$. Given a clock valuation X, let X^{+r} be the function defined by $\mathsf{X}^{+r}(x) = \mathsf{X}(x) + r$ for all $x \in X$. For any stack content $w = \langle a_0, v_0 \rangle \cdots \langle a_n, v_n \rangle$, let w^{+r} be the stack content $\langle a_0, v_0 + r \rangle \cdots \langle a_n, v_n + r \rangle$.

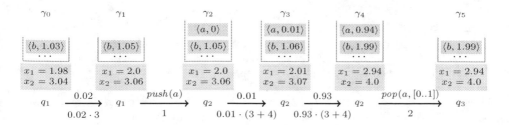

Fig. 1. A fragment of a computation

Given configurations $\gamma = \langle q, \text{X}, w \rangle$ and $\gamma' = \langle q', \text{X}', w' \rangle$, we have $\gamma \xrightarrow{r} \gamma'$ if $q' = q$, $\text{X}' = \text{X}^{+r}$ and $w' = w^{+r}$. This means that a PTPA may perform a timed transition, whereby it advances all clocks and ages of stack symbols by some non-negative real number.

Discrete Transitions. Let $t = \langle q, op, q' \rangle \in \Delta$ be a transition rule. For configurations $\gamma = \langle q, \text{X}, w \rangle$ and $\gamma' = \langle q', \text{X}', w' \rangle$, we have $\gamma \xrightarrow{t} \gamma'$ if either (i) $op = nop$, $w' = w$, and $\text{X}' = \text{X}$, (ii) $op = push(a)$, $w' = w \cdot \langle a, 0 \rangle$, and $\text{X}' = \text{X}$, (iii) $op = pop(a, I)$, $w = w' \cdot \langle a, v \rangle$ for some $v \in I$, and $\text{X}' = \text{X}$, (iv) $op = test(x, I)$, $w' = w$, $\text{X}' = \text{X}$, and $\text{X}(x) \in I$, or (v) $op = reset(x)$, $w' = w$, and $\text{X}' = \text{X}[x \leftarrow 0]$.

A *computation* π to a configuration γ of a PTPA $\mathcal{T}$ is a finite sequence of the form $\gamma_0 \xrightarrow{t_1} \gamma_1 \xrightarrow{t_2} \cdots \xrightarrow{t_n} \gamma_n$, where $\gamma_0 = \gamma_{init}$, $\gamma_n = \gamma$, and for all $1 \leq i \leq n$, $\gamma_i \in Conf(\mathcal{T})$, and either $t_i \in \Delta$ or $t_i \in \mathbb{R}^{\geq 0}$. We define $Comp(\mathcal{T}, q)$ to be the set of computations to a configuration γ such that $\text{State}(\gamma) = q$. If π is a computation to γ, and $\text{State}(\gamma) = q$, we say that π is a computation to q.

Cost of Computation. We will now extend the cost function *Cost*, which we originally defined on stack symbols and transition rules, to transitions and computations. The cost of a discrete transition is given by the cost of the corresponding transition rule, and the cost of a timed transition is the total cost of the stack scaled by the length of the timed transition. Fix a discrete or timed transition $\gamma \xrightarrow{t} \gamma'$, and let $\gamma = \langle q, \text{X}, \langle a_0, v_0 \rangle \cdots \langle a_n, v_n \rangle \rangle$. Formally, $Cost(\gamma \xrightarrow{t} \gamma') = Cost(t)$ if $\gamma \xrightarrow{t} \gamma'$ is discrete, and $Cost(\gamma \xrightarrow{t} \gamma') = t \cdot \sum_{i=0}^{n} Cost(a_i)$ if $\gamma \xrightarrow{t} \gamma'$ is timed. The cost of a computation $\pi = \gamma_0 \xrightarrow{t_1} \gamma_1 \xrightarrow{t_2} \cdots \xrightarrow{t_n} \gamma_n$ is defined as the sum of the costs of its transitions, i.e. $Cost(\pi) = \sum_{i=1}^{n} Cost(\gamma_{i-1} \xrightarrow{t_i} \gamma_i)$. Fig. 1 shows a fragment of a computation of a PTPA where the set of clocks is $\{x_1, x_2\}$ and the stack alphabet is $\{a, b, c\}$ (the stack not shown is filled with c). We assume that the storage costs of a, b and c are 4, 3 and 0, respectively, that the transition cost of $\langle q_1, push(a), q_2 \rangle$ is 1, and that the transition cost of $\langle q_2, pop(a, [0 : 1]), q_3 \rangle$ is 2. The accumulated cost for the example is 9.64.

Cost-Optimality. In this paper, we address the problem of computing the optimal cost required to reach a certain state. The optimal cost is defined as the *infimum* of the set of reachability costs. Formally, we define the optimal cost

$Cost_{opt}(\mathcal{T}, q)$ as $Cost_{opt}(\mathcal{T}, q) = \inf\{Cost(\pi) \mid \pi \in Comp(\mathcal{T}, q)\}$ if $Comp(\mathcal{T}, q) \neq \emptyset$ and $Cost_{opt}(\mathcal{T}, q) = \infty$ otherwise. The *cost-optimal reachability problem* is then, given a PTPA $\mathcal{T}$ and a state q, to compute the optimal cost $Cost_{opt}(\mathcal{T}, q)$.

Theorem 1. *Let $\mathcal{T}$ be* PTPA *and q be a state. Then $Cost_{opt}(\mathcal{T}, q)$ is computable.*

The rest of the paper is devoted to the proof of the above theorem. In Sec. 4, we show that it is sufficient to consider computations in a certain form in order to compute the optimal cost. In Sec. 5, we introduce our symbolic automata (PPA) to which we reduce the cost-optimal reachability problem for PTPA. In Sec. 6, we formally define the region encoding and some related operations. Finally, we construct a PPA that simulates all the computations (in a certain form) of PTPA in Sec. 7.

4   Forms of Computations

Detailed Computations. In order to solve the cost-optimal reachability problem, we will only consider computations that are of a certain form, called *detailed*. This yields a closer correspondence between computations of PTPA and computations in the untimed automaton, defined in section 7. Consider a timed transition $\langle q, \mathtt{X}, \langle a_0, v_0 \rangle \cdots \langle a_n, v_n \rangle \rangle \xrightarrow{r} \langle q', \mathtt{X}', \langle a_0, v_0' \rangle \cdots \langle a_n, v_n' \rangle \rangle$. Let $V = fract(\mathtt{X}(X)) \cup fract(\{v_0, \ldots, v_n\})$ and let $m = \min(V)$ and $d = \max(V)$. In other words, m is the minimal fractional part and d is the maximal fractional part of any value. We can classify the timed transition into two different types:

- *Type 1*: A transition which is taken when no value has fractional part 0, and which may make the values with the largest fractional parts reach the next integer. This is the case when $r > 0$, $m > 0$ and $r \leq 1 - d$.
- *Type 2*: A transition which is taken when some values have fractional parts 0, which makes the fractional parts of those value positive, and which does not change the integral part of any value. This is the case when $r > 0$, $m = 0$ and $r < 1 - d$.

A detailed computation is a computation where all timed transitions are of either type 1 or 2. We use $Comp^{\mathsf{d}}(\mathcal{T}, q)$ to denote all the detailed computations in $Comp(\mathcal{T}, q)$. Since the cost function is linear, considering only detailed computations is not a restriction. Formally, we have:

Lemma 2. $\forall. \pi \in Comp(\mathcal{T}, q), \exists. \pi' \in Comp^{\mathsf{d}}(\mathcal{T}, q)$ *s.t. $Cost(\pi') = Cost(\pi)$.*

Computations in δ-form. A computation is in δ-*form* if the values of all clocks and symbol ages along the computation are strictly within δ of an integer. Formally, given some $\delta : 0 < \delta < \frac{1}{10}$, we say that a configuration $\langle q, \mathtt{X}, \langle a_0, v_0 \rangle \cdots \langle a_m, v_m \rangle \rangle$ is in δ-form if (i) for all $x \in X$, $fract(\mathtt{X}(x)) < \delta$ or $fract(\mathtt{X}(x)) > 1 - \delta$, and (ii) for all $i : 1 \leq i \leq m$, $fract(v_i) < \delta$ or $fract(v_i) > 1 - \delta$. A computation $\pi = \gamma_0 \xrightarrow{t_1} \gamma_1 \xrightarrow{t_2} \cdots \xrightarrow{t_n} \gamma_n$ is in δ-form if for each $i : 1 \leq i \leq n$, the configuration γ_i is in δ-form. Note that we need an upper bound (e.g. $\frac{1}{10}$) on δ to ensure that short and long timed transitions are not mixed. We use $Comp_\delta^d(\mathcal{T}, q)$ to denote all the computations in $Comp^d(\mathcal{T}, q)$ in δ-form.

We use established linear programming techniques, first used for priced timed automata [9] and later for priced timed Petri nets [4], to show that the feasible delays for a set of computations with fixed "structure", i.e. discrete transitions, form a polyhedron with integral vertices. Since the cost-function is linear, its extreme values in the vertices of the polyhedron, which means that the optimal cost is a natural number.

Lemma 3. *For any* PTPA $\mathcal{T}$ *and state* q *of* $\mathcal{T}$, *we have* $Cost_{opt}(\mathcal{T}, q) \in \mathbb{N}$.

Due to strict inequalities in the structure of the computation, the exact delays represented by the vertex which represents the optimal cost might not be feasible, but by choosing the delays arbitrarily close to the vertex, we can get arbitrarily close to the optimal cost.

Lemma 4. *For every* $\pi \in Comp^d(\mathcal{T}, q)$ *and* $\delta : 0 < \delta < \frac{1}{10}$, *there is* $\pi' \in Comp_\delta^d(\mathcal{T}, q)$ *such that* $Cost(\pi') \leq Cost(\pi)$.

From Lemma 4 it follows that in order to find the optimal cost, we only need to consider computations in δ-form for arbitrarily small δ, i.e.

Lemma 5. $\forall \delta : 0 < \delta < \frac{1}{10}$, $Cost_{opt}(\mathcal{T}, q) = \inf\{Cost(\pi) \mid \pi \in Comp_\delta^d(\mathcal{T}, q)\}$.

The fact that a computation is detailed and in δ-form implies that the length of any timed transition is either in $(0 : \delta)$ (a *short* timed transition) or in $(1 - \delta : 1)$ (a *long* timed transition). For example, if $\delta = \frac{1}{11}$, the timed transition between γ_3 and γ_4 in Fig. 1 is long, while the other two are short. We use this to construct a PPA (the *symbolic* automaton), which simulates PTPA computations that are detailed and in δ-form for arbitrarily small δ. Since δ is arbitrarily small, the model simulates long timed transitions with length arbitrarily close to 1, and short timed transition with length arbitrarily close to 0. Therefore, in the discrete model, we pay the full stack cost for long timed transitions and nothing for short timed transitions. The cost for simulating the computation in Fig. 1 would be 10, i.e. slightly higher than the real cost of 9.64. However, the difference between the real and symbolic computations can be made arbitrarily small by decreasing δ. We build on techniques from [1] to handle the simulation of the timed part, and extend the concepts with the necessary information to handle costs.

5   Priced Pushdown Automata

A *priced pushdown automaton* (PPA) $\mathcal{P}$ is a tuple $\langle Q, q_{init}, \Gamma, \Delta_1, \Delta_2, Cost \rangle$, where Q is a finite set of states, $q_{init} \in Q$ is an initial state, Γ is a finite stack alphabet, Δ_1 and Δ_2 are finite sets of transition rules, and $Cost : (\Delta_1 \cup \Gamma) \to \mathbb{N}$ is a cost function assigning costs to both transition rules in Δ_1 and stack symbols in Γ. Intuitively, we separate the transition rules into two different sets. When

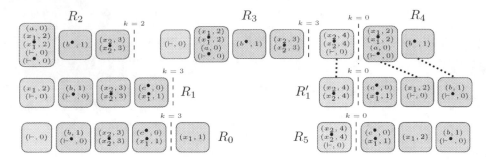

Fig. 2. Example regions used in the simulation of the example in Fig. 1

we perform transitions in Δ_1, we pay the transition cost, and when we perform transitions in Δ_2, we pay for each stack symbol in the stack. The set Δ_1 contains transition rules of the form $\langle q, op, q' \rangle$, where op is one of (i) *nop*, an operation that does not modify the stack, (ii) *push(a)*, which pushes a on top of the stack, or (iii) *pop(a)* which pops a off the stack. The set Δ_2 contains transition rules of the form $\langle q, sc, q' \rangle$, where $q, q' \in Q$ (*sc* stands for "stack cost").

Semantics. A configuration β of $\mathcal{P}$ is a tuple $\langle q, w \rangle$, where $q \in Q$ is a state and w is a word over Γ. The initial configuration β_{init} is the configuration $\langle q_{init}, \epsilon \rangle$. Let $Conf(\mathcal{P})$ denote the set of configurations of $\mathcal{P}$.

For two configurations $\beta = \langle q, w \rangle$ and $\beta' = \langle q', w' \rangle$, we have $\gamma \xrightarrow{t} \gamma'$ if either (i) $w' = w$ and $t = \langle q, nop, q' \rangle \in \Delta_1$, (ii) $w' = w \cdot a$ and $t = \langle q, push(a), q' \rangle \in \Delta_1$ for some $a \in \Gamma$, (iii) $w = w' \cdot a$ and $t = \langle q, pop(a), q' \rangle \in \Delta_1$ for some $a \in \Gamma$, or (iv) $w' = w$ and $t = \langle q, sc, q' \rangle \in \Delta_2$. We define the functions $\texttt{State}(\beta) = q$ and and $\texttt{Stack}(\beta) = w$. A computation π to a configuration β of $\mathcal{P}$ is a finite sequence $\beta_0 \xrightarrow{t_1} \beta_1 \xrightarrow{t_2} \cdots \xrightarrow{t_n} \beta_n$ where $\beta_0 = \beta_{init}$, $\beta_n = \beta$, and for all $1 \leq i \leq n$, $\beta_i \in Conf(\mathcal{P})$ and $t_i \in \Delta_1 \cup \Delta_2$. We define $Comp(\mathcal{P}, q)$ to be the set of computations to a configuration β such that $\texttt{State}(\beta) = q$. In this case, we also say that π is a computation to q.

Cost of Computations. For a transition $\beta \xrightarrow{t} \beta'$, where $\beta = \langle q, a_0 \cdots a_m \rangle$ and $\beta' = \langle q', w \rangle$, we define its cost as $Cost(\beta \xrightarrow{t} \beta') = Cost(t)$ if $t \in \Delta_1$ and $Cost(\beta \xrightarrow{t} \beta') = \sum_{i=0}^{m} Cost(a_i)$ if $t \in \Delta_2$. The cost of a computation $\pi = \beta_0 \xrightarrow{t_1} \beta_1 \xrightarrow{t_2} \cdots \xrightarrow{t_n} \beta_n$ is then defined as $Cost(\pi) = \sum_{i=1}^{n} Cost(\beta_{i-1} \xrightarrow{t_i} \beta_i)$. We define the *cost-optimal reachability problem* for PPA in the same manner as for PTPA. Lemma 6 follows by a reduction to the same problem for *priced discrete-timed pushdown automata* [2]:

Lemma 6. *Let $\mathcal{P}$ be a* PPA *and q be a state. Then $Cost_{opt}(\mathcal{P}, q)$ is computable.*

6   Regions

For timed automata, the proof of reachability is achieved using the classical region encoding [5], which is a finite-state abstraction of an uncountable state space. In [1], we showed that it is possible to extend this encoding to timed pushdown automata. More precisely, we show that given a timed pushdown automaton, it is possible to construct an untimed pushdown automaton that simulates it w.r.t. reachability properties. This works by storing regions in the stack of the untimed automaton. The main difficulty in the construction is that the automaton can only access the top element of the stack. However, we might need to remember relationships between elements that lie arbitrarily far apart in the stack. We showed that it is possible to extend the regions in a finite way to capture all such dependencies.

Let $\mathcal{T} = \langle Q, q_{init}, \Gamma, X, \Delta, Cost \rangle$ be a PTPA. We define the set $Y = X \cup \Gamma \cup \{\vdash\}$ of *plain items* and a corresponding set $Y^\bullet = X^\bullet \cup \Gamma^\bullet \cup \{\vdash^\bullet\}$ of *shadow items*. We then define the set of *items* $Z = Y \cup Y^\bullet$. Intuitively, a shadow item in a region on the stack records the value of its plain counterpart in the region below. Let c_{max} be the largest constant in the definition of $\mathcal{T}$ and $Max = \{0, 1, \ldots, c_{max}, \infty\}$. Here, ∞ is a symbolic value representing anything larger than c_{max}. A *region* R is a tuple $\langle w, k \rangle$, consisting of a word $w = r_0 r_1 \ldots r_n \in (2^{Z \times Max})^+$ and a *boundary position* $k : 0 \le k \le n$, such that w satisfies the following conditions:

- $\sum_{i=0}^{n} |r_i \cap (\Gamma \times Max)| = 1$ and $\sum_{i=0}^{n} |r_i \cap (\Gamma^\bullet \times Max)| = 1$. There is exactly one occurrence of a stack symbol and one occurrence of a shadow stack symbol.
- $\sum_{i=0}^{n} |r_i \cap (\{\vdash\} \times Max)| = 1$ and $\sum_{i=0}^{n} |r_i \cap (\{\vdash^\bullet\} \times Max)| = 1$. There is exactly one occurrence of $\vdash$ and one occurrence of $\vdash^\bullet$.
- For all clocks $x \in X$, $\sum_{i=0}^{n} |r_i \cap (\{x\} \times Max)| = 1$ and $\sum_{i=0}^{n} |r_i \cap (\{x^\bullet\} \times Max)| = 1$. Each plain clock symbol and shadow clock symbol occurs exactly once.
- $r_i \ne \emptyset$ for all $1 \le i \le n$. Only the first set may be empty.

The purpose of the boundary position is to separate the items with low fractional parts from those with high fractional parts. For $z \in Z$, if $\langle z, m \rangle \in r_i$ for some (unique) $m \in Max$ and $i \in \{0, \ldots, n\}$, then let $Val(R)(z) = m$ and $Index(R)(z) = i$. Otherwise, define $Val(R)(z) = \perp$ and $Index(R)(z) = \perp$ (this may only be the case for stack symbols). We define $R^\top = \{z \in Z \mid Index(R)(z) \ne \perp\}$. Note that the set of regions, w.r.t. fixed X, Γ and c_{max}, is finite. As an example, the region R_0 in Fig. 2 is the topmost region in the symbolic stack representing the configuration γ_0 in Fig. 1.

Next, we define a number of operations on regions that we need for the construction of the symbolic automaton.

Satisfiability. Given an item $z \in Z$, an interval $I \in \mathcal{I}$, and a region R such that $z \in R^\top$, we write $R \models (z \in I)$ if one of the following conditions holds:

- $Index(R)(z) = 0$, $Val(R)(z) \ne \infty$ and $Val(R)(z) \in I$. If the fractional part of z is 0, we test if the value of z is in I.
- $Index(R)(z) > 0$, $Val(R)(z) \ne \infty$ and $(Val(R)(z) + v) \in I$ for all $v \in \mathbb{R}^{\ge 0}$ such that $0 < v < 1$. If the fractional part of z is greater than 0, we test if its integral part increased by any real number $v : 0 < v < 1$ is in the interval.
- $Val(R)(z) = \infty$, and I is of the form $(m : \infty)$ or of the form $[m : \infty)$. If the integral part of z is ∞, then the interval cannot have an upper bound.

Adding and Removing Items. For a region $R = \langle r_0 \ldots r_n, k \rangle$, an item $z \in Z$, and $m \in Max$, we define $R \oplus \langle z, m \rangle$ to be the set of regions R' which satisfy one of the following conditions:

- $R' = \langle r_0 \ldots r_{i-1}(r_i \cup \{\langle z, m \rangle\})r_{i+1} \ldots r_n, k \rangle$, where $0 \leq i \leq n$. The item is added into an existing set, in which case k is unchanged.
- $R' = \langle r_0 \ldots r_i \{\langle z, m \rangle\} r_{i+1} \ldots r_n, k + 1 \rangle$, where $1 \leq i \leq k$. The item is added as a new singleton set to the left of k, in which case k is increased by 1.
- $R' = \langle k, r_0 \ldots r_i \{\langle z, m \rangle\} r_{i+1} \ldots r_n, k \rangle$, where $k \leq i \leq n$. The item is added as a new set to the right of k, which is left unchanged.

We define $R \ominus z$ to be the region $R' = \langle r_0 \cdots r_{j-1}(r_j \setminus (\{z\} \times Max))r_{j+1} \ldots r_n, k' \rangle$, where j is the unique value s.t. $r_j \cap (\{z\} \times Max) \neq \emptyset$ and where $k' = k - 1$ if $0 < j \leq k$ and $|r_j| = 1$, and $k' = k$ otherwise. In other words, we delete z from a set, if it exists, and update k accordingly. We extend the definition of $\ominus$ to sets of items by letting $R \ominus \{z_1 \ldots z_n\} = (\cdots((R \ominus z_1) \ominus z_2) \cdots) \ominus z_n$.

Resetting. For a region $R = \langle r_0 \ldots r_n, k \rangle$ and an item $z \in Z$, we define $R[z \leftarrow 0]$ to be the unique region $\langle (r'_0 \cup \{\langle z, 0 \rangle\})r'_1 \cdots r'_n, k \rangle$, where $r'_0 r'_1 \cdots r'_n = R \ominus z$. We delete z from the region and reintroduce it with value 0. This operation is used when we simulate the resetting of clocks.

Pushing New Regions. This operation is used to simulate the pushing of new regions. It takes a region R and a stack symbol $a \in \Gamma$, and creates a new region R' in which the shadow items record the values of the plain items in R and where the value of a is 0. We define $New(R, a)$ to be the region R' such that there are R_1, R_2, R_3 satisfying the following conditions (for example, in Fig. 2, $R_2 = New(R_1, a)$):

- $R_1 = \langle r_0 \cdots r_{n_1}, k \rangle = R \ominus (R^\top \cap Y^\bullet)$. Delete all shadow items from R.
- $R_2 = \langle r'_0 \ldots r'_{n_1}, k \rangle$, where $r'_i = r_i \cup \{\langle y^\bullet, m \rangle \mid \langle y, m \rangle \in r_i\}$ for $0 \leq i \leq n_1$. Add fresh shadow items with the same values and the same indices as their plain counterparts.
- $R_3 = \langle r''_0 \ldots r''_{n_2}, k \rangle = R_2 \ominus (R^\top \cap \Gamma)$. Delete the previous stack symbol.
- $R' = \langle (r''_0 \cup \{\langle a, 0 \rangle\})r''_1 \ldots r''_{n_2}, k \rangle$. Introduce a with value 0.

Passage of Time. Next, we describe operations that simulate the passage of time. Given a pair $\langle z, m \rangle \in Z \times Max$, define $\langle z, m \rangle^\oplus = \langle z, m' \rangle$, where $m' = m + 1$ if $m < c_{max}$, and $m' = \infty$ otherwise. For a set $r \in 2^{Z \times Max}$, define $r^\oplus = \{\langle z, m \rangle^\oplus \mid \langle z, m \rangle \in r\}$. In other words, we increase the integral part of each item by 1, up to c_{max}. For a region $R = \langle r_0 \ldots r_n, k \rangle$, define $R^\oplus = \langle w', k' \rangle$ in the following way:

- If $r_0 \neq \emptyset$, then $w' = \emptyset r_0 \ldots r_n$, and $k' = k + 1$. A small amount of time passes, which results in the first region being "pushed" out.
- If $r_0 = \emptyset$ and $k < n$, then $w' = r_n^\oplus r_1 \ldots r_{n-1}$ and $k' = k$. The items in the last region reach their next integral value but the small fractional parts remain small.
- If $r_0 = \emptyset$ and $k = n$, then $w' = w$ and $k' = 0$. All small fractional parts become large, but no integral part changes.

We denote by $R^{\oplus\oplus}$ the set $\{R, R^{\oplus}, (R^{\oplus})^{\oplus}, \ldots\}$. Note that this set is finite. We define $R_{\vdash}^{\oplus}$ to be the region R' such that there are R_1 and R_2 satisfying the following conditions:

- $R_1 = R^{\oplus} \ominus \vdash$. We remove the item $\vdash$.
- $R_2 \in R_1 \oplus (\vdash, 0)$ and $R_2 \models (\vdash \in [0..0])$. We reintroduce $\vdash$ by placing it in the leftmost set. Note that R_2 is unique.

The operation $R_{\vdash}^{\oplus}$ simulates passage of time while maintaining the value and index of $\vdash$. We define $R_{\vdash}^{\oplus\oplus}$ similarly to $R^{\oplus\oplus}$. In Fig. 2, we have that $R_3 \in R_2{}^{\oplus\oplus}$.

Product. We now define the product operator $\odot$ that merges the information in two regions. This operation is used when simulating *pop* transitions. For regions $R_1 = \langle w_1, k_1 \rangle$ and $R_2 = \langle w_2, k_2 \rangle$, we write $R_1 \preceq R_2$ if there are $i_0 < i_1 < \cdots < i_\ell \leq |w_1|$ and $j_0 < j_1 < \cdots < j_\ell \leq |w_2|$ such that: (1) $i_0 = j_0 = 0$, (2) $\{i_1, \ldots, i_\ell\} \subseteq Index\,(R_1)\,(R_1^\top \cap Y) \subseteq \{i_0, i_1, \ldots, i_\ell\}$, (3) $\{j_1, \ldots, j_\ell\} \subseteq Index\,(R_2)\,\left(R_2{}^\top \cap Y^\bullet\right) \subseteq \{j_0, j_1, \ldots, j_\ell\}$, (4) $(R_2{}^\top \cap Y^\bullet) = \{y^\bullet \mid y \in (R_1^\top \cap Y)\}$, and (5) for every $h : 0 < h \leq \ell$ and $y \in (R_1^\top \cap Y)$, we have: (i) $Index\,(R_1)\,(y) = i_h$ iff $Index\,(R_2)\,(y^\bullet) = j_h$, and (ii) $j_h \leq k_2$ iff $i_h \leq k_1$. In this case we say that R_1 *supports* R_2. Intuitively, this means that the shadow items in R_2 match their plain counterparts in R_1 and that the information in the two regions can be merged. In Fig. 2, we have that $R_1' \preceq R_4$. The matching is illustrated by dotted lines.

Assume that $R_1 \preceq R_2$, $w_1 = r_0 \ldots r_n$ and $w_2 = r_0' \ldots r_{n'}'$. Let $v_h = r_{i_h+1} \cdots r_{i_{h+1}-1}$, $v_h' = r_{i_h+1}' \cdots r_{i_{h+1}-1}'$ for all $h : 0 \leq h < \ell$, $v_\ell = r_{i_\ell+1} \cdots r_n$, and $v_\ell' = r_{i_\ell+1}' \cdots r_{n'}'$. Note that the sequences of indices $i_0, \ldots i_\ell$ and $j_0, \ldots, j_\ell$ are unique. We define $p_h = r_{i_h} \cap (Y^\bullet \cup \Gamma)$ and $p_h' = r_{j_h}' \cap (X \cup \{\vdash\})$ for all $h : 0 < h \leq \ell$. We define $q_0 = p_0 \cup p_0'$ and, for $1 \leq h \leq \ell$, define $q_h = p_h \cup p_h'$ if $p_h \cup p_h' \neq \emptyset$ and $q_h = \epsilon$ otherwise. Then, $w \in w_1 \odot w_2$ if $w = q_0 \cdot w_0' \cdot q_1 \cdots q_\ell \cdot w_\ell'$ and $w_h' \in (v_h \otimes v_h')$ for $h : 0 \leq h \leq \ell$. Intuitively, we take the clocks from w_2, and the shadow items and stack symbol from w_1. For regions $R_1 = \langle w_1, k_1 \rangle$ and $R_2 = \langle w_2, k_2 \rangle$, we have $R = \langle w, k \rangle \in R_1 \odot R_2$ if $w \in w_1 \odot w_2$ and for all $z \in Z$, the following conditions hold (in Fig. 2 we have $R_5 \in R_1' \odot R_4$):

- $Index\,(R)\,(z) \leq k$ iff $Index\,(R_1)\,(z) \leq k_1$, for $z \in \Gamma^\bullet \cup X^\bullet \cup \Gamma$. The boundary should be preserved for the shadow items and the stack symbol from R_1.
- $Index\,(R)\,(z) \leq k$ iff $Index\,(R_2)\,(z) \leq k_2$, for $z \in X$. The boundary should be preserved for the clocks taken from R_2.

7 Simulation

We will describe how, given a PTPA $\mathcal{T} = (Q, q_{init}, \Gamma, X, \Delta, Cost)$, one can construct a priced pushdown automaton PPA $\mathcal{P} = (Q^{\mathcal{P}}, q_{init}^{\mathcal{P}}, \Gamma^{\mathcal{P}}, \Delta_1^{\mathcal{P}}, \Delta_2^{\mathcal{P}}, Cost^{\mathcal{P}})$ that simulates detailed computations in δ-form of $\mathcal{T}$ for arbitrarily small δ.

The states of $\mathcal{P}$ contain both the states of $\mathcal{T}$, which are called *genuine*, and a set of *temporary* states which are used as intermediate states to simulate the transitions of $\mathcal{T}$. More precisely, we use a set $Q_{\tt tmp}$ of temporary states s.t. $Q_{\tt tmp} \cap Q = \emptyset$. We write $\mathtt{tmp}(\ldots)$, $\mathtt{tmp}_1(\ldots)$, $\mathtt{tmp}_2(\ldots)$ and $\mathtt{tmp}_3(\ldots)$ to denote unique elements of Q_{tmp}, where the arguments are used to uniquely identify elements of Q_{tmp}. We also assume that $q_{init}^{\mathcal{P}} \notin Q$ and $q_{init}^{\mathcal{P}} \notin Q_{\text{tmp}}$. Then, $Q^{\mathcal{P}}$ is defined as $Q \cup Q_{\text{tmp}} \cup \{q_{init}^{\mathcal{P}}\}$.

The stack alphabet $\Gamma^{\mathcal{P}}$ of $\mathcal{P}$ is the set of regions over the stack alphabet $\Gamma \cup \{\perp\}$, the set of clocks X and the maximal constant c_{max}, where $\perp \notin \Gamma$ is a special symbol that represents the bottom of the stack. Note that $\Gamma^{\mathcal{P}}$ is finite. We define $Cost^{\mathcal{P}}(\perp) = 0$. The cost of a region R is defined as $Cost^{\mathcal{P}}(R) = Cost(a)$, where a is the unique stack symbol s.t. $R^{\top} \cap \Gamma = \{a\}$.

Let $w_{init} = \{\langle z, 0 \rangle \mid z \in X \cup X^{\bullet} \cup \{\vdash, \vdash^{\bullet}\}\} \cup \{\langle \perp, 0 \rangle, \langle \perp^{\bullet}, 0 \rangle\}$. The symbolic automaton starts in its initial configuration $\langle q_{init}^{\mathcal{P}}, \epsilon \rangle$ and pushes the initial region $R_{init} = \langle w_{init}, 0 \rangle$ on the stack while moving to the initial state of $\mathcal{T}$, i.e. $\Delta_1^{\mathcal{P}}$ contains the rule $\langle q_{init}^{\mathcal{P}}, push(R_{init}), q_{init} \rangle$. We define

$$Cost^{\mathcal{P}}(\langle q_{init}^{\mathcal{P}}, push(R_{init}), q_{init} \rangle) = 0.$$

Then, $\mathcal{P}$ starts the simulation of $\mathcal{T}$. The transitions of $\mathcal{T}$ are simulated in the following way (as an example, Fig. 2 shows the regions involved in the simulation of Fig. 1):

- *Nop.* For every $t = \langle q_1, nop, q_2 \rangle \in \Delta$, the set $\Delta_1^{\mathcal{P}}$ contains $t = \langle q_1, nop, q_2 \rangle$. Define $Cost^{\mathcal{P}}(t) = Cost(t)$.
- *Push.* We need two temporary states to simulate this transition. First, we move to a temporary state and pop the topmost region in order to remember its content. Then, we push back that region, moving to the second temporary state. Finally, we push a new topmost region that we construct using the remembered values. For every $t = \langle q_1, push(a), q_2 \rangle \in \Delta$, and every region R, the set $\Delta_1^{\mathcal{P}}$ contains $t_1 = \langle q_1, pop(R), \text{tmp}_1(t, R) \rangle$, $t_2 = \langle \text{tmp}_1(t, R), push(R), \text{tmp}_2(t, R) \rangle$, and $t_3 = \langle \text{tmp}_2(t, R), push(New(R, a)), q_2 \rangle$. We define $Cost^{\mathcal{P}}(t_1) = 0$, $Cost^{\mathcal{P}}(t_2) = 0$ and $Cost^{\mathcal{P}}(t_3) = Cost(t)$.
- *Pop.* To simulate pop transitions, we use two temporary states. We first pop the topmost region. The new topmost region then needs to be updated to reflect the changes that occurred while it was inaccessible. To do this, we rotate it until it matches the popped region. Then, we merge the information in both regions. In this way, the information about changes "ripples" down the stack as elements are popped. The effect of both these steps is captured by popping the new topmost region and pushing a region which is a product of the two popped regions. Formally, for every $t = \langle q_1, pop(a \in I), q_2 \rangle \in \Delta$, and all regions R_1, R_2 such that $R_2 \models a \in I$, the set $\Delta_1^{\mathcal{P}}$ contains $t_1 = \langle q_1, pop(R_2), \text{tmp}_1(t, R_2) \rangle$ and $t_2 = \langle \text{tmp}_1(t, R_2), pop(R_1), \text{tmp}_2(t, R_2, R_1) \rangle$. Additionally, $\Delta_1^{\mathcal{P}}$ contains $t_{R_3} = \langle \text{tmp}_2(t, R_2, R_1), push(R_3), q_2 \rangle$ for all $R_3 \in \{R_1' \odot R_2 \mid R_1' \in R_1^{\oplus\oplus}, R_1' \preceq R_2\}$. We define their costs as $Cost^{\mathcal{P}}(t_1) = 0$, $Cost^{\mathcal{P}}(t_2) = 0$ and $Cost^{\mathcal{P}}(t_{R_3}) = Cost(t)$ for all the above R_3. In Fig. 2, R_5 is the result of popping R_4 when R_1 is the second topmost region.
- *Test.* We simulate a test transition with two transitions. First, we pop the region if it satisfies the condition, while moving to a temporary state. Next, we push the same region and move to the new genuine state. For every $t = \langle q_1, test(x \in I), q_2 \rangle \in \Delta$, and every region R such that $R \models x \in I$, the set

Δ_1 contains $t_1 = \langle q_1, pop(R), \mathtt{tmp}(t, R) \rangle$ and $t_2 = \langle \mathtt{tmp}(t, R), push(R), q_2 \rangle$. Define $Cost^P(t_1) = 0$ and $Cost^P(t_2) = Cost(t)$.

- *Reset.* We simulate resetting x by popping the topmost region and pushing back a region which is identical, except that x is 0. For every $t = \langle q_1, reset(x), q_2 \rangle \in \Delta$, and every region R, the set of Δ_1^P contains $t_1 = \langle q_1, pop(R), \mathtt{tmp}(t, R) \rangle$ and $t_2 = \langle \mathtt{tmp}(t, R), push(R[x \leftarrow 0]), q_2 \rangle$.

Define $Cost^P(t_1) = 0$ and $Cost^P(t_2) = Cost(t)$.

- *Timed Transitions.* To simulate timed transitions, we pop the topmost region, rotate it, and push it back. Let $R = \langle r_0 \ldots r_n, k \rangle$ be a region.
  - If $r_0 \neq \emptyset$ (resp. $r_0 = \emptyset$ and $k < n$), then we simulate a short timed transition which makes the fractional part of all value positive (resp. the highest fractional part 0). In this case, Δ_1^P contains

$$t_1 = \langle q, pop(R), \mathtt{tmp}_1(time, q, R) \rangle \text{ and}$$

$$t_2 = \langle \mathtt{tmp}_1(time, q, R), push(R_\leftarrow^\oplus), q \rangle .$$

We define $Cost^P(t_1) = Cost^P(t_2) = 0$.
  - If $r_0 = \emptyset$ and $k - n$, then we simulate a long timed transition. In this case, Δ_1^P contains

$$t_1 = \langle q, pop(R), \mathtt{tmp}_1(time, q, R) \rangle \text{ and}$$

$$t_2 = \langle \mathtt{tmp}_1(time, q, R), push(R_\leftarrow^\oplus), \mathtt{tmp}_2(time, q, R) \rangle .$$

Define $Cost^P(t_1) = Cost^P(t_2) = 0$. Since the transition is long, we must pay the stack cost. Therefore, Δ_2^P contains $\langle \mathtt{tmp}_2(time, q, R), sc, q \rangle$.

Correctness. Consider a detailed computation π_T in δ-form of length n in T and its simulation π_P in P. We will give a bound on the difference in cost between each step in π_T compared to its corresponding steps in π_P. The costs of the discrete steps are identical. Now we consider timed transitions. If the timed transition is short, its cost is bounded by δ multiplied by the stack cost at that step, while the cost of the corresponding steps in π_P is equal to 0. On the other hand, if the timed transition is long, the cost is bounded by $(1 - \delta)$ multiplied by the stack cost, while the cost of the corresponding steps is exactly equal to the stack cost. Consequently, the difference is always bounded by δ multiplied by the stack cost. Since the stack cost is bounded by the length of the stack multiplied by the cost of the most expensive symbol, and since the length of the stack is bounded by n, the difference in price at each step is bounded by $\delta \cdot |\pi_T| \cdot \max\{Cost(a) \mid a \in \Gamma\}$. This implies that the total difference in cost between π_T and π_P is bounded by $\delta \cdot |\pi_T|^2 \cdot \max\{Cost(a) \mid a \in \Gamma\}$. This gives the following lemma:

Lemma 7. $\forall \delta : 0 < \delta < \frac{1}{10}$ and $\forall \pi_T . \pi_T \in Comp_\delta^d(T, q)$, $\exists \pi_P . \pi_P \in Comp(P, q)$ s.t. $|Cost^P(\pi_P) - Cost(\pi_T)| \leq \delta \cdot |\pi_T|^2 \cdot \max\{Cost(a) \mid a \in \Gamma\}$.

The other direction can be explained in a similar manner:

Lemma 8. $\forall \pi_{\mathcal{P}}.\ \pi_{\mathcal{P}} \in Comp(\mathcal{P}, q),\ and \forall \delta : 0 < \delta < \frac{1}{10},\ \exists \pi_{\mathcal{T}}.\ \pi_{\mathcal{T}} \in Comp_{\delta}^{\mathsf{d}}(\mathcal{T}, q)$
$s.t.\ |Cost^{\mathcal{P}}(\pi_{\mathcal{P}}) - Cost(\pi_{\mathcal{T}})| \leq \delta \cdot |\pi_{\mathcal{T}}|^2 \cdot \max\{Cost(a) \mid a \in \Gamma\}.$

We now combine Lemma 2, Lemma 4, and Lemmas 8 and 7 to prove the following:

Theorem 9. $Cost_{opt}(\mathcal{T}, q) = Cost_{opt}(\mathcal{P}, q)$ *for any state* $q \in Q$.

Thus, the cost-optimal reachability problem for PTPA reduces to the same problem for PPA, which is computable by Lemma 6. This concludes the proof of Theorem 1.

8   Conclusion

We have studied the cost-optimal reachability problem for priced dense-timed pushdown automata, and shown that this problem can be reduced to the same problem for priced (untimed) pushdown automata, which in turn can be reduced to the reachability problem for ordinary pushdown automata. This yields an algorithm for computing the optimal reachability cost for PTPA. To simplify the exposition, we assumed that push (resp. reset) operations result in the stack symbol having age (resp. affected clock having value) 0. The model still strictly subsumes that of [1] (we can encode the intervals in the stack symbols).

A simple generalization is to make the stack symbol storage cost dependent on the current control-state. Our construction can be trivially extended to handle this case. (We need to consider a similar extension for the priced pushdown automata.)

A challenging problem which we are currently considering is to extend our results to the case of negative costs.

References

1. Abdulla, P.A., Atig, M.F., Stenman, J.: Dense-timed pushdown automata. In: LICS (2012)
2. Abdulla, P.A., Atig, M.F., Stenman, J.: The minimal cost reachability problem in priced timed pushdown systems. In: Dediu, A.-H., Martín-Vide, C. (eds.) LATA 2012. LNCS, vol. 7183, pp. 58–69. Springer, Heidelberg (2012)
3. Abdulla, P.A., Čerāns, K., Jonsson, B., Tsay, Y.K.: General decidability theorems for infinite-state systems. In: LICS, pp. 313–321 (1996)
4. Abdulla, P.A., Mayr, R.: Computing optimal coverability costs in priced timed Petri nets. In: LICS (2011)
5. Alur, R., Dill, D.L.: A theory of timed automata. TCS 126(2), 183–235 (1994)
6. Alur, R., La Torre, S., Pappas, G.J.: Optimal paths in weighted timed automata. TCS 318(3), 297–322 (2004)
7. Behrmann, G., Fehnker, A., Hune, T., Larsen, K.G., Pettersson, P., Romijn, J., Vaandrager, F.W.: Minimum-cost reachability for priced timed automata. In: Di Benedetto, M.D., Sangiovanni-Vincentelli, A.L. (eds.) HSCC 2001. LNCS, vol. 2034, pp. 147–161. Springer, Heidelberg (2001)

8. Bouajjani, A., Esparza, J., Maler, O.: Reachability analysis of pushdown automata: Application to model-checking. In: Mazurkiewicz, A., Winkowski, J. (eds.) CON-CUR 1997. LNCS, vol. 1243, pp. 135–150. Springer, Heidelberg (1997)

9. Bouyer, P., Brihaye, T., Bruyere, V., Raskin, J.F.: On the optimal reachability problem of weighted timed automata. FMSD 31(2), 135–175 (2007)

10. Dang, Z.: Pushdown timed automata: a binary reachability characterization and safety verification. TCS 302(1-3), 93–121 (2003)

11. Karp, R.M., Miller, R.E.: Parallel program schemata. JCSS (1969)

12. Trivedi, A., Wojtczak, D.: Recursive timed automata. In: Bouajjani, A., Chin, W.-N. (eds.) ATVA 2010. LNCS, vol. 6252, pp. 306–324. Springer, Heidelberg (2010)

13. Uezato, Y., Minamide, Y.: Pushdown systems with stack manipulation. In: Van Hung, D., Ogawa, M. (eds.) ATVA 2013. LNCS, vol. 8172, pp. 412–426. Springer, Heidelberg (2013)

Formulae for Polyominoes on Twisted Cylinders

Gadi Aleksandrowicz[1], Andrei Asinowski[2], Gill Barequet[1],
and Ronnie Barequet[3]

[1] Dept. of Computer Science
Technion—Israel Institute of Technology
Haifa 32000, Israel
{gadial,barequet}@cs.technion.ac.il
[2] Institut für Informatik
Freie Universität Berlin
Takustraße 9, 14195 Berlin, Germany
asinowski@gmail.com
[3] Dept. of Computer Science
Tel Aviv University
Tel Aviv 69978, Israel
ronnieba@post.tau.ac.il

Abstract. Polyominoes are edge-connected sets of cells on the square lattice $\mathbb{Z}^2$. We investigate polyominoes on a square lattice embedded on so-called *twisted cylinders* of a bounded width (perimeter) w. We prove that the limit growth rate of polyominoes of the latter type approaches that of polyominoes of the former type, as w tends to infinity. We also prove that for any fixed value of w, the formula enumerating polyominoes on a twisted cylinder of width w satisfies a linear recurrence whose complexity grows exponentially with w. By building the finite automaton that "grows" polyominoes on the twisted cylinder, we obtain the prefix of the sequence enumerating these polyominoes. Then, we recover the recurrence formula by using the Berlekamp-Massey algorithm.

Keywords: Recurrence formula, transfer matrix, generating function.

1 Introduction

Polyominoes (also known as *lattice animals*) of size n are edge-connected sets of n cells on the orthogonal lattice $\mathbb{Z}^2$. *Fixed* polyominoes are considered distinct if they differ in their shapes *or* orientations. The number of fixed polyominoes of size n is usually denoted in the literature by $A(n)$. There are two long-standing open problems related to the study of polyominoes.

1. The enumeration of polyominoes, that is, finding a formula for $A(n)$ or computing $A(n)$ for specific values of n; and
2. Computing $\lim_{n\to\infty} A(n+1)/A(n)$, the asymptotic *growth rate* of polyominoes (also called "Klarner's constant").

A.-H. Dediu et al. (Eds.): LATA 2014, LNCS 8370, pp. 76–87, 2014.

To date, no analytic formula is known for $A(n)$. The best currently-known method (in terms of running time) for counting fixed polyominoes is a transfer-matrix algorithm suggested by Jensen [6]. In a parallel version of this algorithm, Jensen was able to compute $A(n)$ up to $n = 56$ [7]. In a seminal work, Klarner [8] showed that the limit $\lambda := \lim_{n \to \infty} \sqrt[n]{A(n)}$ exists. Only three decades later did Madras [10] show that the limit $\lim_{n \to \infty} A(n+1)/A(n)$ exists and, hence, is equal to λ. At the present time not even a single significant decimal digit of λ is known. The best-known lower [3] and upper [9] bounds on λ are 3.9801 and 4.6496, respectively. It is generally assumed (see, e.g., [5]), as a conclusion from numerical methods applied to the known values of $A(n)$, that $\lambda = 4.06 \pm 0.02$. Jensen [7] refined this analysis, estimating λ at 4.0625696 ± 0.0000005.

The notion of polyominoes on a *twisted cylinder* was introduced in [3]. For a fixed natural number w, the twisted cylinder of *width*[1] w is the surface obtained from the Euclidean plane by identifying all pairs of points of the form (x, y), $(x - kw, y + k)$, for $k \in \mathbb{Z}$. Pictorially, we can imagine this as taking the strip $[0, w] \times \mathbb{R}$ and gluing, for all y, (w, y) with $(0, y + 1)$ (see Figure 1(a)).

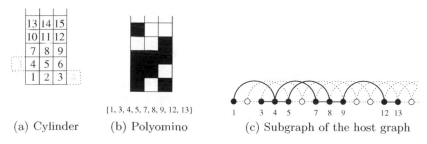

{1, 3, 4, 5, 7, 8, 9, 12, 13}

(a) Cylinder   (b) Polyomino   (c) Subgraph of the host graph

Fig. 1. A polyomino on the twisted cylinder of width 3

A polyomino on the twisted cylinder is a finite edge-connected set of unit integer cells on this cylinder. Two polyominoes obtained from each other by translation are considered identical. In order to have a unique representation, we shall assume without loss of generality that the cell $[0, 1] \times [0, 1]$ is the leftmost cell belonging to all polyominoes. We shall label this cell by 1, and all other cells on the cylinder will lie on the right of 1 so that the cell i will be the right neighbor of the cell $i - 1$. Therefore, the twisted cylinder of width w may be seen as a host graph (dual of the lattice graph) whose vertices are labeled by $\mathbb{N}$, and there is an edge $\{i, j\}$ if $|i - j| = 1$ (a horizontal edge) or $|i - j| = w$ (a vertical edge). Alternatively, we have a half-infinite circulant graph with $\aleph_0$ vertices, in which each vertex $i \in \mathbb{N}$ is connected to vertices $i \pm 1$ and $i \pm w$ (unless the attempted connection is to a non-natural number). In this interpretation, a polyomino P may be regarded as a finite subset of $\mathbb{N}$ such that $1 \in P$ and the corresponding induced subgraph (in the mentioned-above host graph) is connected. Figures 1(b,c) show a sample polyomino on the twisted cylinder of width 3.

---

[1] The term 'width,' and not 'perimeter,' is used for consistency with [6] and [3].

Let $A_w(n)$ denote the number of polyominoes on the twisted cylinder of width w up to translation. It is easy to observe [3] that for $n > w$, we have $A_w(n) < A(n)$. This follows from the fact that one can embed any polyomino on a twisted cylinder in the regular plane, but the opposite operation is not always possible due to self-overlaps of the polyominoes. Furthermore, it is also true [ibid.] that for any fixed value of w, the limit $\lambda_w := \lim_{n \to \infty} A_w(n+1)/A_w(n)$ exists. Moreover, we also have $\lambda_1 \leq \lambda_2 \leq \lambda_3 \leq \cdots$. It was conjectured in [3] that this sequence converges to λ, a claim which we prove in this paper.

It is proven in [3] that λ_w is the only positive eigenvalue of a huge matrix (the size of each of which dimensions is proportional to the wth Motzkin number), and much effort is spent in that work on computing numerically this eigenvalue for as large as possible value of w. The currently best lower bound on λ is achieved by computing $\lambda_{22} \approx 3.9801$. Since $\lim_{w \to \infty} \lambda_w = \lambda$, we believe that finding explicit formulae for the numbers of polyominoes on twisted cylinders of different widths, and understanding the relations between these formulae, will shed light on the real value of λ. In the current paper we investigate the values known of this sequence, and find, by using numerical methods, that it converges to roughly 4.068. This supports the common belief [5] that λ is approximately 4.06, which, as mentioned above, is based on a completely different method.

It is easy to show that $A_2(n) = 2^{n-1}$. This follows immediately from observing the possible edges connecting cells of a polyomino in the host graph. Order the cells according to their ordinal numbers on the cylinder. Then, a cell must be connected to the next cell (according to this order) by either a horizontal edge (if the difference between them is 1) or a vertical edge (if the difference is 2). Since the cylinder is narrow, no "spiraling" is possible.

In this work we obtain enumeration results for polyominoes on twisted cylinders by building finite automata that describe the "growth" of polyominoes on these cylinders, computing their transfer matrices (similarly to [3]), and obtaining from these matrices the generating functions and the recurrence formulae for the number of polyominoes. This method can be generalized to any width. This gives rise to our main result, namely, that the formula enumerating polyominoes on any twisted cylinder obeys a linear recurrence. In practice, the size of the involved matrix increases exponentially (as do Motzkin numbers), and so, the amount of computations, as well as the orders of the computed formulae, becomes prohibitively large (at least, with respect to the computing resources available to us) for width greater than 10.

2   Convergence of Growth Rates

We begin this paper by proving a conjecture made in [3, p. 32], namely, that the asymptotic growth rate of polyominoes on a twisted cylinder of width w approaches that of polyominoes in the plane, as w tends to infinity.

Theorem 1. $\lim_{w \to \infty} \lambda_w = \lambda$.

Proof. By Madras [10] we know that

$$\lambda = \lim_{n \to \infty} A(n+1)/A(n) = \lim_{n \to \infty} \sqrt[n]{A(n)},$$

and that a similar property holds for a twisted cylinder of any width w, that is,

$$\lambda_w = \lim_{n \to \infty} A_w(n+1)/A_w(n) = \lim_{n \to \infty} \sqrt[n]{A_w(n)}.$$

By Barequet et al. [3] we also know that

$$\lambda_1 \le \lambda_2 \le \cdots \le \lambda,$$

thus, the limit $\lambda^* = \lim_{w \to \infty} \lambda_w$ exists and we also have $\lambda^* \le \lambda$. Our goal is to prove that $\lambda^* = \lambda$. We will prove this by showing that for any $\varepsilon > 0$ there exists w_ε such that $\lambda_{w_\varepsilon} \ge \lambda - \varepsilon$.

Since $\lambda = \lim_{n \to \infty} \sqrt[n]{A(n)}$, for any $\varepsilon > 0$ there exists $n_0 = n_0(\varepsilon)$ such that for all $n \ge n_0$ we have $\sqrt[n]{A(n)} \ge \lambda - \varepsilon$, in particular, $\sqrt[n_0]{A(n_0)} \ge \lambda - \varepsilon$. Consider the twisted cylinder of width n_0. For any $k \le n_0$, we obviously have that $A_{n_0}(k) = A(k)$, since the twist of the cylinder has no effect on polyominoes whose size is smaller than the width of the cylinder. In addition, we argue that there exists an infinite monotone increasing subsequence of $(\sqrt[k]{A_{n_0}(k)})$, for $k \ge n_0$.[2] This follows immediately from a standard polyomino concatenation argument. As in the plane, we always have, for any w, that $A_w(n)A_w(m) \le A_w(n+m)$. By fixing $w = n_0$ and $m = n$ we obtain that $A_{n_0}^2(n) \le A_{n_0}(2n)$, that is, $\sqrt[n]{A_{n_0}(n)} \le \sqrt[2n]{A_{n_0}(2n)}$. Thus, $(\sqrt[2^i n_0]{A_{n_0}(2^i n_0)})$ (for $i \ge 0$) is an appropriate subsequence. Since the original sequence converges, so does its subsequence, and since the latter is monotone increasing, it lies entirely below its limit. Therefore,

$$\lambda_{n_0} = \lim_{n \to \infty} \sqrt[n]{A_{n_0}(n)} \ge \sqrt[n_0]{A_{n_0}(n_0)} = \sqrt[n_0]{A(n_0)} \ge \lambda - \varepsilon,$$

hence, choosing $w_\varepsilon = n_0$ completes the proof.

---

[2] A similar claim for polyominoes in the plane is a well-known folklore lemma. In fact, one can prove a slightly stronger claim, that the entire sequence is found below its limit. However, although it is widely believed that the entire sequence in monotone increasing, no proof for this is currently known.

To date, the growth rates λ_w were published [3] up to $w = 22$. Five more values (up to $w = 27$) were recently computed using a supercomputer. Specifically, $(\lambda_w)_{w=1}^{27}$ = (1, 2, 2.65897, 3.06090, 3.31410, 3.48094, 3.59606, 3.67875, 3.74022, 3.78724, 3.82409, 3.85355, 3.87752, 3.89732, 3.91388, 3.92790, 3.93988, 3.95021, 3.95920, 3.96706, 3.97399, 3.98014, 3.98562, 3.99052, 3.99494, 3.99893, 4.00254). Figure 2 plots λ_w as a function of w.

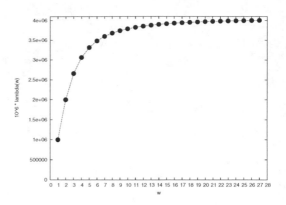

Fig. 2. A plot of λ_w as a function of w

We attempted to estimate the limit of (λ_w) using three methods. First, we represented λ_w as a solution to a linear-programming problem. Assuming that λ_w has a $1/w$-expansion, we approximated it by the sum of the twenty leading terms of this expansion, that is, $\lambda_w \sim f(w) = \sum_{i=0}^{19} c_i/w^i$. The linear system included the constraints $f(w) \leq \lambda_w + \varepsilon$ and $f(w) \geq \lambda_w - \varepsilon$, for $4 \leq w \leq 27$. We excluded $w = 1, 2, 3$ since when trimming the expansion, the smaller w is, the larger the error introduced to the value of the function is. (This is in conjunction with the fact that we trimmed the values of λ_w after the 5th or 6th digit after the decimal point.) The target function was simply ε, and the goal was to minimize it. Using Mathematica, we found that c_0 (the free term in the $1/w$-expansion) was about 4.06714. The other coefficients were $c_1 \approx -0.848309$, $c_2 \approx -27.9819$, $c_3 \approx 116.155$, $c_4 \approx -405.076$, $c_5 \approx 1023.27$, $c_6 \approx -1135.33$, and $c_i = 0$ for $7 \leq i \leq 19$. The obtained error was $\varepsilon \approx 6.49 \cdot 10^{-6}$.

Second, we computed the vector $\bar{x}$ that solves the linear least-squares problem for the matrix equation $B \cdot \bar{x} = \bar{b}$, where $B_{i,j} = 1/i^j$ and $b_i = \lambda_i$ (taking, again, $4 \leq i \leq 27$ and, this time, $0 \leq j \leq 6$). Using Mathematica, we found that $\bar{x}_0$ was about 4.06727. The other values x_i, for $1 \leq i \leq 6$, were approximately -0.856246, -27.7995, 114.068, -392.374, 984.007, and -1087.02, which compares well with c_1 through c_6 from the previous method. These two methods strongly support the widely-believed [5,7] estimate of λ at 4.06 ± 0.02.

Third, Wynn's epsilon algorithm [14] yielded 4.04161 as the estimated limit of the sequence (omitting the first two terms), which is still quite close to the anticipated value. Figure 3 shows the three Mathematica programs used above.

3   Transfer Matrix

Our method for enumerating polyominoes on a twisted cylinder is based on showing that there is a bijection between the set of all these polyominoes and a regular language, and on computing the generating function that enumerates

```
NMinimize[{eps,
    c0+c1/4^1+c2/4^2+c3/4^3+c4/4^4+c5/4^5+c6/4^6+c7/4^7+c8/4^8+c9/4^9+c
10/4^10+c11/4^11+c12/4^12+c13/4^13+c14/4^14+c15/4^15+c16/4^16+c17/4^17
+c18/4^18+c19/4^19+c20/4^20 <= 3.060901+eps &&
    c0+c1/4^1+c2/4^2+c3/4^3+c4/4^4+c5/4^5+c6/4^6+c7/4^7+c8/4^8+c9/4^9+c
10/4^10+c11/4^11+c12/4^12+c13/4^13+c14/4^14+c15/4^15+c16/4^16+c17/4^17
+c18/4^18+c19/4^19+c20/4^20 >= 3.060901-eps &&
    (* ... similarly, two constraints for each value of w ... *)
    c0+c1*1/27+c2/27^2+c3/27^3+c4/27^4+c5/27^5+c6/27^6+c7/27^7+c8/27^8+
c9/27^9+c10/27^10+c11/27^11+c12/27^12+c13/27^13+c14/27^14+c15/27^15+c1
6/27^16+c17/27^17+c18/27^18+c19/27^19+c20/27^20 <= 4.00254+eps &&
    c0+c1*1/27+c2/27^2+c3/27^3+c4/27^4+c5/27^5+c6/27^6+c7/27^7+c8/27^8+
c9/27^9+c10/27^10+c11/27^11+c12/27^12+c13/27^13+c14/27^14+c15/27^15+c1
6/27^16+c17/27^17+c 18/27^18+c19/27^19+c20/27^20 >= 4.00254-eps},
    {c0,c1,c2,c3,c4,c5,c6,c7,c8,c9,c10,c11,c12,c13,c14,c15,c16,c17,c18,
c19,c20,eps}]
```

(a) Linear programming (result: 4.06714)

```
LeastSquares[{
    {1/4^0,1/4^1,1/4^2,1/4^3,1/4^4,1/4^5,1/4^6},
    {1/5^0,1/5^1,1/5^2,1/5^3,1/5^4,1/5^5,1/5^6},
    {1/6^0,1/6^1,1/6^2,1/6^3,1/6^4,1/6^5,1/6^6},
    {1/7^0,1/7^1,1/7^2,1/7^3,1/7^4,1/7^5,1/7^6},
    {1/8^0,1/8^1,1/8^2,1/8^3,1/8^4,1/8^5,1/8^6},
    {1/9^0,1/9^1,1/9^2,1/9^3,1/9^4,1/9^5,1/9^6},
    (* ... lines of the form
    {1/w^0,1/w^1,1/w^2,1/w^3,1/w^4,1/w^5,1/w^6},
        for each value of w ... *)
    {1/23^0,1/23^1,1/23^2,1/23^3,1/23^4,1/23^5,1/23^6},
    {1/24^0,1/24^1,1/24^2,1/24^3,1/24^4,1/24^5,1/24^6},
    {1/25^0,1/25^1,1/25^2,1/25^3,1/25^4,1/25^5,1/25^6},
    {1/26^0,1/26^1,1/26^2,1/26^3,1/26^4,1/26^5,1/26^6},
    {1/27^0,1/27^1,1/27^2,1/27^3,1/27^4,1/27^5,1/27^6}},
{3.060901, 3.314100, 3.480943, 3.596055, 3.678749, 3.740221,
3.787243, 3.824087, 3.853548, 3.877519, 3.897317, 3.913880, 3.92790,
3.93988, 3.95021, 3.95920, 3.96706, 3.97399, 3.98014, 3.98562,
3.99052, 3.99494, 3.99893, 4.00254}]
```

(b) Least squares (result: 4.06727)

```
SequenceLimit[{2.658967, 3.060901, 3.314100, 3.480943, 3.596055,
    3.678749, 3.740221, 3.787243, 3.824087, 3.853548, 3.877519,
    3.897317, 3.913880, 3.92790, 3.93988, 3.95021, 3.95920, 3.96706,
    3.97399, 3.98014, 3.98562, 3.99052, 3.99494, 3.99893, 4.00254
}]
```

(c) Wynn's epsilon algorithm (result: 4.04161)

Fig. 3. Mathematica programs used for estimating the limit of (λ_w)

words in this language. To this aim, we use the *transfer matrix* that encodes the growth of polyominoes on the twisted cylinder of width w.[3]

The first cell of the twisted cylinder is always "occupied." Then, the next cell is made either "occupied" or "empty," and the process continues ad infinitum. The *signature* of a polyomino P characterizes to which connected components the last w cells of P on the cylinder (the "boundary" of P) belong. (The signature encodes both occupied and empty cells.) Obviously, polyominoes with more than one connected component are invalid, but they can become valid later if enough cells are added so as to make the entire polyomino connected. The addition of an empty (resp., occupied) cell to a (possibly disconnected) polyomino P of size n and with a signature q results in a polyomino of size n (resp., $n+1$) with signature q' (resp., q''). The key idea is that one does not need to keep in memory all possible polyominoes, but only all possible signatures. The identity of q, and whether the new cell is empty or occupied, determines unambiguously the size and signature of the new polyomino. This is precisely the information encoded in the transfer matrix. The reader is referred to [3,6,7] for the full details of this method, and, in particular, how signatures encode the possible boundaries, and how to set the transfer rules (between signatures) that define the respective automaton. In what follows we provide a brief description of this automaton, and focus on how to compute from it the function that enumerates the polyominoes.

The transfer matrix encodes a finite automaton defined as follows. The states of the automaton are all possible signatures. The initial state corresponds to the signature that encodes w empty cells. Each state has up to w outgoing edges, labeled '0' through '$w-1$'. Being at state q, and upon reading the input character k (for $0 \le k < w$), the automaton switches to the new state q' that corresponds to concatenating to a polyomino with signature q an occupied cell and then k more empty cells. (Naturally, concatenating w empty cells to any polyomino P will "terminate" it since then any further occupied cells will be disconnected from P.) This guarantees that a sequence of n input characters will correspond to a polyomino of size precisely n. The only accepting state of the automaton, q_a, corresponds to the signature '100...0' that encodes an occupied cell followed by $w-1$ empty cells. (Any valid polyomino can be "normalized" by concatenating to it empty cells as needed.) The connectedness of polyominoes with the signature that corresponds to q_a is guaranteed by the edges of the automaton. Any edge, that manifests the split of the polyomino into components that can never be connected again, leads to a "dead-end" state. Our goal is, thus, to count all words over the alphabet $\{0,1,\dots,w-1\}$ that are recognized by the automaton, that is, correspond to valid (connected) polyominoes.

Figure 4(a) shows the automaton that describes the growth of polyominoes on the twisted cylinder of width 3. States are named according to the notation used for signatures in [3,6,7]. For simplicity of presentation, the original initial state '000' was "united" with the accepting state '100'. (The original initial

---

[3] Jensen [6] invented this method in order to *count* polyominoes on regular (*non-twisted*) cylinders, and his implementation did not actually use the matrix per se. He used a system of "transfer rules" and did not look for any enumerating formula.

state—the empty polyomino—had no incoming edges and had the same outgoing edges as the state '100'.) In addition, the "dead end" state, as well as the edges leading to this state, were omitted from the figure.

We now follow closely the method described by Stanley [13, §4.7]. The transfer matrix B that describes this automaton is of size $k \times k$, where k is the number of different signatures. In fact, $k+1$ is equal to the $(w+1)$st Motzkin number [2,3]. The entry B_{ij} contains the number of edges leading from state q_i to state q_j. (In our case, entries can only be 0 or 1.) Denote by $f_i(x)$ the generating function for the number of words accepted by the automaton if q_i is the initial state, and by $\overline{f}(x)$ the vector of all these generating functions. Since the number of accepting paths of length n starting at q_i is identical to the number of accepting paths of length $n - 1$ starting at states reachable from q_i by one step, we have the relation $f_i = \sum_j B_{ij} x f_j + \delta_i$, where $\delta_i = 1$ if $q_i = q_a$ and 0 otherwise. Therefore, we have $\overline{f} = xB\overline{f} + \overline{v}$ ($xB = Bx$ since this is a commutative ring), where $\overline{v}$ is a binary vector containing 1 in the entries corresponding to accepting states and 0 elsewhere. By solving for $\overline{f}$, we obtain that $\overline{f} = (I - xB)^{-1}\overline{v}$. Our case is simple in the sense that we have only one accepting state that is also identical to the initial state. By using Cramer's rule, one can easily see that the generating function of q_a is given by $\det(C)/\det(M)$, where $M = I - xB$ and C is obtained from M by erasing the row and column corresponding to q_a. (In general, one has to erase the row corresponding to the initial state and the column corresponding to the accepting state.)

Figure 4(b) shows the transfer matrix that corresponds to the growth of polyominoes on the twisted cylinder of width 3. The described method yields the generating function $(x^3+x^2+x-1)/(2x^3+x^2+2x-1)$, which implies immediately [13, p. 202, Th. 4.1.1][4] that the recurrence formula is $a_n = 2a_{n-1} + a_{n-2} + 2a_{n-3}$.

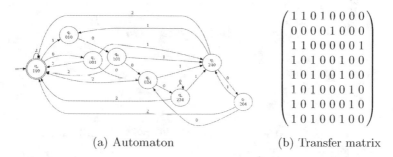

(a) Automaton   (b) Transfer matrix

Fig. 4. An automaton implementing the building of polyominoes on the twisted cylinder of width 3, and the corresponding transfer matrix

---

[4] The cited theorem says that the recurrence is manifested in the denominator of the generating function. Actually, the recurrence can be obtained directly from the transfer matrix (once its minimal polynomial is known), but we find it beneficial to also have the generating function at hand.

An important implication of the fact that a transfer matrix (that is, a finite automaton) controls the growth of polyominoes on a twisted cylinder of *any* width is the following theorem.

Theorem 2. *For any $w \geq 2$, $A_w(n)$ satisfies a linear recurrence.*   □

We next report our results for twisted cylinders of widths up to 10.

3.1 Using a Generating Function

First, we applied the method described above, namely, computing the generating function and deriving from it the recurrence formula. This allowed us to compute the linear recurrences for $w = 4, 5, 6$. (This method exceeded our available computing resources for higher values of w.) As the meaning of a_n should be understood from the context, we omit w from the notation of the recurrence.

For width 4, the obtained generating function is

$$\frac{x^7 - 4x^6 + 2x^5 - 3x^3 + 5x^2 - 4x + 1}{2x^7 - 6x^6 + 2x^5 + 2x^4 - 7x^3 + 8x^2 - 5x + 1},$$

and the minimal linear recurrence is

$$a_n = 5a_{n-1} - 8a_{n-2} + 7a_{n-3} - 2a_{n-4} - 2a_{n-5} + 6a_{n-6} - 2a_{n-7},$$

with $a_1 = 1$, $a_2 = 2$, $a_3 = 6$, $a_4 = 19$, $a_5 = 59$, $a_6 = 181$, and $a_7 = 555$. The growth rate (solution of the characteristic equation) is $\lambda_4 = 3.0609\ldots$.

For width 5, the generating function is found to be $(x^{18} + x^{17} + 7x^{16} - 3x^{15} - 17x^{14} - 13x^{13} + 13x^{12} + 8x^{11} - 10x^{10} - 12x^9 - 10x^8 - 16x^7 + 3x^6 - 2x^4 + 5x^3 - x^2 - 3x + 1)/(2x^{18} + 14x^{16} - x^{15} - 23x^{14} - 22x^{13} + 16x^{12} + 14x^{11} - 10x^{10} - 18x^9 - 9x^8 - 25x^7 - x^6 - 5x^4 + 6x^3 + x^2 - 4x + 1)$. The minimal linear recurrence is

$$a_n = 4a_{n-1} - a_{n-2} - 6a_{n-3} + 5a_{n-4} + a_{n-6} + 25a_{n-7} + 9a_{n-8} + 18a_{n-9}$$
$$+ 10a_{n-10} - 14a_{n-11} - 16a_{n-12} + 22a_{n-13} + 23a_{n-14} + a_{n-15}$$
$$- 14a_{n-16} - 2a_{n-18},$$

with $a_1 = 1$, $a_2 = 2$, $a_3 = 6$, $a_4 = 19$, $a_5 = 63$, $a_6 = 211$, $a_7 = 707$, $a_8 = 2360$, $a_9 = 7853$, $a_{10} = 26070$, $a_{11} = 86434$, $a_{12} = 286416$, $a_{13} = 948991$, $a_{14} = 3144464$, $a_{15} = 10419886$, $a_{16} = 34530671$, $a_{17} = 114435963$, and $a_{18} = 379251561$. The growth rate is $\lambda_5 = 3.3141\ldots$.

For width 6, the computed generating function was $(x^{48} - 6x^{47} + \cdots + 17x^2 - 7x + 1)/(2x^{48} - 9x^{47} + \cdots - 8x + 1)$. The minimal linear recurrence (of degree 48) and its initial values can be found in [1]. The growth rate is $\lambda_6 = 3.4809\ldots$.

3.2 Using the Berlekamp-Massey Algorithm

The procedure described above broke down at $w = 7$ due to an explosion in the running time needed to compute the minimal polynomial of the huge transfer

matrix. Hence, we attempted to compute the linear recurrence directly from the
sequence of numbers of polyominoes obtained by running the automaton.

Our first attempt was to feed enough values of the sequence to the Mathematica command `FindLinearRecurrence`, running this package on a virtual
machine having four vCPU of total 10.1 GHz and 8 GB of RAM, hosted on a
PowerEdge R710. As a testcase, we computed again the linear recurrence for the
number of polyominoes on the twisted cylinder of width 6. This computation required about 1.5 seconds. Repeating the procedure for width 7, it took about 11
minutes to obtain a linear recurrence with 121 terms, whose largest coefficient
in absolute value was about $8.74 \cdot 10^7$ (27 bits). The recurrence is found in [1].

Mathematica was unable to compute the linear recurrence for width 8, so we
turned to another idea. Massey [11] used an algorithm of Berlekamp in order
to find the shortest linear recurrence that satisfies a sequence of elements from
a field. Reeds and Sloane [12] generalized this method to the case in which the
elements of the sequence are integers modulo an arbitrary number whose prime
factorization is known. Observing that a recurrence remains the same if it is
taken modulo some big-enough number N, we followed the same strategy. First,
we computed the leading values of the sequence modulo a few prime numbers
$p_1 = 2, p_2 = 3, p_3 = 5, \ldots, p_k$ and obtained "modulo sequences" $s_1, s_2, s_3, \ldots, s_k$.
Second, we computed, using the Berlekamp-Massey algorithm, the respective
linear recurrence formulae $r_1, \ldots, r_k$ of the sequences $s_1, \ldots, s_k$. Finally, using
the Chinese remainder theorem, we recovered the original recurrence formula
from the recurrences $r_1, \ldots, r_k$.

We only needed to take care that $N =
\Pi_{i=1}^k p_i > 2L$, where $L = L(w)$ is the
largest absolute value of a coefficient of
the original recurrence, in order to guarantee that the latter was recovered correctly. Indeed, this condition implies that
the original non-negative coefficients are
recovered correctly in the range $[0, N/2]$,
while the negative coefficients are recovered uniquely in the range $(N/2, N]$. A
priori we had no clue about the value of
L. From a plot of the values of $\ln \ln L(w)$
as a function of w, for $4 \leq w \leq 10$ (see

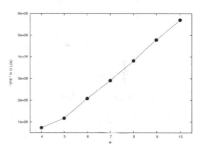

Fig. 5. A plot of $\ln \ln L(w)$ as a
function of w

Figure 5), it seems that $L(w)$ grows in a doubly-exponential manner with respect
to w. A well-known approximation of the primorial $p_n\# = \prod_{k=1}^n p_k$ (the product
of the first n prime numbers) is $e^{(1+o(1))(n \ln n)}$. Hence, $O(\ln L(w)/ \ln \ln L(w)) =
O(e^w/w)$ prime numbers should be sufficient for our purpose. In practice, at least
for small values of w, less than e^w/w primes are needed. Thus, we started with
a nominal number of primes, and whenever the procedure failed, we doubled the
number of primes and repeated the procedure. Checking whether the procedure
succeeded was performed simply by comparing values produced by the recovered

recurrence to the values computed by the automaton, for as many more values as the number of values needed to initialize the recurrence.

If a sequence s satisfies a recurrence R, then the sequence, obtained from s by taking all values modulo a prime p, satisfies the recurrence obtained from R by taking all coefficients modulo p. If we repeat this process for enough prime numbers, the original recurrence can be recovered uniquely since all primes are obviously coprime. This is readily done by using the Chinese remainder theorem, recovering each coefficient independently. In case a recovered coefficient exceeds $N/2$, we simply subtract N from it to obtain the correct value.

Given the sequence (with original values taken modulo p), computing the linear recurrence formula (modulo p) was performed by a C++ program running Massey's algorithm. Currently, we use a simple $O(n^2)$-time implementation of it (where n is the complexity of the recurrence), while the best known upper bound is currently $O(n \log^2 n)$ [4]. The code for recovering the original recurrence formula from all the modulo recurrences was written in Python, which has a built-in capability of handling big numbers.

These two programs were run on a home laptop with a 2.2 GHz processor and 3 GB of main memory. The computation time of the entire procedure for cylinders of up to width 8 was negligible. For width 8, the obtained recurrence included 315 terms, and the largest coefficient in absolute value was about $5.48 \cdot 10^{19}$ (66 bits). For width 9, the computation time was about 20 seconds. The obtained recurrence included 826 terms, and the largest coefficient was about $5.18 \cdot 10^{51}$ (172 bits). The computation of the linear recurrence for the number of polyominoes on a twisted cylinder of size 10 required about 5 minutes, and the obtained recurrence included 2168 terms, with the largest coefficient being about $6.39 \cdot 10^{129}$ (432 bits). The recurrences for widths 8–10 can be found in [1].

Note that the complexities of the recurrence formulae for twisted cylinders of widths 1–10, that is, $1, 1, 3, 7, 18, 48, 121, 315, 826, 2168$, are almost identical to the 2nd through 11th Motzkin numbers, namely, $1, 2, 4, 9, 21, 51, 127, 323, 835, 2188$. This is due to the fact that the respective automata have that many essentially-different states [3], hence, these are the sizes of the corresponding transfer matrices whose minimal polynomials are computed.

4   Conclusion

In this paper we studied the formulae that enumerate fixed polyominoes on twisted cylinders. First, we proved that as w, the width of the cylinder, tends to infinity, λ_w, the growth rate of polyominoes on these cylinders, approaches λ (the growth rate of polyominoes in the plane). By numerical methods we found that the limit of the sequence (λ_w) is roughly 4.06, supporting the common belief. Then, we analyzed the formula enumerating polyominoes on a twisted cylinder by manipulation the transfer matrix that describes the finite automaton that models the growth of the polyominoes. A direct consequence of this method is that the function enumerating polyominoes on a twisted cylinder of *any* width w satisfies a linear recurrence. We recovered these recurrences for up to $w = 10$.

We believe that further investigation of formulae enumerating polyominoes on twisted cylinders may shed more light on the elusive formula for the number of polyominoes in the plane.

Problems for further research include:

- Obtain recurrence formulae for cylinders of width larger than 10.
- Find the rule governing the development of the recurrence formulae.
- Investigate how fast the sequence (λ_w) converges to λ.

References

1. Barequet, G.: Recurrence data for polyominoes on twisted cylinders, http://www.cs.technion.ac.il/~barequet/twisted/
2. Barequet, G., Moffie, M.: On the complexity of Jensen's algorithm for counting fixed polyominoes. J. Discrete Algorithms 5, 348–355 (2007)
3. Barequet, G., Moffie, M., Ribó, A., Rote, G.: Counting polyominoes on twisted cylinders. Integers 6, #A22 (2006)
4. Brent, R.P., Gustavson, F.G., Yun, D.Y.Y.: Fast solution of Toeplitz systems of equations and computation of Padé approximant. J. Algorithms 1, 259–295 (1980)
5. Gaunt, D.S., Sykes, M.F., Ruskin, H.: Percolation processes in d-dimensions. J. Phys. A—Math. Gen. 9, 1899–1911 (1976)
6. Jensen, I.: Enumerations of lattice animals and trees. J. Stat. Phys. 102, 865–881 (2001)
7. Jensen, I.: Counting polyominoes: A parallel implementation for cluster computing. In: Sloot, P.M.A., Abramson, D., Bogdanov, A.V., Gorbachev, Y.E., Dongarra, J., Zomaya, A.Y. (eds.) ICCS 2003, Part III. LNCS, vol. 2659, pp. 203–212. Springer, Heidelberg (2003)
8. Klarner, D.A.: Cell growth problems. Can. J. Math. 19, 851–863 (1967)
9. Klarner, D.A., Rivest, R.L.: A procedure for improving the upper bound for the number of n-ominoes. Can. J. Math. 25, 585–602 (1973)
10. Madras, N.: A pattern theorem for lattice clusters. Ann. Comb. 3, 357–384 (1999)
11. Massey, J.L.: Shift-register synthesis and BCH decoding. IEEE T. Inf. Theory 15, 122–127 (1969)
12. Reeds, J.A., Sloane, N.J.A.: Shift-register synthesis (modulo m). SIAM J. Comput. 14, 505–513 (1985)
13. Stanley, R.: Enumerative Combinatorics, I. Cambridge University Press (1997)
14. Wynn, P.: On a device for computing the $e_m(S_n)$ transformation. Math. Tables and other Aids to Computation 10, 91–96 (1956)

Picture Codes with Finite Deciphering Delay[*]

Marcella Anselmo[1], Dora Giammarresi[2], and Maria Madonia[3]

[1] Dipartimento di Informatica, Università di Salerno
via Giovanni Paolo II, 132-84084 Fisciano (SA), Italy
anselmo@dia.unisa.it
[2] Dipartimento di Matematica, Università Roma "Tor Vergata"
via della Ricerca Scientifica, 00133 Roma, Italy
giammarr@mat.uniroma2.it
[3] Dipartimento di Matematica e Informatica, Università di Catania
Viale Andrea Doria 6/a, 95125 Catania, Italy
madonia@dmi.unict.it

Abstract. A two-dimensional code is defined as a set $X \subseteq \Sigma^{**}$ such that any picture over Σ is tilable in at most one way with pictures in X. The codicity problem is in general undecidable. Very recently in [4] prefix picture codes were introduced as a decidable subclass that generalizes prefix string codes. Finite deciphering delay sets are an interesting class of string codes that coincide with prefix codes in the case of delay equal to 0. An analogous notion is introduced for picture codes and it is proved that they correspond to a bigger class of decidable picture codes that includes interesting examples and special cases.

Keywords: Two-dimensional languages, codes.

1 Introduction

In the theoretical study of formal languages, string codes have been always a relevant subject of research also because of their applications to practical problems. Theoretical results on string codes are related to combinatorics on words, automata theory and semigroup theory. We refer to [8] for complete references. Extensions of classical strings to two dimensions can be done in several ways: in general they bring to the definition of polyominoes, labeled polyominoes, as well as rectangular labeled polyominoes usually referred to as pictures. On the other hand, the notion of codes can be intuitively and naturally transposed to two dimensional objects by exploiting the notion of unique tiling decomposition. A set C of polyominoes is a code if every polyomino that is tilable with (copies of) elements of C, it is so in a unique way. Unfortunately, most of the published results show that in the 2D context we loose important properties. In [7] D. Beauquier and M. Nivat proved that the problem whether a finite set of polyominoes is a

[*] Partially supported by MIUR Projects *"Aspetti matematici e applicazioni emergenti degli automi e dei linguaggi formali"* and *"PRISMA PON04a2 A/F"*, by 60% Projects of University of Catania, Roma "Tor Vergata", Salerno.

A.-H. Dediu et al. (Eds.): LATA 2014, LNCS 8370, pp. 88–100, 2014.

code is undecidable, and that the same result holds also for dominoes. Codes of other variants of polyominoes including bricks (i.e. labelled polyominoes) and pictures are also studied in [1,9,13,14,16] and further undecidability results are proved.

It is worthwhile to remark that all mentioned results consider 2D codes independently from a 2D formal language theory.

Very recently, in [4], a new definition for picture codes was introduced in connection with the family REC of picture languages recognized by finite tiling systems. Remark that finite tiling systems generalize to two dimensions finite state automata for strings and that family REC is considered as the two-dimensional counterpart of regular string languages (see [12]). In [4] codes are defined by using the formal operation of *tiling star* as defined in [17]: the tiling star of a set X is the set X^{**} of all pictures that are tilable (in the polyominoes style) by elements of X. Then X is a code if any picture in X^{**} is tilable in a unique way. Remark that if $X \in$ REC then X^{**} is also in REC. By analogy to the string case, it holds that if X is a finite picture code then, starting from pictures in X we can construct an unambiguous tiling system for X^{**} (see [5] for the definition). Unfortunately, despite this nice connection to the string code theory, it is proved that it is still undecidable whether a given set of pictures is a code. This is actually coherent with the known result of undecidability for unambiguity inside the family REC.

An important and easy-to-construct class of string codes are prefix codes. Recall that a set S of words is called prefix if inside S no word is (left-)prefix of another one. It holds that any prefix set of words is also a code, referred to as a prefix code. Moreover prefix codes have the property that any string can be decoded *on-time*: while reading the string from left-to-right, always only one element in S matches the input.

Looking for decidable subclasses of picture codes, in [4] it is proposed a definition of *prefix code* for pictures. Pictures are then considered with a preferred scanning direction: from top-left corner towards the bottom-right one. Then a picture x is a prefix of a picture p, if x coincides with the top-left portion of p. Unfortunately the mere translation that defines a set X to be prefix by imposing that its pictures are not mutually prefixes is not sufficient to guarantee that X is a code. Then, going from strings to pictures it is maintained the following property: if X is a prefix code, when decoding a picture p starting from top-left corner, it should be uniquely chosen which element in X we can start with.

Notice that, in the generic intermediate step of the computation, we have read only part of the input: this is where the main difficulties arise. In fact the initial part of a string is still a string while a piece of a picture is not in general a picture. The formal definition involves special kind of polyominoes that have straight top border.

Main results in [4] include the proof that it is decidable whether a finite set of pictures is a prefix set and that, as in the 1D case, every prefix set of pictures is a code. Moreover a polynomial time decoding algorithm for finite prefix codes is presented. Some results are also given concerning the generalization to 2D of the notions of *maximal* and *complete* prefix codes. A particular subclass of prefix codes is studied in [3].

In this paper we extend to picture codes the notion of finite deciphering delay for strings, as described in [8,10]. Intuitively, when reading a coded message from left to right, its deciphering can already begin after a *finite lookahead* without waiting for the end of the message. Saying it differently, if S is a string code with deciphering delay d, we need to read a portion of the string corresponding to $d + 1$ code words before actually uniquely decoding the leftmost one.

More formally: if w is a string having two distinct prefixes in S^+ (i.e. they factorize in S) and such that the shorter one is in S^{d+1} then the two prefixes start with the same word in S. In particular it holds that prefix codes have delay equal to zero and for this they are also called *instantaneous* codes (see [8] for all formal definitions).

Extending the concept of finite deciphering delay code to two dimensions seems quite intuitive and natural, but the formalization of such ideas is extremely involved and require to cleverly invent the right definition of polyomino prefix of another polyomino as well as a careful counting of such k code-pictures that play the role of finite scanning lookahead. The main difficulties come, as in the prefix case, from the fact that in the computation the initial part of a picture is a polyomino and not a picture. Moreover we have the further inconvenience that we need to compare two different polyominoes one composed by at least k pictures of the set X.

We give the definition of a set of pictures with finite deciphering delay. As in the string case, a code with delay equal to 0 is a prefix code. Further we exhibit, for each integer k, a set with deciphering delay equal to k and point out a set that has not finite deciphering delay. Moreover, since it is decidable whether a finite set of pictures has deciphering delay equal to k, this notion contributes to enlarge the family of known decidable picture codes.

The paper is organized as follows: Section 2 and 3 report all needed definitions and known results on picture codes; Section 4 includes all new results proposed in this paper. Some conclusions are given in Section 5.

2   Preliminaries

We introduce some definitions about two-dimensional languages (see [12,11]). A *picture* over a finite alphabet Σ is a two-dimensional rectangular array of elements of Σ. Given a picture p, $|p|_{row}$ and $|p|_{col}$ denote the number of rows and columns, respectively while $|p| = (|p|_{row}, |p|_{col})$ denotes the picture *size*. The set of all pictures over Σ is denoted by Σ^{**}. A *two-dimensional language* (or *picture language*) over Σ is a subset of Σ^{**}.

In order to locate a position in a picture, it is necessary to put the picture in a reference system. The set of coordinates $dom(p) = \{1, 2, \ldots, |p|_{row}\} \times \{1, 2, \ldots, |p|_{col}\}$ is referred to as the *domain* of a picture p. We let $p(i, j)$ denote the symbol in p at coordinates (i, j). Positions in $dom(p)$ are ordered following the lexicographic order: $(i, j) < (i', j')$ if either $i < i'$ or $i = i'$ and $j < j'$.

We assume the top-left corner of the picture to be at position $(1,1)$ and this automatically fix the scanning direction for a picture from the top-left corner toward the bottom right one. Moreover, to easily detect border positions of pictures, we use initials of words "top", "bottom", "left" and "right": then, for example the *tl-corner* of p refers to position $(1,1)$.

A *subdomain* of $dom(p)$ is a set d of the form $\{i, i+1, \ldots, i'\} \times \{j, j + 1, \ldots, j'\}$, where $1 \le i \le i' \le |p|_{row}$, $1 \le j \le j' \le |p|_{col}$, also specified by the pair $[(i,j), (i',j')]$. The portion of p corresponding to positions in subdomain $[(i,j), (i',j')]$ is denoted by $p[(i,j), (i',j')]$. Then a picture x is *subpicture of p* if $x = p[(i,j), (i',j')]$, for some $1 \le i \le i' \le |p|_{row}$, $1 \le j \le j' \le |p|_{col}$. In other words, x is a subpicture of p when, after a translation of x that lets its tl-corner be placed on position (i,j) of p, the content of x matches the content of the corresponding portion of p (taking care of considering a common reference for the coordinates of positions). Throughout all the paper, when dealing with subpictures or prefixes of pictures (or in the sequel polyominoes), and a translation of p is needed, it will be implicitly assumed that the positions of the cells will refer to a common reference system (for example, a translation of $dom(p)$ could be considered).

As a special case of subpictures we consider prefixes of pictures. Given pictures x, p, with $|x|_{row} \le |p|_{row}$ and $|x|_{col} \le |p|_{col}$, we say that x is a *prefix* of p if x is a subpicture of p corresponding to its top-left portion, i.e. if $x = p[(1,1), (|x|_{row}, |x|_{col})]$.

In this paper we will consider an interesting star operation for picture language introduced by D. Simplot in [17]: the tiling star. The idea is to compose pictures in a way to cover a rectangular area as, for example, in the following figure.

Definition 1. *The* tiling star *of X, denoted by X^{**}, is the set of pictures p whose domain can be partitioned in disjoint subdomains $\{d_1, d_2, \ldots, d_k\}$ such that any subpicture p_h of p associated with the subdomain d_h belongs to X, for all $h = 1, \ldots, k$.*

Language X^{**} is called the set of all tilings by X in [17]. In the sequel, if $p \in X^{**}$, the partition $t = \{d_1, d_2, \ldots, d_k\}$ of $dom(p)$, together with the corresponding pictures $\{p_1, p_2, \ldots, p_k\}$, is called a *tiling decomposition* of p in X.

In this paper, while dealing with the tiling star of a set X, we will need to manage also non-rectangular "portions" of pictures composed by elements of X: those are actually labeled polyominoes, that we will call polyominoes, for the sake of simplicity.

In order to use consistent notation when dealing with pictures and polyominoes, given a polyomino c we can consider it as plugged in its minimal (rectangular) bounding box. In some sense it can be viewed as a picture where positions not "occupied" by the polyomino contain the blank symbol.

The *domain of a (labeled) polyomino c* is naturally defined as the set of pairs (i, j) corresponding to all occupied positions inside its minimal bounding box, i.e. positions containing a non-blank symbol. As before, to the tl-corner of the minimal bounding box is assigned position $(1, 1)$. Notice that position $(1, 1)$ does not necessarily belong to the domain of a given polyomino (see the examples below).

		a	a
	a	a	a
	b	a	a
a	b	a	a

a	b		
	a		
	b	a	a
a	b	a	a

a	b	a	a
a		a	a
b			

In this paper we restrict our attention to column convex polyominoes, i.e. polyominoes with the property that if two positions in the same column belong to the domain then all the intermediate positions in that column belong to the domain too.

We can also use the notion of a picture x to be *subpicture of a polyomino c* by directly referring to their domains (again actually translation of domains) and comparing the labels of corresponding common positions.

This allows to extend to polyominoes the notion of *tiling decomposition in a set of pictures* X: the domain of a polyomino c is partitioned in a collection of rectangular subdomains in a way that the subpictures corresponding to those subdomains belong to X. Finally, we also define a sort of *polyomino tiling star* that, applied to a set of pictures X, produces the set of all polyominoes that have a tiling decomposition in X. We denote it by X^{p**}. If a polyomino p belongs to the polyomino star of X, we say that p is *tilable* in X.

3   Two-Dimensional Codes

In this section we summarize the results in [4] that introduce the motivations for the main results of this paper given in the next section.

The notion of codes in two dimensions was considered by many authors in the literature: their works defined, in different contexts, polyomino codes, picture codes, and brick codes ([7,9,16]). We directly refer to the definition of code given in [4] where two-dimensional codes are introduced in the setting of the theory of recognizable two-dimensional languages and coherently to the notion of language unambiguity as in [2,5,6].

Definition 2. *Let* Σ *be a finite alphabet.* $X \subseteq \Sigma^{**}$ *is a code iff any* $p \in \Sigma^{**}$ *has at most one tiling decomposition in* X.

Example 3. Let $\Sigma = \{a, b\}$ be the alphabet and let $X = \left\{ \boxed{a\ b}, \begin{array}{c} \boxed{a} \\ \boxed{b} \end{array}, \begin{array}{cc} a & a \\ a & a \end{array} \right\}$. It is easy to see that X is a code. Any picture $p \in X^{**}$ can be decomposed starting at tl-corner and checking the size $(2, 2)$ subpicture $p[(1, 1), (2, 2)]$: it can be univocally decomposed in X. Then, proceed similarly for the next contiguous size $(2, 2)$ subpictures.

The notion of code seems to generalize easily from strings to pictures: the bad new is that the problem whether a given set of pictures is a code is in general undecidable, even for finite sets (and this holds also for other definitions of picture codes found in the literature). A challenging aim is then to find decidable subclasses of codes.

Recall that in one dimension, a set X of strings is prefix if no two strings in X are one prefix of the other one. Moreover, it holds that any prefix set is a code. Unfortunately this cannot be directly generalized to picture world as shown in the following example.

Example 4. Let $X = \left\{ \boxed{a\ b},\ \boxed{b\ a},\ \boxed{\begin{matrix} a \\ a \end{matrix}} \right\}$. Notice that no picture in X is prefix of another picture in X (see the definition in Section 2). Nevertheless X is not a code. Indeed picture $\begin{matrix} a\ b\ a \\ a\ b\ a \end{matrix}$ has the two following different tiling decompositions in X: $t_1 = \boxed{\begin{matrix} a\ b \\ a\ b \end{matrix}\begin{matrix} a \\ a \end{matrix}}$ and $t_2 = \boxed{\begin{matrix} a \\ a \end{matrix}\begin{matrix} b\ a \\ b\ a \end{matrix}}$.

Then a more careful definition of prefix sets of pictures is needed. The basic idea in defining a *prefix code* is to prevent the possibility to start decoding a picture in two different ways (as it is for the prefix string codes). One major difference going from 1D to 2D case is that, while any initial part of a decomposition of a string is still a string, the initial part of a decomposition of a picture has not necessarily a rectangular shape: it is in general a (labeled) polyomino. More specifically, it is a polyomino whose domain contains always the tl-corner position $(1,1)$ that is referred to as *corner polyomino*.

The definition of a picture x prefix of a corner polyomino c is naturally given by referring to their domains as done at the end of Section 2: x corresponds exactly to the top-left portion of c that has the same size of x, i.e $x = c[(1,1), (|x|_{row}, |x|_{col})]$.

Definition 5. *A set $X \subseteq \Sigma^{**}$ is* prefix *if any two different pictures in X cannot be both prefix of the same corner polyomino tilable in X.*

Notice that the set X of Example 3 is prefix. On the contrary, the set X of Example 4 is not prefix: pictures $\boxed{\begin{matrix} a \\ a \end{matrix}}$ and $\boxed{a\ b}$ are both prefixes of the corner polyomino $\boxed{\begin{matrix} a\ b \\ a\ b \end{matrix}}$ tilable in X.

Definition 5 seems a good generalization of prefix sets of strings; in fact it is proved that a prefix set is a code referred to as *prefix code*. Contrarily to the case of all other known classes of 2D codes, the family of finite prefix codes has the important property to be decidable. In [4] it is also given polynomial decoding algorithm for a finite prefix picture code and notions of maximality and completeness are investigated.

4 Codes with Finite Deciphering Delay

In the theory of string codes an important role is played by codes with finite deciphering delay. Informally, they satisfy the property that the decoding of a given string-message can be started after reading a *finite* number of code words. Formally: a set $S \in \Sigma^*$ has *finite deciphering delay* if there is an integer $d \geq 0$ such that for all $x, x' \in S$, $y \in S^d, y' \in S^*$, xy prefix of $x'y'$ implies $x = x'$ (see [8]). They generalize prefix codes that are in fact also called *instantaneous* codes. In this section we afford their generalization to picture codes.

The idea can be translated quite naturally to two dimensions. We fix, as for prefix codes, the top-left to bottom-right scanning direction. Then, deciphering delay equal to k means that we can possibly begin two different candidate decompositions containing up to k code pictures: if we add further code pictures, only one of the attempts of decomposition can survive (and then we can validate the first picture code).

With this concept in mind, we can say that, as in the string case, *prefix picture codes have deciphering delay equal to 0*. To formalize deciphering delay greater than 0 we need to cope with the two dimensions: at each step of decoding process we can proceed in the two different directions rightwards and downwards. Thus an "explored part" composed by k code pictures is not a subpicture but in general it is a polyomino. Remark that, besides using the top-left to bottom-right scanning direction, we will follow the lexicographic order in domain positions: this produces only column-convex polyominoes. Then to extend the formal definition from strings to pictures we need to fix the following notions (as in Definitions 6, 8, 9) :
- prefix in the framework of polyominoes
- "initial" picture of a polyomino decomposition
- polyomino decomposition of k code pictures
- extension of a polyomino decomposition from k to more than k pictures.

To give the definition of polyomino prefix of another one, we need to fix the first position of a polyomino c, that, we call *tl-corner* by uniformity with pictures: it is the minimal position (with respect to the lexicographic order) of $dom(c)$ inside its minimal bounding box.

Definition 6. *Let $c, c' \in \Sigma^{p**}$. Polyomino c is* prefix *of c', we write $c \lhd c'$, if, when using the same coordinates to denote their tl-corners, the following conditions are verified.*
1. *$dom(c) \subseteq dom(c')$*
2. *$c(i, j) = c'(i, j)$ for all $(i, j) \in dom(c)$*
3. *if $(i_0, j_0) \in dom(c)$ then, for all $(i, j) \in dom(c')$ with $i \leq i_0$ and $j \leq j_0$, we have $(i, j) \in dom(c)$.*

Notice that condition 3. in the definition imposes to c a sort of "column prefix-ness": we pretend that each column in c is "prefix" of the corresponding column in c' along the direction from top to bottom. Moreover remark that this definition, when referred to pictures and corner polyominoes, corresponds to the definition of prefix as in [4].

Example 7. In the figure below, polyomino (b) is prefix of polyomino (a), according to the definition. The same polyomino (b), referred to the minimal bounding
box of (a), is shown in (c).

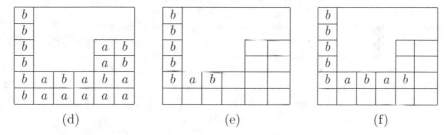

(a)   (b)   (c)

In the figures below polyomino (e) is prefix of polyomino (d), whereas polyomino (f) is not because it does not satisfy condition *3)* of the definition. In
particular position $(5,5)$ is in (f), while positions $(3,5)$ and $(4,5)$ are in (d) but
not in (f).

(d)   (e)   (f)

For the next definition we introduce the notation of "first picture in a decomposition": it is the picture that contains the tl-corner of the domain. More
formally, let $c \subset X^{p**}$ and let $t = \{d_1, d_2, \ldots, d_h\}$ be a tiling decomposition of c
in X, with corresponding pictures $\{p_1, p_2, \ldots, p_h\}$. The picture p_f associated to
the subdomain d_f that contains the tl-corner of c is called the *first picture* in t
and is denoted by $first(t)$.

Another important aspect to consider when generalizing the definition for
strings is "how to count" the k pieces of the decomposition: we are interested
only in the pieces of the decomposition that come "after" the first picture of the
decomposition itself. (Recall we follow a top-left to bottom-right direction and
explore along the lexicographic order of domain positions). This is formalized in
the following definition.

Definition 8. *Let $X \subseteq \Sigma^{**}$, $c \in X^{p**}$. Let t be a tiling decomposition of c in
X and* tl-c$(c) = (1, \bar{\jmath})$. *The decomposition t has k br-components if there are
k subdomains in t, specified by the pairs $[(i_1, j_1), (i'_1, j'_1)], \ldots, [(i_k, j_k), (i'_k, j'_k)]$,
respectively, such that $j_r \geq \bar{\jmath}$ or $i_r > |first(t)|_{row}$, for all $1 \leq r \leq k$.*

Recall that in the string case, when we extend a prefix of k code words, we just
add one code word and consider a prefix with $k + 1$ words. This is not sufficient
in a two-dimensional setting where we need to consider an extension step that
enlarges the prefix along the "frontier". And this is expressed in the following
definition.

Definition 9. Let $X \subseteq \Sigma^{**}$, $c, c', c'' \in X^{p**}$ with $c \lhd c'$, and let t, t'' be tiling decompositions of c, c'', respectively in X. Then t'' extends t consistently with c' if:

- $t \subset t''$
- for any $(i, j) \in dom(c'')$, $c''(i, j) = c'(i, j)$
- $dom(c'')$ contains all positions $(i, j) \in dom(c') \setminus dom(c)$ such that $(i, j - 1)$ or $(i - 1, j - 1)$ or $(i - 1, j)$ is in $dom(c)$.

Finally we are ready to give the definition of picture languages with finite deciphering delay $k \geq 1$. Recall that picture languages with deciphering delay $k = 0$ are prefix codes.

Definition 10. Let $X \subseteq \Sigma^{**}$. X has finite deciphering delay if it is prefix or there is an integer $k \geq 1$ such that, for any $c, c' \in X^{p**}$, with $c \lhd c'$, and for any pair of tiling decompositions in X, t and t', of c and c', respectively, where t has k br-components and $first(t) \neq first(t')$, there is no $c'' \in X^{p**}$ with a tiling decomposition $t'' \neq t'$ that extends t consistently with c'.

If the condition in the previous definition holds for some integer k, then it holds for all $k' \geq k$. If X has finite deciphering delay, then the smallest integer k satisfying the condition in the previous definition, is the *deciphering delay* of X.

The following examples show a picture language with finite deciphering delay and another one that has not finite deciphering delay, respectively.

Example 11. Let $X = \left\{ \boxed{\begin{smallmatrix} a \\ b \end{smallmatrix}}, \boxed{a\ a\ b}, \boxed{\begin{smallmatrix} b\ a \\ b\ b \end{smallmatrix}} \right\}$.

Set X is not prefix; in fact pictures $\boxed{\begin{smallmatrix} a \\ b \end{smallmatrix}}$ and $\boxed{a\ a\ b}$ are both prefixes of the corner polyomino $\boxed{\begin{smallmatrix} a\ a\ b\ a \\ b\ b\ b\ b \end{smallmatrix}}$ tilable in X.

Let us show that X has deciphering delay 1. Indeed, the only possibility for $c, c' \in X^{p**}$ with tiling decompositions t and t' respectively, to have $c \lhd c'$ and $first(t) \neq first(t')$, is that $first(t) = \boxed{\begin{smallmatrix} a \\ b \end{smallmatrix}}$, and $first(t') = \boxed{a\ a\ b}$ (or viceversa).

Therefore the top-left portion of c' must be $\boxed{\begin{smallmatrix} a\ \ a\ \ b \\ b \end{smallmatrix}}$. This fact leds to these two different possible situations: $c = \boxed{a\ a\ b} \in X^{p**}$ prefix of $c' = \boxed{\begin{smallmatrix} a\ a\ b\ a \\ b\ b\ b\ b \end{smallmatrix}} \in X^{p**}$ and $c = \boxed{\begin{smallmatrix} a \\ b \end{smallmatrix}} \in X^{p**}$ prefix of $c' = \boxed{\begin{smallmatrix} a\ a\ b \\ b\ a \\ b\ a \end{smallmatrix}}$ with tiling decompositions in X, t and t' respectively, as in the figure, and $first(t) \neq first(t')$. But in both cases there is no $c'' \in X^{p**}$ with a tiling decomposition that extends t consistently with c'. Remark that t has $k = 1$ br-component.

Example 12. Let $X = \left\{ \begin{array}{c} b \\ a \end{array}, \boxed{b\ a}, \begin{array}{cc} a & a \\ a & a \end{array} \right\}$.

It is easy to see that X is a code but X has an infinite deciphering delay. Indeed, consider for any $k > 0$,

$$
c = \begin{array}{|c|c|c|c|c}
b & a & a & a & \cdots \\
\hline
a & a & a & a & \cdots \\
\hline
a & a \\
\cline{1-2}
\end{array} \in X^{p**} \quad \text{and} \quad c' = \begin{array}{|c|c|c|c|c}
b & a & a & a & \cdots \\
\hline
a & a & a & a & \cdots \\
\hline
a & a \\
\cline{1-2}
a & a \\
\cline{1-2}
\end{array} \in X^{p**}.
$$

Then $c \lhd c'$ and any pair of tiling decompositions in X, t and t' respectively, with $first(t) \neq first(t')$, t with k br-components, must be as those ones given in the figure. But it is easy to see that, for any k, there is always $c'' \in X^{p**}$ with a tiling decomposition t'' that extends t consistently with c'.

Under the hypothesis X finite, the following result can be proved by checking all polyominoes tilable with k pictures in X.

Proposition 13. *Given a finite set of picture $X \subseteq \Sigma^{**}$ and an integer $k \geq 0$, it is decidable whether X has deciphering delay k.*

The next two theorems contain the main results of the paper. We prove first that picture sets with finite deciphering delay form a non trivial hierarchy and second that having finite deciphering delay is a sufficient condition for a picture set to be a code.

Theorem 14. *For any $k \geq 0$ there exists a set $X \subseteq \Sigma^{**}$ that has deciphering delay k.*

Proof. Let $k \geq 0$. Consider the set $X_k = \left\{ \begin{array}{c} a \\ b \end{array}, \boxed{a^k\ b}, \begin{array}{c} b^k\ a \\ b^k\ b \end{array} \right\}$ where a^k (b^k, resp.) denotes the picture with one row and k columns over the alphabet $\{a\}$ ($\{b\}$, resp.). Let us show that, for any $k \geq 0$, X_k has deciphering delay k.

For $k = 0$, X_0 is prefix. For $k > 0$, in order to make the proof easier, let us fix $k = 3$. Set $X_3 = \left\{ \begin{array}{c} a \\ b \end{array}, \boxed{a\ a\ a\ b}, \begin{array}{c} b\ b\ b\ a \\ b\ b\ b\ b \end{array} \right\}$ has deciphering delay greater than 2.

Indeed, consider $c = \begin{array}{|c|c|} a & a \\ \hline b & b \end{array} \in X_3^{p**}$ and $c' = \begin{array}{|c|c|c|c|} a & a & a & b \\ \hline b & b & b & a \\ \hline b & b & b & b \end{array} \in X_3^{p**}$, with the tiling decompositions t and t' in X, respectively, as shown in the figure. Then t has 2 br-components, $first(t) \neq first(t')$ and there is

$$
c'' = \begin{array}{ccccccccc}
 & & & & a & a & a & b & \\
 & & & & b & b & b & a & \\
\hline
a & a & a & b & b & b & b & a \\
 & & & & b & b & b & b \\
\end{array} \in X_3^{p**}
$$

with a tiling decomposition t'' that extends t consistently with c'.

Let us show that X_3 has deciphering delay 3. Note that the only possibility for $c, c' \in X^{p**}$ with tiling decompositions in X t and t' respectively, to have that $c \lhd c'$ and $first(t) \neq first(t')$, is that $first(t) = \begin{array}{|c|} \hline a \\ \hline b \\ \hline \end{array}$ and $first(t') = \begin{array}{|c|c|c|c|} \hline a & a & a & b \\ \hline \end{array}$

(or viceversa). Therefore the top-left portion of c' must be $\begin{array}{cccc} a & a & a & b \\ b & & & \end{array}$. Then

the unique possible scenario is the following: $c = \begin{array}{|c|c|c|} \hline a & a & a \\ \hline b & b & b \\ \hline \end{array} \in X_3^{p**}$ prefix of

$c' = \begin{array}{|c|c|c|c|} \hline a & a & a & b \\ \hline b & b & b & a \\ \hline b & b & b & b \\ \hline \end{array} \in X_3^{p**}$, with tiling decompositions in X, t and t' respectively,

as in the figure, where t has 3 br-components and $first(t) \neq first(t')$. In this case there is no $c'' \in X_3^{p**}$ with a tiling decomposition t'' that extends t consistently with c'.

Theorem 15. *If $X \subseteq \Sigma^{**}$ has finite deciphering delay then X is a code.*

Proof. Let k be the deciphering delay of X. Suppose by contradiction that there exists a picture $u \in \Sigma^{**}$ that admits two different tiling decompositions in X, say t_1 and t_2. Now, let (i_0, j_0) the smallest position (in lexicographic order) of u, where t_1 and t_2 differ. Position (i_0, j_0) corresponds in t_1 to the tl-corner of some $x_1 \in X$, and in t_2 to the tl-corner of some $x_2 \in X$, with $x_1 \neq x_2$. See the figure below, where a dot indicates position (i_0, j_0) of u in t_1 (on the left) and t_2 (on the right), respectively.

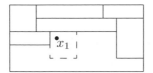   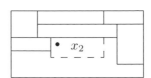

Consider the polyomino $\bar{u}$ obtained by eliminating from u all the pictures whose top-left corner occurs, in the tiling decomposition t_1, in a position (i, j) that is smaller than (i_0, j_0) and consider its corresponding tiling decomposition say $d_{\bar{u}}$. Now let y_1 the picture that covers the bottom-right corner of $\bar{u}$ in $d_{\bar{u}}$ and consider the polyomino $\bar{u}_1$ obtained, from $\bar{u}$, by eliminating the picture y_1 and its corresponding tiling decomposition $d_{\bar{u}_1}$; let y_2 the picture that covers the bottom-right corner of $\bar{u}_1$ in $d_{\bar{u}_1}$ and consider the polyomino $\bar{u}_2$ obtained, from $\bar{u}_1$, by eliminating the picture y_2 and its corresponding tiling decomposition $d_{\bar{u}_2}$

Remark that $\bar{u}_2 \lhd \bar{u}_1$ and that $first(d_{\bar{u}_1}) = x_1 \neq x_2 = first(d_{\bar{u}_2})$. Now, if $\bar{u}_2$ has k br-components, this contradicts the definition of deciphering delay k by choosing $c = \bar{u}_2$, $c' = u$ and $c'' = \bar{u}_1$.

On the contrary, if $\bar{u}_2$ has less than k br-components, an anologous contradiction can be obtained. Indeed, define the pictures u^h as the juxtaposition, row by row, of h copies of u, for an appropriate h, and v as the juxtaposition, row by row, of $\bar{u}$ with u^h.

Then choose c, c' and c'' equal to the picture obtained by gluing, one under the other, v with u, $\bar{u}_1$ and $\bar{u}_2$, respectively. $\square$

Note that the converse of Theorem 15 does not hold as shown by language X in Example 12.

Furthermore the next proposition shows that the parsing problem becomes polynomially solvable for languages that are the tiling star of a finite code with finite deciphering delay. Recall that the parsing problem is NP-complete for finite or, more in general, tiling recognizable languages (see [15]).

Proposition 16. *There exists a polynomial time algorithm that, given a finite code $X \subseteq \Sigma^{**}$ with deciphering delay k, and a picture $p \in \Sigma^{**}$, finds a tiling decomposition of p in X, if it exists, it exits with negative answer, otherwise.*

Proof. (Sketch) The algorithm scans p using a "current position" (i, j) and a "current partial tiling decomposition" $\mathcal{T}$ that contains some of the positions of p grouped as collections of rectangular p subdomains. At the beginning $(i, j) = (1, 1)$. The algorithm finds the pictures in X that are subpictures of p when translated with tl-corner on (i, j) and such that the positions of their subdomains were not yet put in $\mathcal{T}$. If there does not exist a picture as above, it exits with negative answer; if there exists only one picture it adds it to $\mathcal{T}$ and set (i, j) as the minimum position of p not yet covered by $\mathcal{T}$; otherwise for any pair of such pictures x_1 and x_2 in X, the algorithm continues as follows. It tries to extend alternatively two tiling decompositions t_1 and t_2, initially consisting only by $dom(x_1)$, $dom(x_2)$, respectively, in such a way that at each step, t_1 (t_2, resp.) corresponds to a polyomino $c_1 \in X^{p**}$ ($c_2 \in X^{p**}$, resp.) and $c_1 \lhd c_2$ or vice versa. The process continues until one between t_1 or t_2 cannot be anymore extended and this is at most when it has k br-components. At this point the algorithm adds to $\mathcal{T}$ the first picture (x_1 or x_2) of the other decomposition, and set (i, j) as the minimum position of p not yet covered by $\mathcal{T}$. The algorithm stops when it does not find any new position (i, j) and returns $\mathcal{T}$.

All this task can be accomplished by doing a preprocessing that classifies all tiling decompositions with k br-components done with elements of the input code. Then, the algorithm processes the picture p in time polynomial in the size of p and in a function of k and the size of the input code. $\qquad\square$

5   Conclusions

This paper proceeds along the direction started in [4] where prefix picture codes are defined and investigated in relation with the theory of recognizable picture languages (REC family). The general aim is to define subfamilies of decidable picture codes. Here we concentrate on sets of pictures with finite deciphering delay giving a non trivial definition that extends the corresponding definition for strings and that, at the same time, captures meaningful families of sets.

Several research directions are worthwhile to explore as, for instance, the relation between maximal codes with finite deciphering delay and maximal prefix codes. Moreover it will be interesting to investigate possible 2D generalizations of other types of string codes, as for example bifix codes. Finally one can try to remove the finiteness hypothesis and find decidable picture codes belonging to particular sub-families in REC, such as deterministic ones (see [2,6]).

References

1. Aigrain, P., Beauquier, D.: Polyomino tilings, cellular automata and codicity. Theoretical Computer Science 147, 165–180 (1995)
2. Anselmo, M., Giammarresi, D., Madonia, M.: Deterministic and unambiguous families within recognizable two-dimensional languages. Fund. Inform 98(2-3), 143–166 (2010)
3. Anselmo, M., Giammarresi, D., Madonia, M.: Strong prefix codes of pictures. In: Muntean, T., Poulakis, D., Rolland, R. (eds.) CAI 2013. LNCS, vol. 8080, pp. 47–59. Springer, Heidelberg (2013)
4. Anselmo, M., Giammarresi, D., Madonia, M.: Two-dimensional codes of pictures. In: Béal, M.-P., Carton, O. (eds.) DLT 2013. LNCS, vol. 7907, pp. 46–57. Springer, Heidelberg (2013)
5. Anselmo, M., Giammarresi, D., Madonia, M., Restivo, A.: Unambiguous recognizable two-dimensional languages. RAIRO: Theoretical Informatics and Applications 40(2), 227–294 (2006)
6. Anselmo, M., Madonia, M.: Deterministic and unambiguous two-dimensional languages over one-letter alphabet. Theoretical Computer Science 410(16), 1477–1485 (2009)
7. Beauquier, D., Nivat, M.: A codicity undecidable problem in the plane. Theoret. Comp. Sci. 303, 417–430 (2003)
8. Berstel, J., Perrin, D., Reutenauer, C.: Codes and Automata. Cambridge University Press (2009)
9. Bozapalidis, S., Grammatikopoulou, A.: Picture codes. ITA 40(4), 537–550 (2006)
10. Bruyère, V., Wang, L., Zhang, L.: On completion of codes with finite deciphering delay. Europ. J. Combinatorics 11, 513–521 (1990)
11. Giammarresi, D., Restivo, A.: Recognizable picture languages. Int. Journal Pattern Recognition and Artificial Intelligence 6(2-3), 241–256 (1992)
12. Giammarresi, D., Restivo, A.: Two-dimensional languages. In: Rozenberg, G. (ed.) Handbook of Formal Languages, vol. III, pp. 215–268. Springer (1997)
13. Grammatikopoulou, A.: Prefix picture sets and picture codes. In: Procs. CAI 2005, pp. 255–268. Aristotle University of Thessaloniki (2005)
14. Kolarz, M., Moczurad, W.: Multiset, Set and Numerically Decipherable Codes over Directed Figures. In: Smyth, B. (ed.) IWOCA 2012. LNCS, vol. 7643, pp. 224–235. Springer, Heidelberg (2012)
15. Lindgren, K., Moore, C., Nordahl, M.: Complexity of two-dimensional patterns. Journal of Statistical Physics 91(5-6), 909–951 (1998)
16. Moczurad, M., Moczurad, W.: Some open problems in decidability of brick (Labelled polyomino) codes. In: Chwa, K.-Y., Munro, J.I. (eds.) COCOON 2004. LNCS, vol. 3106, pp. 72–81. Springer, Heidelberg (2004)
17. Simplot, D.: A characterization of recognizable picture languages by tilings by finite sets. Theoretical Computer Science 218(2), 297–323 (1991)

Networks of Polarized Evolutionary Processors
Are Computationally Complete

Fernando Arroyo[1], Sandra Gómez Canaval[1], Victor Mitrana[2],
and Ştefan Popescu[3]

[1] Department of Languages, Projects and Computer Information Systems
University School of Informatics, Polytechnic University of Madrid
Crta. de Valencia km. 7 - 28031 Madrid, Spain
{farroyo,sgomez}@eui.upm.es
[2] Department of Organization and Structure of Information
University School of Informatics, Polytechnic University of Madrid
Crta. de Valencia km. 7 - 28031 Madrid, Spain
victor.mitrana@upm.es
[3] Faculty of Mathematics and Computer Science, University of Bucharest
Str. Academiei 14, 010014 Bucharest, Romania
stefipopescu2004@yahoo.com

Abstract. In this paper, we consider the computational power of a new
variant of networks of evolutionary processors which seems to be more
suitable for a software and hardware implementation. Each processor as
well as the data navigating throughout the network are now considered
to be polarized. While the polarization of every processor is predefined,
the data polarization is dynamically computed by means of a valuation
mapping. Consequently, the protocol of communication is naturally defined by means of this polarization. We show that tag systems can be
simulated by these networks with a constant number of nodes, while Turing machines can be simulated, in a time-efficient way, by these networks
with a number of nodes depending linearly on the tape alphabet of the
Turing machine.

1   Introduction

Networks of evolutionary processors (NEP) form a class of highly parallel and
distributed computing models inspired and abstracted from the biological evolution. Informally, a network of evolutionary processor consists of a virtual (complete) graph in which each node hosts a very simple processor called evolutionary
processor. By an evolutionary processor we mean a mathematical construction
which is able to perform very simple operations inspired by the point mutations
in DNA sequences (insertion, deletion or substitution of a single base pair). By
an informal parallelism with the natural process of evolution, each node may be
viewed as a cell having genetic information encoded in DNA sequences which
may evolve by local evolutionary events, that is point mutations. Each node processor, which is specialized just for one of these evolutionary operations, acts on

A.-H. Dediu et al. (Eds.): LATA 2014, LNCS 8370, pp. 101–112, 2014.

the local data and then local data becomes a mobile agent which can navigate in the network following a given protocol. Only that data which is able to pass a filtering process can be communicated. This filtering process may require to satisfy some conditions imposed by the sending processor, by the receiving processor or by both of them. All the nodes send simultaneously their data and the receiving nodes handle also simultaneously all the arriving messages, according to some strategies.

It is worth mentioning that NEPs resemble a pretty common architecture for parallel and distributed symbolic processing, related to the Connection Machine [7] which was defined as a network of microprocessors in the shape of a hypercube. Each microprocessor was very simple, processing one bit per unit time. Also it is closely related to the tissue-like P systems [13] in the membrane computing area [19].

NEPs as language generating devices and problem solvers have been considered in [2] and [14], respectively. They have been further investigated in a series of subsequent works. NEPs as accepting devices and problem solvers have been considered in [12]; later on, a characterization of the complexity classes **NP**, **P**, and **PSPACE** based on accepting NEPs has been reported in [10]. Universal NEPs and some descriptional complexity problems are discussed in [9]. The reader interested in a survey of the main results regarding NEPs is referred to [11].

Software implementations of NEPs have been reported, see, e.g., [3,4,16], most of them in JAVA. They encountered difficulties especially in the implementation of filters. The main idea to simulate the non-deterministic behavior of NEPs has been to consider a safe-thread model of processors, that is to have each rule and filter in a thread, respectively. Clearly the threads corresponding to the filters are much more complicated than those associated with the evolutionary rules. Configuration changes in a NEP are accomplished either by a communication step or by an evolutionary step, but these two steps may be realized in any order. This suggests that evolution or communication may be chosen depending on the thread model of processor [4]. The input and output filters are implemented as threads extending the `Runnable` interface. Therefore a processor is the parent of a set of threads, which use all objects from that processor in a mutual exclusion region. When a processor starts to run, it starts in a cascade way the rule threads and filter threads. As one can see, the filters associated with processors, especially if there are both input and output filters, seem to be hardly implementable. Consequently, it would be of interest to replace the communication based on filters among processors by another protocol. A first attempt was to move filters from each node to the edges between the nodes, see, e.g., [5]. Although this variant seems to be theoretically simpler, the attempts towards an implementation have encountered similar difficulties due to the fact that the filters associated with edges are similar to those associated with nodes.

Work [1] considers a new variant of NEP with the aim of proposing a new type of filtering process and discusses the potential of this variant for solving hard computational problems. The main and completely new feature of this variant is the valuation mapping which assigns to each string an integer value, depending on the values assigned to its symbols. Actually, we are not interested in computing the exact value of a string, but just the sign of this value. By means of this valuation, one may metaphorically say that the strings are electrically polarized. Thus, if the nodes are polarized as well, the strings migration from one node to another through the channel between the two cells seems to be more natural and easier to be implemented.

We consider here a slightly more general variant of networks of polarized evolutionary processors (NPEP) and investigate its computational power. Although the communication protocol based on the polarized processors and the valuation function seems to offer less control, the new variant is still computationally complete. We show that NPEP with a constant number of processors, namely 15, are computationally complete by devising a method for simulating 2-Tag Systems. As a 2-tag system can efficiently simulate any deterministic Turing machine but not nondeterministic ones, we propose a simulation of nondeterministic Turing machines with NPEP which maintains the working time of the Turing machine. That is, every language accepted by a one-tape nondeterministic Turing machine in time $f(n)$ can be accepted by an NPEP in time $O(f(n))$. Unlike the simulation of a 2-tag system, the size of a NPEP simulating an arbitrary Turing machine depends linearly on the number of tape symbols of the Turing machine.

2 Preliminaries

We start by summarizing the notions used throughout this work. An *alphabet* is a finite and nonempty set of symbols. The cardinality of a finite set A is written $card(A)$. Any finite sequence of symbols from an alphabet V is called *word* over V. The set of all words over V is denoted by V^* and the empty word is denoted by ε. The length of a word x is denoted by $|x|$ while $alph(x)$ denotes the minimal alphabet W such that $x \in W^*$. Furthermore, $|x|_a$ denotes the number of occurrences of the symbol a in x.

A homomorphism from the monoid V^* into the monoid (group) of additive integers $\mathbf{Z}$ is called *valuation* of V^* in $\mathbf{Z}$.

We consider here the following definition of 2-tag systems that appears in [20]. This type of tag-system, namely the type $\mathcal{T}_2$ 2-tag-systems that appear in Section 8 of [20], is slightly different but equivalent to those from [18,15]. A 2-tag system $T = (V, \mu)$ consists of a finite alphabet of symbols V, containing a special *halting symbol* H (denoted in [20] with $STOP$) and a finite set of rules $\mu : V \setminus \{H\} \to V^+$ such that $|\mu(x)| \geq 2$ or $\mu(x) = H$. Furthermore, $\mu(x) = H$ for just one $x \in V \setminus \{H\}$. A halting word for the system T is a word that contains the halting symbol H or whose length is less than 2. The transformation t_T (called the tag operation) is defined on the set of non-halting words as follows: if x is the leftmost symbol of a non-halting word w, then $t_T(w)$ is the result of deleting the leftmost 2 symbols of w and then appending the word $\mu(x)$ at the right end of the obtained word. A computation by a 2-tag system as above is a finite sequence of words produced by iterating the transformation t, starting with an initially given non-halting word w and halting when a halting word is produced. A computation is not considered to exist unless a halting word is produced in finitely-many iterations. Note that in [20] the halting words are defined a little bit different, as the words starting with the only symbol y such that $\mu(y) = H$, or the words whose length is less than 2. However, our way of defining halting words is equivalent to that in [20], in the sense that there exists a bijection between the valid computations obtained in each of these two cases. Indeed, if we consider the stopping condition from [20], and obtain in a valid computation a word starting with y, thus a halting word, it is enough to apply once more t_T on this word to obtain a word containing H, a halting word according to our definition, and transform the initial valid computation in a valid computation according to our definition. Conversely, if a word containing H, a halting word for our definition, is obtained in a valid computation, then the halting symbol could not have appeared in that word in other way than by applying t_T on a word starting with y, a halting word for the definition from [20], therefore we have a corresponding valid computation, by that definition. As shown in [20], such restricted 2-tag systems are universal.

A nondeterministic Turing machine is a construct $M = (Q, V, U, \delta, q_0, B, F)$, where Q is a finite set of states, V is the input alphabet, U is the tape alphabet, $V \subset U$, q_0 is the initial state, $B \in U \backslash V$ is the "blank" symbol, $F \subseteq Q$ is the set of final states, and δ is the transition mapping, $\delta : (Q \backslash F) \times U \to 2^{Q \times (U \backslash \{B\}) \times \{R, L\}}$. In this paper, we assume without loss of generality that any Turing machine we consider has a semi-infinite tape (bounded to the left) and makes no stationary moves; the computation of such a machine is described in [21,6,17]. An input word is accepted if and only if after a finite number of moves the Turing machine enters a final state. The language accepted by the Turing machine is a set of all accepted words. We say a Turing machine *decides* a language L if it accepts L and moreover halts on every input.

We say that a rule $a \to b$, with $a, b \in V \cup \{\varepsilon\}$ and $ab \neq \varepsilon$ is a *substitution rule* if both a and b are not ε; it is a *deletion rule* if $a \neq \varepsilon$ and $b = \varepsilon$; it is an *insertion rule* if $a = \varepsilon$ and $b \neq \varepsilon$. The set of all substitution, deletion, and insertion rules over an alphabet V are denoted by Sub_V, Del_V, and Ins_V, respectively. Given a rule σ as above and a word $w \in V^*$, we define the following *actions* of σ on w:

- If $\sigma \equiv a \to b \in Sub_V$, then $\sigma(w) = \begin{cases} \{ubv : \exists u, v \in V^* \ (w = uav)\}, \\ \{w\}, \text{ otherwise.} \end{cases}$
- If $\sigma \equiv a \to \varepsilon \in Del_V$, then

$$\sigma^r(w) = \begin{cases} \{u : w = ua\}, \\ \{w\}, \text{ otherwise} \end{cases} \qquad \sigma^l(w) = \begin{cases} \{v : w = av\}, \\ \{w\}, \text{ otherwise} \end{cases}$$

- If $\sigma \equiv \varepsilon \to a \in Ins_V$, then $\sigma^r(w) = \{wa\}$, $\sigma^l(w) = \{aw\}$.

Note that $\alpha \in \{l, r\}$ expresses the way of applying a deletion or insertion rule to a word, namely in the left ($\alpha = l$), or in the right ($\alpha = r$) end of the word, respectively. It is worth mentioning that the action mode of a substitution rule applied to a word w: it returns the set of all words that may be obtained from w depending on the position in w where the rule was actually applied.

For every evolutionary rule σ, action $\alpha \in \{l, r\}$, (α is missing when σ is a substitution rule) and $L \subseteq V^*$, we define the α-*action of* σ *on* L by $\sigma^\alpha(L) = \bigcup_{w \in L} \sigma^\alpha(w)$. Given a finite set of rules M, we define the α-*action of* M on the word w and the language L by $M^\alpha(w) = \bigcup_{\sigma \in M} \sigma^\alpha(w)$ and $M^\alpha(L) = \bigcup_{w \in L} M^\alpha(w)$, respectively.

Definition 1. A *polarized evolutionary processor* over V is a pair (M, α, π), where:

- M is a set of substitution, deletion or insertion rules over the alphabet V. Formally: $(M \subseteq Sub_V)$ or $(M \subseteq Del_V)$ or $(M \subseteq Ins_V)$. The set M represents the set of evolutionary rules of the processor. As one can see, a processor is "specialized" in one evolutionary operation, only.
- α gives the action mode of the rules of the node. If $M \subseteq Sub_V$, then α is missing.
- $\pi \in \{-, +, 0\}$ is the polarization of the node (negatively or positively charged, or neutral, respectively).

We denote the set of evolutionary processors over V by EP_V. Clearly, the evolutionary processor described here is a mathematical concept similar to that of an evolutionary algorithm, both being inspired from the Darwinian evolution. As compared to evolutionary algorithms, the rewriting operations we have considered here might be interpreted as mutations and the filtering process described above might be viewed as a selection process. Recombination is missing but it was asserted that evolutionary and functional relationships between genes can be captured by taking only local mutations into consideration [22].

Definition 2. A *network of polarized evolutionary processors* (NPEP for short) is a 7-tuple $\Gamma = (V, U, G, \mathcal{R}, \varphi, In, Out)$, where:

- V and U are the input and network alphabet, respectively, $V \subseteq U$.
- $G = (X_G, E_G)$ is an undirected graph without loops with the set of vertices X_G and the set of edges E_G. G is called the *underlying graph* of the network.
- $\mathcal{R} : X_G \longrightarrow EP_U$ is a mapping which associates with each node $x \in X_G$ the polarized evolutionary processor $\mathcal{R}(x) = (M_x, \alpha_x, \pi_x)$.
- φ is a valuation of U^* in $\mathbf{Z}$.
- $In, Out, \in X_G$ are the *input* and the *output* node of Γ, respectively.

We say that $card(X_G)$ is the size of Γ. A *configuration* of a NPEP Γ as above is a mapping $C : X_G \longrightarrow 2^{V^*}$ which associates a set of words with every node of the graph. A configuration may be understood as the sets of words which are present in any node at a given moment. Given a word $w \in V^*$, the initial configuration of Γ on w is defined by $C_0^{(w)}(x_I) = \{w\}$ and $C_0^{(w)}(x) = \emptyset$ for all $x \in X_G \setminus \{x_I\}$.

A configuration can change either by an *evolutionary step* or by a *communication step*. When changing by an evolutionary step, each component $C(x)$ of the configuration C is changed in accordance with the set of evolutionary rules M_x associated with the node x. Formally, we say that the configuration C' is obtained in *one evolutionary step* from the configuration C, written as $C \Longrightarrow C'$, iff

$$C'(x) = M_x^{\alpha_x}(C(x)) \text{ for all } x \in X_G.$$

When changing by a communication step, each node processor $x \in X_G$ sends out copies of all its words but keeping a local copy of the words having the same polarity to that of x only, to all the node processors connected to x and receives a copy of each word sent by any node processor connected with x providing that it has the same polarity as that of x. Note that, for simplicity reasons, we prefer to consider that a word migrate to a node with the same polarity and not an opposed one. Formally, we say that the configuration C' is obtained in *one communication step* from configuration C, written as $C \vdash C'$, iff

$$C'(x) = (C(x) \setminus \{w \in C(x) \mid sign(\varphi(w)) \neq \pi_x\}) \cup$$
$$\bigcup_{\{x,y\} \in E_G} (\{w \in C(y) \mid sign(\varphi(w)) = \pi_x\}),$$

for all $x \in X_G$. Here $sign(m)$ is the sign function which returns $+, 0, -$, provided that m is a positive integer, is 0, or is a negative integer, respectively. Note that all words with a different polarity than that of x are expelled. Further, each expelled word from a node x that cannot enter any node connected to x (no such node has the same polarity as the word has) is lost.

Let Γ be a NPEP, the computation of Γ on the input word $w \in V^*$ is a sequence of configurations $C_0^{(w)}, C_1^{(w)}, C_2^{(w)}, \ldots$, where $C_0^{(w)}$ is the initial configuration of Γ on w, $C_{2i}^{(w)} \Longrightarrow C_{2i+1}^{(w)}$ and $C_{2i+1}^{(w)} \vdash C_{2i+2}^{(w)}$, for all $i \geq 0$. Note that the configurations are changed by alternative steps. By the previous definitions, each configuration $C_i^{(w)}$ is uniquely determined by the configuration $C_{i-1}^{(w)}$. Otherwise stated, each computation in a NPEP is deterministic.

A computation as above *halts*, if there exists a configuration in which the set of words existing in the output node <u>Out</u> is non-empty. Given a NPEP Γ and an input word w, we say that Γ accepts w if the computation of Γ on w halts.

Let Γ be a NPEP with the input alphabet V; the *time complexity* of the finite computation $C_0^{(x)}, C_1^{(x)}, C_2^{(x)}, \ldots C_m^{(x)}$ of Γ on $x \in V^*$ is denoted by $Time_\Gamma(x)$ and equals m. The time complexity of Γ is the function from $\mathbf{N}$ to $\mathbf{N}$,
$$Time_\Gamma(n) = sup\{Time_\Gamma(x) \mid |x| = n\}.$$

3   2-Tag Systems Can Be Simulated by NPEP of Constant Size

In the following we show how a 2-tag system can be simulated by an NPEP. We make use of a similar strategy to that developed in [8].

Theorem 1. *For every 2-tag system $T = (V, \mu)$ there exists a NPEP Γ of size 15 such that $L(\Gamma) = \{w \mid T$ halts on $w\}$.*

Proof. Let $V = \{a_1, a_2, \ldots, a_n, a_{n+1}\}$ be the alphabet of the tag system T with $a_{n+1} = H$ and $V' = V \setminus \{H\}$. Let S be the set of all suffixes of the words in $\{\mu(a) \mid a \in V\}$. We consider the NPEP $\Gamma = (V', U, G, \mathcal{R}, \varphi, 1, 15)$ with the 15 nodes 1, 2, ..., 15. The working alphabet of the network is defined as follows:

$$U = V \cup \{a_0', a_0''\} \cup \{a', \overline{a'}, a'', a^\circ \mid a \in V'\} \cup$$
$$\{[x], \langle x \rangle, \ll x \gg, \overline{\langle x \rangle}, \overline{\ll x \gg}, \langle a_0 x \rangle, \ll a_0 x \gg, \overline{\langle a_0 x \rangle}, \overline{\ll a_0 x \gg} \mid x \in S\}.$$

The processors placed in the 15 nodes of the network are defined in Table 1.

Table 1. The description of the nodes of Γ

Node	M	α	π	Adjacency list
1	$\{a \to [\mu(a)] \mid a \in V'\}$	0		$\{2, 14\}$
2	$\{a \to a^\circ \mid a \in V'\}$	+		$\{1, 3\}$
3	$\{\varepsilon \to a_0'\}$	r	$-$	$\{2, 4, 12\}$
4	$\{a_k' \to a_k'' \mid 0 \leq k \leq n+1\}$	0		$\{3, 5, 9\}$
5	$\{[a_k x] \to \langle a_{k-1} x \rangle \mid 1 \leq k \leq n+1, x \in S\} \cup$ $\{\langle a_k x \rangle \to \langle a_{k-1} x \rangle \mid 1 \leq k \leq n+1, x \in S\}$	+		$\{4, 6\}$
6	$\{\langle x \rangle \to \ll x \gg \mid x \subset S\}$	$-$		$\{5, 7\}$
7	$\{a_{k-1}'' \to \overline{a_k'} \mid 1 \leq k < n+1\}$	0		$\{6, 8\}$
8	$\{\overline{a_k'} \to a_k' \mid 1 \leq k \leq n+1\}$	+		$\{7, 9\}$
9	$\{\ll x \gg \to \langle x \rangle \mid x \subset S\}$	$-$		$\{4, 8, 10\}$
10	$\{\overline{\langle a_0 x \rangle} \to \overline{\ll a_0 x \gg} \mid x \in S\}$	0		$\{9, 11\}$
11	$\{a_k' \to a_k \mid 1 < k \leq n+1\}$	+		$\{10, 12, 15\}$
12	$\{\ll a_0 a_k x \gg \to [a_k x] \mid 1 \leq k \leq n+1, x \in S\} \cup$ $\{\ll a_0 \gg \to [a_0]\}$	0		$\{3, 11, 13\}$
13	$\{[a_0] \to \varepsilon\}$	l	$-$	$\{12, 14\}$
14	$\{u^\circ \to c\}$	l	$-$	$\{1, 13\}$
15	$\emptyset$		$-$	$\{11\}$

The construction of Γ is complete as soon as we define the valuation mapping φ. It is defined as follows:

$$\varphi(a_k) = 0, 1 \leq k \leq n, \qquad \varphi(H) = -10,$$
$$\varphi([x]) = \varphi(\overline{\langle x \rangle}) = \varphi(\overline{\langle a_0 x \rangle}) = 1, x \in S,$$
$$\varphi(a^\circ) = -2, a \in V',$$
$$\varphi(a_k') = 1, 0 \leq k \leq n+1,$$
$$\varphi(a_k'') = 2, 0 \leq k \leq n+1,$$
$$\varphi(\langle x \rangle) = \varphi(\langle a_0 x \rangle) = -1, x \in S,$$
$$\varphi(\ll x \gg) = \varphi(\ll a_0 x \gg) = 0, x \in S,$$
$$\varphi(\overline{a_k'}) = 3, 1 \leq k \leq n+1,$$
$$\varphi(\overline{\ll x \gg}) = \varphi(\overline{\ll a_0 x \gg}) = 2, x \in S.$$

We show that Γ accepts a word w that does not contain H if an only if T eventually halts on w. Let $w = aby, a, b \in V, y \in V^*$ be a word that does not contain H such that T eventually halts on w. We show how w can be accepted by Γ. At the beginning of the computation w is found in node 1, where the first symbol a can be replaced with $[\mu(a)]$. From our further explanations, it will turn out that if the symbol a replaced by $[\mu(a)]$ in node 1 is not the leftmost one, then the computation on this word is blocked. The valuation of the new word, $[\mu(a)]by$, has a positive value such that the word enters node 2 which is positively charged. In node 2, we can rewrite b as b°, getting the new word $[\mu(a)]b^\circ y$. Again, later we infer that if another symbol is replaced, the computation will be blocked. The word $[\mu(a)]b^\circ y$ can only enter node 3 where the symbol a_0' is inserted to its right end obtaining $[\mu(a)]b^\circ y a_0'$. This word can enter node 4 and 12. Note that the copy entering node 12 remains there forever. Therefore, we continue our analysis in node 4. In this node, a_0' is replaced by a_0'' which change the polarization of the word from neutral to positive which makes it to migrate to node 5.

We now assume that $\mu(a) = a_k x$ for some $1 \leq k \leq n + 1$. In node 5, $[a_k x]$ is replaced by $\langle a_{k-1} x \rangle$ and the new word which is negatively charged enters node 6. After its first symbol $\langle a_{k-1} x \rangle$ is substituted with $\ll a_{k-1} x \gg$, the word has a null valuation and enters node 7. The word is successively transformed in $\ll a_{k-1} x \gg b^\circ \overline{y a_1'}$ (in node 7), $\ll a_{k-1} x \gg b^\circ y a_1'$ (in node 8), and $\overline{\langle a_{k-1} x \rangle} b^\circ y a_1'$ (in node 9). If $k > 1$, this string first returns to node 4, resulting in $\langle a_{k-1} x \rangle b^\circ y a_1''$, and then enters node 5, resulting in $\langle a_{k-2} x \rangle b^\circ y a_1''$. This process continues by iteratively passing the sequence of nodes $4, 5, 6, 7, 8, 9$ until a string of the form $\langle a_0 x \rangle b^\circ y a_k'$ is obtained in node 9. Now the current word enters 10 and 11, where it is rewritten into $\overline{\ll a_0 x \gg} b^\circ y a_k'$ and $\overline{\ll a_0 x \gg} b^\circ y a_k$, respectively. If $k = n + 1$, then the word enters the output node 15 and the computation halts. Otherwise, the current string can only enter node 12, where its first symbol $\overline{\ll a_0 x \gg}$ is replaced either by $[a_j x']$, provided that $x = a_j x'$, or by $[a_0]$, if $x = \varepsilon$. In the former case, the whole process described above resumes from the node 3. In the latter, we actually reached a word of the form $[a_0] b^\circ y \mu(a)$. This word can enter node 13, where the symbol $[a_0]$ is deleted, provided that it is the leftmost symbol. A copy of this word remains in 13 forever, but another copy enters 14, where the symbol b° is deleted, provided that it is the leftmost symbol. One can see now that if the symbols substituted in the nodes 1 and 2 were not the right ones, the computation will get stuck. Note that the word obtained in node 14 is exactly $y\mu(a)$, which means that we have correctly simulated the step $a \to \phi(a)$ in the tag system. Now, this word enters 1 where the simulation of the next step in T starts.

By these explanations, we infer that $w \in L(\Gamma)$ if and only if T will eventually halt on w. □

4   Arbitrary Turing Machines Can Be Simulated by NPEP

Although 2-tag systems efficiently simulate deterministic Turing machines, via cyclic tag systems (see, e.g., [23]), the previous result does not allow us to say much about the NPEP accepting in a computationally efficient way all recursively enumerable languages. We now discuss how recursively enumerable languages can be efficiently (from the time complexity point of view) accepted by NPEP by simulating arbitrary Turing machines.

Theorem 2. *For any recursively enumerable language L, accepted in $\mathcal{O}(f(n))$ by a Turing machine with tape alphabet U, there exists an NPEP of size $10\,card(U)$ accepting L in $\mathcal{O}(f(n))$ steps.*

Proof. Let $M = (Q, V, U, \delta, q_0, B, F)$ be a Turing machine with $U \cap Q = \emptyset$, and $U = \{a_1, a_2, \ldots, a_{n+1}\}$, $a_{n+1} = B$. We start the construction of the NPEP Γ accepting the language accepted by M with the definition of its working alphabet:

$$W = U \cup Q \cup \{\overline{q} \mid q \in Q\} \cup \{\overline{a} \mid a \in U\} \cup \{a', a'', \widetilde{a}, \widehat{a} \mid a \in U \setminus \{B\}\} \cup$$
$$\{[q, a, D] \mid q \in Q, a \in U \setminus \{B\}, D \in \{L, R\}\}.$$

We now can define the valuation mapping φ as follows:

$$\varphi(a) = 0, a \in U, \qquad\qquad \varphi(\overline{a}) = 2, a \in U,$$
$$\varphi(a') = -1, a \in U \setminus \{B\}, \qquad \varphi(a'') = 2, a \in U \setminus \{B\},$$
$$\varphi(q) = -1, q \in Q \setminus F \qquad\qquad \varphi(q) = 1, q \in F,$$
$$\varphi(\widetilde{a}_i) = -p_i, 1 \le i \le n, \qquad \varphi(\widehat{a}_i) = p_i, 1 \le i \le n,$$
$$\varphi(\langle s, a_i, L \rangle) = -p_i, s \in Q, 1 \le i \le n, \quad \varphi(\langle s, a_i, R \rangle) = p_i, s \in Q, 1 \le i \le n,$$
$$\varphi([s, a, D]) = -2, \qquad\qquad \varphi(\overline{q}) = 0, q \in Q,$$
$$s \in Q, a \in U \setminus \{B\}, D \in \{L, R\}$$

Here p_i denotes the ith odd prime number. The processors placed in the nodes of the network are defined in Table 2.

We now analyze the computation of Γ on an input word, say w, which is placed in the input node $\underline{In}$. Here the B symbol is added to its right-hand end, yielding wB which has a neutral polarization. Therefore, a copy of this word remains in $\underline{In}$, while another copy migrates to $InsSt$ (Insert State). It will turn out, by our further explanations, that if w is accepted by a computation of M that uses the minimal number t of auxiliary cells (cells that initially contain B), then every word wB^j, with $j < t$, that enters $InsSt$ will be eventually blocked in a node. We assume that wB^j, for some $j \ge t$, enters $InsSt$. It is transformed into $wB^j q_0$ which is negatively charged, such that each node $IdS(a_i)$ (Identify Symbol), $1 \le i \le n+1$, receives one of its copies. Inductively, we may assume that the current word is $yB^k qx$, for some $k \ge 0$, which signifies that the Turing Machine M is in state q and has on its tape the word xy, with its head positioned on the first symbol of y. Assume that $y = a_i z$, for some $1 \le i \le n+1$; note that the copy of $yB^k qx$ that enters $IdS(a_k)$, $k \ne i$, will be further blocked in Del.

Let us follow the copy of $a_i z B^k qx$ that enters $IdS(a_i)$. After an occurrence of a_i, not necessarily the leftmost one, is replaced by $\overline{a}_i$, the word arrives in $ChT(a_i)$ (Choose Transition). Here a symbol $[s, a_m, D]$, such that $(s, a_m, D) \in \delta(q, a_i)$, is adjoined which makes the new word to have a null value through the valuation mapping φ. This word enters Del, where a barred symbol is removed provided that it is the leftmost one. We can see now that, if the occurrence of a_i substituted in the node $IdS(a_i)$ was not the leftmost one, the word remains blocked in Del. Therefore, our current word becomes $zB^k[s, a_m, D]x$, $1 \le m \le n$, $D \in \{L, R\}$. As it is negatively charged, it enters $DetLR$ (Determine Left-Right). In this node, $[s, a_m, D]$ is replaced by $\langle s, a_m, D \rangle$ and the new word is $zB^k\langle s, a_m, D \rangle x$.

Table 2. The definition of the nodes of Γ

Node	M	α	π	Adjacency list
In	$\{\varepsilon \to B\}$	r	0	$\{InsSt\}$
$InsSt$	$\{\varepsilon \to q_0\}$	r	0	$\{In\} \cup$ $\{IdS(a) \mid a \in U\}$
$IdS(a),$ $a \in U$	$\{a \to \overline{a}\}$		$-$	$\{InsSt, RestS, ChT(a)\}$
$ChT(a),$ $a \in U$	$\{q \to [s, b, D] \mid$ $(s, b, D) \in \delta(q, a)\}$		$+$	$\{Del, IdS(a)\}$
Del	$\{\overline{a} \to \varepsilon \mid a \in U\}$	l	0	$\{DetLR\} \cup$ $\{ChT(a) \mid a \in U\}$
$DetLR$	$\{[s, b, D] \to \langle s, b, D \rangle \mid$ $(s, b, D) \in \delta(q, a),$ for some $q \in Q \setminus F, a \in U\}$		$-$	$\{Del\} \cup$ $\{InsL_1(a) \mid a \in U \setminus \{B\}\} \cup$ $\{InsR_1(a) \mid a \in U \setminus \{B\}\} \cup$
$InsR_1(a),$ $a \in U \setminus \{B\}$	$\{\varepsilon \to \widetilde{a}\}$	r	$+$	$\{DetLR, CSR(a)\}$
$CSR(a)$ $a \in U \setminus \{B\}$	$\{\langle s, a, R \rangle \to \overline{s} \mid s \in Q\}$		0	$\{InsR_1(a), InsR_2(a)\}$
$InsR_2(a)$ $a \in U \setminus \{B\}$	$\{\widetilde{a} \to a\}$		$-$	$\{CSR(a), RestS\}$
$InsL_1(a),$ $a \in U \setminus \{B\}$	$\{\varepsilon \to \widehat{a}\}$	l	$-$	$\{DetLR, CSL(a)\}$
$CSL(a)$ $a \in U \setminus \{B\}$	$\{\langle s, a, L \rangle \to \overline{s} \mid s \in Q\}$		0	$\{InsL_1(a), InsL_2(a)\}$
$InsL_2(a)$ $a \in U \setminus \{B\}$	$\{\widehat{a} \to a\}$		$+$	$\{CSL(a), Move\}$
$Move$	$\{a \to a' \mid a \in U \setminus \{B\}\}$		0	$\{InsL_2(a) \mid a \in U \setminus \{B\}\} \cup$ $Ins(a) \mid a \in U \setminus \{B\}$
$Ins(a)$ $a \in U \setminus \{B\}$	$\{\varepsilon \to a''\}$	l	$-$	$\{Move, Del(a)\}$
$Del(a)$ $a \in U \setminus \{B\}$	$\{a' \to \varepsilon\} \cup$	r	$+$	$\{Ins(a), FL\}$
FL	$\{a'' \to a \mid a \in U \setminus \{B\}\}$		$+$	$\{RestS\} \cup$ $\{Del(a) \mid a \in U \setminus \{B\}\}$
$RestS$	$\{\overline{s} \to s \mid s \in Q\}$		0	$\{FL, Out\} \cup$ $\{InsR_2(a) \mid a \in U \setminus \{B\}\} \cup$ $\{IdS(a) \mid a \in U\} \cup$
Out	$\emptyset$		$+$	$\{RestS\}$

The value of this word is either negative, if $D = L$, or positive, if $D = R$. We first consider the case $D = R$; each node $InsR_1(a)$, $a \in U \setminus \{B\}$, receives a copy of the word $zB^k \langle s, a_m, R \rangle x$. If a copy arrives in $InsR_1(a)$, $a \neq a_m$, then it can never be transformed into a word with a null value. After a number of evolutionary steps, each of them inserting a $\widetilde{a}$ at the end, the word gets a negatively value and migrates back to $DetLR$, where it remains forever. The copy of $zB^k \langle s, a_m, R \rangle x$ that enters $InsR_1(a_m)$, is modified into $zB^k \langle s, a_m, R \rangle x\widetilde{a_m}$. As it has a null value, it enters $CSR(a_m)$ (Change State from the Right), where it becomes $zB^k \overline{s} x\widetilde{a_m}$. The word $zB^k \overline{s} x\widetilde{a_m}$ is negatively charged and migrates to $InsR_2(a_m)$, where wta_m is substituted with a_m. Now, the word enters $RestS$ (Restore State), where the word becomes $zB^k sxa_m$, hence we have correctly simulated a move of M to the right.

If $D = L$, then by means of the nodes $InsL_1(a_m)$, $CSL(a_m)$, and $InsL_2(a_m)$, the symbol a_m is inserted in the beginning of the word. Then, by means of the nodes $Move$, $Ins(b)$, $Del(b)$, and FL, the symbol $b \neq B$ is rotated from the end of the word to its beginning. After this rotation, the word enters $RestS$, where it becomes $ba_m zB^k sx_1$, provided that $x = x_1 b$. In conclusion, we have correctly simulated a move of M to the left.

In order to simulate another move, the word enters $IdS(a)$, $a \in U$, and the whole process discussed above resumes. Note that if the state s in a word existing in $RestS$ is a final state, then the word enters $\underline{Out}$ and the computation halts.

Now, the simulation proof is complete. From Table 2, it is easy to see that the size of Γ is exactly $10 card(U)$.   □

An analysis of the proof, reveals that each move of M is simulated by Γ in a constant number of steps. On the other hand, if M accepts in $\mathcal{O}(f(n))$ time, then the number of necessary steps in the beginning of any computation of Γ on a word of length n is at most $\mathcal{O}(f(n))$. Consequently, $Time_\Gamma \in \mathcal{O}(f(n))$.

We finish this work with two *open problems* that naturally arise:

1. Is the size proved in Theorem 1 optimal?
2. Can arbitrary Turing machines be simulated by NPEP of constant size? In the affirmative, is such a simulation still time-efficient?

References

1. Alarcón, P., Arroyo, F., Mitrana, V.: Networks of polarized evolutionary processors as problem solvers. In: Advances in Knowledge-Based and Intelligent Information and Engineering Systems. Frontiers in Artificial Intelligence and Applications, pp. 807–815. IOS Press (2012)
2. Castellanos, J., Martín-Vide, C., Mitrana, V., Sempere, J.M.: Networks of evolutionary processors. Acta Inf. 39, 517–529 (2003)
3. Diaz, M.A., de Mingo, L.F., Gómez, N.: Implementation of massive parallel networks of evolutionary processors (MPNEP): 3-colorability problem. In: Krasnogor, N., Nicosia, G., Pavone, M., Pelta, D. (eds.) NICSO 2007. SCI, vol. 129, pp. 399–408. Springer, Heidelberg (2007)
4. Diaz, M.A., de Mingo, L.F., Gómez, N.: Networks of evolutionary processors: Java Implementation of a threaded processor. International Journal of Information Theories & Applications 15, 37–43 (2008)

5. Drăgoi, C., Manea, F., Mitrana, V.: Accepting networks of evolutionary processors with filtered connections. J. UCS 13, 1598–1614 (2007)
6. Hartmanis, J., Stearns, R.E.: On the computational complexity of algorithms. Trans. Amer. Math. Soc. 117, 533–546 (1965)
7. Hillis, W.D.: The Connection Machine. MIT Press, Cambridge (1979)
8. Loos, R., Manea, F., Mitrana, V.: Small universal accepting hybrid networks of evolutionary processors. Acta Inf. 47, 133–146 (2010)
9. Manea, F., Martín-Vide, C., Mitrana, V.: On the size complexity of universal accepting hybrid networks of evolutionary processors. Mathematical Structures in Computer Science 17, 753–771 (2007)
10. Manea, F., Margenstern, M., Mitrana, V., Pérez-Jiménez, M.J.: A new characterization of NP, P, and PSPACE with accepting hybrid networks of evolutionary processors. Theory Comput. Syst. 46, 174–192 (2010)
11. Manea, F., Martín-Vide, C., Mitrana, V.: Accepting networks of evolutionary word and picture processors: A survey. In: Scientific Applications of Language Methods, Mathematics, Computing, Language, and Life: Frontiers in Mathematical Linguistics and Language Theory, vol. 2, pp. 523–560. World Scientific (2010)
12. Margenstern, M., Mitrana, V., Pérez-Jímenez, M.J.: Accepting Hybrid Networks of Evolutionary Processors. In: Ferretti, C., Mauri, G., Zandron, C. (eds.) DNA 2004. LNCS, vol. 3384, pp. 235–246. Springer, Heidelberg (2005)
13. Martín-Vide, C., Pazos, J., Păun, G., Rodríguez-Patón, A.: A new class of symbolic abstract neural nets: Tissue P systems. In: Ibarra, O.H., Zhang, L. (eds.) COCOON 2002. LNCS, vol. 2387, pp. 290–299. Springer, Heidelberg (2002)
14. Martín-Vide, C., Mitrana, V., Pérez-Jiménez, M.J., Sancho-Caparrini, F.: Hybrid networks of evolutionary processors. In: Cantú-Paz, E., et al. (eds.) GECCO 2003. LNCS, vol. 2723, pp. 401–412. Springer, Heidelberg (2003)
15. Minsky, M.L.: Size and structure of universal Turing machines using tag systems. In: Recursive Function Theory, Symp. in Pure Mathematics, vol. 5, pp. 229–238 (1962)
16. Navarrete, C.B., Echeanda, M., Anguiano, E., Ortega, A., Rojas, J.M.: Parallel simulation of NEPs on clusters. In: Proc. IEEE/WIC/ACM International Joint Conference on Web Intelligence and Intelligent Agent Technology - WI-IAT, pp. 171–174. IEEE Computer Society (2011)
17. Papadimitriou, C.H.: Computational Complexity. Addison-Wesley (1994)
18. Post, E.L.: Formal reductions of the general combinatorial decision problem. Amer. J. Math. 65, 197–215 (1943)
19. Păun, G.: Membrane computing. An introduction. Springer, Berlin (2002)
20. Rogozhin, Y.: Small universal Turing machines. Theoret. Comput. Sci. 168, 215–240 (1996)
21. Rozenberg, G., Salomaa, A. (eds.): Handbook of Formal Languages, vol. I-III. Springer, Berlin (1997)
22. Sankoff, D., et al.: Gene order comparisons for phylogenetic inference:evolution of the mitochondrial genome. Proceedings of the National Academy of Sciences of the United States of America 89, 6575–6579 (1992)
23. Woods, D., Neary, T.: On the time complexity of 2-tag systems and small universal Turing machines. In: 47th Annual IEEE Symposium on Foundations of Computer Science FOCS 2006, pp. 439–448 (2006)

Two Double-Exponential Gaps for Automata with a Limited Pushdown*

Zuzana Bednárová and Viliam Geffert

Dept. Comput. Sci., P. J. Šafárik University
Jesenná 5, 04154 Košice, Slovakia
ivazuzu@eriv.sk,
viliam.geffert@upjs.sk

Abstract. We shall consider nondeterministic and deterministic automata equipped with a limited pushdown (constant height NPDAs and DPDAs) as well as their two-way versions (constant height 2NPDAs and 2DPDAs). We show two double-exponential gaps for these devices, namely, *(i)* for complementing constant height one-way NPDAs and *(ii)* for converting 2NPDAs or 2DPDAs into one-way devices.

Keywords: pushdown automata, finite state automata, regular languages, descriptional complexity.

1 Introduction

Pushdown automata (PDAs) are one of the fundamental models in the formal language theory. They were introduced as a model for solving many practical problems arising from syntax analysis in programming languages and provided a machine counterpart for context-free grammars [3,6,8,11].

In [4], a model of a one-way nondeterministic automaton equipped with a limited pushdown was introduced (constant height NPDA), together with its deterministic version (constant height DPDA). Obviously, such machines can accept regular languages only. However, once a pushdown store is available, even though of a constant height, we may accept some regular languages with machines much smaller than any finite state automata.[1] For some languages, there exists an exponential gap between the size of constant height NPDAs and the size of nondeterministic finite automata (NFAs); the same gap was found between the constant height DPDAs and deterministic finite automata (DFAs) [4].

From that perspective, it makes sense to analyze the blow-up in the size of constant height PDAs incurred by basic Boolean operations such as union, intersection, or complement. The size cost of Boolean operations is a classical problem, see, e.g., [5,7,10,12]. Unlike the PDAs with unrestricted pushdown, both

---

* Supported by the Slovak grant contracts VEGA 1/0479/12 and APVV-0035-10.
[1] For such a machine, a fair *descriptional complexity measure* must take into account all machine's components, i.e., the number of states, the height of the pushdown store, and the size of the pushdown alphabet.

A.-H. Dediu et al. (Eds.): LATA 2014, LNCS 8370, pp. 113–125, 2014.
© Springer International Publishing Switzerland 2014

the constant height NPDAs and DPDAs are closed under all these operations: *(i)* For constant height DPDAs, the costs of union and intersection are single-exponential, but the cost of complement is polynomial [1]. *(ii)* For constant height NPDAs, the cost of union is linear but the cost of intersection is single-exponential [2]. The cost of the given Boolean operation for the constant height version of PDA is at most polynomial if and only if the unrestricted version is closed under the same operation (see e.g. [8]). This holds also for *complementing* constant height NPDAs [2], but here we have a large gap: the upper bound is trivially double-exponential (see also Thm. 1 below), but the proved lower bound is only single-exponential (obtained by application of De Morgan's laws).

The first problem tackled here is complementing of one-way NPDAs. We show that this requires a *double-exponential* blow-up. Namely, for each $c \geq 2$, we provide $\{L_n\}_{n\geq1}$, a family of regular languages, built over the same $(c+2)$-letter alphabet, such that: *(i)* these languages are accepted by one-way NPDAs with $n+O(c)$ states, a pushdown of height n, and c pushdown symbols, but *(ii)* their complements $\{L_n^C\}_{n\geq1}$ cannot be accepted by one-way NPDAs in which both the number of states and the pushdown height are below $2^{1/6 \cdot c^n} - O(n \cdot \log c) \geq 2^{c^{n-O(1)}}$, independently of the used pushdown alphabet.

Second, it turns out that $\{L_n^C\}_{n\geq1}$, for which one-way NPDAs require double-exponential resources, can be accepted by *two-way* machines deterministically even with linear resources, namely, by constant height 2DPDAs with $n+O(c)$ states, a pushdown of height $n+1$, and $c+1$ pushdown symbols. This gives a double-exponential blow-up for converting constant height 2DPDAs (hence, also 2NPDAs) to constant height one-way NPDAs (hence, also DPDAs). For comparison, removing bidirectionality is single-exponential for finite state automata [9], but not possible at all for 2DPDAs or 2NPDAs with unrestricted pushdown, capable of accepting non-context-free languages like $\{a^m b^m c^m : m \geq 0\}$.

2 Preliminaries

We assume the reader is familiar with the standard models of a deterministic and nondeterministic *finite state automaton* (DFA and NFA, for short), see, e.g., [8].

For the given alphabet Σ, let Σ^i denote the set of words of length i, with $\Sigma^0 = \{\varepsilon\}$, $\Sigma^{\leq h} = \bigcup_{i=0}^{h} \Sigma^i$, and $\Sigma^* = \bigcup_{i=0}^{\infty} \Sigma^i$.

For technical reasons, we introduce a *(one-way) nondeterministic pushdown automaton* (NPDA) in the form that distinguishes instructions manipulating the pushdown store from those reading the input tape [4]; as a sextuplet $A = \langle Q, \Sigma, \Gamma, H, q_I, F \rangle$, where Q is the finite set of states, Σ the input alphabet, Γ the pushdown alphabet, $q_I \in Q$ the initial state, $F \subseteq Q$ the set of accepting states, and $H \subseteq Q \times (\{\varepsilon\} \cup \Sigma \cup \{-,+\} \cdot \Gamma) \times Q$ is the *transition relation*, establishing instructions with the following meaning. (q, ε, q'): A gets from the state q to the state q' without using the input tape or the pushdown store. (q, a, q'): if the next input symbol is a, A gets from q to q' by reading a. $(q, -X, q')$: if the symbol on top of the pushdown is X, A gets from q to q' by popping X. $(q, +X, q')$: A gets from q to q' by pushing the symbol X onto the pushdown. An *accepting computation* begins in the state q_I with the empty pushdown store (no initial symbol in the pushdown), and ends by halting in an accepting state $q' \in F$ after reading the entire input. As usual, $L(A)$ denotes the language accepted by A.

A *two-way* nondeterministic pushdown automaton (2NPDA) is defined in the same way as the one-way NPDA, but now A can move in both directions along the input. Transitions in H related to input reading are upgraded as follows. $(q, a\rightarrow, q')$, $(q, a\leftarrow, q')$, $(q, a\circ, q')$: if the current input symbol is a, A gets from q to q' and moves its input head one position to the right, left, or keeps it stationary, respectively. The input in enclosed in between $\vdash, \dashv \notin \Sigma$, called the *left and right endmarkers*, respectively. A starts in q_I with the input head at the left endmarker and accepts by halting in $q' \in F$ anywhere along the input.

A *deterministic* pushdown automaton (DPDA or 2DPDA) is obtained from NPDA or 2NPDA by claiming that the transition relation does not allow executing more than one possible instruction at a time. (See [1] for formal details.)

A *constant height* pushdown automaton (NPDA, 2NPDA, DPDA, or 2DPDA) is $A = \langle Q, \Sigma, \Gamma, H, q_I, F, h \rangle$, where all components are as above, but h is a constant delimiting the *pushdown height*. By definition, if A tries to push a symbol onto the pushdown but h symbols are already piled up, A aborts and rejects.

3   Complementing Constant Height NPDAs

Given a constant height NPDA A, a trivial double-exponential upper bound for an NPDA accepting the complement of $L(A)$ is obtained easily, by converting into a classical DFA for $L(A)^C$. The resulting machine can be viewed as a special case of the constant height NPDA, with the pushdown height equal to zero and not using the power of nondeterminism. This gives:

Theorem 1 ([2, Thm. 9]). *For each constant height NPDA $A = \langle Q, \Sigma, \Gamma, H, q_I, F, h \rangle$, there exists a DFA A' (hence, also a constant height NPDA) accepting the complement of $L(A)$ with at most $2^{\|Q\| \cdot \|\Gamma^{\leq h}\|}$ states.*

We shall now show that the above conversion to DFAs cannot be substantially improved. Let us begin with the family of witness languages. For a given alphabet Σ and $n \geq 1$, let $\mathcal{X} = \Sigma^n$. Clearly, we have $\|\Sigma\|^n$ many different strings in $\mathcal{X}$. We say that a string φ is *well-formed*, if it has the following block structure:

$$\varphi = x_1 \text{¢} x_2 \text{¢} \cdots x_s \text{¢} \$ y_1^R \text{¢} y_2^R \text{¢} \cdots y_r^R \text{¢}, \tag{1}$$

where $\$, \text{¢} \notin \Sigma$ are separator symbols, $s, r \geq 0$, and $x_i, y_j \in \mathcal{X}$, for $i \in \{1, \ldots, s\}$ and $j \in \{1, \ldots, r\}$. We call the string *ill-formed*, if it is not well-formed.

Definition 2. *For the given alphabet Σ and $n \geq 1$, let $L_{\Sigma,n}$ be the language consisting of all strings $\varphi \in (\Sigma \cup \{\$, \text{¢}\})^*$ which are ill-formed, plus all well-formed strings $\varphi \in (\Sigma \cup \{\$, \text{¢}\})^*$, for which $\cup_{i=1}^s \{x_i\} \bigcap \cup_{j=1}^r \{y_j\} \neq \emptyset$. That is, $x_i = y_j$ for some $i \in \{1, \ldots, s\}$ and $j \in \{1, \ldots, r\}$.*

We are going to show that the language $L_{\Sigma,n}$ can be accepted by a constant height NPDA with linear cost, but any constant height NPDA accepting the complement $L^{\mathrm{C}}_{\Sigma,n}$ must use resources that are double-exponential in n:

Lemma 3. *For each alphabet Σ and $n \geq 1$, the language $L_{\Sigma,n}$ can be accepted by a constant height NPDA using $n + 2 \cdot \|\Sigma\| + 8$ states, a pushdown of height n, and $\|\Sigma\|$ pushdown symbols.*

Theorem 4. *Let A_n be any constant height NPDA accepting the language $L^{\mathrm{C}}_{\Sigma,n}$, for some non-unary alphabet Σ and some $n \geq 4$, and let Q_n and h_n be, respectively, the number of states and the pushdown height in A_n. Then*

$$(\|Q_n\|+1)^2 \cdot (h_n+1) > 2^{1/2 \cdot \|\Sigma\|^n - 2n \cdot \log \|\Sigma\| - 2 \cdot \log(n+2) - \log(3/2)}.$$

Consequently, in $\{A_n\}_{n \geq 1}$, either the number of states or else the pushdown height must be above $2^{1/6 \cdot \|\Sigma\|^n - O(n \cdot \log \|\Sigma\|)} \geq 2^{\|\Sigma\|^{n-O(1)}}$.

Proof. Let $A_n = \langle Q_n, \Sigma, \Gamma_n, H_n, q_{1,n}, F_n, h_n \rangle$ be a constant height NPDA for $L^{\mathrm{C}}_{\Sigma,n}$. Assume first that A_n is in the normal form of Lem. 1 in [4]. That is, A_n accepts in the unique state $q_{\mathrm{F},n}$ with empty pushdown at the end of the input, and hence $F_n = \{q_{\mathrm{F},n}\}$. For contradiction, assume also that, for some $n \geq 4$,

$$\|Q_n\|^2 \cdot (h_n+1) \leq 2^{1/2 \cdot \|\Sigma\|^n - 2n \cdot \log \|\Sigma\| - 2 \cdot \log(n+2) - \log(3/2)}$$
$$= 2^{1/2 \cdot \|\Sigma\|^n - \log \|\Sigma\|^{2n} - \log(n+2)^2 - \log(3/2)} = \frac{2^{1/2 \cdot \|\Sigma\|^n} \cdot 2}{\|\Sigma\|^{2n} \cdot (n+2)^2 \cdot 3} . \tag{2}$$

We are now going to fool A_n, i.e., to construct a string $\delta \notin L^{\mathrm{C}}_{\Sigma,n}$ such that A_n accepts δ. First, we partition the blocks in $\mathcal{X} = \Sigma^n$, upon which the languages $L_{\Sigma,n}$ and $L^{\mathrm{C}}_{\Sigma,n}$ are defined, into two disjoint sets

$$\mathcal{U} = \{x_i \in \mathcal{X} : \mathrm{num}(x_i) < \lfloor \tfrac{1}{2} \cdot \|\Sigma\|^n \rfloor\} \text{ and } \mathcal{W} = \{x_i \in \mathcal{X} : \mathrm{num}(x_i) \geq \lfloor \tfrac{1}{2} \cdot \|\Sigma\|^n \rfloor\},$$

where $\mathrm{num}(x_i)$ denotes the integer value corresponding to the string x_i, interpreting the symbols of the alphabet $\Sigma = \{d_0, d_1, \ldots d_{c-1}\}$ as the corresponding digits in the representation of numbers to the base $c = \|\Sigma\|$. It is easy to see that $\|\mathcal{X}\| = \|\mathcal{U}\| + \|\mathcal{W}\|$ and that

$$\|\mathcal{X}\| = \|\Sigma\|^n, \quad \|\mathcal{U}\| = \lfloor \tfrac{1}{2} \cdot \|\Sigma\|^n \rfloor, \quad \|\mathcal{W}\| = \lceil \tfrac{1}{2} \cdot \|\Sigma\|^n \rceil. \tag{3}$$

Define now the following sets of strings composed of these blocks:

$$X_0 = \{x_1 \mathrm{\textcent} \cdots x_s \mathrm{\textcent} : \mathrm{num}(x_1) < \ldots < \mathrm{num}(x_s), x_1, \ldots, x_s \in \mathcal{X}\},$$
$$U_0 = \{u_1 \mathrm{\textcent} \cdots u_s \mathrm{\textcent} : \mathrm{num}(u_1) < \ldots < \mathrm{num}(u_s), u_1, \ldots, u_s \in \mathcal{U}\},$$
$$W_0 = \{w_1 \mathrm{\textcent} \cdots w_s \mathrm{\textcent} : \mathrm{num}(w_1) < \ldots < \mathrm{num}(w_s), w_1, \ldots, w_s \in \mathcal{W}\}.$$

Before passing further, we need some additional notation and terminology. First, for each string $z = z_1 \mathrm{\textcent} \cdots z_s \mathrm{\textcent} \in X_0$, a *complementary string* (with respect to the set $\mathcal{X}$) will be, by definition, the string $z^{\mathrm{C}(\mathcal{X})} = \tilde{z}_1 \mathrm{\textcent} \cdots \tilde{z}_r \mathrm{\textcent}$ consisting of all blocks from $\mathcal{X}$ that do not appear in z, in sorted order. That is, $\{\tilde{z}_1, \ldots, \tilde{z}_r\} =$

$\mathcal{X} \setminus \{z_1, \ldots, z_s\}$, and $\mathrm{num}(\tilde{z}_1) < \ldots < \mathrm{num}(\tilde{z}_r)$. Clearly, $z^{\mathrm{C}(\mathcal{X})}$ is again in X_0. With z, we also associate the set $S_z = \{z_1, \ldots, z_s\}$ consisting of all blocks that appear in z. It is obvious that $S_z \cup S_{z^{\mathrm{C}(\mathcal{X})}} = \mathcal{X}$ and $S_z \cap S_{z^{\mathrm{C}(\mathcal{X})}} = \emptyset$. In the same way, we introduce $z^{\mathrm{C}(\mathcal{U})}, z^{\mathrm{C}(\mathcal{W})}$ for strings $z \in U_0$ or $z \in W_0$, complementing *with respect to $\mathcal{U}$ or $\mathcal{W}$*, so here we have $S_{z^{\mathrm{C}(\mathcal{U})}} = \mathcal{U} \setminus S_z$ and $S_{z^{\mathrm{C}(\mathcal{W})}} = \mathcal{W} \setminus S_z$. (When the base set $\mathcal{B}$ is known from the context, we shall sometimes write z^{C} instead of $z^{\mathrm{C}(\mathcal{B})}$.) Second, we define a *blockwise reversal* of the string $z \in X_0$, slightly different from the usual reversal: by definition, $z^{\overline{\mathrm{R}}} = z_1^{\mathrm{R}} \mathcal{c} \cdots z_s^{\mathrm{R}} \mathcal{c}$, not reversing the whole string, but each block separately in its own place. (For comparison, the classical reversal gives $z^{\mathrm{R}} = \mathcal{c} z_s^{\mathrm{R}} \cdots \mathcal{c} z_1^{\mathrm{R}}$.) Finally, the *blockwise reversal of the complementary string*, with respect to $\mathcal{X}$, will be denoted by $z^{\mathrm{C}(\mathcal{X})\overline{\mathrm{R}}} = (z^{\mathrm{C}(\mathcal{X})})^{\overline{\mathrm{R}}} = \tilde{z}_1^{\mathrm{R}} \mathcal{c} \cdots \tilde{z}_r^{\mathrm{R}} \mathcal{c}$, sometimes simplifying notation to $z^{\mathrm{C}\overline{\mathrm{R}}}$.

Now, we can rewrite the definition of the language $L_{\Sigma,n}^{\mathrm{C}}$. A well-formed string $\varphi = x\$y^{\overline{\mathrm{R}}} = x_1 \mathcal{c} \cdots x_s \mathcal{c} \$ y_1^{\mathrm{R}} \mathcal{c} \cdots y_r^{\mathrm{R}} \mathcal{c}$ is in $L_{\Sigma,n}^{\mathrm{C}}$ if and only if $S_x \cap S_y = \emptyset$.

Now, it is easy to see that X_0 can be expressed as $X_0 = U_0 \cdot W_0$. This follows from the fact that the blocks in each $x = x_1 \mathcal{c} \cdots x_s \mathcal{c} \in X_0$ are sorted. Thus, x can be partitioned into $x = uw$, for some $u \in U_0$ and $w \in W_0$, not excluding the possibility of $u = \varepsilon$ or $w = \varepsilon$. The boundary between u and w is unambiguously given by the position of the first block x_i for which we have $\mathrm{num}(x_i) \geq \lfloor \frac{1}{2} \cdot \|\Sigma\|^n \rfloor$. Because of the one-to-one correspondence between the subsets of $\mathcal{X}, \mathcal{U}, \mathcal{W}$ and the strings in X_0, U_0, W_0, we have, using (3):

$$\|X_0\| = 2^{\|\Sigma\|^n}, \quad \|U_0\| = 2^{\lfloor 1/2 \cdot \|\Sigma\|^n \rfloor}, \quad \|W_0\| = 2^{\lceil 1/2 \cdot \|\Sigma\|^n \rceil}. \tag{4}$$

Consider now an arbitrary input $\varphi_{uw} = uw\$u^{\mathrm{C}(\mathcal{U})\overline{\mathrm{R}}} w^{\mathrm{C}(\mathcal{W})\overline{\mathrm{R}}}$, where $u \in U_0$ and $w \in W_0$. To simplify the notation, we shall write u^{C} and w^{C} instead of $u^{\mathrm{C}(\mathcal{U})}$ and $w^{\mathrm{C}(\mathcal{W})}$, respectively. That is, $\varphi_{uw} = uw\$u^{\mathrm{C}\overline{\mathrm{R}}} w^{\mathrm{C}\overline{\mathrm{R}}}$. This input can also be expresses as $\varphi_x = x\$x^{\mathrm{C}(\mathcal{X})\overline{\mathrm{R}}}$, where x represents an arbitrary string taken from X_0, using decompositions $x = uw$ and $x^{\mathrm{C}(\mathcal{X})\overline{\mathrm{R}}} = u^{\mathrm{C}\overline{\mathrm{R}}} w^{\mathrm{C}\overline{\mathrm{R}}}$.

It is obvious that the string $\varphi_{uw} = uw\$u^{\mathrm{C}\overline{\mathrm{R}}} w^{\mathrm{C}\overline{\mathrm{R}}} = x\$x^{\mathrm{C}(\mathcal{X})\overline{\mathrm{R}}}$ is well-formed and that $S_x \cap S_{x^{\mathrm{C}(\mathcal{X})}} = \emptyset$. This gives that $\varphi_{uw} \in L_{\Sigma,n}^{\mathrm{C}}$ and hence φ_{uw} must be accepted by A_n. In general, φ_{uw} may have more than one accepting computation path but, for each φ_{uw}, we can fix the "leftmost" accepting path (using some lexicographic ordering among the paths). For this fixed path, we define the following parameters (see either side of Fig. 1):

- $y_\ell \in \{0, \ldots, h_n\}$ is the lowest height of pushdown store in the course of reading the string $w\$u^{\mathrm{C}\overline{\mathrm{R}}}$.
- $q_\ell \in Q_n$ is the state in which the height y_ℓ is attained for the last time, along $w\$u^{\mathrm{C}\overline{\mathrm{R}}}$.
- $x_\ell \in \{-(|w|+1), \ldots, -1, +1, \ldots, +(|u^{\mathrm{C}\overline{\mathrm{R}}}|+1)\}$ is the input position in which q_ℓ is entered. This value is relative to the position of the \$-symbol in $w\$u^{\mathrm{C}\overline{\mathrm{R}}}$. Since each block in both w and $u^{\mathrm{C}\overline{\mathrm{R}}}$ is exactly $n+1$ input symbols long (taking also into account the $\mathcal{c}$-symbol at the end of each block) and the same block is not repeated twice, we have $x_\ell \in \{-(\|\mathcal{W}\| \cdot (n+1)+1), \ldots, -1, +1, \ldots, +(\|\mathcal{U}\| \cdot (n+1)+1)\}$. Therefore, by the use of (3), the total number of different values x_ℓ

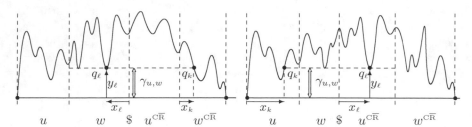

Fig. 1. Parameters and pushdown content along the computation

can be bounded by $\lceil \frac{1}{2} \cdot \|\Sigma\|^n \rceil \cdot (n{+}1) + \lfloor \frac{1}{2} \cdot \|\Sigma\|^n \rfloor \cdot (n{+}1) + 2 = \|\Sigma\|^n \cdot (n{+}1) + 2 \le$ $\|\Sigma\|^n \cdot (n{+}2)$. The last inequality holds for each $\|\Sigma\| \ge 2$ and each $n \ge 4$.
 - $\gamma_{u,w}$ is the pushdown content at the moment when q_ℓ is entered.

The values for the next two parameters, namely, for q_k and x_k, depend on whether $x_\ell \le -1$ or $x_\ell \ge +1$: For $x_\ell \le -1$, that is, if the computation reaches the state q_ℓ before reading the \$-separator (see the left side of Fig. 1), then:

 - $q_k \in Q_n$ is the state at the moment when the pushdown height is equal to y_ℓ for the first time in the course of reading $w^{\mathrm{c}\overline{\mathrm{R}}}$. Such situation must happen because the pushdown height drops down to zero at the end of $w^{\mathrm{c}\overline{\mathrm{R}}}$, starting from at least y_ℓ at the beginning of $w^{\mathrm{c}\overline{\mathrm{R}}}$.
 - $x_k \in \{0, \dots, |w^{\mathrm{c}\overline{\mathrm{R}}}|\}$ is the distance from the beginning of $w^{\mathrm{c}\overline{\mathrm{R}}}$ to the input position in which q_k is entered. Using (3), the total number of different values x_k can be bounded by $\|\mathcal{W}\| \cdot (n{+}1) + 1 = \lceil \frac{1}{2} \cdot \|\Sigma\|^n \rceil \cdot (n{+}1) + 1 \le$ $(\frac{1}{2} \cdot \|\Sigma\|^n {+} 1) \cdot (n{+}1) + 1 \le \frac{1}{2} \cdot \|\Sigma\|^n \cdot (n{+}2)$. The last inequality uses the fact that $n{+}2 \le \frac{1}{2} \cdot \|\Sigma\|^n$ for each $\|\Sigma\| \ge 2$ and each $n \ge 4$.

For $x_\ell \ge +1$, that is, if the computation reaches the state q_ℓ after reading the \$-separator (see the right side of Fig. 1), then:

 - $q_k \in Q_n$ is the state at the moment when the pushdown height is equal to y_ℓ for the last time in the course of reading u. Such situation must happen because the pushdown height shoots up to at least y_ℓ at the end of u, starting from zero at the beginning of u.
 - $x_k \in \{0, \dots, |u|\}$ is the distance from the beginning of u to the input position in which q_k is entered. Here the total number of different values x_k can be bounded by $\|\mathcal{U}\| \cdot (n{+}1) + 1 = \lfloor \frac{1}{2} \cdot \|\Sigma\|^n \rfloor \cdot (n{+}1) + 1 \le \frac{1}{2} \cdot \|\Sigma\|^n \cdot (n{+}2)$.

It is easy to see that, independent of whether $x_\ell \le -1$ or $x_\ell \ge +1$, we have $h_n {+} 1$ different possible values for y_ℓ, $\|Q_n\|$ different possibilities for the state q_ℓ, $\|\Sigma\|^n \cdot (n{+}2)$ possible values for x_ℓ, $\|Q_n\|$ possibilities for the state q_k, and at most $\frac{1}{2} \cdot \|\Sigma\|^n \cdot (n{+}2)$ possibilities for the value x_k. Therefore, the number of different quintuples $[y_\ell, q_\ell, x_\ell, q_k, x_k]$ is bounded by $\|Q_n\|^2 \cdot (h_n{+}1) \cdot \|\Sigma\|^{2n} \cdot (n{+}2)^2 \cdot \frac{1}{2}$.

In conclusion, for each $x = uw \in X_0 = U_0 W_0$, we took the corresponding input $\varphi_{uw} = uw\$u^{\mathrm{c}\overline{\mathrm{R}}} w^{\mathrm{c}\overline{\mathrm{R}}}$ (in total, $\|X_0\|$ inputs of this kind) and, for each of them, we fixed the unique leftmost accepting computation path, which gave the unique

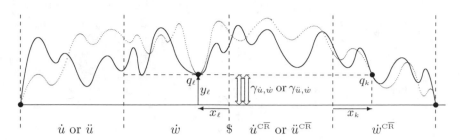

Fig. 2. Computation paths for the inputs $\varphi_{\dot{u}\dot{w}} = \dot{u}\dot{w}\$\dot{u}^{\mathrm{C\overline{R}}}\dot{w}^{\mathrm{C\overline{R}}}$, $\varphi_{\ddot{u}\dot{w}} = \ddot{u}\dot{w}\$\ddot{u}^{\mathrm{C\overline{R}}}\dot{w}^{\mathrm{C\overline{R}}}$, and for $\delta_{\dot{u}\setminus\ddot{u}} = \dot{u}\dot{w}\$\ddot{u}^{\mathrm{C\overline{R}}}\dot{w}^{\mathrm{C\overline{R}}} \notin L^{\mathrm{C}}_{\Sigma,n}$, for the case of $x_\ell \leq -1$

quintuple of parameters $[y_\ell, q_\ell, x_\ell, q_k, x_k]$. Thus, each $x = uw \in X_0$ is associated with exactly one quintuple $[y_\ell, q_\ell, x_\ell, q_k, x_k]$. Hence, a simple pigeonhole argument proves the existence of a set $X_1 \subseteq X_0$, such that all $uw \in X_1$ share the same $[y_\ell, q_\ell, x_\ell, q_k, x_k]$ and, using (1) and (2), the cardinality of such set is

$$\|X_1\| \geq \frac{\|X_0\|}{\|Q_n\|^2 \cdot (h_n+1) \cdot \|\Sigma\|^{2n} \cdot (n+2)^2 \cdot \frac{1}{2}} = \frac{1}{\|Q_n\|^2 \cdot (h_n+1)} \cdot \frac{2^{\|\Sigma\|^n} \cdot 2}{\|\Sigma\|^{2n} \cdot (n+2)^2}$$

$$\geq \frac{\|\Sigma\|^{2n} \cdot (n+2)^2 \cdot 3}{2^{1/2 \cdot \|\Sigma\|^n} \cdot 2} \cdot \frac{2^{\|\Sigma\|^n} \cdot 2}{\|\Sigma\|^{2n} \cdot (n+2)^2} = 3 \cdot 2^{1/2 \cdot \|\Sigma\|^n} = 2^{1/2 \cdot \|\Sigma\|^n} + 2^{1/2 \cdot \|\Sigma\|^n + 1}$$

$$\geq 2^{\lfloor 1/2 \cdot \|\Sigma\|^n \rfloor} + 2^{\lceil 1/2 \cdot \|\Sigma\|^n \rceil} = \|U_0\| + \|W_0\| .$$

Summing up, $X_1 \subseteq X_0 = U_0 \cdot W_0$ and $\|X_1\| \geq \|U_0\| + \|W_0\| - 1$. Therefore, the sets $A = U_0$, $B = W_0$, and $C = X_1$ satisfy the assumptions of Lem. 3 in [2], by which there must exist some strings $\dot{u}, \ddot{u} \in U_0$ and $\dot{w}, \ddot{w} \in W_0$, with $\dot{u} \neq \ddot{u}$ and $\dot{w} \neq \ddot{w}$, such that $\dot{u}\dot{w}$, $\dot{u}\ddot{w}$, and $\ddot{u}\dot{w}$ are all in X_1. Consequently, they all share the same parameters $[y_\ell, q_\ell, x_\ell, q_k, x_k]$ on the corresponding accepting paths. Now we have to distinguish between the two cases, depending on the shared value x_ℓ.

CASE I: $x_\ell \leq -1$. This means that, for $uw \in \{\dot{u}\dot{w}, \dot{u}\ddot{w}, \ddot{u}\dot{w}\}$, all fixed accepting computations for the inputs $\varphi_{uw} = uw\$u^{\mathrm{C\overline{R}}}w^{\mathrm{C\overline{R}}}$ visit the same state $q_\ell \in Q_n$, with the same pushdown height y_ℓ, and at the same position x_ℓ, before reading the \$-separator, in the course of reading w. Thus, for all these inputs, the parameter $q_k \in Q_n$ is taken as the state at the moment when the pushdown height is equal to y_ℓ for the first time along $w^{\mathrm{C\overline{R}}}$, at a position x_k. Also the values q_k and x_k are the same for all these inputs. This situation is depicted in Fig. 2.

Consider now $\varphi_{\dot{u}\dot{w}} = \dot{u}\dot{w}\$\dot{u}^{\mathrm{C\overline{R}}}\dot{w}^{\mathrm{C\overline{R}}}$ and $\varphi_{\ddot{u}\dot{w}} = \ddot{u}\dot{w}\$\ddot{u}^{\mathrm{C\overline{R}}}\dot{w}^{\mathrm{C\overline{R}}}$ (forgetting about $\varphi_{\dot{u}\ddot{w}}$). Both $\varphi_{\dot{u}\dot{w}}$ and $\varphi_{\ddot{u}\dot{w}}$ are in $L^{\mathrm{C}}_{\Sigma,n}$. Since $\dot{u} \neq \ddot{u}$ and $\dot{u}, \ddot{u} \in U_0$, the sets $S_{\dot{u}}, S_{\ddot{u}}$ must differ in at least one block $u_i \in \mathcal{U}$. Assume first that $u_i \in S_{\dot{u}} \setminus S_{\ddot{u}}$. But then $u_i \in S_{\dot{u}} \cap S_{\ddot{u}}^{\mathrm{C}}$, and hence $u_i \in S_{\dot{u}\dot{w}} \cap S_{\ddot{u}^{\mathrm{C}}\dot{w}^{\mathrm{C}}}$. This gives that a hybrid input $\delta_{\dot{u}\setminus\ddot{u}} = \dot{u}\dot{w}\$\ddot{u}^{\mathrm{C\overline{R}}}\dot{w}^{\mathrm{C\overline{R}}}$, which is clearly well-formed, cannot be in $L^{\mathrm{C}}_{\Sigma,n}$.

However, A_n accepts $\delta_{\dot{u}\setminus\ddot{u}}$. First, on the inputs $\varphi_{\dot{u}\dot{w}}$ and $\varphi_{\ddot{u}\dot{w}}$, the pushdown store contains, respectively, the string $\gamma_{\dot{u},\dot{w}}$ or $\gamma_{\ddot{u},\dot{w}}$ at the moment when the machine A_n reaches the state q_ℓ at the position x_ℓ. On both $\varphi_{\dot{u}\dot{w}}$ and $\varphi_{\ddot{u}\dot{w}}$, these deepest y_ℓ symbols will stay unchanged in the pushdown until the moment when A_n reaches the same state q_k at the same position x_k along $\dot{w}^{\mathrm{C\overline{R}}}$.

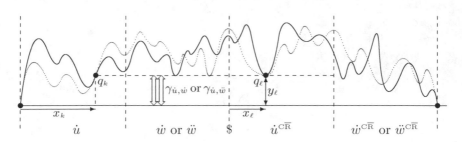

Fig. 3. Computation paths for the inputs $\varphi_{\dot{u}\dot{w}} = \dot{u}\dot{w}\$\dot{u}^{\mathrm{c\overline{R}}}\dot{w}^{\mathrm{c\overline{R}}}$, $\varphi_{\dot{u}\ddot{w}} = \dot{u}\ddot{w}\$\dot{u}^{\mathrm{c\overline{R}}}\ddot{w}^{\mathrm{c\overline{R}}}$ and for $\delta_{\dot{w}\backslash\ddot{w}} = \dot{u}\dot{w}\$\dot{u}^{\mathrm{c\overline{R}}}\ddot{w}^{\mathrm{c\overline{R}}} \notin L^{\mathrm{C}}_{\Sigma,n}$, for the case of $x_\ell \geq +1$

Now, for the input $\delta_{\dot{u}\backslash\ddot{u}}$, one of the possible computations can start by following the trajectory for $\varphi_{\dot{u}\dot{w}}$, reading $\dot{u}$ and the first $|\dot{w}\$| - |x_\ell|$ symbols of $\dot{w}\$$, until it reaches q_ℓ. At this moment, the pushdown store contains $\gamma_{\dot{u},\dot{w}}$. Now the machine switches to the computation path for $\varphi_{\ddot{u}\dot{w}}$, until it gets to q_k. Along this path, the computation does not touch the deepest y_ℓ symbols in the pushdown store, reading the remaining $|x_\ell|$ symbols of $\dot{w}\$$, the entire segment $\ddot{u}^{\mathrm{c\overline{R}}}$, and the first x_k symbols of $\dot{w}^{\mathrm{c\overline{R}}}$. Then the computation on $\delta_{\dot{u}\backslash\ddot{u}}$ switches back to the trajectory for $\varphi_{\dot{u}\dot{w}}$, working with the same content in the pushdown and reading the remaining $|\dot{w}^{\mathrm{c\overline{R}}}| - x_k$ symbols of $\dot{w}^{\mathrm{c\overline{R}}}$. Thus, A_n stops in the unique state $q_{\mathrm{F},n}$ and accepts $\delta_{\dot{u}\backslash\ddot{u}} = \dot{u}\dot{w}\$\ddot{u}^{\mathrm{c\overline{R}}}\dot{w}^{\mathrm{c\overline{R}}} \notin L^{\mathrm{C}}_{\Sigma,n}$, which is a contradiction.

Symmetrically, for $u_i \in S_{\ddot{u}} \setminus S_{\dot{u}}$, the argument switches the roles of $\varphi_{\dot{u}\dot{w}}$ and $\varphi_{\ddot{u}\dot{w}}$, proving that A_n accepts $\delta_{\ddot{u}\backslash\dot{u}} = \ddot{u}\dot{w}\$\dot{u}^{\mathrm{c\overline{R}}}\dot{w}^{\mathrm{c\overline{R}}} \notin L^{\mathrm{C}}_{\Sigma,n}$, a contradiction.

CASE II: $x_\ell \geq +1$. The argument is similar to Case I. Again, the fixed computations on φ_{uw} have the same parameters q_ℓ, y_ℓ, and x_ℓ, this time in the course of reading $u^{\mathrm{c\overline{R}}}$. Hence, q_k is now taken as the state at the moment when the pushdown height equal to y_ℓ for the last time along u, at a position x_k. Also here both q_k and x_k are the same (see Fig. 3). Thus, instead of computations connecting q_ℓ with q_k, we consider segments connecting q_k with q_ℓ, for the inputs $\varphi_{\dot{u}\dot{w}} = \dot{u}\dot{w}\$\dot{u}^{\mathrm{c\overline{R}}}\dot{w}^{\mathrm{c\overline{R}}}$ and $\varphi_{\dot{u}\ddot{w}} = \dot{u}\ddot{w}\$\dot{u}^{\mathrm{c\overline{R}}}\ddot{w}^{\mathrm{c\overline{R}}}$ (forgetting about $\varphi_{\ddot{u}\dot{w}}$). Here we have two subcases: either $S_{\dot{w}} \setminus S_{\ddot{w}} \neq \emptyset$, but then the machine accepts $\delta_{\dot{w}\backslash\ddot{w}} = \dot{u}\dot{w}\$\dot{u}^{\mathrm{c\overline{R}}}\ddot{w}^{\mathrm{c\overline{R}}} \notin L^{\mathrm{C}}_{\Sigma,n}$, or $S_{\ddot{w}} \setminus S_{\dot{w}} \neq \emptyset$, when it accepts $\delta_{\ddot{w}\backslash\dot{w}} = \dot{u}\ddot{w}\$\dot{u}^{\mathrm{c\overline{R}}}\dot{w}^{\mathrm{c\overline{R}}} \notin L^{\mathrm{C}}_{\Sigma,n}$.

In conclusion, we have $\delta \in \{\delta_{\dot{u}\backslash\ddot{u}}, \delta_{\ddot{u}\backslash\dot{u}}, \delta_{\dot{w}\backslash\ddot{w}}, \delta_{\ddot{w}\backslash\dot{w}}\}$ not belonging to $L^{\mathrm{C}}_{\Sigma,n}$ but accepted by A_n. Therefore, the inequality (2) must be reversed. This gives $\|Q_n\|^2 \cdot (h_n+1) > 2^{1/2 \cdot \|\Sigma\|^n - O(n \cdot \log \|\Sigma\|)}$, for $n \geq 4$. However, the argument holds only for NPDAs in "normal form" of Lem. 1 in [4], assuming that A_n accepts with empty pushdown. For general NPDAs, the bound changes to $(\|Q_n\|+1)^2 \cdot (h_n+1) > 2^{1/2 \cdot \|\Sigma\|^n - O(n \cdot \log \|\Sigma\|)}$, since converting into the normal form costs one state, keeping the same pushdown height. Thus, either $\|Q_n\|+1$ or else h_n+1 is above $2^{1/6 \cdot \|\Sigma\|^n - O(n \cdot \log \|\Sigma\|)} \geq 2^{\|\Sigma\|^{n-O(1)}}$. $\square$

By combining Lem. 3 and Thm. 4, for $\|\Sigma\| = c$, we get:

Theorem 5. *For each fixed constant $c \geq 2$, there exists $\{L_n\}_{n\geq 1}$, a family of regular languages built over a fixed $(c+2)$-letter input alphabet, such that:*

(i) *there exists $\{A_n\}_{n\geq 1}$, a sequence of constant height NPDAs accepting these languages with $\|Q_n\| \leq n+O(c)$ states, the pushdown of height $h_n = n$, and c pushdown symbols, but*

(ii) *for any constant height NPDAs $\{A_n^C\}_{n\geq 1}$ accepting the family of their complements $\{L_n^C\}_{n\geq 1}$, either the number of states in A_n^C or else the pushdown height must be above $2^{1/6 \cdot c^n - O(n \cdot \log c)} \geq 2^{c^{n-O(1)}}$, independently of the size of the used pushdown alphabet.*

Finally, by allowing alphabets to grow in n, we can get a slightly wider gap. For example, using $L_n = L_{\Sigma_n,n}$, where $\|\Sigma_n\| = n$, we obtain:

Theorem 6. *There exists $\{L_n\}_{n\geq 1}$, a family of regular languages, such that:*

(i) *there exists $\{A_n\}_{n\geq 1}$, a sequence of constant height NPDAs accepting these languages with $\|Q_n\| \leq 3n+8$ states, the pushdown of height $h_n = n$, and $\|\Gamma_n\| = n$ pushdown symbols, but*

(ii) *for any constant height NPDAs $\{A_n^C\}_{n\geq 1}$ accepting the family of their complements $\{L_n^C\}_{n\geq 1}$, either the number of states in A_n^C or else the pushdown height must be above $2^{1/6 \cdot n^n - O(n \cdot \log n)} \geq 2^{2^{n \cdot \log n - O(1)}}$.*

4 Converting Two-Way PDAs to One-Way Devices

Here we consider the problem of converting into one-way devices. Given a constant height 2NPDA A, a double-exponential upper bound for an equivalent one-way NPDA is obtained as follows. First, we code the pushdown content in the finite state control, with $\|Q'\| \leq \|Q\| \cdot \|\Gamma^{\leq h}\| + 1$ states (cf. [1, Lem. 2.2]). This 2NFA is made one-way, with $\|Q''\| \leq \binom{2 \cdot \|Q'\|}{\|Q'\|+1} \leq 4^{\|Q'\|-\Omega(\log \|Q'\|)} \leq 4^{\|Q\| \cdot \|\Gamma^{\leq h}\|}$ states [9]. The resulting NFA is a special case of a one-way constant height NPDA.

Theorem 7. *For each constant height 2NPDA $A = \langle Q, \Sigma, \Gamma, H, q_1, F, h \rangle$ (hence, also for 2DPDA), there exists an equivalent one-way NFA A' (hence, also a one-way constant height NPDA) with at most $4^{\|Q\| \cdot \|\Gamma^{\leq h}\|}$ states.*

We now show that converting constant height 2DPDAs into one-way devices *must* be paid by a double-exponential blow-up, even if we use nondeterminism as an additional tool. To this aim, we show that the languages $\{L_{\Sigma,n}^C\}_{n\geq 1}$, introduced in Def. 2, are accepted by constant height 2DPDAs with linear resources.

Lemma 8. *For each alphabet Σ and $n \geq 1$, the language $L_{\Sigma,n}^C$ can be accepted by a constant height 2DPDA using $n + 6 \cdot \|\Sigma\| + 21$ states, a pushdown of height $n+1$, and $\|\Sigma\|+1$ pushdown symbols.*

Proof. We present only the main ideas for a 2DPDA A accepting $L_{\Sigma,n}^{\mathrm{C}}$, but all details can be easily filled by the reader. Recall that the input of a two-way machine is enclosed in between two endmarkers, and hence a tape with a well-formed accepted input (see also (1) and Def. 2) looks as follows:

$$\vdash \varphi \dashv \; = \; \vdash x_1 \text{¢} x_2 \text{¢} \cdots x_s \text{¢} \$ y_1^{\mathrm{R}} \text{¢} y_2^{\mathrm{R}} \text{¢} \cdots y_r^{\mathrm{R}} \text{¢} \dashv,$$

with blocks $x_i, y_j \in \Sigma^n$ and $x_i \neq y_j$, for each $i \in \{1, \ldots, s\}$ and $j \in \{1, \ldots, r\}$. Next, we shall use the pushdown alphabet $\Gamma = \Sigma \cup \{X_{\mathrm{I}}\}$, where X_{I} is a new pushdown symbol, to be placed at the bottom at the very beginning. The pushdown is of height $h = n+1$, capable of containing one block $z \in \Sigma^n$, above X_{I}.

After pushing X_{I} onto the pushdown, the machine A verifies whether the input is the well-formed. Traversing the entire tape from left to right, A checks whether the input contains exactly one $\$$-symbol; traversing the tape back, it verifies that each block is exactly of length n. If the input is ill-formed, A halts and rejects. Such procedure uses only $n+O(1)$ states. In addition, if the input is well-formed but it begins with the $\$$-symbol, A halts and accepts. (This reduces the number of states in subsequent phases.)

We are now ready for the main idea. The machine A generates in its pushdown store, one after another, all possible blocks $z \in \Sigma^n = \{d_0, \ldots, d_{c-1}\}^n$, in lexicographic order. In the course of this iteration, each generated block $z \in \Sigma^n$ is compared with the input blocks x_i, for $i = 1, \ldots, s$, and then with y_j, for $j = r, \ldots, 1$, in that order. The input is rejected if and only if, for some $z \in \Sigma^n$ and some i, j, we find out that $z = x_i$ and $z = y_j$.

The iteration is initialized by storing the lexicographically first block $z = d_0^n$ in the pushdown. Pushing n copies of d_0 on top of the pushdown costs only $O(1)$ states, since the input is well-formed and hence we can count from 1 to n just by moving the input head along $\vdash x_1 \text{¢}$.

Now, the current pushdown block z is compared with the input blocks x_i, for $i = 1, \ldots, s$, one after another. Comparing z with x_i is based on ability to read x_i in both directions. That is, the symbols are popped out from the pushdown to the finite state control, after which they are compared against the symbols in x_i, moving the input head to the right. If A finds the first difference, it can restore the pushdown content be reading the input symbols in opposite direction. After that, A can move one input block to the right and compare the restored pushdown block z again, with the next input block x_{i+1}. (When the last block x_s has been compared without a match, A starts the next iteration, for the next pushdown block $z \in \Sigma^n$.) Conversely, if the match is found, that is, if $z = x_i$, the machine A still restores z by reading x_i from right to left but, after that, A moves to the right endmarker $\dashv$, to inspect y_j-blocks. The whole procedure uses $3\|\Sigma\| + O(1)$ states. (Since the input is well-formed, the pushdown need not be tested against overflows or underflows, which saves states.)

If $z = x_i$, for some z and x_i, the current pushdown block z is also compared with $y_r, \ldots, y_1$. This procedure is very similar to that inspecting $x_1, \ldots, x_s$, but the blocks in the right part are written reversely. For this reason, all input head movements are in the opposite direction, starting with y_r and ending by y_1.

The most important difference is that if A finds a match, i.e., if $z = y_j$, for some j, we have found a pushdown block equal to some blocks in both parts of the input, i.e., $x_i = z = y_j$. Therefore, A halts immediately in a rejecting state. (When the last block y_1 has been compared without a match, A starts the next iteration, for the next $z \in \Sigma^n$.) This procedure uses only $2 \cdot \|\Sigma\| + O(1)$ states, instead of $3 \cdot \|\Sigma\| + O(1)$, since A does not have to restore z, if $z = y_j$.

To iterate over all blocks $z \in \Sigma^n = \{d_0, \ldots, d_{c-1}\}^n$, we need also a routine that replaces the current block z in the pushdown by the next block in lexicographic order. Clearly, z can be expressed in the form $\xi d_i d_{c-1}^e$, for some $\xi \in \Sigma^*$ and some $e \in \{0, \ldots, n\}$, and d_i represents the first symbol in z not equal to d_{c-1} from the right. Hence, it is enough to replace z in the pushdown by $\xi d_{i+1} d_0^e$. Thus, in the first phase, A pops all symbols d_{c-1} out from the pushdown, until it finds the first symbol d_i different from d_{c-1}. Then A starts the second phase, pushing the symbol d_{i+1} and the "proper" number of copies of d_0. To avoid counting from 0 to e and then back from e to 0 in the finite state control, we count by input head movement. That is, the first phase starts with the input head at the $\$$-symbol, moving one position to the left per each symbol d_{c-1} popped out. Conversely, in the second phase, we push one copy of d_0 per each move to the right, until we reach the $\$$-symbol. Implemented this way, $\|\Sigma\| + O(1)$ states are sufficient, instead of $\Omega(n \cdot \|\Sigma\|)$. (Since there is at least one block to the left of $\$$, i.e., we have enough room for this input head movement.) There is one special case, an attempt to increment the largest block $z = d_{c-1}^n$. This is detected when A sees the bottom pushdown symbol X_1 in the first phase. This means that the iteration over all possible blocks $z \in \Sigma^n$ has been completed without meeting with the same z on both sides of the input, and hence A halts and accepts.

Carefully implemented, the machine works with $n + 6 \cdot \|\Sigma\| + 21$ states.   □

Combining this with Thm. 5, we get a gap for removing bidirectionality:

Theorem 9. *For each fixed constant $c \geq 2$, there exists $\{L_n^c\}_{n \geq 1}$, a family of regular languages built over a fixed $(c+2)$-letter input alphabet, such that:*

(i) *there exists $\{A_n\}_{n \geq 1}$, a sequence of constant height 2DPDAs (hence, also 2NPDAs) accepting these languages with $\|Q_n\| \leq n + O(c)$ states, the pushdown of height $h_n = n + 1$, and $c + 1$ pushdown symbols, but*

(ii) *for any constant height one-way NPDAs $\{A_n^c\}_{n \geq 1}$ accepting this family (hence, also for one-way DPDAs), either the number of states in A_n^c or else the pushdown height must be above $2^{1/6 \cdot c^n - O(n \cdot \log c)} \geq 2^{c^{n - O(1)}}$, independently of the size of the used pushdown alphabet.*

5   Final Remarks

We tackled two problems for pushdown automata with a limited pushdown height. First, complementing a one-way NPDA A with $\|Q\|$ states and a pushdown of height h, using $\|\Gamma\|$ pushdown symbols, does not cost more than $2^{\|Q\| \cdot \|\Gamma^{\leq h}\|}$

states [2] — actually by a trivial conversion to a classical DFA not using nondeterminism or any pushdown at all (see Thm. 1). More important, in Thms. 5 and 6, we have shown that a double-exponential blow-up is necessary. This reveals that, for one-way constant height NPDAs, the complement is essentially harder than intersection, the cost of which is single-exponential [2].

Second, converting a two-way constant height 2NPDA into a one-way device does not cost more than $4^{\|Q\| \cdot \|\Gamma^{\leq h}\|}$ states — this time by a trivial conversion to a classical NFA, by Thm. 7. Also for removing bidirectionality, a double-exponential blow-up is necessary, as shown in Thm. 9. Moreover, this blow-up is required for converting 2DPDAs into one-way NPDAs or DPDAs as well.

There are not too many computational models with a double-exponential blow-up for such operations. To obtain these gaps, we used some techniques from [2] as a starting point. Nevertheless, several new ideas were required. Namely, in [2], also the *upper* bounds for complementing the used witness languages were only single-exponential, which is not sufficient for our purposes.

However, the *exact* costs are still open; the upper and lower bounds presented here do not match each other exactly. It should also be interesting to investigate the cost of other language operations for constant height PDAs. Finally, we would like to emphasize the interest in studying the power of machines with a limit on some other types of memory (a queue, a stack with a read-only access below the top, and so on). The main reason is that such machines may be viewed as automata with a huge — but finite — number of states, with some replicated patterns in the structure of transitions in the state set.

References

1. Bednárová, Z., Geffert, V., Mereghetti, C., Palano, B.: The size-cost of Boolean operations on constant height deterministic pushdown automata. Theoret. Comput. Sci. 449, 23–36 (2012)
2. Geffert, V., Bednárová, Z., Mereghetti, C., Palano, B.: Boolean language operations on nondeterministic automata with a pushdown of constant height. In: Bulatov, A.A., Shur, A.M. (eds.) CSR 2013. LNCS, vol. 7913, pp. 100–111. Springer, Heidelberg (2013)
3. Chomsky, N.: Context-free grammars and pushdown storage. Quarterly Progress Report 65, Research Lab. Electronics. MIT, Cambridge, Massachusetts (1962)
4. Geffert, V., Mereghetti, C., Palano, B.: More concise representation of regular languages by automata and regular expressions. Inform. & Comput. 208, 385–394 (2010)
5. Gruber, H., Holzer, M.: Language operations with regular expressions of polynomial size. Theoret. Comput. Sci. 410, 3281–3289 (2009)
6. Harrison, M.: Introduction to Formal Language Theory. Addison-Wesley (1978)
7. Holzer, M., Kutrib, M.: Nondeterministic descriptional complexity of regular languages. Internat. J. Found. Comput. Sci. 14, 1087–1102 (2003)
8. Hopcroft, J., Motwani, R., Ullman, J.: Introduction to Automata Theory, Languages, and Computation, 3rd edn. Prentice Hall (2007)
9. Kapoutsis, C.A.: Removing bidirectionality from nondeterministic finite automata. In: Jedrzejowicz, J., Szepietowski, A. (eds.) MFCS 2005. LNCS, vol. 3618, pp. 544–555. Springer, Heidelberg (2005)

10. Kutrib, M., Malcher, A., Wotschke, D.: The Boolean closure of linear context-free languages. Acta Inform. 45, 177–191 (2008)
11. Schützenberger, M.: On context-free languages and pushdown automata. Inform.& Control 6, 246–264 (1963)
12. Yu, S., Zhuang, Q., Salomaa, K.: The state complexities of some basic operations on regular languages. Theoret. Comput. Sci. 125, 315–328 (1994)

Covering Pairs in Directed Acyclic Graphs

Niko Beerenwinkel[1], Stefano Beretta[2,4], Paola Bonizzoni[2],
Riccardo Dondi[3], and Yuri Pirola[2]

[1] Dept. of Biosystems Science and Engineering, ETH Zurich, Basel, Switzerland
niko.beerenwinkel@bsse.ethz.ch
[2] DISCo, Univ. degli Studi di Milano-Bicocca, Milan, Italy
{beretta,bonizzoni,pirola}@disco.unimib.it
[3] Dip. di Scienze Umane e Sociali, Univ. degli Studi di Bergamo, Bergamo, Italy
riccardo.dondi@unibg.it
[4] Inst. for Biomedical Technologies, National Research Council, Segrate, Italy

Abstract. The Minimum Path Cover problem on directed acyclic graphs
(DAGs) is a classical problem that provides a clear and simple mathematical formulation for several applications in different areas and that
has an efficient algorithmic solution. In this paper, we study the computational complexity of two constrained variants of Minimum Path Cover
motivated by the recent introduction of next-generation sequencing technologies in bioinformatics. The first problem (MinPCRP), given a DAG
and a set of pairs of vertices, asks for a minimum cardinality set of paths
"covering" all the vertices such that both vertices of each pair belong to
the same path. For this problem, we show that, while it is NP-hard to
compute if there exists a solution consisting of at most three paths, it is
possible to decide in polynomial time whether a solution consisting of at
most two paths exists. The second problem (MaxRPSP), given a DAG
and a set of pairs of vertices, asks for a single path containing the maximum number of the given pairs of vertices. We show its NP-hardness
and also its W[1]-hardness when parametrized by the number of covered
pairs. On the positive side, we give a fixed-parameter algorithm when
the parameter is the maximum overlapping degree, a natural parameter
in the bioinformatics applications of the problem.

1 Introduction

The *Minimum Path Cover* (MinPC) problem is a well-known problem in graph
theory. Given a *directed acyclic graph* (DAG), MinPC asks for a minimum-
cardinality set Π of paths such that each vertex of G belongs to at least one
path of Π. The problem can be solved in polynomial time with an algorithm
based on a proof of the well-known Dilworth's theorem for partially ordered sets
which allows to relate the size of a minimum path cover to that of a maximum
matching in a bipartite graph obtained from the input DAG [6].

The Minimum Path Cover problem has important applications in several fields
ranging from bioinformatics [1,5,11] to software testing [7,10]. In particular, in
bioinformatics the Minimum Path Cover problem is applied to the reconstruction of a set of highly-similar sequences starting from a large set of their short

A.-H. Dediu et al. (Eds.): LATA 2014, LNCS 8370, pp. 126–137, 2014.

fragments (called *short reads*) [5, 11]. More precisely, each fragment is represented by a single vertex and two vertices are connected if the alignments of the corresponding reads on the genomic sequence overlap. In [11], the paths on such a graph represent putative transcripts and a minimum-cardinality set of paths "covering" all the vertices represents a set of protein isoforms which are likely to originate from the observed reads. On the other hand, in [5] the paths on such a graph represent the genomes of putative viral haplotypes and a minimum-cardinality set of paths covering the whole graph represents the likely structure of a viral population.

Recently, different constraints have motivated the definition of new variants of the minimum path cover problem. In [1], given a DAG D and a set P of required paths, the proposed problem asks for a minimum cardinality set of paths such that: (1) each vertex of the graph belongs to some path, and (2) each path in P is a subpath of a path of the solution. The authors have described a polynomial-time algorithm to solve this problem by collapsing each required path into a single vertex and then finding a minimum path cover on the resulting graph. Other constrained problems related to minimum path cover have been proposed in the context of social network analysis and, given an edge-colored graph, ask for the maximum number of vertex-disjoint uni-color paths that cover the vertices of the given graph [2, 12].

Some constrained variants of the minimum path cover problem have been introduced in the past by Ntafos and Hakimi in the context of software testing [10] and appear to be relevant for some sequence reconstruction problems of recent interest in bioinformatics. More precisely, in software testing each procedure to be tested is modeled by a graph where vertices correspond to single instructions and two vertices are connected if the corresponding instructions are executed sequentially. The test of the procedure should check each instruction at least once, hence a minimum path cover of the graph represents a minimum set of execution flows that allows to test all the instructions. Clearly, not all the execution flows are possible. For this reason, Ntafos and Hakimi proposed the concept of required pairs, which are pairs of vertices that a feasible solution must include in a path, and that of impossible pairs, which are pairs of vertices that a feasible solution must not include in the same path. In particular, one of the problems introduced by Ntafos and Hakimi is the *Minimum Required Pairs Cover* (Min-RPC) problem where, given a DAG and a set of required pairs, the goal is to compute a minimum set of paths *covering* all the required pairs, *i.e.*, a minimum set of paths such that, for each required pair, at least one path contains both vertices of the pair.

The concept of required pairs is also relevant for sequence reconstruction problems in bioinformatics, as short reads are often sequenced in pairs (*paired-end reads*) and these pairs of reads must align to a single genetic sequence. As a consequence, each pair of vertices corresponding to paired-end reads must belong to the same path of the cover. Paired-end reads provide valuable information that, in principle, could greatly improve the accuracy of the reconstruction. However, they are often used only to filter out the reconstructed sequences that do not meet such constraints, instead of directly exploiting them during the reconstruction process. Notice that MinRPC asks for a solution that covers only the required pairs, while in bioinformatics we are also interested in covering all the vertices. For this reason, we consider a variant of the Minimum Path Cover problem, called *Minimum Path Cover with Required Pairs* (MinPCRP), that, given a DAG and a set of required pairs, asks for a minimum set of paths covering all the vertices and all the required pairs. Clearly, MinPCRP is closely related to MinRPC. In fact, as we show in Section 2, the same reduction used in [10] to prove the NP-hardness of MinRPC can be applied to our problem, leading to its intractability.

In this paper, we continue the analysis of [10] by studying the complexity of path covering problems with required pairs. More precisely, we study how the complexity of these problems is influenced by two parameters relevant for the sequence reconstruction applications in bioinformatics: (1) the minimum number of paths covering all the vertices and all the required pairs and (2) the maximum *overlapping degree* (defined later). In the bioinformatics applications we discussed, the first parameter—the number of covering paths—is often small, thus an algorithm exponential in the size of the solution could be of interest. The second parameter we consider in this paper, the maximum overlapping degree, can be informally defined as follows. Two required pairs overlap when there exists a path that connects the vertices of the pairs, and the path cannot be split in two disjoint subpaths that separately connect the vertices of the two pairs. Then, the overlapping degree of a required pair is the number of required pairs that overlap with it. In the sequence reconstruction applications, as the distance between two paired-end reads is fixed, the maximum overlapping degree is small compared to the number of vertices, hence it is a natural parameter for investigating the computational complexity of the problem.

First, we investigate how the computational complexity of MinPCRP is influenced by the first parameter. In this paper we prove that it is NP-complete to decide if there exists a solution of MinPCRP consisting of at most three paths (via a reduction from the 3-Coloring problem). We complement this result by giving a polynomial-time algorithm for computing a solution with at most 2 paths, thus establishing a sharp tractability borderline for MinPCRP when parameterized by the size of the solution. These results significantly improve the hardness result that Ntafos and Hakimi [10] presented for MinRPC (and that holds also for MinPCRP), where the solution contains a number of paths which is polynomial in the size of the input.

Then, we investigate how the computational complexity of MinPCRP is influenced by the second parameter, the overlapping degree. Unfortunately, Min-PCRP is NP-hard even if the maximum overlapping degree is 0. In fact, this can be easily obtained by modifying the reduction presented in [10] to hold also for restricted instances of MinPCRP with no overlapping required pairs.

A natural heuristic approach for solving MinPCRP is the one which computes a solution by iteratively adding a path that covers a maximum set of required pairs not yet covered by a path of the solution. This approach leads to a natural combinatorial problem, the *Maximum Required Pairs with Single Path* (MaxRPSP) problem, that, given a DAG and a set of required pairs, asks for a path that covers the maximum number of required pairs. We investigate the complexity of MaxRPSP and we show that it is not only NP-hard, but also W[1]-hard when the parameter is the number of covered required pairs. This result shows that it is unlikley that the problem is fixed-parameter tractable, when parameterized by the number of required pairs covered by a single path. We refer the reader to [3,9] for an in-depth presentation of the theory of fixed-parameter complexity. We consider also the MaxRPSP problem parameterized by the maximum overlapping degree but, differently from MinPCRP, we give a fixed-parameter algorithm for this case. This positive result shows a gap between the complexity of MaxRPSP and the complexity of MinPCRP when parameterized by the maximum overlapping degree.

The rest of the paper is organized as follows. First, in Section 2 we give some preliminary notions and we introduce the formal definitions of the two problems. In Section 3, we investigate the computational complexity of MinPCRP when the solution consists of a constant number of paths: we show that it is NP-complete to decide if there exists a solution of MinPCRP consisting of at most three paths, while the existence of a solution consisting of at most two paths can be computed in polynomial time. In Section 4, we investigate the computational complexity of MaxRPSP: we prove its W[1]-hardness when the parameter is the number of required pairs covered by the path (Section 4.1) and we give a fixed-parameter algorithm when the parameter is the maximum overlapping degree (Section 4.2).

Due to the page limit, some proofs are omitted. A complete version of this paper, containing all the proofs, is available at http://arxiv.org/abs/1310.5037.

2 Preliminaries

In this section, we introduce the basic notions used in the rest of the paper and we formally define the two combinatorial problems we are interested in.

While our problems deal with directed graphs, we consider both directed and undirected graphs. We denote an *undirected graph* as $G = (V, E)$ where V is the set of vertices and E is the set of (undirected) edges, and a *directed graph* as $D = (N, A)$ where N is the set of vertices and A is the set of (directed) arcs. We denote an edge of $G = (V, E)$ as $\{v, u\} \in E$ where $v, u \in V$. Moreover, we denote an arc of $D = (N, A)$ as $(v, u) \in A$ where $v, u \in N$.

Given a directed graph $D = (N, A)$, a *path* π from vertex v to vertex u, denoted as vu-path, is a sequence of vertices $\langle v_1, \ldots, v_n \rangle$ such that $(v_i, v_{i+1}) \in A$, $v = v_1$ and $u = v_n$. We say that a vertex v *belongs to* a path $\pi = \langle v_1, \ldots, v_n \rangle$, denoted as $v \in \pi$, if $v = v_i$, for some $1 \leq i \leq n$. Given a path $\pi = \langle v_1, \ldots, v_n \rangle$, we say that a path $\pi' = \langle v_i, v_{i+1}, \ldots, v_{j-1}, v_j \rangle$, with $1 \leq i \leq j \leq n$, is a subpath of π. Given a set $N' \subseteq N$ of vertices, a path π *covers* N' if every vertex of N' belongs to π.

In the paper, we consider a set R of pairs of vertices in N. We denote each pair as $[v_i, v_j]$, to avoid ambiguity with the notations of edges and arcs.

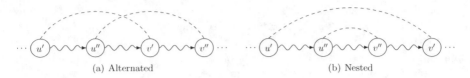

Fig. 1. Examples of two overlapping required pairs $[u', v']$ and $[u'', v'']$. In (a) the required pairs are *alternated*, while in (b) they are *nested*.

Now, we are able to give the definitions of the combinatorial problems we are interested in.

Problem 1. Minimum Path Cover with Required Pairs (MinPCRP)
Input: a directed acyclic graph $D = (N, A)$, a source $s \in N$, a sink $t \in N$, and a set $R = \{[v_x, v_y] \mid v_x, v_y \in N, v_x \neq v_y\}$ of required pairs.
Output: a minimum cardinality set $\Pi = \{\pi_1, \ldots, \pi_n\}$ of directed st-paths such that every vertex $v \in N$ belongs to at least one st-path $\pi_i \in \Pi$ and every required pair $[v_x, v_y] \in R$ belongs to at least one st-path $\pi_i \in \Pi$, i.e. v_x, v_y belongs to π_i.

Problem 2. Maximum Required Pairs with Single Path (MaxRPSP)
Input: a directed acyclic graph $D = (N, A)$, a source $s \in N$, a sink $t \in N$ and a set $R = \{[v_x, v_y] \mid v_x, v_y \in N, v_x \neq v_y\}$ of required pairs.
Output: an st-path π that covers a set $R' = \{[v_x, v_y] \mid v_x, v_y \in \pi\} \subseteq R$ of maximum cardinality.

Two required pairs $[u', v']$ and $[u'', v'']$ in R *overlap* if there exists a path π in D such that the four vertices appear in π in one of the following orders (assuming that the vertex u' appears before u'' in π), where v' and u'' are two distinct vertices of G (see Fig. 1):

- $\langle u', u'', v', v'' \rangle$ (the two required pairs are *alternated*);
- $\langle u', u'', v'', v' \rangle$ (the required pair $[u'', v'']$ is *nested* in $[u', v']$).

Notice that, from this definition, the required pairs $[x, y]$ and $[y, z]$ do not overlap.

Finally, consider a required pair $[u', v']$ of R. We define the *overlapping degree* of $[u', v']$ as the number of required pairs in R that overlap with $[u', v']$.

Hardness of MinPCRP. As we mentioned in the introduction, MinPCRP is related to a combinatorial problem which has been studied in the context of program testing [10], where it is shown to be NP-hard. More precisely, given a directed acyclic graph $D = (N, A)$, a source $s \in N$, a sink $t \in N$ and a set $R = \{[v_x, v_y] \mid v_x, v_y \in N, v_x \neq v_y\}$ of required pairs, the *Minimum Required Pairs Cover* (MinRPC) problem asks for a minimum cardinality set $\Pi = \{\pi_1, \ldots, \pi_n\}$ of directed st-paths such that every required pair $[v_x, v_y] \in R$ belongs to at least one st-path $\pi_i \in \Pi$, i.e. $v_x, v_y \in \pi_i$.

MinRPC can be easily reduced to MinPCRP due to the following property: each vertex of the graph D (input of MinRPC) must belong to at least one required pair. Indeed, if this condition does not hold for some vertex v, we can modify the graph D by contracting v (that is removing v and adding an edge (u, z) to A, for each $u, z \in N$ such that $(u, v), (v, z) \in A$). This implies that, since in an instance of MinRPC all the resulting vertices belong to some required pair, a feasible solution of that problem must cover every vertex of the graph. Then, a solution of MinRPC is also a solution of MinPCRP, which implies that MinPCRP is NP-hard.

MinPCRP on *directed* graphs (not necessarily acyclic) is as hard as MinPCRP on DAGs. In fact, since each strongly connected component can be covered with a single path, we can replace them with single vertices, obtaining a DAG and without changing the size of the solution. Clearly, MinPCRP on general graphs and requiring that the covering paths are simple is as hard as the Hamiltonian path problem, which is NP-complete.

3   A Sharp Tractability Borderline for MinPCRP

In this section, we investigate the computational complexity of MinPCRP and we give a sharp tractability borderline for k-PCRP, the restriction of MinPCRP where we ask whether there exist k paths that cover all the vertices of the graph and all the set of required pairs. First, we show (Sect. 3.1) that 3-PCRP is NP-complete. This result implies that k-PCRP does not belong to the class XP [1], so it is probably hopeless to look for an algorithm having complexity $O(n^k)$, and hence for a fixed-parameter algorithm in k. We complement this result by giving (Sect. 3.2) a polynomial time algorithm for 2-PCRP, thus defining a sharp borderline between tractable and intractable instances of MinPCRP.

3.1   Hardness of 3-PCRP

In this section we show that 3-PCRP is NP-complete. We prove this result via a reduction from the well-known 3-Coloring (3C) problem which, given an undirected (connected) graph $G = (V, E)$, asks for a coloring $c : V \rightarrow \{c_1, c_2, c_3\}$ of the vertices of G with exactly 3 colors, such that, for every $\{v_i, v_j\} \in E$, we have $c(v_i) \neq c(v_j)$.

Starting from an undirected graph $G = (V, E)$ (instance of 3C), we construct a corresponding instance $\langle D = (N, A), R \rangle$ of 3-PCRP as follows. For every subset $\{v_i, v_j\}$ of cardinality 2 of V, we define a graph $D_{i,j} = (N_{i,j}, A_{i,j})$ (in the following we assume that, for each $D_{i,j}$ associated with set $\{v_i, v_j\}$, $i < j$). The vertex set $N_{i,j}$ is $\{s^{i,j}, n_i^{i,j}, n_j^{i,j}, f^{i,j}, t^{i,j}\}$. The set $A_{i,j}$ of arcs connecting the vertices of $N_{i,j}$ can have two possible configurations, depending on the fact that $\{v_i, v_j\}$ belongs or does not belong to E. In the former case, that is $\{v_i, v_j\} \in E$, $D_{i,j}$ is in *configuration (1)* (see Fig. 2 (a)) and:

$$A_{i,j} = \{(s^{i,j}, n_i^{i,j}), (s^{i,j} n_j^{i,j}), (s^{i,j}, f^{i,j}), (n_i^{i,j}, t^{i,j}), (n_j^{i,j}, t^{i,j}), (f^{i,j}, t^{i,j})\}$$

---

[1] We recall that the class XP contains those problems that, given a parameter k, can be solved in time $O(n^{f(k)})$.

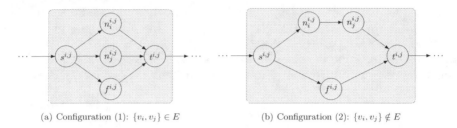

(a) Configuration (1): $\{v_i, v_j\} \in E$ (b) Configuration (2): $\{v_i, v_j\} \notin E$

Fig. 2. Example of the two configurations of subgraph $D_{i,j} = (N_{i,j}, A_{i,j})$ associated with a pair $\{v_i, v_j\}$ of vertices of a graph $G = (V, E)$

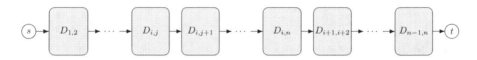

Fig. 3. Example of graph $D = (N, A)$ associated with graph $G = (V, E)$. Grey boxes represent subgraphs $D_{i,j}$ in one of the two possible configurations of Fig. 2.

In the latter case, that is $\{v_i, v_j\} \notin E$, $D_{i,j}$ is in *configuration (2)* (see Fig. 2 (b)) and:

$$A_{i,j} = \{(s^{i,j}, n_i^{i,j}), (s^{i,j}, f^{i,j}), (n_i^{i,j}, n_j^{i,j}), (n_j^{i,j}, t^{i,j}), (f^{i,j}, t^{i,j})\}$$

The whole graph $D = (N, A)$ is constructed by concatenating the graphs $D_{i,j}$ (for all $1 \leq i < j \leq n$) according to the lexicographic order of their indices i, j. The sink $t^{i,j}$ of each graph $D_{i,j}$ is connected to the source $s^{i',j'}$ of the graph $D_{i',j'}$ which immediately follows $D_{i',j'}$. A distinguished vertex s is connected to the source of $D_{1,2}$ (*i.e.*, the first subgraph), while the sink of $D_{n-1,n}$ (*i.e.*, the last subgraph) is connected to a second distinguished vertex t. Fig. 3 depicts such a construction.

The set R of required pairs is defined as follows.

$$R = \{[s, f^{i,j}] \mid \{v_i, v_j\} \in E\} \cup \bigcup_{1 \leq i \leq n} R_i \quad \text{where } R_i = \{[n_i^{i,j}, n_i^{i,h}] \mid 1 \leq j \leq h \leq n\}$$

The following lemmas prove the correctness of the reduction.

Lemma 3. *Let $G = (V, E)$ be an undirected (connected) graph and let $\langle D = (N, A), R \rangle$ be the corresponding instance of 3-PCRP. Then, given a 3-coloring of G we can compute in polynomial time three paths of D that cover all its vertices and every required pair in R.*

Lemma 4. *Let $G = (V, E)$ be an undirected graph and let $\langle D = (N, A), R \rangle$ be the corresponding instance of 3-PCRP. Then, given three paths in D that cover all its vertices and every required pair in R we can compute in polynomial time a 3-coloring of G.*

Since 3-PCRP is clearly in NP, the following result is a consequence of the previous lemmas and of the NP-hardness of 3C [8].

Theorem 5. *3-PCRP is NP-complete.*

3.2 A Polynomial Time Algorithm for 2-PCRP

In this section we give a polynomial time algorithm for computing a solution of 2-PCRP. Notice that 1-PCRP can be easily solved in polynomial time, as there exists a solution of 1-PCRP if and only if the reachability relation of the vertices of the input graph is a total order.

The algorithm for solving 2-PCRP is based on a polynomial-time reduction to the 2-Clique Partition problem, which, given an undirected graph $G = (V, E)$, asks whether there exists a partition of V in two sets V_1, V_2 both inducing a clique in G. Computing the existence of a 2-Clique Partition over a graph G is equivalent to computing if there exists a 2-Coloring of the complement graph G' (hence deciding if G' is bipartite), which is well-known to be solvable in polynomial time [8, probl. GT15]. To perform this reduction we assume that given $\langle D = (N, A), R \rangle$, instance of 2-PCRP, every vertex of the graph D belongs to at least one required pair in R. Otherwise, we add to R the required pairs $[s, v_i]$ for all $v_i \in N$ that do not belong to any required pair. Therefore, a solution that covers all the required pairs in R covers also all the vertices, hence it is a feasible solution of 2-PCRP. Moreover, notice that this transformation does not affect the solution of 2-PCRP, since all the paths start from s and cover all the nodes of the graph, including the additional required pairs.

The algorithm, starting from an instance $\langle D = (N, A), R \rangle$ of 2-PCRP, computes in polynomial time a corresponding undirected graph $G = (V, E)$ where $V = \{v_c \mid c \in R\}$ and $E = \{\{v_{c_i}, v_{c_j}\} \mid \exists$ a path in D that covers both c_i and $c_j\}$.

Given a set of required pairs $R' \subseteq R$, we denote by $V(R')$ the corresponding set of vertices of G (*i.e.*, $V(R') = \{v_c \mid c \in R'\}$).

The algorithm is based on the following fundamental property.

Lemma 6. *Given an instance $\langle D = (N, A), R \rangle$ of 2-PCRP and the corresponding graph $G = (V, E)$, then there exists a path π that covers a set R' of required pairs if and only if $V(R')$ is a clique of G.*

From Lemma 6, it follows that, in order to compute the existence of a solution of 2-PCRP over the instance $\langle D = (N, A), R \rangle$ (in which every vertex of D belongs to at least one required pair in R), we have to compute if there exists a 2-Clique Partition of the corresponding graph G. Since the 2-Clique Partition problem can be solved in polynomial-time [8, probl. GT15], we can conclude that 2-PCRP can be decided in polynomial time.

4 Parameterized Complexity of MaxRPSP

In this section, we consider the parameterized complexity of MaxRPSP. We show that, although MaxRPSP is W[1]-hard (hence unlikely fixed-parameter tractable) when parameterized by the number of required pairs covered by a single path (Section 4.1), the problem becomes fixed-parameter tractable if the maximum overlapping degree is a parameter (Section 4.2).

4.1   W[1]-Hardness of MaxRPSP Parameterized by the Optimum

In this section, we investigate the parameterized complexity of MaxRPSP when parameterized by the size of the solution, that is the maximum number of required pairs covered by a single path, and we prove that the problem is W[1]-hard (notice that this result implies the NP-hardness of MaxRPSP). This result shows that it is unlikley that the problem is fixed-parameter tractable, when parameterized by the number of required pairs covered by a single path. For details on the theory of fixed-parameter complexity, we refer the reader to [3,9].

We prove this result via a parameterized reduction from the h-Clique problem to the decision version of MaxRPSP (k-RPSP), parameterized by the sizes of the respective solutions. Given an undirected graph $G = (V, E)$ and an integer h, h-Clique asks to decide if there exists a clique $C \subseteq V$ of size h. On the other hand, given a DAG D, a set R of required pairs, and an integer k, the k-RPSP problem consists of deciding if there exists a path in D that "covers" k required pairs. We recall that h-Clique is known to be W[1]-hard [4].

First, we start by showing how to construct an instance of k-RPSP starting from an instance of h-Clique. Given an (undirected) graph $G = (V, E)$ with n vertices $v_1, \ldots, v_n$, we construct the associated directed acyclic graph $D = (N, A)$ as follows. The set N of vertices is defined as:

$$N = \{v_i^z \mid v_i \in V, 1 \leq z \leq h\} \cup \{s, t\}$$

Informally, N consists of two distinguished vertices s, t and of h copies $v_i^1, \ldots, v_i^h$ of every vertex v_i of G.

The set of arcs A is defined as:

$$A = \{(v_i^z, v_j^{z+1}) \mid \{v_i, v_j\} \in E, 1 \leq z \leq h-1\} \cup \{(s, v_i^1), (v_i^h, t) \mid v_i \in V\}$$

Informally, we connect every two consecutive copies associated with vertices that are adjacent in G, the source vertex s to all the vertices v_i^1, with $1 \leq i \leq n$, and all the vertices v_i^h, with $1 \leq i \leq n$, to the sink vertex t.

The set R of required pairs is defined as:

$$R = \{[v_i^x, v_j^y] \mid \{v_i, v_j\} \in E, 1 \leq x < y \leq h\}$$

Informally, for each edge $\{v_i, v_j\}$ of G there is a required pair $[v_i^x, v_j^y]$, $1 \leq x < y \leq h$, between every two different copies associated with v_i, v_j.

By construction, the vertices in N (except for s and t) are partitioned into h *independent sets* $I_z = \{v_i^z \mid 1 \leq i \leq n\}$, with $1 \leq z \leq h$, each one containing a copy of every vertex of V. Moreover, the arcs of A only connect two vertices of consecutive subsets I_z and I_{z+1}, with $1 \leq z \leq h-1$. Figure 4 presents an example of directed graph D associated with an undirected graph G.

Now, we are able to prove the main properties of the reduction.

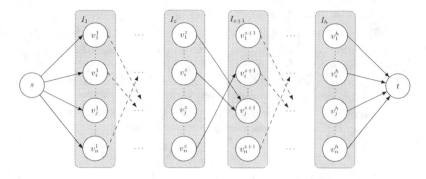

Fig. 4. Example of directed acyclic graph $D = (N, A)$ associated with an instance $G = (V, E)$ of the h-Clique problem. Each gray box highlight an independent set I_z composed of one copy of the vertices in V. Edges (v_1^z, v_j^{z+1}), (v_i^z, v_j^{z+1}), and (v_n^z, v_i^{z+1}) are some of the directed edges in A associated with edges $\{v_1, v_j\}, \{v_i, v_j\}, \{v_i, v_n\} \in E$.

Lemma 7. *Let $G = (V, E)$ be an undirected graph and $\langle D = (N, A), R \rangle$ be the associated instance of k-RPSP. Then: (1) starting from an h-clique in G we can compute in polynomial time an st-path π in D that covers $\binom{h}{2}$ required pairs of R; (2) starting from an st-path π in D that covers $\binom{h}{2}$ required pairs we can compute in polynomial time an h-clique in G.*

The W[1]-hardness of k-RPSP easily follows from Lemma 7 and from the W[1]-hardness of h-Clique when parameterized by h [4].

Theorem 8. *k-RPSP is W[1]-hard when parameterized by the number of required pairs covered by an st-path.*

4.2   An FPT Algorithm for MaxRPSP Parameterized by the Maximum Overlapping Degree

In this section we propose a fixed parameter algorithm (FPT) for the MaxRPSP problem, where the parameter is the maximum overlapping degree of the required pairs in R. For the rest of the section, let $\langle D = (N, A), R \rangle$ be an instance of the MaxRPSP problem.

For ease of exposition, we fix an order of the required pairs in R and we represent the i-th required pair of the ordering as $[v_i^1, v_i^2]$. Whenever no confusion arises, we will refer to that required pair as i-*pair*. Intuitively, we want that the order of the required pairs is "compatible" with the topological order of the vertices. More formally, in this ordering, for any two distinct required pairs $[v_j^1, v_j^2]$ and $[v_i^1, v_i^2]$ with $j < i$, pair $[v_j^1, v_j^2]$ is nested in $[v_i^1, v_i^2]$ or there does not exist a path π from s to v_j^2 that covers both the required pairs (that is, π passes through v_j^2 before v_i^2). Clearly, an order that satisfies this condition can be easily computed from the topological order of the vertices.

We present a parameterized algorithm based on dynamic programming for the MaxRPSP problem when the parameter p is the maximum number of overlapping required pairs. In fact, we can decompose a path π, starting in s, ending in a vertex v, and covering k required pairs, into two subpaths: the first one—π_1—starts in s, ends in a vertex v', and covers k_1 required pairs, while the other one—π_2—starts in v', ends in v, and covers the remaining $k_2 = k - k_1$ required pairs (possibly using vertices of π_1). The key point to define the recurrence is that, for each required pair p, we keep track the set of required pairs overlapping p and covered by the path. To this aim, for each required pair $[v_i^1, v_i^2]$, we define the set $OP([v_i^1, v_i^2])$ as the set of vertices v such that v belongs to a required pair that overlaps $[v_i^1, v_i^2]$ and such that v_i^2 is reachable from v. By a slightly abuse of the notation, we consider that $OP([v_i^1, v_i^2])$ always contains vertex v_i^1.

The recurrence relies on the following observation. Let π be a path covering a set P of required pairs and let $N(P)$ be the set of vertices belonging to the required pairs in P. Consider two required pairs $[v_i^1, v_i^2]$ and $[v_j^1, v_j^2]$ in P, with $j < i$. Then, either $[v_j^1, v_j^2]$ is nested in $[v_i^1, v_i^2]$ (hence the fact that π covers the pair $[v_j^1, v_j^2]$ can be checked by the recurrence looking only at the required pairs that overlap with $[v_i^1, v_i^2]$) or pairs $[v_i^1, v_i^2]$ and $[v_j^1, v_j^2]$ are alternated. In the latter case, since $[v_i^1, v_i^2]$ is in P, we only have to consider the vertices in the set $N(P) \cap OP([v_i^1, v_i^2]) \cap OP([v_j^1, v_j^2])$. Moreover, let p_i be the number of required pairs that overlap the required pair $[v_i^1, v_i^2]$, then $|OP([v_i^1, v_i^2])|$ is at most $2p_i$. Hence, the cardinality of set $N(P) \cap OP([v_i^1, v_i^2]) \cap OP([v_j^1, v_j^2])$ is bounded by $2\max(p_i, p_j)$. Moreover, given two sets S and S' of vertices such that $S \subseteq OP([v_i^1, v_i^2])$ and $S' \subseteq OP([v_j^1, v_j^2])$, we say that S is in *agreement* with S' if $S \cap (OP([v_i^1, v_i^2]) \cap OP([v_j^1, v_j^2])) = S' \cap (OP([v_i^1, v_i^2]) \cap OP([v_j^1, v_j^2]))$. Informally, when S and S' are in agreement, they must contain the same subset of vertices of $OP([v_i^1, v_i^2]) \cap OP([v_j^1, v_j^2])$.

Let $P([v_i^1, v_i^2], S)$ denote the maximum number of required pairs covered by a path π ending in vertex v_i^2 and such that the set $S \subseteq OP([v_i^1, v_i^2])$ is covered by π. In the following we present the recurrence to compute $P([v_i^1, v_i^2], S)$. For ease of exposition we only focus on vertices that appear as second vertices of the required pairs. In fact, paths that do not end in such vertices are not able to cover new required pairs. Furthermore, for simplicity, we consider the source s as the second vertex of a fictitious required pair (with index 0) $[\perp, s]$ which does not overlap any other required pair. Such a fictitious required pair does not contribute to the total number of required pairs covered by the path.

The recurrence is:

$$P([v_i^1, v_i^2], S) = \max_{\substack{[v_j^1, v_j^2]\text{ not nested in }[v_i^1, v_i^2]\text{ and } j < i; \\ S'\text{ in agreement with } S; \\ \exists\text{ a path from } v_j^2 \text{ to } v_i^2 \\ \text{covering all vertices in } S \setminus S';}} \left\{ P([v_j^1, v_j^2], S') + |Ov([v_i^1, v_i^2], S, S')| \right\} \quad (1)$$

where $Ov([v_i^1, v_i^2], S, S') = \{[v_h^1, v_h^2] \mid [v_h^1, v_h^2] \text{ is nested in } [v_i^1, v_i^2] \wedge v_h^1 \in S \wedge v_h^2 \in S \setminus S'\}$. Notice that each required pair is assumed to be nested in itself.

The base case of the recurrence is $P([\perp, s], \varnothing) = 0$.

The correctness of the recurrence derives from the following two lemmas.

Lemma 9. *If $P([v_i^1, v_i^2], S) = k$, then there exists a path π in D ending in v_i^2, such that every vertex in S belongs to π and the number of required pairs covered by π is k.*

Lemma 10. *Let π be a path in D ending in v_i^2 and covering k required pairs. Let S be the set of all the vertices belonging to required pairs covered by π and overlapping $[v_i^1, v_i^2]$. Then $P([v_i^1, v_i^2], S) \geq k$.*

Let p be the maximum number of overlapping required pairs in D (that is, $p = \max_i\{p_i\}$). It follows that the number of possible subsets S is bounded by $O(2^p)$. Then, each entry $P[v_i^2, S]$ requires time $O(2^p n)$ to be computed, and, since there exist $O(2^p n)$ entries, the recurrence requires time $O(4^p n^2)$. From Lemma 9 and Lemma 10, it follows that an optimal solution for MaxRPSP can be obtained by looking for the maximum of the values $P[v_i^2, S]$. Hence, the overall time complexity of the algorithm is bounded by $O(4^p n^2)$.

Acknowledgements. SB, PB, and YP are supported by the Fondo di Ateneo 2011 grant "Metodi algoritmici per l'analisi di strutture combinatorie in bioinformatica". PB, RD, and YP are supported by the MIUR PRIN 2010-2011 grant "Automi e Linguaggi Formali: Aspetti Matematici e Applicativi", code H41J12000190001.

References

1. Bao, E., Jiang, T., Girke, T.: BRANCH: Boosting RNA-Seq assemblies with partial or related genomic sequences. Bioinformatics 29(10), 1250–1259 (2013)
2. Bonizzoni, P., Dondi, R., Pirola, Y.: Maximum disjoint paths on edge-colored graphs: Approximability and tractability. Algorithms 6(1), 1–11 (2013)
3. Downey, R., Fellows, M.: Parameterized complexity. Springer (1999)
4. Downey, R., Fellows, M.: Fixed-parameter tractability and completeness II: On completeness for W[1]. Theoretical Computer Science 141(1&2), 109–131 (1995)
5. Eriksson, N., Pachter, L., Mitsuya, Y., Rhee, S., Wang, C., Gharizadeh, B., Ronaghi, M., Shafer, R., Beerenwinkel, N.: Viral population estimation using pyrosequencing. PLoS Computational Biology 4(5), e1000074 (2008)
6. Fulkerson, D.R.: Note on Dilworth's decomposition theorem for partially ordered sets. Proc. American Mathematical Society 7, 701–702 (1956)
7. Gabow, H., Maheshwari, S., Osterweil, L.: On two problems in the generation of program test paths. IEEE Trans.on Software Engineering 2(3), 227–231 (1976)
8. Garey, M., Johnson, D.: Computer and intractability: A guide to the theory of NP-completeness. W. H. Freeman (1979)
9. Niedermeier, R.: Invitation to fixed-parameter algorithms. Oxford Univ. Press (2006)
10. Ntafos, S., Hakimi, S.: On path cover problems in digraphs and applications to program testing. IEEE Trans. on Software Engineering 5(5), 520–529 (1979)
11. Trapnell, C., Williams, B., Pertea, G., Mortazavi, A., Kwan, G., van Baren, M., Salzberg, S., Wold, B., Pachter, L.: Transcript assembly and quantification by RNA-Seq reveals unannotated transcripts and isoform switching during cell differentiation. Nature Biotechnology 28(5), 516–520 (2010)
12. Wu, B.Y.: On the maximum disjoint paths problem on edge-colored graphs. Discrete Optimization 9(1), 50–57 (2012)

Efficient List-Based Computation of the String Subsequence Kernel*

Slimane Bellaouar[1], Hadda Cherroun[1], and Djelloul Ziadi[2]

[1] Laboratoire LIM, Université Amar Telidji, Laghouat, Algérie
{s.bellaouar,hadda_cherroun}@mail.lagh-univ.dz
[2] Laboratoire LITIS - EA 4108, Université de Rouen, Rouen, France
djelloul.ziadi@univ-rouen.fr

Abstract. Kernel methods are powerful tools in machine learning. They have to be computationally efficient. In this paper, we present a novel list-based approach to compute efficiently the string subsequence kernel (SSK). Our main idea is that our list-based SSK reduces to range query problem. We started by the construction of a *match list* $L(s,t) = \{(i,j) : s_i = t_j\}$ where s and t are the strings to be compared; such match list contains only the required data that contribute to the result. To do some intermediate processing efficiently, we constructed a *layered range tree* and applied the corresponding computational geometry algorithms. Moreover, we extended our match list to be a *list of lists* in order to improve the computation efficiency of the SSK. The whole process takes $O(|L| \log |L| + pK)$ time and $O(|L| \log |L| + K)$ space, where $|L|$ is the size of the match list, p is the length of the SSK and K is the total reported points by range queries over all the entries of the list.

Keywords: string kernel, computational geometry, layered range tree, range query.

1 Introduction

Kernel methods [6] offer an alternative solution to the limitation of traditional machine learning algorithms, applied solely on linear separable problems. They map data into a high dimensional feature space where we can apply linear learning machines based on algebra, geometry and statistics. Hence, we may discover non-linear relations. Moreover, kernel methods enable other data type processings (biosequences, images, graphs, ...).

Strings are among the important data types. Therefore, machine learning community devotes a great effort of research to string kernels, which are widely used in the fields of bioinformatics and natural language processing. The philosophy of all string kernels can be reduced to different ways to count common substrings or subsequences that occur in both strings to be compared, say s and t.

In the literature, there are two main approaches to improve the computation of the SSK. The first one is based on dynamic programming; Lodhi et al. [8] apply

---

* This work is supported by the MESRS - Algeria under Project 8/U03/7015.

A.-H. Dediu et al. (Eds.): LATA 2014, LNCS 8370, pp. 138–148, 2014.

dynamic programming paradigm to the suffix version of the SSK. They achieve a complexity of $O(p|s||t|)$, where p is the length of the SSK. Later, Rousu and Shawe-Taylor [9] propose an improvement to the dynamic programming approach based on the observation that most entries of the dynamic programming matrix (DP) do not really contribute to the result. They use a set of match lists combined with a sum range tree. They achieve a complexity of $O(p|L|\log\min(|s|,|t|))$, where L is the set of matches of characters in the two strings. Beyond the dynamic programming paradigm, the trie-based approach [9,7,11] is based on depth first traversal on an implicit trie data structure. The idea is that each node in the trie corresponds to a co-occurrence between strings. But the number of gaps is restricted, so the computation is approximate.

Motivated by the efficiency of the computation, a key property of kernel methods, in this paper we focus on improving the SSK computation. Our main idea consists to map a machine learning problem on a computational geometry one. Precisely, the list-based SSK computation reduces to 2-dimensional range queries on a layered range tree (a range tree enhanced by the fractional cascading technique). We started by the construction of a *match list* $L(s,t) = \{(i,j) : s_i = t_j\}$ where s and t are the strings to be compared; such match list contains only the required data that contribute to the result. To do some intermediate processing efficiently, we constructed a *layered range tree* and applied the corresponding computational geometry algorithms. Moreover, we extended our match list to be a *list of lists* in order to improve the computation efficiency of the SSK. The overall time complexity is $O(|L|\log|L| + pK)$, where $|L|$ is the size of the match list and K is the total reported points by range queries over all the entries of the match list.

The rest of this paper is organized as follows. Section 2 deals with some concept definitions used in the other sections. In section 3, we recall formally the SSK computation. We also review three efficient computations of the SSK, namely, dynamic programming, trie-based and sparse dynamic programming approaches. Section 4 is devoted to the presentation of our contribution. Section 5 includes conclusions and discussion.

2 Preliminaries

Let Σ be an alphabet of a finite set of symbols. We denote the number of symbols in Σ by $|\Sigma|$. A string $s = s_1...s_{|s|}$ is a finite sequence of symbols of length $|s|$ where s_i denotes the ith element of s. We use Σ^n to denote the set of all finite strings of length n and Σ^* to denote the set of all strings. The notation $[s = t]$ is a boolean function that returns

$$\begin{cases} 1 & \text{if } s \text{ and } t \text{ are identical;} \\ 0 & \text{otherwise.} \end{cases}$$

The string $s(i : j)$ denotes the substring $s_i s_{i+1}...s_j$ of s. Accordingly, a string t is a substring of a string s if there are strings u and v such that $s = utv$ (u and v can be empty). The substrings of length n are referred to as n-grams (or n-mers).

The string t is a subsequence of s if there exists an increasing sequence of indices $I = (i_1, ..., i_{|t|})$ in s, $(1 \leq i_1 < ... < i_{|t|} \leq |s|)$ such that $t_j = s_{i_j}$, for $j = 1, ..., |t|$. In the literature, we use $t = s(I)$ if t is a subsequence of s in the positions given by I. The empty string ϵ is indexed by the empty tuple. The absolute value $|t|$ denotes the length of the subsequence t which is the number of indices ($|I|$), while $l(I) = i_{|t|} - i_1 + 1$ refers to the number of characters of s covered by the subsequence t.

3   String Subsequence Kernels

The SSK adopts a new weighting method that reflects the degree of contiguity of a subsequence in the string. In order to measure the distance of non contiguous elements of the subsequence, a gap penalty $\lambda \in]0, 1]$ is introduced. Formally, the mapping function $\phi^p(s)$ in the feature space F can be defined as follows:

$$\phi_u^p(s) = \sum_{I:u=s(I)} \lambda^{l(I)}, \; u \in \Sigma^p.$$

The associated kernel can be written as:

$$K_p(s, t) = \langle \phi^p(s), \phi^p(t) \rangle = \sum_{u \in \Sigma^p} \sum_{I:u=s(I)} \sum_{J:u=t(J)} \lambda^{l(I)+l(J)}.$$

A suffix kernel is defined to assist in the computation of the SSK. The associated embedding is given by:

$$\phi_u^{p,S}(s) = \sum_{I \in I_p^{|s|}:u=s(I)} \lambda^{l(I)}, u \in \Sigma^p,$$

where I_p^k denotes the set of p-tuples of indices I with $i_p = k$.

The associated kernel can be defined as follows:

$$K_p^S(s, t) = \langle \phi^{p,S}(s), \phi^{p,S}(t) \rangle$$
$$= \sum_{u \in \Sigma^p} \phi_u^{p,S}(s).\phi_u^{p,S}(t).$$

The SSK can be expressed in terms of its suffix version as follows:

$$K_p(s, t) = \sum_{i=1}^{|s|} \sum_{j=1}^{|t|} K_p^S(s(1 : i), t(1 : j)), \tag{1}$$

with $K_1^S(s, t) = [s_{|s|} = t_{|t|}] \lambda^2$.

3.1   Naive Implementation

The computation of the similarity of two strings (sa and tb) is conditioned by their final symbols. In the case where $a = b$, we have to sum kernels of all prefixes of s and t. Hence, a recursion has to be devised:

$$K_p^S(sa, tb) = [a = b] \sum_{i=1}^{|s|} \sum_{j=1}^{|t|} \lambda^{2+|s|-i+|t|-j} K_{p-1}^S(s(1:i), t(1:j)). \tag{2}$$

This computation leads to a complexity of $O(p(|s|^2|t|^2))$.

3.2   Efficient Implementations

We will present three methods that compute the SSK efficiently, namely the dynamic programming [8], the trie-based [9,7,11] and the sparse dynamic programming approaches [9].

Dynamic Programming Approach. The starting point of the dynamic programming approach is the suffix recursion given by equation (2). From this equation, we can consider a separate dynamic programming table DP_p for storing the double sum:

$$DP_p(k, l) = \sum_{i=1}^{k} \sum_{j=1}^{l} \lambda^{k-i+l-j} \, K_{p-1}^S(s(1:i), t(1:j)). \tag{3}$$

It is easy to see that: $K_p^S(sa, tb) = [a = b] \lambda^2 \, DP_p(|s|, |t|)$.
  Computing ordinary DP_p for each (k, l) would be inefficient. So we can devise a recursive version of equation (3) with a simple counting device:

$$DP_p(k, l) = K_{p-1}^S(s(1:k), t(1:l)) + \lambda DP_p(k-1, l) +$$
$$\lambda DP_p(k, l-1) - \lambda^2 DP_p(k-1, l-1).$$

Consequently, the complexity of the SSK becomes $O(p|s||t|)$.

Trie-Based Approach. This approach is based on search trees known as tries, introduced by E. Fredkin in 1960. The key idea of the trie-based approach is that leaves play the role of the feature space indexed by the set Σ^p. In the literature, there are variants of trie-based string subsequence kernels. For instance the (p, m)-mismatch string kernel [7] and restricted SSK [11].
  In the present section, we try to describe a trie-based SSK presented in [9] that slightly differ from those cited above [7,11]. Given that each node in the trie corresponds to a co-occurrence between strings, the algorithm stores all matches $s(I) = u_1 \cdots u_q$, $I = i_1 \cdots i_q$ in such node. In parallel, it maintains a list of alive matches $L_s(u, g)$ that records the last index i_q (g is the number of gaps in the

occurences). Notice that in the same list we are able to record many occurrences with different gaps. Similarly, the algorithm is applied to the string t. The process will continue until achieving the depth p where the kernel is evaluated as follows:

$$K_p(s,t) = \sum_{u \in \Sigma^p} \phi_u^p(s)\phi_u^p(t) = \sum_{u \in \Sigma^p} \sum_{g_s,g_t} \lambda^{g_s+p}|L_s(u,g_s)| \cdot \lambda^{g_t+p}|L_t(u,g_t)|.$$

Given that, there exists $\binom{p+g_{max}}{g_{max}}$ different entries at leaf nodes, the worst-case time complexity of the algorithm is $O(\binom{p+g_{max}}{g_{max}}(|s|+|t|))$.

Sparse Dynamic Programming Approach. It is built on the fact that in many cases, most of the entries of the DP matrix are zero and do not contribute to the result. Rousu and Shawe-Taylor [9] have proposed a solution using two data structures. The first one is a set of match lists instead of K_p^S matrix. The second one is a range sum tree, which is a B-tree, that replaces the DP_p matrix. It is used to return the sum of n values within an interval in $O(\log n)$ time. Their algorithm runs in $O(p|L|\log\min(|s|,|t|))$, where $L = \{(i,j)|s_i = t_j\}$ is the set of matches of characters in both strings.

4   List and Layered Range Tree Based Approach

Looking forward to improving the complexity of SSK, our approach is based on two observations. The first one concerns the computation of $K_p^S(s,t)$ that is required only when $s_{|s|} = t_{|t|}$. Hence, we have kept only a list of index pairs of these entries rather than the entire suffix table, $L(s,t) = \{(i,j) : s_i = t_j\}$. If we consider the example which computes $K_p(gatta, cata)$, the list generated is

$$L(gatta, cata) = \{(2,2),(5,2),(3,3),(4,3),(2,4),(5,4)\}.$$

In the rest of the paper, while measuring the complexity of different computations, we will consider, $|L|$, the size of the match list $L(s,t)$ as the parameter indicating the size of the input data.

The complexity of the naive implementation of the list version is $O(p|L|^2)$, and it seems not obvious to compute $K_p^S(s,t)$ efficiently on a list data structure. In order to address this problem, we have made a second observation that the suffix table can be represented as a 2-dimensional space (plane) and the entries where $s_{|i|} = t_{|j|}$ as points in this plane.

At the light of this observation, the computation of $K_p^S(s,t)$ can be interpreted as an orthogonal range query. In the literature, there are several data structures that are used in computational geometry. We have examined a spatial data structure known as Kd-tree [1,4,10]. It records a total time cost of $O(p(|L|\sqrt{|L|}+K))$ for computing the SSK, where K is the total of the reported points. It is clear that this relative amelioration is not sufficiently satisfactory. So we adopted another spatial data structure, called range tree [4,10,2,3], which has better query time for rectangular range queries. We will describe such data structure and its relationship with SSK in the following subsections.

4.1   Suffix Table Representation

The entries (k, l) in $L(s, t)$ correspond to a set S of points in the plane, where the index pairs (k, l) play the role of the point coordinates. The set S is represented by a 2-dimensional range tree, where nodes represent points in the plane. Thereby, representing the suffix table tend to be the construction of a 2-dimensional range tree. A range tree, denoted by $\mathcal{RT}$ is primarily a balanced binary search tree (BBST) augmented with an associated data structure. In order to build such data structure, first, we consider the set S_x of the first coordinate (x-coordinate) values of all the points in S. Thereafter, a BBST called x-$\mathcal{RT}$ is constructed with points of S_x in the leaves. Both internal and leaf nodes v of x-$\mathcal{RT}$ are augmented by a 1-dimensional range tree, it can be a BBST or a sorted array, of a canonical subset $P(v)$ on y-coordinates, denoted by y-$\mathcal{RT}$. The subset $P(v)$ is the points stored in the leaves of the sub tree rooted at the node v. Figure 1 illustrates the construction process of a 2-dimensional range tree.

In the case where two points have the same x or y-coordinate, we have to define a total order by using a lexicographic one. It consists to replace the real number by a composite-number space [4]. The composite number of two reals x and y is denoted by $x|y$, so for two points, we have:

$$(x|y) < (x'|y') \Leftrightarrow x < x' \vee (x = x' \wedge y < y').$$

In such situation, we have to transform the range query $[x_1 : x_2] \times [y_1 : y_2]$ related to a set of points in the plane to the range query $[(x_1| - \infty) : (x_2| + \infty)] \times [(y_1| - \infty) : (y_2| + \infty)]$ related to the composite space.

Based on the analysis of computational geometry algorithms, our 2-dimensional range tree requires $O(|L| \log |L|)$ storage and can be constructed in $O(|L| \log |L|)$ time. This leads to the following lemma.

Lemma 1. *Let s and t be two strings and $L(s, t) = \{(i, j) : s_i = t_j\}$ the match list associated to the suffix version of the SSK. A range tree for $L(s, t)$ requires $O(|L| \log |L|)$ storage and takes $O(|L| \log |L|)$ construction time.*

4.2   Location of Points in a Range

We recall that computing the recursion for the SSK given by the equation (2) can be interpreted as the evaluation of a 2-dimensional range query applied to a 2-dimensional range tree. Such evaluation locates all points that lie in the specified range.

A useful idea, in terms of efficiency, consists on treating a rectangular range query as a two nested 1-dimensional queries. In other words, let $[x_1 : x_2] \times [y_1 : y_2]$ be a 2-dimensional range query, we first ask for the points with x-coordinates in the given 1-dimensional range query $[x_1 : x_2]$. Consequently, we select a collection of $O(\log |L|)$ subtrees. We consider only the canonical subset of the resulted subtrees, which contains, exactly, the points that lies in the x-range $[x_1 : x_2]$. At the next step, we will only consider the points that fall in the y-range $[y_1 : y_2]$. The total task of a range query can be performed in $O(\log^2 |L| + k)$ time, where

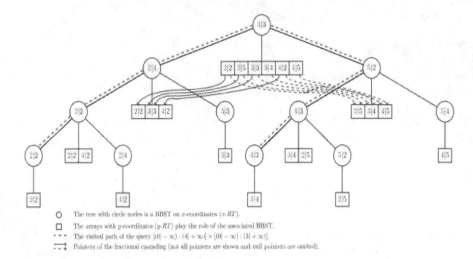

Fig. 1. Layered range tree $\mathcal{RT}$ related to K_p^S(gatta, cata)

k is the number of points that are in the range. We can improve it by enhancing the 2-dimensional range tree with the fractional cascading technique which is described in the following subsection.

4.3   Fractional Cascading

The key observation made during the invocation of a rectangular range query is that we have to search the same range $[y_1 : y_2]$ in the associated structures y-$\mathcal{RT}$ of $O(\log |L|)$ nodes found while querying the x-$\mathcal{RT}$ by the range query $[x_1 : x_2]$. Moreover, there exists an inclusion relationship between these associated structures. The goal of the fractional cascading consists on executing the binary search only once and use the result to speed up other searches without expanding the storage by more than a constant factor.

The application of the fractional cascading technique introduced by [5] on a range tree creates a new data structure so called *layered range tree*. We illustrate such technique through an example of SSK computing in Fig. 1.

Using this technique, the rectangular search query time becomes $O(\log |L| + k)$, where k is the number of reported points. For the computation of $K_p^S(s,t)$ we have to consider $|L|$ entries of the match list. The process iterates p times, therefore, we get a time complexity of $O(p|L| \log |L| + K)$ for evaluating the SSK, where K is the total of reported points over all the entries of $L(s,t)$. This result combined to that of Lemma. 1 lead to the following lemma:

Lemma 2. *Let s and t be two strings and $L(s,t) = \{(i,j) : s_i = t_j\}$ the match list associated to the suffix version of the SSK. A layered range tree for $L(s,t)$ uses $O(|L| \log |L|)$ storage and it can be constructed in $O(|L| \log |L|)$ time. With this layered range tree, the SSK of length p can be computed in $O(p(|L| \log |L| + K))$, where K is the total number of reported points over all the entries of $L(s,t)$.*

4.4   List of Lists Building and SSK Computation

Another observation leads us to pursue our improvement of the SSK computation complexity. It is obvious to state that point coordinates, in our case, in the plane remain unchanged during the entire process. So instead of invoking the 2-dimensional range query multiple times according to the evolution of the parameter p, it is more beneficial if we do the computation only once. Accordingly, in this phase, we extended our match list to be a list of lists (Fig. 2), where each entry (k, l) points to a list that contains all the points that lie in the corresponding range. Algorithm 1 builds this list of lists. The complexity of the construction of the list of lists is the complexity of invoking the 2-dimensional range query over all the entries of the match list. This leads to $O(|L| \log |L| + K)$ time complexity.

Algorithm 1. List of Lists Creation

Input: match list $L(s,t)$ and Layered Range Tree $\mathcal{RT}$
Output: List of Lists $LL(s,t)$: The match list augmented by lists
            containing reported points

1 **foreach** *entry* $(k,l) \in L(s,t)$ **do**
2   *Preparing the range query*
3      $x_1 \leftarrow 0$
4      $y_1 \leftarrow 0$
5      $x_2 \leftarrow k - 1$
6      $y_2 \leftarrow l - 1$
7   relatedpoints $\leftarrow$ 2D-RANGE-
       QUERY($\mathcal{RT}$, $[(x_1| - \infty) : (x_2| + \infty)] \times [(y_1| - \infty) : (y_2| + \infty)]$)

8   **while** *There exists* $(i,j) \in relatedpoints$ **do**
9   add (i,j) to (k,l)-list

Once the list of lists constructed, the SSK computation will sum over all the reported points stored on it. The process is described in Algorithm 2. The cost of this computation is $O(K)$. Since we will evaluate the SSK for $p \in [1.. \min(|s|, |t|)]$, this leads to a complexity of $O(pK)$. So the over all complexity is $O(|L| \log |L| + pK)$ which include the construction of the list of lists and the computation of SSK in the strict sense. This leads to the following theorem that summarizes the result for the computation of the SSK.

Theorem 3. *Let s and t be two strings and $L(s,t) = \{(i,j) : s_i = t_j\}$ the match list associated to the suffix version of the SSK. A layered range tree and a list of lists for $L(s,t)$ require $O(|L| \log |L| + K)$ storage and they can be constructed in $O(|L| \log |L| + K)$ time. With these data structures, the SSK of length p can be computed in $O(|L| \log |L| + pK)$, where K is the total number of reported points over all the entries of $L(s,t)$.*

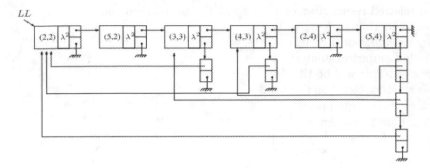

Fig. 2. List of lists inherent to K_1^S(gatta,cata)

Algorithm 2. SSK computation

Input: List of Lists $LL(s,t)$, subsequence length p and penalty coefficient
λ

Output: Kernel values $K_q(s,t) = K(q) : q = 1, \ldots, p$

1 **for** $q{=}1{:}p$ **do**

2 | *Initialization*

3 | $K(q) \leftarrow 0$

4 | $KPS(1 : |max|) \leftarrow 0$

5 | **foreach** *entry* $(k,l) \in LL(s,t)$ **do**

6 | | **foreach** *entry* $r \in (k,l) - list$ **do**

7 | | | $(k,l) - list$ is a list associated to the entry (k,l)

8 | | | $(i,j) \leftarrow r.Key$

9 | | | $KPS_{(i,j)} \leftarrow r.Value$

10 | | | $KPS(k,l) \leftarrow KPS(k,l) + \lambda^{k-i+l-j} \; KPS_{(i,j)}$

11 | | $K(q) \leftarrow K(q) + KPS(k,l))$

12 | *Preparing $LL(s,t)$ For the next computation*

13 | **foreach** *entry* $(k,l) \in KPS$ **do**

14 | | Update $LL(k,l)$ with $KPS(k,l)$

---

5   Conclusions

We have presented a novel algorithm that efficiently computes the string subsequence kernel (SSK). Our approach is refined over three phases. We started by the construction of a match list $L(s,t)$ that contains, only, the information that contributes in the result. In order to locate, efficiently, the related positions for each entry of the match list, we have constructed a layered range tree. At last, we have built a list of lists to compute efficiently the SSK. The Whole task takes $O(|L| \log |L| + pK)$ time and $O(|L| \log |L| + K)$ space, where p is the length of the SSK and K is the total number of reported points.

The reached result gives evidence of an asymptotic complexity improvement compared to that of a naive implementation of the list version $O(p|L|^2)$. On the other hand, our algorithm is output sensitive. Such property dictates to us to conduct empirical analysis in order to compare our contribution with other approaches. This will be the subject of a future research.

Nevertheless, based on the asymptotic complexities of the different approaches and the experiments presented in [9], we make some discussions. The dynamic programming approach is faster when the DP_p table is dense. This case is achieved on long strings if the alphabet is small and on short strings. The trie-based approach is faster on medium-sized alphabets but it suffers from gap number restriction. Furthermore, recall that our approach and the sparse dynamic programming one are proposed in the context where the most of the entries of the DP_p table are zero. This case occurred for large-sized alphabets. From the asymptotic complexity of the sparse approach, $O(p|L|\log\min(|s|,|t|))$, it is clear that its efficiency depends on the size of the strings. For our approach it depends only on the number of common subsequences. Under these conditions our approach outperforms for long strings. These discussions will be validated by a future empirical study.

A noteworthy advantage is that our approach separates the process of required data location from the strict computation one. This separation limits the impact of the length of the SSK on the computation. It have influence, only, on the strict computation process. Moreover, such separation property can be favorable if we assume that the problem is multi-dimensional, e.g. *the multiple string comparison* problem, one of the most active research in biological sequence analysis. We believe that our approach open a new vision for attacking this problem. In terms of complexity, this can have influence, only, on the location process by a logarithmic factor. Indeed, the layered range tree can report points that lies in a rectangular range query in $O(\log^{d-1}|L| + k)$, in a d-dimensional space.

At the implementation level, great programming effort is supported by well-studied and ready to use computational geometry algorithms. Hence, the emphasis is shifted to a variant of string kernel computations that can be easily adapted.

References

1. Bentley, J.L.: Multidimensional binary search trees used for associative searching. Commun. ACM 18(9), 509–517 (1975), http://doi.acm.org/10.1145/361002.361007
2. Bentley, J.L.: Decomposable searching problems. Inf. Process. Lett. 8(5), 244–251 (1979), http://dblp.uni-trier.de/db/journals/ipl/ipl8.html#Bentley79
3. Bentley, J.L., Maurer, H.A.: Efficient worst-case data structures for range searching. Acta Inf. 13, 155–168 (1980), http://dblp.uni-trier.de/db/journals/acta/acta13.html#BentleyM80
4. Berg, M.D., Cheong, O., Kreveld, M.V., Overmars, M.: Computational Geometry: Algorithms and Applications, 3rd edn. Springer-Verlag TELOS, Santa Clara (2008)
5. Chazelle, B., Guibas, L.J.: Fractional cascading: I. a data structuring technique. Algorithmica 1(2), 133–162 (1986)

6. Cristianini, N., Shawe-Taylor, J.: An introduction to support Vector Machines: and other kernel-based learning methods. Cambridge University Press, New York (2000)

7. Leslie, C., Eskin, E., Noble, W.: Mismatch String Kernels for SVM Protein Classification. In: Neural Information Processing Systems, vol. 15, pp. 1441–1448 (2003), http://citeseerx.ist.psu.edu/viewdoc/summary?doi=10.1.1.58.4737

8. Lodhi, H., Saunders, C., Shawe-Taylor, J., Cristianini, N., Watkins, C.: Text classification using string kernels. J. Mach. Learn. Res. 2, 419–444 (2002), http://dx.doi.org/10.1162/153244302760200687

9. Rousu, J., Shawe-Taylor, J.: Efficient computation of gapped substring kernels on large alphabets. J. Mach. Learn. Res. 6, 1323–1344 (2005), http://dl.acm.org/citation.cfm?id=1046920.1088717

10. Samet, H.: The design and analysis of spatial data structures. Addison-Wesley Longman Publishing Co., Inc., Boston (1990)

11. Shawe-Taylor, J., Cristianini, N.: Kernel Methods for Pattern Analysis. Cambridge University Press, New York (2004)

Channel Synthesis Revisited

Béatrice Bérard[1,*] and Olivier Carton[2,**]

[1] Université Pierre & Marie Curie, LIP6/MoVe, CNRS UMR 7606, BC 169
4 Place Jussieu, 75252 Paris Cedex 05, France
[2] Université Paris Diderot, LIAFA, CNRS UMR 7089, Case 7014
75205 Paris Cedex 13, France
Beatrice.Berard@lip6.fr, Olivier.Carton@liafa.univ-paris-diderot.fr

Abstract. Given a system modeled by a rational relation R, a channel is a pair (E, D) of rational relations that respectively encode and decode binary messages, and such that the composition ERD is the identity relation. This means that the message between E and D has been perfectly transmitted through R. Investigating the links between channels and the growth of rational sets of words, we give new characterizations for relations with channels. In the particular case where the relation is given as a union of functions, we obtain as a consequence the decidability of the synthesis problem with a linear complexity.

Keywords: Distributed synthesis, channels, rational relations, transducers, security.

1   Introduction

Channel synthesis. The problem of channel synthesis was introduced in [1,2] as a special case of the general distributed synthesis problem: Given an architecture defined by processes and communication links between them or with the environment, and a specification on the messages transmitted over these links, this general problem aims at deciding the existence of local programs, one for each process, that together meet the specification, whatever the environment does.

In the asynchronous setting, this problem is undecidable for LTL specifications as soon as there are two processes [3]. It was then proposed in [1,2] to consider two processes modeled by finite transducers, that respectively encode and decode finite binary messages. They communicate asynchronously through a medium, acting as noise over the link between them and also described by a fixed non deterministic finite transducer. Moreover, a particular basic external specification expresses faithful communication: the message received is equal to the message emitted. Such an encoder/decoder pair was called a *channel*. The *channel synthesis problem* then asks if, given the noisy process, the encoder and decoder can be synthesized. This question is related to security properties: when

---

* Supported by a grant from Coopération France-Québec, Service Coopération et Action Culturelle 2012/26/SCAC.
** Supported by ANR FREC.

A.-H. Dediu et al. (Eds.): LATA 2014, LNCS 8370, pp. 149–160, 2014.

the noisy process describes some protocol, the existence of a channel may lead to possibly illegal communication [4,5,6]. The problem was proved undecidable (Σ_1^0-complete) for rational relations, but decidable in polynomial time for a rational function. When a channel exists for such a function, it can be effectively computed.

Contribution. We revisit here this notion of channel and show that it has strong links with rational bijections [7], hence it is also related to the growth of languages. Given a language L, the growth function associates with an integer $n \geq 0$ the number of words in L of length less than or equal to n. We introduce the notion of *patterns*, which generate typical languages of exponential growth, and establish some of their properties. Then, combining these properties with results on rational bijections, we prove a new characterization for bounded relations with channels: If R is a bounded rational relation, given as a union of rational functions $h_1 + \cdots + h_n$, then the following conditions are equivalent: (1) R has a channel, (2) at least one of the h_is has a channel, (3) the range of R has an exponential growth. We obtain as a corollary that the channel synthesis problem is decidable in linear time for a finite union of functions. The latter result was already stated in [8] (with a polynomial time complexity), but the proof was not satisfactory. We believe that the notion of exponential growth is central to the study of channels, although it was often implicit in previous works.

2   Definitions and Notations

The set of natural numbers is denoted by $\mathbb{N}$ and the set of *words* over a finite alphabet A is denoted by A^*, with ε for the empty word and $A^+ = A^* \setminus \{\varepsilon\}$. The binary alphabet $\{0,1\}$ is denoted by $\mathbb{B}$. The length of a word u is written $|u|$ and for $n \in \mathbb{N}$, we denote by $A^{\leq n}$ (respectively A^n) the subset of A^* of words of length less than or equal to n (respectively equal to n). A *language* is a subset of A^*. We denote also by $|L|$ the cardinality of a language L.

For two words u and v, v is a *prefix* of u, written $v \preccurlyeq u$, if there is some word w such that $u = vw$. Two words w and w' are called *prefix compatible* if either w is a prefix of w' or w' is a prefix of w. If they are not prefix compatible, there exist two different letters a and b and three words u, v and v' such that $w = uav$ and $w' = ubv'$. The word u is, by definition, the *longest common prefix* of w and w'. The notions of *suffix compatible* words and *longest common suffix* are defined similarly.

A subset X of A^* is a code if any word in X^* admits a unique decomposition over X. A set of two words $X = \{u, v\}$ is not a code if and only if u and v commute ($uv = vu$) ([9]).

Finite Automata. A finite automaton, or automaton for short, is a tuple $\mathcal{A} = \langle Q, I, Lab, \Delta, F \rangle$, where Q is a finite set of states, $I \subseteq Q$ is the subset of initial states, Lab is a finite set of labels, $\Delta \subseteq Q \times Lab \times Q$ is a finite transition relation and $F \subseteq S$ is a of final states. Note that Lab can be an alphabet but also a (subset of a) monoid. Given two states $q, q' \in Q$, a *path* from q to q' with *label* u, written as $q \xrightarrow{u} q'$, is a sequence of transitions $q \xrightarrow{a_1} q_1, q_1 \xrightarrow{a_2} q_2, \cdots q_{n-1} \xrightarrow{a_n} q'$, with $a_i \in Lab$ and $q_i \in Q$, for $1 \leq i \leq n-1$ such that $u = a_1 \cdots a_n$. The path is *accepting* if $q \in I$ and $q' \in F$, and the language of $\mathcal{A}$, denoted by $\mathcal{L}(\mathcal{A})$, is the set of labels of accepting paths. A state $q \in Q$ is *useful* if it occurs in some accepting run. Since the accepted language is the same when removing non useful states, we assume in the sequel that the set Q contains only useful states, in which case $\mathcal{A}$ is called *trim*. A regular language over an alphabet A is a subset of A^* accepted by a finite automaton with set of labels $Lab = A$. The regular languages over A are also the rational sets of A^*.

Finite Transducers. A finite transducer (or transducer for short) is a finite automaton $\mathcal{T}$ with set of labels $Lab \subseteq A^* \times B^*$ for two alphabets A and B. A label $(u, v) \in A^* \times B^*$ is also written as $u|v$. The subset $\mathcal{L}(\mathcal{T})$ of $A^* \times B^*$ is a *rational relation* [10] from A^* to B^*. The transducer $\mathcal{T}$ is said to realize the relation $\mathcal{L}(\mathcal{T})$.

Given a rational relation R, we write $R(u) - \{v \in B^* \mid (u, v) \in R\}$ for the image of $u \in A^*$, $R^{-1}(v) = \{u \in A^* \mid (u, v) \in R\}$ for the inverse image of $v \in B^*$, possibly extended to subsets of A^* or B^* respectively, $\mathrm{dom}(R) = \{u \in A^* \mid \exists v \in B^*, (u, v) \in R\}$ for the domain of R and $\mathrm{rg}(R) = \{v \in B^* \mid \exists u \in A^*, (u, v) \in R\}$ for the range of R.

For a subset P of A^*, the identity relation $\{(u, u) \mid u \in P\}$ on $A^* \times A^*$ is denoted by Id_P. The composition of rational relations R_1 on $A^* \times B^*$ and R_2 on $B^* \times C^*$, denoted by $R_1 R_2$ (from left to right) or by $R_2 \circ R_1$ (from right to left), is the rational relation on $A^* \times C^*$ defined by $\{(u, w) \mid \exists v \ (u, v) \in R_1 \wedge (v, w) \in R_2\}$ ([11]). Moreover, the image and inverse image of a regular language by a rational relation are regular languages [10].

The relation R is bounded if there exists $k \in \mathbb{N}$ such that for each word $u \subset A^*$, $|R(u)| \leq k$. It is a function if $k - 1$. We often identify a function f with its graph $\{(x, f(x) \mid x \in \mathrm{dom}(f)\}$ and we write $f \subseteq R$ (resp. $f \subset f'$) to mean that its graph is contained in R (resp. in the graph of f'). We also write $R + f$ to mean the relation which is the union of R and the graph of f. Since functions play a central role in the rest of the paper, we recall below two powerful and useful results. The first one states that it is always possible to extract a rational function from a rational relation. The second one gives a representation of a bounded rational relation as a union of rational functions.

Theorem 1 (Uniformization [10]). *For any rational relation R, there exists a rational function $f \subseteq R$ with the same domain. Furthermore, a transducer realizing f can be effectively computed from a transducer realizing R.*

Theorem 2 (Bounded relations [12,13]). *A rational relation R is bounded by k if and only if there exist k rational functions $f_1, \ldots, f_k$ such that $R = f_1 + \cdots + f_k$.*

Growth. The growth of a language L is the function mapping each integer $n \in \mathbb{N}$ to the number of words in L of length less than or equal to n. The growth of L is *polynomial* if it is bounded by some polynomial, that is $|L \cap A^{\leq n}| = O(n^k)$

for some $k \in \mathbb{N}$. For instance, the growth of $P_k = (0^*1)^k0^*$ is polynomial since $|P_k \cap \mathbb{B}^{\leq n}| = O(n^{k+1})$. The growth of the set $L = (0 + 10)^*$ is not polynomial.

The growth of a set L is *exponential* if it is greater that some exponential, that is if $\theta^n = O(|L \cap A^{\leq n}|)$ for some real number $\theta > 1$. The growth of the set $L = (0 + 10)^*$ is, for instance, exponential. Note that the growth cannot be more than exponential since $|A^{\leq n}| = (|A|^{n+1} - 1)/(|A| - 1)$, for $|A| \geq 2$. The following proposition states that, for rational sets, there is a gap:

Proposition 1 ([7]). *The growth of a rational set of words is either polynomial or exponential. Moreover, for a finite automaton $\mathcal{A}$, the language $\mathcal{L}(\mathcal{A})$ has an exponential growth if and only if there exist words $u, v, \bar{v}, w$ with $|v| = |\bar{v}|$ and $v \neq \bar{v}$, and a state q of $\mathcal{A}$ such that $i \xrightarrow{u} q \xrightarrow{v} q \xrightarrow{w} f$ and $q \xrightarrow{\bar{v}} q$ in $\mathcal{A}$ where i is an initial state and f is a final state.*

This result suggests the notion of patterns studied in details in Section 4, and defined as tuples of words $(u, v, \bar{v}, w)$ with $|v| = |\bar{v}|$ and $v \neq \bar{v}$.

An automaton with ε-transitions is an automaton in which any transition has either the form $p \xrightarrow{a} q$ for a letter $a \in A$ or the form $p \xrightarrow{\varepsilon} q$. It is well-known [14] that ε-transitions can be removed and that any automaton with ε-transitions is equivalent to an automaton without ε-transition. This latter transformation may however introduce a quadratic blow-up of the number of transitions. The following proposition states that it can be directly checked, without removing ε-transitions, whether the language accepted by an automaton has an exponential growth (the result is more or less part of folklore but a proof is given in [15]).

Proposition 2. *It can be checked in linear time whether the language accepted by an automaton with ε-transitions has an exponential growth.*

The *degree* $\deg(L)$ of a set L with a polynomial growth is the least integer k such that $|L \cap A^{\leq n}| = O(n^k)$. The degree of the set $P_k = (0^*1)^k0^*$ is, for instance, $\deg(P_k) = k + 1$. Note that a rational set L is finite whenever $\deg(L) = 0$. If L is exponential, we set $\deg(L) = \infty$. The following theorem characterizes the existence of a rational bijection between two rational sets.

Theorem 3 ([7]). *There exists a rational bijection between two rational sets L and L' if and only if either they are both finite, (that is $\deg(L) = \deg(L') = 0$) and they have the same cardinality or they are both infinite and $\deg(L) = \deg(L')$.*

When the languages L and L' are infinite, the relation $\deg(L) = \deg(L')$ should be understood as either their growths are both exponential, that is $\deg(L) = \deg(L') = \infty$, or their growth are both polynomial with the same degree $\deg(L) = \deg(L') < \infty$.

3   Channels

A channel is a way to achieve reliable communication between two processes, an encoder and a decoder, via a noisy medium modeled by a non deterministic transducer with labels in $A^* \times B^*$. The encoder E reads binary input and produces an output in A^*, while the decoder D reads words in B^* and produces a binary word. The pair (E, D) is a channel if the binary message is correctly transmitted:

Definition 1. *Let $R \subseteq A^* \times B^*$ be a rational relation. A* rational channel *(or* channel *for short) for R is a pair (E, D) of rational relations in $\mathbb{B}^* \times A^*$ and $B^* \times \mathbb{B}^*$ respectively such that $ERD = \mathrm{Id}_{\mathbb{B}^*}$.*

As a consequence of this definition, a rational relation R has a channel if and only if R^{-1} has a channel.

Examples. We set here $A = B = \mathbb{B}$.

The prefix relation $\mathrm{Pref} = \{(u, v) \mid u \preccurlyeq v\}$ has a rational channel with $E = \{(0, 00), (1, 11)\}^*(\varepsilon, 01)$ and $D = \{(00, 0), (11, 1)\}^*(01, \varepsilon)$, as illustrated in Figure 1.

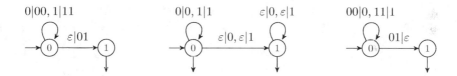

Fig. 1. From left to right: E, Pref and D

The relation $\mathrm{Diff}_1 = \{(u, v) \mid u$ and v differ by at most 1 bit$\}$ has a rational channel where 0 and 1 are encoded respectively by 000 and 111 in E. A first decoder D consists in a majority choice (this would be a subcase of channels with substitution in [2]), associating 0 with 000, 001, 010 and 100, and similarly for 1. Another decoder can also be simply the inverse of E, hence a "sub-decoder" of D, ignoring the substitution.

Note that in the definition above, E and D must be rational relations. However, there exist some relations without rational channels but with a channel satisfying the relation $ERD = \mathrm{Id}_{\mathbb{B}^*}$. Let f be a bijection from $\mathbb{B}^*$ onto $\mathbb{N}$ and let R be the relation defined by $R = \{(u, v) \mid |u| = |v|\}$. Then for $E = \{(u, 0^{f(u)}) \mid u \in \mathbb{B}^*\}$ and $D = \{(0^{f(u)}, u) \mid u \in \mathbb{B}^*\}$, the inverse of E, (E, D) satisfies $ERD = \mathrm{Id}_{\mathbb{B}^*}$ but the characterization proved in Section 5 shows that R has no rational channel.

In the sequel, we only consider rational channels. The main result of this paper is the following:

Theorem 4. *Let $R = h_1 + \cdots + h_n$ be a bounded relation where each h_i is a rational function. The following statements are equivalent:*

1. *R has a channel,*
2. *At least one function h_i has a channel,*
3. *$\mathrm{rg}(R)$ has an exponential growth.*

Using the equivalence between 1 and 3 and Proposition 2, we obtain the decidability of the channel synthesis problem:

Corollary 1. *The channel existence problem for bounded relations, given as a union of functions $h_1 + \cdots + h_n$, is decidable in linear time. When it exists, the channel can be effectively computed.*

4 Patterns

It can be observed in the examples above that, when the encoder/decoder pair (E, D) exists, E can be chosen as a bijection from $\mathbb{B}^*$ onto some language L of the form $u(v + \bar{v})^*w$, with D the inverse of E. In order to generalize this observation, and in relation with Proposition 1, we now introduce the notion of pattern:

Definition 2. *A pattern is a 4-tuple $s = (u, v, \bar{v}, w)$ of words such that $|v| = |\bar{v}|$ and $v \neq \bar{v}$. The language associated with s is $L_s = u(v + \bar{v})^*w$. A sub-pattern of s is a pattern of the form $(ux, y, \bar{y}, zw)$ where $x, y, \bar{y}, z \in (v + \bar{v})^*$. Two patterns s and s' of the form $s = (u, xv, x\bar{v}, xw)$ and $s' = (ux, vx, \bar{v}x, w)$ (or $s' = (ux, \bar{v}x, vx, w)$) are called conjugated.*

Note that in a pattern $s = (u, v, \bar{v}, w)$, the set $\{v, \bar{v}\}$ is a code hence, the notion of pattern can be seen as the basic element for the canonical channels and encoding states or nodes defined in [5,6,2].

If s' is a sub-pattern of s, then the inclusion $L_{s'} \subseteq L_s$ holds. Moreover, if the two patterns s and s' are conjugated, then the languages L_s and $L_{s'}$ are equal. For a pattern $s = (u, v, \bar{v}, w)$, we denote by μ_s the morphism from $\mathbb{B}^*$ to $(v + \bar{v})^*$ which maps 0 to v and 1 to $\bar{v}$. If $t = (x, y, \bar{y}, z)$ is a pattern over the alphabet $\mathbb{B}$, then $(u\mu_s(x), \mu_s(y), \mu_s(\bar{y}), \mu_s(z)w)$ is a sub-pattern of s which we denote by $s \diamond t$. Note that this composition of patterns is associative. Indeed, if $t = (x, y, \bar{y}, z)$ and $t' = (x', y', \bar{y}', z')$ are two patterns over the alphabet $\mathbb{B}$, then $s \diamond (t \diamond t') = (s \diamond t) \diamond t'$.

The proofs of the next lemmas are omitted and can be found in [15]. These proofs are based on a pumping argument.

Lemma 1. *Let s be a pattern and L a rational set of words. There exists a sub-pattern s' of s such that either $L_{s'} \subseteq L$ or $L_{s'} \cap L = \emptyset$.*

Lemma 2. *Let s be a pattern and let L be a rational set accepted by an automaton $\mathcal{A}$. If $L_s \subseteq L$, there exist a sub-pattern $s' = (u', v', \bar{v}', w')$ of s and paths $i \xrightarrow{u'} q \xrightarrow{v'} q \xrightarrow{w'} f$ and $q \xrightarrow{\bar{v}'} q$ in $\mathcal{A}$ where i is an initial state and f is a final state.*

The following proposition shows that if two patterns are not conjugated, one of them can be replaced by one of its sub-patterns to make the associated languages disjoint. In particular, two patterns are conjugated if and only if their associated languages are equal.

Proposition 3. *If the patterns s and s' are not conjugated, then either there exists a sub-pattern s'' of s such that $L_{s''} \cap L_{s'} = \emptyset$ or there exists a sub-pattern s'' of s' such that $L_{s''} \cap L_s = \emptyset$.*

Proof. Let s and s' be the two patterns $(u, v, \bar{v}, w)$ and $(u', v', \bar{v}', w')$. Let k, k', m and m' be the integers defined by $k = |uw|$, $k' = |u'w'|$, $m = |v| = |\bar{v}|$ and $m' = |v'| = |\bar{v}'|$. Let M_s and $M_{s'}$ be the sets $\{|x| \mid x \in L_s\}$ and $\{|x| \mid x \in L_{s'}\}$ of lengths of words in L_s and $L_{s'}$. These two sets M_s and $M_{s'}$ are respectively contained in the sets $k + m\mathbb{N}$ and $k' + m'\mathbb{N}$.

We first suppose that $m \neq m'$ and by symmetry we can assume that $m < m'$. We then consider two sub-cases depending on whether $k \equiv k' \mod m'$ or not. We first suppose that $k \not\equiv k' \mod m'$. The two sets $k + mm'\mathbb{N}$ and $k' + m'\mathbb{N}$ are then disjoint. It follows that the sub-pattern $s'' = (u, v^{m'}, \bar{v}^{m'}, w)$ of s satisfies $L_{s''} \cap L_{s'} = \emptyset$. We now suppose $k \equiv k' \mod m'$. Since $m < m'$, $k + m \not\equiv k' \mod m'$ holds and the two sets $k + m + mm'\mathbb{N}$ and $k' + m'\mathbb{N}$ are then disjoint. It follows that the sub-pattern $s'' = (uv, v^{m'}, \bar{v}^{m'}, w)$ of s satisfies $L_{s''} \cap L_{s'} = \emptyset$.

From now on, we suppose that $m = m'$. If $k \not\equiv k' \mod m$, the two sets M_s and $M_{s'}$ are already disjoint and we can set $s'' = s$. From now on, we also suppose that $k \equiv k' \mod m$.

If uv and u' are not prefix compatible, the sub-pattern $s'' = (uv, v, \bar{v}, w)$ satisfies $L_{s''} \cap L_{s'} = \emptyset$. A similar solution can be found if $u\bar{v}$ and u' are not prefix compatible or if w and $v'w'$ are not suffix compatible. We can now suppose that uv and u' (resp. $u\bar{v}$ and u', u and $u'v'$, u and $u'\bar{v}'$) are prefix compatible and that vw and w' (resp. $\bar{v}w$ and w', w and $v'w'$, w and $\bar{v}'w'$) are suffix compatible.

Since u and $u'v'$ are compatible and u and $u'\bar{v}'$ are also compatible, the word u is a prefix of $u'z$ where z is the longest common prefix of v' and $\bar{v}'$ and thus $|u| \leq |u'| + |z|$. By a similar argument, the word w satisfies $|w| \leq |w'| + |z'|$ where z' is the longest common suffix of v' and $\bar{v}'$. Combining these two relations gives $k = |uw| \leq k' + |zz'| < k' + m$ where the relation $|zz'| < m$ follows from $v' \neq \bar{v}'$. By symmetry the relation $k' \leq k + m$ also holds and this implies $k - k'$ since $k \equiv k' \mod m$.

We now have $m = m'$ and $k = k'$. We can assume by symmetry that $|u| \leq |u'|$ and thus $|w'| \leq |w|$. Since u and u' are prefix compatible, the word u is a prefix of u'. There exists a word x such that $u' = ux$. By symmetry, the word w' is a suffix of w and there exists a word x' such that $w = x'w'$. Since u' and uv are prefix compatible and u' and $u\bar{v}$ are also prefix compatible, the word x is a prefix of v and $\bar{v}$. There exist two words z and $\bar{z}$ such that $v = xz$ and $\bar{v} = x\bar{z}$. By symmetry, the word x' is a suffix of v' and $\bar{v}'$ and there exist two words z' and $\bar{z}'$ such that $v' = z'x'$ and $\bar{v}' = \bar{z}'x'$. Note that $|x| = |x'| = |u'| - |u| = |w| - |w'|$ and that $|z| = |\bar{z}| = |z'| = |\bar{z}'| = m - |x|$.

We first suppose that $x \neq x'$. Let s'' be the sub-pattern $(u'v'^2, v', \bar{v}', w')$ of s'. Any word in $L_{s''}$ starts with $u'v'^2 = uxz'x'z'x'$ whereas any word of L_s is either shorter or starts with a prefix in $u(v + \bar{v})^2 = ux(z + \bar{z})x(z + \bar{z})$. This shows that $L_{s''} \cap L_s = \emptyset$. We now suppose that $x = x'$. If the two sets $\{z, \bar{z}\}$ and $\{z', \bar{z}'\}$ are equal, the two patterns s and s' are conjugated and this a contradiction with the hypothesis. If these two sets are different, we can assume by symmetry that $z \notin \{z', \bar{z}'\}$. The sub-pattern $(uv, v, \bar{v}, w)$ of s satisfies then $L_{s''} \cap L_{s'} = \emptyset$.

5   Channel Characterizations

Recall that in both examples of section 3, the encoder and decoder can be chosen as rational bijections. The following characterization generalizes this observation.

Proposition 4. *There is a channel for a relation R if and only there exist two rational sets L_0 and L_1 with exponential growth such that $R \cap (L_0 \times L_1)$ is a bijection between L_0 and L_1.*

Note that it is assumed, in the previous proposition, that both L_0 and L_1 have an exponential growth. It is actually sufficient to assume that only one of them has. Theorem 3 and the fact that $R \cap (L_0 \times L_1)$ is a bijection between L_0 and L_1 ensure that the other one also has an exponential growth.

Proof. We first prove that the condition is sufficient. Suppose that $R \cap (L_0 \times L_1)$ is a bijection between L_0 and L_1. Since L_0 has an exponential growth, there exists, by Theorem 3, a rational bijection E between $\{0,1\}^*$ and L_0. The relation ER is thus a bijection between $\{0,1\}^*$ and L_1. Set $D = (ER)^{-1}$. It is then clear than $ERD = \mathrm{Id}_{\mathbb{B}^*}$.

Suppose now that there are two rational relations E and D such that $ERD = \mathrm{Id}_{\mathbb{B}^*}$. We first claim that there exists a function $E' \subseteq E$ from $\{0,1\}^*$ to A^* and another function $D' \subseteq D$ from B^* to $\{0,1\}^*$ such that $E'RD' = \mathrm{Id}_{\mathbb{B}^*}$. Let K_1 be the rational set $\mathrm{rg}(ER) = \{v \mid \exists u \in \{0,1\}^* \ (u,v) \in ER\}$ and D' the restriction $D' = D \cap (K_1 \times \{0,1\}^*)$. It is clear that $ERD' = \mathrm{Id}_{\mathbb{B}^*}$ and that D' must be functional. Let K_0 be the set $\mathrm{dom}(RD') = \{u \mid \exists v \in \{0,1\}^* \ (u,v) \in RD'\}$ and let E'' be the restriction $E'' = E \cap (\{0,1\}^* \times K_1)$. The relation E'' might not be functional but there exists by Theorem 1 a rational function $E' \subseteq E''$. It is also clear that $E'RD' = \mathrm{Id}_{\mathbb{B}^*}$.

We now suppose that E and D are two functions. Applying the reasoning to E^{-1} and D^{-1} there are two relations $E' \subseteq E$ and $D' \subseteq D$ such that E'^{-1} and D'^{-1} are functions and $E'RD' = \mathrm{Id}_{\mathbb{B}^*}$. Let L_0 and L_1 be the sets $L_0 = \mathrm{dom}(RD') = \{u \mid \exists v \in \{0,1\}^* \ (u,v) \in RD'\}$ and $L_1 = \mathrm{rg}(E'R) = \{v \mid \exists u \in \{0,1\}^* \ (u,v) \in E'R\}$. The relation E' is then a bijection between $\{0,1\}^*$ and L_0 and the relation D' is a bijection between L_1 and $\{0,1\}^*$. It follows that $R \cap (L_0 \times L_1)$ must be a bijection between L_0 and L_1.

For two patterns $s = (u, v, \bar{v}, w)$ and $s' = (u', v', \bar{v}', w')$, we denote by $h_{s,s'}$ the function whose graph is the rational relation $(u, u')((v, v') + (\bar{v}, \bar{v}'))^*(w, w')$. Note that this function is a bijection from L_s to $L_{s'}$ and that the inverse function $h_{s,s'}^{-1}$ is actually the function $h_{s',s}$. Let s_0 be the pattern $(\varepsilon, 0, 1, \varepsilon)$. The set L_{s_0} is then the set $\mathbb{B}^*$ and $h_{s_0,s}$ is a bijection from $\mathbb{B}^*$ to L_s. Note finally that if t is a pattern over the alphabet $\mathbb{B}$ the restriction of the function $h_{s,s'}$ to the domain $L_{s \diamond t}$ is the function $h_{s \diamond t, s' \diamond t}$ from $L_{s \diamond t}$ to $L_{s' \diamond t}$.

Lemma 3. *Let h be a rational function such that $\mathrm{rg}(h)$ has an exponential growth. Then there exist two patterns s and s' such that $h_{s,s'} \subseteq h$.*

Proof. Let $\mathcal{T}$ be a transducer realizing the function h. Let $\mathcal{A}$ be the automaton obtained by ignoring the input label of each transition of $\mathcal{T}$ and taking the output label as the label. This automaton accepts the set $\mathrm{rg}(h)$. By Proposition 1 there exists a pattern $s' = (u', v', \bar{v}', w')$ and paths $i \xrightarrow{u'} q \xrightarrow{v'} q \xrightarrow{w'} f$ and $q \xrightarrow{\bar{v}'} q$ in $\mathcal{A}$ where i is an initial state and f is a final state. Since these paths come from paths in $\mathcal{T}$, there are words u, v_0, $\bar{v}_0$ and w and paths $i \xrightarrow{u|u'} q \xrightarrow{v_0|v'} q \xrightarrow{w|w'} f$ and $q \xrightarrow{\bar{v}_0|\bar{v}'} q$ in $\mathcal{T}$. Note however that the words v_0 and $\bar{v}_0$ may not have the same length. Let v and $\bar{v}$ be the words $v_0\bar{v}_0$ and $\bar{v}_0v_0$. These words have the same length but they are different. Otherwise h maps the single word $uv_0\bar{v}_0$ to the two differents words $u'v'\bar{v}'w'$ and $u'\bar{v}'v'w'$. The function $h_{s,s''}$ where $s = (u, v, \bar{v}, w)$ and $s'' = (u', v'\bar{v}', \bar{v}'v', w')$ satisfies then $h_{s,s''} \subseteq h$. $\qed$

The proof of Theorem 4 proceeds by induction on the number of functions, so we first establish the result for a single function.

Proposition 5. *If h is a function, then h has a channel if and only if $\mathrm{rg}(h)$ has an exponential growth.*

Proof. By Proposition 4, the condition is necessary. Indeed, if h has a channel, there exist two languages L_0 and L_1 with an exponential growth such that $h \cap (L_0 \times L_1)$ is a bijection from L_0 to L_1. The set L_1 is thus contained in $\mathrm{rg}(h)$ and $\mathrm{rg}(h)$ has a exponential growth.

By Lemma 3, there are two patterns s and s' such that $h_{s,s'} \subseteq h$. Since h is a function, $h \cap (L_s \times L_{s'}) = h_{s,s'}$. By Proposition 4, the function h has a channel. $\qed$

The next lemma, which is one of the key ingredients for the main result, was present in [8] (with an unsatisfactory proof).

Lemma 4. *Let R be a rational relation and let h be a rational function. If R has a channel, then $R + h$ also has a channel.*

Proof. If R has a channel, by Proposition 4, there exist two languages L_0 and L_1 with an exponential growth such that $g = R \cap (L_0 \times L_1)$ is a bijection from L_0 to L_1. Applying Lemma 3 to g, there exist two patterns $s = (u, v, \bar{v}, w)$ and $s' = (u', v', \bar{v}', w')$ such that $h_{s,s'} \subseteq g$. Let L be the domain of the function h. By Lemma 1, s can be replaced by one of its sub-patterns such that either $L_s \cap L = \emptyset$ or $L_s \subseteq L$. Like in the previous proof, $g \cap (L_s \times L_{s'}) = h_{s,s'}$. If $L_s \cap L = \emptyset$ then $(R + h) \cap (L_s \times L_{s'}) = h_{s,s'}$ and the function $h_{s,s'}$ provides a channel for $R + h$.

We now suppose that s satisfies $L_s \subseteq L$. Let $\mathcal{T}$ be a transducer realizing the function h and let $\mathcal{A}$ be the automaton obtained by ignoring the output label of each transition of $\mathcal{T}$ and taking the input label as the label. This automaton accepts the set $L = \mathrm{dom}(h)$. By Lemma 2, s can be replaced by one its sub-patterns such that there are paths $i \xrightarrow{u} q \xrightarrow{v} q \xrightarrow{w} f$ and $q \xrightarrow{\bar{v}} q$ in $\mathcal{A}$ where i is an initial state and f is a final state. Since these paths come from paths in $\mathcal{T}$, we obtain a pattern $s'' = (u'', v'', \bar{v}'', w'')$ such that $i \xrightarrow{u|u''} q \xrightarrow{v|v''} q \xrightarrow{w|w''} f$ and $q \xrightarrow{\bar{v}|\bar{v}''} q$ in $\mathcal{T}$.

We distinguish two cases depending on whether the two patterns s' and s'' are conjugated or not. If they are not conjugated, it can be assumed, without loss of generality, that $L_{s'} \cap L_{s''} = \emptyset$ by Proposition 3. In this latter case, the function $h_{s,s'}$ provides a channel for $R + h$. If they are conjugated, the two functions $h_{s,s'}$ and $h_{s,s''}$ are equal and therefore again provide a channel for $R + h$.

We are now ready to prove Theorem 4.

Proof (Proof of Theorem 4). Let $R = h_1 + \ldots + h_n$ be a bounded relation, where $h_1, \ldots, h_n$ are rational functions. If R has a channel, then by Proposition 4, $\mathrm{rg}(R)$ has an exponential growth, hence (1) implies (3). If $\mathrm{rg}(R)$ has an exponential growth, then one of the images $\mathrm{rg}(h_i)$ has an an exponential growth. By Proposition 5, the function h_i has a channel, hence (3) implies (2). Finally, let us assume that h_1 has a channel. Using the previous lemma, it can be proved by induction on i, that each relation $h_1 + \ldots + h_i$ has a channel, hence R has a channel. Therefore (2) implies (1) which concludes the proof.

Example of Channel Synthesis
We finally illustrate the channel construction of Corollary 1 on an example. With $A = B = \mathbb{B}$, Figure 2 depicts a transducer realizing a relation R which is the union $R = h_1 + h_2$ of two functions h_1 and h_2 from $\mathbb{B}^*$ to $\mathbb{B}^*$. The function h_1 (on the left side) is the morphism which maps the symbols 0 and 1 to 01 and 1 respectively. The function h_2 (on the right side) maps each word $a_1 a_2 \ldots a_n$ to $a_1 a_3 \ldots$ keeping only symbols at odd positions.

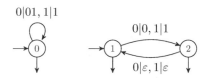

Fig. 2. A union of 2 functions

The function h_1 has obviously a channel since it is one-to-one from $\mathbb{B}^*$ onto $(01 + 1)^*$. We show how the proof of Lemma 4 extracts a channel for the relation R.

1. Since the range $\mathrm{rg}(h_1) = (01 + 1)^*$ has an exponential growth, we obtain the pattern $s_1 = (\varepsilon, 011, 101, \varepsilon)$ such that $L_{s_1} \subseteq (01 + 1)^*$. The corresponding pattern of the inputs for h_1 is $s = (\varepsilon, 01, 10, \varepsilon)$, hence the function h_{s,s_1} provides a channel for h_1.
2. Examining the part of the transducer realizing h_2, we observe that the input pattern s induces the corresponding output pattern $s_2 = (\varepsilon, 0, 1, \varepsilon)$, but the two patterns s_1 and s_2 satisfy $L_{s_1} \cap L_{s_2} \neq \emptyset$, so h_{s,s_1} does not correspond to a channel for R itself.
3. Since s_1 and s_2 and are not conjugated, it is possible to find a sub-pattern $s_3 = (\varepsilon, 00, 11, \varepsilon)$ of s_2 such that $L_{s_1} \cap L_{s_3} = \emptyset$. The corresponding input pattern for h_2 is then the pattern $s' = (\varepsilon, 0101, 1010, \varepsilon)$ obtained from s.

4. The channel in $R = h_1 + h_2$ is then built from the function h_{s',s_3}. It consists of the pair (E, D) where E is the morphism which maps 0 and 1 to 0101 and 1010 respectively, and D maps 00 and 11 to 0 and 1 respectively.

Besides Corollary 1, we also obtain:

Corollary 2. *Let $R = h_1 + \cdots + h_n$ be a relation where each h_i is either a function or the inverse of a function. The following statements are equivalent:*

1. *R has a channel,*
2. *At least one relation h_i has a channel.*

This last result has to be compared with Theorem 4. When R is the finite union of functions and inverses of functions, it may have a domain and a range with exponential growth without having a channel as shown by the following example. Let A and B the alphabet $\mathbb{B}$. Let R be given by $R = h_1 + h_2$ where h_1 is the function from $\mathbb{B}^*$ to $\mathbb{B}^*$ which maps each word to the empty word and h_2 is the inverse of h_1. The domain and the range of R are both equal to $\mathbb{B}^*$ but R has no channel. Hence the decidability result cannot apply in this case.

Proof. Let us recall that a relation R has a channel if and only if R^{-1} has a channel. If one relation h_i has a channel, then R has a channel by Lemma 4.

Suppose now that R has a channel. By Proposition 4, there exist two languages L_0 and L_1 with an exponential growth such that $R' = R \cap (L_0 \times L_1)$ is a bijection between L_0 and L_1. For each $1 \le i \le n$, let us denonte by h_i' the restriction $h_i \cap (L_0 \times L_1)$. Since $h_i' \subseteq R'$, and R' is a bijection, each relation h_i' is also a bijection. We may then suppose that each h_i' is a function. It follows then from Thereom 4 that R' has a channel and that R has also a channel.

6   Conclusion

We proved a new characterization of bounded relations with channels, linked to the growth of their image. We conjecture that this characterization could be extended to relations R for which there exists a polynomial P such that for each word u, $|R(u)| \le P(|u|)$. We also plan to investigate more powerful channels described by (subclasses of) two-way transducers instead of simple transducers.

References

1. Bérard, B., Benattar, G., Lime, D., Mullins, J., Roux, O.H., Sassolas, M.: Channel synthesis for finite transducers. In: Dömösi, P., I.S. (eds.) Proceedings of the 13th International Conference on Automata and Formal Languages (AFL 2011), pp. 79–92 (August 2011)
2. Benattar, G., Bérard, B., Lime, D., Mullins, J., Roux, O.H., Sassolas, M.: Channel Synthesis for Finite Transducers. International Journal of Foundations of Computer Science 23(6), 1241–1260 (2012)
3. Schewe, S., Finkbeiner, B.: Synthesis of asynchronous systems. In: Puebla, G. (ed.) LOPSTR 2006. LNCS, vol. 4407, pp. 127–142. Springer, Heidelberg (2007)

4. Millen, J.K.: 20 years of covert channel modeling and analysis. In: Proc. of the 1999 IEEE Symposium on Security and Privacy, pp. 113–114 (May 1999)
5. Hélouet, L., Zeitoun, M., Degorre, A.: Scenarios and Covert channels: another game. In: de Alfaro, L. (ed.) Proc. of Games in Design and Verification (GDV 2004). ENTCS, vol. 119, pp. 93–116. Elsevier (2005)
6. Hélouët, L., Roumy, A.: Covert channel detection using information theory. In: Chatzikokolakis, K., Cortier, V. (eds.) Proc. of the 8th Int. Workshop on Security Issues in Concurrency (SecCo 2010) (August 2010)
7. Maurer, A., Nivat, M.: Rational bijection of rational sets. Acta Inf. 13, 365–378 (1980)
8. Benattar, G.: Synthèse de systèmes informatiques temporisés non interférents. PhD thesis, Université de Nantes (2011)
9. Lothaire, M.: Combinatorics on words. Encyclopedia of Mathematics, vol. 17. Addison-Wesley, Reading (1983)
10. Sakarovitch, J.: Elements of automata theory. Cambridge University Press (2009)
11. Elgot, C.C., Mezei, J.E.: On relations defined by generalized finite automata. IBM Journal Res. Develop. 9, 47–68 (1965)
12. Weber, A.: Decomposing a k-valued transducer into k unambiguous ones. RAIRO Theoretical Informatics and Applications 30(5), 379–413 (1996)
13. Sakarovitch, J., de Souza, R.: On the decomposition of k-valued rational relations. In: Albers, S., Weil, P. (eds.) Proceedings of the 25th Symposium on Theoretical Aspects of Computer Science (STACS 2008), pp. 621–632 (2008)
14. Harrison, M.A.: Introduction to formal language theory. Addison-Wesley (1978)
15. Bérard, B., Carton, O.: Channel synthesis revisited. Technical report, LIP6 (2013), `http://pagesperso-systeme.lip6.fr/Beatrice.Berard/PDF/rr-channels-BBOC.pdf`

Characterisation of the State Spaces
of Live and Bounded Marked Graph Petri Nets

Eike Best[1,*] and Raymond Devillers[2]

[1] Department of Computing Science, Carl von Ossietzky Universität Oldenburg
26111 Oldenburg, Germany
eike.best@informatik.uni-oldenburg.de
[2] Département d'Informatique, Université Libre de Bruxelles
Boulevard du Triomphe - C.P. 212, 1050 Bruxelles, Belgium
rdevil@ulb.ac.be

Abstract. The structure of the reachability graph of a live and bounded marked graph Petri net is fully characterised. A dedicated synthesis procedure is presented which allows the net and its bounds to be computed from its reachability graph.

Keywords: Petri nets, region theory, system synthesis, transition systems.

1 Introduction

Deducing behavioural properties from structural properties is one of the major objectives of the analysis of systems. In this paper, a similar question about system synthesis is addressed: given regular behaviour, can one find a generating system that is well-structured? An answer will be given for marked graph Petri nets [7,8], leading to a full characterisation of their state spaces.

Petri net region theory [1,2] investigates general conditions under which an edge-labelled directed graph (or a labelled transition system) is the reachability graph of a Petri net. However, not much is implied about the structure of the net, if it exists. This paper shows that if a labelled transition system exhibits a characteristically uniform cyclic structure, then it can be generated by a marked graph, and the marking bounds may easily be deduced from some paths. Such cyclic behaviour arises, for instance, in the context of persistent Petri nets [3,11], or in the context of signal transition graphs [10].

Labelled transition systems and Petri nets are defined in sections 2 and 3, respectively. The cyclic (and other) behavioural properties studied in this paper are introduced at the end of section 3. The synthesis procedure and its application to marked graphs are described in sections 4 and 5, respectively. Section 6 concludes and describes ideas for future work. Proofs of some auxiliary results have been moved to Appendix A.

---

* The first author gratefully acknowledges the support of Université d'Évry-Val-d'Essonne and Université Paris-Est Créteil Val-de-Marne.

A.-H. Dediu et al. (Eds.): LATA 2014, LNCS 8370, pp. 161–172, 2014.

2   Labelled Transition Systems

Definition 1. LTS, REVERSE LTS, REACHABILITY, PARIKH VECTORS, CYCLES

A labelled transition system with initial state, abbreviated lts, is a quadruple $(S, \rightarrow, T, s_0)$ where S is a set of *states*, T is a set of *labels* with $S \cap T = \emptyset$, $\rightarrow \subseteq (S \times T \times S)$ is the *transition relation*, and $s_0 \in S$ is an *initial state*. The *reverse lts* is $(S, \leftarrow, T, s_0)$ with $(s, t, s') \in \leftarrow$ iff $(s', t, s) \in \rightarrow$. A label t is *enabled* in a state s, denoted by $s[t\rangle$, if there is some state s' such that $(s, t, s') \in \rightarrow$. For $s \in S$, let $s^\bullet = \{t \in T \mid s[t\rangle\}$. For $t \in T$, $s[t\rangle s'$ iff $(s, t, s') \in \rightarrow$, meaning that s' is *reachable* from s through the execution of t. The definitions of enabledness and of the reachability relation are extended to sequences $\sigma \in T^*$:

$s[\varepsilon\rangle$ and $s[\varepsilon\rangle s$ are always true;
$s[\sigma t\rangle$ $(s[\sigma t\rangle s')$ iff there is some s'' with $s[\sigma\rangle s''$ and $s''[t\rangle$ $(s''[t\rangle s'$, respectively).

A state s' is reachable from state s if there exists a label sequence σ such that $s[\sigma\rangle s'$. By $[s\rangle$, we denote the set of states reachable from s. For a finite sequence $\sigma \in T^*$ of labels, the *Parikh vector* $\Psi(\sigma)$ is a T-vector (i.e., a vector of natural numbers with index set T), where $\Psi(\sigma)(t)$ denotes the number of occurrences of t in σ. $s[\sigma\rangle s'$ is called a *cycle*, or more precisely a *cycle at state s*, if $s = s'$. The cycle is *nontrivial* if $\sigma \neq \varepsilon$. An lts is called *acyclic* if it has no nontrivial cycles. A nontrivial cycle $s[\sigma\rangle s$ around a reachable state $s \in [s_0\rangle$ is called *small* if there is no nontrivial cycle $s'[\sigma'\rangle s'$ with $s' \in [s_0\rangle$ and $\Psi(\sigma') \lneqq \Psi(\sigma)$.   □

Definition 2. BASIC PROPERTIES OF AN LTS

A labelled transition system $(S, \rightarrow, T, s_0)$ is called

- *totally reachable* if $[s_0\rangle = S$ (i.e., every state is reachable from s_0);
- *finite* if S and T (hence also $\rightarrow$) are finite sets;
- *(super-)deterministic*, if for any states $s, s', s'' \in [s_0\rangle$ and sequences $\sigma, \tau \in T^*$ with $\Psi(\sigma) = \Psi(\tau)$: $(s[\sigma\rangle s' \wedge s[\tau\rangle s'') \Rightarrow s' = s''$ and $(s'[\sigma\rangle s \wedge s''[\tau\rangle s) \Rightarrow s' = s''$ (i.e., from any one state, Parikh-equivalent sequences may not lead to two different successor states, nor come from two different predecessor states);
- *reversible* if $\forall s \in [s_0\rangle$: $s_0 \in [s\rangle$ (i.e., s_0 always remains reachable);
- *persistent* if for all reachable states s and labels t, u, if $s[t\rangle$ and $s[u\rangle$ with $t \neq u$, then there is some state $r \in S$ such that both $s[tu\rangle r$ and $s[ut\rangle r$ (i.e., once two different labels are both enabled, neither can disable the other, and executing both, in any order, leads to the same state);
- *backward persistent* if for all reachable states s, s', s'', and labels t, u, if $s'[t\rangle s$ and $s''[u\rangle s$ and $t \neq u$, then there is some reachable state $r \in S$ such that both $r[u\rangle s'$ and $r[t\rangle s''$ (i.e., persistency in backward direction).   □

If the lts is totally reachable, reversibility is the same as strong connectedness in the graph-theoretical sense. If the lts is strongly connected, backward persistency is the same as persistency in the reverse lts. The lts depicted in Figure 1 satisfies all properties given in Definition 2.

Fig. 1. A transition system (l.h.s.) and a Petri net solving it (r.h.s.)

3   Petri Nets

Definition 3. PETRI NETS, MARKINGS, REACHABILITY GRAPHS

A (finite, initially marked, place-transition, arc-weighted) Petri net is a tuple $N = (P, T, F, M_0)$ such that P is a finite set of *places*, T is a finite set of *transitions*, with $P \cap T = \emptyset$, F is a *flow* function $F \colon ((P \times T) \cup (T \times P)) \to \mathbb{N}$, M_0 is the *initial marking*, where a *marking* is a mapping $M \colon P \to \mathbb{N}$. A transition $t \in T$ is *enabled by* a marking M, denoted by $M[t\rangle$, if for all places $p \in P$, $M(p) \geq F(p, t)$. If t is enabled at M, then t can *occur* (or *fire*) in M, leading to the marking M' defined by $M'(p) = M(p) - F(p, t) + F(t, p)$ (noted $M[t\rangle M'$). The set of markings reachable from M is denoted $[M\rangle$. The *reachability graph of N* is the labelled transition system with the set of vertices $[M_0\rangle$ and set of edges $\{(M, t, M') \mid M, M' \in [M_0\rangle \land M[t\rangle M'\}$. If an lts TS is isomorphic to the reachability graph of a Petri net N, we will also say that N *solves TS*.   □

Definition 4. BASIC STRUCTURAL PROPERTIES OF PETRI NETS

For a place p of a Petri net $N = (P, T, F, M_0)$, let ${}^\bullet p = \{t \in T \mid F(t, p) > 0\}$ and $p^\bullet = \{t \in T \mid F(p, t) > 0\}$. N is called *connected* if it is weakly connected as a graph; *plain* if $cod(F) \subseteq \{0, 1\}$; *pure* or *side-condition free* if $p^\bullet \cap {}^\bullet p = \emptyset$ for all places $p \in P$; ON (*place-output-nonbranching*) if $|p^\bullet| \leq 1$ for all places $p \in P$; a *marked graph* if N is plain and $|p^\bullet| \leq 1$ and $|{}^\bullet p| \leq 1$ for all places $p \in P$.   □

Definition 5. BASIC BEHAVIOURAL PROPERTIES OF PETRI NETS

A Petri net $N = (P, T, F, M_0)$ is *weakly live* if $\forall t \in T \exists M \in [M_0\rangle \colon M[t\rangle$ (i.e., there are no unfireable transitions); *k-bounded* for some fixed $k \in \mathbb{N}$, if $\forall M \in [M_0\rangle \forall p \in P \colon M(p) \leq k$ (i.e., the number of tokens on any place never exceeds k); *bounded* if $\exists k \in \mathbb{N} \colon N$ is k-bounded; *persistent* (*backward persistent*, *reversible*) if its reachability graph is persistent (backward persistent, reversible, respectively); and *live* if $\forall t \in T \forall M \in [M_0\rangle \exists M' \in [M\rangle \colon M'[t\rangle$ (i.e., no transition can be made unfireable).   □

Proposition 6. PROPERTIES OF PETRI NET REACHABILITY GRAPHS

The reachability graph RG of a Petri net N is totally reachable and deterministic. N is bounded iff RG is finite.   □

This paper focusses on the basic finite situation, on lts generated by Petri nets, and on systems without superfluous transitions. Therefore, we shall assume that

All transition systems are finite, totally reachable, and deterministic.
All Petri nets are connected, weakly live, and bounded.

In the next definition, ρ mimicks the notion of a Petri net place in terms of an lts. $\mathbb{R}$ corresponds to the marking of this place at the various states; and $\mathbb{B}$ ($\mathbb{F}$) correspond to its outgoing (incoming, respectively) transitions.

Definition 7. REGIONS OF LTS

A triple $\rho = (\mathbb{R}, \mathbb{B}, \mathbb{F}) \in (S \to \mathbb{N}, T \to \mathbb{N}, T \to \mathbb{N})$ is a *region* of an lts $(S, \to, T, s_0)$ if for all $s[t\rangle s'$ with $s \in [s_0\rangle$, $\mathbb{R}(s) \geq \mathbb{B}(t)$ and $\mathbb{R}(s') = \mathbb{R}(s) - \mathbb{B}(t) + \mathbb{F}(t)$. □

An lts $(S, \to, T, s_0)$ satisfies SSP (state separation property) iff

$$\forall s, s' \in [s_0\rangle \colon s \neq s' \;\Rightarrow\; \exists \text{ region } \rho = (\mathbb{R}, \mathbb{B}, \mathbb{F}) \text{ with } \mathbb{R}(s) \neq \mathbb{R}(s')$$

and ESSP (event/state separation property) iff

$$\forall s \in [s_0\rangle \, \forall t \in T \colon (\neg s[t\rangle) \;\Rightarrow\; \exists \text{ region } \rho = (\mathbb{R}, \mathbb{B}, \mathbb{F}) \text{ with } \mathbb{R}(s) < \mathbb{B}(t).$$

Theorem 8. BASIC REGION THEOREM FOR PLACE/TRANSITION NETS [2]

A (finite, totally reachable, deterministic) lts is the reachability graph of a (possibly non-plain, or non-pure) Petri net iff it satisfies SSP and ESSP. □

Let $\Upsilon \colon T \to \mathbb{N} \backslash \{0\}$ be a fixed Parikh vector with no zero entries. The principal properties of any lts TS studied in this paper are the ones listed below.

b   : TS is finite, totally reachable, and deterministic.
rp   : TS is reversible and persistent.
PΥ   : The Parikh vector of any small cycle in TS equals Υ.
bp   : TS is backward persistent.

For example, the lts shown in Figure 1 satisfies all four requirements. Figure 2 violates **P1** (i.e.: **PΥ** with constant Parikh vector 1) but satisfies **P2** as well as all other properties – **b**, **rp**, and **bp**. The lts shown in Figure 3 satisfies all properties **b** to **P1**, but not **bp**. Two solutions are also depicted: a plain non-ON one in the middle of the figure, and a non-plain ON one on the right-hand side.

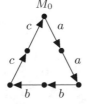
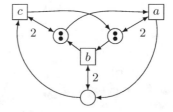

Fig. 2. An lts satisfying all properties but **P1**. The Petri net shown on the right-hand side solves it. However, there is no ON Petri net, much less a marked graph, solution.

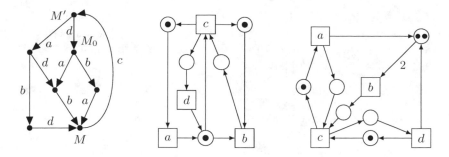

Fig. 3. An lts that cannot be solved by a marked graph, and two solutions

Theorem 9. PROPERTIES OF LIVE MARKED GRAPHS [7,8]

The reachability graph of a connected, live and bounded marked graph is finite and satisfies **b**, **rp**, **P**1, *and* **bp**.   □

Theorem 9 implies that the lts shown in Figure 3 cannot be solved by a marked graph. Consider state M: it has incoming arrows a and d which violate **bp**.

4   Solving an lts, Using rp, P1, and bp

Let $TS = (S, \rightarrow, T, s_0)$ satisfy properties **b** (basic), **rp** (reversible and persistent), **P**1 (constant Parikh vector 1 of small cycles), and **bp** (backward persistent). We present an algorithm that produces a Petri net with isomorphic reachability graph. We shall assume that TS is nontrivial, in the sense that $|S| \geq 2$ and $|T| \geq 2$. Otherwise TS can be solved trivially.

For $s, s' \in S$, let a path $s[\tau\rangle s'$ be called *short* if $|\tau| \leq |\tau'|$ for every path $s[\tau'\rangle s'$, where $|\tau|$ denotes the length of τ. Also, let the *distance* $\Delta_{s,s'}: T \rightarrow \mathbb{N}$ be defined as $\Delta_{s,s'} = \Psi(\tau)$, where $s[\tau\rangle s'$ is any short path. By Lemmata 22 and 24 in the appendix, $\Delta_{s,s'}$ is well-defined for any two states s, s'.

Fix a label $x \in T$. Let $TS\text{-}x$ be defined from TS by erasing every arrow labelled with x, as illustrated in Figure 4. The resulting lts has state set S and label set $T \setminus \{x\}$. By Lemma 21, the paths of $TS\text{-}x$ are precisely the short paths of TS not containing x.

Lemma 10. PROPERTIES OF $TS\text{-}x$

$TS\text{-}x$ *is acyclic, has a unique maximal state s_x, a unique minimal state r_x, and is weakly connected.*

Proof: Acyclicity arises from the fact that every nontrivial cycle must contain at least one x by property **P**1. The existence of s_x follows from Lemma 25. By Lemma 26, there is a short directed path not containing x from any state into s_x. Hence, connectedness (between s and s') results from going forward from s to s_x and then backward from s_x to s'. The existence of r_x also follows from Lemma 25, applied to the reverse lts (which is allowed because the assumed properties are, as a whole, preserved by reversal).   □

These properties depend heavily on $\mathbf{P}\Upsilon$ with $\Upsilon = 1$. For instance, if all a-arrows are erased in Figure 2, the resulting lts is not weakly connected.

Let $Seq(x)$ be the set of *sequentialising states w.r.t.* x in which, by definition, x is not enabled but in all of whose immediate successor states, x is enabled:

$$Seq(x) \quad = \quad \{s \in S \mid \neg s[x\rangle \wedge \forall a \in T\colon s[a\rangle \Rightarrow s[ax\rangle\}$$

The terminology is motivated in [6] for ON nets. E.g., in Figure 3, $M' \in Seq(b)$. The ON solution shown on the right-hand side contains a "sequentialising place" having a and d as input transitions and b as an output transition.

In general, the set S is partitioned into $X \cup\!\!\!\!\bullet (S\backslash X)$ where X is the set of states enabling x. $S \setminus X$ includes r_x and $Seq(x)$, as well as all states in between. The latter is implied by persistency. X includes all states between $Seq(x)$ (exclusively) and s_x (inclusively). In Figure 4, X is represented by slim nodes, while $S \setminus X$ is represented by fat nodes. It is an easy consequence of our basic assumptions that all sets are nonempty.

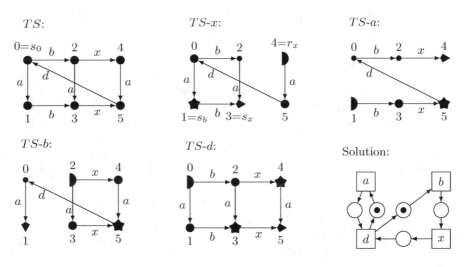

Fig. 4. A fully worked, simple example. *Legend:* r_x is represented by a semicircle; s_x is represented by a kite symbol; elements in $Seq(x)$ are represented as stars; the five places of the solution correspond to the five stars.

Let x be fixed as before and pick, in addition, a state s in $\max(S\backslash X) = Seq(x)$.

Lemma 11. PROPERTIES OF $\Delta_{r_x,s}$
   $\Delta_{r_x,s}$ *has exactly two entries that are zero; all other entries are positive.*

Proof: $\Delta_{r_x,s}(x) = 0$, by persistency and because s does not enable x.
   Assume that all other entries of $\Delta_{r_x,s}$ are positive; there is a path $r_x[\alpha\rangle s$ with $\Psi(\alpha) = \Delta_{r_x,s}$. By Lemma 20 and $\mathbf{P}1$, there is a cycle $r_x[\beta\rangle r_x$ with $\Psi(\beta) = 1$, hence β contains x. Thus, by Keller's theorem (cf. Appendix), $r_x[\alpha\rangle s[\beta \overset{\bullet}{-} \alpha\rangle$, so that $s[x\rangle$, contradicting $s \in S \setminus X$. Therefore, $\Delta_{r_x,s}$ has at least two entries 0.

Assume that $s[a\rangle q[x\rangle q'$. This is possible by $s \in Seq(x)$. By Lemma 20, there is a cycle $s[ax\rangle q'[\gamma\rangle s$ where every letter except a and x occurs in γ. Let $r_x[\delta\rangle q'$ be any short path (not containing x). Then $r_x[\delta\gamma\rangle s$ is a path from r_x to s not containing x, and therefore short, but containing all transitions in TS-x except a. Therefore, $\Delta_{r_x,s}$ has at most two entries 0.   □

This proof implies that (i): a label a with $s[a\rangle$ is uniquely determined by the choice of x and s, and (ii): $s = s_a$, the unique state enabling only a.

Next, we define a function $\mathbb{R}^{s,x}: S \to \mathbb{N}$, also depending on s and x. Let a be the unique label with $s = s_a$. For any state $q \in S$, define $\mathbb{R}^{s,x}(q) = \Delta_{r_x,q}(a)$. For example, let the initial state on the top left-hand corner of Figure 4 be $s_0 = 0$. Then with TS-x and $s = s_b = 1$, $\mathbb{R}^{s,x}(s_0) = 0$, because on any path from $r_x = 4$ to $s_0 = 0$, no b occurs.

A net will now be assembled from $TS = (S, \to, T, s_0)$ by the following algorithm.

for every label $x \in T$ **do for** every state $s \in Seq(x)$ **do**
    determine $a \in T$ for which $s = s_a$;
    define a place $p=p^{s,x}$ with $^\bullet p=\{a\}$, $F(a,p)=1$ and $p^\bullet=\{x\}$, $F(p,x)=1$; (1)
    compute $\mathbb{R}^{s,x}$ as above and put $M_0(p^{s,x}) - \mathbb{R}^{s,x}(s_0)$ tokens on $p^{s,x}$
end for end for

In the net so constructed, every place $p^{s,x}$ has exactly one input transition, viz. a, and exactly one output transition, viz. x, and the net is plain. So, it is a marked graph, and moreover, it is side-condition-free because $a \neq x$.

Lemma 12. $\mathbb{R}^{s,x}$ "DISABLES" x IN s AND "ENABLES" x IN ALL STATES IN X
$\mathbb{R}^{s,x}(s) - 0$, and $\mathbb{R}^{s,x}(q) \geq 1$ for every state $q \in X$.

Proof: $\mathbb{R}^{s,x}(s) = 0$ because a does not occur on any path from r_x to s.

Every $q \in X$ is above some $s' \in Seq(x)$, i.e. $s'[a'\alpha\rangle q$ for some a' and some α. As shown in the proof of Lemma 11, every label except a' and x occurs on a short path from r_x to s', so that $\mathbb{R}^{s,x}(q) \geq 1$ by the definition of $\mathbb{R}^{s,x}$, independently of whether $a = a'$ or $a \neq a'$.   □

Let a be determined from x and s, as before, and define

$$\mathbb{B}(t) = \begin{cases} 1 \text{ if } t = x \\ 0 \text{ if } t \neq x \end{cases} \text{ and } \mathbb{F}(t) = \begin{cases} 0 \text{ if } t \neq a \\ 1 \text{ if } t = a \end{cases} \quad (2)$$

Lemma 13. $(\mathbb{R}^{s,x}, \mathbb{B}, \mathbb{F})$ IS A REGION
The triple $\rho^{s,x} = (\mathbb{R}^{s,x}, \mathbb{B}, \mathbb{F})$, as constructed above, is a region in TS.

Proof: Suppose $s_1[t\rangle s_2$. $\mathbb{R}^{s,x}(s_1) \geq \mathbb{B}(t)$ follows from the second claim of Lemma 12 if $t = x$ and from $\mathbb{B}(t) = 0$ and the semipositiveness of $\mathbb{R}^{s,x}$ if $t \neq x$. $\mathbb{R}^{s,x}(s_2) = \mathbb{R}^{s,x}(s_1) + \mathbb{F}(t) - \mathbb{B}(t)$ follows from the first line of (2) if $t = x$, and from the second line of (2) if $t \neq x$.   □

Theorem 14. Isomorphism of TS and $RG(N, M_0)$

*Let a labelled transition system $TS = (S, \rightarrow, T, s_0)$ with properties **b**, **rp**, **P**1, and **bp** be given. Let N with initial marking M_0 be the Petri net constructed according to the above procedure. Then TS and the reachability graph $RG(N, M_0)$ of (N, M_0) are isomorphic.*

Proof: Lemma 12 implies that the set of regions constructed above satisfy ESSP, which ensures that TS and $RG(N, M_0)$ are language-equivalent. To see that SSP is also satisfied, assume that s_1 and s_2 in TS are mapped to the same marking M reachable in (N, M_0). By the strong connectedness of TS, there is a sequence $s_1[\sigma\rangle s_2$. Since $M[\sigma\rangle$ by language equivalence, and because s_2 is mapped to M, there is also a sequence $s_2[\sigma\rangle s_3$. Using the finiteness of TS, we get $s_i[\sigma^\ell\rangle s_i$ for some $i, \ell \geq 1$. Because this is a cycle, property **P**1 implies that every letter occurs equally often in σ^ℓ, and hence also equally often in σ. Thus σ is itself cyclic, entailing $s_1 = s_2$. The claim follows by Theorem 8. $\square$

Note that N has no isolated places. Hence it is connected, because otherwise, each connected component generates small cycles which do not satisfy **P**1.

5   Marked Graphs, and Place Bounds

Theorem 15. Live and bounded marked graph reachability graphs

*A labelled transition system satisfying **b** is isomorphic to the reachability graph of a connected live and bounded marked graph iff it satisfies the properties **rp**, **P**1 and **bp**.*

Proof: For ($\Rightarrow$), see Theorem 9. For ($\Leftarrow$), see Theorem 14. $\square$

Theorem 15 characterises the structure of the reachability graph of a connected, live and bounded marked graph. Let us now look more carefully at this bound.

Lemma 16. Exact bound

*Assume that $TS = (S, \rightarrow, T, s_0)$ satisfies **b**, **rp**, **P**1, and **bp**. The bound of the marked graph constructed by (1) is $\max\{\Delta_{s_a, s_x}(a) \mid x \in T, s_a \in \max(S \setminus X)\}$.*

Proof: We already saw that $M_r(p^{s,x}) = \Delta_{r_x, r}(a)$ for each $x \in T$, $s = s_a \in \max(S \setminus X)\}$ and $r \in S$, and $M_s(p^{s,x}) = 0$. Hence, the maximum marking for that place is $M_{s_x}(p^{s,x}) = \Delta_{r_x, s_x}(a)$ so that, if $s = s_a$, $M_{s_x}(p^{s,x}) = \Delta_{r_x, s_x}(a) = \Delta_{r_x, s_a}(a) + \Delta_{s_a, s_x}(a) = \Delta_{s_a, s_x}(a)$, and this is the maximal marking of that place. The claimed bound results. $\square$

Lemma 17. Minimality

*Assume that $TS = (S, \rightarrow, T, s_0)$ satisfies **b**, **rp**, **P**1, and **bp**. Any marked graph solution of TS contains (a copy of) the net constructed by (1).*

Proof: Let us consider some $x \in T$ and $s_a \in \max(S \setminus X)$ as above. There must be a place $p_{x,a}$ in the solution that excludes x at s_a, that is $M_{s_a}(p_{x,a}) = 0$ since the net is a marked graph, hence plain. Let us assume that it is a place from b to x. For any state $r \in S$ we must also have $M_r(p_{x,a}) = M_{r_x}(p_{x,a}) + \Delta_{r_x,r}(b)$, so that $M_{r_x}(p_{x,a}) = 0 = \Delta_{r_x,s_a}(b)$ as well. Therefore, there is no label b between r_x and s_a. But since $s_a \in Seq(x)$, from Lemma 11 and **P1**, the only missing labels between r_x and s_a are a and x, so that $p_{x,a} = p^{a,s}$, with the same initial marking. The property results.   □

Corollary 18. LIVE AND k-BOUNDED MARKED GRAPH REACHABILITY GRAPHS
   *Assume that $TS = (S, \rightarrow, T, s_0)$ satisfies **b**, **rp**, **P1**, and **bp**.*
Let $K = \max\{\Delta_{s_a,s_x}(a) \mid x \in T, s_a \in \max(S \setminus X)\}$.
(a): *If $k \geq K$, then TS is (isomorphic to) the reachability graph of a connected, live, k-bounded marked graph.* **(b):** *If $k < K$, then no marked graph whose reachability graph is isomorphic to TS is k-bounded.*

Thus K is the tightest possible bound for a marked graph realising TS: this results from Lemmata 16 and 17. As a consequence, the constructed marked graph is not only minimal, but also unique. Moreover, if an lts satifying all properties **b**, **rp**, **P1**, **bp** is reduced by fusing the endpoints of all x-labelled edges, one gets a well-defined new lts (with one transition less) which also satisfies all properties, and thus corresponds again to a marked graph.

6   Concluding Remarks

In this paper, we have proved that every labelled transition system satisfying some basic properties as well as reversibility, persistency, backward persistency, and a Parikh 1 property of small cycles, is isomorphic to the reachability graph of a live and bounded marked graph. This result, and the corresponding one for k-bounded marked graphs, seem to be novel, even though marked graphs enjoy a long history of being studied.

We would like to emphasise the key role of backward persistency. If **bp** is not true, then the state r_x of Lemma 10 cannot be used, as the set $S \setminus X$ may have more than one minimum; also, Lemma 11 fails. If **bp** is dropped but all other properties are kept, one can find examples which cannot be solved by ON Petri nets even if arbitrary arc weights and arbitrary side-conditions are allowed, disproving a conjecture of [5]. Such examples are rather complex; they are described in [6].

Future work might be concerned with the following issues:

- Extending the characterisations to non-live and/or unbounded marked graphs, while relaxing the plainness and pureness assumptions [12].
- Checking whether nets (N, M_0) which satisfy **rp** and whose initial marking satisfies $\gcd\{M_0(p) \mid p \in P\} > 1$ are backward persistent. (A positive answer would settle a question left open in [4].)

Acknowledgments. We are grateful to Philippe Darondeau, Hanna Klaudel, and Elisabeth Pelz for discussions. It was Philippe, in particular, who promoted the general idea of trying to characterise the state spaces of (distributable) Petri nets.

Appendix

A   Auxiliary Results

Let $TS = (S, \to, T, s_0)$ be an lts satisfying **b**, **rp**, and **PΥ** with some positive Υ. For sequences $\sigma, \tau \in T^*$, $\tau \overset{\bullet}{-} \sigma$ denotes the *residue* of τ w.r.t σ, i.e. the sequence left after cancelling successively in τ the leftmost occurrences of all symbols from σ, read from left to right. Formally and inductively: for $t \in T$, $\tau \overset{\bullet}{-} t = \tau$ if $\Psi(\tau)(t) = 0$; $\tau \overset{\bullet}{-} t =$ the sequence obtained by erasing the leftmost t in τ if $\Psi(\tau)(t) \neq 0$; $\tau \overset{\bullet}{-} \varepsilon = \varepsilon$; and $\tau \overset{\bullet}{-} (t\sigma) = (\tau \overset{\bullet}{-} t) \overset{\bullet}{-} \sigma$.

Theorem 19. KELLER'S THEOREM [9]
If $s[\tau\rangle$ and $s[\sigma\rangle$ for some $s \in [s_0\rangle$, then $s[\tau(\sigma \overset{\bullet}{-} \tau)\rangle s'$ and $s[\sigma(\tau \overset{\bullet}{-} \sigma)\rangle s''$ as well as $\Psi(\tau(\sigma \overset{\bullet}{-} \tau)) = \Psi(\sigma(\tau \overset{\bullet}{-} \sigma))$ and $s' = s''$.   □

Lemma 20. CYCLIC EXTENSIONS
Suppose $s[\alpha\rangle$ with $\alpha \in T^*$ and $\Psi(\alpha) \leq \Upsilon$. Then there is a small cycle $s[\kappa\rangle s$ such that α is a prefix of κ.

Proof: Let $\widetilde{\alpha}$ be such that $s[\widetilde{\alpha}\rangle s$ and $\Upsilon = \Psi(\widetilde{\alpha})$. Such a sequence $\widetilde{\alpha}$ exists by persistency, reversibility, and because small cycles can be pushed to all states (cf. Corollary 4 of [3]). Suppose $s[\alpha\rangle s'$. By Keller's theorem, $s[\alpha\rangle s'[\widetilde{\alpha} \overset{\bullet}{-} \alpha\rangle s''$. By $\Psi(\alpha) \leq \Upsilon = \Psi(\widetilde{\alpha})$, $\Psi(\widetilde{\alpha}) = \Psi(\alpha(\widetilde{\alpha} \overset{\bullet}{-} \alpha))$. By the cyclicity of $\widetilde{\alpha}$, $s'' = s$. Choosing $\kappa = \alpha(\widetilde{\alpha} \overset{\bullet}{-} \alpha)$ proves the lemma.   □

Lemma 21. CHARACTERISATION OF SHORT PATHS
Suppose that $s[\tau\rangle s'$. Then $s[\tau\rangle s'$ is short iff $\neg(\Upsilon \leq \Psi(\tau))$.

Proof: ($\Rightarrow$): By contraposition. Suppose that $s[\tau\rangle s'$ and that $\Upsilon \leq \Psi(\tau)$. There is some cycle $s[\kappa\rangle s$ with $\Psi(\kappa) = \Upsilon$. By Keller's theorem, $s[\kappa\rangle s[\tau \overset{\bullet}{-} \kappa\rangle s''$. By $\Psi(\kappa) = \Upsilon \leq \Psi(\tau)$, $\Psi(\tau) = \Psi(\kappa(\tau \overset{\bullet}{-} \kappa))$, and therefore, $s' = s''$ (by determinacy, which holds by property **b**). Since neither κ nor τ is the empty sequence, and by the fact that κ contains every transition at least once, $|\tau \overset{\bullet}{-} \kappa| < |\tau|$. Hence $s[\tau\rangle s'$ is not short.

($\Leftarrow$): Suppose that $s[\tau\rangle s'$ and $\neg(\Upsilon \leq \Psi(\tau))$. Consider any other path $s[\tau'\rangle s'$ from s to s'. By reversibility, there is some path ρ from s' to s. Hence both $s'[\rho\tau\rangle s'$ and $s'[\rho\tau'\rangle s'$ are cycles at s'. By Keller's theorem, $s'[\rho\tau'\rangle s'[(\rho\tau) \overset{\bullet}{-} (\rho\tau')\rangle s'$. Hence $s'[\tau \overset{\bullet}{-} \tau'\rangle s'$, and since this is a cycle, $\Psi(\tau \overset{\bullet}{-} \tau')$ is a multiple of Υ. In view of $\neg(\Upsilon \leq \Psi(\tau))$ and $1 \leq \Upsilon$, this can only be the case if $\Psi(\tau \overset{\bullet}{-} \tau') = 0$, i.e., $\tau \overset{\bullet}{-} \tau' = \varepsilon$. This implies, in particular, that $\Psi(\tau) \leq \Psi(\tau')$ and that $|\tau| \leq |\tau'|$, and therefore, $s[\tau\rangle s'$ is short.   □

Lemma 22. UNIQUENESS OF SHORT PARIKH VECTORS
Suppose that $s[\tau\rangle s'$ and $s[\tau'\rangle s'$ are both short. Then $\Psi(\tau) = \Psi(\tau')$.

Proof: By Lemma 21, both $\neg(\Upsilon \leq \Psi(\tau))$ and $\neg(\Upsilon \leq \Psi(\tau'))$. As in the second part of the previous proof, we may conclude, using some suitable (in fact any) path $s'[\rho\rangle s$, both $s'[\tau \overset{\bullet}{-} \tau'\rangle s'$ and $s'[\tau' \overset{\bullet}{-} \tau\rangle s'$. Therefore, both $\Psi(\tau) \leq \Psi(\tau')$ and $\Psi(\tau') \leq \Psi(\tau)$, implying $\Psi(\tau) = \Psi(\tau')$. $\qquad\square$

Lemma 23. CHARACTERISATION OF PARIKH VECTORS OF PATHS
Suppose that $s[\tau\rangle s'$. Then $\Psi(\tau) = \Psi(\tau') + m\cdot\Upsilon$, with some number $m \in \mathbb{N}$, where $s[\tau'\rangle s'$ is any short path.

Proof: Assume that $s[\tau\rangle s'$. Let m be the maximal number in $\mathbb{N}$ such that $\Psi(m\cdot\Upsilon) \leq \Psi(\tau)$. Let $s[\kappa\rangle s$ be some cycle with $\Psi(\kappa) = \Upsilon$. Then also $s[\kappa^m\rangle s$, with $\Psi(\kappa^m) = m\cdot\Upsilon$. By Keller's theorem, $s[\kappa^m\rangle s[\tau'\rangle s'$, with $\tau' = \tau \overset{\bullet}{-} \kappa^m$. By the maximality of m, $s[\tau'\rangle s'$ is short, and by $\Psi(\kappa^m) \leq \Psi(\tau)$, $\Psi(\tau)$ can be written as $\Psi(\tau) = \Psi(\tau') + \Psi(\kappa^m)$. By Lemma 22, the choice of τ' is arbitrary. $\qquad\square$

Lemma 24. EXISTENCE OF SHORT PATHS
Suppose that s, s' are states. There is a short path from s to s'.

Proof: By reversibility, $s[\tau\rangle s'$ for some τ. Just take the path $s[\tau'\rangle s'$ from the proof of Lemma 23. $\qquad\square$

So far, only **PΥ** was needed, but the remaining Lemmata depend on **P1**.

Lemma 25. EVERY LABEL HAS A UNIQUE SINGULAR ENABLING STATE
For every $x \in T$ there is a unique state s_x on which only x is enabled.

Proof: There must be at least one such state, because otherwise one can create a cycle without any x, by bypassing every outgoing edge labelled x on every state and using the finiteness of the lts, eventually contradicting property **P1**.

Suppose s_x and s'_x are two such states and let $s_x[x\rangle s$. By **P1**, we can find $s[\alpha x\rangle s$, without any x in α. Let $s[\beta\rangle s'_x$ be a short path, which exists by Lemma 24. By Keller's theorem, $s'_x[\alpha x \overset{\bullet}{-} \beta\rangle$. Hence all of α are wiped out by β because s'_x enables only x. Therefore, and because β is short, every letter except x occurs at least once in β. Similarly, if $s'_x[x\rangle s'[\beta'\rangle s_x$ (with a short β'), then every letter except x occurs at least once in β'. Now consider the cycle $s_x[x\rangle s[\beta\rangle s'_x[x\rangle s'[\beta'\rangle s_x$. It has every letter exactly twice, because x occurs exactly twice in it, and because of **P1**. Therefore, β has every letter except x exactly once, which implies $s_x = s'_x$, again by **P1**. $\qquad\square$

Lemma 26. LABELS ON SHORT PATHS INTO s_x
On any short path into s_x, there is no label x.

Proof: Assume that $r[x\alpha\rangle s_x$ is a short path such that α has no label x. (Other short paths into s_x containing x can be reduced to this case by taking suffixes.) Also, let $r[x\delta\rangle r$ be a cycle where δ contains no x but every other letter once. By Keller's theorem, $s_x[x\delta \overset{\bullet}{-} x\alpha\rangle$ which cannot be empty (because otherwise $r[x\alpha\rangle s_x$ is not short) but also does not start with an x; contradiction. $\qquad\square$

References

1. Badouel, E., Bernardinello, L., Darondeau, P.: Theory of Regions (to appear)
2. Badouel, E., Darondeau, P.: Theory of Regions. In: Reisig, W., Rozenberg, G. (eds.) APN 1998. LNCS, vol. 1491, pp. 529–586. Springer, Heidelberg (1998)
3. Best, E., Darondeau, P.: A Decomposition Theorem for Finite Persistent Transition Systems. Acta Informatica 46, 237–254 (2009)
4. Best, E., Darondeau, P.: Separability in Persistent Petri Nets. Fundamenta Informaticae 112, 1–25 (2011)
5. Best, E., Darondeau, P.: Petri Net Distributability. In: Clarke, E., Virbitskaite, I., Voronkov, A. (eds.) PSI 2011. LNCS, vol. 7162, pp. 1–18. Springer, Heidelberg (2012)
6. Best, E., Devillers, R.: Solving LTS with Parikh-unique Cycles. Technical Report (2013)
7. Commoner, F., Holt, A.W., Even, S., Pnueli, A.: Marked Directed Graphs. J. Comput. Syst. Sci. 5(5), 511–523 (1971)
8. Genrich, H.J., Lautenbach, K.: Synchronisationsgraphen. Acta Inf. 2, 143–161 (1973)
9. Keller, R.M.: A Fundamental Theorem of Asynchronous Parallel Computation. In: Tse-Yun, F. (ed.) Parallel Processing. LNCS, vol. 24, pp. 102–112. Springer, Heidelberg (1975)
10. Kondratyev, A., Cortadella, J., Kishinevsky, M., Pastor, E., Roig, O., Yakovlev, A.: Checking Signal Transition Graph Implementability by Symbolic BDD Traversal. In: Proc. European Design and Test Conference, Paris, France, pp. 325–332 (1995)
11. Landweber, L.H., Robertson, E.L.: Properties of Conflict-Free and Persistent Petri Nets. JACM 25(3), 352–364 (1978)
12. Teruel, E.: On Weighted T-systems. In: Jensen, K. (ed.) ICATPN 1992. LNCS, vol. 616, pp. 348–367. Springer, Heidelberg (1992)

Computing Depths of Patterns[*]

Francine Blanchet-Sadri[1], Andrew Lohr[2], Sean Simmons[3],
and Brent Woodhouse[4]

[1] Department of Computer Science, University of North Carolina
P.O. Box 26170, Greensboro, NC 27402–6170, USA
blanchet@uncg.edu
[2] Department of Mathematics, Mathematics Building
University of Maryland, College Park, MD 20742, USA
alohr1@umd.edu
[3] Department of Mathematics, Massachusetts Institute of Technology
77 Massachusetts Avenue, Cambridge, MA 02139–4307, USA
seanken@mit.edu
[4] Department of Mathematics, Purdue University
150 N. University Street, West Lafayette, IN 47907–2067, USA
bwoodhou@purdue.edu

Abstract. Pattern avoidance is an important research topic in combinatorics on words which dates back to Thue's construction of an infinite word over three letters that avoids squares, i.e., a sequence with no two adjacent identical factors. This result finds applications in various algebraic contexts where more general patterns than squares are considered. A more general form of pattern avoidance has recently emerged to allow for undefined positions in sequences. New concepts on patterns such as depth have been introduced and a number of questions have been raised, some of them we answer. In the process, we prove a *strict* bound on the number of square occurrences in an unavoidable pattern, and consequently, any pattern with more square occurrences than distinct variables is avoidable over three letters. We also prove a *strict* bound on the length of an avoidable pattern with at least four distinct variables. We finally provide an algorithm that determines whether a given pattern is of bounded depth, and if so, computes its depth.

1 Introduction

A *pattern p*, i.e., a word over an alphabet Δ of *variables* denoted by $A, B, C, \ldots$, is *avoidable* over some finite alphabet Σ if there exists an infinite word (or sequence) over Σ with no occurrence of p. The terminology of avoidable pattern, although studied by Thue at the beginning of the twentieth century, was introduced much later by Bean et al. [1] and by Zimin [12] who described a simple procedure to decide avoidability. The problem of deciding whether a pattern is k-avoidable, i.e., avoidable over k letters, has however remained open. Thus the problem of

---

[*] This material is based upon work supported by the National Science Foundation under Grant Nos. DMS–0754154 and DMS–1060775.

A.-H. Dediu et al. (Eds.): LATA 2014, LNCS 8370, pp. 173–185, 2014.

classifying the avoidability indices of all patterns over a fixed number of variables has become subject of investigation [8, 11] (the smallest k such that a pattern is k-avoidable is its *avoidability index*). Chapter 3 of [9] provides background on avoidable patterns.

A more general form of pattern avoidance has recently emerged to allow for undefined positions. In this context, *partial words* are sequences that may have such positions, called don't care symbols or holes, that match any letter of the alphabet (partial words without holes are *full words*). The occurrences of the same variable in a pattern are replaced with pairwise "compatible" partial words. For example, an occurrence of the pattern AAA has the form uvw where u is compatible with both v and w, and v is compatible with w. Constructing an infinite partial word with infinitely many holes that avoids a given pattern amounts to constructing an infinite set of infinite words that avoid the pattern. New research topics are being developed such as pattern avoidance with respect to hole sparsity [2], abelian pattern avoidance [6], etc.

Clearly AA is unavoidable due to occurrences of *trivial* squares of the form $a\diamond$ or $\diamond a$, where a is a letter and $\diamond$ is the hole symbol. In [10], it was shown that there exists a partial word with infinitely many holes over two letters that avoids the pattern A^n, $n \geq 3$, and so its avoidability index in partial words is 2. Ref. [3–5] provide, using "division" of patterns, the avoidability indices of *all* binary patterns, those over A and B, and *almost all* ternary patterns, those over A, B and C, except for four patterns whose avoidability index was shown to be between 2 and 5. To calculate the avoidability index of a pattern p, the lower bound is usually computed using backtracking. For the upper bound, a HD0L system is built that consists of an inner morphism ϕ and of an outer morphism ψ. Then $\psi(\phi^\omega(a))$ is shown to avoid p, for some letter a.

In the process of classifying the ternary patterns with respect to partial word avoidability, new concepts such as depth and shallowness, were introduced and a number of questions were raised [4]. Among them are the following:

1. If p is k-shallow and p_1 and p_2 are (h_1, k)-deep and (h_2, k)-deep respectively, is $p_1 A p_2$ $(h_1 + h_2, k)$-deep? In general, what relation does the depth of $p_1 A p_2$ have with the depth of p_1 and p_2? (Concepts are recalled in Section 5.)
2. Can every unavoidable pattern be written in the form of [4, Corollary 9]? (Corollary 9 is recalled in Section 5.)

In relation to 1, it was mentioned that the classification of the depths of patterns may give insight; this classification was completed in [4] though the problem remained open. Among other things, we answer these questions. In Section 2, we review a few basic concepts and notations. In Section 3, we prove, in particular, a *strict* bound on the number of square occurrences in a pattern that is partial word unavoidable, and consequently, any pattern with more square occurrences than distinct variables is 3-avoidable in partial words. We also prove a *strict* bound on the length of a pattern with at least four variables that is partial word avoidable. In Section 4, we exhibit an unavoidable pattern that cannot be written in the form of [4, Corollary 9], negatively answering 2 above. In Section 5, we provide an algorithm that determines if a given pattern has bounded depth, and if so, outputs its depth. Finally in Section 6, we conclude with some remarks.

2   Basic Concepts and Notations

Let Σ be a finite alphabet of letters. Define $\Sigma_\diamond = \Sigma \cup \{\diamond\}$, where $\diamond \notin \Sigma$ represents an undefined position or a hole. Partial words over Σ are sequences over $\Sigma_\diamond$ while full words over Σ are partial words over Σ with no $\diamond$'s. The empty word is denoted by ε. The set of all full words (resp., non-empty full words) over Σ is denoted by Σ^* (resp., Σ^+), while the set of all partial words (resp., non-empty partial words) over Σ by $\Sigma_\diamond^*$ (resp., $\Sigma_\diamond^+$). If u, v are partial words over Σ of equal length, then u is *contained in* v, denoted $u \subset v$, if $u[i] = v[i]$ for all i such that $u[i] \in \Sigma$; u, v are *compatible*, denoted $u \uparrow v$, if $u[i] = v[i]$ for all i such that $u[i], v[i] \in \Sigma$. If u, v are non-empty and compatible, then uv is a *square*. A partial word u is a *factor* of a partial word v if there exist x, y such that $v = xuy$. A full word is a *subword* of v if it is compatible with a factor of v. A *completion* of a partial word is a full word compatible with it. We denote by $v[i..j]$ (resp., $v[i..j)$) the factor $v[i] \cdots v[j]$ (resp., $v[i] \cdots v[j-1]$).

Let Δ be an alphabet of variables, $\Sigma \cap \Delta = \emptyset$, and let $p = A_0 \cdots A_{n-1}$, where $A_i \in \Delta$, be a pattern. The set of distinct variables that occur in p is denoted by $\alpha(p)$. If a variable occurs only once in p, it is a *singleton* variable. Define an *occurrence* of p in a partial word w over an alphabet Σ as a factor $u_0 \cdots u_{n-1}$ of w, where for all i, $u_i \neq \varepsilon$, and for all i, j, if $A_i = A_j$, then $u_i \uparrow u_j$. In other words, $u_0 \cdots u_{n-1} \subset \varphi(p)$, where φ is any non-erasing morphism from Δ^* to Σ^*. The partial word w *meets* the pattern p, or p *occurs* in w, if for some factorization $w = xuy$, we have that u is an occurrence of p in w; otherwise, w *avoids* p or w is *p-free*. For instance, $ab\diamond ba\diamond bba$ meets $ABBA$ (take the morphism $\varphi(A) = bb$ and $\varphi(B) = a$), while $\diamond babbbaaab$ avoids $ABBA$. These definitions also apply to (one-sided) infinite partial words over Σ, which are functions from $\mathbb{N}$ to $\Sigma_\diamond$.

A pattern $p \in \Delta^*$ is *k-avoidable in partial words* if there are infinitely many partial words in $\Sigma_\diamond^*$ with h holes, for any integer $h > 0$, that avoid p, where Σ is any alphabet of size k. If there is a partial word over Σ with infinitely many holes that avoids p, then p is obviously k-avoidable. On the other hand, if, for some integer $h \geq 0$, every long enough partial word in $\Sigma_\diamond^*$ with h holes meets p, then p is *k-unavoidable* (it is unavoidable over Σ). Finally, a pattern which is k-avoidable for some k is *avoidable*, and a pattern which is k-unavoidable for every k is *unavoidable*. The *avoidability index* of a pattern p is the smallest integer k such that p is k-avoidable, or is ∞ if p is unavoidable. Note that k-avoidability implies $(k+1)$-avoidability.

If a pattern p occurs in a pattern q, then p *divides* q and denote this by $p \mid q$; for instance, $AA \nmid ABA$ but $AA \mid ABAB$. Note that if $p \mid q$ and an infinite word avoids p then it also avoids q, and so the avoidability index of q is less than or equal to the avoidability index of p.

3   Avoiding Patterns

Any infinite partial word with at least one hole must meet A^2, so A^2 is clearly unavoidable in partial words. The theorem below addresses the avoidability of all other patterns where each variable occurs at least twice.

Theorem 1. *Let p be a pattern with $|p| > 2$ such that each variable in p occurs at least twice. Then p can be avoided by an infinite full word over k letters, for some k, and there exists a partial word with infinitely many holes over an alphabet of size $k + 5$ that avoids p. Moreover, if there are no squares of length two in p, there exists a partial word with infinitely many holes over an alphabet of size $k + 3$ that avoids p.*

Proof. By [9, Corollary 3.2.10], p can be avoided by an infinite full word if each of its variables occurs at least twice. Therefore, let w be an infinite word over an alphabet Σ of cardinality k, such that w avoids p. Take $m = 1$ if there are no squares of length two in p and $m = 2$ otherwise. There exist some $a_0, a_1, \ldots, a_{2m} \in \Sigma$ (not necessarily distinct) such that $a_0 a_1 \cdots a_{2m}$ occurs infinitely often as a factor of w. We create a sequence of integers $\{k_j\}$ as follows. Let k_0 be the smallest positive integer where $a_0 a_1 \cdots a_{2m} = w[k_0 - m]w[k_0 - (m-1)] \cdots w[k_0 + m]$. Define k_j recursively so that k_{j+1} is the smallest integer with $k_{j+1} > 4k_j$ and $a_0 a_1 \cdots a_{2m} = w[k_{j+1} - m]w[k_{j+1} - (m-1)] \cdots w[k_{j+1} + m]$.

Define the alphabet $\Sigma' = \Sigma \cup \{b_0, b_1, \ldots, b_{2m}\}$, where $b_i \notin \Sigma$ for all i. We define the partial word v as follows. If $j \equiv 0 \mod 6|p|$, for $0 \le i \le m - 1$ let $v[k_j + i + 1] = b_{m+i+1}$ and $v[k_j - i - 1] = b_{m-i-1}$; also define $v[k_j] = \diamond$. If $j \not\equiv 0 \mod 6|p|$, let $v[i] = b_m$ if $i = k_j$, and in all other cases let $v[i] = w[i]$. Note that v is basically w, except the factor $b_0 \cdots b_{m-1} \diamond b_{m+1} \cdots b_{2m}$ is inserted infinitely often, and between each two occurrences of this factor, there are $6|p|-1$ instances of b_m, where the distance between any two such instances is greater than or equal to the distance from the first b_m to the beginning of the word. This construction also guarantees that for any i with $v[i] = b_0$, we must have $v[i + m] = \diamond$. Likewise, for any i with $v[i] = b_{2m}$, $v[i - m] = \diamond$. Thus b_0 and b_{2m} can be viewed as "sentinel" letters on the left and right of the holes in v.

The partial word v is well-defined, and its letters come from an alphabet of size $k+2m+1$. We show that v avoids p by assuming that v meets p and reaching a contradiction. Set $p = A_0 \cdots A_{n-1}$, where each A_i is a variable in Δ. Define j_0 and j_1 so that $u = u_0 \cdots u_{n-1} = v[j_0..j_1]$ is a factor of v such that if $A_i = A_j$ then $u_i \uparrow u_j$, i.e., u is an occurrence of p in v.

Two occurrences of the same variable A in p, say A_i and $A_{i'}$, where $i < i'$, correspond to partial words $u_i, u_{i'}$ such that $u_i \uparrow u_{i'}$. Moreover, there exist s, t, and ℓ, $s \le s + \ell < t \le t + \ell$, so that $u_i = v[s] \cdots v[s+\ell]$ and $u_{i'} = v[t] \cdots v[t+\ell]$. Let $J_1 = \{j \mid s \le k_j \le s + \ell\}$ and $J_2 = \{j \mid t \le k_j \le t + \ell\}$. We show that $|J_2| \le 1$, thus $|J_1| \le 2$. Assume for the sake of contradiction that $|J_2| > 1$, so there exists $j \in J_2$ such that $j + 1 \in J_2$. However,

$$\ell = t + \ell - t \ge k_{j+1} - k_j > k_j > s + \ell \ge \ell,$$

$|J_2| > 1$, therefore $|J_1| \leq 2$.

Since each variable in p occurs at least twice, $|J_1| \leq 2$ and $|J_2| \leq 1$ imply there are at most $2|p|$ non-negative integers j with $j_0 \leq k_j \leq j_1$. By construction of v, there are $6|p| - 1$ integers j such that $v[k_j] = b_m$ between any two holes in v. Thus u contains at most one hole. We can show that u actually contains no holes. Thus $u_i \neq \diamond$ for all u_i in u. Define $\varphi : (\Sigma')^* \to \Sigma^*$ with $\varphi(a) = a$ if $a \in \Sigma$ and $\varphi(b_i) = a_i$ for $0 \leq i \leq 2m$. By construction $\varphi(u)$ is a factor of w that represents an occurrence of p, contradicting the fact that w avoids p. $\qquad\square$

It was shown in [3, 4] that the maximum length unary unavoidable pattern is A^2, length 2, the maximum length binary unavoidable pattern is A^2BA^2, length 5, and the maximum length ternary unavoidable pattern is $A^2BA^2CA^2$, length 8. Now, let A_m, for $m \in \mathbb{N}$ be different variables. Let $Z_0 = \varepsilon$, and for all $m \in \mathbb{N}$, $Z_{m+1} = Z_m A_{m+1} Z_m$, the Z_m's being the Zimin words known to be unavoidable in full words [9]. Referring to [7], all the Zimin words are unavoidable in partial words. Note that Z_m is over m distinct variables and $|Z_m| = 2^m - 1$. Therefore, the following corollary provides a *strict* avoidability bound for patterns with at least four variables. It also extends [9, Corollary 3.2.11].

Corollary 2. *Let p be a pattern with $n \geq 4$ distinct variables. If $|p| \geq 2^n$, then p is avoidable in partial words.*

The next theorem provides a bound on the number of square occurrences in a pattern that is partial word unavoidable. This bound cannot be improved. For a variable alphabet of size n, the pattern

$$A_0 A_0 A_1 A_0 A_0 A_2 A_0 A_0 \cdots A_0 A_0 A_{n-1} A_0 A_0$$

has n square occurrences, and is unavoidable in partial words (there has to be a factor "$a\diamond$" that occurs infinitely often for some letter a).

Theorem 3. *The number of square occurrences in a pattern that is partial word unavoidable is less than or equal to the number of distinct variables used. Consequently, any pattern with more square occurrences than distinct variables is 3-avoidable.*

Proof. If any square occurrence in the pattern is of length greater than two, it is divisible by $ABAB$, which is 3-avoidable, so, we restrict to only when we have a single variable squared. Also, no square occurrences are adjacent, otherwise the pattern would be divisible by $AABB$, which is 3-avoidable. Lastly, no square occurrences overlap, because an overlap of two length two square occurrences is an occurrence of AAA, which is 2-avoidable.

Suppose that p is an unavoidable pattern over an alphabet of n variables Δ. Proceeding by contradiction, write $p = A_1 A_1 u_1 A_2 A_2 \cdots A_n A_n u_n A_{n+1} A_{n+1}$

$(, \ldots, , \;)$ such that for every $A \in \alpha(u_i) \setminus S, A \in \alpha(u_{f(A)})$.
Put another way, every variable that occurs in u_i has to either appear in a square occurrence or at some point further left in p. We know such an i exists because with each m, u_m can either use a variable from $\Delta \setminus S$ that has not been used before, or it can only use variables that occurred before. Because there are more u_m than variables in $\Delta \setminus S$ to introduce for the first time, at least one of them has to not introduce any new variables from $\Delta \setminus S$, that is the u_i we want.

Let $T = \alpha(u_i) \setminus S$ and let $g : \Delta \to \mathbb{N}$ map A to the number of times A appears in u_i. We can prove that the partial word w' with infinitely many holes over three letters constructed below avoids p. Let $\Sigma = \{a, b, c\}$ and let $\theta : \Sigma^* \to \Sigma^*$ be the morphism defined by $\theta(a) = abc$, $\theta(b) = ac$, and $\theta(c) = b$. Define the morphism $\phi : \Sigma^* \to \Sigma_\diamond^*$ as θ^3 with the factor bab of $\theta^3(a)$ changed to $b \diamond b$, i.e.,

$$\phi(\ell) = \begin{cases} abcacb \diamond bcbac, & \text{if } \ell = a; \\ abcacbac, & \text{if } \ell = b; \\ abcb, & \text{if } \ell = c. \end{cases}$$

Let $w = \phi \circ \theta^\omega(a)$ and let $\langle i_m \rangle$ be the sequence of indices of holes of w, i.e., $w[i] = \diamond$ if and only if $i \in \langle i_m \rangle$. Let $\langle j_m \rangle$ be any subsequence of $\langle i_m \rangle$ such that

$$j_{m+1} > (1 + \sum_{A \in T} g(A)) j_m + 4 + 2 \sum_{A \in S} g(A).$$

We construct w' from w by replacing, for all i_m, $w[i_m - 1..i_m + 1]$ with $b \diamond b$ if $i_m \in \langle j_m \rangle$ or bab if not.

Since p is unavoidable, we have a meeting morphism h. Note that, for every $A \in T$, $u_{f(A)}$ contains A means $|h(u_{f(A)})| \geq |h(A)|$. Then because each square occurrence in p has its ends within two positions of a $\diamond$, there is a function gap mapping u_m, $m \leq i$, to $j_{m'} - j_{m''}$ where the two terms $j_{m'}$ and $j_{m''}$ of $\langle j_m \rangle$ selected are the positions of the two holes that the ends of u_m are near. This means $\text{gap}(u_i) \geq |h(u_i)| \geq \text{gap}(u_i) - 4$. So, for every $A \in T$, if $f(A) \neq i - 1$,

$$|h(A)| \leq |h(u_{f(A)})| \leq \text{gap}(u_{f(A)}) < \text{gap}(u_{i-1}) - 4 \leq |h(u_{i-1})|.$$

Then,

$$|h(u_i)| > \left(\sum_{A \in T} g(A)\right) |h(u_{i-1})| + 2 \sum_{A \in S} g(A)$$
$$\geq \left(\sum_{A \in T} g(A)|h(A)|\right) + 2 \sum_{A \in S} g(A),$$

which contradicts the fact that

$$|h(u_i)| = \sum_{A \in \Delta} g(A)|h(A)| \leq \left(\sum_{A \in T} g(A)|h(A)|\right) + 2 \sum_{A \in S} g(A).$$

$\square$

The next theorem will be useful for computing the depth of a given pattern in the next section.

Theorem 4. *If p is a pattern with no squares that is k-avoidable in full words, then, for every positive integers m, h, there is an infinite avoiding partial word over $k + 4h$ letters with h holes each at least m positions away from each other and the beginning of the word.*

Proof. We do induction on the number of variables in the pattern p. For the basis, the smallest alphabet size over which it is possible to have a square-free pattern that is avoidable by infinite full words is three. The only such ternary pattern with a singleton is $ABACBAB$ which, by [4], can be avoided with a partial word having infinitely many holes over only three letters, using the HD0L system given by $\phi(\theta^\omega(a))$ where

$$\theta(\ell) = \begin{cases} ad, & \text{if } \ell = a; \\ ab, & \text{if } \ell = b; \\ db, & \text{if } \ell = c; \\ c, & \text{if } \ell = d; \end{cases} \quad \text{and} \quad \phi(\ell) = \begin{cases} bb, & \text{if } \ell = a; \\ caabc, & \text{if } \ell = b; \\ aab\diamond acbaabc, & \text{if } \ell = c; \\ ac, & \text{if } \ell = d. \end{cases}$$

Then, we just fill in all but h of the holes that are far enough apart. In the case of no singletons, by Theorem 1, p is avoidable over $k + 3$ letters, and we have our base case with $\alpha(p) = 3$.

For the inductive step, let w be an infinite full word over k letters that avoids p. Let w' be a length five factor of w that occurs infinitely often. Let $S_{w'} = \{w_0, w_1, \ldots, w_{h-1}\}$ be a collection of h disjoint occurrences of w' in w that each occurs at least m positions apart and m positions from the ends of the word. Let $\{a_i, b_i, c_i, d_i\}_{0 \le i < h}$ be distinct letters that do not appear in w. Define $w'_i = a_i b_i \diamond c_i d_i$. Then, for each $w_i \in S_{w'}$, replace w_i with w'_i, call the resulting word W. Our claim is that W avoids p because under the mapping $\theta(v) = w'[0]$ if $v = a_i$ for some i; $w'[1]$ if $v = b_i$ for some i; $w'[2]$ if $v = \diamond$; $w'[3]$ if $v = c_i$ for some i; $w'[4]$ if $v = d_i$ for some i; and v otherwise, $\theta(W) = w$ avoids p. Assume to the contrary that W meets p with pattern occurrence $u_1 \cdots u_{|p|}$. Then there must be some i such that $\diamond$ is a factor of u_i.

Suppose towards a contradiction that for every i such that $\diamond$ is a factor of u_i, the variable corresponding to u_i is a singleton. Then, if u_i corresponds to a singleton, u_i is not required to be compatible with any u_j, $j \ne i$, so, the pattern occurrence is preserved under applying θ to W, contradicting the fact that w is p-free. We are also able to require that at least one of the non-singleton hole containing factors does not correspond to the factor $\diamond$ of W. So let A be a non-singleton corresponding to some u_j that contains a hole.

Suppose towards a contradiction that $\diamond$ is at some position other than either the first or last position of u_j. Then there is an i such that $b_i \diamond c_i$ is a factor of u_j, but any possible completion of $b_i \diamond c_i$ is a subword that only appears in u_j, contradicting the fact that u_j corresponds to a non-singleton. So, u_j either starts or ends with a $\diamond$.

Suppose towards a contradiction that $|u_j| > 2$. Then, there is an i such that u_j has either $a_i b_i \diamond$ or $\diamond c_i d_i$ as a suffix or prefix respectively. Any completion of either however is not a subword that appears anywhere else. So, u_j is $\diamond$ (Case 1), $\diamond c_i$ (Case 2), or $b_i \diamond$ (Case 3). Note that Case 3 is symmetric to Case 2. Letting the function f map u_n to the variable that corresponds to u_n, the rest of the proof is based on the concept of "holeboundedness".   □

4   Answering a Conjecture

It has been conjectured that every unavoidable pattern may be written in the form of [4, Corollary 9] which is stated as follows. Let p be a pattern of only distinct variables over Δ and let $0 \leq i < |p|$ be such that $p_0, p_1, \ldots, p_n \in \Delta^*$ are compatible with factors of $\mathrm{dig}_i(p)$, where $\mathrm{dig}_i(p)$ is defined by $\mathrm{dig}_i(p)[i] = \diamond$ and $\mathrm{dig}_i(p)[j] = p[j]$ if $j \neq i$. If $A_1, \ldots, A_n$ are distinct variables not in Δ, then $p_0 A_1 p_1 \cdots A_n p_n$ is unavoidable. We answer negatively this conjecture.

Theorem 5. *The above mentioned conjecture is false.*

Proof. It suffices to provide a pattern that is unavoidable and that is neither full word unavoidable nor of the form of [4, Corollary 9]. The pattern

$$p = Z_4 EFF = ABACABADABACABAEFF$$

satisfies such property. It is clearly not full word unavoidable because it is divisible by AA. It is also not of the form of Corollary 9, as the only possibilities for $A_1, \ldots, A_n$ are $A_1 = D$ and $A_2 = E$ as they are the only variables occurring only once, meaning that $p_0 = p_1 = ABACABA$ and $p_2 = FF$. However, since p_0 has A occurring more than twice, it cannot be compatible with a factor of some $\mathrm{dig}_i(p)$, pattern formed from at most one hole and some number of distinct variables, as was the restriction on p.

To see that p is unavoidable, first note that it is unavoidable even in full words. Thus, there must be some hole occurring at least two letters to the right of an occurrence of Z_4. Let F map to the letter occurring immediately to the right of the hole, and E be the factor of the word between the occurrence of Z_4 and the hole.   □

5   Computing Depths of Patterns

Recall the definitions of *depth* and *shallowness* from [4]. A k-unavoidable pattern p is (h, k)-*deep* if there exists some $m \in \mathbb{N}$ such that every partial word w over a k-letter alphabet meets p whenever w has at least h holes separated pairwise from each other and from the first and final position of w by factors of length m or greater. A function $\delta : \mathbb{N} \setminus \{0, 1\} \to \mathbb{N}$ is the depth-function of an unavoidable pattern p if for all k the pattern p is $(\delta(k), k)$-deep and p is not (j, k)-deep for any $j < \delta(k)$. When the depth function of p is bounded, its supremum d is the *depth* of p and p is d-deep. On the other hand, a pattern p is k-*shallow* if p is

$(0, k)$-deep or $(1, k)$-deep. If p is k-shallow for all k, p is *shallow*. The pattern p is *k-non-shallow* if it is k-unavoidable but not k-shallow. Shallow patterns have some properties in common with full word unavoidable patterns that higher-depth patterns do not have.

A use of shallowness from [4] states that if p_1, p_2 are k-unavoidable patterns over an alphabet of variables Δ and A is a variable which does not appear in p_1 or p_2, i.e., $A \in \Delta \setminus (\alpha(p_1) \cup \alpha(p_2))$, then the pattern $p_1 A p_2$ is k-unavoidable if there exists some k-shallow pattern p such that p_1 and p_2 are factors of p. Note that it is also much easier to check that a given pattern is shallow for a given k than to check that it has higher depth. This is done just by starting with a hole then trying to add a letter on each side, backtracking if no letter works. If there are only finitely many such words, then the pattern is unavoidable with depth 1. This does not work as easily for higher depths because if the backtracking came up finite, then it could be that the two holes starting the backtracking were not far enough apart.

The classification of the depths of the 2-unavoidable binary patterns has been completed.

Theorem 6 ([4]). *The 2-unavoidable binary patterns fall into five categories with respect to depth:*

1. *The patterns ε, A, AB, and ABA are shallow with depth 0;*
2. *The patterns AA and AAB are $(0, 2)$ and $(1, k)$-deep for all $k \geq 3$;*
3. *The pattern $AABA$ is $(0, 2)$, $(1, 3)$, and $(2, k)$-deep for all $k \geq 4$;*
4. *The pattern $AABAA$ has depth function δ satisfying $\delta(2) = 0$ and, for all $k \geq 3$, $\delta(k) = k + 1$;*
5. *The patterns $AABAB$, $AABB$, $ABAB$, $ABBA$ are $(0, 2)$-deep.*

For example, consider the depth of $AABAA$. To have an occurrence, the same square must occur twice, separated by at least one letter. If we have $k + 1$ holes, at least one letter occurs next to two distinct holes. Then we have the same trivial square introduced twice. This means the word must meet the pattern, so there can be no avoiding word with $k + 1$ holes. If we have $k - 1$ holes, we can surround them like $a_k a_1 \diamond a_1 a_k, a_k a_2 \diamond a_2 a_k, \ldots, a_k a_{k-1} \diamond a_{k-1} a_k$, which avoids. To increase the lower bound on the depth to $k + 1$, we can construct an avoiding word with k holes starting with the fixed point at a of the morphism mapping a to abc, b to ac, and c to b.

The next theorem describes the form of all 1-deep patterns, knowing that the variable that appears squared cannot appear anywhere else, and the variables appearing around the square occurrence must be singletons. The rest of the pattern must be 0-deep, once the square surrounded by singletons is replaced with a single singleton.

Theorem 7. *The only patterns that are 1-deep have exactly one square occurrence.*

The proof, which can be done by induction on the number of variables in the patterns, relies on the following two lemmas.

Lemma 8. *If the patterns p_1 and p_2 are (h_1, k) and (h_2, k)-deep respectively, then $p = p_1 A p_2$ is not (h, k)-deep with $h < h_1 + h_2$.*

This implies that p is either k-avoidable or (h, k)-deep with $h \geq h_1 + h_2$. We do not necessarily have $h = h_1 + h_2$, as demonstrated by $ABACAA$ which is 3-deep even though ABA has depth 0 and AA has depth 1.

Lemma 9. *If the patterns p_1 and p_2 are both 0-deep, then, taking A to be a variable not appearing in either p_1 or p_2, we have $p = p_1 A p_2$ is not h-deep for any $h > 0$.*

We now have the necessary machinery to describe Algorithm 1, which finds the depth of an arbitrary pattern p. We say a variable A of p is *holebound* if all but a single occurrence of A must map to a $\diamond$ in any meeting morphism. This is used in the proof of Theorem 10 in which we insert arbitrarily many factors of the form $a_1 \cdots a_{|p|} \diamond a_{|p|+1} \cdots a_{2|p|}$, where each of the a_i's that are used are unique to each hole.

Algorithm 1. Determine if a pattern has bounded depth, if so, find its depth

---

Require: p is a pattern
Ensure: the depth of p if p has bounded depth, FALSE otherwise
1:   $V \leftarrow \emptyset$
2:   $S \leftarrow \emptyset$
3:   $S_f \leftarrow \emptyset$
4:   **for** variables A that appear in a square occurrence in p **do**
5:      **if** A has two or more square occurrences in p **then**
6:         return FALSE
7:      $S \leftarrow \{\text{all maximal occurrences of powers of } A\} \cup S$
8:      $V \leftarrow \{A\} \cup V$
9:   **while** $S \neq \emptyset$ **do**
10:      remove an occurrence O from S
11:      $S_f \leftarrow \{O\} \cup S_f$
12:      **for** occurrences O_A of variables A between O and either the side of the word or a singleton **do**
13:         **if** $A \in V$ **then**
14:            return FALSE
15:         $V \leftarrow \{A\} \cup V$
16:         $S \leftarrow \{\text{all occurrences of } A \text{ other than } O_A\} \cup S$
17:   $f \leftarrow$ the formula obtained by removing all the occurrences in S_f from p
18:   **if** f is full word avoidable (using Zimin's procedure) **then**
19:      return FALSE
20:   return $|S_f|$

To illustrate Algorithm 1, suppose that we want to determine if the pattern $AABCECDFBGD$ has bounded depth. Build the table:

S	S_f	V	
$\{AA\}$	$\emptyset$	$\{A\}$	$AABCECDFBGD$
$\{B\}$	$\{AA\}$	$\{A,B\}$	$AABCECDFBGD$
$\{B,C\}$	$\{AA\}$	$\{A,B,C\}$	$AABCECDFBGD$
$\{B,C\}$	$\{AA\}$	$\{A,B,C\}$	$AABCECDF\underline{B}GD$
$\{C\}$	$\{AA,B\}$	$\{A,B,C\}$	$AABCE\underline{C}DFBGD$
$\{D\}$	$\{AA,B,C\}$	$\{A,B,C,D\}$	$AABCE\underline{CD}FBGD$
$\emptyset$	$\{AA,B,C,D\}$	$\{A,B,C,D\}$	$AABCECDFBG\underline{D}$

We then have that the given pattern is unavoidable with depth four.

Theorem 10. *Given as input a pattern p, Algorithm 1 determines if p has depth that is bounded, and if so, it outputs its depth; otherwise, it returns FALSE.*

Proof. If p is full word unavoidable, then it has bounded depth 0. Now suppose that it is full word avoidable over k letters. Construct the word over $k + 2h|p|$ letters by arbitrarily far apart inserting h holes surrounded by $2|p|$ unused letters extending $|p|$ on each side. Call this word w. By Theorem 4, if p does not have at least one square occurrence, then p cannot have bounded depth, because it cannot be $(h, k + 4h)$-deep. Also, if the same variable appears squared twice, then p is divisible by $AABAA$ meaning that its depth function is unbounded. Any square occurrence must correspond to a hole and a letter adjacent to a hole, meaning that variable is holebound because any letter that appears adjacent to a hole never appears again. Here V serves to keep track of exactly those variables which have been holebound within two positions of hole-containing occurrences of variables already considered, i.e., already in either S or S_f. This means that if a variable that is in V ever appears again when considering some different hole-containing occurrence, then that variable is holebound in two different locations, a contradiction. So, the word w would avoid p.

Every time we take an occurrence O from S, it either corresponds exactly to a hole, or, if it came from a square in the first for loop, a hole and one letter that is adjacent. Because each of the letters as we go out on either side are distinct, the neighbors are either holebound or singletons. To see this, note that no subword of length greater than two ever appears again. Holebound variables only have length one meaning that we must eventually reach a singleton or the end of the pattern. For the non-singletons considered before reaching a singleton, their other occurrences must correspond to holes, so they go onto S. Note then, that splitting p on its singletons, each term must consist entirely of variables in V, or have no variables in V. After removing all such chunks of the holebound variables, the pattern we are left with is square-free, and, if it is full word avoidable, we can get an avoiding word over $k + 4h$ letters, meaning that the pattern is not of bounded depth. On the other hand, if what we are left with is full word unavoidable, then there is a way of spacing the holes that each corresponded to some occurrence that is in S_f far enough apart so that the terms whose variables were disjoint from V must appear between the holes. Because the only holes that were used were for occurrences in S_f, and only one each, the depth is $|S_f|$.   □

If the pattern, after deleting variables occurring in V, is entirely composed of singletons, then we are in the interesting case where p is unavoidable even over an infinite alphabet so long as there are $|S_f|$ holes spaced far enough apart.

6 Conclusion

Dealing with unavoidable patterns with unbounded depth functions is much more complicated than dealing with patterns with bounded ones because letters around holes must be reused at some point. Because Algorithm 1 uses a construction that introduces $2|p|$ new letters per hole, every depth function is either bounded or is in $\Omega(k)$ where k is the alphabet size.

For any m, the pattern $p = A_0 A_0 A_1 A_2 \cdots A_{m-2} A_{m-1} A_0 A_0 A_1 A_2 \cdots A_{m-2}$ over m variables has depth function in $\Theta(k^{m-1})$. In fact, it has depth function at least $(k-3)^{m-1} + 1$. To see this, p has avoidability index 3 in full words, and by starting with a square-free word over three letters, $A_0 A_0$ has to line up with a square. By using the $k-3$ letters left over, we can fill in the $m-1$ positions to the right of the hole in $(k-3)^{m-1}$ different ways. By exponentially spacing the holes, for every $0 \leq i < m-1$, A_i must have an image of length one, but each of the length $m-1$ factors to the right of the hole is distinct. So, this word over k letters with $(k-3)^{m-1}$ holes avoids p. There may be patterns with more quickly growing depths than p.

References

1. Bean, D.R., Ehrenfeucht, A., McNulty, G.: Avoidable patterns in strings of symbols. Pacific Journal of Mathematics 85, 261–294 (1979)
2. Blanchet-Sadri, F., Black, K., Zemke, A.: Unary pattern avoidance in partial words dense with holes. In: Dediu, A.-H., Inenaga, S., Martín-Vide, C. (eds.) LATA 2011. LNCS, vol. 6638, pp. 155–166. Springer, Heidelberg (2011)
3. Blanchet-Sadri, F., Lohr, A., Scott, S.: Computing the partial word avoidability indices of binary patterns. Journal of Discrete Algorithms 23, 113–118 (2013)
4. Blanchet-Sadri, F., Lohr, A., Scott, S.: Computing the partial word avoidability indices of ternary patterns. Journal of Discrete Algorithms 23, 119–142 (2013)
5. Blanchet-Sadri, F., Mercaş, R., Simmons, S., Weissenstein, E.: Avoidable binary patterns in partial words. Acta Informatica 48, 25–41 (2011)
6. Blanchet-Sadri, F., Simmons, S.: Abelian pattern avoidance in partial words. In: Rovan, B., Sassone, V., Widmayer, P. (eds.) MFCS 2012. LNCS, vol. 7464, pp. 210–221. Springer, Heidelberg (2012)
7. Blanchet-Sadri, F., Woodhouse, B.: Strict bounds for pattern avoidance. Theoretical Computer Science 506, 17–28 (2013)
8. Cassaigne, J.: Motifs évitables et régularités dans les mots. Ph.D. thesis, Paris VI (1994)

9. Lothaire, M.: Algebraic Combinatorics on Words. Cambridge University Press, Cambridge (2002)
10. Manea, F., Mercaş, R.: Freeness of partial words. Theoretical Computer Science 389, 265–277 (2007)
11. Ochem, P.: A generator of morphisms for infinite words. RAIRO-Theoretical Informatics and Applications 40, 427–441 (2006)
12. Zimin, A.I.: Blocking sets of terms. Mathematics of the USSR-Sbornik 47, 353–364 (1984)

Solving Equations on Words
with Morphisms and Antimorphisms

Alexandre Blondin Massé, Sébastien Gaboury, Sylvain Hallé,
and Michaël Larouche

Laboratoire d'informatique formelle
Université du Québec à Chicoutimi
Chicoutimi (Québec) G7H 2B1, Canada
{ablondin,s1gabour,shalle}@uqac.ca

Abstract. Word equations are combinatorial equalities between strings of symbols, variables and functions, which can be used to model problems in a wide range of domains. While some complexity results for the solving of specific classes of equations are known, currently there does not exist any equation solver publicly available. Recently, we have proposed the implementation of such a solver based on Boolean satisfiability that leverages existing SAT solvers for this purpose. In this paper, we propose a new representation of equations on words having fixed length, by using an enriched graph data structure. We discuss the implementation as well as experimental results obtained on a sample of equations.

1 Introduction

Combinatorics on words has known increasing popularity in the last decades and has proven useful in a large variety of situations, such as computational linguistics [12], bioinformatics [7], automated testing [8] and discrete geometry [3]. A subfield having been considerably developped in parallel is that of equations involving words.

Equations on words have been studied from both theoretical and practical perspectives. Literature on combinatorics often refers to the so-called "Makanin's algorithm" [13], used to demonstrate that the existential theory of equations over free monoids is decidable. It was, however, probably not meant to be used to actually *solve* equations on words, and rather appears as a proof device by which decidability is deduced. Nevertheless, a LISP implementation of Makanin's algorithm was discussed in [1], although this implementation is neither available nor has been experimentally evaluated. Moreover, Makanin's result only applies to word equations with variables and constants; even if an implementation were available, it would not handle morphisms (i.e. substitutions) and antimorphisms (i.e. substitutions reversing the letters' order).

More recent tools have considered the solving of string constraints using regular expressions, such as HAMPI [9], Omega [6] and STRANGER [15]. However, while regular expressions and word equations present some overlap, morphisms and antimorphisms are outside their range of expressiveness. This entails that

A.-H. Dediu et al. (Eds.): LATA 2014, LNCS 8370, pp. 186–197, 2014.

many problems (such as those presented in Section 4) are not within reach of known algorithms and tools.

A recent approach followed by the authors of this paper consists in translating a set of word equations into an equivalent instance of the Boolean satisfiability (SAT) problem [10]. This problem is then sent to an off-the-shelf SAT solver called Minisat [4], whose answer is then converted back into appropriate values for each variable appearing in the equations. However, the SAT approach, while successful at enumerating all solutions of a given system of equations, provides little insight into the structure of this set of solutions. Our main objective here is to present an alternate way of solving the same problem addressing this particular issue.

The sections are divided as follows. As usual, Section 2 contains the definitions and notation used in the next sections. Section 3 describes two domains of application of equations involving words. Section 4 deals with theoretical aspects and introduce the data structure used to solve particular types of equations. Finally, Section 6 discusses an implementation of a solver with two paradigms (Boolean constraints satisfiability and based on a graph data structure). Section 7 concludes briefly.

2   Definitions and Notation

The usual terminology and notation on words is found in [11]. An *alphabet* A is a finite set whose elements are called *letters* or *symbols*. A *word* is a (finite or infinite) sequence of elements of A. Given an alphabet A, the *free monoid* A^* is the set of all finite words on A. The *length* of a word w, denoted by $|w|$, is the number of letters it contains. The *empty word*, usually denoted by ε is the only word having length 0. The basic operation on words is *catenation*, writtent uv or $u \cdot v$ for words u and v. A practical notation is x^k to denote $xx \cdots x$ (k times).

Given two alphabets A and B, a *morphism* (or *substitution*) is a function $\varphi : A^* \to B^*$ compatible with catenation, i.e. $\varphi(uv) = \varphi(u)\varphi(v)$ whenever u and v are words of A. A *k-uniform morphism* φ (or simply *uniform*) is a morphism such that $|\varphi(a)| = k$ for all letters $a \in \Sigma$. If φ is k-uniform, we set $|\varphi| = k$.

Similarly, an *antimorphism* is a function $\varphi : A \to B$ such that $\varphi(uv) = \varphi(v)\varphi(u)$. For instance, the reversal operator $\sim$ is an antimorphism since $\widetilde{uv} = \widetilde{v}\widetilde{u}$. It is easy to verify that antimorphisms are the same as morphisms up to the reversal operator. More precisely, if φ is antimorphism, then there exists a (unique) morphism φ' such that $\varphi = \varphi' \circ \sim$. Moreover, every morphism or antimorphism is completely determined by the image of each letter of the alphabet. Uniform morphisms φ are of particular interest since the length of $\varphi(u)$ can be determined unequivocally from the length of u.

In general, equations on words are studied only with constants and variables. In our case, we are interested to extend the theory to equations on words involving morphisms and the reversal operator. For this purpose, we introduce the following four sets of symbols: (i) Σ is called the set of *constants*; (ii) $\mathcal{V}$ is the set of *variables*; (iii) $\mathcal{M}$ is the set of *morphisms*; (iv) $\mathcal{A}$ is the set of *antimorphisms*.

Usually, elements of Σ are denoted by the first letters a, b, c, ... and elements of $\mathcal{V}$ by the last letters x, y, z, t, ... of the latin alphabet. Elements of $\mathcal{M}$ and $\mathcal{A}$ are identified with capital letters A, B, ... or greek letters φ, μ, ...

To introduce formally what is an *equation on words*, we first define *word expressions*. We say that u is a *word expression* (or simply *expression*) if

   (i) $u = \varepsilon$, $u \in \Sigma$ or $u \in \mathcal{V}$;
  (ii) $u = (v_1, v_2, \ldots, v_k)$, where v_i is an expression, for $i = 1, 2, \ldots, k$;
 (iii) $u = (\varphi, v)$, where $\varphi \in \mathcal{M} \cup \mathcal{A}$ and v is itself an expression.

The set of all expressions is denoted by $\mathcal{E}$. An *equation on words* (or simply *equation*) is a couple (L, R), where $L = (u_1, u_2, \ldots, u_\ell)$ and $R = (v_1, v_2, \ldots, v_r)$ are expressions. Often, we write $L = u_1 u_2 \cdots u_\ell$, $R = v_1 v_2 \ldots v_r$ and $L = R$ for simplifying the notation. For instance, let $\Sigma = \{a, b\}$, $\mathcal{V} = \{x, y\}$, $\varphi : a \mapsto b, b \mapsto a$, $\mathcal{M} = \{\varphi\}$ and $\mathcal{A} = \{\widetilde{}\}$. Then

$$ab\widetilde{x}y = \varphi(x)yab, \tag{1}$$

is an equation on words with respect to these four sets.

An *assignment* is a map $I : \mathcal{V} \to \Sigma^*$. Any assignment can naturally be extended to arbitrary expressions by applying the following recursive rules:

   (i) $I(\varepsilon) = \varepsilon$;
  (ii) If $u \in \Sigma$, then $I(u) = u$;
 (iii) If $u = (v_1, v_2, \ldots, v_k)$, where v_i is an expression, for $i = 1, 2, \ldots, k$, then $I(u) = I(v_1)I(v_2) \cdots I(v_k)$;
 (iv) If $u = (\varphi, v)$, where $\varphi \in \mathcal{M} \cup \mathcal{A}$ and v is an expression, then $I(u) = \varphi(I(v))$.

A *solution* is an assignment $S : \mathcal{V} \to \Sigma^*$ whose extension verifies $S(L) = S(R)$. As an example, the assignment $S(x) = ba$ and $S(y) = \varepsilon$ is a solution to Equation (1) since $S(L) = abab = S(R)$.

Equations for which the lengths of all variables are known will be of special interest. Consider a map $\lambda : \mathcal{V} \to \mathbb{N}$ assigning to each variable a fixed length and suppose that all morphisms and antimorphisms are uniform. As for assignment maps, λ can naturally be extended to any expression:

   (i) $\lambda(\varepsilon) = 0$;
  (ii) If $u \in \Sigma$, then $\lambda(u) = 1$;
 (iii) If $u = (v_1, v_2, \ldots, v_k)$, where v_i is an expression, for $i = 1, 2, \ldots, k$, then $\lambda(u) = \lambda(v_1) + \lambda(v_2) + \ldots + \lambda(v_k)$;
 (iv) If $u = (\varphi, v)$, where $\varphi \in \mathcal{M} \cup \mathcal{A}$ and v is an expression, then $\lambda(u) = |\varphi| \cdot \lambda(v)$.

3   Motivation

Equipped with the previous definitions, it now becomes possible to formulate a range of mathematical problems as equations on words.

3.1   Distinct Square Conjecture

A first example comes from pure combinatorics on words. In [5], one may read the following conjecture in the concluding section:

Conjecture 1 ([5]). A word w of length n contains less than n distinct squares.

For instance, the eight square factors of the word $w = 0000110110101$ are

$$00, 11, 0000, 0101, 1010, 011011, 101101, 110110$$

and it can be shown by exhaustive enumeration that no word of length 13 can achieve more than 8 squares. An intuitive argument in favor of Conjecture 1 is that if a word contains many squares, then it is strongly periodic, so that the squares cannot all be distinct.

Now, although there are many ways of generating examples with many squares, an alternate approach would be to model the problem with an equation and to enumerate all possible solutions. More precisely, in order to compute a word with k squares, one could solve the equations

$$x_1 u_1^2 y_1 = x_2 u_2^2 y_2 = \ldots = x_k u_k^2 y_k$$

with various distinct lengths for the words x_i, $i = 1, 2, \ldots, k$. Obviously, this would not provide a proof to Conjecture 1, but it could be a simple way to generate candidates for words that are rich in distinct squares.

3.2   Tilings

Another interesting example comes from the field of discrete geometry. Consider the alphabet $\mathcal{F} = \{0, 1, 2, 3\}$, where the symbols in $\mathcal{F}$ represent unit-length moves in each of the four directions of a square grid, starting from **0** (meaning "north") and rotating 90° counter-clockwise at each successive symbol. Some words of $\mathcal{F}$, called *boundary words* represent a sequence of moves surrounding a closed and connected region of the plane, called a *polyomino*.

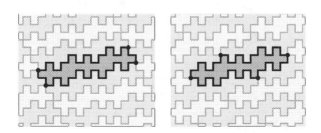

Fig. 1. A 2-square tile and the two tilings it induces. The two factorizations (**010301010301030101030 10**, **121**, **232123232123212323 21232**, **303**) (on the left) and (**03010301010301030**, **101030101**, **21232123232123212**, **323212323**) (on the right) are represented by black dots.

Consider then the (unique) antimorphism on $\mathcal{F}$ such that $\mathbf{0} \mapsto \mathbf{2}$, $\mathbf{1} \mapsto \mathbf{3}$, $\mathbf{2} \mapsto \mathbf{0}$ and $\mathbf{3} \mapsto \mathbf{1}$. Intuitively, the application of $\hat{}$ on some word produces the "mirror" path of its input, i.e. the same path traveled in the opposite direction. A 2-*square tile* is a polyomino whose boundary word w verifies the equations

$$w = xy\widehat{x}\widehat{y} \tag{2}$$

$$ww = pzt\,\widehat{z}\,\widehat{t}s, \tag{3}$$

where x, y, z and t are nonempty words on $\mathcal{F}$ that describe two distinct factorizations of w (see Figure 1) and p and s are some prefix and suffix of w. It turns out that the set of all solutions of Equations (2)–(3) is quite complex to describe and it is only thanks to computer exploration that the authors of [2] were able to compute it.

4   Solving Equations on Words

The examples given in the previous section illustrate how a number of mathematical objects can be specified as solutions of equations on words. In this section, we briefly describe the problem complexity and next, we focus our attention on a particular subset of equations on words.

The exact complexity class for the problem of solving word equations is currently unknown. When the length of equations is not known in advance, the current best upper bound is due to Plandowski, who showed that the problem is in PSPACE [14]. However, if we restrict ourselves to equations of a known length, we can show that the problem becomes NP-complete.

Theorem 1. *Let Σ, $\mathcal{V}$, $\mathcal{M}$ and $\mathcal{A}$ be some sets of constants, variables, morphisms and antimorphisms. Assume that all morphisms and antimorphisms are uniform and consider an equation $L = R$ on these four sets, where $|L| = |R| = n$. Then the problem of finding some solution $S : \mathcal{V} \to \Sigma^*$ for $L = R$ is NP-complete.*

Proof. The problem is clearly in NP, since a given solution S can be verified by replacing all variables with their values and then performing a symbol-by-symbol comparison of both sides of the equality, even if some morphisms and antimorphisms are involved.

It remains to show that the problem is NP-hard. We do so by reducing 3-SAT to our problem. An instance of the 3-SAT problem is a set of Boolean variables $B = \{b_1, b_2, \ldots, b_n\}$ and a set $\{\{x_{1,1}, x_{1,2}, x_{1,3}\}, \{x_{2,1}, x_{2,2}, x_{2,3}\}, \ldots, \{x_{m,1}, x_{m,2}, x_{m,3}\}\}$, where each triplet $\{x_{i,1}, x_{i,2}, x_{i,3}\}$ is called a clause and each $x_{i,j}$ is either b_i or $\neg b_i$ for some $b_i \in B$. Let

$$\Sigma = \{000, 001, 010, 011, 100, 101, 110, 111\}$$

be the set of constants, $\mathcal{V} = \{s_{i,j} \mid 1 \le i \le m, j = 1, 2, 3\}$ the set of variables, $\mathcal{M} = \{E, \sigma\}$ and $\mathcal{A} = \emptyset$, where E and σ are defined by

$$E(x) = \overline{x} \quad \text{and} \quad \sigma(x) = \begin{cases} 000, & \text{if } x = 000; \\ 111, & \text{otherwise.} \end{cases}$$

for any $x \in \Sigma$, where $\bar{x}$ is the symbol in Σ with the 0's and 1's inverted in x. We then associate a given 3-SAT instance with the word equation:

$$\sigma(s_{1,1} \cdot s_{1,2} \cdot s_{1,3} \cdot s_{2,1} \cdot s_{2,2} \cdots s_{m,3}) = (111)^{3m} \tag{4}$$

where $s_{i,j} = b_i$ if $x_{i,j} = b_i$, and $s_{i,j} = \overline{b_{i,j}}$ if $x_{i,j} = \neg b_{i,j}$, for $b_{i,j}$ the i-th variable of the j-th clause of the 3-SAT instance. For example the 3-SAT instance $\{\{b_1, \neg b_2, b_3\}, \{\neg b_1, \neg b_3, b_4\}\}$ translates into the equation $\sigma(b_1 \cdot \overline{b_2} \cdot b_3 \cdot \overline{b_1} \cdot \overline{b_3} \cdot b_4) = (111)^6$.

A solution to the 3-SAT problem is an assignment $\nu : B \to \{\top, \bot\}$ such that each clause contains at least one $x_{i,j}$ that is assigned to the value true ($\top$). This happens exactly when each triplet of the left-hand side of Equation 4 contains at least one occurrence of the symbol "1", which in turn entails that the application of σ to this triplet will produce symbol "111". Hence one can observe that this instance admits a solution if and only if Equation 4 does. $\qquad\square$

It follows from Theorem 1 that, unless P = NP, one cannot expect to design a polynomial time algorithm solving any equation on words, even with fixed length. Nevertheless, it is possible to achieve reasonable performance by using convenient data structures that allow one to decide whether such a solution exists, and even exhaustive enumeration of all solutions.

5 Data Structure

We now describe the main data structure behind our word equations solver.

5.1 Two Examples

Suppose that we wish to find a word of length 20 having square prefixes of length 8 and 12. What form would have that word? This can easily be represented by the equation $uuv = wwx$, where $|u| = 4$, $|v| = 12$, $|w| = 6$ and $|x| = 8$. Checking the result by hand takes some time and it seems more convenient to use a computer to solve it. The idea is to represent each letter of the variables and of the equation by a vertex and add an edge between two vertices whenever they host equal letters. Then we merge the connected components to obtain a final graph that describes the value of every letter (see Figure 2).

Since we are interested in handling equations with morphisms as well, let us next consider an equation over the alphabet $\{R, L, F\}$ involving the antimorphism $\widehat{\cdot}$ defined by $\widehat{u} = \widetilde{\bar{u}}$, where $\bar{\cdot}$ is the morphism defined by $R \leftrightarrow L$ and $F \leftrightarrow F$. We wish to find the solution to the equation

$$(xLyL\widehat{x}L\widehat{y}L)^2 = uzLtL\widehat{z}L\widehat{t}Lv, \tag{5}$$

where $|x| = 3$, $|y| = 4$, $|z| = 3$, $|t| = 4$, $|u| = 3$ and $|v| = 15$. As in the first example, we add vertices x_1, x_2, x_3 for the variable x, vertices $\widehat{x}_1$, $\widehat{x}_2$, $\widehat{x}_3$ for the variable $\widehat{x}$, and so on for the other variables. Since u and v are not interesting

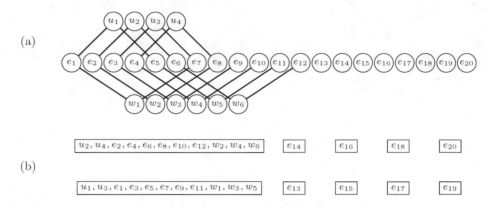

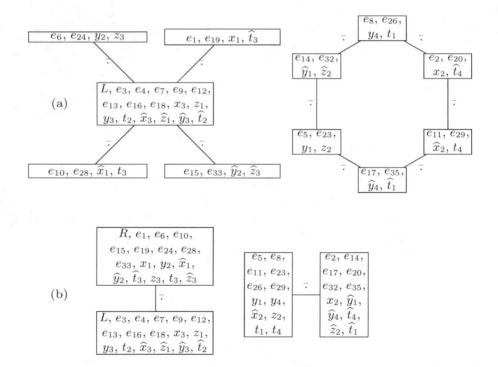

Fig. 2. The relation between letters for the equation $uuv = wwx$. (a) The initial graph. (b) The resulting graph whose connected components have been merged.

Fig. 3. (a) The merged graph obtained from Equation (5). It has two connected components. Each label is identified with the morphism $\bar{\cdot}$. (b) Resulting graph after having merged further the vertices with the information deduced from the involutory morphism $\bar{\cdot}$.

for the problem (they may be considered as slack variables), we do not show them in the graph. After having merged the connected components, we obtain a graph with 11 vertices (see Figure 3). Next, we connect vertex i to vertex j with label ℓ if $\ell(i) = j$. In this case, the label is the morphism $\bar{\cdot}$.

Let us denote by $[v]$ the vertex containing the value v. We can see from the graph of Figure 3 that additional information can be deduced. For instance, since L belongs to the vertex $[e_3]$, this implies that $[e_6] = [e_1] = [e_{10}] = [e_{15}] = R$. In the same spirit, since $\bar{\cdot}$ is involutory, the cycle with 6 vertices to the right could be reduced to a cycle with 2 vertices since $[e_2] = [e_{17}] = [e_{14}]$ and $[e_5] = [e_8] = [e_{11}]$. Therefore, there are exactly three solutions to Equation (5), when the lengths are fixed as above, according to whether $[e_5] = R$, $[e_5] = L$ or $[e_5] = F$.

5.2   The General Case

The two examples of Subsection 5.1 are easily generalized to arbitrary equations with known lengths. To construct the graph, we proceed as follows. Let

$$(u_1, u_2, \ldots, u_\ell) = (v_1, v_2, \ldots, v_r)$$

be an equation of length n with fixed lengths and uniform morphisms and antimorphisms. First, we create a vertex for every constant occuring in the equation. Also, we create a vertex e_i for $i = 1, 2, \ldots, n$. Next, for every variable v having length k, we create a vertex v_i for $i = 1, 2, \ldots, k$. Finally, for every variable v such that $\varphi(v)$ occurs in the equation, we introduced the vertex $\varphi(v)_i$, for $i = 1, 2, \ldots, |v| \cdot |\varphi|$.

It remains to describe how to construct the arcs. Clearly, if some variable v occurs at position i in the equation, then we add the non directed edge $\{v_{1+j}, e_{i+j}\}$ for $j = 0, 1, \ldots, |v| - 1$. The same applies for constants occurring in the equation, as well as *morphic expressions*, i.e. expressions of the form $\varphi(v)$ for some expression v. It only remains to establish a link between a variable v and its image under some morphism φ. If φ is 1-uniform, then it suffices to add an arc $(v_i, \varphi(v)_i)$ for $i = 1, 2, \ldots, |v|$. On the other hand, if φ is k-uniform for $k > 1$, then we add edges of the form $(v_i, \varphi(v)_{|\varphi| \cdot (i-1)+j})$ for $i = 1, 2, \ldots, |v|$ and $j = 1, 2, \ldots, |\varphi|$. Notice that the direction of the arc is important. In particular, whenever the restriction of the morphism is invertible, it is important to add an arc in the opposite direction labelled with the inverse. Figures 2 and 3 illustrate the construction.

When the graph is constructed, it is possible to reduce it according to different simple rules:

(1) We can merge every pair of vertices linked by an unlabelled edge, since they correspond to equal values.
(2) If some vertex v contains a constant letter a, then we can derive the value of any other vertex u connected to v by some morphism φ. It suffices to apply φ to a and to merge the result with the vertex u.
(3) If two arcs with the same label leave some vertex v to vertices u_1 and u_2, then we can merge vertices u_1 and u_2.

(4) If $\varphi^n = \varphi$ for some positive integer n, then we merge any pair of vertices to the ends of directed paths composed of n arcs labelled φ.

(5) If $\varphi(a) = a$ for a unique letter $a \in \Sigma$ and if there is a loop labelled φ from vertex u to itself, then we can merge u and a.

In the first example, only Rule (1) has been applied, since there is no morphism/antimorphism involved. On the other hand, in the second example, we used Rule (1) to get the graph of Figure 3(a), and then we used Rules (2) and (4) to reduce the graph further.

It is also worth mentioning that in a concrete implementation, vertices are not actually merged, but are rather assigned a common *representant* as for the *union-find* datastructure, thus guaranteeing a better theoretical complexity. Therefore, to solve any equation on words, it suffices to try every possible lengths for the variable, construct the graph described above and then compute either one or all solutions by applying Rules (1)–(5).

6   Implementation

To assess the practical feasibility of this representation, we implemented it into a generic word equation solver made of 1,800 lines of Python code, which is publicly available under an open source license.[1] The solver provides a uniform interface for the solving of word equations through two different methods. The first method is the graph technique presented in this paper; the second method converts the input data into an equivalent Boolean satisfiability problem (SAT), as was mentioned earlier [10].

Figure 4 shows a sample input file for the solver. The first line must be the enumeration of the alphabet; the second statement lists the word variables used in the equations, which must differ from the alphabet symbols. Each word must have at least one length; either a single value (first line), a list or a range of lengths. Morphisms are then defined on alphabet symbols, and the set of equations ends the file. The token * stands for concatenation and ~ is the reversal operator. The solver return any combination of words that meets all equations.

```
Alphabet [a,b,c];
Word u,v,w;
|u| = 2;
|v| = [1,2,4];
|w| = [3-6];
Morphism sigma : a->b, b->c, c->a;
u*w=ab*v*ba;
sigma(u)*v=abba*~(v);
```

Fig. 4. Sample input file for the word equation solver

---

[1] http://github.com/ablondin/solver

The solver was run on a batch of four equations corresponding to four classes of problems: 1) Only variables and constants: $uwba = wabv$; 2) Variables, constants and mirror operator: $\widetilde{u}baw = ab\widetilde{w}v$; 3) Variables, constants and morphisms: $\sigma(u)wba = \sigma(w)abv$; 4) Variables, constants, mirror operator and morphisms: $\widetilde{u}abw = \sigma(w)bav$. For each of these equations, we ran the solver by increasing either the size of the alphabet or the length of the solutions to find. The running times for both experiments are shown respectively in Figures 5 and 6. The computed times were obtained on an Intel Xeon W3520 CPU running at 2.67 GHz. For the sake of completeness, we ran the solver on each problem for both the graph- and the SAT-based method and plotted their results on the same figure.

We observed that, while both alphabet size and word length progressively increase the solver's running time, the main contributing factor, for the equations we studied, is word length. Indeed, using variables of length 10 yields a solving time of more than 400 seconds for most equations, irrespective of the algorithm used, while increasing the size of the alphabet up to 25 requires a solving time of less than 200 seconds even for the worst case measured.

The use of morphisms in the graph-based method takes its toll on solving time. This can be clearly seen in Figure 5, where the growth of solving time for both equations involving morphisms exceeds that of all other problem instances. The same trend can be observed in Figure 6. However, we can also observe that, for equations that do not involve morphisms, the use of the graph technique outperforms the SAT-based approach for all problem instances. This is best illustrated in Figure 6, where the graph-based algorithm beats the SAT solver by a large margin; for example with word length fixed to 10, equation (2) is solved

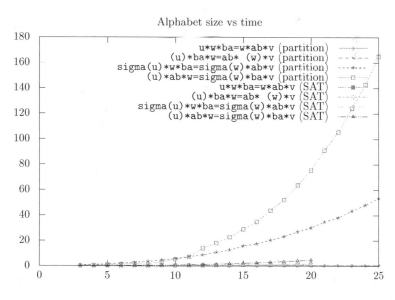

Fig. 5. Solving time of various equations for increasing alphabet size (word length fixed to 5)

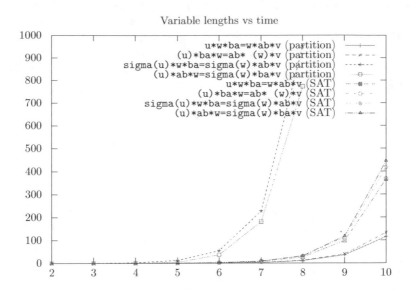

Fig. 6. Solving time of various equations for increasing word length (alphabet size fixed to 2)

in 130 seconds by the graph solver, while the SAT solver takes 420 seconds to solve the *same* problem instance —an almost fourfold increase. This gap further widens as the word length increases.

This finding is especially pleasing, since SAT solvers are generally regarded as heavily-optimized software for most NP-complete problems. The fact that the graph-based method, implemented in an interpreted language such as Python, still surpasses SAT solvers for this class of problems is a telling sign of its adequacy. As for the use of morphisms, an initial profiling of the graph solver indicates that there is ample room for improvement through the use of more optimized data structures to handle the merging of vertices in a graph.

7   Conclusion

The resolution of equations on words is a fertile concept that is being increasingly used for the modelling of many mathematical problems. However, the absence of a dedicated and efficient solver for word equations slows down its adoption by researchers; in that respect, the encoding of word equations as graph structures, as presented in this paper, turns out to be a promising way to compute solutions to a system of word equations.

The implementation of a general-purpose solver reveals that solutions to complex equations can be computed efficiently; in addition, the output returned by the solver does not merely enumerate solutions, but provides the user with a graph revealing the structure of these solutions. Additions and future work on this topic includes the optimization of node-merging code, the relaxation of bound conditions on variables and the handling of non-uniform morphisms. In time, the widespread use of such a solver can prove instrumental in extending the reach of word equations to other mathematical problems of importance.

References

1. Abdulrab, H.: Implementation of Makanin's algorithm. In: Schulz, K.U. (ed.) IWWERT 1990. LNCS, vol. 572, pp. 61–84. Springer, Heidelberg (1992)
2. Blondin Massé, A., Garon, A., Labbé, S.: Generation of double square tiles. Theoretical Computer Science (2012) (to appear)
3. Brlek, S.: Interactions between digital geometry and combinatorics on words. In: Ambroz, P., Holub, S., Masáková, Z. (eds.) WORDS. EPTCS, vol. 63, pp. 1–12 (2011)
4. Eén, N., Sörensson, N.: An extensible SAT-solver. In: Giunchiglia, E., Tacchella, A. (eds.) SAT 2003. LNCS, vol. 2919, pp. 502–518. Springer, Heidelberg (2004)
5. Fraenkel, A.S., Simpson, J.: How many squares can a string contain? Journal of Combinatorial Theory, Series A 82(1), 112 – 120 (1998),
   http://www.sciencedirect.com/science/article/pii/S0097316597928430
6. Fu, X., Li, C.-C.: A string constraint solver for detecting web application vulnerability. In: SEKE, pp. 535–542. Knowledge Systems Institute Graduate School (2010)
7. Gusfield, D.: Algorithms on strings, trees, and sequences: computer science and computational biology. Cambridge Univ. Press (January 2007),
   http://www.amazon.com/exec/obidos/
   redirect?tag=citeulike07-20&path=ASIN/0521585198
8. Kiezun, A., Ganesh, V., Artzi, S., Guo, P., Hooimeijer, P., Ernst, M.: HAMPI: A solver for word equations over strings, regular expressions and context-free grammars. ACM Trans. on Software Engineering and Methodology 21(4) (2012) (to appear)
9. Kiezun, A., Ganesh, V., Guo, P.J., Hooimeijer, P., Ernst, M.D.: Hampi: a solver for string constraints. In: Rothermel, G., Dillon, L.K. (eds.) ISSTA, pp. 105–116. ACM (2009)
10. Larouche, M., Blondin Mass, A., Gaboury, S., Hall, S.: Solving equations on words through Boolean satisfiability. In: Maldonado, J.C., Shin, S.Y. (eds.) SAC, pp. 104–106. ACM (2013)
11. Lothaire, M.: Algebraic Combinatorics on Words. Cambridge University Press (2002)
12. Lothaire, M.: Applied Combinatorics on Words (Encyclopedia of Mathematics and its Applications). Cambridge University Press, New York (2005)
13. Makanin, G.: The problem of solvability of equations in a free semigroup. Mathematics of the USSR-Sbornik 32(2), 129 (1977)
14. Plandowski, W.: Satisfiability of word equations with constants is in PSPACE. In: FOCS, pp. 495–500. IEEE Computer Society (1999)
15. Yu, F., Bultan, T., Ibarra, O.H.: Relational string verification using multi-track automata. Int. J. Found. Comput. Sci. 22(8), 1909–1924 (2011)

On the Arithmetics of Discrete Figures[*]

Alexandre Blondin Massé[1,2,**], Amadou Makhtar Tall[1], and Hugo Tremblay[1,2]

[1] Laboratoire d'informatique formelle, Université du Québec à Chicoutimi
555, boul. de l'Université, Chicoutimi, G7H 2B1, Canada
{ablondin,amadou-makhtar.tall1}@uqac.ca
[2] Laboratoire de combinatoire et d'informatique mathématique
Université du Québec à Montréal
CP 8888, Succ. Centre-ville, Montral (Qubec) H3C 3P8
hugo.tremblay@lacim.ca
http://lif.uqac.ca

Abstract. Discrete figures (or polyominoes) are fundamental objects in combinatorics and discrete geometry, having been studied in many contexts, ranging from game theory to tiling problems. In 2008, Provençal introduced the concept of prime and composed polyominoes, which arises naturally from a *composition* operator acting on these discrete figures. Our goal is to study further polyomino composition and, in particular, factorization of polyominoes as a product of prime ones. We provide a polynomial time (with respect to the perimeter of the polyomino) algorithm that allows one to compute such a factorization. As a consequence, primality of polyominoes can be decided in polynomial time.

Keywords: Discrete figures, polyominoes, boundary words, primality, tiling, morphism.

1 Introduction

Although polyominoes are known since antiquity, it is only in 1953 that the word was coined by S.W. Golomb and was later popularized by M. Gardner, who was very active in recreational mathematics for a large part of the 20th century [6]. In fact, polyominoes are well-known from the general public: One only needs to think about the very popular Tetris video game, which consists in filling lines with tetrominoes (i.e. polyominoes composed of four unit cells). Polyominoes are also well-known and useful in combinatorics and theoretical computer science. For instance, one of the applications of the famous dancing links algorithm proposed by D. Knuth in 2000 consists in solving polyomino tiling puzzles efficiently [7] (an implementation of such a solver is found in [8]).

Polyominoes have also been fundamental objects in the study of tilings. It is known since 1970 that the problem of tiling the plane with free polyominoes (polyominoes that can be rotated and reflected) picked from a finite set is

---

[*] This research is supported by the Natural Sciences and Engineering Research Council of Canada (NSERC).

[**] Corresponding author.

A.-H. Dediu et al. (Eds.): LATA 2014, LNCS 8370, pp. 198–209, 2014.

Fig. 1. Two double parallelogram tiles and the tilings they induce. The black dots indicate the points shared by four copies of the tile. (a) A Christoffel tile. (b) A Fibonacci tile (see [10] for more details).

undecidable [5]. When restricted to only one tile, it is not known if it is decidable or not. Also, the problem of deciding if a given polyomino tiles another polyomino is known to be NP-complete [13]. In the case where no rotation and no reflection is allowed, the problem becomes much simpler. Indeed, not only is it decidable, but it can also be determined in polynomial time (a $\mathcal{O}(n^2)$ bound is proved in [4], which was reduced to $\mathcal{O}(n)$ for tilings whose boundary has bounded local periodicity [2]) by using results from [1,16]. The basic idea comes from D. Beauquier and M. Nivat, who characterized such objects by the shape of their boundary: Indeed, they are polyominoes whose boundary can be divided into four or six pieces that are pairwise parallel [1]. Pseudo-square tiles have been extensively studied (see for instance [3] and [12]).

In [11], Blondin Massé et al. considered the problem of enumerating polyominoes called *double parallelogram tiles*, i.e. polyominoes yielding two distinct parallelogram tilings (see Figure 1). In order to prove one of their main results, they had to rely on the concept of *prime* and *composed* polyominoes, introduced by Provençal in his Ph.D thesis [15]. However, very little is known about that classification.

This article is divided as follows. In Section 2, we introduce the usual definitions and notation. Section 3 is devoted to the link between discrete paths and words. Composition, prime and composed polyominoes are defined in Section 4. We provide the main algorithms of this article in Section 5 and we briefly conclude in Section 6.

2   Definitions and Notation

The *square grid* is the set $\mathbb{Z}^2$. A *cell* is a unit square in $\mathbb{R}^2$ whose corners are points of $\mathbb{Z}^2$. We shall denote by $c(i,j)$ the cell

$$c(i,j) = \{(x,y) \in \mathbb{R}^2 \mid i \leq x \leq i+1, j \leq y \leq j+1\},$$

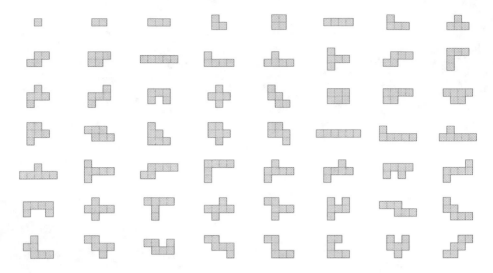

Fig. 2. Free polyominoes without hole having area at most 6

i.e. (i, j) is the bottom left corner of the cell. The set of all cells is denoted by $\mathcal{C}$. Two cells $c, d \in \mathcal{C}$ are called 4-*neighbors* if they have exactly one side in common. Clearly, every cell $c(i, j)$ has exactly four 4-neighbors: $c(i + 1, j)$, $c(i, j + 1)$, $c(i - 1, j)$ and $c(i, j - 1)$. A 4-*connected region of* $\mathbb{Z}^2$ is any subset R of $\mathbb{R}^2$ such that for every $c, d \in R$, there exist cells $c_1, c_2, \ldots, c_k$ such that $c = c_1$, $d = c_k$, and c_i, c_{i+1} are 4-neighbors for $i = 1, 2, \ldots, k - 1$.

The usual isometries (translation, rotation and reflections) are naturally defined on 4-connected regions. In some cases, it is convenient to consider two regions equivalent up to some isometries. For instance, the relation $R \sim_\theta S$ defined by "R is a translated copy of S" is an equivalence relation: A *fixed polyomino* (or simply *polyomino*) is any equivalence class of $\sim_\theta$. In the same spirit, if the relation $R \sim_\rho S$ is defined by "R is a translated or rotated copy of S", then a *one-sided polyomino* is any equivalence class of $\sim_\rho$. Finally, a *free polyomino* is any equivalence class of the relation $R \sim_\sigma S$ defined by "R is a translated, rotated, reflected or glided reflected copy of S".

Another notion of interest is that of *holes*. Given a region R, let $\overline{R}$ denote its *complement*, i.e. $\overline{R} = \mathcal{C} - R$. We say that R is *without hole* if $\overline{R}$ is a 4-connected region. All free polyominoes of c cells, $c = 1, 2, 3, 4, 5, 6$, are illustrated in Figure 2.

In many situations, it is convenient to represent a polyomino by its boundary, which in turn is easily represented by a word on a 4-letter alphabet encoding the elementary steps $\rightarrow$ (east), $\uparrow$ (north), $\leftarrow$ (west) and $\downarrow$ (south). In the following, we recall basic definitions from combinatorics on words [9].

An *alphabet* $\mathcal{A}$ is a finite set whose elements are *letters*. A finite word w is a function $w : \{1, 2, \ldots, n\} \to \mathcal{A}$, where w_i (also denoted by $w[i]$) is the i-th letter, $1 \leq i \leq n$. The *length* of w, denoted by $|w|$, is the integer n. The *empty word* ε is the unique word having length 0.

The *free monoid* $\mathcal{A}^*$ is the set of all finite words over $\mathcal{A}$. The *reversal* of $w = w_1 w_2 \cdots w_n$ is the word $\widetilde{w} = w_n w_{n-1} \cdots w_1$. Given a nonempty word w, let $\mathrm{FST}(w) = w_1$ and $\mathrm{LST}(w) = w_n$ denote respectively the first and last letter of the word w. A word u is a *factor* of another word w if there exist $x, y \in \mathcal{A}^*$ such that $w = xuy$. A *proper factor* of w is any factor u such that $u \neq \varepsilon$ and $u \neq w$. The factor u of w starting at position i and ending with position j is denoted by $w[i:j]$ and the integer i is called an *occurrence of u in w*. Moreover, if $x = \varepsilon$, then u is called *prefix* and if $y = \varepsilon$, it is called a *suffix* of w. We denote by $|w|_u$ the number of occurrences of u in w. Two words u and v are *conjugate*, written $u \equiv v$ or sometimes $u \equiv_{|x|} v$, when x, y are such that $u = xy$ and $v = yx$. Conjugacy is an equivalence relation and the class of a word w is denoted by $[w]$. A *power* of a word u is a word of the form u^k for some integer $k \in \mathbb{N}$. It is convenient to set $u^0 = \varepsilon$ for each word u.

Given two alphabets $\mathcal{A}$ and $\mathcal{B}$, a *morphism* is a function $\varphi : \mathcal{A}^* \to \mathcal{B}^*$ compatible with concatenation, that is, $\varphi(uv) = \varphi(u)\varphi(v)$ for any $u, v \in \mathcal{A}^*$. It is clear that a morphism is completely defined by its action on the letters of $\mathcal{A}$. In the same spirit, an *antimorphism* is a function $\varphi : \mathcal{A}^* \to \mathcal{B}^*$ such that $\varphi(uv) = \varphi(v)\varphi(u)$ whenever $u, v \in \mathcal{A}^*$. The reversal $\widetilde{\cdot}$ is an antimorphism and it is not difficult to show that for any antimorphism φ, we have $\varphi = \widetilde{\cdot} \circ \varphi'$, where φ' is a morphism, i.e. φ can be expressed as the composition of a morphism and the reversal operator.

3   Paths as Words

The Freeman chain code $\mathcal{F} = \{\mathbf{0}, \mathbf{1}, \mathbf{2}, \mathbf{3}\}$ is considered as the additive group of integers modulo 4. To distinguish the number (for instance 0) from the letter (for instance $\mathbf{0}$), we shall denote the latter with bold font. Basic transformations on $\mathcal{F}$ are rotations $\rho^i : x \mapsto x + i$ and reflections $\sigma_i : x \mapsto i - x$, which extend uniquely to morphisms on $\mathcal{F}^*$. Two other useful functions for our purpose are the morphism $\overline{\cdot}$ defined by $\mathbf{0} \leftrightarrow \mathbf{2}$ and $\mathbf{1} \leftrightarrow \mathbf{3}$ and the antimorphism $\widehat{\cdot} = \overline{\cdot} \circ \widetilde{\cdot}$.

From now on, words are considered over $\mathcal{F}$ and are called *paths* to emphasize their geometrical nature. A path w is *closed* if it satisfies $|w|_{\mathbf{0}} = |w|_{\mathbf{2}}$ and $|w|_{\mathbf{1}} = |w|_{\mathbf{3}}$, and it is *simple* if no proper factor of w is closed. A *boundary word* is a simple and closed path. It is convenient to represent each closed path w by its conjugacy class $[w]$, also called *circular word*. It follows from this definition that counter-clockwise circular boundary words and the polyominoes without hole they describe are in bijection.

In [1], D. Beauquier and M. Nivat proved that a polyomino P tiles the plane by translation if and only if it admits a boundary word

$$w = xyz\widehat{x}\widehat{y}\widehat{z},$$

with at most one empty word among x, y and z. If x, y or z is empty, then P is called *parallelogram polyomino* (*pseudo-square* in [2]). Otherwise, it is called

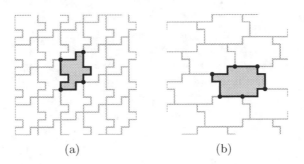

(a)　　　　　　　　　　　　(b)

Fig. 3. (a) A parallelogram polyomino and the tiling it induces (one of its boundary words admits a BN-factorization **0010·101211·2322·330323**). The black dots indicate the factorization points of the boundary word and they clearly form a parallelogram. (b) An hexagon polyomino having boundary word **000·0010·1121·222·2322·3033** and the corresponding tiling.

hexagon polyomino (pseudo-hexagon in [2]). It should be noted that parallelogram polyominoes were introduced more than 40 years ago with a completely different meaning [14], but we still prefer to use the word "parallelogram" in this case as it seems the most appropriate word for the concept (see Figure 3).

The following simple fact, proven in [11], is one of the key idea for designing Algorithm 1. Roughly speaking, it states that the factors of any parallelogram polyomino always start and end with the same elementary step, and that all four letters are exactly covered once:

Proposition 1 ([11]). *Let* $w \equiv xy\widehat{x}\widehat{y}$ *be a boundary word of a parallelogram polyomino. Then* $\mathrm{FST}(x) = \mathrm{LST}(x)$, $\mathrm{FST}(y) = \mathrm{LST}(y)$ *and the first letter of* x, $\widehat{x}$, y, $\widehat{y}$ *are mutually distinct, that is,*

$$\{\mathrm{FST}(x), \mathrm{FST}(\widehat{x}), \mathrm{FST}(y), \mathrm{FST}(\widehat{y})\} = \{\mathbf{0}, \mathbf{1}, \mathbf{2}, \mathbf{3}\}.$$

4   Prime and Composed Polyominoes

On the Freeman alphabet, a class of morphisms is of particular interest for our purpose:

Definition 1 ([11,15]). *A morphism* $\varphi : \mathcal{F}^* \to \mathcal{F}^*$ *is called* homologous *if* $\varphi(a) = \widehat{\varphi(\overline{a})}$ *for every* $a \in \mathcal{F}$.

Roughly speaking, homologous morphisms replace the two horizontal elementary steps by an arbitrary path ($\varphi(\mathbf{0})$ is traveled in the opposite direction with respect to $\varphi(\mathbf{2})$) and the same idea applies to vertical steps. One proves easily that homologous morphisms satisfy $\widehat{\varphi(w)} = \varphi(\widehat{w})$ for any $w \in \mathcal{F}^*$.

As polyominoes without hole have simple boundary word, an additional condition on homologous morphism is necessary in order to define prime and composed polyominoes without hole.

Definition 2. *Let φ be an homologous morphism. We say that φ is a* parallelogram morphism *if*

(i) $\varphi(\mathbf{0123})$ *is the boundary word of a parallelogram polyomino;*
(ii) $\mathrm{FST}(\varphi(a)) = a$ *for every $a \in \mathcal{F}$.*

The purpose of condition (ii) of Definition 2 is justified by the following extension of Proposition 1:

Proposition 2. *Let φ be a parallelogram morphism. Then for every $a \in \mathcal{F}$, $\mathrm{FST}(\varphi(a)) = a = \mathrm{LST}(\varphi(a))$.*

Let $\mathcal{M}$ be the set of morphisms on $\mathcal{F}$, $\mathcal{H}$ the set of homologous morphisms and $\mathcal{P}$ the set of parallelogram morphisms.

Proposition 3. *$\mathcal{H}$ is a submonoid of $\mathcal{M}$ and $\mathcal{P}$ is a submonoid of $\mathcal{H}$ with respect to the concatenation.*

Proof. The identity morphism Id is both an homologous and parallelogram morphism. Now, let $\varphi, \varphi' \in \mathcal{H}$. Then for every letter $a \in \mathcal{F}$,

$$(\widehat{\varphi \circ \varphi'})(\overline{a}) = \widehat{\varphi(\varphi'(\overline{a}))} = \varphi(\widehat{\varphi'(\overline{a})}) = \varphi(\varphi'(a)) = (\varphi \circ \varphi')(a),$$

so that $\varphi \circ \varphi' \in \mathcal{H}$ as well. Similarly, on one hand, if $\varphi, \varphi' \in \mathcal{P}$, then

$$\mathrm{FST}((\varphi \circ \varphi')(a)) = \mathrm{FST}((\varphi(\varphi'(a))) = \mathrm{FST}(\varphi'(a)) = a.$$

On the other hand $(\varphi \circ \varphi')(\mathbf{0123})$ is closed. The proof that $(\varphi \circ \varphi')(\mathbf{0123})$ is simple is more technical and is ommitted due to lack of space.   □

Every parallelogram morphism φ is associated with a unique polyomino, denoted by $\mathrm{POLY}(\varphi)$. Conversely, one might be tempted to state that each parallelogram polyomino is uniquely represented by a parallelogram morphism. Indeed, condition (i) in Definition 2 ensures that the boundary word is traveled counterclockwise and that only one of its conjugate is chosen. However, even when taking those restrictions into account, there exist parallelogram polyominoes with two distinct parallelogram factorizations:

Example 1. The parallelogram morphisms φ defined by $\varphi(0) = 010, \varphi(1) = 121$ and φ' defined by $\varphi'(0) = 030, \varphi'(1) = 101$ both yield the X pentamino (see the polyomino in the third row, fourth column of Fig. 2).

In fact, every double parallelogram tile admits two nontrivial distinct factorizations [11].

We are now ready to define prime and composed polyominoes. It shall be noted that prime polyominoes were defined in [15], but the (equivalent) definition below relies on parallelogram morphisms:

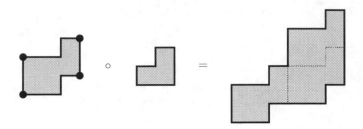

Fig. 4. Composition of a polyomino with a parallelogram polyomino. The parallelogram factorization is illustrated with black dots. The resulting composed polyomino can be tiled by copies of the parallelogram polyomino.

Definition 3 ([15]). *A polyomino P distinct from the one cell polyomino is called* composed *if there exists some boundary word u and some parallelogram morphism φ such that*

(i) $\mathrm{POLY}(\varphi(u)) = P$;
(ii) $\mathrm{POLY}(u)$ *is not the unit square;*
(iii) $\varphi \neq \mathrm{Id}$.

Otherwise, P is called prime.

In other words, a polyomino is prime if and only if it cannot be decomposed as a product of some parallelogram polyomino and some other polyomino. It is worth mentioning that the problem of deciding if a polyomino is prime is as least as hard as the problem of deciding if a number is prime. Indeed, one notices that the rectangle $1 \times n$ having boundary word $\mathbf{0}^n\mathbf{1}\mathbf{2}^n\mathbf{3}$ is prime if and only if n is prime. In the same spirit, a boundary word u is called *prime* if $\mathrm{POLY}(u)$ is a prime polyomino and a parallelogram morphism φ is called *prime* if $\mathrm{POLY}(\varphi(\mathbf{0123}))$ is a prime polyomino.

Following Definition 3, one is naturally led to ask whether the fundamental theorem of arithmetic can be extended to polyominoes. There are two conditions to verify: the existence of a prime factorization and the unicity of this factorization. The former is easy to prove:

Theorem 1. *Let P be any non-unit polyomino without hole. Then either P is a prime polyomino or there exist prime parallelogram morphisms $\varphi_1, \varphi_2, \ldots, \varphi_n$ and a prime boundary word u such that $P = \mathrm{POLY}((\varphi_1 \circ \varphi_2 \circ \ldots \circ \varphi_n)(u))$.*

Proof. By induction on the perimeter of P. If P is prime, then there is nothing to prove. Otherwise, there exist a parallelogram morphism $\varphi \neq \mathrm{Id}$ and a boundary word u, with $\mathrm{POLY}(u)$ different for the unit square, such that $P = \mathrm{POLY}(\varphi(u))$. By induction applied to $\mathrm{POLY}(u)$, there exist parallelogram morphisms $\varphi_1, \varphi_2, \ldots, \varphi_n$ and a prime boundary word v such that

$$u = (\varphi_1 \circ \varphi_2 \circ \ldots \circ \varphi_n)(v),$$

so that $P = \text{POLY}((\varphi \circ \varphi_1 \circ \varphi_2 \circ \ldots \circ \varphi_n)(v))$. If φ is prime, then the claim is proved. Otherwise, using induction, one shows that $\varphi = \varphi'_1 \circ \varphi'_2 \circ \ldots \circ \varphi'_k$ for some prime parallelogram morphisms $\varphi'_1, \varphi'_1, \ldots, \varphi'_k$, which concludes the proof. □

The proof of unicity seems rather involved and is discussed in the last section (see Conjecture 1).

5   Algorithms

In this section, we shift our attention to the algorithmic aspects surrounding composed and prime polyominoes. Basically, to determine if a polyomino P is composed, we need to find a parallelogram morphism $\varphi \neq \text{Id}$ and a boundary word $u \notin [\mathbf{0123}]$ such that $\varphi(u)$ is a boundary word of P. If no such morphism exists, then we may conclude that the polyomino is prime.

5.1   Naive Version

The simplest straightforward approach consists of trying every possible factorization until either one is found or all have been exhausted. To reduce the number of cases to be considered, by Proposition 2, we may restrict ourselves to factors starting and ending with the same letter. More precisely, let P be any polyomino and w one of its boundary word. For every $a \in \mathcal{F}$, let F_a be the set of factors of w^2 starting and ending with a. The following steps can be used to factorize P:

1. Compute $F_{\mathbf{0}}$ and $F_{\mathbf{1}}$;
2. Let $u \in F_{\mathbf{0}}$, $v \in F_{\mathbf{1}}$;
3. Let φ be the parallelogram morphism induced by u and v;
4. If there is some conjugate w' of w such that $w = \varphi(w')$, then return (φ, u).
5. Otherwise, repeat steps 2–4 until all possible u and v have been exhausted.

The complexity of the previous algorithm is clearly polynomial.

Theorem 2. *Any polyomino P may be factorized as a product of prime polyominoes in $\mathcal{O}(n^6)$, where n is the perimeter of P.*

Proof. Let w be any boundary word of P. Step 1 is done in $\mathcal{O}(n^2)$, as there are $\mathcal{O}(n^2)$ factors starting and ending with some letter in any word on $\mathcal{F}$. Steps 2–4 are then repeated $\mathcal{O}(n^4)$. The construction of φ in Step 3 is done in constant time, while Step 4 takes $\mathcal{O}(n)$ since it must be performed for every conjugate of w. Therefore, decomposing P as $\varphi(u)$, for some parallelogram morphism φ and some boundary word u is done in $\mathcal{O}(n^5)$. Since it must be repeated as long as either φ or u is not prime, the overall complexity is $\mathcal{O}(n^6)$. □

As a consequence:

Corollary 1. *Given a polyomino P having perimeter n, it can be decided in polynomial time with respect to n whether P is prime or composed.* □

The algorithm was implemented in Python and tested for polyominoes having number of cells between 1 and 10. Figure 5 contains the result. As there are many more prime than composed polyominoes, only the composed ones are illustrated.

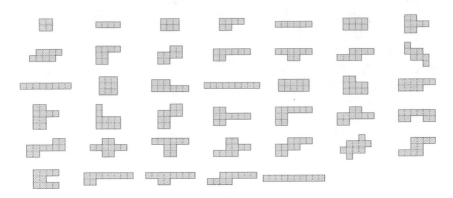

Fig. 5. Free polyominoes having area at most 10 that are composed

5.2   Improving the Naive Algorithm

An upper bound of $\mathcal{O}(n^6)$ is rather crude and one should expect to reduce it. More precisely, instead of enumerating all possible factors u in F_0 and v in F_1, the choice should be sensible about whether u and v occur contiguously or with overlap. This yields an algorithm such as the following one. As a first step, choose any boundary word w starting with $\mathbf{0}$. The first idea consists in trying to divide w into blocks starting and ending with the same letter, in virtue of Proposition 2. It suffices to look at every occurrence of $\mathbf{0}$ for the ending of $\varphi(\mathbf{0})$. When such a block is chosen, then $\varphi(\mathbf{0})$ and $\varphi(\mathbf{2})$ are completely determined.

The next step consists in checking the letter following the first block. If it is $\mathbf{0}$ or $\mathbf{2}$, then we check if the following letters match $\varphi(\mathbf{0})$. If it is not the case, then we have to try with another choice for $\varphi(\mathbf{0})$. On the other hand, if there is a match, then we go on to the next block. We repeat the previous steps until we reach either the letter $\mathbf{1}$ or $\mathbf{3}$. In the same manner as for the letter $\mathbf{0}$, we then try every possible block for either $\mathbf{1}$ or $\mathbf{3}$ (and then the image of the last letter under φ is uniquely determined). When the four images of φ over single letters are chosen, it only remains to verify if the boundary word may be factorized as a product of the four blocks (this step is called the *decoding step*).

Based on the previous paragraphs, one might design a branch-and-bound algorithm for factorizing any polyomino. The pseudocode is found in Algorithm 1 and it was also implemented in Python.

Theorem 3. *Any polyomino P may be factorized as a product of prime polyominoes in $\mathcal{O}(n^5)$, where n is the perimeter of P.*

Proof. Let w be any boundary word of P starting with $\mathbf{0}$. Choosing each occurrence of $\mathbf{0}$ in w to construct $\varphi(\mathbf{0})$, there are at most n possible values (and then $\varphi(\mathbf{2})$ is determined). Once $\varphi(\mathbf{0})$ is chosen, there are at most n possible values for $\varphi(\mathbf{1})$ (and then $\varphi(\mathbf{3})$ is determined). Finally, when $\varphi(a)$ is known for each $a \in \mathcal{F}$, it remains to verify if w might be decoded from φ, which is done in linear time at most. Therefore, it can be decided in $\mathcal{O}(n^3)$ whether w is decodable according

---

Algorithm 1. Improved factorization of polyomino

```
 1  Require: w is simple;
 2  Function Factorize(w : boundary word)
 3  begin
 4    │  for i ∈ Occurrences(0, w) do
 5    │  │    /* We try to factorize every conjugate starting with 0  */
 6    │  │    u ← Conjugate(w, i)
 7    │  │    φ ← FactorizeRec(w, ∅, 0)
 8    │  │    if φ ≠ ∅ then
 9    │  │    └    return φ
10    │  return ∅

11  Function FactorizeRec(w : boundary word, φ : morphism, i : integer)
12  begin
13    │  if i ≥ |w| then
14    │  │    /* We first check if the decoding is complete        */
15    │  │    if φ is completely defined and non-trivial then
16    │  │    └    return φ
17    │  │    else
18    │  │    └    return ∅
19    │  else
20    │  │    ℓ ← w[i]
21    │  │    if φ(ℓ) is defined then
22    │  │    │    /* The next block should match φ(ℓ)              */
23    │  │    │    k ← |φ(ℓ)|
24    │  │    │    if k > |w| − i or w[i : i + k] then
25    │  │    │    └    return ∅
26    │  │    │    else
27    │  │    │    └    return FactorizeRec(w[i + k : |w| − 1], φ, i + k)
28    │  │    else
29    │  │    │    /* We consider constructions of the next block   */
30    │  │    │    for j ∈ Occurrences(ℓ, w[i : |w| − 1]) do
31    │  │    │    │    φ(ℓ) ← w[i : j]
32    │  │    │    │    φ(ℓ̄) ← ŵ[i : j]
33    │  │    │    │    φ ← FactorizeRec(w[j + 1 : |w| − 1], φ, j + 1)
34    │  │    │    │    if φ is not trivial then
35    │  │    │    │    └    return φ
36    │  │    │    │    else
37    │  │    │    │    └    return ∅
38    │  │    └    return ∅
```

to some parallelogram morphism φ. Since the test must be performed for every conjugate of w starting with $\mathbf{0}$, we obtain a bound of $\mathcal{O}(n^4)$. Finally, repeating this reduction until a prime decomposition is obtained yields the claimed $\mathcal{O}(n^5)$ complexity. $\square$

5.3   Improving the Upper Bound

The $\mathcal{O}(n^5)$ bound seems rather high and it is not clear whether it can be realized. To support this impression, let us study Algorithm 1 on the polyomino P having boundary word

$$w = \mathbf{0}^k \mathbf{1}^k \mathbf{2}^{k-1} \mathbf{3} \mathbf{2} \mathbf{3}^{k-1}.$$

Let $n = |w| = 4k$. Since P is a square $k \times k$ minus one corner cell, one might prove that it is prime. For the algorithm to take as much time as possible, assume that k has d divisors. Then the algorithm will try the k conjugate of w starting with 0 and will construct d images for $\varphi(\mathbf{0})$. Similarly, we will have to consider d possible images for $\varphi(\mathbf{1})$. The decoding being performed in linear time, we get an overall bound of $\mathcal{O}(kd^2 n) = \mathcal{O}(n^2 d^2)$. But d is in general much smaller than k: It is easy to see for instance that $d \leq \sqrt{k}$ (tighter bounds from number theory can be derived). Therefore, we obtain in that case a bound of $\mathcal{O}(n^3)$.

We believe that a significant improvement could reduce the theoretical bound to $\mathcal{O}(n^4)$ or even $\mathcal{O}(n^3)$ by taking into account the *repetitions* of some factors. For instance, when we try to factorize the polyomino P of the previous paragraph with $\varphi(\mathbf{0}) = \mathbf{0}$ and we read the factor $\mathbf{0}^k$, it would be more efficient to keep in memory the fact that $\varphi(\mathbf{0})$ can be set only to powers of $\mathbf{0}$ that divide k.

6   Concluding Remarks

In this paper, we have provided an algorithm to express any polyomino as a product of prime polyominoes in polynomial time. As a consequence, it follows that we can decide if a polyomino is prime or composed in polynomial time as well. Another result worth mentioning is Theorem 1, which guarantees the existence of a prime factorization. However, it seems more difficult to verify if such a factorization is unique. Hence, we are led to propose the following conjecture:

Conjecture 1. Let P be some composed polyomino. Then there exist unique prime parallelogram morphisms φ_1, φ_2, ..., φ_n and a unique prime boundary word u such that $P = \text{POLY}((\varphi_1 \circ \varphi_2 \circ \ldots \circ \varphi_n)(u))$.

Indeed, actual computational explorations have not succeeded in providing a counter-example to this conjecture for polyominoes having area at most 10.

As mentioned above, the $\mathcal{O}(n^5)$ bound is rather crude and it would not be surprising to design more efficient algorithms for solving the factorization problem.

The reader interested in toying with an implementation of Algorithm 1 is invited to look at the publicly available source code hosted on Github[1], which only depends on a basic Python installation to be run.

References

1. Beauquier, D., Nivat, M.: On translating one polyomino to tile the plane. Discrete Comput. Geom. 6, 575–592 (1991)
2. Brlek, S., Provençal, X.: On the problem of deciding if a polyomino tiles the plane by translation. In: Holub, J., Žďárek, J. (eds.) Proceedings of the Prague Stringology Conference 2006, Czech Technical University in Prague, Prague, Czech Republic, August 28-30, pp. 65–76 (2006) ISBN80-01-03533-6
3. Brlek, S., Frosini, A., Rinaldi, S., Vuillon, L.: Tilings by translation: enumeration by a rational language approach. Electronic Journal of Combinatorics 13, 15 (2006)
4. Gambini, I., Vuillon, L.: An algorithm for deciding if a polyomino tiles the plane by translations. Theoret. Informatics Appl. 41, 147–155 (2007)
5. Golomb, S.W.: Tiling with sets of polyominoes. Journal of Combinatorial Theory 9(1), 60–71 (1970),
http://www.sciencedirect.com/science/article/pii/S0021980070800552
6. Golomb, S.W.: Polyominoes: Puzzles, Patterns, Problems, and Packings. Princeton Academic Press, Princeton (1996)
7. Knuth, D.E.: Dancing links (2000), http://arxiv.org/abs/cs/0011047
8. Labbé, S.: Tiling solver in Sage (2011), http://www.sagemath.org/doc/reference/combinat/sage/combinat/tiling.html
9. Lothaire, M.: Combinatorics on Words. Cambridge University Press, Cambridge (1997)
10. Blondin-Massé, A., Brlek, S., Garon, A., Labbé, S.: Christoffel and Fibonacci tiles. In: Brlek, S., Reutenauer, C., Provençal, X. (eds.) DGCI 2009. LNCS, vol. 5810, pp. 67–78. Springer, Heidelberg (2009)
11. Blondin Massé, A., Garon, A., Labbé, S.: Combinatorial properties of double square tiles. Theoretical Computer Science (2012),
http://www.sciencedirect.com/science/article/pii/S0304397512009723,
http://www.sciencedirect.com/science/article/pii/S0304397512009723
12. Massé, A.B., Frosini, A., Rinaldi, S., Vuillon, L.: On the shape of permutomino tiles. Discrete Applied Mathematics 161(15), 2316–2327 (2013),
http://www.sciencedirect.com/science/article/pii/S0166218X12003344;
advances in Discrete Geometry: 16th International Conference on Discrete Geometry for Computer Imagery
13. Moore, C., Michael, J.: Hard tiling problems with simple tiles (2000),
http://arxiv.org/abs/math/0003039
14. Polyá, G.: On the number of certain lattice polygons. Journal of Combinatorial Theory 6(1), 102–105 (1969),
http://www.sciencedirect.com/science/article/pii/S0021980069801134
15. Provençal, X.: Combinatoire des mots, géométrie discrète et pavages. Ph.D. thesis, D1715, Université du Québec à Montréal (2008)
16. Wijshoff, H., van Leeuven, J.: Arbitrary versus periodic storage schemes and tessellations of the plane using one type of polyomino. Inform. Control 62, 1–25 (1984)

---

[1] http://github.com/ablondin/prime-polyominoes

On the List Update Problem with Advice

Joan Boyar[1,*], Shahin Kamali[2,**], Kim S. Larsen[1,*],
and Alejandro López-Ortiz[2]

[1] University of Southern Denmark
Department of Mathematics and Computer Science
Campusvej 55, 5230 Odense M, Denmark
{joan,kslarsen}@imada.sdu.dk
[2] University of Waterloo, School of Computer Science
200 University Avenue West, Waterloo, ON N2L 3G1, Canada
{s3kamali,alopez-o}@cs.uwaterloo.ca

Abstract. We study the online list update problem under the advice model of computation. Under this model, an online algorithm receives partial information about the unknown parts of the input in the form of some bits of advice generated by a benevolent offline oracle. We show that advice of linear size is required and sufficient for a deterministic algorithm to achieve an optimal solution or even a competitive ratio better than $15/14$. On the other hand, we show that surprisingly two bits of advice is sufficient to break the lower bound of 2 on the competitive ratio of deterministic online algorithms and achieve a deterministic algorithm with a competitive ratio of $1.\bar{6}$. In this upper-bound argument, the bits of advice determine the algorithm with smaller cost among three classical online algorithms.

1 Introduction

List update is a well-studied problem in the context of online algorithms. The input is a sequence of requests to items of a list; the requests appear in a sequential and online manner, i.e., while serving a request an algorithm cannot look at the incoming requests. A request involves accessing an item in the list[1]. To access an item, an algorithm should linearly probe the list; each probe has a cost of 1, and accessing an item in the ith position results in a cost of i. The goal is to maintain the list in a way to minimize the total cost. An algorithm can make a *free exchange* to move an accessed item somewhere closer to the front of the list. Further, it can make any number of *paid exchanges*, each having a cost of 1, to swap the positions of any two consecutive items in the list.

---

* Supported in part by the Danish Council for Independent Research and the Villum Foundation.
** Supported in part by Dr. Derick Wood Graduate Scholarship in Computer Science.
[1] Similar to other works, we consider the *static* list update problem in which there is no insertion or deletion.

A.-H. Dediu et al. (Eds.): LATA 2014, LNCS 8370, pp. 210–221, 2014.

Similar to other online problems, the standard method for comparing online list update algorithms is competitive analysis. The competitive ratio of an online algorithm $\mathbb{A}$ is the maximum ratio between the cost of $\mathbb{A}$ for serving any sequence and the cost of OPT for serving the same sequence. Here, OPT is an optimal offline algorithm. It is known that, for a list of length l, no deterministic online algorithm can achieve a competitive ratio better than $2l/(l+1)$ (reported in [13]); this converges to 2 for large lists. There are 2-competitive (hence optimal) algorithms for the problem; these include Move-To-Front (MTF) [20] and TIMESTAMP [1].

Although competitive analysis has been accepted as the standard tool for comparing online algorithms, there are objections to it. One relevant objection in this context is that assuming a total lack of information about the future is unrealistic in many applications. This is particularly the case for the list update problem when it is used as a method for compression [4]. In this application, each character of a text is treated as an item in the list, and the text as the input sequence which is parsed (revealed) in a sequential manner. A compression algorithm can be devised from a list update algorithm $\mathbb{A}$ by writing the access cost of $\mathbb{A}$ for serving each character in unary[2]. Hence, the size of the compressed file is roughly equal to the access cost of the list update algorithm. In this application, it is possible to include some partial information about the structure of the sequence (text) in the compressed file, for example, which of three algorithms was used to do the compression. This partial information could potentially be stored using very little space compared to the subsequent savings in the size of the compressed file compared with the original file, due to the availability of the partial information.

Advice complexity provides an alternative for the analysis of online problems. Under the advice model, the online algorithm is provided with some bits of advice, generated by a benevolent offline oracle with infinite computational power. This reduces the power of the adversary relative to the online algorithm. Variant models are proposed and studied for the advice complexity model [10,11,7,6]. Here, we use a natural model from [7,6] that assumes advice bits are written on a tape, and the online algorithm can access the tape at any time. The advice complexity of an algorithm is then the length of the shortest prefix of the tape that includes all accessed bits. Since its introduction, many online problems have been studied under the advice model. These include classical online problems such as paging [7,12,15], k-server [11,6,19], and bin packing [9].

1.1 Contribution

When studying an online problem under the advice model, the first question to answer is how many bits of advice are required to achieve an optimal solution. We show that advice of size OPT(σ) is sufficient to optimally serve a sequence σ, where OPT(σ) is the cost of an optimal offline algorithm for serving σ, and it is linear in the length of the sequence, assuming that the length of the list is

---

[2] Encodings other than unary correspond to other cost models for list update.

a constant. We further show that advice of linear size is required to achieve a deterministic algorithm with a competitive ratio better than $15/14$.

Another important question is how many bits of advice are required to break the lower bound on the competitive ratio of any deterministic algorithm. We answer this question by introducing a deterministic algorithm that receives two bits of advice and achieves a competitive ratio of $1.\bar{6}$. The advice bit for a sequence σ simply indicates the best option between three online algorithms for serving σ. These three algorithms are TIMESTAMP, MTF-Odd (MTFO) and MTF-Even (MTFE). TIMESTAMP inserts an accessed item x in front of the first item y (from the front of the list) that precedes x in the list and was accessed at most once since the last access to x. If there is no such item y or x is accessed for the first time, no items are moved. MTFO (resp. MTFE) moves a requested item x to the front on every odd (resp. even) request to x.

2   Optimal Solution

In this section, we provide upper and lower bounds on the number of advice bits required to optimally serve a sequence. We start with an upper bound:

Theorem 1. *Under the advice model,* $\text{OPT}(\sigma) - n$ *bits of advice are sufficient to achieve an optimal solution for any sequence* σ *of length* n, *where* $\text{OPT}(\sigma)$ *is the cost of an optimal algorithm for serving* σ.

Proof. It is known that there is an optimal algorithm that moves items using only a family of paid exchanges called *subset transfer* [16]. In a subset transfer, after serving a request to an item x, a subset S of items preceding x in the list is moved (using paid exchanges) to just after x in the list, so that the relative order of items in S among themselves remains unchanged. Consider an optimal algorithm OPT which only moves items via subset transfer. After a request to x at index i, an online algorithm can read $i-1$ bits from the advice tape, indicating (bit vector style) the subset which should be moved to after x. Provided with this, the algorithm can always maintain the same list as OPT. The total number of bits read by the algorithm will be equal to $\text{OPT}(\sigma) - n$.   ☐

The above theorem implies that for lists of constant size, advice of linear size is sufficient to optimally serve a sequence. We show that advice of linear size is also required to achieve any competitive ratio smaller than $15/14$.

Consider instances of the list update problem on a list of two items x and y which are defined as follows. Assume the list is ordered as $[x, y]$ before the first request. Also, to make explanation easier, assume that the length of the sequence, n, is divisible by 5. Consider an arbitrary bitstring B, of size $n/5$, which we refer to as the *defining bitstring*. Let σ denote the list update sequence defined from B in the following manner: For each bit in B, there are five requests in σ, which we refer to as a *round*. We say that a round in σ is of type 0 (resp. 1) if the bit associated with it in B is 0 (resp. 1). For a round of type 0, σ will contain the requests $yyyxx$, and for a round of type 1, the requests $yxxxx$. For example, if $B = 011\ldots$, we will have $\sigma = \langle yyyxx, yxxxx, yxxxx, \ldots \rangle$.

Since the last two requests in a round are to the same item x, it makes sense for an online algorithm to move x to the front after the first access. This is formalized in the following lemma, which is easy to prove.

Lemma 2. *For any online list update algorithm $\mathbb{A}$ serving a sequence σ created from a defining bitstring, there is another algorithm whose cost is not more than $\mathbb{A}$'s cost for serving σ and that ends each round with the list in the order $[x, y]$.*

Provided with the above lemma, we can restrict our attention to algorithms that maintain the ordering $[x, y]$ at the end of each round. In what follows, by an 'online algorithm' we mean an online algorithm with this property.

Lemma 3. *The cost of an optimal algorithm for serving a sequence of length n, where the sequence is created from a defining bitstring, is at most $7n/5$.*

Proof. Since there are $n/5$ rounds, it is sufficient to show that there is an algorithm which incurs a cost of at most 7 for each round. Consider an algorithm that works as follows: For a round of type 0, the algorithm moves y to the front after the first access to y. It also moves x to the front after the first access to x. Hence, it incurs a cost $2+1+1+2+1 = 7$. For a round of type 1, the algorithm does not move any item and incurs a cost of $2+1+1+1+1 = 6$. In both cases, the list ordering is $[x, y]$ at the end of the round and the same argument can be repeated for the next rounds.   □

For a round of type 0 (with requests to $yyyxx$), if an online algorithm $\mathbb{A}$ moves each of x and y to the front after the first accesses, it has cost 7. If it does not move y immediately, it has cost at least 8. For a round of type 1 (i.e., a round of requests to $yxxxx$), if an algorithm does no rearrangement, its cost will be 6; otherwise its cost is at least 7. To summarize, an online algorithm should 'guess' the type of each round and act accordingly after accessing the first request of the round. If the algorithm makes a wrong guess, it incurs a 'penalty' of at least 1 unit. This relates our problem to the binary guessing problem, defined in [11,5].

Definition 4 ([5]). *The* Binary String Guessing Problem with known history *(2-SGKH) is the following online problem. The input is a bitstring of length m, and the bits are revealed one by one. For each bit b_t, the online algorithm $\mathbb{A}$ must guess if it is a 0 or a 1. After the algorithm has made a guess, the value of b_t is revealed to the algorithm.*

Lemma 5 ([5]). *On an input of length m, any deterministic algorithm for 2-SGKH that is guaranteed to guess correctly on more than αm bits, for $1/2 \leq \alpha < 1$, needs to read at least $(1 + (1 - \alpha) \log(1 - \alpha) + \alpha \log \alpha)m$ bits of advice.* [3]

We reduce the 2-SGKH problem to the list update problem:

---

[3] In this paper we use $\log n$ to denote $\log_2(n)$.

Theorem 6. *On an input of size n, any algorithm for the list update problem which achieves a competitive ratio of γ $(1 < \gamma \leq 15/14)$ needs to read at least $(1 + (7\gamma - 7)\log(7\gamma - 7) + (8 - 7\gamma)\log(8 - 7\gamma))/5 \times n$ bits of advice.*

Proof. Consider the 2-SGKH problem for an arbitrary bitstring B. Given an online algorithm $\mathbb{A}$ for the list update problem, define an algorithm for 2-SGKH as follows: Consider an instance σ of the list update problem on a list of length 2 where σ has B as its defining bitstring, and run $\mathbb{A}$ to serve σ. For the first request y in each round in σ, $\mathbb{A}$ should decide whether to move it to the front or not. The algorithm for the 2-SGKH problem guesses a bit as being 0 (resp. 1) if, after accessing the first item requested in the round associated with the bit in B, $\mathbb{A}$ moves it to front (resp. keeps it at its position). As mentioned earlier, for each incorrect guess $\mathbb{A}$ incurs a penalty of at least 1 unit, i.e., $\mathbb{A} \geq \text{OPT} + w$, where w is the number of wrong guesses for critical requests. Since $\mathbb{A}$ has a competitive ratio of γ, we have $\mathbb{A} \leq \gamma \text{OPT}$. Consequently, we have $w \leq (\gamma - 1)\text{OPT}(\sigma)$ and by Lemma 3, $w \leq 7(\gamma - 1)/5 \times n$. This implies that if $\mathbb{A}$ has a competitive ratio of γ, the 2-SGKH algorithm makes at most $7(\gamma - 1)/5 \times n$ mistakes for an input bitstring B of size $n/5$, i.e., at least $n/5 - 7(\gamma-1)/5 \times n = (8 - 7\gamma) \times n/5$ correct guesses. Define $\alpha = 8 - 7\gamma$, and note that α is in the range $[1/2, 1)$ when γ is in the range stated in the lemma. By Lemma 5, at least $(1 + (1 - \alpha)\log(1 - \alpha) + \alpha\log\alpha)n/5$ bits of advice are required by such a 2-SGKH algorithm. Replacing α with $8 - 7\gamma$ completes the proof. □

Thus, to obtain a competitive ratio better than $15/14$, a linear number of bits of advice is required. For example, to achieve a competitive ratio of 1.01, at least $0.12n$ bits of advice are required. Theorems 1 and 6 imply the following corollary.

Corollary 7. *For any fixed list, $\Theta(n)$ bits of advice are required and sufficient to achieve an optimal solution for the list update problem. Also, $\Theta(n)$ bits of advice are required and sufficient to achieve a 1-competitive algorithm.*

3   An Algorithm with Two Bits of Advice

In this section we show that two bits of advice are sufficient to break the lower bound of 2 on the competitive ratio of deterministic algorithms and achieve a deterministic online algorithm with a competitive ratio of $1.\bar{6}$. The two bits of advice for a sequence σ indicate which of the three algorithms TIMESTAMP, MTF-Odd (MTFO) and MTF-Even (MTFE), have the lower cost for serving σ. Recall that MTFO (resp. MTFE) moves a requested item x to the front on every odd (resp. even) request to x. We prove the following theorem:

Theorem 8. *For any sequence σ, we have either $\text{TIMESTAMP}(\sigma) \leq 1.\bar{6}\,\text{OPT}(\sigma)$, $\text{MTFO}(\sigma) \leq 1.\bar{6}\,\text{OPT}(\sigma)$, or $\text{MTFE}(\sigma) \leq 1.\bar{6}\,\text{OPT}(\sigma)$.*

To prove the theorem, we show that for any sequence σ, $\text{TIMESTAMP}(\sigma) + \text{MTFO}(\sigma) + \text{MTFE}(\sigma) \leq 5\,\text{OPT}(\sigma)$. We note that all three algorithms have the *projective property*, meaning that the relative order of any two items only depends on the requests to those items and their initial order in the list (and not on the requests to other items). MTFO (resp. MTFE) is projective since in its list an item y precedes x if and only if the last odd (resp. even) access to y is more recent than the last odd (resp. even) access to x. In the lists maintained by TIMESTAMP, item y precedes item x if and only if in the projected sequence on x and y, y was requested twice after the second to last request to x or the most recent request was to y and x has been requested at most once. Hence, TIMESTAMP also has the projective property.

Similar to most other work for analysis of projective algorithms,[4] we consider the *partial cost model*, in which accessing an item in position i is defined to have cost $i-1$. We say an algorithm is *cost independent* if its decisions are independent of the cost it has paid for previous requests. The cost of any cost independent algorithm for serving a sequence of length n decreases n units under the partial cost model when compared to the *full* cost model. Hence, any upper bound for the competitive ratio of a cost independent algorithm under the partial cost model can be extended to the full cost model.

To prove an upper bound on the competitive ratio of a projective algorithm under the partial cost model, it is sufficient to prove that the claim holds for lists of size 2. The reduction to lists of size two is done by applying a *factoring lemma* which ensures that the total cost of a projective algorithm $\mathbb{A}$ for serving a sequence σ can be formulated as the sum of the costs of $\mathbb{A}$ for serving projected sequences of two items. A projected sequence of σ on two items x and y is a copy of σ in which all items except x and y are removed. We refer the reader to [8, p. 16] for details on the factoring lemma. Since MTFO, MTFE, and TIMESTAMP are projective and cost-independent, to prove Theorem 8, it suffices to prove the following lemma:

Lemma 9. *Under the partial cost model, for any sequence σ_{xy} of two items, we have* $\mathrm{MTFO}(\sigma_{xy}) + \mathrm{MTFE}(\sigma_{xy}) + \mathrm{TIMESTAMP}(\sigma_{xy}) \leq 5 \times \mathrm{OPT}(\sigma_{xy})$.

Before proving the above lemma, we study the aggregated cost of MTFO and MTFE on certain subsequences of two items. One way to think of these algorithms is to imagine they maintain a bit for each item. On each request, the bit of the item is flipped; if it becomes '0', the item is moved to the front. Note that the bits of MTFO and MTFE are complements of each other. Thus, we can think of them as one algorithm started on complementary bit sequences. We say a list is in state $[ab]_{(i,j)}$ if item a precedes b in the list and the bits maintained for a and b are i and j ($i, j \in \{0, 1\}$), respectively. To study the value of $\mathrm{OPT}(\sigma_{xy})$, we consider an offline algorithm which uses a free exchange to move an accessed item from the second position to the front of the list if and only if the following request is to the same item. It is known that this algorithm is optimal for lists of two items [17].

---

[4] Almost all existing algorithms for the list update problem are projective; the only exceptions are TRANSPOSE, Move-Fraction [20], and SPLIT [13]; see [14] for a survey.

Table 1. Assuming the initial ordering of items is $[ab]$, the cost of a both MTFO and MTFE for serving subsequence $\langle baba \rangle$ is at most 3 (under the partial cost model). The final ordering of the items will be $[ab]$ in three of the cases.

Bits for (a,b)	Cost for $\langle baba \rangle$	Orders before accessing items	Final order
$(0,0)$	$1+0+1+1=3$	$[ab]\ [ab]\ [ab]\ [ba]$	$[ab]$
$(0,1)$	$1+1+0+1=3$	$[ab]\ [ba]\ [ba]\ [ba]$	$[ab]$
$(1,0)$	$1+0+1+1=3$	$[ab]\ [ab]\ [ab]\ [ba]$	$[ba]$
$(1,1)$	$1+1+1+0=3$	$[ab]\ [ba]\ [ab]\ [ab]$	$[ab]$

Lemma 10. *Consider a subsequence of two items a and b of the form $(ba)^{2i}$, i.e., i repetitions of $\langle baba \rangle$. Assume the initial ordering is $[ab]$. The cost of each of MTFO and MTFE for serving the subsequence is at most $3i$ (under the partial cost model). Moreover, at the end of serving the subsequence, the ordering of items in the list maintained by at least one of the algorithms is $[ab]$.*

Proof. We refer to repetition of $baba$ as a *round*. We show that MTFO and MTFE have a cost of at most 3 for serving each round. Assume the bits associated with both items are '0' before serving $baba$. The first request has a cost of 1 and b remains in the second position, the second request has cost 0, and the remaining requests each have a cost of 1. In total, the cost of the algorithm is 3. The other cases (when items have different bits) are handled similarly. Table 1 includes a summary of all cases. As illustrated in the table, if the bits maintained for a and b before serving $baba$ are $(0,0)$, $(0,1)$, or $(1,1)$, the list order will be $[ab]$ after serving the round. Since both a and b are requested twice, the bits will be also the same after serving $baba$. Hence, in these three cases, the same argument can be repeated to conclude that the list order will be $[ab]$ at the end of serving $(ba)^{2i}$. Since the bits maintained for the items are complements in MTFE and MTFO, at least one of them starts with bits $(0,0)$, $(0,1)$, or $(1,1)$ for a and b; consequently, at least one algorithm ends up with state $[ab]$ at the end.   □

Lemma 11. *Consider a subsequence of two items a and b which has form $\langle baa \rangle$. The total cost that MTFE and MTFO incur together for serving this subsequence is less than or equal to 4 (under the partial cost model).*

Proof. If the initial order of a and b is $[ba]$, the first request has no cost, and each algorithm incurs a total cost of at most 2 for the other two requests of the sequence. Hence, the aggregated cost of the two algorithms is 4. Next, assume the initial order is $[ab]$. Assume the bits maintained by one of the algorithms for a and b are $(1,0)$, respectively. As illustrated in Table 2, this algorithm incurs a cost of 1 for serving baa; the other algorithm incurs a cost of 3. In total, the algorithms incur a cost of 4. In the other case, when bits maintained for a and b are both '0' in one algorithm (consequently, both are '1' in the other algorithm), the total cost of the algorithms for serving $\langle baa \rangle$ is 3.   □

Table 2. The total cost of MTFO and MTFE for serving a sequence $\langle baa \rangle$ is at most 4 (under the partial cost model). Note that the bits of these algorithms for each item are complements of each other.

Initial order	Bits for (a,b)	Cost for $\langle baa \rangle$	Orders before accessing items	Bits and Costs (other algorithm)	Total cost (both algs.)
$[ab]$	$(0,0)$	$1+0+0=1$	$[ab]\ [ab]\ [ab]$	$(1,1) \to 2$	$1+2=3$
$[ab]$	$(0,1)$	$1+1+1=3$	$[ab]\ [ba]\ [ba]$	$(1,0) \to 1$	$3+1=4$
$[ab]$	$(1,0)$	$1+0+0=1$	$[ab]\ [ab]\ [ab]$	$(0,1) \to 3$	$1+3=4$
$[ab]$	$(1,1)$	$1+1+0=2$	$[ab]\ [ba]\ [ab]$	$(0,0) \to 1$	$2+1=3$
$[ba]$	$(0,0)\ (0,1)$ $(1,0)\ (1,1)$	$\leq 0+1+1=2$	-	≤ 2	$2+2=4$

Using Lemmas 10 and 11, we are ready to prove Lemma 9:

Proof (Lemma 9, and consequently Theorem 8).

Consider a sequence σ_{xy} of two items x and y. We use the *phase partitioning technique* as discussed in [8]. We partition σ_{xy} into *phases* which are defined inductively as follows. Assume we have defined phases up until, but not including, the tth request ($t \geq 1$) and the relative order of the two items is $[xy]$ before the tth request. Then the next phase is of *type 1* and is of one of the following forms ($j \geq 0$ and $k \geq 1$):

$$(a)\ x^j yy \quad (b)\ x^j (yx)^k yy \quad (c)\ x^j (yx)^k x$$

In case the relative order of the items is $[yx]$ before the tth request, the phase has type 2 and its form is exactly the same as above with x and y interchanged. Note that, after two consecutive requests to an item, TIMESTAMP, MTFO and MTFE all have that item in the front of the list. So, after serving each phase, the relative order of items is the same for all three algorithms. This implies that σ_{xy} is partitioned in the same way for all three algorithms. To prove the lemma, we show that its statement holds for every phase.

Table 3 shows the costs incurred by all three algorithms as well as OPT for each phase. Note that phases of the form (b) and (c) are divided into two cases, depending on whether k is even or odd. We discuss the different phases of type 1 separately. Similar analyses, with x and y interchanged, apply to the phases of type 2. Note that before serving a phase of type 1, the list is ordered as $[xy]$ and the first j requests to x have no cost.

Consider phases of form (a), $x^j yy$. MTFO and MTFE incur a total cost of 3 for serving yy (one of them moves y to the front after the first request, while the other keeps it in the second position). TIMESTAMP incurs a cost of 2 for serving yy (it does not move it to the front after the first request). So, in total, the three algorithms incur an aggregated cost of 5. On the other hand, OPT incurs a cost of 1 for the phase. So, the ratio between the sum of the costs of the algorithms and the cost of OPT is 5.

Next, consider phases of the form (b). TIMESTAMP incurs a cost of $2k$ for serving the phase; it incurs a cost of 1 for all requests in $(yx)^{2i}$ except the very first one, and a cost of 1 for serving the second to last request to y. Assume k is even and we have $k = 2i$ for some $i \geq 1$, so the phase looks like $x^j(yx)^kyy$. By Lemma 10, the cost incurred by MTFO and MTFE is at most $3i$ for serving $(yx)^{2i}$. We show that for the remaining two requests to y, MTFO and MTFO incur an aggregated cost of at most 3. If the list maintained by any of the algorithms is ordered as $[yx]$ before serving yy, that algorithm incurs a cost of 0 while the other algorithm incurs a cost of at most 2 for these requests; in total, the cost of both algorithms for serving yy will be at most 2. If the lists of both algorithms are ordered as $[xy]$, one of the algorithms incurs a cost of 1 and the other incurs a cost of 2 (depending on the bit they keep for y). In conclusion, MTFO and MTFE incur a total cost of at most $6i + 3$. TIMESTAMP incurs a cost of $2k = 4i$, while OPT incurs a cost of $2i + 1$ for the phase. To conclude, the aggregated cost of all algorithms is at most $10i + 3$ compared to $2i + 1$ for OPT, and the ratio between them is less than 5.

Next, assume k is odd and we have $k = 2i - 1$, i.e., the phase has the form $x^j(yx)^{2i-2}yxyy$. The total cost of MTFO and MTFE for $(yx)^{2i-2}$ is at most $2 \times (3(i - 1))$ (Lemma 10), the total cost for the next request to y is at most 2, and the total cost for subsequent xyy is at most 4 (Lemma 11). In total, MTFO and MTFE incur a cost of at most $6i$ for the phase. On the other hand, TIMESTAMP incurs a cost of $4i - 2$ for the phase. The aggregated cost of the three algorithms is at most $10i - 2$ for the phase, while OPT incurs a cost of $2i$. So, the ratio between sum of the costs of the algorithms and OPT is less than 5.

Next, consider phases of type 1 and form (c). TIMESTAMP incurs a cost of $2k - 1$ in this case. Assume k is even, i.e., the phase has the form $x^j(yx)^{2i}x$. By Lemma 10, MTFO and MTFE each incur a total cost of at most $3i$ for $(yx)^{2i}$. Moreover, after this, the list maintained for at least one of the algorithms is ordered as $[xy]$. Hence, the aggregated cost of algorithms for the next request to x is at most 1. Consequently, the total cost of MTFE and MTFO is at most $6i + 1$ for the round. Adding the cost $2k - 1 = 4i - 1$ of TIMESTAMP, the total cost of all three algorithms is at most $10i$. On the other hand, OPT incurs a cost of $2i$ for the phase. So, the ratio between the aggregated cost of all three algorithms and the cost of OPT is at most 5. Finally, assume k is odd, i.e., the phase has form $x^j(yx)^{2i-2}yxx$. By Lemma 10, MTFO and MTFE together incur a total cost of $2 \times 3(i - 1)$ for $x^j(yx)^{2i-2}$. By Lemma 11, they incur a total cost of at most 4 for yxx. In total, they incur a cost of at most $6(i - 1) + 4$ for the phase. TIMESTAMP incurs a cost of $4i - 3$; this sums up to $10i - 5$ for all three algorithms. In this case, OPT incurs a cost of $2i - 1$. Hence, the ratio between the sum of the costs of all three algorithms and OPT is at most 5.

In fact, the upper bound provided in Theorem 3 for the competitive ratio of the better algorithm among TIMESTAMP, MTFO and MTFE is tight under the partial cost model. To show this, we make use of the following lemma.

Lemma 12. *Consider a sequence $\sigma_\alpha = x(yxxx\ yxxx)^k$, i.e., a single request to x, followed by k repetitions of $(yxxx\ yxxx)$. Assume the list is initially ordered as $[xy]$. We have $\text{MTFO}(\sigma) = \text{MTFE}(\sigma) = 4k$ while $\text{OPT}(\sigma) = 2k$ (under the partial cost model).*

Table 3. The costs of MTFO, MTFE, and TIMESTAMP for a phase of type 1 (the phase has type 1, i.e., the initial ordering of items is xy). The ratio between the aggregated cost of algorithms and the cost of OPT for each phase is at most 5. ALGMIN (resp. ALGMAX) is the algorithm among MTFO and MTFE, which incurs less (resp. more) cost for the phase. Note that the costs are under the partial cost model.

Phase	ALGMIN	ALGMAX	TIMESTAMP	Sum (ALGMIN + ALGMAX + TIMESTAMP)	OPT'	$\frac{\text{Sum}}{\text{OPT'}}$
$x^j yy$	1	2	2	5	1	5
$x^j(yx)^{2i}yy$	$\le 3i+1$	$\le 3i+2$	$2\times 2i = 4i$	$\le 10i+3$	$2i+1$	<5
$x^j(yx)^{2i-2}yxyy$	$\le 3(i-1)+1$ $+$ ALGMIN($\langle xyy\rangle$)	$\le 3(i-1)+1$ $+$ ALGMAX($\langle xyy\rangle$)	$2\times(2i-1)$ $=4i-2$	$\le 6(i-1)+2+4$ $+(4i-2)=10i-2$	$2i$	<5
$x^j(yx)^{2i}x$	$\le 3i$	$\le 3i+1$	$2\times 2i-1$ $=4i-1$	$\le (6i+1)+(4i-1)$ $=10i$	$2i$	≤ 5
$x^j(yx)^{2i-2}yxx$	$\le 3(i-1)$ $+$ ALGMIN($\langle yxx\rangle$)	$\le 3(i-1)$ $+$ ALGMAX($\langle yxx\rangle$)	$2\times(2i-1)-1$ $=4i-3$	$\le 6(i-1)+4$ $+(4i-3)=10i-5$	$2i-1$	≤ 5

Proof. We refer to each repetition of $(yxxx\ yxxx)$ as a round. Initially, the bits maintained by MTFO (resp. MTFE) for x,y are $(1,1)$ (resp. $(0,0)$). After the first request to x, the bits of MTFO (resp. MTFE) change to $(0,1)$ (resp. $(1,0)$) for x,y. MTFO incurs a cost of 3 for the first half of each round; it incurs a cost of 1 for all requests except the last request to x. MTFE incurs a cost of 1 for serving the first half of a round; it only incurs a cost of 1 on the first requests y. After serving the first half, the list for each algorithm will be ordered as $[xy]$ and the bits maintained by MTFO (resp. MTFE) for x,y will be $(1,0)$ (resp. $(0,1)$). Using a symmetric argument, the costs of MTFO and MTFE for the second half of a round are respectively 1 and 3. In total, both MTFO and MTFE incur a cost of 4 for each round. After serving the round, the list maintained by both algorithms will be ordered as $[xy]$ and the bits associated with the items will be the same as at the start of the first round. Thus, MTFO and MTFE each have a total cost of $4k$ on σ_α. An optimal algorithm OPT never changes the ordering of the list and has a cost of 2 for the whole round, giving a cost of $2k$ for σ_α. □

Theorem 13. *There are sequences for which the costs of all of* TIMESTAMP, *MTFE, and* MTFO *are* $1.\bar{6}$ *times that of* OPT *(under the partial cost model).*

Proof. Consider a sequence $\sigma = \sigma_\alpha\sigma_\beta$ where $\sigma_\alpha = x(yxxx\ yxxx)^{k_\alpha}$ and $\sigma_\beta = (yyxx)^{k_\beta}$. Here, k_α is an arbitrary large integer and $k_\beta = 2k_\alpha$. By Lemma 12, we have $\text{MTFO}(\sigma_\alpha) = \text{MTFE}(\sigma_\alpha) = 4k_\alpha$ while $\text{OPT}(\sigma_\alpha) = 2k_\alpha$. We have $\text{TIMESTAMP}(\sigma_\alpha) = 2k_\alpha$, because it does not move y from the second position.

Next, we study the cost of MTFO and MTFE for serving σ_β. Note that after serving σ_α, the lists maintained by these algorithms is ordered as $[xy]$ and the bits associated with x and y are respectively $(0,1)$ for MTFO and $(1,0)$ for MTFE (see the proof of Lemma 12).We show that for each round $yyxx$ of σ_β, the cost of each algorithm is 3. On the first request to y, MTFO moves it to the front (since the bit maintained for y is 1); so it incurs a cost of 1 for the first requests to y. On the first request to x, MTFO keeps x in the second position; hence it incurs a cost of 2 for the requests to x. In total, it has a cost of 3 for the round. With a similar argument, MTFE incurs a cost of 2 for the requests to y and a cost of 1 for the requests to x and a total cost of 3. The list order and bits maintained for the items will be the same at the end of the round as at the start. Hence, the same argument can be extended to other rounds to conclude that the cost of both MTFE and MTFO for serving σ_β is $3k_\beta$. On the other hand, TIMESTAMP incurs a cost of 4 on each round as it moves items to the front on the second consecutive request to them; hence, the cost of TIMESTAMP for serving σ_β is $4k_\beta$. An algorithm that moves items in front on the first of two consecutive request to them will incur a cost of 2 on each round; hence the cost of OPT for serving σ_β is at most $2k_\beta$.

To summarize, the cost of each of MTFO and MTFE for serving σ is $4k_\alpha + 3k_\beta = 10k_\alpha$ while the cost of TIMESTAMP is $2k_\alpha + 4k_\beta = 10k_\alpha$, and the cost of OPT is $2k_\alpha + 2k_\beta = 6k_\alpha$. As a consequence, all three algorithms have a cost which is $10/6 = 1.\bar{6}$ times that of OPT.   □

4   Concluding Remarks

It is generally assumed that the offline oracle that generates advice bits has unbounded computational power. We used this assumption when we showed that $\text{OPT}(\sigma)$ bits are sufficient to achieve an optimal solution in Section 2. However, for the algorithm introduced in Section 3, the advice bits can be generated in polynomial time. The offline version of the list update problem is known to be NP-hard [3]. In this sense, our algorithm can be seen as a linear-time approximation algorithm with an approximation ratio of $1.\bar{6}$; this is, to the best of our knowledge, the best deterministic offline algorithm for the problem. It should be mentioned that there is a randomized online algorithm BIT which also has a competitive ratio of $1.\bar{6}$ against an oblivious adversary [18]. BIT maintains a bit for each item and flips the bit on each access; whenever the bit becomes '0' it moves the item to the front. The bits are initially set uniformly at random; hence, BIT uses l bits of advice for lists of length l. COMB is another randomized algorithm which makes use of a linear number of random bits and improves the competitive ratio to 1.6 [2]. We can conclude that there are online algorithms which achieve a competitive ratio of at most 1.6 when provided a linear (in the length of the list) number of advice bits. However, from a practical point of view, it is not clear how an offline oracle can smartly generate such bits of advice.

We proved that with two bits of advice, one can achieve a (deterministic) algorithm with a competitive ratio of at most $1.\bar{6}$. This bound is tight under the partial cost model (Theorem 13); however, the lower bound argument for the competitive ratio of this algorithm does not extend to the full cost mode, i.e., the upper bound of $1.\bar{6}$ might be overly pessimistic. All studied projective algorithms have the same competitive ratio under partial and full cost models; our algorithm might be distinctive in this sense.

While two bits of advice can break the lower bound of 2 on the competitive ratio of online algorithms, it remains open whether this can be done with one bit of advice. Regardless, it is not hard to see that any algorithm with one bit of advice has a competitive ratio of at least 1.5. We conjecture that this lower bound can be improved and leave it as future work.

References

1. Albers, S.: Improved randomized on-line algorithms for the list update problem. SIAM J. Comput. 27(3), 682–693 (1998)
2. Albers, S., von Stengel, B., Werchner, R.: A combined BIT and TIMESTAMP algorithm for the list update problem. Inf. Proc. Lett. 56, 135–139 (1995)
3. Ambühl, C.: Offline list update is NP-hard. In: Paterson, M. (ed.) ESA 2000. LNCS, vol. 1879, pp. 42–51. Springer, Heidelberg (2000)
4. Bentley, J.L., Sleator, D., Tarjan, R.E., Wei, V.K.: A locally adaptive data compression scheme. Commun. ACM 29, 320–330 (1986)
5. Böckenhauer, H.-J., Hromkovič, J., Komm, D., Krug, S., Smula, J., Sprock, A.: The string guessing problem as a method to prove lower bounds on the advice complexity. In: Du, D.-Z., Zhang, G. (eds.) COCOON 2013. LNCS, vol. 7936, pp. 493–505. Springer, Heidelberg (2013)
6. Böckenhauer, H.-J., Komm, D., Královič, R., Královič, R.: On the advice complexity of the k-server problem. In: Aceto, L., Henzinger, M., Sgall, J. (eds.) ICALP 2011, Part I. LNCS, vol. 6755, pp. 207–218. Springer, Heidelberg (2011)
7. Böckenhauer, H., Komm, D., Královič, R., Královič, R., Mömke, T.: On the advice complexity of online problems. In: Dong, Y., Du, D.-Z., Ibarra, O. (eds.) ISAAC 2009. LNCS, vol. 5878, pp. 331–340. Springer, Heidelberg (2009)
8. Borodin, A., El-Yaniv, R.: Online Computation and Competitive Analysis. Cambridge University Press (1998)
9. Boyar, J., Kamali, S., Larsen, K.S., López-Ortiz, A.: Online bin packing with advice. CoRR abs/1212.4016 (2012)
10. Dobrev, S., Královič, R., Pardubská, D.: Measuring the problem relevant information in input. RAIRO Inform. Theor. Appl. 43(3), 585–613 (2009)
11. Emek, Y., Fraigniaud, P., Korman, A., Rosén, A.: Online computation with advice. Theoret. Comput. Sci. 412(24), 2642–2656 (2011)
12. Hromkovič, J., Královič, R., Královič, R.: Information complexity of online problems. In: Hliněný, P., Kučera, A. (eds.) MFCS 2010. LNCS, vol. 6281, pp. 24–36. Springer, Heidelberg (2010)
13. Irani, S.: Two results on the list update problem. Inf. Proc. Lett. 38, 301–306 (1991)
14. Kamali, S., López-Ortiz, A.: A survey of algorithms and models for list update. In: Brodnik, A., López-Ortiz, A., Raman, V., Viola, A. (eds.) Ianfest-66. LNCS, vol. 8066, pp. 251–266. Springer, Heidelberg (2013)
15. Komm, D., Královič, R.: Advice complexity and barely random algorithms. RAIRO Inform. Theor. Appl. 45(2), 249–267 (2011)
16. Reingold, N., Westbrook, J.: Optimum off-line algorithms for the list update problem. Tech. Rep. YALEU/DCS/TR-805, Yale University (1990)
17. Reingold, N., Westbrook, J.: Off-line algorithms for the list update problem. Inf. Proc. Lett. 60(2), 75–80 (1996)
18. Reingold, N., Westbrook, J., Sleator, D.D.: Randomized competitive algorithms for the list update problem. Algorithmica 11, 15–32 (1994)
19. Renault, M.P., Rosén, A.: On online algorithms with advice for the k-server problem. In: Solis-Oba, R., Persiano, G. (eds.) WAOA 2011. LNCS, vol. 7164, pp. 198–210. Springer, Heidelberg (2012)
20. Sleator, D., Tarjan, R.E.: Amortized efficiency of list update and paging rules. Commun. ACM 28, 202–208 (1985)

Shift-Reduce Parsers for Transition Networks*

Luca Breveglieri, Stefano Crespi Reghizzi, and Angelo Morzenti**

Dip. di Elettronica, Informazione e Bioingegneria (*DEIB*)
Politecnico di Milano, Piazza Leonardo Da Vinci n. 32, 20133, Milano, Italy
{luca.breveglieri,stefano.crespireghizzi,angelo.morzenti}@polimi.it

Abstract. We give a new direct construction of the shift-reduce $ELR(1)$ parsers for recursive Transition Networks (*TN*), which is suitable for languages specified by Extended *BNF* grammars (*EBNF*). Such parsers are characterized by their absence of conflicts, not just the classical shift-reduce and reduce-reduce types, but also a new type named *convergence* conflict. Such a condition is proved correct and is more general than the past proposed conditions for the shift-reduce parsing of *EBNF* grammars or *TN*'s. The corresponding parser is smaller than a classical one, without any extra bookkeeping. A constraint on *TN*'s is mentioned, which enables top-down deterministic $ELL(1)$ analysis.

Keywords: extended grammar, *EBNF*, *LR* syntax analysis, bottom-up parser.

1 Introduction

Extended Backus-Naur Form grammars (*EBNF*) improve the readability of context-free (*BNF*) grammars and are widely used for language specification. They often appear in the graphical form of Transition Networks (*TN*), pioneered by [5]. Our contribution is a new method for constructing parsers starting from a *TN*. Here we focus on the deterministic bottom-up or shift-reduce parsers, corresponding to the $LR(1)$ approach [12] widely used and supported by parser generation tools. The Knuth's $LR(1)$ condition exactly characterizes the deterministic *BNF* languages and his classical mapping from a grammar to a shift-reduce parser – i.e., a (deterministic) pushdown automaton *(D)PDA* – will be referred to as the *standard* method.

Though any *EBNF* grammar or *TN* can be converted into an equivalent *BNF* one by replacing each regular expression (r.e.) or finite automaton (*FA*) with a *BNF* subgrammar, the proposed methods for directly checking determinism are difficult or overly restricted, and none of them has reached consensus. The broad and accurate recent survey in [10] says: "What has been published about *LR*-like parsing theory is so complex that not many feel tempted to use it; ... Of course, such a simplistic resumé does not do justice to the great efforts that have gone

---

* Work partially supported by *PRIN* "Automi e Linguaggi Formali", and by *CNR* - *IEIIT*.
** We thank two anonymous referees for their accurate reading and insightful comments.

A.-H. Dediu et al. (Eds.): LATA 2014, LNCS 8370, pp. 222–235, 2014.

into research and implementation of the methods described. But it is a striking phenomenon that the ideas behind recursive descent parsing of *ECFG*'s [i.e., *EBNF*] can be grasped and applied immediately, whereas most of the literature on *LR*-like parsing of *RRPG*'s [i.e., *TN*] is very difficult to access. Given the developments in computing power and software engineering, and the practical importance of *ECFG*'s and *RRPG*'s, a uniform and coherent treatment of the subject seems in order." This is precisely our objective: offering a practical and general construction for such parsers. Most proposed methods for the construction of deterministic parsers when *TN*'s have bifurcating and circular paths as well as recursive invocations, add to the standard $LR(1)$ method a few complex bookkeeping operations to manage the reduction of strings of unbounded length. We contribute instead a novel mapping from a general *TN* to a shift-reduce parser *DPDA* with no extra bookkeeping. We also give a rigorous $ELR(1)$ condition for the mapping to produce a *DPDA*: in addition to shift-reduce *SR* and reduce-reduce *RR* conflicts, there should not be any *convergence* conflict, to be later defined. We compare our directly obtained $ELR(1)$ parser with the $LR(1)$ ones indirectly obtained by translating the *TN* to a *BNF* grammar and by applying the standard method. Our direct method is proved as general as the indirect one, and the (descriptive) state complexity of the indirect $LR(1)$ parsers always exceeds that of our direct $ELR(1)$ parsers.

Sect. 2 sets terminology and notation. Sect. 3 contains the $ELR(1)$ condition for *TN*'s, the main theoretical property and its proof, and the direct construction of the parser. Sect. 4 compares the direct and indirect parsers. Sect. 5 discusses related work and gives conclusions.

2   Preliminaries

The terminal alphabet is Σ and the empty string is ε. A *BNF* (context-free) *grammar* G consists of a 4-tuple (Σ, V, P, S). A *grammar symbol* is an element of the union alphabet $\Sigma \cup V$. In a rule $A \to \alpha$ the symbols A and α are the left and right part, respectively. Alternative rules such as $A \to \alpha$ and $A \to \beta$ may be joint into one rule $A \to \alpha \mid \beta$. A grammar is *right-linear* (*RL*) if every rule has the form $A \to a\,B$ or $A \to \varepsilon$, with $a \in \Sigma$ and $B \in V$.

In an *EBNF* grammar G, for each nonterminal $A \in V$ there is exactly one rule $A \to \alpha$ where the *regular expression* (r.e.) α over the union alphabet $\Sigma \cup V$ uses union, catenation and Kleene star; the regular language associated to A is denoted as $R(\alpha)$ or R_A $(\neq \emptyset)$.

An immediate *derivation* is denoted by $u\,A\,v \Rightarrow u\,w\,v$, with strings u, w and v possibly empty, and with $w \in R(\alpha)$. A derivation is *rightmost* if string v does not contain any nonterminal. A derivation $A \overset{*}{\Rightarrow} A\,v$, with $v \neq \varepsilon$, is called *left-recursive*. For a derivation $u \Rightarrow v$ the reverse relation is named *reduction* and is denoted as $v \leadsto u$.

The language of grammar G is $L(G) = \{\, x \in \Sigma^* \mid \quad S \overset{*}{\Rightarrow} x \,\}$. A sentence in $L(G)$ is unambiguous if it has only one syntax tree (if grammar G is *EBNF* a tree node can have unboundedly many child nodes), and grammar G is unambiguous if every sentence in $L(G)$ is so.

We represent each rule $A \to \alpha$ as a (deterministic) *(D)FA* M_A (named "machine") that recognizes language $R(\alpha)$. A set $\mathcal{M}$ of such *FA*'s for all nonterminals is named *transition network* (*TN*) and represents the grammar. Actually the same net represents any grammar with the same nonterminals and the same associated regular languages.

Definition 1. *Let G be an EBNF grammar $S \to \sigma$, $A \to \alpha$, ... The transition net $\mathcal{M} = \{\, M_S, M_A, \dots \,\}$ is a set of DFA's that accept the regular languages R_S, R_A, ... To prevent confusion, the state set of machine M_A is denoted as $Q_A = \{\, 0_A, \dots, q_A, \dots \,\}$, its initial state is 0_A and its set of final states is $F_A \subseteq Q_A$. The union of all the states of the net is $Q = \bigcup_{M_A \in \mathcal{M}} Q_A$. The transition function of every machine is denoted as δ, at no risk of confusion since their state sets are disjoint. We assume that all the machines are non-reentrant, i.e., no edges enter their initial states.*

For a state $q_A \in Q_A$ we denote as $R(M_A, q_A) = R(q_A)$ the regular language, over the union alphabet $\Sigma \cup V$, accepted by machine M_A starting from q_A and ending in any final state.

The terminal language defined by machine M_A when starting from state q_A, is denoted as $L(M_A, q_A) = L(q_A) = \{\, y \in \Sigma^* \mid \eta \in R(q_A) \wedge \eta \overset{*}{\Rightarrow} y \,\} \neq \emptyset.$ □

For a non-reentrant machine, a computation that stays inside the machine graph, never revisits the initial state. Clearly any machine can be so normalized with a negligible overhead, by adding one state and a few edges. Such an adjustment, though minor, greatly simplifies the reduction moves of the parser. We remark that any two structurally equivalent *EBNF* grammars (i.e., such that their syntax trees are identical up to an isomorphism of nonterminal names) can be represented by the same *TN*. Therefore it is meaningful to refer to all such grammars as being *associated* to a given *TN*.

Example 2. Grammar G in Fig. 1.a and its *TN* in Fig. 1.b will be used as a running example. Notice that grammar G features a null rule, union, star, union under star, and self-nested derivations. Machine M_S features a null path, loops, alternative paths, and recursion. □

Next we introduce a special *BNF* grammar that mirrors the structure of a *TN*. Since an *FA* is equivalent to a (*BNF*) *RL* grammar, which encodes *FA* edges as *RL* rules, consider such a grammar $\hat{G}_A$ for machine M_A, with $R(M_A, 0_A) = L(\hat{G}_A) \subseteq (\Sigma \cup V)^*$. The nonterminals of $\hat{G}_A$ are the states of Q_A (axiom 0_A); there is a rule $p_A \to X\, r_A$ if an edge $p_A \overset{X}{\longrightarrow} r_A$ is in δ; and there is the empty rule $p_A \to \varepsilon$ if p_A is a final state. Notice that a rule of $\hat{G}_A$ with the form $p_A \to B\, r_A$, where symbol B is a nonterminal of the original *EBNF* grammar, is still *RL* since symbol B is viewed as a "terminal" symbol for $\hat{G}_A$.

Then for every machine M_A and grammar $\hat{G}_A$, and for every rule $p_A \to B\, r_A$, replace the nonterminal symbol $B \in V$ by 0_B, and so obtain a rule $p_A \to 0_B\, r_A$. The resulting *BNF* grammar is denoted as $\hat{G}$, named the *right-linearized* grammar (*RLZG*) of the net: it has terminal alphabet Σ, nonterminal set Q and

(a) *EBNF grammar*

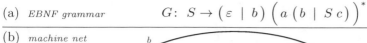

$$G: S \to \big(\varepsilon \mid b\big)\big(a\,(b \mid S\,c)\big)^*$$

(b) *machine net*

$$\mathcal{M} = \{\, M_S \,\}$$

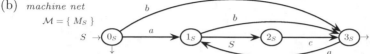

(c) *ELR pilot graph*

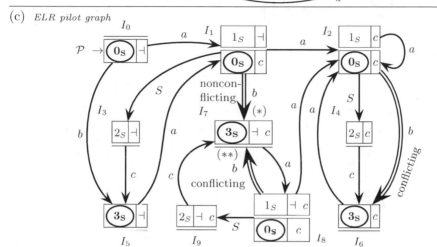

(d) *bottom-up analysis simulation of "abaacc" - stack contents shown before each reduction*

input tape (with cell numbering and end-of-text ⊣; the gap is for the null reduction):

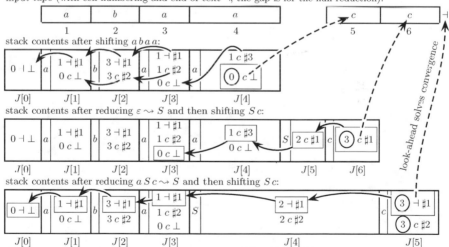

the parser reduces $a\,b\,a\,S\,c \rightsquigarrow S$, the stack contains $J[0] = \{\,\langle\,0 \dashv \perp\,\rangle\,\}$ and the parser accepts

The parser is a pushdown automaton (*PDA*). The superimposed arrows are for visualization. Solid arcs show links between stack items: all the *iid* chains in the 1st stack, but only the reduction handle in the 2nd and 3rd stack. Dashed arcs show look-ahead matching with the input at reduction. The items in the reduction handle are framed. The final states on the stack top, which trigger a reduction move if their look-ahead matches the input, are encircled. The subscript S of the machine states and the item brackets are omitted, e.g., item $\langle\,0_S \dashv \perp\,\rangle$ is shortened as $0 \dashv \perp$, and so on the others.

Fig. 1. *EBNF* gram. G (a), net $\mathcal{M}$ (b), pilot $\mathcal{P}$ (c), and analysis with convergence (d).

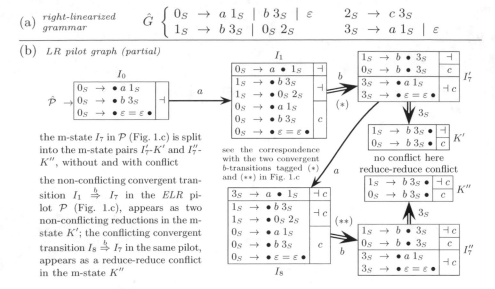

Fig. 2. *LR* pilot $\hat{\mathcal{P}}$ (b) of the *RLZG* $\hat{G}$ (a): dotted rules are used instead of *TN* states

axiom 0_S. Its right parts have length zero or two, and may contain two non-terminal symbols; thus grammar $\hat{G}$ is not *RL*. Grammar $\hat{G}$ generates language $L(G)$. See the running example in Fig. 2.a.

The classical notions of $LR(1)$ *look-ahead set*, dotted rule and item are rephrased for *TN*'s. Dotted rules are replaced by machine states, e.g., $S \to \bullet b$ becomes 0_S. An $ELR(1)$ *item* is a pair $\langle q_B, a \rangle$ in $Q \times (\Sigma \cup \{\dashv\})$, where $\dashv$ is the end-of-text: this says that a is a legal look-ahead token for the current activation of machine M_B in the state q_B. The standard closure function [12] is adjusted for the legal items of a *TN*.

Closure function. It is the smallest set that contains a given set C of items:

$$closure(C) = C \cup \left\{ \langle 0_B, b \rangle \,\middle|\, \begin{array}{l} \exists\, \langle q, a \rangle \in closure(C) \,\wedge\, \exists\, \left(q \xrightarrow{B} r \right) \in \mathcal{M} \\ \wedge\, b \in Ini\,(\,L(r) \cdot a\,) \end{array} \right\}$$

For simplicity we group together the items that have the same state, in this way: $\langle q, \{a_1, \ldots, a_k\} \rangle \equiv \{\langle q, a_1 \rangle, \ldots, \langle q, a_k \rangle\}$, where the *look-ahead set* $\{a_1, a_2, \ldots, a_k\}$ is never empty.

3   From *TN* to Shift-Reduce Parser

For brevity we only comment the passages that essentially depart from the standard ones. Given a *TN* $\mathcal{M}$, we present the new direct $ELR(1)$ construction of the finite state-transition function, denoted as ϑ, of the *DFA* that controls the parser (*DPDA*). This *DFA* is named *pilot* to shorten its traditional name "recognizer of viable $LR(1)$ prefixes". Its states, named *macro-states* (*m-states*) to distinguish them from *TN* states, are sets of items.

The construction has three phases: (i) from $\mathcal{M}$ we construct the *pilot*; (ii) we check the forthcoming $ELR(1)$ condition, which involves the new *convergence* conflict, besides the standard shift-reduce (SR) and reduce-reduce (RR) ones; and (iii) if the test is passed, we construct the $DPDA$. For an item $\langle p_A, \rho \rangle$ and a grammar symbol X, the *shift* (or "go to") function (qualified as *terminal / nonterminal* according to X) is $\vartheta\left(\langle p_A, \rho \rangle, X\right) = \langle q_A, \rho \rangle$ if the edge $p_A \overset{X}{\to} q_A$ exists, or the empty set otherwise. For a set C of items, the shift under X is the union of the shifts of all the items in C.

Algorithm 3. *Let the pilot DFA be $\mathcal{P} = (\Sigma \cup V, R, \vartheta, I_0, F = R)$ and let its initial state be $I_0 = closure\left(\{\langle 0_S, \dashv \rangle\}\right)$. The pilot state set $R = \{I_0, I_1, \dots\}$ and its transition function $\vartheta \colon R \times (\Sigma \cup V) \to R$ are computed by the next steps:*

$R' := \{I_0\}$
repeat $R := R'$
    for each m-state $I \in R$ and grammar symbol $X \in \Sigma \cup V$ do
        $I' := closure\left(\vartheta(I, X)\right)$ $R' := R \cup \{I'\}$
        add the edge $I \overset{X}{\to} I'$ to the transition function ϑ
until $R' = R$ □

For a pilot (e.g., Fig. 1.c) every m-state I is partitioned into two subsets. The *base* contains the non-initial items: $I_{|base} = \{\langle q, \pi \rangle \in I \mid q \text{ is not initial state}\}$; the *closure* contains the remaining ones: $I_{|closure} = \{\langle q, \pi \rangle \in I \mid q \text{ is initial state}\}$. All the m-states, but the initial one I_0, have a non-empty base, while their closures may be empty. The *kernel* of a m-state I is the projection of all the items contained in I on their first component (which is a TN state): $I_{|kernel} = \{q \in Q \mid \langle q, \pi \rangle \in I\} \subseteq Q$.

When in the same m-state I, it happens that two items shift under the same grammar symbol, then determinism may be defeated, as next explained.

Definition 4. *A m-state I has the multiple transition property (MTP) if it includes two items $\langle q, \pi \rangle$ and $\langle r, \rho \rangle$, such that for some grammar symbol $X \in \Sigma \cup V$ both TN transitions $\delta(q, X)$ and $\delta(r, X)$ are defined. Moreover a pilot transition $\vartheta(I, X)$ is convergent if $\delta(q, X) = \delta(r, X)$. The pilot transition has a convergence conflict if the look-ahead sets overlap, i.e., $\pi \cap \rho \neq \emptyset$.* □

We formalize the conditions for all the decisions of a parser to be deterministic.

Definition 5. $ELR(1)$ *condition for a TN.*

1. *For no m-state I there is a shift-reduce (SR) or a reduce-reduce (RR) conflict:*

$$(SR) \quad \forall \text{ item } \langle q, \pi \rangle \in I \text{ s.t. } q \text{ is final and } \forall \text{ edge } I \overset{a}{\to} I' : a \notin \pi \quad (1)$$
$$(RR) \quad \forall \text{ items } \langle q, \pi \rangle, \langle r, \rho \rangle \in I \text{ s.t. } q \text{ and } r \text{ are final} : \pi \cap \rho = \emptyset \quad (2)$$

2. *No transition of the pilot has any convergence conflict.* □

In the pilot $\mathcal{P}$ shown in Fig. 1.c, the MTP m-states I_1, I_2 and I_8 have outgoing convergent edges, two of which are conflicting. The presence of a convergent non-conflicting transition and of a convergent conflicting one that go into the same m-state I_7, shows that the convergence conflict is a property of the transition, not of the destination m-state.

ELR (1) *versus Standard LR (1) Conditions.* A BNF grammar, since the number of alternative rules $A \rightarrow \alpha \mid \beta \mid \ldots$ is known and the right parts do not contain any star or union operations, has a natural equivalent FA N_A, which in general is nondeterministic; the graph of N_A is acyclic and shaped as a tree. Clearly N_A satisfies the no-reentrance hypothesis for the initial state; it is often not minimal because each graph path ends in a distinct but undistinguishable final state. No m-state in the standard pilot has the multiple transition property, and the only possible conflicts are SR and RR.

Our machine M_A may differ from N_A in two ways. First, machine M_A is deterministic, though out of convenience, not of necessity. To illustrate, consider the alternatives $C \rightarrow$ **if** E **then** I | **if** E **then** I **else** I. Determinizing has the effect of left-factoring the longest common prefix, i.e., of using the $EBNF$ rule $C \rightarrow$ **if** E **then** I (ε | **else** I).

Second, the graph of M_A typically has fewer states than that of N_A: the final states (and maybe others) of N_A coalesce if they are equivalent for M_A. Hence some pilot edges may get convergent, thus the RR conflicts in the pilot of a BNF grammar may turn into *convergence* ones in the pilot of an equivalent $EBNF$ grammar (Fig. 1.c, 2.b). The correspondence between convergence conflicts and RR conflicts is not so evident. To have a grasp, notice how the convergence conflict in the m-state I_7 of Fig. 1.c matches the RR conflict in the m-state I_7'' of Fig. 2.b (read also the included comments).

The next central result supports the view that in a sense our ELR (1) condition is the most general possible for a TN. We believe this theorem is essential because early proposals to extend LR (1) concepts to $EBNF$ grammars or to TN's, since they omitted formal analysis, were later found to be flawed or not general enough (see Sect. 5).

Theorem 6. *Let $\mathcal{M}$ be a TN and $\hat{G}$ be the equivalent $RLZG$. Net $\mathcal{M}$ meets the ELR (1) condition if, and only if, grammar $\hat{G}$ meets the LR (1) condition.*   □

Proof. The proof shows that (if-part) if the ELR (1) condition is violated by the pilot $\mathcal{P}$ of net $\mathcal{M}$, then the pilot $\hat{\mathcal{P}}$ of grammar $\hat{G}$ violates the LR (1) condition, and viceversa (only-if part) that an LR (1) conflict in $\hat{\mathcal{P}}$ entails an ELR (1) one in $\mathcal{P}$. We study the correspondence between pilots $\mathcal{P}$ and $\hat{\mathcal{P}}$, namely between their transition functions ϑ and $\hat{\vartheta}$, and their m-states I and $\hat{I}$. It helps to compare the pilots in Fig. 1.c, 2.b.

We observe that since grammar $\hat{G}$ is BNF, all the edges of pilot $\hat{\mathcal{P}}$ that enter the same m-state, have identical labels. This does not hold for pilot $\mathcal{P}$, and so a m-state of $\mathcal{P}$ may be split into several m-states of $\hat{\mathcal{P}}$. We also notice, for any non-empty rule $X \rightarrow Y Z$ of $\hat{G}$, that it holds $X \in Q, Y \in \Sigma \cup \{ 0_A \mid \ 0_A \in Q \}$ and $Z \in Q \setminus \{ 0_A \mid \ 0_A \in Q \}$. Due to the special form of the rules of $\hat{G}$, the m-states $\hat{I}$ of $\hat{\mathcal{P}}$ are divided into three sets: the set $\{\ \hat{I}_0\ \}$ of the *initial* m-state; the set of the *intermediate* m-states, i.e., those such that every item of $\hat{I}_{|base}$ has the form $p_A \to Y \bullet q_A$; and the set of the *sink reduction* m-states, i.e., those such that every item has the form $p_A \to Y\, q_A \bullet$.

We say that an item $\langle q_X, \lambda \rangle$ of pilot $\mathcal{P}$ *corresponds* to an item of pilot $\hat{\mathcal{P}}$ of the form $\langle p_X \to s \bullet q_X, \rho \rangle$, if it holds $\lambda = \rho$. Then two m-states I of $\mathcal{P}$ and $\hat{I}$ of $\hat{\mathcal{P}}$ are *correspondent* if the items in $I_{|base}$ and in $\hat{I}_{|base}$ correspond to each other; moreover the initial m-states I_0 and $\hat{I}_0$ are defined as correspondent.

The following properties of correspondent m-states (Lemma 7) are straightforward and can be easily proved, see [2].

Lemma 7. *The mapping defined by the correspondence relation from the set containing the m-state $\hat{I}_0$ and the intermediate m-states of pilot $\hat{\mathcal{P}}$, to the set of the m-states of pilot $\mathcal{P}$, is total, many-to-one and onto (surjective). The following statements hold:*

1. *For any grammar symbol s, and any correspondent m-states I and $\hat{I}$, it holds: transition $\vartheta\,(I,\, s) = I'$ is defined $\Longleftrightarrow$ transition $\vartheta\,(\hat{I},\, s) = \hat{I}'$ is defined and m-state $\hat{I}'$ is intermediate. Moreover m-states I' and $\hat{I}'$ are correspondent.*
2. *Let a final state f_A be non-initial. Item $\langle f_A, \lambda \rangle$ is in m-state I (in $I_{|base}$) $\Longleftrightarrow$ a correspondent m-state $\hat{I}$ contains both items $\langle p_A \to s \bullet f_A, \lambda \rangle$ and $\langle f_A \to \varepsilon \bullet, \lambda \rangle$.*
3. *Let the initial state 0_A be final too, but with $A \neq S$. Item $\langle 0_A, \pi \rangle$ is in m-state I $\Longleftrightarrow$ a correspondent m-state $\hat{I}$ contains item $\langle 0_A \to \varepsilon \bullet, \pi \rangle$.*
4. *Let the axiomatic initial state 0_S be final too. Item $\langle 0_S, \pi \rangle$ is in m-state I_0 $\Longleftrightarrow$ item $\langle 0_S \to \varepsilon \bullet, \pi \rangle$ is in m-state $\hat{I}_0$.*
5. *For any pair of correspondent m-states I and $\hat{I}$, and any initial state 0_A, it holds: $0_A \in I_{|closure} \Longleftrightarrow 0_A \to \bullet\alpha \in \hat{I}_{|closure}$ for every alternative rule $0_A \to \alpha$ of 0_A.* $\quad\square$

Part "if". Consider the three conflict types that may occur in the pilot $\mathcal{P}$.

SR conflict. Consider a conflict in m-state $I \ni \langle f_B, \{a\} \rangle$, where f_B is final and non-initial and $\vartheta\,(I,\, a)$ is defined. By Lemma 7 (points (1) and (2)) there exists a correspondent m-state $\hat{I}$ such that $\vartheta\,(\hat{I},\, a)$ is defined and $\langle f_B \to \varepsilon\bullet, \{a\} \rangle \in \hat{I}$, hence the same conflict is in $\hat{\mathcal{P}}$. A similar reasoning, by exploiting (1) and (3), applies to a conflict in m-state $I \ni \langle 0_B, \{a\} \rangle$, where 0_B is final and initial, and $\vartheta\,(I,\, a)$ is defined.

RR conflict. If a conflict is present in m-state $I \supseteq \{\langle f_A, \{a\} \rangle, \langle f_B, \{a\} \rangle\}$, where f_A and f_B are final and non-initial, then from (2) the same conflict exists in one or more m-states $\hat{I} \supseteq \{\langle f_A \to \varepsilon\bullet, \{a\} \rangle, \langle f_B \to \varepsilon\bullet, \{a\} \rangle\}$. Similar reasonings apply to the cases where one final state or both final states in the items are initial.

Convergence conflict. Consider a conflicting transition $I \xrightarrow{X} I'$, where it holds $I \supseteq \{\langle p_A, \{a\} \rangle, \langle q_A, \{a\} \rangle\}$, $\delta\,(p_A, X) = \delta\,(q_A, X) = r_A$ and $I'_{|base} \ni \langle r_A, \{a\} \rangle$. If neither p_A nor q_A are initial, both items are in the base of I. By (1) there are correspondent intermediate m-states and a transition $\hat{I} \xrightarrow{X} \hat{I}'$ with $\hat{I}' \supseteq$

$\{\langle p_A \to X \bullet r_A, \{a\}\rangle, \langle q_A \to X \bullet r_A, \{a\}\rangle\}$, and the sink reduction m-state $\hat{\vartheta}(\hat{I}', r_A)$ has an RR conflict. Similarly, if (arbitrarily) $q_A = 0_A$, then $\langle q_A, \{a\}\rangle \in I_{|closure}$ from the non-reentrance hypothesis for M_A, and so $I_{|base}$ necessarily contains an item $C = \langle s, \rho\rangle$ such that $closure(C) = \langle 0_A, \{a\}\rangle$. Therefore for some t and Y, there exists a m-state $\hat{I}$ correspondent of I such that $\langle t \to Y \bullet s, \rho\rangle \in \hat{I}_{|base}$ and $\langle 0_A \to \bullet X r_A, \{a\}\rangle \in \hat{I}_{|closure}$, hence it holds $\hat{\vartheta}(\hat{I}, X) = \hat{I}'$ and $\langle 0_A \to X \bullet r_A, \{a\}\rangle \in \hat{I}'$, whence $\hat{\vartheta}(\hat{I}', r_A)$ has an RR conflict.

Part "only if". Consider the two conflict types that may occur in the pilot $\hat{\mathcal{P}}$.

SR conflict. The conflict occurs in m-state $\hat{I}$ such that $\langle f_B \to \varepsilon\bullet, \{a\}\rangle \in \hat{I}$ and $\hat{\vartheta}(\hat{I}, a)$ is defined. By Lemma 7 (points (1) and (2) (or (3))), the correspondent m-state I contains $\langle f_B, \{a\}\rangle$ and the move $\vartheta(I, a)$ is defined, thus resulting in the same conflict.

RR conflict. Consider m-state $\hat{I}$ s.t. $\{\langle f_A \to \varepsilon\bullet, \{a\}\rangle, \langle f_B \to \varepsilon\bullet, \{a\}\rangle\} \subseteq \hat{I}_{|closure}$, where f_A and f_B are final non-initial. By (2), the correspondent m-state I contains the items $\langle f_A, \{a\}\rangle, \langle f_B, \{a\}\rangle$ and has the same conflict. A similar reasoning, based on points (2) and (3), applies if either state f_A or f_B is initial. Finally consider an RR conflict in a sink reduction m-state $\hat{I}$ s.t. $\{\langle p_A \to X r_A\bullet, \{a\}\rangle, \langle q_A \to X r_A\bullet, \{a\}\rangle\} \subseteq \hat{I}$. There exist m-states $\hat{I}'$, $\hat{I}''$ and transitions $\hat{I}'' \overset{X}{\to} \hat{I}' \overset{r_A}{\to} \hat{I}$ s.t. $\hat{I}'$ contains items $\langle p_A \to X \bullet r_A, \{a\}\rangle$ and $\langle q_A \to X \bullet r_A, \{a\}\rangle$, the correspondent m-state I' contains item $\langle r_A, \{a\}\rangle$, and $\{\langle p_A \to \bullet X r_A, \{a\}\rangle, \langle q_A \to \bullet X r_A, \{a\}\rangle\} \subseteq \hat{I}''_{|closure}$. Since $\hat{I}''_{|closure} \neq \emptyset$, $\hat{I}''$ is not a sink reduction m-state. Let I'' be its correspondent m-state. Then: if p_A is initial, $\langle p_A, \{a\}\rangle \in I''_{|closure}$ by (5); if p_A is not initial, there exists an item $\langle t_A \to Z \bullet p_A, \{a\}\rangle \in \hat{I}''_{|base}$, and $\langle p_A, \{a\}\rangle \in I''$. A similar reasoning applies to state q_A, hence $\langle q_A, \{a\}\rangle \in I''$, and $I'' \overset{X}{\to} I'$ has a convergence conflict. $\qquad\square$

We address a possible criticism to the significance of Theorem 6: that, starting from an *EBNF* grammar, several equivalent *BNF* grammars can be obtained by removing the r.e. operations in different ways. Such grammars may or may not be $LR(1)$, a fact that would seem to make somewhat arbitrary the choice of the left-linearized form in our definition of $ELR(1)$. We defend the significance and generality of our choice on two grounds. First, our original grammar specification is not a set of r.e.'s, but a *TN* set of *DFA*'s, and the choice to transform the *DFA* into a right-linear grammar is not only natural but also opportunistic because, as shown by [9], the other natural form - left-linear - would exhibit conflicts in most cases. Second, the same author shows that this definition of $ELR(1)$ grammar dominates the alternative definitions available at his time (and we believe also the later definitions). We illustrate by the case of an $ELR(1)$ *TN* (by Theorem 6 the right-linearized grammar $\hat{G}$ is $LR(1)$) where an equivalent *BNF* grammar obtained by a natural transformation has conflicts.

Example 8. The language structure is $E(sE)^*$, where substring E has the (context-free) form $b^+ b^n e^n$ or $b^n e^n e$, with $n \geq 0$. It is defined by the *TN* below, which meets the $ELR(1)$ condition.

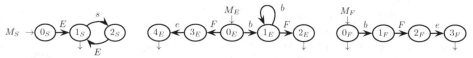

On the contrary, in the equivalent *BNF* grammar with nonterminals S, E, F and B, and with rules $S \to E\,s\,S \mid E$, $E \to B\,F \mid F\,e$, $F \to b\,F\,e \mid \varepsilon$ and $B \to b\,B \mid b$, there is an *SR* conflict. The *RLZG* postpones any reduction decision as late as possible and so avoids the conflicts.   □

$ELR\,(1)$ Parser. Given the pilot of an $ELR\,(1)$ *TN*, we specify the *DPDA* that recognizes and parses the sentences. In the *DPDA* stack, elements of two types alternate: grammar symbols and *stack m-states* (*sms*). Since for a given *TN* $\mathcal{M}$ there are finitely many different items, the number of m-states is bounded, and the number of items in any m-state is bounded by $C_{Max} = |Q| \times (|\Sigma| + 1)$. An *sms*, denoted as J, is an *ordered* set of triples named *stack items*, which are items with one more field named *item identifier* (*iid*). A stack item has the form $\langle q_A, \pi, iid \rangle$, where $q_A \in Q$, $\pi \subseteq \Sigma \cup \{\dashv\}$, and $1 \leq iid \leq C_{Max}$ or $iid = \bot$. For readability the *iid* values are prefixed by a $\sharp$ marker.

The parser uses a surjective mapping μ from the set of *sms* to that of m-states: it holds $\mu\,(J) = I$ if, and only if, by dropping their *iid*'s, the items in J equal those in I. For explaining the parsing algorithm, we stipulate that identically indexed symbols $J[l]$ and I_l are related by $\mu\,(J[l]) = I_l$.

Algorithm 9. *$ELR\,(1)$ parser as a DPDA. Let $J[0]$ or $J[0]\,a_1\,J[1]\,a_2 \ldots a_k\,J[k]$ be the current stack contents, where a_i $(1 \leq i \leq k)$ is any grammar symbol (terminal or nonterminal) and the sms on the stack top is $J[k]$ (see Fig. 1.d).*

Initialization. *The stack contents are set to $J[0] = \{\,s \mid s = \langle q, \pi, \bot \rangle$ for every item $\langle q, \pi \rangle \in I_0\,\}$, where $\mu\,(J[0]) = I_0$ is the initial m-state of the pilot.*

Shift move. *The top sms is J, with $\mu\,(J) = I$, the current input token is $a \in \Sigma$, and it holds $\vartheta\,(I, a) = I'$. Suppose that by inspecting I, the pilot decides to shift. The shift move performs two steps:*
1. *pushes on stack the token a and gets the next token*
2. *pushes on stack the sms J' computed as follows:*

$$J' = \{\, \langle q'_A, \rho, \sharp i \rangle \mid \langle q_A, \rho, \sharp j \rangle \text{ is at position } i \text{ in } J \wedge \left(q_A \xrightarrow{a} q'_A \right) \in \delta \} \quad (3)$$
$$\cup \{\, \langle 0_B, \sigma, \bot \rangle \mid \langle 0_B, \sigma \rangle \in I'_{|closure}\,\} \qquad \text{(see Fig. 3)} \qquad (4)$$

so that $\mu\,(J') = I'$. Notice the last condition in (3) implies it holds $q'_A \in I'_{|base}$.

Reduction move (non-initial state). *The current stack contents are the sequence $J[0]\,a_1\,J[1]\,a_2 \ldots a_k\,J[k]$ and the corresponding m-states are $\mu\,(J[l]) = I_l$, with $0 \leq l \leq k$. Suppose that by inspecting I_k, the pilot chooses to reduce an item $\langle q_A, \pi \rangle \in I_k$, where q_A is a final yet non-initial state. Let $t_k = \langle q_A, \rho, \sharp i_k \rangle \in J[k]$ be the (only) sms such that the current token is $a \in \rho$. An iid chain starts from $\sharp i_k$, which links t_k to a stack item $t_{k-1} = \langle p_A, \rho, \sharp i_{k-1} \rangle \in J[k-1]$, and so on as far as it reaches a stack item $t_h \in J[h]$ that has a null iid (thus its state is initial): $t_h = \langle 0_A, \rho, \bot \rangle$. The reduction move performs three steps:*

1. *grows the syntax (sub)tree list by applying reduction $a_{h+1}\, a_{h+2} \ldots a_k \rightsquigarrow A$*
2. *pops the stack symbols in this order: $J[k]\, a_k\, J[k-1]\, a_{k-1} \ldots J[h+1]\, a_{h+1}$*
3. *executes the nonterminal shift move $\vartheta\,(I[h],\, A)$ (see below)*

Reduction move (initial state). *It differs from the preceding case (non-initial state) in that the chosen item is $\langle 0_A,\, \pi,\, \bot \rangle$. The parser move grows the syntax (sub)tree list by the null reduction $\varepsilon \rightsquigarrow A$ and performs the nonterminal shift move corresponding to $\vartheta\,(I[k],\, A)$.*

Nonterminal shift move. *It is the same as a shift move, except that the shifted symbol A is a nonterminal and the parser does not read the next input token as it does instead in the step 1 of the shift move.*

Acceptance. *The parser halts and accepts when the stack contains only $J[0]$ and the current input token is $\dashv$ (end-of-text).*   □

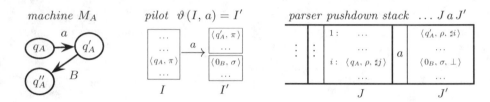

Fig. 3. Shift move scheme, with machine, pilot and stack ($\mu\,(J) = I$ and $\mu\,(J') = I'$).

For a shift move we notice that an *sms* J' computed by Alg. 9, may contain multiple items that have the same state. This happens whenever an edge $I \to \theta\,(I,\,a) = I'$ of the pilot graph is convergent. The scheme of a shift move (eq.s (3) and (4)) is shown in Fig. 3.

Stack items are linked through *iid* chains, e.g., item $\langle q'_A,\, \rho,\, \sharp i \rangle$ is linked to $\langle q_A,\, \rho,\, \sharp j \rangle$ via $\sharp i$. Every stack item is mapped by function μ onto a pilot m-state item that has the same machine state. In general the look-ahead set π of such a pilot item is a superset of the look-ahead set ρ in the stack item, due to the possible presence of convergent transitions. The two sets ρ and π coincide if at parsing time the pilot does not take a convergent transition. We omit the straightforward correctness proof of Alg. 9 and we move to an example.

Example 10. Fig. 1.d shows the parse trace of string $a\,b\,a\,a\,c\,c$. The stack alternates grammar symbols $X \in \Sigma \cup V$ and *sms*'s $J[l]$, and starts from *sms* $J[0]$. Each pair $(X,\, J)$ spans the recognized (series of) token(s). In an *sms* $J[l]$ each item is numbered from 1 (top item). An *iid* $\sharp i$ in an item of $J[l+1]$ refers back to the i-th item in $J[l]$. The trace shows from top to bottom: a null reduction, a non-null one of an acyclic path in M_S, and a final non-null cyclic one. The *sms*'s J differ from the m-states I of $\mathcal{P}$: *sms* $J[3]$ has three items $\langle 1_S,\, \dashv,\, \sharp 1 \rangle$, $\langle 1_S,\, c,\, \sharp 2 \rangle$ and $\langle 0_S,\, c,\, \bot \rangle$, which come from m-state I_8 (i.e., $\mu\,(J[3[) = I_8$) by splitting the look-ahead set of the item $\langle 1_S,\, \dashv c \rangle$ that results from convergence; and *sms* $J[2]$ is a splitting of m-state I_7 (i.e., $\mu\,(J[2[) = I_7$), with two *sms*'s coming from one m-state item. More comments are in Fig. 1 (all parts).   □

4   Standard $LR(1)$ versus $ELR(1)$ Parser

First, we argue that the direct pilot has fewer m-states than the indirect standard pilot of an equivalent *BNF* grammar. Consider a family of regular languages L_n, where integer n is the *star height*, and their one-rule *EBNF* grammars $S \to L_n$ (with $n \geq 1$). We show a simpler family with an alphabet growing with n, but others exist with a bounded alphabet. Take these languages: $L_1 = a^*$, $L_2 = (a^* b)^*$, $L_3 = ((a^* b)^* c)^*$, etc. The minimal (non-reentrant) *DFA* that recognizes L_n, has $n + 1$ states and is isomorphic to the direct pilot of $S \to L_n$. Converting such a grammar to *BNF* needs n nonterminals and its standard pilot has a number of m-states that is lower-bounded by $2n$.

Second, we examine how the typical transformations of a *BNF* grammar into an *EBNF* one affect the pilot size. Space prevents a complete analysis and we consider two main cases only. If a left-recursive (or right-recursive) *BNF* rule, such as $A \to A u \mid v$, is replaced by an *EBNF* rule $A \to v u^*$ that features a star operation, then the pilot size shrinks. If a language substitution is applied, e.g., by replacing rules $A \to a B b$ and $B \to u \mid v$ with one rule $A \to a (u \mid v) b$, then the number of nonterminals and *FA*'s decreases, and so also does the pilot size. Therefore the repeated application of left-recursion removal and substitution yields a *TN* with a smaller pilot.

At last, we compare the run-time memory needed for the pilot stacks. At a first glance the direct pilot stack contains twice the number of entries of the indirect pilot stack, because the latter pilot does not need to store the grammar symbols between *sms*'s (all of which are m-states), whereas the former one needs them to build the syntax tree when performing a reduction (see the trace in Fig. 1.d). Yet it would be straightforward to take the grammar symbol X shifted by a direct pilot transition $I_1 \overset{X}{\dashrightarrow} I_2$, and encode it into the *sms* associated to I_2 by μ. This would make equal the stack lengths of the two parsers. Although such a coding enlarges the stack alphabet size, it does not penalize run-time efficiency. To sum up, the direct parser has a smaller pilot and the same stack memory occupation at run-time as the standard indirect parser. Experimentation on real grammars is obviously needed to assess the practicality of the direct method.

5   Related Work and Conclusion

Many early authors have proposed to extend the standard *LR* method to *EBNF* grammars, each proposal purporting to improve over previous attempts, but no clear-cut optimal solution has surfaced. The following discussion particularly draws from the later papers [10,11,16]. The first dichotomy concerns the source language specification: either an *EBNF* grammar or a *TN*. Since it is now perfectly clear that r.e.'s and finite automata are interchangeable notations, the distinction is moot.

Some authors impose restrictions on the r.e.'s, for instance by limiting the star depth to one or forbidding common subexpressions. Although the original motivation to simplify parser construction has since vanished, it is fair to say that the r.e.'s used in the language manuals are typically quite simple, for the reason of avoiding obscurity.

Others, including ourselves, use TN's and specify the right parts by DFA's (notably [16]). Notice that the use of NFA's would make little difference for readability, yet it could be easily accommodated within our parser generation framework. Our inexpensive normalization of DFA's that disallows reentering the initial state (Def. 1), greatly simplifies the parser construction. For TN specifications two approaches to parser construction exist: (A) eliminates star and union (i.e., converts $EBNF$ into BNF) and applies the standard $LR(1)$ construction; (B) directly constructs the parser. In [3] a systematic transformation from $EBNF$ to BNF is used to obtain an $ELR(1)$ parser that simulates the classical Knuth's one for BNF. It is generally agreed [16] that approach (B) is superior, because transforming to BNF adds inefficiency and obscures the semantic structure. Moreover we have seen in Sect. 4 that the size of the directly produced parser is smaller. The only advantage of approach (A) is to leverage on existing parser generators, e.g., Bison.

The major difficulty with approach (B) is to identify the left end of the reduction handle, since its length is unpredictable and possibly unbounded. A list of proposed solutions is in the cited surveys. Some algorithms use a special shift move, sometimes called *stack-shift*, to record into the stack the left end of the handle when a new computation on a machine is started. But if the initial state is reentered, then a conflict between stack-shift and normal shift is unavoidable, and various complicated devices have been invented to arbitrate the conflict. Some authors add read-back states to control how deep the parser should dig into the stack [4,14], others (e.g., [17]) use counters for the same purpose, let alone further proposed devices. Unfortunately it was shown in [8,11] that several proposals do not precisely characterize the grammars they apply to, and in certain cases they may fall into unexpected errors. Motivated by the flaws of past attempts, paper [15] offers a characterization of the $LR(k)$ property for TN's. Although their definition is intended to ensure that the languages "can be parsed from left to right with a look-ahead of k symbols", the authors admit that "the subject of efficient techniques for locating the left end of a handle is beyond the scope of this paper".

We mention a link (proved in [2]) with the classical top-down deterministic $ELL(1)$ TN's. These are a special case of $ELR(1)$ TN's meeting two conditions: there are no left-recursive derivations and every m-state base has at most one item. The second condition rules out convergent transitions and conflicts. The two conditions generalize the Beatty's characterization [1] of $LL(1)$ BNF grammars. If an $ELR(1)$ pilot meets these conditions, a few steps permit to transform the shift-reduce parser into a predictive one.

After a long history of moderately successful attempts, our $ELR(1)$ condition and the naturally corresponding parser hopefully offer a definitive solution to this long-standing problem. Our approach is more or equally general as any previous proposal known to us, and is simple: it just adds the treatment of convergent edges to the Knuth's definition. The technical difficulties were understood since long, and we have combined and improved existing ideas into a practical and demonstrably correct solution. In our teaching experience [6], the new *TN* parser construction method economically unifies bottom-up and top-down approaches, usually taught as independent methods. Moreover such a unification is promising for building heterogeneous parsers [7] based on grammar partition [13]. For real application a parser generation tool has to be developed, which we estimate to be about as complex as the existing $LR(1)$ tools.

References

1. Beatty, J.C.: On the relationship between the LL(1) and LR(1) grammars. JACM 29(4), 1007–1022 (1982)
2. Breveglieri, L., Crespi Reghizzi, S., Morzenti, A.: Parsing methods streamlined. arXiv:1309.7584 [cs.FL], pp. 1–64 (2013)
3. Celentano, A.: LR parsing technique for extended context-free grammars. Comput. Lang. 6(2), 95–107 (1981)
4. Chapman, N.P.: LALR(1,1) parser generation for regular right part grammars. Acta Inform. 21, 29–45 (1984), http://dx.doi.org/10.1007/BF00289138
5. Conway, M.E.: Design of a separable transition-diagram compiler. Comm. ACM 6(7), 396–408 (1963)
6. Crespi Reghizzi, S., Breveglieri, L., Morzenti, A.: Formal languages and compilation, 2nd edn. Springer, London (2013)
7. Crespi Reghizzi, S., Psaila, G.: Grammar partitioning and modular deterministic parsing. Comput. Lang. 24(4), 197–227 (1998)
8. Gálvez, J.F.: A note on a proposed LALR parser for extended context-free grammars. Inf. Process. Lett. 50(6), 303–305 (1994), http://dx.doi.org/10.1016/0020-0190(94)00051-4
9. Heilbrunner, S.: On the definition of ELR(k) and ELL(k) grammars. Acta Inform. 11, 169–176 (1979)
10. Hemerik, K.: Towards a taxonomy for ECFG and RRPG parsing. In: Dediu, A.H., Ionescu, A.M., Martín-Vide, C. (eds.) LATA 2009. LNCS, vol. 5457, pp. 410–421. Springer, Heidelberg (2009), http://dx.doi.org/10.1007/978-3-642-00982-2
11. Kannapinn, S.: Reconstructing LR theory to eliminate redundance, with an application to the construction of ELR parsers (in German). Ph.D. thesis, Tech. Univ. Berlin (2001)
12. Knuth, D.E.: On the translation of languages from left to right. Information and Control 8, 607–639 (1965)
13. Korenjak, A.J.: A practical method for constructing LR(k) processors. Commun. ACM 12(11), 613–623 (1969)
14. LaLonde, W.R.: Constructing LR parsers for regular right part grammars. Acta Inform. 11, 177–193 (1979), http://dx.doi.org/10.1007/BF00264024
15. Lee, G.O., Kim, D.H.: Characterization of extended LR(k) grammars. Inf. Process. Lett. 64(2), 75–82 (1997), http://dx.doi.org/10.1016/S0020-0190(97)00152-X
16. Morimoto, S., Sassa, M.: Yet another generation of LALR parsers for regular right part grammars. Acta Inform. 37, 671–697 (2001)
17. Sassa, M., Nakata, I.: A simple realization of LR-parsers for regular right part grammars. Inf. Process. Lett. 24(2), 113–120 (1987), http://dx.doi.org/10.1016/0020-0190(87)90104-9

Optimal Sorting Networks

Daniel Bundala and Jakub Závodný

Department of Computer Science, University of Oxford
Wolfson Building, Parks Road, Oxford, OX1 3QD, UK
{daniel.bundala,jakub.zavodny}@cs.ox.ac.uk

Abstract. This paper settles the optimality of sorting networks given in
The Art of Computer Programming vol. 3 more than 40 years ago. The
book lists efficient sorting networks with $n \leq 16$ inputs. In this paper we
give general combinatorial arguments showing that if a sorting network
with a given depth exists then there exists one with a special form. We
then construct propositional formulas whose satisfiability is necessary for
the existence of such a network. Using a SAT solver we conclude that the
listed networks have optimal depth. For $n \leq 10$ inputs where optimality
was known previously, our algorithm is four orders of magnitude faster
than those in prior work.

1 Introduction

In their celebrated result, Ajtai, Komlós and Szemerédi (AKS) [1], gave an optimal oblivious sorting algorithm with $O(n \log n)$ comparisons in $O(\log n)$ parallel
steps. An oblivious sorting algorithm is one in which the order of comparisons is
fixed and depends only on the number of inputs but not their values. Compare
this with standard algorithms such as MergeSort or QuickSort where the order
of comparisons crucially depends on the input values.

A popular model of oblivious sorting algorithms are so-called sorting networks,
which specify a sequence of swap-comparisons on a set of inputs, and whose depth
models the number of parallel steps required. Even though the AKS network
has asymptotically optimal depth, it is infamous for the large constant hidden
in the big O bound; recursively constructed networks of depth $O(\log^2 n)$ [2]
prove superior to the AKS network for all practical values of n. Small networks
for small numbers of inputs serve as base cases for these recursive methods.
However, constructing networks of optimal depth has proved extremely difficult
(e.g., [5,7]) and is an open problem even for very small number of inputs. We
address this problem in this paper.

Already in the fifties and sixties various constructions appeared for small
sorting networks on few inputs. In 1973 in The Art of Computer Programming
vol. 3 [4], Knuth listed the best sorting networks with $n \leq 16$ inputs known at
the time. It was further shown in [3] that these networks have optimal depth for
$n \leq 8$. No progress had been made on the problem until 1989 when Parberry [7]
showed that the networks listed in [4] are optimal for $n = 9$ and $n = 10$. The result was obtained by implementing an exhaustive search with pruning based on

A.-H. Dediu et al. (Eds.): LATA 2014, LNCS 8370, pp. 236–247, 2014.
© Springer International Publishing Switzerland 2014

symmetries in the first two parallel steps in the sorting networks, and executing the algorithm on a supercomputer (Cray-2). Despite the great increase in available computational power in the 24 years since, the algorithm would still not be able to handle the case $n = 11$. Recently there were attempts [5] at solving the case $n = 11$ but we are not aware of any successful one.

Forty years after the publication of the list of small sorting networks by Knuth [4], we finally settle their optimality for the remaining cases $n = 11$ up to and including 16. We give general combinatorial arguments showing that if a small-depth sorting network exists then there exists one with a special form. We then construct propositional formulas whose satisfiability is necessary for the existence of such a network. By checking the satisfiability of the formulas using a SAT solver we conclude that no smaller networks than those listed exist.

We obtained all our results using an off-the-shelf SAT solver running on a standard desktop computer. It is noteworthy that our algorithm required less than a second to prove the optimality of networks with $n \leq 10$ inputs whereas the algorithm in [7] was estimated to take hundreds of hours on a supercomputer and that in [5] took more than three weeks on a desktop computer.

2   Sorting Networks

A **comparator network** C with n channels and depth d is defined as a tuple $C = \langle L_1, \ldots, L_d \rangle$ of **layers** $L_1, \ldots, L_d$. Each layer consists of **comparators** $\langle i, j \rangle$ for pairs of channels $i < j$. Every channel i is required to occur at most once in each layer L_k, i.e., $|\{j \mid \langle i, j \rangle \in L_k \vee \langle j, i \rangle \in L_k\}| \leq 1$. A layer L is called **maximal** if no more comparators can be added into L, i.e., $|L| = \lfloor \frac{n}{2} \rfloor$.

An input to a comparator network is a sequence of numbers applied to channels in the first layer. The numbers are propagated through the network; each comparator $\langle i, j \rangle$ takes the values from channels i and j and outputs the smaller value on channel i and the larger value on channel j. For an input sequence $x_1, \ldots, x_n$ define the value $V(k, i)$ of channel $1 \leq i \leq n$ at layer $k = 0$ (input) to be $V(0, i) = x_i$ and at layer $1 \leq k \leq d$ to be:

$$V(k, i) = \begin{cases} \min(V(k-1, i), V(k-1, j)) & \text{if } \langle i, j \rangle \in L_k, \\ \max(V(k-1, i), V(k-1, j)) & \text{if } \langle j, i \rangle \in L_k, \\ V(k-1, i) & \text{otherwise.} \end{cases}$$

The **output** $C(x)$ of C on x is the sequence $\langle V(d, 1), V(d, 2), \ldots, V(d, n) \rangle$. See Fig. 1 for an example of a network and its evaluation on an input.

Each comparator permutes the values on two channels and hence the output of a comparator network is always a permutation of the input. A comparator network is called a **sorting network** if the output $C(x)$ is sorted (ascendingly) for every possible input $x \in \mathbb{Z}^n$. We denote the set of all sorting networks with n channels and depth d by $\boldsymbol{S(n, d)}$.

In this work, we are interested in finding the optimal-depth sorting networks for small values of n. That is, given n, what is the least value of d, denoted by $\boldsymbol{V(n)}$, such that $S(n, d)$ is nonempty?

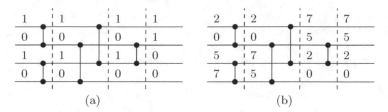

(a)   (b)

Fig. 1. A comparator network $(L_1 = \{\langle 1, 2\rangle, \langle 3, 4\rangle\}, L_2 = \{\langle 1, 3\rangle, \langle 2, 4\rangle\}, L_3 = \{\langle 2, 3\rangle\})$ with 4 channels, 5 comparators, and depth 3. The channels go from left to right, the first channel is at the bottom, the dashed lines separate the layers. The network on the left is evaluated on the input $\langle 0, 1, 0, 1\rangle$ and the network on the right on $\langle 7, 5, 0, 2\rangle$. Diagram shows the values on channels after each layer.

Observe that the function $V(n)$ is non-decreasing. Let C be a sorting network with n channels, and construct a network D from C by removing the last channel and all comparators attached to it. Then D is a sorting network with $n - 1$ channels: its behaviour on any input is simulated by the first $n - 1$ channels of C if the input to the last channel is set larger than all other inputs ($C(x\infty)$ is $D(x)\infty$, and $C(x\infty)$ is sorted so $D(x)$ is also sorted).

2.1   Known Bounds on $V(n)$

Fig. 2 summarises the best bounds on $V(n)$ for $n \leq 16$ channels known before our work. See [7] for lower bounds on $V(9)$ and $V(10)$, all other numbers appeared already in [4]. The main contribution of this paper is that $S(11, 7)$ and $S(13, 8)$ are empty. Thus we improve the lower bounds for $n = 11, 12$ and $13 \leq n \leq 16$ to 8 and 9, respectively, thereby matching the respective upper bounds.

n	1	2	3	4	5	6	7	8	9	10	11	12	13	14	15	16
Upper bound	0	1	3	3	5	5	6	6	7	7	8	8	9	9	9	9
Lower bound	0	1	3	3	5	5	6	6	7	7	7*	7*	7*	7*	7*	7*

Fig. 2. Table summarising the best lower and upper bounds known before our work. We improve the starred lower bounds to match the corresponding upper bound.

One can think of a layer of a comparator network as a matching on n elements: a comparator joins two distinct elements. The number of matchings on n elements grows exponentially in n. (See Fig. 3 for values for $n \leq 13$.) In particular, there are 35696 matchings on 11 elements, so to establish the lower bound $V(11) \geq 8$ we have to show that none of the $35696^7 \geq 10^{31}$ comparator networks with 11 channels and depth 7 is a sorting one. Similarly, to establish $V(13) \geq 9$ we have to consider $568504^8 \geq 10^{46}$ candidate networks. These numbers immediately make any exhaustive search approach infeasible. In the next section we present techniques to reduce the search space of possible sorting networks, and in Section 4 we show how to explore this space using a SAT solver.

3   Search Space Reduction

In the previous section we showed that the number of comparator networks grows very quickly. In this section we study general properties of sorting networks with arbitrary numbers of channels and depth, and we show that if $S(n, d)$ is non-empty then it contains a sorting network of a particular form, thus restricting the set of possible candidate networks. For example, for $n = 13$ this restricts the set of $568504^2 \geq 3 \cdot 10^{11}$ possible first-two layers to only 212 candidates.

Our arguments build upon and extend those from [7], and are based on four technical lemmas given in the following subsections. We make use of the following notation. The set of all layers on n channels is denoted as $\boldsymbol{G_n}$. For two networks $C = \langle L_1, \ldots, L_p \rangle$ and $D = \langle M_1, \ldots, M_q \rangle$ with the same number of channels, the **composition** $C \, \mathring{,} \, D$ of C and D is the network $\langle L_1, \ldots, L_p, M_1, \ldots, M_q \rangle$. That is, we first apply C and then D; for any input $x \in \mathbb{B}^n$ we have $(C \, \mathring{,} \, D)(x) = D(C(x))$. A **prefix** of a network C is a network P such that $C = P \, \mathring{,} \, Q$ for some network Q. If L is a single layer, we abuse the notation, treat L as a comparator network of depth 1, and write $L(x)$ for the application of the layer L on input x.

3.1   A Sufficient Sorting Condition

Before we even start looking for sorting networks it seems necessary to check infinitely many inputs (every $x \in \mathbb{Z}^n$) just to determine whether a comparator network is a sorting one. However, a standard result restricts the set of sufficient inputs to the Boolean ones. Denote $\mathbb{B} = \{0, 1\}$.

Lemma 1 ([4]). *Let C be a comparator network. Then C is a sorting network if and only if C sorts every Boolean input (every $x \in \mathbb{B}^n$).*

3.2   Output-Minimal Networks

When looking for a sorting network $C = \langle L_1, \ldots, L_d \rangle$, we can assume without loss of generality that the first layer L_1 is maximal, since by adding comparators to the first layer we can only restrict the set of its possible outputs. We cannot assume that all layers are maximal, but we can assume that the individual prefixes are maximally sorting in the following sense.

By $\boldsymbol{outputs(C)} = \{C(x) \mid x \in \mathbb{B}^n\}$ we denote the set of all possible outputs of a comparator network C on Boolean inputs. The following lemma states that it suffices to consider prefixes P with minimal $outputs(P)$.

Lemma 2. *Let $C = P \, \mathring{,} \, S$ be a sorting network of depth d and Q be a comparator network such that $depth(P) = depth(Q)$ and $outputs(Q) \subseteq outputs(P)$. Then $Q \, \mathring{,} \, S$ is a sorting network of depth d.*

Proof. Since $depth(P) = depth(Q)$ we have $depth(Q \, \mathring{,} \, S) = depth(P \, \mathring{,} \, S) = d$.

Let $x \in \mathbb{B}^n$ be an arbitrary input. Then $Q(x) \in outputs(Q) \subseteq outputs(P)$. Hence, there is $y \in \mathbb{B}^n$ such that $Q(x) = P(y)$. Thus, $(Q \, \mathring{,} \, S)(x) = S(Q(x)) = S(P(y)) = (P \, \mathring{,} \, S)(y) = C(y)$, which is sorted since C is a sorting network.   □

3.3   Generalised Sorting Networks and Symmetry

We further restrict the set of candidate sorting networks by exploiting their symmetry. To facilitate such arguments, we introduce so-called generalised comparator networks [4] where we lift the condition that the min-channel of a comparator is the one with a smaller index.

Formally, a **generalised comparator network** C with n channels and depth d is a tuple $C = \langle L_1, \ldots, L_d \rangle$ whose layers $L_1, \ldots, L_d$ consists of comparators $\langle i, j \rangle$ for channels $i \neq j$, such that each channel occurs at most once in each layer. A comparator $\langle i, j \rangle$ is called a **min-max comparator** if $i < j$ and a **max-min comparator** otherwise. Channel i receives the minimum and channel j receives the maximum of the values on channels i and j.

A generalised comparator network can move smaller values to the channel with larger index; we adapt the definition of a sorting network to reflect this. A generalised comparator network C is a **generalised sorting network** if there exists a permutation π_C such that for every $x \in \mathbb{B}^n$ the value of $C(x)$ is sorted after applying π_C. That is, if $C(x) = (y_1, \ldots, y_n)$ then we require $(y_{\pi_C(1)}, \ldots, y_{\pi_C(n)})$ to be sorted. It is well known [6,4] that a generalised sorting network can always be untangled into an "ordinary" sorting network of the same dimensions. Furthermore, this operation preserves the "ordinary" prefix:

Lemma 3 ([6,4]). *If G is a generalised sorting network of depth d then there exist a sorting network C of depth d. Furthermore, if $G = P \,\mathring{,}\, H$ where P is a comparator network then $C = P \,\mathring{,}\, I$ where I is a comparator network.*

Let π be a permutation on n elements. For a comparator $\langle i, j \rangle$ we define the comparator $\pi(\langle i, j \rangle) = \langle \pi(i), \pi(j) \rangle$, and we extend the action of π to layers and networks: $\pi(L) = \{\pi(C_1), \ldots, \pi(C_k)\}$ and $\pi(C) = \langle \pi(L_1), \ldots, \pi(L_d) \rangle$. Intuitively, applying π to a comparator network is equivalent to permuting the channels according to π; possibly flipping min-max and max-min comparators. Since a generalised sorting network sorts all inputs up to a fixed permutation (π_C) of the output, so do its permutations $\pi(C)$ (up to the permutation $\pi_C \circ \pi^{-1}$).

Lemma 4 ([7]). *Let C be a generalised sorting network with n channels and π be any permutation on n elements. Then $\pi(C)$ is a generalised sorting network.*

Lemmas 1 and 2 also hold for generalised comparator networks.

3.4   First Layer

We showed in Section 3.2 that if there is a sorting network in $S(n, d)$, then there is one whose first layer is maximal. Now we show that for *any* maximal layer L, there exists a sorting network in $S(n, d)$ whose first layer is L.

Lemma 5 ([7]). *Let L be a maximal layer on n inputs. If there is a sorting network in $S(n, d)$ there is a sorting network in $S(n, d)$ whose first layer is L.*

Proof. Let $C = L_1 \, \mathring{,} \, N$ be a sorting network with L_1 its first layer. By Section 3.2, if $L_1^+ \supseteq L_1$ is a maximal layer, then $C^+ = L_1^+ \, \mathring{,} \, N$ is also a sorting network. Since L_1^+ and L are both maximal, there is a permutation π such that $\pi(L_1^+) = L$. Then, $\pi(C^+)$ is a generalised sorting network by Lemma 4. Now, $\pi(C^+) = \pi(L_1^+) \, \mathring{,} \, \pi(N) = L \, \mathring{,} \, \pi(N)$, and by Lemma 3 there is a comparator network I such that $L \, \mathring{,} \, I$ is a sorting network and $depth(L \, \mathring{,} \, I) = depth(C^+) = depth(C)$. □

Lemma 5 allows us to consider only networks with a given maximal first layer. For networks on n inputs we fix the first layer to

$$F_n = \{\langle i, \lceil \tfrac{n}{2} \rceil + i \rangle \mid 1 \leq i \leq \lfloor \tfrac{n}{2} \rfloor\}.$$

3.5   Second Layer

Next we reduce the possibilities for the second layer[1], not to a single candidate but to a small set of candidate second layers. For $n = 13$ we arrive at 212 candidates out of the possible 568504 second layers.

As for the first layer, we can consider second layers modulo permutations of channels. However, we must take into account that the first layer is already fixed to F_n, and only consider permutations that leave the first layer intact.

Lemma 6 ([7]). *Let π be a permutation such that $\pi(F_n) = F_n$ and let L be a layer on n channels such that $\pi(L)$ is a layer. If $S(n, d)$ contains a network with first layer F_n and second layer L, it also contains a network with first layer F_n and second layer $\pi(L)$.*

Denote by $\boldsymbol{H_n}$ the group of permutations on n elements that fix F_n. Two layers L and L' are equivalent under H_n if $L' = \pi(L)$ for some $\pi \in H_n$. For any set S of layers, denote by $\boldsymbol{R(S)}$ a set of (lexicographically smallest) representatives of S equivalent under H_n. Lemma 6 then implies that it suffices to consider networks with second layers from $R(G_n)$.

Recall from Lemma 2 that it is enough to consider prefixes of comparator networks with minimal sets of possible outputs. We apply a symmetry argument similar to Lemma 6 to the sets of possible outputs, and observe that it extends to all permutations on n channels. In particular we show that it is enough to consider second layers whose sets of possible outputs are minimal up to any permutation of channels.

Lemma 7. *Let L_a and L_b be layers on n channels such that $outputs(F_n \, \mathring{,} \, L_b) \subseteq \pi(outputs(F_n \, \mathring{,} \, L_a))$ for some permutation π on n channels. If $S(n, d)$ contains a network with first layer F_n and second layer L_a, it also contains a network with first layer F_n and second layer L_b.*

Proof. Let $C = F_n \, \mathring{,} \, L_a \, \mathring{,} \, N$ be a sorting network of depth d. Then $\pi(C) = \pi(F_n \, \mathring{,} \, L_a) \, \mathring{,} \, \pi(N)$ is a generalised sorting network. Since $outputs(F_n \, \mathring{,} \, L_b) \subseteq \pi(outputs(F_n \, \mathring{,} \, L_a)) = outputs(\pi(F_n \, \mathring{,} \, L_a))$, Lemma 2 implies that $F_n \, \mathring{,} \, L_b \, \mathring{,} \, \pi(N)$ is also a generalised sorting network. Then, by Lemma 3, there exists a comparator network I such that $F_n \, \mathring{,} \, L_b \, \mathring{,} \, I$ is a sorting network of depth d. □

---

[1] We assume that $n > 2$ so that the first layer F_n is not yet a sorting network.

If $outputs(F_n \,\mathring{,}\, L_b) \subseteq \pi(outputs(F_n \,\mathring{,}\, L_a))$ for some permutation π, we write $L_b \leq_{po} L_a$ where po stands for *permuted outputs*. For a set S of layers, denote by $\boldsymbol{R_{po}(S)}$ a minimal set of representatives from S such that for each $s \in S$, there is a representative $r \in R_{po}(S)$ such that $r \leq_{po} s$. Lemma 7 implies that it suffices to consider second layers from $R_{po}(G_n)$. Fig. 3 compares numbers of candidate layers $|R_{po}(G_n)|$ and $|R(G_n)|$ with $|G_n|$ for various n.

Computing the Representatives $\boldsymbol{R_{po}(G_n)}$. Although we can speed up the search for sorting networks dramatically by only considering second layers from $R_{po}(G_n)$ instead of G_n, computing $R_{po}(G_n)$ is a non-trivial task even for $n = 13$.

Just establishing the inequality $L_a \geq_{po} L_b$ for two layers L_a and L_b involves the comparison of sets $outputs(F_n \,\mathring{,}\, L_b)$ and $\pi(outputs(F_n \,\mathring{,}\, L_a))$, both of size up to 2^n, for all permutations π. A naive algorithm comparing all sets of outputs for all pairs of layers thus takes time $O(|G_n|^2 \cdot n! \cdot 2^n)$, and is infeasible for $n = 13$. We present three techniques to speed up the computation of $R_{po}(G_n)$.

First we note that in the second layer it is useless to repeat a comparator from the first layer, and in most other cases adding a comparator to the second layer decreases the set of its possible outputs. Call a layer L **saturated** if it contains no comparator from F_n, and its unused channels are either all min-channels, or all max-channels of comparators from F_n. Let S_n be the set of all saturated layers on n channels.

Lemma 8. *Let n be odd and let L be a layer on n channels. There exists a saturated layer S such that $S \leq_{po} L$.*

Proof. Let L be any layer on n channels. First construct L^0 by removing from L all comparators that also appear in F_n. For any input, L and L^0 give the same output, so $outputs(F_n \,\mathring{,}\, L) = outputs(F_n \,\mathring{,}\, L^0)$. Next, suppose that L^0 is not saturated. Then one of the following holds.

- We can add a comparator between a channel $i \leq \lfloor \frac{n}{2} \rfloor$, which is a min-channel in F_n, and a channel $j \geq \lceil \frac{n}{2} \rceil + 1$, which is a max-channel in F_n such that $\langle i, j \rangle$ is not a comparator from F_n. (If n is odd and L^0 is not saturated, there are at least 3 unused channels, and we can always choose a pair which is not in F_n, not a pair of min-channels, and not a pair of max-channels from F_n.) Denote $L^1 = L^0 \cup \langle i, j \rangle$ and consider the output of $F_n \,\mathring{,}\, L^1$ on some input $x \in \mathbb{B}^n$. We will show that $(F_n \,\mathring{,}\, L^1)(x)$ can also arise as the output of $F_n \,\mathring{,}\, L^0$. If $(F_n \,\mathring{,}\, L^1)(x) \neq (F_n \,\mathring{,}\, L^0)(x)$, then the output of $(F_n \,\mathring{,}\, L^0)(x)$ must be 1 on channel i and 0 on channel j, and the added comparator $\langle i, j \rangle$ flips these values in the output of $F_n \,\mathring{,}\, L^1$. Since channel i is the min-channel of the comparator $\langle i, i + \lceil \frac{n}{2} \rceil \rangle$ in F_n, both channels i and $i + \lceil \frac{n}{2} \rceil$ must carry the value 1 in the input x. Similarly, since channel j is the max-channel of the comparator $\langle j - \lceil \frac{n}{2} \rceil, j \rangle$ of F_n, both channels j and $j - \lceil \frac{n}{2} \rceil$ must carry the value 0 in the input x. By changing the value of channel i to 0 and the value of channel j to 1, these changes propagate to the output in $F_n \,\mathring{,}\, L^0$, and yield the same output as that of $F_n \,\mathring{,}\, L^1$ on x. It follows that of $outputs(F_n \,\mathring{,}\, L^1) \subseteq outputs(F_n \,\mathring{,}\, L^0)$.

– We can add a comparator between some channel i and channel $j = \lceil \frac{n}{2} \rceil$, which is unused in F_n, obtaining a layer L^1. Similarly as in the previous case we can prove that $outputs(F_n \,\mathring{,}\, L^1) \subseteq outputs(F_n \,\mathring{,}\, L^0)$.

By induction, we obtain layers $L^1, L^2, \ldots$, until some L^k is saturated. Then $outputs(F_n \,\mathring{,}\, L^k) \subseteq outputs(F_n \,\mathring{,}\, L^0) = outputs(F_n \,\mathring{,}\, L)$, so $L^k \leq_{po} L$. □

Second we note that if two networks are the same up to a permutation π, then their sets of outputs are also the same up to π. In particular, $L \leq_{po} \pi(L)$ for any layer L and any $\pi \in H_n$. This observation and the above lemma together imply that it suffices to consider representatives of saturated layers up to permutations from H_n before computing the representatives with respect to $\leq_{po}$.

Lemma 9. For odd n, we have $R_{po}(G_n) = R_{po}(R(S_n))$.

Checking whether a layer is saturated only takes time $O(n^2)$ and computing $R(\cdot)$ involves checking only $\lfloor \frac{n}{2} \rfloor!$ permutations compared to all $n!$ for $R_{po}(\cdot)$. Instead of computing $R_{po}(G_n)$ directly, we first compute $R(S_n)$ and only on this much smaller set we compute the most expensive reduction operation R_{po}. Figure 3 summarises the number of layers, saturated layers, representatives and representatives modulo rotation for different n.

Finally we show how to compute representatives R_{po}. Recall that $L_b \leq_{po} L_a$ iff $outputs(F_n \,\mathring{,}\, L_b) \subseteq \pi(outputs(F_n \,\mathring{,}\, L_a))$ for some permutation π. A necessary condition for $outputs(F_n \,\mathring{,}\, L_b) \subseteq \pi(outputs(F_n \,\mathring{,}\, L_a))$ is that the number of outputs of $(F_n \,\mathring{,}\, L_a)$ where channel i is set to 1 is at least the number of outputs of $(F_n \,\mathring{,}\, L_b)$ where channel $\pi(i)$ is set to 1. We obtain a similar necessary condition by considering only outputs with value 1 on exactly k channels. For each $i = 1, \ldots, n$ and each $k = 0, \ldots, n$ we obtain a necessary condition on π for $outputs(F_n \,\mathring{,}\, L_b) \subseteq \pi(outputs(F_n \,\mathring{,}\, L_a))$ to hold. These conditions are fast to check and significantly prune the space of possible permutations π, thereby making the check $L_b <_{po} L_a$ feasible for any two layers L_b and L_a. For $n = 13$ we were able to compute $R(S_n)$ in 2 seconds and subsequently $R_{po}(R(S_n))$ in 32 minutes.

n	3	4	5	6	7	8	9	10	11	12	13
$|G_n|$	4	10	26	76	232	764	2620	9496	35696	140152	568504
$|S_n|$	2	7	10	51	74	513	700	6345	8174	93255	113008
$|R(G_n)|$	4	8	18	28	74	101	295	350	1134	1236	4288
$|R(S_n)|$	2	-	8	-	29	-	100	-	341	-	1155
$|R_{po}(G_n)|$	2	2	6	6	14	15	37	27	88	70	212

Fig. 3. Number of candidates for second layer on n channels. Candidate sets are: G_n = set of all layers, S_n = set of saturated layers, $R(S)$ = set of representatives of S under permutations fixing the first layer, $R_{po}(S)$ = set of representatives of S under permuted outputs. Note that $R(S_n)$ is used to compute $R_{po}(G_n)$ only for odd n.

4 Propositional Encoding of Sorting Networks

In the previous section we showed how to restrict the set of possible first two layers of sorting networks. In this section we describe how to reduce the existence of such a sorting network to the satisfiability of a set of propositional formulas. We then employ the power of modern SAT solvers to determine the satisfiability of the obtained formulas.

Recall that to check whether a comparator network is a sorting one it suffices to consider only its outputs on Boolean inputs (Lemma 1). Now, for Boolean values $x, y \in \mathbb{B}$ a min-max comparator reduces to: $\min(x, y) = x \wedge y$ and $\max(x, y) = x \vee y$. The authors of [5] observed that a comparator network of a given size can be represented by a propositional formula, and the existence of a sorting network in $S(n, d)$ is equivalent to its satisfiability. We improve upon the work of [5] and give a more natural translation to propositional formulas.

We represent a comparator network with n channels and depth d by Boolean variables $C_n^d = \{g_{i,j}^k\}$ for $1 \leq i < j \leq n$ and $1 \leq k \leq d$. The variable $g_{i,j}^k$ indicates whether the comparator $\langle i, j \rangle$ occurs in layer k. We then define

$$once_i^k(C_n^d) = \bigwedge_{1 \leq i \neq j \neq l \leq n} (\neg g_{\min(i,j),\max(i,j)}^k \vee \neg g_{\min(i,l),\max(i,l)}^k) \qquad \text{and}$$

$$valid(C) = \bigwedge_{1 \leq k \leq d,\, 1 \leq i \leq n} once_i^k(C_n^d),$$

where $once_i^k(C_n^d)$ enforces that channel i is used at most once in layer k, and $valid(C_n^d)$ enforces that this constraint holds for each channel in every layer, i.e., that C represents a valid comparator network.

Let $x = \langle x_1, \ldots, x_n \rangle \in \mathbb{B}^n$ be a Boolean input and $y = \langle y_1, \ldots, y_n \rangle$ be the sequence obtained by sorting x. To evaluate the network C_n^d on an input x we introduce variables v_i^k for $0 \leq k \leq d$ and $1 \leq i \leq n$ denoting $V(k, i)$–the value of channel i after layer k. The correct value of v_i^k is enforced by $update_i^k(C_n^d)$ which implements the recursive formula for $V(k, i)$ from Section 2:

$$update_i^k(C_n^d) = (\neg used_i^k(C_n^d) \implies (v_i^k \leftrightarrow v_i^{k-1})) \wedge$$
$$\bigwedge_{1 \leq j < i} [g_{j,i}^k \implies (v_i^k \leftrightarrow (v_j^{k-1} \vee v_i^{k-1}))] \wedge$$
$$\bigwedge_{i < j \leq n} [g_{i,j}^k \implies (v_i^k \leftrightarrow (v_j^{k-1} \wedge v_i^{k-1}))] \qquad \text{and}$$
$$used_i^k(C_n^d) = \bigvee_{j < i} g_{j,i}^k \vee \bigvee_{i < j} g_{i,j}^k,$$

where the formula $used_i^k(C_n^d)$ denotes whether channel i is used in layer k. We can express the predicate "$C_n^d(x)$ is sorted" as:

$$sorts(C_n^d, x) = \bigwedge_{1 \leq i \leq n} (v_i^0 \leftrightarrow x_i) \wedge \bigwedge_{\substack{1 \leq k \leq d, \\ 1 \leq i \leq n}} update_i^k(C_n^d) \wedge \bigwedge_{1 \leq i \leq n} (v_i^d \leftrightarrow y_i)$$

where the first term ensures that we start with the input x, the second term that the v_i^k update appropriately, and the last term that the output is sorted.

Lemma 10. *A sorting network with n channels and depth d exists if and only if $valid(C_n^d) \wedge \bigwedge_{x \in \mathbb{B}^n} sorts(C_n^d, x)$ is satisfiable.*

Further, for inputs of the form $x = 0^p y 1^q$, we hard-wire the variables v_i^k in the formula $sorts(C_n^d, x)$ to false for $1 \leq i \leq p$ and to true for $n - q < i \leq n$. These values are implied by the $update_i^k(C_n^d)$ formulas (see also Example 11). However, we find that hard-wiring these values speeds up the SAT solver approximately by a factor of 4 for $n \leq 12$ as the SAT solver is not able to discover them directly by unit propagation.

In Section 3 we showed that it suffices to consider sorting networks with first layer F_n and second layer $S \in R_{po}(G_n)$. We can incorporate such restriction into the propositional formula easily. For the first layer, let $T = outputs(F_n)$ be the set of possible outputs, then a sorting network with n channels, depth d, and first layer F_n exists if and only if $valid(C_n^{d-1}) \wedge \bigwedge_{x \in T} sorts(C_n^{d-1}, x)$ is satisfiable. A similar adaptation works for fixing the first two layers; we produce one formula for each $S \in R_{po}(G_n)$ and check the satisfiability of each of them.

Instantiating these SAT formulas and checking their satisfiability was sufficient to establish $V(n)$ for $n \leq 12$ in less than 2 minutes in each case (see Fig. 4). A further optimisation substantially reduced the time to establish $V(13)$.

4.1   Existence of Subnetworks: A Necessary Condition

Our final optimisation in showing the nonexistence of sorting network is restricting attention to inputs of the form $0^p y 1^q$. This optimisation is based on the idea that if a comparator network sorts its input, its subnetworks must also sort their respective subinputs. Consider the following example.

Example 11. Consider the evaluation of a sorting network C on input $0x$ where $x \in \mathbb{B}^{n-1}$. Since C consists of min-max comparators, the value on the first channel is always 0. Hence, also the output of the first channel is 0. (See also Fig. 1.) Let D be the comparator network obtained from C by removing the first channel and all comparators attached to it. Then $C(0x) = 0D(x)$ for all x. Requiring that $C(0x)$ is sorted for all $x \in \mathbb{B}^{n-1}$ is the same as requiring that D is a sorting network. A similar argument can be made for inputs of the form $y1$ for $y \in \mathbb{B}^{n-1}$, and in general for $0^p y 1^q$ for $y \in \mathbb{B}^{n-p-q}$.

Let $T^{p,q} = \{t = 0^p x 1^q \mid t \in T, x \in \mathbb{B}^{n-p-q}\} \subseteq T$ be the set of all inputs from T beginning with p zeros and ending with q ones. Intuitively, evaluating a network C on inputs from $T^{p,q}$ exercises only the subnetwork obtained by removing first p and last q channels from C.

For subnetwork size $m < n$ let $T_m = \bigcup_{p+q=n-m} T^{p,q}$. Then $T_m \subseteq T$ and so if network C sorts all inputs from T then C sorts all inputs from T_m. Therefore, a necessary condition for the existence of a network on n channels and depth d sorting inputs T is the satisfiability of the formula

$$subnets(n, d, m, T) = valid(C_n^d) \wedge \bigwedge_{x \in T_m} sorts(C_n^d, x).$$

Empirically, we were always able to find m with $m < n$ such that the resulting formula $subnets(n, d, m, T)$ was unsatisfiable. Furthermore, the SAT solver established unsatisfiability of this formula significantly faster than for the original formula (see Fig. 4).

5   Experimental Results

In this section we present an experimental evaluation of the described techniques, and show how we used them to obtain bounds on $V(n)$ for $n \leq 16$. We instantiated propositional formulas encoding the existence of a sorting network for various values of n and d and various stages of optimisation as presented in the previous sections.[2] We checked their satisfiability using an off-the-shelf propositional SAT solver[3] running on a standard desktop computer[4]. The times taken by the SAT solver are reported in Fig. 4.

n	5	6	7	8	9	10	11	12	13
d	4	4	5	5	6	6	7	7	8
SAT	0.02s	0.05s	1.79s	1.93s	864s	1738s	$> 10^5$s	$> 10^5$s	-
Fix-1	0s	0s	0s	0.02s	0.5s	0.5s	314s	452s	-
Fix-1 + subnet	0s	0s	0s	0.01s	0.27s	0.26s	112s	143s	-
Fix-2	0s	0s	0.03s	0.07s	0.93s	1.13s	63s	87s	22h23m
Fix-2 + subnet	0s	0s	0.02s	0.05s	0.77s	0.78s	49s	48s	13h1m
d	5	5	6	6	7	7	8	8	9
SAT	0s	0.04s	0.13s	1.12s	59.7s	949s	1294s	$> 10^5$s	-
Fix-1	0s	0s	0s	0.01s	0.20s	3.6s	24s	172s	1h40m

Fig. 4. Time required by a SAT solver to solve particular instances of n and d using different variants of propositional formulas: the basic formula from Lemma 10 (SAT), a formula fixing the first layer to F_n (Fix-1), formulas fixing the first two layers to $F_n \, \overset{\circ}{\,}\, S$ for each $S \in R_{po}(G_n)$ (Fix-2), and the *subnets*(n, d, m) versions thereof for appropriate values of m (subnet). The top series corresponds to $d = V(n) - 1$, the largest depth for which no sorting network exists and the formulas are unsatisfiable, the bottom series corresponds to $d = V(n)$ and the formulas are satisfiable. A missing value indicates that the SAT solver ran out of available memory.

Our computations confirm the known values of $V(n)$ for $n \leq 10$. Noteworthy is the case $n = 9$ where we establish the nonexistence of a sorting network of depth 6 in less than a second. The specially crafted and low-level optimised program of [7] was estimated to take 200 hours on the supercomputer Cray-2. Recent work [5] also expressed the existence of such a network as a propositional formula, but their technique by compilation from a higher-level language yields an unnecessarily complicated formula whose SAT checking took over 16 hours.

After 5 minutes of computation when fixing the first layer (2 minutes with the subnetwork optimisation and 1 minute with fixed second layers), we found that $S(11, 7)$ is empty. Since $V(11), V(12) \leq 8$ (see Fig. 2), we have:

Theorem 12. *The optimal depth of a sorting network with $n = 11$ or 12 channels is eight.*

---

[2] Code is available at http://www.cs.ox.ac.uk/people/daniel.bundala/networks/
[3] MiniSAT version 2.2.0
[4] Linux, CPU: 2.83GHz, Memory: 3.7GiB. All reported times are using a single CPU.

Note from Fig. 4 that checking all Fix-2 formulas for all candidate first-two layers is already faster than checking the single Fix-1 formula; despite the drawback that the SAT solver is restarted for each different second layer. Furthermore, checking the Fix-1 formula requires much more memory, and for the case $n = 13$, the SAT solver consumed all available memory (4GB) before finishing. Checking a Fix-2 formula is well within available memory, and different instances for different second layers can be distributed to different computers. This also allows us to start with a small subnetwork size in the subnetwork optimisation and increase it only in instances (second layers) where it yields a satisfiable formula.

For $n = 13$ for each of the 212 depth-two prefixes $F_{13} \, \S \, L$ we generated a formula $subnets(13, 6, 10, T)$ with subnetwork size $m = 10$ and determined that all of them are unsatisfiable in cumulative computation time of 13 hours. Hence, none of the 212 candidate second layers can be extended to a sorting network.

Theorem 13. *The optimal depth of a sorting network with $n = 13, 14, 15$ or 16 channels is nine.*

Even though we were able to compute lower bounds for $11 \leq n \leq 16$, the case $n = 17$ is beyond the scope of current techniques. We leave the depth of the optimal sorting network on 17 channels as the main open problem of this paper.

Acknowledgments. We would like to thank Donald E. Knuth for valuable comments on an earlier draft of this paper which led to strengthening of Lemma 7, reformulation of the *subnets* criterion, and inclusion of the hard-wiring optimisation. He also observed that a top-to-bottom reflection of a sorting network is a sorting network, reducing the set of candidate second layers to only 118 in the case $n = 13$.

References

1. Ajtai, M., Komlós, J., Szemerédi, E.: An O(n log n) sorting network. In: Proceedings of the Fifteenth Annual ACM Symposium on Theory of Computing, STOC 1983, pp. 1–9 (1983), http://doi.acm.org/10.1145/800061.808726
2. Batcher, K.E.: Sorting networks and their applications. In: Proceedings of the Spring Joint Computer Conference, AFIPS 1968, pp. 307–314. ACM (1968) (Spring)
3. Floyd, R.W., Knuth, D.E.: The Bose-Nelson sorting problem. J. ACM 9, 282–296 (1962)
4. Knuth, D.E.: The art of computer programming, volume 3: sorting and searching. Addison Wesley, Reading (1973)
5. Morgenstern, A., Schneider, K.: Synthesis of parallel sorting networks using SAT solvers. In: MBMV, OFFIS-Institut für Informatik, pp. 71–80 (2011)
6. Parberry, I.: Parallel complexity theory. Research notes in theoretical computer science. Pitman (1987)
7. Parberry, I.: A computer-assisted optimal depth lower bound for nine-inputs sorting networks. In: Proceedings of the 1989 ACM/IEEE Conference on Supercomputing (1989)

Satisfiability for MTL and TPTL over Non-monotonic Data Words

Claudia Carapelle*, Shiguang Feng*, Oliver Fernández Gil*, and Karin Quaas**

Institut für Informatik, Universität Leipzig,
04109 Leipzig, Germany

Abstract. In the context of real-time systems, Metric Temporal Logic (MTL) and Timed Propositional Temporal Logic (TPTL) are prominent and widely used extensions of Linear Temporal Logic. In this paper, we examine the possibility of using MTL and TPTL to specify properties about classes of non-monotonic data languages over the natural numbers. Words in this class may model the behaviour of, *e.g.*, one-counter machines. We proved, however, that the satisfiability problem for many reasonably expressive fragments of MTL and TPTL is undecidable, and thus the use of these logics is rather limited. On the positive side we prove that satisfiability for the existential fragment of TPTL is NP-complete.

1 Introduction

Recently, verification and analysis of sets of *data words* have gained a lot of interest [6, 18, 11, 10, 3–5]. A data word is a sequence over $\Sigma \times D$, where Σ is a finite set of labels, and D is a (potentially infinite) set of *data values*. In this paper, we consider data words as behavioural models of one-counter machines. In this regard, we let the data domain be the set of natural numbers. Note that the sequence of data values of a word may be non-monotonic.

For reasoning about data words, one may use extensions of linear temporal logic (LTL). One of these is FreezeLTL, which extends LTL with a *freeze quantifier* that stores the current data value in a register variable. The registers can be used to test for equality of data values at different positions of a data word. In spite of this limited access to data values, the satisfiability problem for FreezeLTL is undecidable [10]. However, over *finite* data words, and if the logic is restricted to a single register, then the satisfiability problem is decidable, albeit not primitive recursive [10]. This lower bound has been confirmed for satisfiability for the fragment of FreezeLTL where the only temporal modality is the *finally* modality [12].

Originally, the freeze quantifier was introduced in *Timed Propositional Temporal Logic* (TPTL, for short) [2]. With TPTL, in addition to FreezeLTL, one can compare data values of a data word using linear inequations of the form, *e.g.*, $x - y \leq c$. Another widely used logic in the context of real-time systems is *Metric*

---

* The author is supported by *Deutsche Forschungsgemeinschaft (DFG).*, GRK 1763
** The author is supported by *DFG*, project QU 316/1-1.

A.-H. Dediu et al. (Eds.): LATA 2014, LNCS 8370, pp. 248–259, 2014.

Temporal Logic (MTL, for short). MTL extends LTL by constraining the temporal operators with intervals of the non-negative reals. Both logics, however, have not gained much attention in the specification of *non-monotonic* data words, albeit they can express many interesting properties.

FreezeLTL is a fragment of TPTL, and thus it is clear that one cannot find better decidability results for TPTL than for FreezeLTL. In fact it is well known that the satisfiability problem for TPTL over non-monotonic data words is undecidable [2]. In the context of monotonic data words over the natural numbers, MTL and TPTL are equally expressive, and the satisfiability problem for both logics is EXPSPACE-complete [1, 2]. However, over timed words, TPTL has been proved strictly more expressive than MTL [7], and while satisfiability for both logics is undecidable over infinite timed words [1, 15], there is a difference in the finite words case: TPTL has an undecidable satisfiability problem [1], while satisfiability for MTL is decidable (but not primitive recursive) [16]. We recently proved in [8] that also for *non-monotonic* data words over the natural numbers TPTL is strictly more expressive than MTL, and indeed, there are properties which can be expressed in FreezeLTL, but cannot be expressed in MTL. Hence there was the possibility that MTL would have a better complexity for the satisfiability problem.

However, as a main result we prove that the satisfiability problem for MTL over non-monotonic finite data words is undecidable. This is even the case if we do not allow for propositional variables.

We then investigate the *unary fragments* of MTL and TPTL where the only allowed temporal modalities are unary. We show that the satisfiability problem over finite data words is undecidable for both logics, and for TPTL it is undecidable even if we restrict the formulae to contain at most one register variable and no next modality. This is opposed to the decidability result for FreezeLTL with one register variable evaluated over finite data words [10].

After that we consider another syntactic restriction of the logics, namely we restrict the negation operator to propositional variables, which results in what we call the *positive* fragments of our logics. This excludes the globally modality, which is used in most of the undecidability proofs. However, we prove that this restriction does not lead to any changes in the results for the satisfiability problem compared to the full logics.

Last but not least, we prove that for the *unary positive* fragment (called *existential* fragment in [7]), the satisfiability problem for TPTL is NP-complete.

The main insight of this paper is that both MTL and TPTL have a very limited use in specifying properties over non-monotonic data languages. This adds an important piece to complete the picture about decidability of satisfiability problems for data-relevant extensions of temporal logics.

2   Preliminaries

We use $\mathbb{Z}$ and $\mathbb{N}$ to denote the set of integers and the non-negative integers, respectively. We let P be a finite set of propositional variables.

A *data word* over P is a finite or infinite sequence $(P_0, d_0)(P_1, d_1) \ldots$ of pairs in $2^P \times \mathbb{N}$. We use $(2^P \times \mathbb{N})^*$ and $(2^P \times \mathbb{N})^\omega$, respectively, to denote the set of finite and infinite, respectively, data words over P. The *length* of a data word w is denoted by $|w|$, where we set $|w| = \infty$ if w is an infinite data word.

A *two-counter machine* $\mathcal{M}$ is a finite sequence $(\mathcal{I}_j)_{j=1}^n$ of instructions operating on two counters denoted by C_1 and C_2, where $\mathcal{I}_j$ is one of the following instructions (with $i \in \{1, 2\}$ and $j, k, m \in \{1, \ldots, n\}$):

increment	$\mathcal{I}_j : C_i := C_i + 1;$ go to $\mathcal{I}_k$
zero test/decrement	$\mathcal{I}_j :$ if $C_i = 0$ then go to $\mathcal{I}_k$ else $C_i := C_i - 1;$ go to $\mathcal{I}_m$
halt	$\mathcal{I}_j :$ halt

A *configuration* of a two-counter machine $\mathcal{M}$ is a triple $\gamma = (J, c, d) \in \{\mathcal{I}_1, \ldots, \mathcal{I}_n\} \times \mathbb{N} \times \mathbb{N}$, where J indicates the current instruction, and c and d are the current values of the counters C_1 and C_2, respectively. A *computation* of $\mathcal{M}$ is a finite or infinite sequence $(\gamma_i)_{i \geq 0}$ of configurations, such that $\gamma_0 = (\mathcal{I}_1, 0, 0)$ and γ_{i+1} is the result of executing the instruction $\mathcal{I}_i$ on γ_i for each $i \geq 0$. Without loss of generality, we assume that $\mathcal{I}_n$ is the only instruction of the form halt. The *halting problem* for two-counter machines asks, given a two-counter machine $\mathcal{M}$, whether the (unique) computation of $\mathcal{M}$ reaches a configuration with instruction $\mathcal{I}_n$, *i.e.*, the halting instruction. This problem is Σ_1^0-complete [14]. The *recurrent state problem* for two-counter machines asks, given a two-counter machine $\mathcal{M}$, whether the (unique) computation of $\mathcal{M}$ visits instruction $\mathcal{I}_1$ infinitely often. This problem is Σ_1^1-complete [2]. We will use reductions of these problems to show lower bounds of the satisfiability problem for some fragments of MTL and TPTL .

3 Extensions of Linear Temporal Logic

3.1 Metric Temporal Logic

The set of formulae of MTL is built up from P by boolean connectives and a constraining version of the *until* modality:

$$\varphi ::= p \mid \neg\varphi \mid \varphi_1 \wedge \varphi_2 \mid \varphi_1 \mathsf{U}_I \varphi_2$$

where $p \in P$ and $I \subseteq \mathbb{Z}$ is an interval with endpoints in $\mathbb{Z} \cup \{-\infty, +\infty\}$. We use pseudo-arithmetic expressions to denote intervals, like, *e.g.*, ≥ 1 to denote $[1, \infty)$. If $I = \mathbb{Z}$, then we may omit the annotation I on U_I.

Formulae in MTL are interpreted over data words. Let $w = (P_0, d_0)(P_1, d_1) \ldots$ be a data word, and let $i \leq |w|$. We define the *satisfaction relation for* MTL, denoted by $\models_{\mathsf{MTL}}$, inductively as follows:

$$(w, i) \models_{\mathsf{MTL}} p \Leftrightarrow p \in P_i, \quad (w, i) \models_{\mathsf{MTL}} \neg\varphi \Leftrightarrow (w, i) \not\models_{\mathsf{MTL}} \varphi,$$

$$(w, i) \models_{\mathsf{MTL}} \varphi_1 \wedge \varphi_2 \Leftrightarrow (w, i) \models_{\mathsf{MTL}} \varphi_1 \text{ and } (w, i) \models_{\mathsf{MTL}} \varphi_2,$$

$$(w, i) \models_{\mathsf{MTL}} \varphi_1 \mathsf{U}_I \varphi_2 \Leftrightarrow \exists j.i < j \leq |w| : (w, j) \models_{\mathsf{MTL}} \varphi_2 \text{ and } d_j - d_i \in I, \text{ and}$$

$$\forall k.i < k < j : (w, k) \models_{\mathsf{MTL}} \varphi_1.$$

We say that a data word *satisfies* an MTL formula φ, written $w \models_{\mathsf{MTL}} \varphi$, if $(w, 0) \models_{\mathsf{MTL}} \varphi$. We use the following syntactical abbreviations: $\varphi_1 \vee \varphi_2 := \neg(\neg\varphi_1 \wedge \neg\varphi_2)$, $\varphi_1 \rightarrow \varphi_2 := \neg\varphi_1 \vee \varphi_2$, $\mathtt{true} := p \vee \neg p$, $\mathtt{false} := \neg\mathtt{true}$, $\mathsf{X}_I \varphi := \mathtt{false}\mathsf{U}_I\varphi$, $\mathsf{F}_I\varphi := \mathtt{true}\mathsf{U}_I\varphi$, $\mathsf{G}_I\varphi := \neg\mathsf{F}_I\neg\varphi$. Note that the use of the *strict* semantics for the until modality is essential to derive the next modality.

We define the *length* of a formula ψ, denoted by $|\psi|$, as the number of symbols occurring in ψ where all integer constants in ψ are given in a binary encoding.

In the following, we define some fragments of MTL. A unaryMTL formula is built from propositional variables, using the boolean connectives, and the unary temporal modalities X and F. We use positiveMTL to denote the subset of MTL where negation is restricted to propositional variables. A posUnaryMTL formula is a positiveMTL formula where the only allowed modalities are the F and X modalities.

3.2 Timed Propositional Temporal Logic

Next we define formulae of TPTL. For this, let X be a countable set of *register variables*. Formulae in TPTL are defined by the following grammar:

$$\varphi ::= p \mid x \sim c \mid \neg\varphi \mid \varphi_1 \wedge \varphi_2 \mid \varphi_1\mathsf{U}\varphi_2 \mid x.\varphi$$

where $p \in \mathsf{P}$, $x \in X$, $c \in \mathbb{Z}$, and $\sim \in \{<, \leq, =, \geq, >\}$.

Formulae in TPTL are interpreted over data words. A *register valuation* ν is a function from X to $\mathbb{N}$. Let $w = (P_0, d_0)(P_1, d_1)\dots$ be a data word, let ν be a register valuation, and let $i \in \mathbb{N}$. The satisfaction relation for TPTL, denoted by $\models_{\mathsf{TPTL}}$, is inductively defined in a similar way as for MTL; we only give the definitions for the new formulae:

$$(w, i, \nu) \models_{\mathsf{TPTL}} \varphi_1\mathsf{U}\varphi_2 \Leftrightarrow \exists i.i < j \leq |w|.(w, j, \nu) \models_{\mathsf{TPTL}} \varphi_2,$$
$$\text{and } \forall k.i < k < j.(w, k, \nu) \models_{\mathsf{TPTL}} \varphi_1,$$
$$(w, i, \nu) \models_{\mathsf{TPTL}} x \sim c \Leftrightarrow d_i - \nu(x) \sim c,$$
$$(w, i, \nu) \models_{\mathsf{TPTL}} x.\varphi \Leftrightarrow (w, \nu[x \mapsto d_i], i) \models_{\mathsf{TPTL}} \varphi.$$

Here, $\nu[x \mapsto d_i]$ is the valuation that agrees with ν on all $y \in X\backslash\{x\}$, and maps x to d_i. We say that a data word w satisfies a TPTL formula φ, written $w \models_{\mathsf{TPTL}} \varphi$, if $(w, 0, \bar{0}) \models_{\mathsf{TPTL}} \varphi$, where $\bar{0}$ denotes the valuation that maps all variables to the initial data value of the word, *i.e.* to d_0.

We use the same syntactical abbreviations as for MTL. The *length* of TPTL formulae is also defined as for MTL formulae. We define the fragments unaryTPTL, positiveTPTL, and posUnaryTPTL like the corresponding fragments in MTL. Additionally, we define FreezeLTL to be the subset of TPTL formulae φ where $\sim c$ is of the form $= 0$ whenever φ contains the subformula $x \sim c$. Given $n \geq 0$ and a TPTL fragment $\mathcal{L}$, we use $\mathcal{L}^n$ to denote the subset of $\mathcal{L}$ that corresponds to the set of $\mathcal{L}$ formulae that use at most n different register variables.

4   The Satisfiability Problem

In this paper, we are interested in infinitary and finitary versions of the *satisfiability problem* (SAT, for short): given a formula φ in a logic $\mathcal{L}$, is there some infinite (finite, respectively) data word w with $w \models_{\mathcal{L}} \varphi$?

In the table below, we summarize the complexity results for the satisfiability problem for different fragments of TPTL and MTL. The results shaded in grey are new and presented in this paper.

		full	unary	unary without X	positive	posUnary
Finitary	MTL	Σ_1^0-cpl.	Σ_1^0-cpl.	?	Σ_1^0-cpl.	NP-cpl.
	TPTL^1	Σ_1^0-cpl.	Σ_1^0-cpl.	Σ_1^0-cpl.	Σ_1^0-cpl.	NP-cpl.
	FreezeLTL^1	not pr. rec. [10]	not pr. rec. [10]	not pr. rec. [12]	not pr. rec.	NP-cpl.
Infinitary	MTL	Σ_1^1-cpl.	Σ_1^1-cpl.	?	Σ_1^0-cpl.	NP-cpl.
	TPTL^1	Σ_1^1-cpl. [2]	Σ_1^1-cpl.	Σ_1^1-cpl.	Σ_1^0-cpl.	NP-cpl.
	FreezeLTL^1	Π_1^0-cpl. [10]	Π_1^0-cpl. [10]	Π_1^0-cpl. [12]	not pr. rec.	NP-cpl.

5   Results for Full and Unary Fragments

Alur and Henzinger proved already 20 years ago that infinitary SAT for TPTL is undecidable, even if one does not allow for propositional variables [2]. The proof in the cited paper is by reduction of the recurrent state problem for two-counter machines. In the reduction more than one register variable is used, however, one can easily adapt the proof and strenghten the result.

Theorem 1. *For* TPTL^1, *finitary SAT is* Σ_1^0-*complete, even for the fragment that does not allow propositional variables.*

Proof. Σ_1^0-hardness of finitary SAT can be proved in a similar way to Σ_1^1-hardness of infinitary SAT using a reduction of the halting problem for two-counter machines. It remains to show that finitary SAT is in Σ_1^0. For this, we note that given a TPTL^1 formula ψ and a finite data word w with $w \models_{\mathsf{TPTL}} \psi$, there exists a finite data word w' such that: $|w'| = |w| = n$, $w' \models_{\mathsf{TPTL}} \psi$ and the data values occurring in w' are bounded by $n * \max(|\,min_c\,|, |\,max_c\,|)$, where min_c and max_c are the smallest and greatest constants occurring in ψ. Based on this, the satisfiability of ψ can be characterized by a Σ_1^0 sentence $\exists n \phi(n)$ where $\phi(n)$ has only bounded quantifiers. Any data word can be encoded by using a unary predicate for each propositional variable p in ψ and a binary predicate to encode pairs of position numbers and associated data values. The variable n represents the length of a finite data word and $\phi(n)$ expresses whether ψ is satisfied by a data word of length n. $\qquad\square$

It is well known that every formula in MTL can effectively be translated into a TPTL^1 formula defining the same language of data words. Hence the upper bounds of SAT for TPTL^1 also apply to SAT for MTL. However, we have recently proved that TPTL^1 over non-monotonic data words is *strictly more expressive* than MTL [8]. It is thus natural to consider the exact complexity of SAT for MTL, in particular, as it is further known that finitary SAT for MTL is decidable (albeit with non-primitive recursive complexity) for *timed words* [16], and EXPSPACE-complete for *monotonic* data words [1]. However, we prove that the undecidability of SAT for TPTL^1 also applies to MTL, even if we do not allow for propositional variables.

Theorem 2. *For* MTL, *finitary SAT is* Σ_1^0-*complete, and infinitary SAT is* Σ_1^1-*complete, even for the fragment that does not allow propositional variables.*

Proof. (Sketch) The upper bounds follow from Th. 1 and the result by Alur and Henzinger [2]. For the lower bounds of finitary, respectively, infinitary SAT, we reduce the halting problem, respectively the recurrent state problem for two-counter machines to SAT for MTL: given a two-counter machine $\mathcal{M}$, we define an MTL formula $\varphi_{\mathcal{M}}$ that is satisfiable if, and only if, $\mathcal{M}$ reaches the halting instruction, respectively visits the first instruction infinitely often. In order to avoid the usage of propositional variables, we use an idea similar to the one in the proof of Lemma 2 in [11]. Each instruction $\mathcal{I}_i$ is encoded by a sequence of data values, starting with value 3, which is followed by n positions with data values in $\{1, 2\}$. The value 2 occurs at the ith position after the value 3. After n positions, there is a position for encoding the value of C_1 plus 4, and after that there is a further position for encoding the value of C_2 plus 4. The first position with data value 3 can be used to identify an instruction of $\mathcal{M}$. For example, the configuration $(\mathcal{I}_3, 1, 2)$ of a two-counter machine with 4 instructions is encoded by s $s+3$ s $s+1$ s $s+1$ s $s+2$ s $s+1$ s $s+4$ s $s+5$ s. Here, $s \in \mathbb{N}$ is some arbitrary initial data value. $\qquad\square$

In [10], non-primitive recursive complexity for finitary SAT for $\mathsf{unaryFreezeLTL}^1$ is proved. This result was strengthened to SAT for $\mathsf{unaryFreezeLTL}^1$ without the X modality [12]. Unfortunately, if we extend $\mathsf{unaryFreezeLTL}^1$ to $\mathsf{unaryTPTL}^1$, we already obtain undecidability of SAT. We also prove undecidability of SAT for $\mathsf{unaryMTL}$, however, it is an open problem whether undecidability also holds for the $\mathsf{unaryMTL}$ fragment in which the X modality is not allowed.

Theorem 3. *For* $\mathsf{unaryMTL}$, *finitary SAT is* Σ_1^0-*complete, and infinitary SAT is* Σ_1^1-*complete. For* $\mathsf{unaryTPTL}^1$, *this is the case even if we do not allow for the* X *modality.*

Proof. The upper bounds follow from Theorems 1 and 2. For the lower bounds, we reduce the halting problem for two-counter machines to SAT for $\mathsf{unaryMTL}$ and to SAT for $\mathsf{unaryTPTL}^1$ without the X modality, respectively. Let $\mathcal{M}$ be a two-counter machine with n instructions. Recall that $\mathcal{I}_n$ is the only halting instruction. Define $\mathsf{P} = \{\#, \mathcal{I}_1, \ldots, \mathcal{I}_n, r_1, r_2\}$. We define a formula $\varphi_{\mathcal{M}}$ over P that is satisfiable if, and only if, $\mathcal{M}$ has a halting computation.

Let $\gamma = (J_0, c_0, d_0)(J_1, c_1, d_1) \ldots$ be a computation of $\mathcal{M}$, where $(J_0, c_0, d_0) = (\mathcal{I}_1, 0, 0)$. We encode γ as a data word over 2^P as follows:

$$(\{\#\}, s)(\{J_0, r_1\}, s + c_0)(\{J_0, r_2\}, s + d_0)(\{\#\}, s + 1)(\{J_1, r_1\}, s + 1 + c_1) \ldots$$

i.e., for each $i \geq 0$, the ith configuration of π is encoded by the data word

$$(\{\#\}, s + i)(\{J_i, r_1\}, s + i + c_i)(\{J_i, r_2\}, s + i + d_i).$$

Here $s \in \mathbb{N}$ is again an arbitrary initial data value.

The crucial point in this encoding is that the sequence of data values at positions where $\#$ holds is strictly monotonically increasing by exactly 1. In all of the unaryTPTL^1 formulae (and some of the unaryMTL formulae, respectively) defined below, we can exploit this fact and determine when the encoding of a new configuration starts without using the X modality.

Next we define some unaryMTL formulae. The conjunction of all these formulae is satisfied by a data word w, if and only if, w encodes a *halting* computation of $\mathcal{M}$. Recall that every unaryMTL formula without the X modality can effectively be translated into a unaryTPTL^1 formula without the X modality. For the reduction to SAT for unaryTPTL^1 without the X modality, we give extra formulae only in the case that the unaryMTL formula uses the X modality. We further remark that, due to the strict semantics of our logics, some of the formulae using the G modality have to be additionally stated for the initial position of the data word, but have been omitted here due to lack of space. We start by defining the auxiliary formula $\varphi_{\mathsf{idz}} = \bigvee_{i \in \{1, \ldots, n-1\}} \mathcal{I}_i$ (*i.e.*, the disjunction of all instructions without the halting instruction $\mathcal{I}_n$).

(1) At each position in the data word, exactly one of the following subsets of P must occur: $\{\#\}$, $\{\mathcal{I}_i, r_1\}$ and $\{\mathcal{I}_i, r_2\}$, for some $i \in \{1, \ldots, n\}$. No other propositional variables are allowed. This can be expressed in unaryMTL without the X modality in a straightforward way.

(2) There are two consecutive positions in the data word where $\mathcal{I}_n$ holds. The data word ends after the second occurrence of $\mathcal{I}_n$.
 - $\mathsf{F}(\mathcal{I}_n \wedge \mathsf{F}(\mathcal{I}_n))$ (There are two different positions where $\mathcal{I}_n$ holds.)
 - $\mathsf{G}(\mathcal{I}_n \rightarrow \mathsf{G}(\mathcal{I}_n \rightarrow \mathsf{Gfalse}))$ (After the second occurrence of $\mathcal{I}_n$ the data word ends.)
 - $\mathsf{G}(\mathcal{I}_n \rightarrow \neg\mathsf{F}(\# \vee r_1 \vee \varphi_{\mathsf{idz}}))$ (After the first occurrence of $\mathcal{I}_n$, the symbols $\#, r_1, \mathcal{I}_1, \ldots, \mathcal{I}_{n-1}$ should never occur again. Hence, by (1) the only propositional variables that may occur after the first occurrence of $\mathcal{I}_n$ are $\mathcal{I}_n$ and r_2. This implies that the two occurrences of $\mathcal{I}_n$ (whose existences are guaranteed by the first formula, are consecutive.)

The following formulae express important conditions on the data values.

(3) The data values in the positions where $\#$ holds are strictly monotonic and increase progressively by exactly 1.
 - $\mathsf{G}(\# \rightarrow \neg\mathsf{F}_{\leq 0}\#)$ (The data value at a position where $\#$ holds can never be smaller than or equal to the data value at a preceding position where $\#$ holds.)

- $\mathsf{G}((\# \wedge \mathsf{F}\varphi_{\mathsf{idz}}) \to \mathsf{F}_{=1}\#)$ (If after the $\#$ symbol with data value $d_\#$ an instruction different from $\mathcal{I}_n$ is occurring, *i.e.*, by (2) we have not reached the last configuration, then there will finally be a further $\#$ with data value $d_\# + 1$. This and the first formula imply that the *next* occurrence of $\#$ has data value $d_\# + 1$.)

(4) The data values at positions where r_1 holds are weakly monotonic. (Similarly for r_2.)
  - $\mathsf{G}(r_1 \to \neg \mathsf{F}_{<0}r_1)$

(5) The data values at positions where $\#$ holds serve as a reference value for 0. Hence the data values at positions where r_1 hold should always be greater than or equal to this value. (Similarly for r_2.)
  - $\mathsf{G}(\# \to \neg \mathsf{F}_{<0}r_1)$.

Using these conditions, we can express the remaining details of the structure of a data word encoding a computation of $\mathcal{M}$.

(6) The data word should start with the prefix $(\#, s)(\{\mathcal{I}_1, r_1\}, s)(\{\mathcal{I}_1, r_2\}, s)$. The unaryMTL formula is of the form $\# \wedge \mathsf{X}_{=0}(\mathcal{I}_1 \wedge r_1 \wedge \mathsf{X}_{=0}(\mathcal{I}_1 \wedge r_2))$. In order to express this condition in unaryTPTL[1] without the X modality, we have to consider more elaborate formulae:
  - $\# \wedge x.\mathsf{F}(\mathcal{I}_1 \wedge r_1 \wedge x = 0 \wedge \mathsf{F}(\mathcal{I}_1 \wedge r_2 \wedge x - 0))$ (After the first $\#$ with data value $d_\#$, there will be some $\{\mathcal{I}_1, r_1\}$ and data value $d_\#$, followed by some $\{\mathcal{I}_1, r_2\}$ with data value $d_\#$.)
  - For $i = 1, 2$, we define $\# \wedge x.\mathsf{G}(r_i \to \neg \mathsf{F}(r_i \wedge \mathsf{F}(\# \wedge x = 1)))$ (After the first $\#$ with data value $d_\#$, there cannot be two different r_i before another $\#$ with data value $d_\# + 1$ occurs.)

The second formula and (3) express that after the first $\#$ there is at most one r_i before the next $\#$ occurs. By (5), each r_i following symbol $\#$ must have data value at least as big as that for $\#$. Hence, the r_i with data value $d_\#$ whose existence is enforced by the first formula must occur before the second occurrence of $\#$.

(7) In the remaining data word, the symbol $\#$ is followed by $\{r_1, \mathcal{I}_j\}$, which is followed by $\{r_2, \mathcal{I}_j\}$ for some $j \in \{1, \ldots, n\}$. This is repeated until the end of the word. In unaryMTL, this can be expressed by the formula $\mathsf{G}[\# \to \bigvee_{1 \le i \le n}(\mathsf{X}(\mathcal{I}_i \wedge r_1) \wedge \mathsf{XX}(\mathcal{I}_i \wedge r_2 \wedge (\mathsf{X}\# \vee \mathsf{G}\mathsf{false})))]$. In unaryTPTL[1] without X modality, we define
  - $\mathsf{G}[\# \wedge \mathsf{F}\# \to x. \bigvee_{1 \le i \le n-1} \mathsf{F}(\mathcal{I}_i \wedge r_1 \wedge \mathsf{F}(\mathcal{I}_i \wedge r_2 \wedge \mathsf{F}(\# \wedge x = 1)))]$
    (Together with (3) this guarantees that after $\#$, the symbol r_1 followed by symbol r_2 occur *before the next* occurrence of $\#$.)
  - $\mathsf{G}[\# \wedge \neg \mathsf{F}\# \to \neg \mathsf{F}\varphi_{\mathsf{idz}} \wedge \mathsf{F}(\mathcal{I}_n \wedge r_1 \wedge \mathsf{F}(\mathcal{I}_n \wedge r_2))]$

And for $i = 1, 2$, we define
  - $\mathsf{G}[\# \wedge \mathsf{F}\# \to x.\mathsf{G}(r_i \to \neg \mathsf{F}(r_i \wedge \mathsf{F}(\# \wedge x = 1)))]$
    (There cannot be two different r_i between two $\#$.)

(8) We define the correct encoding of an increment instruction of the form $\mathcal{I}_j : C_1 := C_1 + 1$; go to $\mathcal{I}_k$. Note that incrementing the first counter by 1 corresponds to incrementing the data value of r_1 by exactly 2. The value

of the second counter should not be changed, and this corresponds to incrementing the data value of r_2 by exactly 1. With unaryMTL, this can be expressed by the following formulae:

- $G\langle(\mathcal{I}_j \wedge r_1) \rightarrow (\neg F_{<2}r_1 \wedge F_{=2}r_1)\rangle$ (Together with (4) this implies that the data value at the next occurrence of r_1 is incremented by 2.)
- $G\langle(\mathcal{I}_j \wedge r_2) \rightarrow (\neg F_{<1}r_2 \wedge F_{=1}r_2)\rangle$ (Similarly, this and (4) imply that the data value at the next occurrence of r_2 is incremented by 1.)
- For $i = 1, 2$, define $G\langle(\mathcal{I}_j \wedge r_i) \rightarrow XXX(\mathcal{I}_k \wedge r_i)\rangle$.

For unaryTPTL[1] without X modality, we define

- $G\langle(\mathcal{I}_j \wedge r_1) \rightarrow (x.G(r_1 \rightarrow x \geq 2) \wedge x.F(r_1 \wedge x = 2))\rangle$
- $G\langle(\mathcal{I}_j \wedge r_2) \rightarrow (x.G(r_2 \rightarrow x \geq 1) \wedge x.F(r_2 \wedge x = 1))\rangle$
- $G\langle\# \rightarrow x.G((\mathcal{I}_j \wedge r_1 \wedge F(\# \wedge x = 1)) \rightarrow (\varphi_1 \vee \varphi_2))\rangle$, where
  - $\varphi_1 = F(\# \wedge x = 1 \wedge F(\mathcal{I}_k \wedge F(\# \wedge x = 2)))$, and
  - $\varphi_2 = F(\# \wedge x = 1 \wedge \neg F\# \wedge F\mathcal{I}_k)$.

(9) We define the correct encoding of instructions of the form $\mathcal{I}_j$: if $C_1 = 0$ then go to $\mathcal{I}_k$ else $C_1 := C_1 - 1$; go to $\mathcal{I}_m$: Recall that the data value at $\#$ serves as a reference value for 0. A successful zero test of the first counter (and no change in the value of the first counter) can thus be defined in unaryMTL as follows:

- $G((\# \wedge F_{=0}(\mathcal{I}_j \wedge r_1)) \rightarrow F_{=1}(\mathcal{I}_k \wedge r_1))$ (Note that (3) to (5) guarantee that the position where $\mathcal{I}_k \wedge r_1$ holds with data value incremented by 1 is directly after the next $\#$.)

The negative zero test and decrement instruction is similar. Note that decrementing the value of the first counter corresponds to not changing the data value at r_1.

- $G\langle(\# \wedge X_{>0}(\mathcal{I}_j \wedge r_1)) \rightarrow X(\mathcal{I}_j \wedge r_1 \wedge F_{=0}r_1 \wedge XXX\mathcal{I}_m)\rangle$

We further can use the same formulae as in (8) to express that the value of the other counter does not change. The unaryTPTL[1] formula without the X modality for expressing the negative zero test is a bit more elaborate. Again, (3) to (5) are crucial for ensuring that the correct positions in the data word are defined.

- $G\langle\varphi_1 \rightarrow (\varphi_2 \wedge \varphi_3)\rangle$, where
  - $\varphi_1 = \# \wedge x.F(\mathcal{I}_j \wedge r_1 \wedge F(\# \wedge x = 1)) \wedge x.G(\mathcal{I}_j \wedge r_1 \rightarrow x > 0)$, and
  - $\varphi_2 = x.F\langle\mathcal{I}_j \wedge r_1 \wedge F(\# \wedge x = 1) \wedge x.F(r_1 \wedge x = 0)\rangle$, and
  - $\varphi_3 = x.F\langle\# \wedge x = 1 \wedge F(\mathcal{I}_m \wedge (F(\# \wedge x = 2) \vee \neg F\#))\rangle$.

This finishes the hardness proof for finitary SAT. For infinitary SAT, we do not need (2) and instead use the usual approach and define a formula which expresses that $\mathcal{I}_1$ is visited infinitely often: $GF\mathcal{I}_1$. We can further simplify some of the formulae, as we do not have to check whether we have reached the halting instruction.   □

Remark 4. Using Ehrenfeucht-Fraïssé-Games defined in [8], one can prove that unaryTPTL[1]-formulae of the form $x.F(b \wedge F(c \wedge x = 0))$ cannot be expressed in MTL. We remark that it is exactly this kind of formulae that we use in the unaryTPTL[1] formulae without X modality in (6) to (9). It is an open problem whether we can express the conditions stated there in unaryMTL without using the X modality.

6   Results for Positive Fragments

Next we consider the fragment of MTL and TPTL, in which negation is restricted to propositional variables. Note that this excludes the globally operator, which seems to be crucial in the proofs for lower bounds of SAT for the mentioned logics. It also allows us to prove the following interesting property:

Theorem 5 (Finite model property). *Let* $\varphi \in$ positiveTPTL. *Then* φ *is satisfiable by a data word if, and only if, it is satisfiable by a finite data word.*

This implies that infinitary and finitary SAT are equivalent problems. It turns out that the restriction of negation to propositional variables does not change anything about the complexity status of SAT compared to *finitary* SAT for the corresponding full logics.

Theorem 6. *For* positiveMTL *and* positiveTPTL^1, *SAT is* Σ_1^0-*complete. For* positiveFreezeLTL^1, *SAT is not primitive recursive.*

Proof. For positiveMTL, we reduce the halting problem for two-counter machines to SAT. Given a two-counter machine $\mathcal{M}$ with n instructions, we define a positiveMTL formula $\varphi_\mathcal{M}$ over $\mathsf{P} = \{\mathcal{I}_1, \ldots, \mathcal{I}_n, r_1, r_2\}$ that is satisfiable if, and only if, $\mathcal{M}$ has a halting computation. A computation $(J_0, c_0, d_0)(J_1, c_1, d_1) \ldots$ is encoded by a data word of the form

$$(\{J_0\}, s)(\{r_1\}, s + c_0)(\{r_2\}, s + d_0)(\{J_1\}, s)(\{r_1\}, s + c_1)(\{r_2\}, s + d_1) \ldots$$

Using this structure, we can avoid using the G modality.

For positiveTPTL^1 and positiveFreezeLTL^1, the proof is by reduction of SAT for the corresponding unary fragment to SAT for the corresponding positive fragment without the X modality. Let φ be, *e.g.*, a formula in unaryTPTL^1. For this we may assume without loss of generality that φ is in *negation normal form*, where the application of negation is restricted to propositional variables. For the reduction to work, we must assume that every data word contains a special symbol halt marking the end of the *finite* data word. The idea is to exploit the fact that for finite data words, the formula $G\varphi$ means that φ must hold *until* the symbol halt marks the end of the word. We define a function h mapping unaryTPTL^1 formulae in negation normal form into positiveTPTL^1 formulae. The definition is by induction on the construction of a formula, we only give the interesting cases: $h(G\varphi) := (h(\varphi) \wedge \neg\mathsf{halt})\mathsf{U}\mathsf{halt}$, $h(F\varphi) := (\mathbf{true} \wedge \neg\mathsf{halt})\mathsf{U}(h(\varphi) \wedge \neg\mathsf{halt})$. We have φ is satisfiable if, and only if, $h(\varphi)$ is satisfiable.

Note that this proof idea is not trivially applicable to positiveMTL, because it is not clear how to express the semantics of the $G_{[a,b]}$-modality for $[a, b] \neq (-\infty, +\infty)$. $\qquad\square$

Last but not least, we consider the unary fragment of positiveTPTL^1, in which the only allowed modalities are the F and X modalities. This fragment has also been considered for MTL and TPTL over monotonic timed words [7]. In this setting, SAT for both logics is NP-complete. Here, we show that this applies also to the setting of non-monotonic data words.

Theorem 7. *For* posUnaryTPTL, *SAT is* NP-*complete.*

Proof. (Sketch) The lower bound follows by reduction from SAT for propositional logic. Now we prove that the problem is in NP. Let ψ be a posUnaryTPTL formula. We denote by $\Gamma(\psi)$ the set containing all those formulae that can be obtained from ψ by resolving the non-determinism induced by the occurrences of disjunctions. Formulae in $\Gamma(\psi)$ are is satisfiable if, and only if, there exists a satisfiable formula ψ^* in $\Gamma(\psi)$.

Now, consider a formula $\psi^* \in \Gamma(\psi)$ that is satisfied by a data word w. One can see that only $n \leq |\psi^*|$ points in w are relevant for a successful model checking of w on ψ^*. This is because new further points are only required by subformulae of the form $\mathsf{F}\phi$ or $\mathsf{X}\phi$ and negation only occurs in front of atoms. Therefore, a word w' of length n can be obtained from w such that $w' \models \psi^*$:

Lemma 8. *Let ψ be a* posUnaryTPTL *formula. If ψ is satisfiable, then there exists a data word w such that $w \models \psi$ and $|w| \leq |\psi|$.*

The difference to Th. 5 is that we are able to additionally provide a bound to the length of the possible finite witness of satisfiability.

Based on this, we can decide satisfiability in polynomial non-deterministic time: guess a formula ψ^* in $\Gamma(\psi)$; then guess a data word $w = (P_0, 0) \ldots (P_{n-1}, 0)$ of length $n = |\psi^*|$; last, for each subformula ϕ of ψ^* guess a position from $\{0 \ldots n - 1\}$ and verify that w model checks ψ^* (without considering data value constraints) with respect to the guessed positions. Finally, solve the set of linear inequalities $\mathcal{C}$ built up from the position assigned to each subformula $x \sim c$ and its corresponding less outer subformula $x.\phi$ (in case it does not exists, then ψ^* is used). All the guesses are independent and the verifications can be done in polynomial time. Each inequality in $\mathcal{C}$ belongs to the class of *difference constraints* and a system of such a class of constraints can be solved in polynomial time [9]. Thus, the problem is in NP. □

Corollary 9. *For* posUnaryMTL *and* posUnaryFreezeLTL, *SAT is* NP-*complete.*

7   Conclusion and Open Problems

The main open problem of this paper is the decidability status for the unary fragment of MTL without the X modality. While the X modality can be avoided in reductions for showing undecidability of SAT for unaryTPTL, it seems to be fundamental in all reductions we have looked at for showing undecidability of the unary fragment of MTL. At the same time it seems surprising that the only absence of the X modality could be enough to change the decidability status of the SAT problem.

We are also surprised that the decidability of SAT for unaryFreezeLTL does neither apply to unaryMTL nor to unaryTPTL. This is opposed to a recent extension of a decidability result for FreezeLTL[1]-model checking *deterministic* one-counter automata [11] to MTL and TPTL[1] [17].

Note that our undecidability results for SAT of different fragments of MTL and TPTL, also imply the undecidability of the existential model checking and (apart from the positive fragments) of the universal model checking problem for one-counter machines and the corresponding logics.

References

1. Alur, R., Henzinger, T.A.: Real-time logics: Complexity and expressiveness. Inf. Comput. 104(1), 35–77 (1993)
2. Alur, R., Henzinger, T.A.: A really temporal logic. J. ACM 41(1), 181–204 (1994)
3. Bojanczyk, M., David, C., Muscholl, A., Schwentick, T., Segoufin, L.: Two-variable logic on data words. ACM Trans. Comput. Log. 12(4), 27 (2011)
4. Bollig, B.: An automaton over data words that captures EMSO logic. In: Katoen, J.-P., König, B. (eds.) CONCUR 2011. LNCS, vol. 6901, pp. 171–186. Springer, Heidelberg (2011)
5. Bollig, B., Cyriac, A., Gastin, P., Narayan Kumar, K.: Model checking languages of data words. In: Birkedal, L. (ed.) FOSSACS 2012. LNCS, vol. 7213, pp. 391–405. Springer, Heidelberg (2012)
6. Bouyer, P.: A logical characterization of data languages. Inf. Process. Lett. 84(2), 75–85 (2002)
7. Bouyer, P., Chevalier, F., Markey, N.: On the expressiveness of TPTL and MTL. Inf. Comput. 208(2), 97–116 (2010)
8. Carapelle, C., Feng, S., Fernandez Gil, O., Quaas, K.: Ehrenfeucht-Fraïssé games for TPTL and MTL over data words, http://arxiv.org/abs/1311.6250
9. Cormen, T.H., Leiserson, C.E., Rivest, R.L., Stein, C.: Introduction to Algorithms, Second Edition, 2nd edn. The MIT Press and McGraw-Hill Book Company (2001)
10. Demri, S., Lazic, R.: LTL with the freeze quantifier and register automata. ACM Trans. Comput. Log. 10(3) (2009)
11. Demri, S., Lazić, R.S., Sangnier, A.: Model checking Freeze LTL over one-counter automata. In: Amadio, R.M. (ed.) FOSSACS 2008. LNCS, vol. 4962, pp. 490–504. Springer, Heidelberg (2008)
12. Figueira, D., Segoufin, L.: Future-looking logics on data words and trees. In: Královič, R., Niwiński, D. (eds.) MFCS 2009. LNCS, vol. 5734, pp. 331–343. Springer, Heidelberg (2009)
13. Jaffar, J., Maher, M.J., Stuckey, P.J., Yap, R.H.C.: Beyond finite domains. In: Borning, A. (ed.) PPCP 1994. LNCS, vol. 874, pp. 86–94. Springer, Heidelberg (1994)
14. Minsky, M.L.: Recursive unsolvability of Post's problem of "tag" and other topics in theory of Turing machines. Annals of Mathematics 74(3), 437–455 (1961)
15. Ouaknine, J., Worrell, J.B.: On metric temporal logic and faulty Turing machines. In: Aceto, L., Ingólfsdóttir, A. (eds.) FOSSACS 2006. LNCS, vol. 3921, pp. 217–230. Springer, Heidelberg (2006)
16. Ouaknine, J., Worrell, J.: On the decidability and complexity of metric temporal logic over finite words. Logical Methods in Computer Science 3(1) (2007)
17. Quaas, K.: Model checking metric temporal logic over automata with one counter. In: Dediu, A.-H., Martín-Vide, C., Truthe, B. (eds.) LATA 2013. LNCS, vol. 7810, pp. 468–479. Springer, Heidelberg (2013)
18. Segoufin, L.: Automata and logics for words and trees over an infinite alphabet. In: Ésik, Z. (ed.) CSL 2006. LNCS, vol. 4207, pp. 41–57. Springer, Heidelberg (2006)

(k,l)-Unambiguity and Quasi-Deterministic Structures: An Alternative for the Determinization

Pascal Caron[1], Marianne Flouret[2], and Ludovic Mignot[1]

[1] LITIS, Université de Rouen, 76801 Saint-Étienne du Rouvray Cedex, France
{pascal.caron,ludovic.mignot}@univ-rouen.fr
[2] LITIS, Université du Havre, 76058 Le Havre Cedex, France
marianne.flouret@univ-lehavre.fr

Abstract. We focus on the family of (k, l)-unambiguous automata that encompasses the one of deterministic k-lookahead automata introduced by Han and Wood. We show that this family presents nice theoretical properties that allow us to compute quasi-deterministic structures. These structures are smaller than DFAs and can be used to solve the membership problem faster than NFAs.

1 Introduction

One of the best known automata construction is the position construction [8]. If a regular expression has n occurrences of symbols, then the corresponding position automaton, which is not necessarily deterministic, has exactly $n + 1$ states. The 1-unambiguous regular languages have been defined by Brüggemann-Klein and Wood [2] as languages denoted by regular expressions the position automata of which are deterministic. They have also shown that there exist regular languages that are not 1-unambiguous. This property has practical implication, since it models a property needed in XML DTDs [1]. Indeed, XML DTDs are defined as an extension of classical context-free grammars in which the right hand side of any production is a 1-unambiguous regular expression. Consequently, characterization of such languages, that has been considered *via* the deterministic minimal automaton, is very important, since it proves that not all the regular languages can be used in XML DTDs. The computation of a small deterministic recognizer is also technically important since it allows a reduction of the time and of the space needed to solve the membership problem (to determine whether or not a given word belongs to a language). As a consequence, one may wonder whether there exists a family of languages encompassing the 1-unambiguous one that can be recognized by a polynomial-size deterministic family of recognizers.

On the one hand, numerous extensions of 1-unambiguity have been considered, like k-block determinism [7], k-lookahead determinism [9] or weak 1-unambiguity [4]. All of these extensions, likely to the notion of 1-unambiguity, are expression-based properties. A regular language is 1-unambiguous (resp. k-block deterministic, k-lookahead deterministic, weakly 1-unambiguous) if it is

A.-H. Dediu et al. (Eds.): LATA 2014, LNCS 8370, pp. 260–272, 2014.

denoted by a 1-unambiguous (resp. k-block deterministic, k-lookahead deterministic, weakly 1-unambiguous) regular expression. And all of these three properties
are defined through a recognizer construction.

On the other hand, the concept of lookahead delegation, introduced in [5], handles determinism without computing a deterministic recognizer; the determinism
is simulated by a fixed number of input symbols read ahead, in order to select
the *right* transition in the NFA. This concept arose in a formal study of web-
services composition and its practical applications [6]. Questions about complexity and decidability of lookahead delegation have been answered by Ravikumar
and Santean in [13]. Finally, defining predictable semiautomata, Brzozowski and
Santean [3] improved complexity of determining whether an automaton admits
a lookahead delegator.

The notion of (k, l)-unambiguity for automata is the first step of the study of
the (k, l)-unambiguity for languages. In this paper, we define the notion of (k, l)-
unambiguity for automata, leading to the computation of quasi-deterministic
structures, that are smaller than DFAs and that can be used to solve the membership problem faster than NFAs. Next step is to study the (k, l)-unambiguous
languages, that are languages denoted by some regular expressions the position
automaton of which is (k, l)-unambiguous; Having such a regular expression
allows us to directly compute a quasi-deterministic structures to solve the membership problem.

In Section 3, after defining the (k, l)-unambiguity as an extension of k-
lookahead determinism, we characterize it making use of the *square automaton*.
In Section 4, we define quasi-deterministic structures that allow us to perform a
constant space membership test. Section 5 is devoted to the computation of the
quasi-deterministic structure associated with a (k, l)-unambiguous automaton.

2   Preliminaries

Let ε be *the empty word*. An *alphabet* Σ is a finite set of distinct symbols. The
usual concatenation of symbols is denoted by $\cdot$, and ε is its identity element. We
denote by Σ^* the smallest set containing $\Sigma \cup \{\varepsilon\}$ and closed under $\cdot$. Any subset
of Σ^* is called a *language over* Σ. Any element of Σ^* is called a *word*. The length
of a word w, noted $|w|$, is the number of symbols in Σ it is the concatenation of
(*e.g.* $|\varepsilon| = 0$). For a given integer k, we denote by Σ^k the set of words of length k
and by $\Sigma^{\leq k}$ the set $\bigcup_{k' \leq k} \Sigma^{k'}$. Let $w = a_1 \cdots a_{|w|}$ be a word in Σ^* such that for
any k in $\{1, \ldots, |w|\}$, a_k is a symbol in Σ. Let i and j be two integers such that
$i \leq j \leq |w|$. We denote by $w[i, j]$ the subword $a_i \cdots a_j$ of w starting at position
i and ending at the position j and by $w[i]$ the i-th symbol a_i of w.

A *nondeterministic finite automaton* (**NFA**) A is a 5-tuple $(\Sigma, Q, I, F, \delta)$
where Σ is an alphabet, Q is a *set of states*, $I \subset Q$ is a *set of initial states*,
$F \subset Q$ is a *set of final states* and δ is a *transition function* defined from $Q \times \Sigma$
to 2^Q. The function δ can be interpreted as a subset of $Q \times \Sigma \times Q$ defined by
$q' \in \delta(q, a) \Leftrightarrow (q, a, q') \in \delta$. The domain of δ is extended to $2^Q \times \Sigma^*$ as follows: for
any symbol a in Σ, for any state q in Q, for any subset P of Q, for any word w in

Σ^*: $\delta(P, \varepsilon) = P$, $\delta(P, a) = \bigcup_{p \in P} \delta(p, a)$, $\delta(P, aw) = \delta(\delta(P, a), w)$. The automaton A is said to be *deterministic* if the two following properties hold: $\mathrm{Card}(I) = 1$ and $\forall (q, a) \in Q \times \Sigma$, $\mathrm{Card}(\delta(q, a)) \leq 1$. The automaton A is *accessible* if for any state q in Q, there exists a word w in Σ^* such that $q \in \delta(I, w)$.

Given a word w and an n-state automaton A, the membership test [10], *i.e.* deciding whether w belongs to $L(A)$ can be performed in time $O(n^2 \times |w|)$ and in space $O(n)$. Let us suppose that A' is the determinized n'-state automaton of A (computed as the classical accessible part of the powerset automaton of A). The membership test can be performed in time $O(|w|)$ and in space $O(1)$. But n' can be exponentially greater than n.

Glushkov [8] and McNaughton and Yamada [12] have independently defined the construction of the *Glushkov automaton* or *position automaton* G_E of a regular expression E. The number of states of G_E is $n + 1$, where n is the number of occurrences of symbols of the alphabet in E. The automaton G_E is a $(|E| + 1)$-state finite automaton that recognizes $L(E)$.

A regular expression E is *deterministic* if and only if its Glushkov automaton is. A language is said to be 1-*unambiguous* if there exists a deterministic expression to denote it. Brüggemann-Klein and Wood [2] have shown that determining whether a regular language is 1-unambiguous or not is a decidable problem. Furthermore, they proposed a characterization and showed that both 1-unambiguous languages and non 1-unambiguous regular languages exist. The notion of k-lookahead determinism [9] extends the one of 1-unambiguity of expressions. In that purpose, Han and Wood define the k-*lookahead deterministic position automaton* of an expression.

Definition 1 ([9]). *Let $A = (\Sigma, Q, I, F, \delta)$ be a position automaton of an expression. Then A is a deterministic k-lookahead automaton if for any state q_0 in Q, where (q_0, a_0, q_0), (q_0, a_1, q_1), ..., (q_0, a_m, q_m) are the out-transitions of q_0, with $q_i \neq q_j$ for $0 \leq i, j \leq m$, it holds: $a_i \cdot \mathbb{F}_{k-1}(q_i) \cap a_j \cdot \mathbb{F}_{k-1}(q_j) = \emptyset$, where $0 \leq i < j \leq m$ and $\mathbb{F}_{k-1}(q_i)$ is the set of words of length $k-1$ that labels a path starting at q_i.*

Notice that this definition can be extended to any automaton that is not a position one. Informally, an automaton is k-lookahead deterministic if and only if for any state q, for any word $w = a_1 \cdots a_k$ of length k, all the w-labelled paths starting at q share the same successor q_1^1 after a step of length 1 (see Figure 1).

Brzozowski and Santean [3] introduced the notion of predictability for an automaton and linked it to the one of lookahead determinism: as far as an automaton admits a unique initial state, it is k-predictable if and only if it is $(k + 1)$-lookahead deterministic.

In order to decide whether a given automaton is predictable, they make use of the square automaton defined as follows: let $A = (\Sigma, Q, I, F, \delta)$. The *square automaton* s_A of A is the automaton $(\Sigma, Q \times Q, I \times I, F \times F, \delta')$ where for any pair (q_1, q_2) of states in Q, for any symbol a in Σ, $\delta'((q_1, q_2), a) = \delta(q_1, a) \times \delta(q_2, a)$.

Finally, from the square automaton, they define the pair automata of critical subsets of Q (the set of initial states and the sets of successors of a fork).

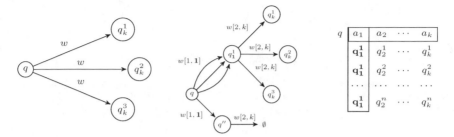

Fig. 1. The k-lookahead determinism

An automaton is predictable if and only if its pair automata admit no cycle. A closely related method has already been applied in comparable settings for Moore machines [11].

3   The (k,l)-Unambiguity

The definition of k-lookahead determinism can be extended by the introduction of an additional parameter l. The maximal length of ambiguity in two distinct paths from the same state and labelled by a same word is bounded by this parameter. Hence an automaton is said to be (k, l)-unambiguous if and only if for any state q, for any word $w = a_1 \cdots a_k$ of length k, there exists an integer $i \le l$ such that all the w-labelled paths starting at q share the same successor q_i^1 after a step of length i. In other words, for any path $c_j = (q_1^j, \ldots, q_k^j)$ from q labelled by w, there exists an integer $i \le l$ such that for any j', $q_i^{j'} = q_i^1$ (see Figure 2).

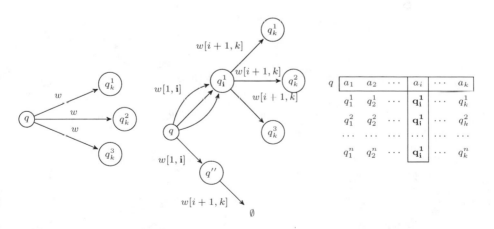

Fig. 2. The (k, l)-unambiguity

Definition 2. *Let k and l be two integers such that $l \leq k$. A finite automaton $A = (\Sigma, Q, I, F, \delta)$ is (k,l)-unambiguous if $\mathrm{Card}(I) = 1$ and if for any state q in Q, for any word w in Σ^k, there exists an integer $1 \leq i \leq l$ such that:*
$$\mathrm{Card}(\{q' \in Q \mid q' \in \delta(q, w[1, i]) \ \wedge \ \delta(q', w[i+1, k]) \neq \emptyset\}) \leq 1.$$

As a direct consequence of this definition, it holds that any (k,l)-unambiguous automaton is also a $(k, l+1)$-unambiguous automaton whenever $l < k$.

The following example enlightens the notion of (k,l)-unambiguity while illustrating the difference between (k,l)-unambiguity and k-lookahead determinism.

Example 3. Let us consider the automaton $A = (\Sigma, Q, I, F, \delta)$ in Figure 3. Let us notice that $\mathrm{Card}(\delta(0, a)) = \mathrm{Card}(\delta(0, ab)) = \mathrm{Card}(\delta(0, aba)) = 2$. As a consequence, the automaton is not $(3,3)$-unambiguous since it is not possible to define an integer $i \leq 3$ such that $\mathrm{Card}(\delta(0, aba[1, i])) \leq 1$. Increasing the length k of the window allows us to avoid this ambiguity. Indeed, for any word w of length 4, $\mathrm{Card}(\delta(0, w)) \leq 1$. Hence A is $(4,4)$-unambiguous. Furthermore, A is also $(4,3)$-unambiguous since 5 is the only state q reached from 0 by aba such that $\delta(q, b)$ is not empty, and since 6 is the only state q reached from 0 by aba such that $\delta(q, a)$ and $\delta(q, c)$ are not empty. However, since the states 3 and 4 are states q reached from 0 by ab such that $\delta(q, ac)$ is not empty, the automaton A is not $(4,2)$-unambiguous. Finally, let us notice that this automaton is not k-lookahead deterministic for any integer k since for any integer j and for any prefix $w = aw'$ of $(abaa)^j$, $\delta(0, a) = \{1, 2\}$ and $w' \in \mathbb{F}_{|w'|}(1) \cap \mathbb{F}_{|w'|}(2)$.

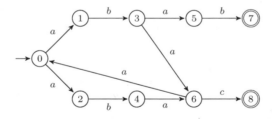

Fig. 3. The automaton of Example 3

Proposition 4. *An automaton is deterministic k-lookahead if and only if it is a $(k,1)$-unambiguous automaton.*

Proposition 5. *There exist (k,l)-unambiguous automata that are not k'-lookahead deterministic for any integer k'.*

Proof. A counterexample is given in Example 3.

Whenever an automaton is not (k,l)-unambiguous for any couple (k,l) of integers, there exists a state from which it cannot be decided without ambiguity which successor will appear during the run. Hence there exists an infinite hesitation between two paths, that can be decided *via* the square automaton.

Theorem 6. *Let A be an accessible automaton and P be the accessible part of its square-automaton. The two following propositions are equivalent:*

1. *there exists a couple (k, l) such that A is (k, l)-unambiguous,*
2. *every cycle in P contains a pair (p, p) for some p in Q.*

In order to prove Theorem 6, let us first state the following definitions and lemmas. Let $A = (\Sigma, Q, I, F, \delta)$ be an automaton, k be an integer and w be a word in Σ^k. A *path p labelled by w* is a finite sequence $p = (p_j)_{0 \leq j \leq k}$ such that for any integer $0 \leq j < k$, $p_{j+1} \in \delta(p_j, w[j+1])$. The path p *starts* with p_0. Two paths $p = (p_j)_{0 \leq j \leq k}$ and $p' = (p'_j)_{0 \leq j \leq k}$ labelled by w are *totally distinct* if for any integer $0 < j \leq k$, $p_j \neq p'_j$.

Lemma 7. *Let $A = (\Sigma, Q, I, F, \delta)$ be an automaton. If there exists a word w in Σ^+ and a state q in Q such that for any integer $1 \leq k \leq |w|$, $\mathrm{Card}(\{q' \in \delta(q, w[1, k]) \mid k < |w| \Rightarrow \delta(q', w[k+1]) \neq \emptyset\}) \geq 2$, then there exists at least two totally distinct paths labelled by w that starts with q.*

Lemma 8. *Let A be an automaton and P be its square-automaton. Let w be a word in Σ^* and q be a state in Q. If there exists two totally distinct paths labelled by w that starts with q in A, then there exists a path $p = (p_j)_{0 \leq j \leq k}$ in P labelled by w starting with (q, q) such that for any integer $1 \leq j \leq k$, $p_j = (c, c')$ with $c \neq c'$.*

Lemma 9. *Let $A = (\Sigma, Q, I, F, \delta)$ be an automaton and $P = (\Sigma, Q', I', F', \delta')$ be its square-automaton. Let w be a word in Σ^* and q_1 and q_2 be two states in Q. If $q_2 \in \delta(q_1, w)$ then $(q_2, q_2) \in \delta'((q_1, q_1), w)$.*

Proof (Theorem 6). Let us set $A = (\Sigma, Q, \{i\}, F, \delta)$ and $P = (\Sigma, Q', I', F', \delta')$.

$(1 \Rightarrow 2)$ Let us suppose that there exists a cycle C in P that does not contain any pair (p, p) for all state p in Q. As a consequence, there exists a path R from (i, i) to a state $s = (c, c')$ in C such that any predecessor of the first occurrence of s does not belong to C. Let q be the state in Q such that **(a)** (q, q) appears on the path R from (i, i) to the first occurrence of (c, c') and **(b)** there exists no state p in Q such that (p, p) appears on the path R between (q, q) and the first occurrence of (c, c'). Notice that q exists since i satisfies the previous propositions. Hence for any integer $k \geq 1$, there exists a word w in Σ^k such that $\delta'((q, q), w) \neq \emptyset$ and then $\mathrm{Card}(\delta(q, w)) \geq 2$. Consequently, there exists no couple (k, l) such that A is (k, l)-unambiguous.

$(1 \Leftarrow 2)$ Let us suppose that for every integer k, there exists a word w in Σ^k and a state q in Q such that for any integer $i \leq k$, $\mathrm{Card}(\{q' \mid q' \in \delta(q, w[1, i]) \wedge \delta(q', w[i+1, k]) \neq \emptyset\}) > 1$. Hence according to Lemma 7, for any integer k, there exists a word in Σ^k such that there exists at least two totally distinct paths labelled by w that starts with q. Since q is reachable from i, then it holds from Lemma 9 that (q, q) belongs to Q' since it is reachable from (i, i). According to Lemma 8, for any integer k, there exists a word in Σ^k such that there exists a path $p = (p_j)_{0 \leq j \leq k}$ in P labelled by w starting with (q, q) such that for any integer $1 \leq j \leq k$, $p_j = (c, c')$ with $c \neq c'$. Finally, whenever

$k \geq \mathrm{Card}(Q) \times (\mathrm{Card}(Q) - 1)$, there exists two integer $1 \leq k_1 < k_2 \leq k$ such that $p_{k_1} = p_{k_2}$. Consequently there exists a cycle in P that contains no pair (p, p) for any p in Q. $\square$

Notice that Theorem 6 defines a polynomial decision procedure to test if, for a given NFA A, there exists a couple (k, l) of integer such that A is (k, l)-unambiguous.

   The next section is devoted to the definition of quasi-deterministic structures. These structures allow us to solve the membership problem with the same complexity as deterministic automata while being possibly exponentially smaller. Finally, we show in Section 5 how to convert a (k, l)-unambiguous NFA into a quasi-deterministic structure.

4   The Quasi-Deterministic Structure

A quasi-deterministic structure is a structure derived from an automaton: it embeds a second transition function that is used to shift the input window (of a fixed length) while reading a word (see Example 11). In the following, the symbol $\perp$ is used to represent undefined states and transitions.

Definition 10. *A quasi-deterministic structure (QDS) is a 6-tuple $S = (\Sigma, \mathcal{Q}, i, F, \delta, \gamma)$ where:*

- *Σ is an alphabet,*
- *$\mathcal{Q} = (Q_j)_{j \in \{1,\ldots,m\}}$ is a family of m disjoint set of states,*
- *$i \in Q_1$ is the initial state,*
- *$F \subset \bigcup_{j \in \{1,\ldots,m\}} Q_j$ is the set of final states,*
- *δ is a function from $Q_j \times \Sigma$ to $Q_{j+1} \cup \{\perp\}$ for $j \in \{1,\ldots,m-1\}$,*
- *γ is a function from Q_m to $Q_1 \times \{1,\ldots,m\}$.*

   The function δ can be extended for any state q in $\bigcup_{j \in \{1,\ldots,m\}} Q_j$, for any state q' in Q_m, for any word w in Σ^* and for any symbol a in Σ to $\delta(q, \varepsilon) = q$, $\delta(q', a) = \perp$, $\delta(\perp, a) = \perp$, $\delta(q, aw) = \delta(\delta(q, a), w)$.

Example 11. The quasi-deterministic structure $S = (\Sigma, \mathcal{Q}, i, F, \delta, \gamma)$ represented in Figure 4 is defined by $\Sigma = \{a, b\}$, $\mathcal{Q} = (\{1, 6\}, \{2, 3, 7\}, \{4, 5, 8\})$, $i = 1$, $F = \{2, 7\}$, $\delta = \{(1, a, 2), (1, b, 3), (2, b, 4), (2, a, 5), (3, a, 5), (3, b, 5), (6, a, 7), (6, b, 7), (7, a, 8)\}$, and $\gamma = \{(5, 2, 1), (4, 1, 6), (8, 2, 6)\}$.

   Such a structure can be used as a recognizer: a word w is recognized if it labels a path from an initial state to a final one. However, the notions of path and of label are different in a quasi-deterministic structure. Indeed, some factors of the word can be repeated all along the path. As a consequence, there exists a path from a state q to a state q' labelled by w if and only if $q' = \Delta(q, w)$ where Δ is the function defined as follows:

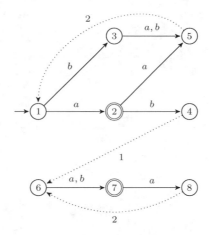

Fig. 4. A Quasi-Deterministic Structure

Definition 12. *Let* $S = (\Sigma, (Q_j)_{j \in \{1,\dots,m\}}, i, F, \delta, \gamma)$ *be a QDS. The extended transition function of S is the function Δ from $Q_1 \times \Sigma^*$ to $(\bigcup_{j \in \{1,\dots,m\}} Q_j) \cup \{\bot\}$ defined for any pair (q, w) in $Q_1 \times \Sigma^*$ by:*

$$\Delta(q, w) = \begin{cases} \delta(q, w) & \text{if } |w| \le m - 1, \\ \Delta(q', w[j+1, |w|]) & \text{if } |w| \ge m \ \wedge \ \gamma(\delta(q, w[1, m-1])) = (q', j) \ \wedge \ q' \ne \bot, \\ \bot & \text{otherwise.} \end{cases}$$

Definition 13. *Let $S = (\Sigma, Q, i, F, \delta, \gamma)$ be a quasi-deterministic structure. The language of S is the language $L(S)$ defined by:*
$$L(S) = \{w \in \Sigma^* \mid \Delta(i, w) \in F\}.$$

Finally, let us show how to determine whether a given word is recognized by a given quasi-deterministic structure (see Example 15).

Require: $S = (\Sigma, Q, i, F, \delta, \gamma)$ a quasi deterministic structure, w in Σ^*
Ensure: Returns $w \in L(S)$
1: $k \leftarrow \text{Card}(Q) - 1$
2: **if** $|w| \le k$ **then**
3:     **return** $\delta(i, w) \in F$
4: **end if**
5: $c \leftarrow i$
6: $w' \leftarrow w$
7: **while** $|w'| > k \ \wedge \ c \ne \bot$ **do**
8:     $(c, j) \leftarrow \gamma(\delta(c, w'[1, k]))$
9:     $w' \leftarrow w'[j+1, |w'|]$
10: **end while**
11: **return** $c \ne \bot \ \wedge \ \delta(c, w') \in F$
  Algorithm 1. Membership Test for Quasi-Deterministic Structure

Proposition 14. *Algorithm 1 returns TRUE if and only if $w \in L(S)$. Further-more, its execution always halts, and is performed in time $O(k \times \frac{|w|}{s})$, where $k = \mathrm{Card}(\mathcal{Q}) - 1$, $s = \min\{j \mid \exists q \in Q_{k+1}, \gamma(q) = (p, j)\}$ and in space $O(1)$.*

Example 15. Let us consider the structure S defined in Example 11. Let $w = bbbaabab$. Following computation illustrates that $\Delta(1, w) = 7$, and since $7 \in F$, it holds that $w \in L(S)$.

$$
\begin{array}{llllllll}
1\;\boxed{b\;\;b} & b & a & a & b & a & b & \\[4pt]
\boxed{b\;\;b}\;\mathbf{5}\;b & a & a & b & a & b & & (\delta(1, bb) = 5) \\[4pt]
b\;\;b\;\;1\;\boxed{b\;\;a} & a & b & a & b & & & (\gamma(5) = (1, 2)) \\[4pt]
b\;\;b\;\;\boxed{b\;\;a}\;\mathbf{5}\;a & b & a & b & & & & (\delta(1, ba) = 5) \\[4pt]
b\;\;b\;\;b\;\;a\;\;1\;\boxed{a\;\;\;\;b} & a & b & & & & & (\gamma(5) = (1, 2)) \\[4pt]
b\;\;b\;\;b\;\;a\;\;\boxed{a\;\;\;\;b}\;\mathbf{4}\;a & b & & & & & & (\delta(1, ab) = 4) \\[4pt]
b\;\;b\;\;b\;\;a\;\;a\;\;\mathbf{6}\;\boxed{b\;\;\;\;a} & b & & & & & & (\gamma(4) = (6, 1)) \\[4pt]
b\;\;b\;\;b\;\;a\;\;a\;\;\boxed{b\;\;\;\;a}\;\mathbf{8}\;b & & & & & & & (\delta(6, ba) = 8) \\[4pt]
b\;\;b\;\;b\;\;a\;\;a\;\;b\;\;a\;\;\mathbf{6}\;b & & & & & & & (\gamma(8) = (6, 2)) \\[4pt]
b\;\;b\;\;b\;\;a\;\;a\;\;b\;\;a\;\;b\;\;\mathbf{7} & & & & & & & (7 \in F \Rightarrow w \in L(S))
\end{array}
$$

Next section is devoted to the conversion of a (k, l)-unambiguous NFA into a quasi-deterministic structure.

5 Quasi-Determinization of a (k,l)-Unambiguous NFA

For any (k, l)-unambiguous automaton, given a state q and a word w of length k, there exists an integer $i \leq l$ such that there exists at most one state q' in $\delta(q, w[1, i])$ such that $\delta(q', w[i+1, k])$ is not empty. Quasi-deterministic structures can be used in order to simulate each run in a unique way. For any pair (q, w), the integer i and the state q' can be precomputed; then the run can restart in q' with a word w' that is a suffix of w. The integer i is the called *step index of q w.r.t. w*, and the state q' the *step successor of q w.r.t. w*.

Definition 16. *Let $A = (\Sigma, Q, \{i\}, F, \delta)$ be a (k, l)-unambiguous automaton, q be a state in Q and w be a word in Σ^k. The step index of q w.r.t. w, denoted by $\mathrm{StepIndex}_w(q)$ is the biggest integer $j \leq l$ satisfying:*

$$\text{Card}(\{q' \in Q \mid q' \in \delta(q, w[1, j]) \ \wedge \ \delta(q', w[j + 1, k]) \neq \emptyset\}) \leq 1.$$

Definition 17. *Let $A = (\Sigma, Q, \{i\}, F, \delta)$ be a (k,l)-unambiguous automaton, q be a state in Q and w be a word in Σ^k. The step successor of q w.r.t. w, denoted by $\text{StepSucc}_w(q)$, is defined by:*

$$\text{StepSucc}_w(q) = \begin{cases} \bot & \text{if } \left\{ \begin{array}{l} q' \in Q \mid \quad q' \in \delta(q, w[1, \text{StepIndex}_w(q)]) \\ \wedge \ \delta(q', w[\text{StepIndex}_w(q) + 1, k]) \neq \emptyset \end{array} \right\} = \emptyset, \\ p & \text{if } \left\{ \begin{array}{l} q' \in Q \mid \quad q' \in \delta(q, w[1, \text{StepIndex}_w(q)]) \\ \wedge \ \delta(q', w[\text{StepIndex}_w(q) + 1, k]) \neq \emptyset \end{array} \right\} = \{p\}. \end{cases}$$

Example 18. Let $\Sigma = \{a, b\}$. Let A be the automaton in Figure 5 that denotes the language $\Sigma^* \cdot \{a\} \cdot \Sigma$. It can be shown that the automaton A is a $(3,1)$-unambiguous NFA. As an example let us consider the state 1: For any word w in Σ^3:

$$\text{Card}(\{q' \in Q \mid q' \in \delta(1, w[1, 1]) \ \wedge \ \delta(q', w[2, 3]) \neq \emptyset\}) \leq 1,$$

i.e. for any word w in Σ^3, $\text{StepIndex}_w(1) = 1$ and $\text{StepSucc}_w(1) = 1$.

Moreover, the automaton is also $(3,3)$-unambiguous, since (k,l)-unambiguous $\Rightarrow (k, l+1)$-unambiguous. As a consequence, the step index can be increased. As an example let us consider the state 1: For any word w in Σ^3, there exists an integer $1 \leq i \leq 3$ such that:

$$\text{Card}(\{q' \in Q \mid q' \in \delta(1, w[1, i]) \ \wedge \ \delta(q', w[i + 1, 3]) \neq \emptyset\}) \leq 1.$$

For any word w in $\Sigma\{a\}\Sigma$, $\text{StepIndex}_w(1) = 1$ and $\text{StepSucc}_w(1) = 1$.
For any word w in $\Sigma\{ba\}$, $\text{StepIndex}_w(1) = 2$ and $\text{StepSucc}_w(1) = 1$.
For any word w in $\Sigma\{bb\}$, $\text{StepIndex}_w(1) = 3$ and $\text{StepSucc}_w(1) = 1$.

Fig. 5. The automaton A

The computation of the pairs $(\text{StepIndex}_w(q), \text{StepSucc}_w(q))$ for any pair of a state and a word (q, w) is sufficient to compute a quasi-deterministic structure. The quasi-deterministic structure is exponentially bigger than the automaton, but with respect to the size of the alphabet, that has to be compared with the exponential growth with respect to the number of states in the classical determinization.

Definition 19. *Let $A = (\Sigma, Q, \{i\}, F, \delta)$ be a (k,l)-unambiguous automaton. The quasi-deterministic structure associated with A is $S = (\Sigma, (Q_j)_{j \in \{1, \ldots, k+1\}}, i', F', \delta', \gamma')$ where:*

- $\forall j \in \{1, \ldots, k+1\}$, $Q_j = \{(q, w) \mid q \in Q \ \wedge \ w \in \Sigma^{j-1}\}$,
- $i' = (i, \varepsilon)$,

$-\ F' = \{(q,w) \mid \delta(q,w) \cap F \neq \emptyset\},$

$-\ \forall a \in \Sigma, \forall (q,w) \in (Q_j)_{j\in\{1,\dots,k\}},$

$$\delta'((q,w),a) = \begin{cases} (q, w \cdot a) & \text{if } (q, w \cdot a) \in Q_{j+1}, \\ \bot & \text{otherwise}, \end{cases}$$

$-\ \forall (q,w) \in Q_{k+1},\ \gamma'((q,w)) = ((\text{StepSucc}_w(q), \varepsilon), \text{StepIndex}_w(q)).$

Proposition 20. *Let $A = (\Sigma, Q, I, F, \delta)$ be a (k,l)-unambiguous automaton and S be the quasi-deterministic structure associated with A. Then the number of states of S is $\text{Card}(Q) \times \frac{\text{Card}(\Sigma)^{k+1}-1}{\text{Card}(\Sigma)-1}$.*

Proposition 21. *Let A be a (k,l)-unambiguous automaton and S be the quasi-deterministic structure associated with A. Then:*
$$L(S) = L(A).$$

Example 22. Let us consider the automaton A defined in Example 18. After removing unreachable states, the quasi-deterministic structure associated with A is given in Figure 6.

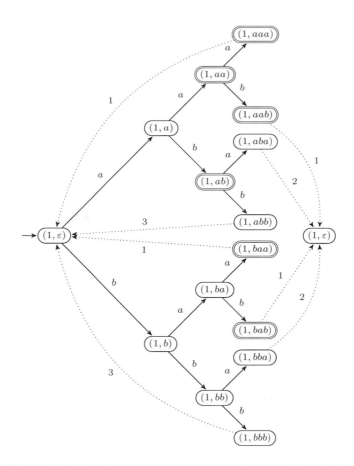

Fig. 6. The Quasi-Deterministic Structure Associated with A

6   Conclusion and Perspectives

Quasi-deterministic structures are an alternative to the computation of a deterministic automaton since they can be used as recognizers, while reducing the space needed to solve the membership problem once computed.

Theorem 23. *There exists a QDS that recognizes* $L_k = \{a,b\}^*\{a\}\{a,b\}^k$ *which is exponentially smaller than the minimal DFA associated with* L_k.

A regular language is (k,l)-unambiguous if it is denoted by some regular expression the position automaton of which is (k,l)-unambiguous. Similar extensions were already defined for deterministic automata (1-unambiguity). Denoting a language by such an expression allows us to directly compute a quasi-deterministic structure in order to solve the membership problem.

One may wonder whether every regular language admits a (k,l)-unambiguous position automaton recognizing it. If the answer is negative, then a second question arises: Is it possible to characterize languages having a (k,l)-unambiguous position automaton, as Brüggemann-Klein and Wood did for languages having a deterministic position automaton ?

References

1. Bray, T., Paoli, J., Queen, C.S.M., Maler, E., Yergeau, F.: Extensible Markup Language (XML) 1.0, 4th edn. (2006),
   http://www.w3.org/TR/2006/REC-xml-20060816
2. Brüggemann-Klein, A., Wood, D.: One-unambiguous regular languages. Inform. Comput. 140, 229–253 (1998)
3. Brzozowski, J.A., Santean, N.: Predictable semiautomata. Theor. Comput. Sci. 410(35), 3236–3249 (2009)
4. Caron, P., Han, Y.-S., Mignot, L.: Generalized one-unambiguity. In: Mauri, G., Leporati, A. (eds.) DLT 2011. LNCS, vol. 6795, pp. 129–140. Springer, Heidelberg (2011)
5. Dang, Z., Ibarra, O.H., Su, J.: Composability of infinite-state activity automata. In: Fleischer, R., Trippen, G. (eds.) ISAAC 2004. LNCS, vol. 3341, pp. 377–388. Springer, Heidelberg (2004)
6. Gerede, C.E., Hull, R., Ibarra, O.H., Su, J.: Automated composition of e-services: lookaheads. In: Aiello, M., Aoyama, M., Curbera, F., Papazoglou, M.P. (eds.) ICSOC, pp. 252–262. ACM (2004)
7. Giammarresi, D., Montalbano, R., Wood, D.: Block-deterministic regular languages. In: Restivo, A., Ronchi Della Rocca, S., Roversi, L. (eds.) ICTCS 2001. LNCS, vol. 2202, pp. 184–196. Springer, Heidelberg (2001)
8. Glushkov, V.M.: The abstract theory of automata. Russian Mathematical Surveys 16, 1–53 (1961)
9. Han, Y.S., Wood, D.: Generalizations of 1-deterministic regular languages. Inf. Comput. 206(9-10), 1117–1125 (2008)

10. Hopcroft, J.E., Motwani, R., Ullman, J.D.: Introduction to Automata Theory, Languages and Computation, 3rd edn. Pearson Addison-Wesley, Upper Saddle River (2007)
11. Kohavi, Z.: Switching and Finite Automata Theory. Computer Science Series, 2nd edn. McGraw-Hill Higher Education (1990)
12. McNaughton, R.F., Yamada, H.: Regular expressions and state graphs for automata. IEEE Transactions on Electronic Computers 9, 39–57 (1960)
13. Ravikumar, B., Santean, N.: On the existence of lookahead delegators for nfa. Int. J. Found. Comput. Sci. 18(5), 949–973 (2007)

Solutions to the Multi-dimensional Equal Powers Problem Constructed by Composition of Rectangular Morphisms

Anton Černý

Department of Information Science
College of Computing Sciences and Engineering
Kuwait University, Kuwait
`anton.cerny@ku.edu.kw`

Abstract. Based on the original approach of Eugène Prouhet, using composition of morphism-like array-words mappings, we provide a construction of solutions to the multi-dimensional Prouhet-Tarry-Escott problem.

Keywords: Prouhet-Tarry-Escott problem, array word, spectrum, symbol position.

1 Introduction

The French mathematician Eugène Prouhet is considered to be one of the fathers of the discipline of combinatorics on words (see [3]). In his paper [18], he provided a solution to the equal powers problem, which later, after his work has been rediscovered, started to be called the Prouhet-Tarry-Escott problem. Recently, the multi-dimensional version of the Prouhet-Tarry-Escott problem was investigated in [2]. The problem can be described as follows:

Problem 1 (PTE$_k^d$). For a given integer $d \geq 1$ (the dimension of the problem) and $k \geq 1$ (the degree of the problem) find two distinct multisets A, B of integer d-tuples satisfying, for all $0 \leq r_0, \ldots r_{d-1}$ such that $r_0 + \cdots r_{d-1} < k$, the equality

$$\sum_{\langle a_0, \ldots a_{d-1} \rangle \in A} A\left(\langle u_0, \ldots a_{d-1}\rangle\right) a_1^{r_0} \cdot \ldots \cdot a_{d-1}^{r_{d-1}}$$
$$= \sum_{\langle b_0, \ldots b_{d-1} \rangle \in B} B\left(\langle b_0, \ldots b_{d-1}\rangle\right) b_1^{r_0} \cdot \ldots \cdot b_{d-1}^{r_{d-1}}$$

Here $A\left(\langle a_1, \ldots a_d \rangle\right)$ denotes the multiplicity of containment of the tuple $\langle a_1, \ldots a_d \rangle$ in the multiset A. Let us note that the degree k, as introduced in Problem 1, is greater by 1 compared to the degree considered in [2]. Prouhet constructed a solution to the Problem 1 for $d = 1$. He actually provided a solution for $d = 1$ of the following more general problem.

Problem 2 (PTE$_{n,k}^d$). For a given integer $d \geq 1$ (the dimension of the problem), $k \geq 1$ (the degree of the problem) and $n \geq 2$, find a n-tuple of multisets $\{A\}_{i=0}^{n-1}$

A.-H. Dediu et al. (Eds.): LATA 2014, LNCS 8370, pp. 273–284, 2014.

of integer d-tuples such that, for all $0 \leq r_0, \ldots r_{d-1}$ satisfying $r_0 + \cdots r_{d-1} < k$, the value

$$\sum_{\langle a_0, \ldots a_{d-1}\rangle \in A_i} A_i \left(\langle a_0, \ldots a_{d-1}\rangle\right) a_0^{r_0} \cdot \cdots \cdot a_{d-1}^{r_{d-1}} \tag{1}$$

does not depend on i.

In the Prouhet's solution for $d = 1$, all the multisets are sets and form a partition of the set $\{0, \ldots, n^k - 1\}$. The solution is based on the structure of words, which can be obtained by iteration of a uniform morphism. These words, in the case $n = 2$, form the building blocks of the well-known sequence of Thue-Morse (which is sometimes called the sequence of Prouhet-Thue-Morse).

In [2], the authors described an exponential-size solution to Problem 1, which they (not quite properly) call the generalization of the Prouhet's theorem. The ideas of the construction are based on geometrical approach entirely different from that of Prouhet. In the current paper we will show, that a more genuine generalization of the Prouhet's construction provides a solution to the Problem 2 for any dimension d. All our considerations will be done for the case $d = 2$, there is no principal obstacle to extend them, in a straightforward way, to any dimension $d > 2$. We use the (rectangular) array words as a generalization of the concept of (linear) words.

2   Basic Notions

We will use angle brackets to denote ordered tuples (alternatively, they may be referred to as finite sequences). Let $\mathbb{N} = \{0, 1, \ldots\}$ denote the set of all natural numbers. For $m \in \mathbb{N}$, let $[\![m]\!] = \{0, 1, \ldots, m - 1\}$. By default, for $i \in [\![m]\!]$, α_i will denote the i-th element of an ordered m-tuple α, thus $\alpha = \langle \alpha_0, \ldots, \alpha_{m-1}\rangle$. We will consider the *alphabetical order* on $[\![m]\!] \times [\![n]\!]$, $m, n \in \mathbb{N}$, where $\langle i_1, j_1 \rangle < \langle i_2, j_2 \rangle$ if either $i_1 + j_1 < i_2 + j_2$ or $i_1 + j_1 = i_2 + j_2$ and $i_1 < j_1$.

Besides the usual radix-n notation for natural numbers, we will deal with mixed-radix notation (see, for example, [12, p. 192]). Assume integers $p \geq 1$, $m_0, \ldots, m_{p-1} \geq 2$.

A *multiset* on a set U is a mapping $A : U \to \mathbb{N}$. We will identify a set $B \subseteq U$ with the multiset defined by $B(u) = 1$ if $u \in B$ and $B(u) = 0$ otherwise. For $u \in U$, we will write $u \in A$ if $A(u) > 0$, otherwise we will write $u \notin A$. The multiset A is *finite* if $u \in A$ for finitely many $u \in U$.

The definition of the basic terms of the (one-dimensional) formal languages theory can be found, for example in [15]. We will concentrate here on definitions for the two-dimensional case. An *alphabet* is any non-empty set; elements of an alphabet are called *symbols*. We will mostly consider alphabets in the form $[\![m]\!], m \in \mathbb{N}$. Thus numbers will be used as symbols. For $m, n \in \mathbb{N}$, a $\langle m, n \rangle$-*array word* ([1,4,7,17,10,19]) is a mapping $\alpha : [\![m]\!] \times [\![n]\!] \to \Sigma$. The pair $\langle m, n \rangle$ denotes the *order* of the array word αT. We denote $\alpha_{i,j} = \alpha(i, j)$. Alternatively, an array word may be viewed as a matrix of symbols. The i-th *row* and the j-th *column* of α are the (linear) words $\alpha_{i,0} \cdots \alpha_{i,n-1}$ and $\alpha_{0,j} \cdots \alpha_{m-1,j}$, respectively. The set of all array words over the alphabet Σ and the set of all

$\langle m, n \rangle$-words will be denoted as Σ^{**} and $\Sigma^{m,n}$, respectively. A (scattered) array subword of an array word $\alpha \in \Sigma^{m,n}$, $m, n \in N$, is an array word obtained by deleting zero or more rows and zero or more columns from α. More formally, let $I = \{i_0 < i_1 < \cdots < i_{r-1}\} \subseteq [\![m]\!]$, $J = \{j_0 < j_1 < \cdots < j_{s-1}\} \in [\![m]\!]$, $\langle r, s \rangle \in [\![m]\!] \times [\![n]\!]$. The (scattered) array subword occurring at position $I \times J$ in α is the $\langle |I|, |J| \rangle$-array word $\alpha_{I \times J}$ satisfying, for $\langle p, q \rangle \in I \times J$, $(\alpha_{I \times J})_{p,q} = \alpha_{i_p, j_q}$. An array subword may occur in the original word at different positions - after deleting different sets of rows and/or columns. The number of occurrences of the array subword γ in an array subword α is denoted as $|\alpha|_\gamma$.

For $r, s \in \mathbb{N}$, we define a $\langle r, s \rangle$-array *morphism*, as a 2-dimensional analogue to the uniform morphism[1]. It is the extension of a mapping $h : \Sigma \to \Gamma^{r,s}$, to $h : \Sigma^{**} \to \Gamma^{**}$ described as follows. For $\alpha \in \Sigma^{m,n}$, $\langle i, j \rangle \in [\![m]\!] \times [\![n]\!]$ and $\langle u, v \rangle \in [\![r]\!] \times [\![s]\!]$, the image $h(\alpha)$ is the $\langle rm, sn \rangle$-word satisfying $h(\alpha)_{ri+u, sj+v} = h(\alpha_{i,j})_{u,v}$.

In the rest of the paper, we will omit the adjective "array". Hence by "word", "subword", "morphism" we will mean "array word", "array subword", "array morphism", respectively. We will specifically point out when we mean the one-dimensional case.

Example 3. Consider the $\langle 3, 2 \rangle$-word

$$\alpha = \begin{matrix} 0 & 1 \\ 1 & 1 \\ 1 & 0 \end{matrix}$$

over the alphabet $\{0, 1\}$ and the $\langle 2, 2 \rangle$-morphism $h : \{0, 1\}^{**} \to \{0, 1, 2\}^{**}$ given as

$$h(0) = \begin{matrix} 0 & 1 \\ 2 & 0 \end{matrix} \qquad h(1) = \begin{matrix} 1 & 2 \\ 0 & 1 \end{matrix}.$$

Then $h(\alpha)$ is the $\langle 6, 4 \rangle$-word

$$h(\alpha) = \begin{matrix} 0 & 1 & 1 & 2 \\ 2 & 0 & 0 & 1 \\ 1 & 2 & 1 & 2 \\ 0 & 1 & 0 & 1 \\ 1 & 2 & 0 & 1 \\ 0 & 1 & 2 & 0 \end{matrix}.$$

The bold symbols in $h(\alpha)$ denote the $\langle 3, 2 \rangle$-subword $\begin{matrix} 0 & 1 \\ 1 & 1 \\ 2 & 1 \end{matrix}$ occurring at position $\{1, 3, 4\} \times \{1, 3\}$ (the subword is obtained by removing from $h(\alpha)$ rows $0, 2, 5$ and columns $0, 2$.)

---

[1] We call the mapping "morphism", though it is not a true morphism in algebraic sense.

3   Spectra of Array Words and the 2-Dimensional Prouhet-Tarry-Escott Problem

The spectrum of a linear word is the multiset of its subwords (with the multiplicity given by the number of distinct occurrences of the subword under consideration) of length up to the given value. The question of characterization of words by their spectra has been widely investigated ([11,8,16,9]). In the 2-dimensional case, we analogously define the k-spectrum of a word $\alpha \in \Sigma^{m,n}$, $m, n \in \mathbb{N}$, $k \in \mathbb{N}$, to be the multiset of array words $S_\alpha^{(k)} = \cup_{p+q \leq k+1} D_\alpha^{\langle p,q \rangle}$, where, for $r, s \in \mathbb{N}$, $D_\alpha^{\langle p,q \rangle}$ is the $\langle p, q \rangle$-deck of α, being the multiset on $\Sigma^{p,q}$, satisfying, for $\gamma \in \Sigma^{p,q}$, $D_\alpha^{\langle p,q \rangle}(\gamma) = |\alpha|_\gamma$.[2]

Let us first mention, that to check the equality of spectra of two words, it is enough to compare the number of occurrences of the "largest" subwords, as follows from a straightforward generalization of the property stated in [9], [6]:

Proposition 4. *Let and* $\alpha, \beta \in \Sigma^{m,n}, m, n \geq 1$. *Let* $\langle r, s \rangle \in [\![m]\!] \times [\![n]\!]$. *If* $D_\alpha^{\langle r,s \rangle} = D_\beta^{\langle r,s \rangle}$ *then, for* $\langle t, u \rangle \in [\![r]\!] \times [\![s]\!], D_\alpha^{\langle t,u \rangle} = D_\beta^{\langle t,u \rangle}$.

The property follows from the equality (valid for all words $\nu \in \Sigma^{t,u}$)

$$\binom{m-t}{r-t}\binom{n-u}{s-u} |\alpha|_\nu = \sum_{\gamma \in \Sigma^{r,s}} |\gamma|_\nu \, |\alpha|_\gamma. \tag{2}$$

The left-hand side of (2) expresses the fact that an occurrence of ν in α appears within $\binom{m-t}{r-t}\binom{n-u}{s-u}$ distinct occurrences of $\langle r, s \rangle$-words in α, while the right-hand side is the count of occurrences of $\langle r, s \rangle$-words γ in α containing ν, each counted as many times as the number of occurrences of ν it contains.

It is proved in [13,5] that positions of the same symbol in two distinct words with identical k-spectrum, $k \geq 1$, determine a solution of the Prouhet-Tarry-Escott problem of degree k. We will show here, by following the ideas of the proof from [13], that analogous assertion is true in the 2-dimensional case.

Assume $\alpha \in \Sigma^{m,n}$, $m, n \in \mathbb{N}$. Let $\langle p, q \rangle \in [\![m+1]\!] \times [\![n+1]\!]$. Consider a symbol $c \in \Sigma$ and $\langle i, j \rangle \in [\![p]\!] \times [\![q]\!]$. Then the number of array words in $D_\alpha^{\langle p,q \rangle}$ (taking in consideration the multiplicity of their occurrences in α) containing the symbol $c \in \Sigma$ at position $\langle i, j \rangle$ is

$$\sigma_\alpha^c(i,j) = \sum_{\substack{\gamma \in D_\alpha^{\langle p,q \rangle} \\ \gamma_{i,j}=c}} D_\alpha^{\langle p,q \rangle}(\gamma)$$

$$= \sum_{\substack{\langle t,u \rangle \in [\![m]\!] \times [\![n]\!] \\ \alpha_{t,u}=c}} \binom{t}{i}\binom{m-t-1}{p-i-1}\binom{u}{j}\binom{n-u-1}{q-j-1} \tag{3}$$

---

[2] The inequality "$p + q \leq k + 1$" in the definition of $S_\alpha^{(k)}$ will translate to "$p + q \leq k + d - 1$" if d-dimensional (instead of 2-dimensional) case is considered.

since if a $\langle p, q \rangle$-subword containing c at position $\langle i, j \rangle$ occurs in α if its $\langle i, j \rangle$-th position coincides with some $\langle t, u \rangle$-th position in α containing c. In each such subword γ, the first i rows of γ occur within the first t rows of α and the remaining $p - i - 1$ rows occur within its last $m - t - 1$ rows. The first j columns of γ occur within the first u columns of α and the last $q - j - 1$ columns occur within its last $n - u - 1$ columns.

We will say that a non-zero bivariate polynomial $f(x, y)$ has bi-order $\langle p, q \rangle$ (for $p, q \geq 1$) if all degrees of x, y occurring in f are from $[\![p]\!] \times [\![q]\!]$, respectively. (Hence the degrees of x, y in f are at most $p-1, q-1$, respectively.) For $m, n \geq 1$ and $\langle p, q \rangle \in [\![m+1]\!] \times [\![n+1]\!]$, consider, for $\langle i, j \rangle \in [\![p]\!] \times [\![q]\!]$, the following bivariate polynomials of bi-order $\langle p, q \rangle$:

$$f_{i,j}^{m,n,p,q}(x, y) = \binom{x}{i}\binom{m-x-1}{p-i-1}\binom{y}{j}\binom{n-y-1}{q-j-1}. \tag{4}$$

Lemma 5. *For fixed integers $m, n \in \mathbb{N}$, $\langle p, q \rangle \in [\![m + 1]\!] \times [\![n + 1]\!]$, the set (4) forms a base of the vector space of all bivariate polynomials of bi-order $\langle p, q \rangle$.*

Proof. It is sufficient to show the linear independence of the set. Assume in contrary that there exist coefficients $\lambda_{i,j}$, not all of them equal to zero, such that

$$\varphi(x, y) = \sum_{\langle i,j \rangle \in [\![p]\!] \times [\![q]\!]} \lambda_{i,j} f_{i,j}^{m,n,p,q}(x, y)$$

is the zero polynomial. Let $\lambda_{t,u}$ be the first of the coefficients in alphabetical order of indices, which is not equal to zero. Then $\varphi(t, u) = \lambda_{t,u}\binom{m\ t\ 1}{p-t-1}\binom{n\ u\ 1}{q-u-1}$, since for $\langle i, j \rangle > \langle t, u \rangle$, either $\binom{t}{i} = 0$ or $\binom{u}{j} = 0$ and, consequently, $f_{i,j}^{m,n,p,q}(t, u) = 0$. We arrive to a contradiction, since, apparently, $\varphi(t, u) \neq 0$.

We will first that two array words with identical decks induce solution to an instance of the equal-power problem. This result will imply that two array words with identical spectra induce a solution to the 2-dimensional Prouhet-Tarry-Escott problem.

Lemma 6. *Let $\alpha, \beta \in \Sigma^{m,n}$, $m, n \in \mathbb{N}$, and let $\langle p, q \rangle \in [\![m + 1]\!] \times [\![n + 1]\!]$. If $D_\alpha^{(p,q)} = D_\beta^{(p,q)}$ then, for each $c \in \Sigma$, and $\langle r, s \rangle \in [\![p]\!] \times [\![q]\!]$*

$$\sum_{\substack{\langle a_1, a_2 \rangle \in [\![m]\!] \times [\![n]\!] \\ \alpha_{a_1,a_2} = c}} a_1^r a_2^s = \sum_{\substack{\langle b_1, b_2 \rangle \in [\![m]\!] \times [\![n]\!] \\ \beta_{b_1,b_2} = c}} b_1^r b_2^s.$$

Proof. Let $c \in \Sigma$ and $\langle r, s \rangle \in [\![p]\!] \times [\![q]\!]$. The equality (3) implies, for $\langle i, j \rangle \in [\![p]\!] \times [\![q]\!]$, $\sigma_\alpha^c(i, j) = \sigma_\beta^c(i, j)$. For $\langle t, u \rangle \in [\![m]\!] \times [\![n]\!]$, denote $\delta_{t,u}^c = 1$ if $\alpha_{t,u} = c$ and $\beta_{t,u} \neq c$, $\delta_{t,u}^c = -1$ if $\alpha_{t,u} \neq c$ and $\beta_{t,u} = c$, and $\delta_{t,u}^c = 0$ otherwise. Then, observing 3, we obtain, for $a \in \Sigma$,

$$\sum_{\langle t,u \rangle \in [\![m]\!] \times [\![n]\!]} \delta^c_{t,u} f^{m,n,p,q}_{i,j}(t,u) =$$

$$= \sum_{\langle t,u \rangle \in [\![m]\!] \times [\![n]\!]} \delta^c_{t,u} \binom{t}{i}\binom{m-t-1}{p-i-1}\binom{u}{j}\binom{n-u-1}{q-j-1} \quad (5)$$

$$= \sigma^c_\alpha(i,j) - \sigma^c_\beta(i,j)$$

$$= 0 \quad (6)$$

Take now the following polynomial expressed as linear combination of the polynomials in base (4).

$$x^r y^s = \sum_{\langle i,j \rangle \in [\![o]\!] \times [\![q]\!]} \mu_{i,j} f^{m,n,p,q}_{i,j}(x,y)$$

Applying (6), we obtain

$$\sum_{\substack{\langle t,u \rangle \in [\![m]\!] \times [\![n]\!] \\ \alpha_{t,u}=c}} t^r u^s - \sum_{\substack{\langle t,u \rangle \in [\![m]\!] \times [\![n]\!] \\ \beta_{t,u}=c}} t^r u^s =$$

$$= \sum_{\langle t,u \rangle \in [\![m]\!] \times [\![n]\!]} \delta^{ca}_{t,u} t^r u^s$$

$$= \sum_{\langle t,u \rangle \in [\![m]\!] \times [\![n]\!]} \delta^c_{t,u} \sum_{\langle i,j \rangle \in [\![o]\!] \times [\![q]\!]} \mu_{i,j} f^{m,n,p,q}_{i,j}(t,u)$$

$$= \sum_{\langle i,j \rangle \in [\![o]\!] \times [\![q]\!]} \mu_{i,j} \sum_{\langle t,u \rangle \in [\![m]\!] \times [\![n]\!]} \delta^a_{t,u} f^{m,n,p,q}_{i,j}(t,u)$$

$$= 0$$

As a direct corollary, we obtain the needed relationship between the pairs array words with identical spectra and solutions of the 2-dimensional Prouhet-Tarry-Escott problem.

Theorem 7. *Let* $k,m,n \in \mathbb{N}$, $c \in \Sigma$, *and* $\alpha, \beta \in \Sigma^{m,n}$, *such that* $S^{(k)}_\alpha = S^{(k)}_\beta$. *Then, for each* $c \in \Sigma$,

$$\sum_{\substack{\langle a_1,a_2 \rangle \in [\![m,n]\!] \\ \alpha_{a_1,a_2}=c}} a^r_1 a^s_2 = \sum_{\substack{\langle b_1,b_2 \rangle \in [\![m,n]\!] \\ \beta_{b_1,b_2}=c}} b^r_1 b^s_2. \quad (7)$$

for all $0 \le r,s$ *such that* $r+s < k$.

Proof. Let $0 \le r,s$ such that $r+s < k$. Then $(r+1)+(s+1) \le k+1$ and $D^{(r+1,s+1)}_\alpha = D^{(r+1,s+1)}_\beta$. Lemma 6 implies (7).

4 Permutation Array Morphisms

In [6], in the 1-dimensional case, uniform morphisms where the images of any two symbols have the same 1-spectra have been considered. As a partial result, it has been proved that application of such morphism to a pair of words of equal length with identical k-spectra results in a pair of words with identical $(k+1)$-spectra. The same seems not to be true in two dimensions, as shown in the following Example 8.

Example 8. Consider the words $\alpha = \begin{smallmatrix} 0\,0 \\ 1\,1 \end{smallmatrix}$, $\beta = \begin{smallmatrix} 0\,1 \\ 0\,1 \end{smallmatrix}$ having the same 2-spectra (since they contain the same number of occurrences of single symbols being $\langle 1,1 \rangle$-words and 1 occurrence of λ being a $\langle 2,0 \rangle$-word, $\langle 1,0 \rangle$-word, $\langle 0,0 \rangle$-word, $\langle 0,1 \rangle$-word and $\langle 0,2 \rangle$-word in the same time), and the morphism $h\,(0) = \begin{smallmatrix} 1\,0 \\ 0\,1 \end{smallmatrix}$, $h\,(1) = \begin{smallmatrix} 1\,1 \\ 0\,0 \end{smallmatrix}$. Then

$$h\,(\alpha) = \begin{smallmatrix} 1\,0\,1\,0 \\ 0\,1\,0\,1 \\ 1\,1\,1\,1 \\ 0\,0\,0\,0 \end{smallmatrix}, \, h\,(\beta) = \begin{smallmatrix} 1\,0\,1\,1 \\ 0\,1\,0\,0 \\ 1\,0\,1\,1 \\ 0\,1\,0\,0 \end{smallmatrix}.$$

Despite the fact that both $h\,(0)$ and $h\,(1)$ consist of two symbols 0 and two symbols 1, the 3-spectra of the words $h\,(\alpha)$ and $h\,(\beta)$ are different, since $h\,(\alpha)$ contains 8 occurrences of the $\langle 1,2 \rangle$-word $\begin{smallmatrix} 0\,0 \end{smallmatrix}$ while $h\,(\beta)$ contains only 6 its occurrences.

We will restrict ourselves to a rather narrow class of morphisms where all the symbol images have the same 1-spectrum. We will call them permutation morphisms. In this section, we will use the following convention. For two words $\alpha, \beta \in [\![m]\!]^{p,q}$, $m,p,q \geq 1$, and for integer c, we will write $\beta = (\alpha + c)\,\mathrm{mod}\,m$ if $\beta_{r,s} = (\alpha_{r,s} + c)\,\mathrm{mod}\,m$ for each $\langle r,s \rangle \in [\![p]\!] \times [\![q]\!]$.

Definition 9. *A $\langle p,q \rangle$-morphism $h : [\![m]\!]^{**} \to [\![m]\!]^{**}$, $m,p,q \geq 1$, is a permutation morphism, if*

1. *there exist a permutation π_1 of the set $[\![p]\!]$ and a permutation π_2 of the set $[\![q]\!]$ such that, for $\langle r,s \rangle \in [\![p]\!] \times [\![q]\!]$,*

$$h\,(0)_{\pi_1(r),\pi_2(s)} = \left(h\,(0)_{r,s} + 1 \right) \mathrm{mod}\,m,$$

2. *for $i \in [\![m]\!]$,*

$$h\,(i) = (h\,(0) + i)\,\mathrm{mod}\,m.$$

Proposition 10. *If h is a permutation morphism, then (using the notation from Definition 9),*

1. *the length of each cycle of the permutation $\langle \pi_1, \pi_2 \rangle$ of the set $[\![p]\!] \times [\![q]\!]$ is a multiple of m*

2. *the least common multiple of the lengths of any cycle in π_1 and any cycle in π_2 is a multiple of m*
3. *pq is a multiple of m.*

Example 11. Let $m = 3, p = 2, q = 3$,

$$h(0) = \begin{smallmatrix} 1 & 0 & 2 \\ 2 & 1 & 0 \end{smallmatrix}, \qquad h(1) = \begin{smallmatrix} 2 & 1 & 0 \\ 0 & 2 & 1 \end{smallmatrix}.$$

Then h is a permutation morphism (choose $\pi_1 : 0 \mapsto 0, 1 \mapsto 1$, $\pi_2 : 0 \mapsto 2, 1 \mapsto 0, 2 \mapsto 1$).

Theorem 12. *Let $m, t, u \geq 1, k \geq 0, z \in \llbracket m \rrbracket$, and let $\alpha, \beta \in \llbracket m \rrbracket^{t,u}$ such that $S_\alpha^{(k)} = S_\beta^{(k)}$ and $(\beta = \alpha + z) \bmod m$. Let $h : \llbracket m \rrbracket^{**} \to \llbracket m \rrbracket^{**}$ be a permutation morphism. Then $S_{h(\alpha)}^{(k+1)} = S_{h(\beta)}^{(k+1)}$ and $h(\beta) = (h(\alpha) + z) \bmod m$ h*

Proof. The latter assertion follows from the fact that

$$
\begin{aligned}
h(\beta_{i,j}) &= (h(\alpha_{i,j} + z) \bmod m) \bmod m \\
&= ((h(\alpha_{i,j}) + z) \bmod m) \bmod m \\
&= (h(\alpha_{i,j}) + z) \bmod m.
\end{aligned}
$$

To prove the latter, we will assume that h is a $\langle p, q \rangle$-morphism, $p, q \geq 1$, and use the notation from Definition 9. There is a bijection ϕ matching subword occurrences in α of words from $S_\alpha^{(k)}$ to subword occurrences of the same words in β. We will define a bijection ϑ matching the occurrences of subwords of $h(\alpha)$ of order not greater than $k + 1$ to array subword occurrences of the same words in $h(\beta)$. Assume an occurrence of a $\langle r, s \rangle$-word γ, where $r + s \leq k + 1$ in $h(\alpha)$ at some position $I_\gamma \times J_\gamma$. For $\langle i, j \rangle \in \llbracket t \rrbracket \times \llbracket u \rrbracket$, let $I_{\gamma,i} = (ip + \llbracket p \rrbracket) \cap I_\gamma$, $J_{\gamma,j} = (jq + \llbracket q \rrbracket) \cap J_\gamma$, and $[i, j, \gamma] = I_{\gamma,i} \times J_{\gamma,j}$. Then $[i, j, \gamma]$ is the position of that part of the occurrence of the word γ, which belongs to the image $h(\alpha_{i,j})$. Let

$$
\begin{aligned}
I &= \{i' \in \llbracket t \rrbracket; [i', j', \gamma] \neq \varnothing \text{ for some } j' \in \llbracket u \rrbracket\}, \\
J &= \{j' \in \llbracket u \rrbracket; [i', j', \gamma] \neq \varnothing \text{ for some } i' \in \llbracket t \rrbracket\}.
\end{aligned}
$$

Observe that $[i', j', \gamma] \neq \varnothing$ for each $\langle i', j' \rangle \in I \times J$, since, otherwise, either $I = \varnothing$ or $J = \varnothing$. We will consider two cases.

1. There is at least one pair $\langle i, j \rangle$ such that $|[i, j, \gamma]| \geq 2$. Then $[I] + [J] \leq k$ and ϕ maps the position $I \times J$ of the occurrence of the word $\alpha_{I \times J} \in S_\alpha^{(k)}$ to some position $\phi(I \times J)$ of an occurrence of the word $\beta_{\phi(I \times J)'} = \alpha_{I \times J}$ in β. That means, each position $\langle i', j' \rangle \in \llbracket t, u \rrbracket$ such that $h(\alpha)_{[i', j', \gamma]} \neq \varnothing$ is mapped to a position $\langle i'', j'' \rangle \in \llbracket t, u \rrbracket$. Since the symbols at the matching positions of the words $\alpha_{I \times J}$ and $\beta_{\phi(I \times J)'}$ are identical, the symbols in $h(\alpha_{i', j'})$ belonging to γ can be found in precisely same positions within $h(\beta)_{i'', j''}$. Now the ϑ-image of the occurrence of γ in β is assembled following, in the reversed way, the decomposition of the occurrence of γ in α to the parts belonging to $h(\alpha_{i', j'})$.

2. For each pair $\langle i, j \rangle$, $\|[i, j, \gamma]\| \leq 1$. Take $\langle i, j \rangle \in I_\gamma \times J_\gamma$ and denote (by slightly overusing the notation of ϑ), $\vartheta(i) = i - (i \bmod p) + \pi_1^{-z}(i \bmod p)$, $\vartheta(j) = j - (j \bmod q) + \pi_2^{-z}(j \bmod q)$; $I'_\gamma = \{\vartheta(i) ; i \in I_\gamma\}$, $J'_\gamma = \{\vartheta(j) ; j \in J_\gamma\}$. Then we set $\vartheta(I_\gamma \times J_\gamma) = I'_\gamma \times J'_\gamma$. This mapping determines a position of some $\langle r, s \rangle$-word in β. We will show, that the subword occurring at this position is γ. Indeed,

$$\beta_{\vartheta(i),\vartheta(j)} = h\left(\beta_{i \operatorname{div} p, j \operatorname{div} q}\right)_{\pi_1^{-z}(i \bmod p), \pi_2^{-z}(j \bmod q)}$$
$$= h\left((\alpha_{i \operatorname{div} p, j \operatorname{div} q} + z) \bmod m\right)_{\pi_1^{-z}(i \bmod p), \pi_2^{-z}(j \bmod q)}$$
$$= (h\left((\alpha_{i \operatorname{div} p, j \operatorname{div} q} + z) \bmod m\right) - z) \bmod m$$
$$= h\left(\alpha_{i \operatorname{div} p, j \operatorname{div} q}\right)$$

The mapping ϑ is clearly a bijection, since, swapping the role of α and β one obtains the inverse mapping.

Theorem 13. *Let* $m \geq 1$ *and let* $h_0, \ldots, h_{k-1}$ *be a sequence of permutation morphisms,* $h_i : [\![m]\!]^{**} \to [\![m]\!]^{**}$, $i \in [\![k]\!]$. *Then*

$$S^k_{h_{k-1}(h_{k-2}(\cdots h_0(0)))} = S^k_{h_{k-1}(h_{k-2}(\cdots h_0(j)))}$$

for each $j \in [\![m]\!]$ *and*

$$h_{k-1}(h_{k-2}(\cdots h_0(j))) = (h_{k-1}(h_{k-2}(\cdots h_0(0))) + j) \bmod m.$$

Corollary 14. *Let* $h : [\![m]\!]^{**} \to [\![m]\!]^{**}$, $m, p, q \geq 1$, *be a permutation morphism,* $k \geq 0$. *Then* $S^k_{h^k(i)} = S^k_{h^k(j)}$.

Example 15. Consider the following three morphisms on $\{0, 1\}$.

$$h_0(0) = \begin{matrix} 0\ 1 \\ 1\ 0 \end{matrix}, \ h_0(1) = \begin{matrix} 1\ 0 \\ 0\ 1 \end{matrix}$$

$$h_1(0) = 0\ 1, \ h_1(1) = 1\ 0$$

$$h_2(0) = \begin{matrix} 1\ 0 \\ 1\ 0 \\ 0\ 1 \end{matrix}, \ h_2(1) = \begin{matrix} 0\ 1 \\ 0\ 1 \\ 1\ 0 \end{matrix}$$

Then

$$h_2(h_1(h_0(0))) = \begin{matrix} 1\ 0\ 0\ 1\ 0\ 1\ 1\ 0 \\ 1\ 0\ 0\ 1\ 0\ 1\ 1\ 0 \\ 0\ 1\ 1\ 0\ 1\ 0\ 0\ 1 \\ 0\ 1\ 1\ 0\ 1\ 0\ 0\ 1 \\ 0\ 1\ 1\ 0\ 1\ 0\ 0\ 1 \\ 1\ 0\ 0\ 1\ 0\ 1\ 1\ 0 \end{matrix}$$

The image of the symbol 1 is obtained by replacing 0 by 1 and 1 by 0. One can check (in rather painstaking way) that 3-spectra of these two words are identical. For example, both of them contain 36 occurrences of the $\langle 2, 2 \rangle$-word $\begin{matrix} 0\ 0 \\ 0\ 0 \end{matrix}$ and neither of them contains an occurrence of $\begin{matrix} 0\ 0 \\ 0\ 1 \end{matrix}$.

Let us note that the morphism h_0 in Example 15 may be considered to be a 2-dimensional generalization of the morphism of Thue-Morse.

The latter assertion of Theorem 13 implies that $h_{k-1}(h_{k-2}(\cdots h_0(j)))$ contains at some position the symbol 0 if and only if $h_{k-1}(h_{k-2}(\cdots h_0(0)))$ contains $(-j) \bmod m$ at the very same position. This fact, together with Theorem 7 and Theorem 13 yield.

Theorem 16. *Let* $m, p_0, \ldots, p_{k-1}, q_0, \ldots, q_{k-1} \geq 1$ *and let* $h_0, \ldots, h_{k-1}$ *be a sequence of permutation morphisms, where* $h_i : [\![m]\!]^{**} \to [\![m]\!]^{**}$, $i \in [\![k]\!]$, *is a* $\langle p_i, q_i \rangle$-*morphism. Then the partition of the set* $[\![p_0 \cdots p_{k-1}]\!] \times [\![q_0 \cdots q_{k-1}]\!]$ *into* m *parts, where the* j-*th part,* $j \in [\![m]\!]$, *consists of all positions in the word* $h_{k-1}(h_{k-2}(\cdots h_0(0)))$ *containing the symbol* j, *is a solution to the problem* $PTE^2_{m,k}$.

Example 17. Consider the sequence of morphisms from Example 15. The word $h_2(h_1(h_0(0)))$ induces the following solution to $PTE^2_{2,3}$.

$$A_0 = \{\, \langle 0,1 \rangle, \langle 0,2 \rangle, \langle 0,4 \rangle, \langle 0,7 \rangle, \langle 1,1 \rangle, \langle 1,2 \rangle, \langle 1,4 \rangle, \langle 1,7 \rangle,$$
$$\langle 2,0 \rangle, \langle 2,3 \rangle, \langle 2,5 \rangle, \langle 2,6 \rangle, \langle 3,0 \rangle, \langle 3,3 \rangle, \langle 3,5 \rangle, \langle 3,6 \rangle,$$
$$\langle 4,0 \rangle, \langle 4,3 \rangle, \langle 4,5 \rangle, \langle 4,6 \rangle, \langle 5,1 \rangle, \langle 5,2 \rangle, \langle 5,4 \rangle, \langle 5,7 \rangle \}$$

$$A_1 = \{\, \langle 0,0 \rangle, \langle 0,3 \rangle, \langle 0,5 \rangle, \langle 0,6 \rangle, \langle 1,0 \rangle, \langle 1,3 \rangle, \langle 1,5 \rangle, \langle 1,6 \rangle,$$
$$\langle 2,1 \rangle, \langle 2,2 \rangle, \langle 2,4 \rangle, \langle 2,7 \rangle, \langle 3,1 \rangle, \langle 3,2 \rangle, \langle 3,4 \rangle, \langle 3,7 \rangle,$$
$$\langle 4,1 \rangle, \langle 4,2 \rangle, \langle 4,4 \rangle, \langle 4,7 \rangle, \langle 5,0 \rangle, \langle 5,3 \rangle, \langle 5,5 \rangle, \langle 5,6 \rangle \}$$

with the sums

$$\sum_{\langle a_0, a_1 \rangle \in A_i} a_0^0 a_1^0 = 24, \sum_{\langle a_0, a_1 \rangle \in A_i} a_0^0 a_1^1 = 84,$$
$$\sum_{\langle a_0, a_1 \rangle \in A_i} a_0^1 a_1^0 = 60, \sum_{\langle a_0, a_1 \rangle \in A_i} a_0^1 a_1^1 = 210.$$

Example 18. Consider the following three morphisms on $\{0, 1, 2\}$.

$$h_0(0) = 2\,1\,0\ , \ h_0(1) = 0\,2\,1\ , \ h_0(2) = 1\,0\,2$$

$$h_1(0) = \begin{matrix} 1\,0 \\ 0\,2 \\ 2\,1 \end{matrix}, \ h_1(1) = \begin{matrix} 2\,1 \\ 1\,0 \\ 0\,2 \end{matrix}, \ h_1(2) = \begin{matrix} 0\,2 \\ 2\,1 \\ 1\,0 \end{matrix}$$

$$h_2 = h_0$$

Then

$$h_2(h_1(h_0(0))) = \begin{matrix} 2\,1\,0\,1\,0\,2\,1\,0\,2\,0\,2\,1\,0\,2\,1\,2\,1\,0 \\ 1\,0\,2\,0\,2\,1\,0\,2\,1\,2\,1\,0\,2\,1\,0\,1\,0\,2 \\ 0\,2\,1\,2\,1\,0\,2\,1\,0\,1\,0\,2\,1\,0\,2\,0\,2\,1 \end{matrix}$$

The word $h_2(h_1(h_0(0)))$ induces the following solution to $PTE^2_{2,3}$.

$$A_0 = \{\, \langle 0,2 \rangle, \langle 0,4 \rangle, \langle 0,7 \rangle, \langle 0,9 \rangle, \langle 0,12 \rangle, \langle 0,17 \rangle, \langle 1,1 \rangle, \langle 1,3 \rangle, \langle 1,6 \rangle,$$
$$\langle 1,11 \rangle, \langle 1,14 \rangle, \langle 1,16 \rangle, \langle 2,0 \rangle, \langle 2,5 \rangle, \langle 2,8 \rangle, \langle 2,10 \rangle, \langle 2,13 \rangle, \langle 2,15 \rangle \}$$

$$A_1 = \{\,\langle 0, 1\rangle\,, \langle 0, 3\rangle\,, \langle 0, 6\rangle\,, \langle 0, 11\rangle\,, \langle 0, 14\rangle\,, \langle 0, 16\rangle\,, \langle 1, 0\rangle\,, \langle 1, 5\rangle\,, \langle 1, 8\rangle\,,$$
$$\langle 1, 10\rangle\,, \langle 1, 13\rangle\,, \langle 1, 15\rangle\,, \langle 2, 2\rangle\,, \langle 2, 4\rangle\,, \langle 2, 7\rangle\,, \langle 2, 9\rangle\,, \langle 2, 12\rangle\,, \langle 2, 17\rangle\}$$

$$A_2 = \{\,\langle 0, 0\rangle\,, \langle 0, 5\rangle\,, \langle 0, 8\rangle\,, \langle 0, 10\rangle\,, \langle 0, 13\rangle\,, \langle 0, 15\rangle\,, \langle 1, 2\rangle\,, \langle 1, 4\rangle\,, \langle 1, 7\rangle\,,$$
$$\langle 1, 9\rangle\,, \langle 1, 12\rangle\,, \langle 1, 17\rangle\,, \langle 2, 1\rangle\,, \langle 2, 3\rangle\,, \langle 2, 6\rangle\,, \langle 2, 11\rangle\,, \langle 2, 14\rangle\,, \langle 2, 16\rangle\}$$

with the sums

$$\sum_{\langle a_0, a_1\rangle \in A_i} a_0^0 a_1^0 = 18, \sum_{\langle a_0, a_1\rangle \in A_i} a_0^0 a_1^1 = 153,$$
$$\sum_{\langle a_0, a_1\rangle \in A_i} a_0^1 a_1^0 = 18, \sum_{\langle a_0, a_1\rangle \in A_i} a_0^1 a_1^1 = 153.$$

5   Conclusion

We provided a construction based on composition of array morphisms leading to solutions to the two-dimensional Prouhet-Tarry-Escott problem. The construction can be easily generalized to higher dimensions. Though in the 1-dimensional case quite a wide class of morphisms can be used in such construction, in our approach highly regular morphisms had to be applied. We do not know whether this requirement can be relaxed. One may also wonder, whether some more analogy to the 1-dimensional case can be drawn. One such analogy is closure of the solution space under linear transformations.

Theorem 19. *Let $\{A_i\}_{i \in [\![n]\!]}$ be a solution to the Problem $PTE^2_{n,k}$, i.e.,*

$$\sum_{\langle a, b\rangle \in A_i} A_i\,(\langle a, b\rangle)\,a^r b^s,$$

for $0 \le r + s < k$, does not depend on i. Let $P_0, Q_0, P_1, Q_1 \in \mathbb{N}$. Let B_i, $i \in [\![n]\!]$ be the multiset, such that

$$B_i\,(\langle c, d\rangle) = \text{if } \langle c, d\rangle = \langle P_1 a + Q_1, P_2 b + Q_2\rangle \text{ for some } (a, b) \in A_i$$
$$\text{then } A_i\,(a, b) \text{ else } 0.$$

Then $\{B_i\}_{i \in [\![n]\!]}$ is a solution to the Problem $PTE^2_{n,k}$.

Proof.

$$\sum_{(c,d) \in A_i} B_i\,(\langle c, d\rangle) c^r d^s = \sum_{(a,b) \in A_i} A_i\,(\langle a, b\rangle)\,(P_1 a + Q_1)^r (P_2 b + Q_2)^s$$

$$= \sum_{t \in [\![r]\!]} \sum_{u \in [\![s]\!]} \binom{r}{t}\binom{s}{u} P_1^t Q_1^{r-t} P_2^u Q_q^{s-u} \sum_{\langle a,b\rangle \in A_i} A_i\,(\langle a, b\rangle)\,a^t b^u,$$

for $0 \le r + s < k$, which does not depend on i, since $0 \le t + u \le r + s < k$.

We believe that construction of other solutions, based on the radix notation of the elements of the solution obtained here may be possible, in analogy to the Theorem of Lehmer [14],[6]. Solutions to the multidimensional Prouhet-Tarry-Escott problem within an arbitrary semiring may be considered, as well.

References

1. Allouche, J.P., Shallit, J.O.: Automatic sequences - theory, applications, generalizations. Cambridge University Press (2003)
2. Alpers, A., Tijdeman, R.: The two-dimensional Prouhet-Tarry-Escott problem. J. Number Theory 123(2), 403–412 (2007)
3. Berstel, J., Perrin, D.: The origins of combinatorics on words. Eur. J. Comb. 28(3), 996–1022 (2007)
4. Černý, A., Gruska, J.: Modular trellises. In: The Book of L, pp. 45–61. Springer, Heidelberg (1986)
5. Černý, A.: Generalizations of Parikh mappings. RAIRO Theor. Inform. Appl. 44(2), 209–228 (2010)
6. Černý, A.: On Prouhet's solution to the equal powers problem. Theoretical Computer Science 491(17), 33–46 (2013)
7. Černý, A., Gruska, J.: Modular real-time trellis automata. Fundamenta Informaticae IX, 253–282 (1986)
8. Choffrut, C., Karhumäki, J.: Combinatorics of words. In: Rozenberg, G., Salomaa, A. (eds.) Handbook of Formal Languages, pp. 329–438. Springer (1997)
9. Dudík, M., Schulman, L.J.: Reconstruction from subsequences. J. Comb. Theory Ser. A 103, 337–348 (2003)
10. Frougny, C., Vuillon, L.: Coding of two-dimensional constraints of finite type by substitutions. Journal of Automata, Languages and Combinatorics 10(4), 465–482 (2005)
11. Kalashnik, L.: The reconstruction of a word from fragments. In: Numerical Mathematics and Computer Technology. Akad. Nauk Ukrain. SSR Inst. Mat., pp. 56–57 (1973) (in Russian)
12. Knuth, D.E.: The art of computer programming, 2nd edn. Seminumerical algorithms, vol. 2. Addison-Wesley (1981)
13. Krasikov, I., Roditty, Y.: On a reconstruction problem for sequences. J. Comb. Theory, Ser. A 77(2), 344–348 (1997)
14. Lehmer, D.H.: The Tarry-Escott problem. Scripta Mathematica 13, 37–41 (1947)
15. Lothaire, M.: Combinatorics on words. Cambridge University Press (1997)
16. Milner, R.: The spectra of words. In: Middeldorp, A., van Oostrom, V., van Raamsdonk, F., de Vrijer, R. (eds.) Processes, Terms and Cycles: Steps on the Road to Infinity. LNCS, vol. 3838, pp. 1–5. Springer, Heidelberg (2005)
17. Mozes, S.: Tilings, substitution systems and dynamical systems generated by them. Journal d'Analyse Mathamatique 53, 139–186 (1989)
18. Prouhet, E.: Mémoire sur quelques relations entre les puissances des nombres. C.R. Acad. Sci. Paris 33, 255 (1851)
19. Rigo, M., Maes, A.: More on generalized automatic sequences. Journal of Automata, Languages and Combinatorics 7(3), 351–376 (2002)

Succinct Encodings of Graph Isomorphism

Bireswar Das[1], Patrick Scharpfenecker[2], and Jacobo Torán[2]

[1] Indian Institute of Technology
Gandhinagar, India
bireswar@iitgn.ac.in
[2] Institute of Theoretical Computer Science
University of Ulm, Germany
{patrick.scharpfenecker,jacobo.toran}@uni-ulm.de

Abstract. It is well known that problems encoded with circuits or formulas generally gain an exponential complexity blow-up compared to their original complexity.

We introduce a new way for encoding graph problems, based on CNF or DNF formulas. We show that contrary to the other existing succinct models, there are examples of problems whose complexity does not increase when encoded in the new form, or increases to an intermediate complexity class less powerful than the exponential blow up.

We also study the complexity of the succinct versions of the Graph Isomorphism problem. We show that all the versions are hard for PSPACE. Although the exact complexity of these problems is not known, we show that under most existing succinct models the different versions of the problem are equivalent. We also give an algorithm for the DNF encoded version of GI whose running time depends only on the size of the succinct representation.

Keywords: Complexity, Succinct, Graphisomorphism, CNF, DNF.

1 Introduction

In many applications, like VLSI design or computer aided verification, graphs and other combinatorial structures present many regularities allowing to encode them in a compact way, much more succinctly than for example the usual adjacency matrices or lists. Galperin and Wigderson [8] studied for the first time the complexity of several graph decision problems when the adjacency relation is presented as a Boolean circuit. The hope was that the succinctly encoded instances contain enough structure to make the considered problem easier to solve. They observed however an exponential blow-up in the complexity of these problems, showing that the regularities that allow a succinct representation, do not really help in order to solve the problem. Several extensions to this work [11,2,15,16] showed that the exponential complexity blow-up is the general behavior, by proving upgrading theorems for several reducibilities: if a problem is complete with respect to certain low level reducibility for a complexity class, then the succinct version of the problem with its instances encoded as a Boolean

A.-H. Dediu et al. (Eds.): LATA 2014, LNCS 8370, pp. 285–296, 2014.

circuit is complete for the corresponding exponentially higher class, with respect to polynomial time reducibilities. Other extensions of the original work concentrated in more restricted encodings of the input instances, like Boolean formulas [15] or ordered binary decision diagrams (OBDDs) [4],[16]. Even if these representation models are more restricted than the Boolean circuits, in these works the same exponential blow-up in the complexity of the succinct version of the problem is shown.

Very recently, the class of NC^0 circuits was considered as a succinct model for encoding problem instances. In [9] the authors prove a blow-up result for the complexity of the Satisfiability problem encoded this way.

We introduce here the use of Boolean formulas in conjunctive (CNF) or disjunctive normal form (DNF) in order to encode graphs. This is a considerable restriction with respect to the model based on general Boolean formulas from Veith [16]. The size of these representations is closely related to the bi-clique decomposition of the graphs [5]. We show that in this limited model an upgrade theorem is not possible since there are examples of problems whose complexity does not increase when the input graphs are given in the form of a DNF formula, while it presents an exponential blow-up when the dual representation is considered. In other cases, the complexity of the succinct version of the problem is neither that of the original problem nor exponentially higher. For example the Dominating Set problem becomes PP-complete when the input graphs are encoded as a DNF formula and it becomes complete for NEXP when a CNF formula is considered, while the connectivity problem for directed graphs remains NL complete with a DNF encoding and becomes complete for PSPACE with a CNF encoding. This is a new phenomenon in the area of succinct representations, since in all the existing models and examples the complexity of the problem blows up exponentially in the succinct version.

The graph isomorphism problem, GI, asks whether there is a bijection between the nodes of two given graphs preserving the adjacency relationship. The problem has been extensively studied (see e.g [10]) because of its graph theoretic importance, but also because it is one of the few problems in NP whose exact complexity is unknown: the problem is not known to be solvable in polynomial time but also it is not expected to be NP-complete. We study here the complexity of the succinct version of this problem considering several input representations. The motivation for this is twofold. On the one hand the complexity of the succinct version of GI might shed some light to the complexity of the standard version of the problem. On the other hand, algorithms for succinct GI that go beyond the trivial decoding of the graph and then the application of an algorithm for the standard problem to it, would be very useful in areas like computer aided verification. Concerning the first goal, since in all the considered models the graph encoded in the input is at most exponentially larger that the input itself, the succinct version of GI is in NEXP. We obtain in Section 6 for all the succinct encoding methods discussed here a hardness result for succinct-GI for the class PSPACE. This is done with the help of the existing upgrade theorems and the hardness results for the standard version of GI.

Although the exact complexity of the succinct versions of GI is still unknown, we prove that in most of the encoding models there is no difference in the complexity of the succinct problem. We show that cir(GI), cnf(GI) and dnf(GI) are equally powerful by giving polynomial time reductions between all these problems. Additionally, we reduce obdd(GI) to dnf(GI). This contrasts with the computational power of the models, since it is well known that OBDDs and formulas can be exponentially larger than Boolean circuits computing certain functions, and that CNF formulas and OBDD's cannot simulate each other without super-polynomially increasing their size.

Concerning the second point in the motivation, we give an algorithm for cnf(GI) (or for dnf(GI)) whose complexity depends solely on the size of the encoding formulas. Based on kernelization methods from the area of parametrized algorithms [5], we present an algorithm that on input two CNF formulas F_1 and F_2 with $2n$ variables each and at most s clauses, encoding graphs of size 2^n, decides whether the encoded graphs are isomorphic in time $O(2^{\sqrt{s2^{O(s)}}})$. This presents an improvement over the straightforward method decoding the graphs and then applying the fastest known isomorphism algorithm to them, in the cases in which s is smaller than n.

The rest of this paper is organized as follows: after a preliminary section explaining our notation and basic definitions, we introduce in Section 3 the CNF and DNF succinct encodings and show some examples on how the complexity of the succinct versions can vary. Section 4 shows the equivalence for encodings in case of the GI problem. Based on these results, we present in Section 5 an algorithm for graphs encoded with DNF formulas. Finally, we show in Section 6, that all the considered succinct versions of GI, obdd(GI), cnf(GI), dnf(GI) and cir(GI) are hard for PSPACE.

2   Preliminaries

We refer the reader to standard textbooks for basic definitions and notation in complexity theory including complexity classes, reductions and graphs. Building on that, we present some further definitions.

$\leq_m^p$ denotes the polynomial time many-one reduction. With $\leq^{LT}$, we denote a many-one reduction defined by a function f so that computing the i-th bit of $f(x)$ (denoted by $f(x)_i$) can be done in logarithmic time in $|x|$. For this we consider a machine model with direct access to the input. This kind of machine has a special query tape, which on input a position, outputs the bit written in the corresponding input position. This is a very weak type of reduction since it has very limited access to the input x [2]. $\leq^{qfr}$ denotes the quantifier free reduction. For space reasons we refer the reader to [16] for a description of this type of reduction.

A Boolean circuit or formula $C(x, y)$ (where x and y are variable vectors) can be interpreted as a succinct encoding of a graph $G = (V, E)$ (or any other structure). If $|x| = |y| = n$, we consider that $C(x, y)$ represents a directed graph on the set of vertices $V = [2^n]$ and edges defined by $C(x, y) = 1$ iff $(x, y) \in E$. Note that the encoded graph has exponential size in n. We say that C encodes the graph G_C. Also formulas in CNF and DNF can be encoded in a similar way and we consider F_C, to be the CNF formula with $2^{|x|}$ literals and $2^{|y|}$ clauses with $C(x, y) = 1$ iff literal x is in clause y (similarly for formulas in DNF).

We will consider the input model in which C is a CNF or DNF formula. Note that every graph can be encoded with a polynomially sized DNF formula with each edge encoded as a single implicant. A CNF encoding is similar with each non-edge encoded as a single clause. This trivial observation shows that the problems encoded as CNF or DNF formulas cannot be easier than the original problems, for example, the Dominating Set problem for graphs encoded as CNF formulas is hard for NP.

Looking at graphs encoded as DNF or CNF formulas, we note that in the first case, each implicant or term adds a directed biclique to the total graph (a biclique is a complete bipartite graph). The source side of the biclique is defined by the set of vertices whose labels satisfy the x part of the implicant, while the target side of the biclique is defined by the vertices satisfying the part of the y variables. Therefore a graph encoded by a DNF with m implicants can be decomposed as the union m bicliques.

In the case of a CNF formula, each clause subtracts a directed biclique from the complete graph K_{2^n}. Every clause removes all edges between the set of x and y nodes falsifying the clause.

Another class of succinct encodings considered in the literature [4,16] is that defined by ordered binary decision diagrams, OBDDs. A binary decision diagram is a directed, acyclic, rooted graph. Vertices are labeled with input variables x_i and edges are labeled with 0 or 1. Every vertex (except the two sinks) has two outgoing edges and they have different labels. There are two special vertices without successors, denoted 0 and 1, denoting the Boolean values true and false. A BDD O describes a Boolean function f over a set of variables $x_1, \ldots, x_n$ as follows: beginning at the root, in each vertex labeled with a variable the outgoing edge labeled with the value of this variable is followed, until the 0 or the 1 vertex is reached. This represents the value of the function. O is an OBDD if there is a permutation π on the set of variables such that in all the paths from the root to one of the special vertices the variables are consistent with the order defined by π.

Two graphs G, H are isomorphic, (represented by $G \cong H$) iff there is a bijection $\pi : V_G \to V_H$ such that for all pairs of vertices $u, v \in V_G$, $(u, v) \in E_G$ iff $((\pi(u), \pi(v)) \in E_H$ holds. The decision problem, given two graphs, determine if they are isomorphic, is denoted as GI.

cir(GI), cnf(GI), dnf(GI) and obdd(GI) are the succinct versions of the isomorphism problem when the graphs are encoded respectively by circuits, CNF, DNF formulas or OBDDs.

3   Succinct CNF and DNF Encodings

We present in this section CNF and DNF encodings of several well known problems, showing that in some cases the inputs given in this way change the complexity of the problems while in others the complexity of the original problem is preserved. Figure 1 summarizes our results. For space reasons we omit some of the proofs in this version of the paper.

Problem	CNF encoding	DNF encoding
Dominating Set	NEXP-complete	PP-complete
CNF-SAT	NEXP-complete	NP-complete
DNF-TAUT	coNEXP-complete	coNP-complete
STCONN	PSPACE-complete	NL-complete

Fig. 1. Complexity differences between DNF and CNF encodings

Theorem 1. *dnf(Dominating Set) is PP-complete.*

Proof. The hardness proof follows by standard methods and it is omitted. We show that dnf(Dominating Set) is in the class PP. For this we sketch an algorithm that on input a DNF formula F and a number k decides if the graph encoded by the formula contains a dominating set of size $\leq k$. The algorithm works in polynomial time with the help of non-adaptive queries to PP. Since PP is closed under truth-table reductions [6], this proves the result. Because we are dealing with directed graphs, a dominating set consists of all vertices with in-degree 0 together with a subset of the vertices dominating all vertices with in-degree greater than 0. The key observation is that this second subset cannot be larger than m, the number of implicants in F. Recall that an implicant contains an x and y part where the vertices consistent with the x literals are the source of the defined edges. Therefore taking at most one vertex satisfying the x part in each of the m implicants is enough to dominate all vertices with in-degree greater than 0. We consider the list of all possible pairs (i, j) with $0 \leq j \leq m$ and $i = k - j$ and ask two queries for each pair: 1) is the number of vertices with in-degree 0 at most j? and 2) is there a subset of vertices of in-degree at least 1, of size $\leq i$ and such that the vertices defined by the y part in every implicant is dominated by a vertex in the subset or by a vertex with in-degree 0? It is not hard to see that both are PP queries. We can conclude that there is a dominating set of size at most k if and only if, for at least one of the m pairs both queries are answered positively. This defines a truth-table reduction.   □

Using the the next result for cnf(CNF-SAT) and the standard reduction from SAT to Dominating Set, it is not hard to see that cnf(Dominating Set) is also NEXP-complete.

Theorem 2. *cnf(CNF-SAT) is NEXP-complete.*

Proof. It is clear that the problem is in NEXP. For showing the hardness we use a recent result from Jahanjou, Miles and Viola [9]. There it is shown that the satisfiability problem for formulas in 3-CNF, when encoded with polynomially many NC^0 functions is NEXP-complete. In their setting the formula F being tested for satisfiability is encoded in a way in which an NC^0 function f_k gets as input a clause index from F and computes the k-th bit of the three literals contained in that clause. We reduce this problem to cnf(CNF-SAT). For this we make a transformation between both types of encodings. We use a single formula encoding the literal-clause relation of F instead of a polynomial amount of NC^0 functions computing the bit encoding of a clause. For this we first transform each of the circuits computing the f_k functions into three NC^0 circuits, with one output bit each, computing $f_{i,j}$ for the j-th bit of the i-th literal ($1 \leq i \leq 3$). Each such function depends only on constant many input bits and can therefore be computed by a constant size CNF. The following circuit computes the literal-clause relation for a clause encoded in the x variables and a literal encoded in the y variables.

$$\bigvee_{i \in \{1,2,3\}} \bigwedge_{j \leq |y|} f_{i,j}(x) = y_j$$

Since $f_{i,j}$ can be represented by a constant size formula, $f_{i,j} \leftrightarrow y_j$ can also be expressed by a constant size CNF formula. The conjunction of these $|y|$ formulas (one for each j) is still a CNF. Finally transforming the disjunction of the 3 resulting formulas (one for each i) into CNF has size $O(|y|^3) = O(poly(n))$. This completes the reduction. □

4   Succinct Encodings of GI

In this and the following sections we concentrate on the complexity of the different succinct versions of GI. We start by investigating the relation between DNF, CNF, circuit- and OBDD encodings for this problem. Theorem 3 states our results. As described in the preliminaries, we have chosen to use encodings for directed graphs. This is not a restriction for studying the complexity of GI since both versions of the problem, for directed or undirected graphs, are equivalent.

Theorem 3. *dnf(GI) $\equiv^p_m$ cnf(GI) $\equiv^p_m$ cir(GI) and obdd(GI) $\leq^p_m$ cnf(GI).*

Proof. Obviously, cnf(GI) $\leq^p_m$ cir(GI), since a Boolean formula is a restricted version of a circuit. The equivalence between dnf(GI) and cnf(GI) follows from the observation, that two graphs are isomorphic iff their complementary graphs are isomorphic. Given a CNF (DNF) formula encoding a graph, the negation of the formula can be easily written as a DNF (CNF) formula and encodes the complementary graph. More interesting is the proof of cir(GI) $\leq^p_m$ cnf(GI).

Given two circuits $C(x,y)$ and $C'(x,y)$ with $|x| = |y| = n$, encoding two graphs $G = G_C$ and $H = G_{C'}$ on the vertex set $V = [2^n]$, we create two formulas F and F' encoding two new graphs $G' = G_F$ and $H' = G_{F'}$ such that

$G \cong H \Leftrightarrow G' \cong H'$. We give first an intuitive description of our construction of F. We use the so called Tseitin transformation from circuits to satisfiability equivalent CNF formulas (see for example [12]). In this transformation a Boolean circuit C is transformed into a satifiability equivalent formula F_C by introducing for each each gate g in C a new variable z_g and a small set of clauses expressing the value of the gate for a certain input. For example if gate g is an OR gate with the input gates e and f we add clauses expressing the subformula $z_g \leftrightarrow (z_e \vee z_f)$. The transformation also adds the single variable clause z_{output} for the output gate of the circuit. It should be clear that C is satisfiable if an only if F_C is satisfiable.

Going back to our construction, we can get from $C(x, y)$ a formula $F_1(x, y, z)$ which encodes the structure and evaluation of the gates of C on input x and y. Variable vector z contains a bit for each gate in C. The satisfying assignments for $F_1(x, y, z)$ consists of values for the x and y variables satisfying $C(x, y)$ plus further assignment values for the z variables with the value of the corresponding gate in C on input x, y. z contains $2n$ additional bits (at the beginning) containing a copy of x and y. This implies that for each satisfying assignment x, y for C, there is a unique z such that $F(x, y, z) = 1$. Moreover, no other satisfying assignment x', y' shares this particular z. If C has s gates, the new formula has linear size in s, is in 3-CNF, and can be interpreted as an encoding for a hypergraph G_1 on the set of vertices $[2^{2n+s}]$ that contains only hyperedges of degree 3. In a second step, these hypergraphs are transformed into standard graphs.

We now give a more detailed construction of F. F includes a constant number of clauses for each gate i in C (represented as D_i in the following formula, encoding the evaluation of this gate. We also separate the original vertices encoded in the x and y assignments from the new z vertices by forcing the x, y vertices to begin with 0 and the z vertices with 1. This can be done with one additional variable. The succinct formula model assumes that all vertices are encoded with the same number of bits. To achieve this, we pad the x and y vectors with $(s + 2n + 1) - (n + 1) = s + n$ zero bits. The last line of the following formula enforces that in each satisfying assignment, the first $2n$ bits of z contain the values of x and y at the beginning, making it unique.

$$F(x, y, z) = D_1 \wedge \ldots \wedge D_s$$
$$\wedge (x_0 = 0) \wedge (y_0 = 0) \wedge (z_0 = 1)$$
$$\wedge \bigwedge_{i=n+1}^{s+2n+1} (x_i = y_i = 0)$$
$$\wedge \bigwedge_{i=1}^{n} (x_i = z_i) \wedge \bigwedge_{i=2n+1}^{2n} (y_i = z_i)$$

Let G_1 be the hypergraph encoded by $F(x, y, z)$ (analogously for H_1). We claim:

Claim: $G \cong H \Leftrightarrow G_1 \cong H_1$

Proof: Clearly, if $G \cong H$, then $G_1 \cong H_1$ since an isomorphism between G and H can be extended to map the z vertices according to the unique hyperdedge they belong to.

Conversely, suppose $G_1 \cong H_1$ via an isomorphism ρ. Observe that all (non isolated) z vertices belong to exactly one hyperedge of 3 nodes. If ρ maps z

vertices to other z vertices then ρ defines an isomorphism between G and H. Suppose that there is a z vertex being mapped to one of the x vertices. Then x belongs to a unique hyperedge. We look at two cases:

- Both neighbors of x belong to a unique hyperedge. Then these three vertices define an isolated hyperedge. ρ can be easily modified to another isomorphism respecting the z vertices, by mapping z to the z neighbor of x. This also defines an isomorphism between G and H.
- Only one of the neighbors of x belongs to a unique hyperedge. Then x is connected to z and y of degree > 2. In the original graphs, x only had one neighbor (y). Again, ρ defines an isomorphism between G and H by swapping the roles of x and z.

$\square$

In a second step we create two standard graphs G_2 and H_2 such that $G_1 \cong H_1 \Leftrightarrow G_2 \cong H_2$. The vertex set of these graphs is the set of assignments for the variable vectors u, x, y, z with $|u| = 1$. Each such vertex encodes a hyperedge (x, y, z) in the previous construction and an additional bit. The formula $F^*(x^*, y^*) = F'(x^*, y^*) \vee F''(x^*, y^*)$, made of the following two subformulas, implements this transformation.

$$F'(uxyz, vx'y'z') = F(x, y, z) \wedge (u = 0) \wedge (v = 1)$$
$$\wedge \bigwedge_{i=0}^{|x'|-1} x'_i = y'_i = 0$$
$$\wedge (z' = y \vee z' = z)$$

$$F''(vx'y'z', uxyz) = F(x, y, z) \wedge (u = 0) \wedge (v = 1)$$
$$\wedge \bigwedge_{i=0}^{|x'|-1} x'_i = y'_i = 0$$
$$\wedge (z' = x \vee z' = z)$$

Note that u and v are single bit variables. F^* encodes a graph that is the union of two vertex sets. The first one, encoded with a leading $u = 0$ encodes the set of all degree 3 hyperedges. The second one is the set of all vertices in G_F. These are forced to begin with 1 and are padded with zeroes.

F' adds edges from a hyperedge vertex $\{x, y, z\}$ to the vertices y and z. Similarly F'' adds edges from x and z to the hyperedge $\{x, y, z\}$. This encodes the directed edge (x, y) in G_F.

F^* is a CNF formula. The whole construction from C to F^* only needs polynomial time. Applying this transformation to C and C' (using some additional dummy gates to get equal circuit sizes) gives us the formulas F and F' satisfying for $G' = G_F$ and $H' = G_{F'}$

$$G \cong H \Leftrightarrow G' \cong H'$$

The statement $\mathrm{obdd(GI)} \leq_m^p \mathrm{cnf(GI)}$ follows from the fact that an OBDD for a Boolean function can be transformed into a polynomial size Boolean circuit for the function, with the above reductions from $\mathrm{cir(GI)}$ to $\mathrm{cnf(GI)}$. $\square$

5   An Algorithm for dnf(GI)

Using the characterization for DNF encoded graphs as the union of bicliques, we give an algorithm for the succinct version of GI when the input graphs are encoded as DNF formulas. The running time of the algorithm depends on the number of implicants in the input.

To achieve this, we extend a kernelization technique explained in [5]. Kernelization is a common method in fixed parametrized algorithms. It basically consists in reducing problem instances to a small part of it from which the complexity of the input can still be recovered. This is called a kernel. In some cases algorithms can be designed having polynomial running time for performing the kernelization plus an exponential running time on the size of the kernel.

Looking at graph isomorphism, our kernelization step consists in merging together all the vertices that have exactly the same set of neighbors. We consider these nodes to be equivalent. Doing this in both graphs has to be done with some care, because some information might be lost in this way, namely the number of merged nodes. For example, if one graph contains one vertex connected to vertices a, b and c and the second graph contains two such vertices, the graphs may seem isomorphic after merging these nodes. To avoid this problem, we add some coloring or labeling encoding the number of vertices that are merged. In the previous example, we would give the vertex in the kernel in the first graph a certain color and in the second graph a different one. These colors can be replaced with gadgets. Such a kernelization preserves isomorphism and can be constructed in time polynomial in the size of the given graphs.

If our graphs are unions of s bicliques, we assign to every vertex v a vector $b(v)$ in $\{0, 1, 2, 3\}^s$ where $b(v)_i$ is 1 if v is in the source side of the i-th biclique, meaning that v has outgoing edges to all vertices in the target side of the i-th DNF term. Similar, $b(v)_i = 2$ if v is in the target side of i, $b(v)_i = 3$ if it is both sides and $b(v)_i = 0$ if it is in none of them. Note that, in contrast to [5], we have four cases since we allow a node to be in the source and target side. There are at most 4^s possible vectors. It should be clear that after this first kernelization step the vertices with the same vector are equivalent. There may still be however equivalent vertices left with different vectors, which can create problems when dealing with the isomorphism of two graphs. Consider for example the case in which two different terms of a DNF formula define two different sets of source vertices but with the same target set. The source sets would have different vectors although they might be equivalent. But this and similar problems are easy to detect. For all pairs of different vertices remaining after the first part of the kernelization, we check if their in- and out-going edges are the same. This is done comparing the union of all relevant source sides and the union of the relevant target sides. If they are the same, we merge these two vectors. This comes at a cost of $(4^s)^2 = 4^{O(s)}$ steps but only decreases the size of the kernel. At last, the kernelization merges all vertices with the same vector and colors this class of nodes with its size.

The kernelization together with an algorithm for GI can be used to compute dnf(GI).

Theorem 4. *Given two DNF formulas F_1 and F_2 with $2n$ inputs each and with at most s implicants, encoding graphs G_1 and G_2 on 2^n vertices, isomorphism for these graphs can be tested in time $2^{\sqrt{s2^{O(s)}}}$.*

Proof. We describe the algorithm and prove its running time and correctness. On input two DNF formulas we apply the explained kernelization to them obtaining two graphs of size at most $4^s = 2^{O(s)}$ in time $2^{O(s)}$. Using the isomorphism test of Babai and Luks (see [1]), which runs in time $2^{\sqrt{n}\log n}$ on graphs with n vertices, it can be tested whether the kernels are isomorphic. This clearly gives a correct result. The running time is $2^{O(s)}$ for the kernelization as well as $2^{\sqrt{s2^{O(s)}}}$ for the isomorphism test. $\qquad\square$

The same algorithm works for CNF formulas by first negating the formula and then applying this algorithm to the resulting DNF formula. If $s \in o(n)$, this provides a better upper bound than obtaining an explicit representation of the graphs (in time $2^n poly(n)$) and applying then the algorithm from Babai and Luks.

Note that this algorithm can be also transformed into an algorithm for cir(GI) by first using our transformation from cir(GI) to dnf(GI) and then applying the kernelization and the isomorphism test. But since the size of the computed DNF formula is $O(c^2)$ where $c \geq n$ is the circuit size, the algorithm would have a worse running time than decoding the graphs from their succinct representations and applying then an isomorphism algorithm to them.

6   Hardness Results

We show in this section that the circuit, DNF, CNF and OBDD succinct versions of GI are hard for PSPACE under polynomial time many-one reducibilities. For this we use the following Conversion Lemma relating the complexity of standard and succinct encodings of the same problem. The Lemma for circuits appeared in [2] and was improved in [3]. The version for OBDD's is from [16].

Lemma 5. *Let $A, B \subseteq \{0,1\}^*$. If $A \leq^{LT} B$, then $cir(A) \leq^p_m cir(B)$. If $A \leq^{qfr} B$ then $obdd(A) \leq^p_m obdd(B)$.*

Consider cir(USTCONN), the succinct version of the undirected reachability problem. It is known that this problem is PSPACE-complete under polynomial time many-one reducibilities [2]. Moreover, USTCONN is AC^0 reducible to the complement of GI [13,14]. We show here that in fact, this reduction can be done in logarithmic time.

Theorem 6. *USTCONN $\leq^{LT} \overline{GI}$.*

Proof. Let $G = (V, E)$ be an undirected graph with $|V| = n$ and two designated vertices $s, t \in V$. Consider a graph $G' = G_1 \cup G_2$ where G_1 and G_2 are two copies of G, and for a vertex $v \in V$ let us call v_1 and v_2 the copies of v in G_1 and G_2 respectively. Furthermore, G' is defined to have vertex t_1 labeled with color 1, (the rest of the vertices have color 0). It is not hard to see that there are not any paths from s to t in G if and only if there is an automorphism φ in G' mapping s_1 to s_2. The question of whether there is an automorphism in G' with the mentioned properties, can in turn be reduced to GI by considering the pair of graphs $(\hat{G}, \hat{H})$ where $\hat{G}$ are $\hat{H}$ are copies of G' but with s_1 marked with a new color 2 in $\hat{G}$ and s_2 marked with the same color in $\hat{H}$. It holds that $G \in \text{USTCONN}$ iff $(\hat{G}, \hat{H}) \in \overline{\text{GI}}$. It is only left to show that the constructions of the graphs $\hat{G}$ and $\hat{H}$ as well as the color labels in the reduction can be done in logarithmic time. A way to do this, is to consider that graph $\hat{G}$ has $4n$ vertices (the construction of $\hat{H}$ is completely analogous). For each vertex v in V we consider the four vertices abv with $a, b \in \{0, 1\}$. The vertices $00v$ and $11v$ define two exact copies of G, while the vertices $01v$ and $10v$ are used for the color labels of s_1 and t_1. For this we can add edges connecting all the $01v$ vertices between themselves forming a clique and to $00s_1$ and connecting all the $10v$ vertices to $00t_1$ (and not between themselves to distinguish them from the 01 vertices). With this construction, one bit of the adjacency matrix on $\hat{G}$ can be computed in logarithmic time (on input G and the position in the matrix) since at most one position in the adjacency matrix of G is needed for this.   □

This result, together with Lemma 5 imply that cir(GI) is PSPACE-hard with respect to the polynomial time many-one reducibility. Theorem 3 implies that even DNF encoded GI is hard for PSPACE.

Corollary 7. *dnf(GI) is hard for PSPACE.*

Theorem 6 can in fact be strengthened for the case of a quantifier free reduction [16] (for space reasons, this is omitted in this version). Using the second part of Lemma 5 this proves that also the OBDD version of the problem is hard for PSPACE.

Corollary 8. *obdd(GI) is hard for PSPACE.*

There are stronger hardness results for GI than the one used here [13]. However, applying the Conversion Lemma to these we do not obtain better hardness results for cir(GI). The translation of these results would imply hardness for the class #PSPACE, but this class is known to coincide with FPSPACE [7].

7   Conclusions and Open Problems

We introduced the new CNF and DNF models for encoding problems succinctly. We showed that contrary to the other existing models, there are examples for which the complexity of succinct version of the problem does not blow up exponentially. The size of the graph encoding in the new models are related to certain graph decompositions. It would be interesting to study further examples of graph problems encoded in these models, trying to obtain algorithms acting directly on the succinct versions as we did for the case of GI.

We also studied the complexity of succinct-GI in the different models and proved that although the complexity of this problem is not well understood yet, the versions for GI under DNF, CNF and even circuit encodings are all equivalent. A question that remains open is whether the OBDD version of GI is also equivalent to the other versions or easier.

References

1. Babai, L., Luks, E.M.: Canonical labeling of graphs. In: Proceedings of the Fifteenth Annual ACM Symposium on Theory of Computing, STOC 1983, pp. 171–183. ACM Press, New York (1983)
2. Balcázar, J.L., Lozano, A., Torán, J.: The Complexity of Algorithmic Problems on Succinct Instances. Computer Science, Research and Applications. Springer US (1992)
3. Eiter, T., Gottlob, G., Mannila, H.: Adding disjunction to datalog (extended abstract). In: Proceedings of the Thirteenth ACM SIGACT-SIGMOD-SIGART Symposium on Principles of Database Systems, PODS 1994, pp. 267–278. ACM Press, New York (1994)
4. Feigenbaum, J., Kannan, S., Vardi, M.Y., Viswanathan, M.: Complexity of Problems on Graphs Represented as OBDDs. In: Meinel, C., Morvan, M. (eds.) STACS 1998. LNCS, vol. 1373, pp. 216–226. Springer, Heidelberg (1998)
5. Fleischner, H., Mujuni, E., Paulusma, D., Szeider, S.: Covering Graphs with Few Complete Bipartite Subgraphs. In: Arvind, V., Prasad, S. (eds.) FSTTCS 2007. LNCS, vol. 4855, pp. 340–351. Springer, Heidelberg (2007)
6. Fortnow, L., Reingold, N.: PP Is Closed under Truth-Table Reductions. Information and Computation 124(1), 1–6 (1996)
7. Galota, M., Vollmer, H.: Functions computable in polynomial space. Information and Computation 198(1), 56–70 (2005)
8. Galperin, H., Wigderson, A.: Succinct representations of graphs. Information and Control 56(3), 183–198 (1983)
9. Jahanjou, H., Miles, E., Viola, E.: Local reductions (2013)
10. Köbler, J., Schöning, U., Torán, J.: The graph isomorphism problem: its structural complexity. Birkhauser (August 1994)
11. Papadimitriou, C.H., Yannakakis, M.: A note on succinct representations of graphs. Information and Control 71(3), 181–185 (1986)
12. Schöning, U., Torán, J.: The Satisfiability Problem: Algorithms and Analyses. Lehmanns Media (2013)
13. Toran, J.: On the hardness of graph isomorphism. In: Proceedings of the 41st Annual Symposium on Foundations of Computer Science, pp. 180–186 (2000)
14. Torán, J.: Reductions to Graph Isomorphism. Theory of Computing Systems 47(1), 288–299 (2008)
15. Veith, H.: Languages represented by Boolean formulas. Information Processing Letters 63(5), 251–256 (1997)
16. Veith, H.: How to encode a logical structure by an OBDD. In: Proceedings of the 13th IEEE Conference on Computational Complexity, pp. 122–131. IEEE Comput. Soc. (1998)

Extremal Combinatorics of Reaction Systems[*]

Alberto Dennunzio[1], Enrico Formenti[2], and Luca Manzoni[2]

[1] Dipartimento di Informatica, Sistemistica e Comunicazione
Università degli Studi di Milano - Bicocca
Viale Sarca 336, 20126 Milano Italy
dennunzio@disco.unimib.it
[2] Laboratoire I3S, Université Nice Sophia Antipolis
CS 40121, 06903 Sophia Antipolis CEDEX, France
enrico.formenti@unice.fr,
luca.manzoni@i3s.unice.fr

Abstract. Extremal combinatorics is the study of the size that a certain collection of objects must have in order to certainly satisfy a property. Reaction systems are a recent formalism for computation inspired by chemical reactions. This work is a first contribution to the study of the behaviour of large reaction systems by means of extremal combinatorics. We defined several different properties that capture some basic behaviour of a reaction system and we prove that they must necessarily be satisfied by large enough systems. Explicit bounds and formulae are also provided.

Keywords: Reaction systems, extremal combinatorics.

1 Introduction

Reaction systems (RS) are a formalism introduced by Ehrenfeucht and Rozenberg in 2004 [2] which is inspired by chemical reactions. Similarly to their real counterparts, RS reactions need a set of chemicals (the *reactants*) to act on and either can be inhibited by other chemicals (the *inhibitors*) or produce some *products*. Formally, a *reaction* consists of these three sets, each of which is coded as a set of symbols and a RS is a set of reactions over a common set of symbols (the *entities*). The size of such a system is simply defined as the number of reactions of which it is composed.

This work provides a first study on the behaviour of large RS. For the first time (at our knowledge) concepts from extremal combinatorics are applied to the field of RS. Other works on RS dynamics have also been carried on adopting a completely different point of view [1]. Indeed, the authors investigated the properties of random generated RS. Conversely, here we focus on some property P and the goal is to find minimal bounds on the size of RS such that all RS of larger size exhibit P.

---

[*] This work has been partially supported by the French National Research Agency project EMC (ANR-09-BLAN-0164).

Combinatorial statements arise almost naturally in many fields. Particularly important are those about the properties that a certain structure can exhibit when its size increases. A famous statement of this kind is, for example, the *Ramsey theorem*, that states that for every $k \in \mathbb{N}$, any large enough complete two-colored graph has a monochromatic subgraph of k nodes. Other examples come from the most different mathematical structures. These statements can be presented in many ways. One can either point out the existence of a size bound after which a certain property holds (a more *Ramsey*-like view), or try to find the exact value of this bound or, at least, obtain some information on its order of magnitude (an extremal combinatorics point of view). To prove statements of extremal combinatorics a large body of work regarding proof techniques has been produced, ranging from various counting techniques to combinatorial proofs and linear algebra methods. Since it is impossible to give a detailed account of all the results and the techniques, we refer the reader to specific books (see, for example, [5] for extremal combinatorics and [4] for Ramsey theory).

This paper studies the minimal size after which a reaction system:

- necessarily exhibits a non-sequential behaviour.
- always exhibits a non-sequential behaviour.
- includes two reactions such that the reactants of one of them inhibit the other.
- necessarily has a reaction that produces the inhibitors of another reaction.
- can be substituted by a smaller reaction system with the same dynamics.
- cannot be written as the union of two reaction systems with a disjoint set of entities.

These properties help to clarify the limits in size and in parallelism that are inherent in the definition of RS. Many of the results obtained in this paper can also be restated as Ramsey-like statements (i.e., in terms of presence of an "ordered" substructure inside a large enough structure).

The paper is structured as follows. The next section recall basic notions about RS. Section 3 contains the main results. Further remarks and discussion are provided in Section 4.

2   Reaction Systems

In this section, we give the basic notions about reaction systems, including the concepts of dynamics and equivalence. Notation is taken from [3].

2.1   Basics of Reaction Systems

We first recall the main concept of a reaction. Inspired by real chemical reactions, it comprises three sets corresponding to the reactants, the inhibitors, and the products of the reaction.

Definition 1. *A reaction* $a = (R_a, I_a, P_a)$ *is a triple of three non-empty sets such that* $R_a \cap I_a = \emptyset$. R_a *is called the set of* reactants, I_a *the set of* inhibitors, *and* P_a *the set of* products, *respectively. For any set S, we say that a is a reaction in S if* $R_a \subseteq S$, $I_a \subseteq S$, *and* $P_a \subseteq S$.

Let S be a finite set. The *set of all reactions in S* is denoted by $rac(S)$. For any $T \subseteq S$ and any reaction $a \in rac(S)$, we say that *a is enabled by T* if $R_a \subseteq T$ and $I_a \cap T = \emptyset$. The *result $res_a(T)$ of a reaction $a \in rac(S)$ on a set $T \subseteq S$* is defined as $res_a(T) = P_a$ if a is enabled by T, and $res_a(T) = \emptyset$ otherwise. The previous notion extend naturally to sets of reactions: the *result set of A on a subset $T \subseteq S$* is $res_A(T) = \bigcup_{a \in A} res_a(T)$. For any $A \subseteq rac(S)$, the *reactant, inhibitor, and product set of A* are $R_A = \bigcup_{a \in A} R_a$, $I_A = \bigcup_{a \in A} I_a$, and $P_A = \bigcup_{a \in A} P_a$, respectively. Furthermore, we say that *A is enabled by T* if every reaction $a \in A$ is enabled by T.

Definition 2. *A Reaction System (RS) is a pair $\mathcal{A} = (S, A)$, where S is a finite set of symbols and $A \subseteq rac(S)$. The set S is called the* background *of $\mathcal{A}$.*

The *result set of a reaction system $\mathcal{A} = (S, A)$ on a subset $T \subseteq S$* is $res_A(T) = res_A(T)$. The *T-activity of A*, denoted by $en_A(T)$, is the set of all reactions of $\mathcal{A}$ enabled by T. A reaction system $\mathcal{B} = (S, B)$ is a *subsystem* of $\mathcal{A} = (S, A)$ if $B \subseteq A$.

Every RS $\mathcal{A} = (S, A)$ defines a finite dynamical system where $\mathcal{P}(S)$ (the subsets of S) are the possible states and the next state function is res_A.

The following notion is crucial to our study.

Definition 3. *The size of a RS $\mathcal{A} = (S, A)$ is $|A|$.*

Note that the above definition does not take into account of $|S|$ simply because $|A|$ can be exponentially larger than $|S|$.

2.2   Equivalence of Reaction Systems

In order to obtain lower bounds on the structure of RS having a given property, one needs the notion of equivalence between RS.

Let S be a set of entities, two reactions $a, b \in rac(S)$ are *functionally equivalent* (denoted by $a \sim b$) if for all $T \subset S$, $res_a(T) = res_b(T)$. Ehrenfeucht and Rozenberg found the necessary and sufficient conditions to establish if two reactions are functionally equivalent [3]. Indeed, they proved that two reactions $a, b \in rac(S)$ are functionally equivalent iff $R_a = R_b$, $I_a = I_b$ and $P_a = P_b$. The notion of functional equivalence can also be extended to sets of reactions. Two sets $A, B \subseteq rac(S)$ of reactions are *functionally equivalent* (denoted by $A \sim B$) iff for all $T \subseteq S$, $res_A(T) = res_B(T)$. It has been proved that the problem of the functional equivalence between sets of reactions is coNP-complete [3]. Two RS are *functionally equivalent* if their sets of reactions are functionally equivalent.

In [3], also a partial order between reactions has been introduced. For any pair of reactions $a, b \in rac(S)$, *a covers b* (denoted by $a \geq b$) iff for all $T \subseteq S$, $res_a(T) \supseteq res_b(T)$. It is immediate that $a \sim b \Leftrightarrow a \geq b \wedge b \geq a$. It has been proved that $a \geq b$ if and only if $R_a \subseteq R_b$, $I_a \subseteq I_b$ and $P_b \subseteq P_a$ [3].

3   Properties and Bounds of Reaction Systems

An important insight on the behaviour of a large group of RS can be obtained by studying the properties that any RS with a large enough size has to exhibit. In this section we will define several properties for RS and we will prove that all these hold for RS with large enough size. In particular, for most of properties the minimal RS size making them emerge will turn out to be asymptotically smaller than the number of all the possible RS. Therefore, we can state that these properties are verified by the large majority of RS.

We now introduce the concept of threshold property, i.e., a property for RS expressed in terms of a predicate over the class of all RS which is true for those RS with size greater than a certain threshold.

Definition 4. *Let P be a predicate over the set of all possible RS. P is said to be a* threshold *property for RS if for every $n \in \mathbb{N}_+$ there exists a natural $k \leq (2^n - 1)(3^n - 2^{n+1} + 1)$ such that $P(\mathcal{A})$ is true for any RS $\mathcal{A} = (S, A)$ with $|S| = n$ and size $|A| \geq k$. Depending on P and $n \in \mathbb{N}_+$, the integer k is denoted by $R(P, n)$ and called the* threshold *for P.*

The following lemma gives the number of all possible reactions in any set of a given number of entities. It will be very useful in the sequel to compute bounds for the threshold properties.

Lemma 5. *For any $n \in \mathbb{N}$ and any set S of symbols with $|S| = n$, it holds that $|rac(S)| = (2^n - 1)(3^n - 2^{n+1} + 1)$.*

Proof. Let $n \in \mathbb{N}$ and let S be any set with $|S| = n$. The number of reactions in S is the number of possible combinations of reactants, inhibitors and products. The number of possible combinations of reactants and inhibitors is:

$$\sum_{i=1}^{n-1} \binom{n}{i} \sum_{j=1}^{n-i} \binom{n-i}{j} = \sum_{i=1}^{n-1} \binom{n}{i}(2^{n-i} - 1) =$$
$$\sum_{i=1}^{n-1} \binom{n}{i} 2^{n-i} - \sum_{i=1}^{n-1} \binom{n}{i} = \sum_{i=1}^{n-1} \binom{n}{i} 2^{n-i} - 2^n + 2$$

By the properties of binomial coefficient one finds:

$$\sum_{i=1}^{n-1} \binom{n}{i} 2^{n-i} = -2^n - 1 + \sum_{i=0}^{n} \binom{n}{i} 2^{n-i} = -2^n - 1 + (2+1)^n$$

Then, we get $3^n - 2^{n+1} + 1$ combinations of reactants and inhibitors. The multiplication with the number of all possible product sets gives a total of $(2^n - 1)(3^n - 2^{n+1} + 1)$ reactions.   □

Remark that, by Lemma 5, a RS with n entities and at least $(2^{n-1}-1)(3^{n-1} - 2^n + 1) + 1$ reactions necessarily has each entity used in at least one reaction.

The following proposition is an immediate consequence of Definition 4 and gives a necessary and sufficient condition for recognizing threshold properties.

Proposition 6. *A predicate P is a threshold property for RS if and only if for any set S of entities, $P(\mathcal{A})$ is true for the RS $\mathcal{A} = (S, rac(S))$.*

The following example shows that not all the properties on the RS dynamics are threshold properties.

Example 7. Compare the following two predicates defined for any RS $\mathcal{A} = (S, A)$ as:

1. $\mathcal{A}$ has a cycle of period 2, i.e., there exists $T \subseteq S$ such that $res_\mathcal{A}(res_\mathcal{A}(T)) = T$ and $res_\mathcal{A}(T) \neq T$.
2. $\mathcal{A}$ has a fixed point, i.e., there exists $T \subseteq S$ such that $res_\mathcal{A}(T) = T$.

Even if they might seem similar, only the second predicate is a threshold property. Indeed, it is clearly true for any $\mathcal{A} = (S, rac(S))$ since $res_\mathcal{A}(\emptyset) = \emptyset$. Consider now the evaluation of the first predicate on $(S, rac(S))$. Since for every $T \subseteq S$ with $T \neq \emptyset$ and $T \neq S$ the reaction $(T, S \setminus T, S)$ is enabled, it holds that $res_\mathcal{A}(T) = S$. Furthermore, $res_\mathcal{A}(S) = res_\mathcal{A}(\emptyset) = \emptyset$. Hence, the first predicate is not a threshold property.

Each RS property we are going to consider will be actually expressed in terms of predicates over the class of all RS. When no confusion is possible, we will identify the term property with the associated predicate.

The first property we deal with is totality. It holds for those RS that produce a non-empty result set on every proper subset of entities. In some sense, it is a kind of liveliness property for RS. More formally, totality is defined for any RS $\mathcal{A} = (S, A)$ by means of $res_\mathcal{A}$ viewed as a function on the subsets of S.

Definition 8. *For any RS $\mathcal{A} = (S, A)$, the function $res_\mathcal{A}$ is said to be* total *if for all $T \subseteq S$ with $T \neq \emptyset$ and $T \neq S$, $res_\mathcal{A}(T) \neq \emptyset$. A RS has property* Tot *if its result function is total.*

In [6], the problem of determining if the result function of a RS is total was proved to be coNP-complete. By Proposition 6, it is clear that Tot is a threshold property. The following proves how large in size a RS should be to define a total result function.

Proposition 9. *The threshold for property* Tot *is*

$$R(Tot, n) = (2^n - 1)(3^n - 3 \cdot 2^n + 2^{\lceil \frac{n}{2} \rceil} + 2^{\lfloor \frac{n}{2} \rfloor}) + 1.$$

Proof. For every $n \in \mathbb{N}_+$, let S be any set with $|S| = n$ and let T be any proper subset of S with $|T| = k$ for some $0 < k < n$. The number of all possible reactions in S enabled by T is $(2^k - 1)(2^{n-k} - 1)(2^n - 1)$ since any non-empty subset of T, resp., $S \setminus T$, resp., S can be selected as set of reactants, resp., inhibitors, resp., products.

By Lemma 5, a RS with $(2^n - 1)(3^n - 2^{n+1} + 1) - (2^k - 1)(2^{n-k} - 1)(2^n - 1)$ reactions can be defined with no reaction enabled in T. This value is maximized for $k = \lceil \frac{n}{2} \rceil$ or $k = \lfloor \frac{n}{2} \rfloor$. Hence,

$$R(\text{Tot}, n) = (2^n - 1)((3^n - 2^{n+1} + 1) - (2^{\lceil \frac{n}{2} \rceil} - 1)(2^{\lfloor \frac{n}{2} \rfloor} - 1)) + 1$$
$$= (2^n - 1)(3^n - 3 \cdot 2^n + 2^{\lceil \frac{n}{2} \rceil} + 2^{\lfloor \frac{n}{2} \rfloor}) + 1$$

that proves the statement. □

We now consider Minimal-Concurrency (MC). It is that property exhibited by non-sequential RS, i.e., with at least two reactions that can be executed concurrently.

Definition 10. *A RS $\mathcal{A} = (S, A)$, $\mathcal{A}$ has the Minimal-Concurrency (MC) property if there exist two reactions $a, b \in A$ and a set $T \subseteq S$ such that $a, b \in en_A(T)$.*

Clearly, MC is a threshold property.

Proposition 11. *The threshold for MC property is $R(\text{MC}, n) = 2^n - 1$.*

Proof. Let $n \in \mathbb{N}_+$. Any RS with n entities and at least $2^n - 1$ reactions necessarily has two reactions with the same reactant set, since only $2^n - 2$ distinct reactant sets are possible. Hence, it has at least two reactions that can be enabled at the same time, i.e., $R(\text{MC}, n) \leq 2^n - 1$. To prove that this bound is strict, consider the RS $\mathcal{A} = (S, A)$ where $|S| = n$ and A contains all the possible reactions in the form $a = (R_a, S \setminus R_a, P)$ for some common $P \subseteq S$. Clearly, $|A| = 2^n - 2$. Since each $a \in A$ is enabled only by $T = R_a$, and such a T is never the same for distinct reactions, $\mathcal{A}$ does not have the MC property. Therefore, $R(\text{MC}, n) = 2^n - 1$. □

We now bring to the extreme the MC property, that is, we consider a fully parallel behavior for RS.

Definition 12. *A RS $\mathcal{A} = (S, A)$ has the Always-Parallel (AP) property if for every set $T \subseteq S$ and every reaction $a \in en_A(T)$ there exists a reaction $b \in A \setminus \{a\}$ such that $b \in en_A(T)$.*

By Proposition 6, AP is a threshold property.

Proposition 13. *The threshold for AP property is*

$$R(\text{AP}, n) = (2^n - 1)\left(3^n - 5 \cdot 2^{n-1} + 2\right) + 2.$$

Proof. For every $n \in \mathbb{N}_+$, let S be any set with $|S| = n$. We count the number of reactions in S enabled by each subset $T \subseteq S$, i.e., those reactions $c = (R_c, I_c, P_c)$ with $R_c \subseteq T$, $I_c \cap T = \emptyset$, and an arbitrary P_c. This number is $(2^n - 1)(2^{|T|} - 1)(2^{n-|T|} - 1)$ that is minimized when $|T| = n - 1$ or, symmetrically, $|T| = 1$. Thus, the minimum number of reactions enabled by any subset of S is $(2^n - 1)(2^{n-1} - 1)$. Let $k = (2^n - 1)(3^n - 2^{n+1} + 1 - 2^{n-1} + 1) + 2$. By Lemma 5, for any RS (S, A) with $|A| \geq k$ every subset $T \subseteq S$ enables at least two distinct reactions.

On the other hand, consider the RS $\mathcal{A} = (S, A)$ where A is made of the reaction $a = (\{s\}, S \setminus \{s\}, P)$ for some fixed $s \in S$ and $P \subseteq S$ and all the ones with a reactant set different from $\{s\}$. $\mathcal{A}$ has exactly $k - 1$ reactions but AP does not hold for $\mathcal{A}$, since $a \in en_{\mathcal{A}}(T)$ for $T = \{s\}$ but no reaction different from a is enabled by T. Therefore, $R(\mathsf{AP}, n) = k$.   □

As immediate generalization of the AP property, we can introduce the (threshold) AP_m property with $2 \leq m \leq (2^n - 1)(2^{n-1} - 1)$, that is, the predicate which is true for any RS such that every subset of its entities enables either 0 or at least m reactions. Clearly, $\mathsf{AP} = \mathsf{AP}_2$ and the following fact holds.

Corollary 14. *For each $2 \leq m \leq (2^n - 1)(2^{n-1} - 1)$, the threshold for the AP_m property is*
$$R(\mathsf{AP}_m, n) = (2^n - 1)\left(3^n - 5 \cdot 2^{n-1} + 2\right) + m.$$

No-Concurrency (NC) is the property exhibited by those RS with at least two reactions that can never be executed at the same time step. In some sense, it is an "upper" limit on the concurrency that a RS can achieve.

Definition 15. *A RS $\mathcal{A} = (S, A)$ has the No-Concurrency (NC) property if there exist $a, b \in A$ such that for all $T \subseteq S$, either $a \notin en_{\mathcal{A}}(T)$ or $b \notin en_{\mathcal{A}}(T)$.*

Equivalently, any RS (S, A) has the NC property if and only if it admits two reactions $a = (R_a, I_a, P_a)$ and $b = (R_b, I_b, P_b)$ such that $R_a \cap I_b \neq \emptyset$, i.e., if and only if $R_A \cap I_A \neq \emptyset$. Then, it follows from Proposition 6 that NC is a threshold property.

Proposition 16. *The threshold for NC property is*
$$R(\mathsf{NC}, n) = (2^n - 1)\left(2^n - 2^{\lceil \frac{n}{2} \rceil} - 2^{\lfloor \frac{n}{2} \rfloor} + 1\right) + 1.$$

Proof. For every $n \in \mathbb{N}_+$, let S be any set with $|S| = n$. We count the largest number of possible reactions in any reaction set A such that $R_A \cap I_A = \emptyset$ and $|R_A|$ is fixed. Putting $h = |R_A|$, this number is $(2^h - 1)(2^{n-h} - 1)(2^n - 1) = (2^n - 1)(2^n - 2^h - 2^{n-h} + 1)$ which is maximized by either $h = \lfloor \frac{n}{2} \rfloor$ or $\lceil \frac{n}{2} \rceil$. Thus, the maximum number of reactions of any reaction set A such that $R_A \cap I_A = \emptyset$ is $(2^n - 1)\left(2^n - 2^{\lceil \frac{n}{2} \rceil} - 2^{\lfloor \frac{n}{2} \rfloor} + 1\right)$ and this concludes the proof.   □

On the other hand, we also consider RS with at least one reaction that produces at least one inhibitor of another reaction. Clearly, the behavior of RS without this feature is determined only by reactants and products since inhibitors play no role.

Definition 17. *A RS $\mathcal{A} = (S, A)$ has property Inh if there exist two reactions $a = (R_a, I_a, P_a), b = (R_b, I_b, P_b) \in A$ such that $P_a \cap I_b \neq \emptyset$.*

It is clear that Inh is a threshold property.

Proposition 18. *The threshold for* Inh *property is bounded as follows:*

$$\left(2^{n-1} - 1\right)^2 < R(Inh, n) < (2^n - 1)\left(2^n - 2^{\lceil \frac{n}{2} \rceil} - 2^{\lfloor \frac{n}{2} \rfloor} + 1\right).$$

Proof. For every $n \in \mathbb{N}_+$, let S be any set with $|S| = n$. To obtain the lower bound, we compute the maximal cardinality of any reaction set A such that $P_A \cap I_A = \emptyset \wedge R_A \cap I_A = \emptyset$. Once $|I_A|$ has been fixed to some $n - h$ with $1 \le h \le n - 1$, the largest cardinality of any reaction set A satisfying the above conjunction condition is $\left(2^h - 1\right)^2 \left(2^{n-h} - 1\right)$ since the cardinalities of P_A and R_A can be at most h. Since this value is maximum for $h = n - 1$ and the conjunction condition is stronger than the statement $[\forall a \in A, R_a \cap I_a = \emptyset]$, the maximal cardinality of any reaction set A satisfying this latter statement is at most $\left(2^{n-1} - 1\right)^2$. This proves the lower bound for $R(Inh, n)$.

As to the upper bound, we compute the cardinality of any set A of triples $a = (R_a, I_a, P_a)$ such that $P_A \cap I_a = \emptyset$ for all $a \in A$ without any restriction on $R_a \cap I_a$ (i.e., we also count triples that are not reactions). When $|P_A|$ is fixed to some h with $0 < h < n$, it holds that $|I_A| = n - h$. Since R_A is not constrained, the cardinality of such a set A with $|P_A| = h$ is $(2^n - 1)(2^{n-h} - 1)(2^h - 1)$, a value that is maximum when $h = \lceil \frac{n}{2} \rceil$. As a consequence, the maximal cardinality of any such a set A of triples is $k = (2^n - 1)\left(2^n - 2^{\lceil \frac{n}{2} \rceil} - 2^{\lfloor \frac{n}{2} \rfloor} + 1\right)$ and, hence, $R(Inh, n) \le k$. To prove that the inequality is strict, just remark that $a = (I_a, I_a, P_a) \in A$, but $a \notin rac(S)$. □

Proposition 18 does not provide an exact threshold value for Inh. However, the ratio between the lower and upper bound is relatively small, in the sense that these are asymptotically proportional. Indeed,

$$\lim_{n \to +\infty} \frac{\left(2^{n-1} - 1\right)^2}{(2^n - 1)\left(2^n - 2^{\lfloor \frac{n}{2} \rfloor} - 2^{\lceil \frac{n}{2} \rceil} + 1\right)} = \frac{1}{4}.$$

We now introduce a notion of redundancy for RS.

Definition 19. *A RS $\mathcal{A} = (S, A)$ is* Red *(redundant) if there exists $C \subset A$ such that $A \sim C$.*

In other words, a RS is Red if its set of reactions contains some redundancy, i.e., reactions that do not contribute to the dynamics of the system. As an example of redundant system consider $(S, rac(S))$. Indeed, it is equivalent to $(S, \emptyset)$. Hence, by Proposition 6, Red is a threshold property. The following property is very useful to provide a sufficient condition assuring that a RS has property Red.

Definition 20. *A RS $\mathcal{A} = (S, A)$ has property* Comp *if for every $a \in rac(S)$, there exists $b \in A$ such that either $a \ge b$ or $b \ge a$.*

In other words, any RS has the Comp property if it contains a maximal antichain of reactions.

Proposition 21. *Let $\mathcal{A} = (S, A)$ be a RS. If there exists $B \subset A$ such that the RS $\mathcal{B} = (S, B)$ has property* Comp, *then $\mathcal{A}$ has property* Red.

Proof. Suppose that for some $B \subset A$ the RS (S, B) has the Comp property and let $C \subseteq A$ be the set of maximal elements of A. We are going to prove that the RS $\mathcal{C} = (S, C)$ is functionally equivalent to $\mathcal{A}$ and $C \subset A$. Remark that any RS with two reactions $a, b \in A$ with $a \geq b$ is functionally equivalent to the one obtained by removing b from the former. Thus, $\mathcal{C} \sim \mathcal{A}$. Since (S, B) has the Comp property, there exist $a \in A \setminus B$ and $b \in B$ such that either $a \geq b$ or $b \geq a$. Because of that, at least one between a and b cannot be an element of C and, hence, $C \subset A$. $\square$

The property Comp holds for $(S, rac(S))$ and thus it is a threshold property. It is difficult to find an exact threshold for Comp property. However, using Proposition 22 convenient bounds can be given.

Proposition 22. *The threshold for property* Comp *is bounded as follows:*

$$\frac{n!(2^n - 2)}{\lceil \frac{n}{2} \rceil! \lfloor \frac{n}{2} \rfloor!} \leq R(\mathsf{Comp}, n) \leq \frac{n!(3^n - 2^{n+1} + 1)}{\lceil \frac{n}{2} \rceil! \lfloor \frac{n}{2} \rfloor!}.$$

Proof. For every $n \in \mathbb{N}_+$, let S be any set with $|S| = n$. Denote by l and u the lower and upper bound for $R(\mathsf{Comp}, n)$, respectively, in the statement of this Proposition.

To prove that $R(\mathsf{Comp}, n) \geq l$, it is enough to provide a reaction set in S containing an antichain of l reactions. Let A be the set of all the reactions of form $(R, S \setminus R, P)$ with $|P| = \lceil \frac{n}{2} \rceil$. All reactions from A are pairwise incomparable. Indeed, consider any two distinct reactions $a, b \in A$. If $P_a \neq P_b$, then they are not comparable since their product sets have the same cardinality. Otherwise, it holds that either $R_a \neq R_b$ or $I_a \neq I_b$, and so necessarily a and b are not comparable. We now compute $|A|$. There are $\binom{n}{k}$ subsets of S with k elements and this value is maximized for either $k = \lceil \frac{n}{2} \rceil$ or $k = \lfloor \frac{n}{2} \rfloor$. Hence, there are $\binom{n}{\lceil \frac{n}{2} \rceil}$ possible product sets for reactions in A. Since there are $2^n - 2$ possible reactant sets and the inhibitor sets are completely determined by them, it follows that $|A| = l$.

On the other hand, to prove that $R(\mathsf{Comp}, n) \leq u$ we exhibit a set B of reactions in S such that the RS (S, B) has the property Comp, $|B| = u$, and any antichain of reactions in S has smaller cardinality. Let B be the set of all reactions of form (R, I, P) with $|P| = \lceil \frac{n}{2} \rceil$. Every reaction $a = (R_a, I_a, P_a) \in rac(S) \setminus B$ is comparable with $(R_a, I_a, P) \in B$ for some $P \subset P_a$ or $P \supset P_a$. Thus, (S, B) has the property Comp. Furthermore, by the same arguments used as above to compute $|A|$ and in the proof of Lemma 5, we get $|B| = u$. Consider now any antichain C of reactions in S. For each $a \in C$, define $C_a = \{(R, I, P) \in C : R = R_a \text{ and } I = I_a\}$ and $B_a = \{(R, I, P) \in B : R = R_a \text{ and } I = I_a\}$. Let π_3 be the projection mapping any reaction to its third component. Since each C_a is an antichain, each $\pi_3(C_a)$ is also an antichain (in $\mathcal{P}(S)$ ordered by the inclusion relation). Moreover, each $\pi_3(B_a)$ is a maximal antichain in $\mathcal{P}(S)$ and then $|C_a| \leq |B_a|$ for every $a \in C$. Since it is possible to extract from $\{C_a\}_{a \in C}$ a partition Γ of C and, consequently, from $\{B_a\}_{a \in C}$ a collection of disjoint subsets of B, it follows that $|C| = \sum_{C_a \in \Gamma} |C_a| \leq \sum_{C_a \in \Gamma} |B_a| \leq |B|$. $\qquad \square$

Knowing if a RS has property Comp is useful to find RS with property Red. In fact, Proposition 21 shows that the existence of a subsystem with property Comp is a sufficient condition for a RS to contain redundant reactions. Hence, by Propositions 21 and 22, the bounds for property Comp immediately give bounds for Red.

The previous proposition can also be reformulated into a Ramsey-like statement as follows.

Proposition 23. *Every RS of large enough size contains a maximal antichain of reactions.*

Equivalently, if one wants to put more emphasis on the property of containing redundant reactions, Proposition 22 can be turned into the following.

Proposition 24. *Every RS of large enough size contains an equivalent subsystem.*

Another interesting property concerns the "decomposability", i.e., the possibility that a RS can be decomposed as the union of two or more smaller RS.

Definition 25. *A RS $\mathcal{A} = (S, A)$ is decomposable if there exist two RS $\mathcal{A}_1 = (S_1, A_1)$ and $\mathcal{A}_2 = (S_2, A_2)$ such that $S_1, S_2 \neq \varnothing$, $S_1 \cup S_2 = S$, $S_1 \cap S_1 = \varnothing$, $A_1 \cup A_2 = A$, and $A_1 \cap A_2 = \varnothing$. A system is non-decomposable (or has property NC) if it is not decomposable.*

There are pretty simple systems which are not decomposable. As an example, the RS $\mathcal{A} = (\{a, b\}, \{(\{a\}, \{b\}, \{a\})\})$ cannot be decomposed. The same holds for $(S, rac(S))$. Therefore, it is natural to search for the threshold assuring that any RS of bigger size is non-decomposable.

Proposition 26. *The threshold for property ND is*

$$R(\mathsf{ND}, n) = (2^{n-1} + 1)(3^{n-1} - 2^n + 1) + 1.$$

Proof. For every $n \in \mathbb{N}_+$, let S be any set of n entities. Suppose that a RS $\mathcal{A} = (S, A)$ can be decomposed into two RS $\mathcal{A}_1 = (S_1, A_1)$ and $\mathcal{A}_2 = (S_2, A_2)$ with $|S_1| = h$ for some $h \in \{1, \ldots, n-1\}$. By Lemma 5, $|A_1| \leq (2^h - 1)(3^h - 2^{h+1} + 1)$ and $|A_2| \leq (2^{n-h} - 1)(3^{n-h} - 2^{n-h+1} + 1)$. Since $|A| = |A_1| + |A_2|$, we can find the value of h that maximizes $|A|$. This value is $h = 1$ or, symmetrically, $h = n - 1$. Hence, if $\mathcal{A}$ can be decomposed, then A has at most $k = (2^{n-1} + 1)(3^{n-1} - 2^n + 1)$ reactions. Furthermore, if $|A|$ exceeds h, then $\mathcal{A}$ cannot be decomposed. Hence, $R(\mathsf{ND}, n) = k + 1$. $\qquad \square$

As generalization of the ND property, for $1 \leq m \leq \lceil \frac{n}{2} \rceil$ the (threshold) ND_m property is the predicate which is true for any RS decomposable in two RS both with a background of at least m entities.

Corollary 27. *For each $1 \leq m \leq \lfloor \frac{n}{2} \rfloor$, the threshold for the ND_m property is*

$$R(\mathsf{ND}_m, n) = (2^{n-m} - 1)(3^{n-m} - 2^{n-m+1} + 1) + (2^m - 1)(3^m - 2^{m+1} + 1) + 1.$$

Proof. The value $R(\mathsf{ND}_m, n) - 1$ is the maximum size of a RS decomposable into two RS both with background of at least m entities. It is obtained as done in the proof of Proposition 26 for RS decomposed into two RS with background of m and $n - m$ entities. $\qquad \square$

4 Conclusion

This work provides a first application of extremal combinatorics to the study of large RS. The results concern some basic properties of RS. In particular, we deal with questions about concurrency, full parallelism, decomposability. All these are fundamental structural properties of RS. Indeed, assume to endow the set of RS having n entities with the uniform probability. Then, extracting a RS with the property MC, or NC, or Inh, or Comp is an event whose probability goes to one as n goes to infinity. It would be interesting to understand if this also holds for property ND, since our result proves that the probability is asymptotically lower bounded by a positive constant. However, we do not know if this probability goes to one or not.

Future works should aim at enlarging the set of threshold properties and build up a corpus of tools favouring the transfer of the extremal combinatorics results. We have studied bounds that involved only one application of the result function. It would be interesting to find natural threshold properties on the dynamics of RS.

References

1. Ehrenfeucht, A., Main, M., Rozenberg, G.: Combinatorics of life and death for reaction systems. International Journal of Foundations of Computer Science 21, 345–356 (2010)
2. Ehrenfeucht, A., Rozenberg, G.: Basic notions of reaction systems. In: Calude, C.S., Calude, E., Dinneen, M.J. (eds.) DLT 2004. LNCS, vol. 3340, pp. 27–29. Springer, Heidelberg (2004)
3. Ehrenfeucht, A., Rozenberg, G.: Reaction systems. Fundamenta Informaticae 75, 263–280 (2007)
4. Graham, R.L., Rothschild, B.L., Spencer, J.H.: Ramsey Theory. Wiley-Interscience Series in Discrete Mathematics and Optimization Advisory. Wiley-Interscience (1990)
5. Jukna, S.: Extremal combinatorics: with applications in computer science. Springer (2001)
6. Salomaa, A.: Functional constructions between reaction systems and propositional logic. International Journal of Foundations of Computer Science 24(1), 147–159 (2013)

Stochastic k-Tree Grammar and Its Application in Biomolecular Structure Modeling

Liang Ding[1], Abdul Samad[1], Xingran Xue[1], Xiuzhen Huang[4],
Russell L. Malmberg[2,3], and Liming Cai[1,2,*]

[1] Department of Computer Science
[2] Institute of Bioinformatics
[3] Department of Plant Biology
University of Georgia, GA 30602, USA
{lding,samad,xrxue,russell,cai}@uga.edu
[4] Dept. of Computer Science, Arkansas State University
Jonesboro, AR 72467, USA
xhuang@astate.edu

Abstract. Stochastic context-free grammar (SCFG) has been successful in modeling biomolecular structures, typically RNA secondary structure, for statistical analysis and structure prediction. Context-free grammar rules specify parallel and nested co-occurren-ces of terminals, and thus are ideal for modeling nucleotide canonical base pairs that constitute the RNA secondary structure. Stochastic grammars have been sought, which may adequately model biomolecular tertiary structures that are beyond context-free. Some of the existing linguistic grammars, developed mostly for natural language processing, appear insufficient to account for crossing relationships incurred by distant interactions of bio-residues, while others are overly powerful and cause excessive computational complexity.

This paper introduces a novel stochastic grammar, called *stochastic k-tree grammar* (SkTG), for the analysis of context-sensitive languages. With the new grammar rules, co-occurrences of distant terminals are characterized and recursively organized into k-tree graphs. The new grammar offers a viable approach to modeling context-sensitive interactions between bioresidues because such relationships are often constrained by k-trees, for small values of k, as demonstrated by earlier investigations. In this paper it is shown, for the first time, that probabilistic analysis of k-trees over strings are computable in polynomial time $n^{O(k)}$. Hence, SkTG permits not only modeling of biomolecular tertiary structures but also efficient analysis and prediction of such structures.

Keywords: stochastic grammar, context-sensitive language, k-tree, dynamic programming, biomolecule, RNA tertiary structure.

1 Introduction

Stochastic formal language systems, typically the stochastic context-free grammar (SCFG), have been significantly valuable to various applications. Such a

---

* Correspondent author.

A.-H. Dediu et al. (Eds.): LATA 2014, LNCS 8370, pp. 308–322, 2014.
© Springer International Publishing Switzerland 2014

system essentially consists of a finite set of rules that syntactically dictate generation of strings for a desired language. Any generation process of a language string is a series of Chomsky rewriting rule applications and thus yields a syntactic structure associated with (the terminal occurrences in) the string. Because syntactic rules often are nondeterministic, there may be more than one syntactic process to generate the same string [26,9]. Stochastic versions of such formal systems may be established by associating a probability distribution with the rules. Compounding the probabilities of rules used in a generation process of a string gives rise to the probability for the corresponding syntactic structure admitted by the string [28,8]. Therefore, a stochastic language system defines a probability space for all the syntactic structures admitted by the string. At the same time, it also defines a probability space for all the strings in the language.

In addition to the apparent wide application in natural language processing [18,15,35,16,27,2], SCFG has also been extensively adopted for statistical analysis of biomolecular structures [25,8,4,5,29]. A biomolecule consists of a string of linearly arranged residues that can spatially interact to fold the string into a 3D structure of biological significance. Interactions between residues are interpreted as co-occurrences of lexical objects in each parsing of the string. SCFG can conveniently model nested and parallel relationships of the interacting residues on a biomolecule. Figure 1 shows an RNA molecule with parallel and nested canonical base parings (in gray, lighter lines) between nucleotides, which is context-free. Indeed, SCFG has enabled the development of a number of effective computer programs for the prediction of RNA secondary structure [21,39,17,1,24]. Such programs are also computationally efficient by taking the advantage of dynamic programming algorithms permitted by context-free rules.

Nevertheless, SCFG cannot account for crossing interactions of a context-sensitive nature, e.g., the interactions in Figure 1 denoted by both gray (lighter) and pink (darker) lines. Since crossing, distant interactions are the signature of a biomolecule forming a tertiary (3D) structure, adequate modeling of such interactions with a stochastic grammar would have the potential for effective analysis and even prediction of biomolecular tertiary structures. Modeling context-sensitive languages with Chomsky context-sensitive grammars can be inconvenient and may incur computational intractability [19,9]. Previous work in more constrained languages has studied mildly context-sensitive grammars, typically the Tree-Adjoining Grammar [13] and its equivalent variants [14,34], to model limited cross-serial dependencies arising in natural language processing. There has been limited success in the applications of such grammars in biomolecular structure modeling [33,29,4]; they were mostly used for the characterization of local, secondary structures. The global structure of a biomolecule involving cross relationships between arbitrarily distant residues may be beyond limited cross-serial dependencies.

In this paper, we introduce a novel stochastic grammar called *stochastic k-tree grammar* (SkTG), for the analysis of context-sensitive languages. With succinct grammar rules, co-occurrences of distant terminals are recursively characterized as *k-trees*. A k-tree is a chordal graph that does not contain cliques

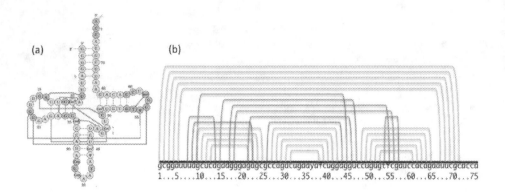

Fig. 1. A single RNA molecule can fold back on itself to form secondary and tertiary structures through bio-residue interactions. (a) The secondary structure of tRNA (Phe of yeast, PDB id: 1EHZ)) consists of parallel and nested canonical base parings (gray, lighter connections) between nucleotides, which is context-free. The tertiary structure formed with additional non-canonical tertiary interactions (pink, darker connections) between nucleotides is context-sensitive. (b) Illustration of the bio-residues interactions of the tRNA molecule in terms of co-occurrences of terminals on a language string.

of size more than $k + 1$ as a graph minor [23,3]. For small values of k, k-trees are tree-like graphs; they are adopted in this work to constrain crossing relationships of terminal occurrences on language strings. Such constrained context-sensitivity has been discovered in biomolecular structures; recent studies have revealed that graphs describing bio-residue interactions found in resolved biomolecular 3D structures are actually (subgraphs of) k-trees, typically for $k \leq 4$ [36,31,37,11,10]. Therefore, the new grammar SkTG offers a viable approach to statistical modeling, analysis, and prediction of biomolecular tertiary structures.

Previous studies showed that statistical analysis problems over general k-trees are extremely difficult, in particular, NP-hard even for $k = 2$, excluding the possibility to feasibly implement such a framework [32,40,30]. However, with the linear chain of vertices constrained on k-trees, we are able to show, for the first time, that the k-tree parsing problem is solvable in polynomial-time for every fixed value of k. In particular, we will show that SkTG makes it possible to define a probability space for all k-tree structures admitted by any given language string. We will demonstrate efficient dynamic programming algorithms for computing the most probable k-tree structure for any given string. In this paper, we will also discuss the application in the prediction of biomolecular tertiary structures that has motivated this work.

2 k-Trees and the k-Tree Grammar

Definition 1. [23] Let integer $k \geq 1$. The class of k-trees are defined with the following inductive steps:

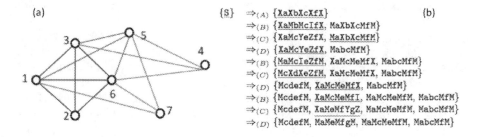

Fig. 2. (a) A generation of a 3-tree of 7 vertices by Definition 1. (b) A derivation of string `abcdefg` with 3-tree grammar rules introduced in Definition 3, with the types of applied grammar rules shown and the LHS of every applied rule underscored. The derivation also results in an induced 3-tree, the same graph shown in (a).

1. A k-tree of $k+1$ vertices is a clique of $k+1$ vertices;
2. A k-tree of n vertices, for $n > k+1$, is a graph consisting of a k-tree G of $n-1$ vertices and a vertex v, which does not occur in G, such that v forms a new $(k+1)$-clique with some size-k clique already in G.

Figure 2 (a) shows of a 3-tree with seven vertices. By Definition 1, the order in which 4-cliques formed is: initially $\{1, 2, 3, 6\}$ (black edges), vertex 5 and blue edges added, then vertex 7 and red edges added, and finally vertex 4 and green edges added.

2.1   The k-Tree Grammar

Chomsky grammars derive a language sentence by series of rewritings on a single symbolic string. Instead, our new grammar derives a language sentence by rewritings on *multiple* symbolic strings, thus resulting in multiple symbolic strings. The language sentence generated in such a derivation consists of the terminal symbols that occur in the resulting multiple symbolic strings; the positional ordering of the derived terminals is completely determined by the derivation.

Let Σ be an alphabet, $\mathcal{N}$ be the set of non-terminals, and ϵ be the empty string. We call a symbolic string an *m-alternating string*, if it has the format $X_0 a_1 X_1 \cdots a_m X_m$ for some $m \geq 0$, such that $X_i \in \mathcal{N} \cup \{\epsilon\}$ for all $0 \leq i \leq m$ and $a_i \in \Sigma$ for all $1 \leq i \leq m$.

Definition 2. Let $\alpha = X_0 a_1 X_1 \cdots a_m X_m$ be an m-alternating string for some $m \geq 0$. Let ω be the substring $X_i a_{i+1} \cdots X_j$ in α, for some $0 \leq i \leq j \leq m$, and $\sigma \in (\mathcal{N} \cup \Sigma)^*$ be a string. Then $\alpha|_\sigma^\omega$ is the string obtained from α with the substring ω being substituted by σ.

For two non-overlapping substrings ω_1 and ω_2 in α, we use $\alpha|_{\sigma_1, \sigma_2}^{\omega_1, \omega_2}$ to denote the string obtained from α with ω_1 being replaced by σ_1 and ω_2 being replaced by σ_2 at the same time. We also allow aggregation $\forall i$ to denote multiple simultaneous substitutions involving all applicable indexes i. In particular, $\alpha|_{Y_i}^{\forall i X_i}$ is the string obtained from α by replacing X_i with Y_i for every i

Definition 3. *Let* $k \geq 2$ *be a fixed integer. A* k-*tree grammar is a 6-tuple* $\Gamma = (\Sigma, \mathcal{N}, \mathcal{R}, M, I, S)$, *where* Σ *is a finite alphabet,* $\mathcal{N}$ *is a set of nonterminals,* S, I, *and* M *are the* starting, importing *and* masking *nonterminals in* $\mathcal{N}$, *respectively, and* $\mathcal{R}$ *is a set of grammar rules. Each rule has the format of* $\alpha \to \mathcal{A}$, *where* α *is either* S *or a* $(k+1)$-*alternating string and* $\mathcal{A}$ *is a subset of* $(k+1)$-*alternating strings, and has one of the following four types. (In the following we assume* $\alpha = X_0 a_1 X_1 \cdots a_{k+1} X_{k+1}$, *where* $\forall i = 1, \cdots, k+1$, $a_i \in \Sigma$, *and* $\forall j = 0, \cdots, k+1$, $X_j \in \mathcal{N}$.)

(A) $S \to \{\beta\}$, for $\beta = Y_0 b_1 Y_1 \cdots b_{k+1} Y_{k+1}$, where $\forall i = 1, \cdots, k+1$, $b_i \in \Sigma$, and $\forall j = 0, 1, \cdots, k+1$, $Y_j \in \mathcal{N} - \{M, I\}$.

(B) $\alpha \to \{\beta, \gamma\}$, where $\exists s, 0 \leq s \leq k+1$, $X_s \neq M$, such that
  (1) $\beta = \alpha|_{Y_i}^{\forall i X_i}$, $\gamma = \alpha|_{Z_i}^{\forall i X_i}$, $Y_s = I$, and $Z_s = M$.
  (2) $\forall i = 0, 1, \cdots, k+1$, if $X_i = M$ then $Y_i = Z_i = M$; else either $Y_i = X_i$ and $Z_i = M$, or $Y_i = M$ and $Z_i = X_i$.

(C) $\alpha \to \{\beta\}$, where $\exists s, 0 \leq s \leq k+1$, $X_s = I$, and $\exists t, 0 \leq t \leq k$, $t - s \geq 1$ or $s - t > 1$, $X_t = X_{t+1} = M$, such that $\beta = \alpha|_{YaZ}^{X_s}|_M^{X_t a_{t+1} X_{t+1}}$, for some $Y, Z \in \mathcal{N} - \{M, I\}$ and some $a \in \Sigma$.

(D) $\alpha \to \{\beta\}$, such that $\beta|_{Y_i}^{\forall i X_i}$ and $\forall i = 0, 1, \cdots, k+1$, if $X_i = M$ then $Y_i = M$; else $Y_i = \epsilon$.

We note that rules of types (B) and (C) are tightly related by the importing nonterminal I. In particular, a rule of type (C) can be used if and only if a related rule of type (B) has been used.

Definition 4. *Let* $\Gamma = (\Sigma, \mathcal{N}, \mathcal{R}, M, I, S)$ *be a* k-*tree grammar. Let set* $\mathcal{T} \subseteq (\Sigma \cup \mathcal{N})^+$. *Let* $\alpha \in \mathcal{T}$, $\alpha \to \mathcal{A} \in \mathcal{R}$, *and define* $\mathcal{T}' = \mathcal{T} - \{\alpha\} \cup \mathcal{A}$. *We say that* $\mathcal{T}$ derives $\mathcal{T}'$ *with rule* $\alpha \to \mathcal{A}$ *and denote it by* $\mathcal{T} \Rightarrow_{\alpha \to \mathcal{A}} \mathcal{T}'$ *(or simply* $\mathcal{T} \Rightarrow \mathcal{T}'$ *when the used rule is clear in the context).*

We call $\mathcal{T} \Rightarrow^* \mathcal{T}'$ a *derivation* if and only if either $\mathcal{T} = \mathcal{T}'$ or there are $\mathcal{T}''$ and $\alpha \to \mathcal{A}$ such that $\mathcal{T} \Rightarrow_{\alpha \to \mathcal{A}} \mathcal{T}''$ and $\mathcal{T}'' \Rightarrow^* \mathcal{T}'$ is a derivation.

Let $\mathcal{T} \subseteq (\Sigma \cup \mathcal{N})^+$ be a subset. A terminal *occurs in* $\mathcal{T}$ if it occurs in some string contained in $\mathcal{T}$. Binary relation $\preceq$ on the set of all terminal occurrences in $\mathcal{T}$ is such that, for any two terminal occurrences a_i and a_j in $\mathcal{T}$, $a_i \preceq a_j$ if and only if (a) $a_i = a_j$, or (b) a_i occurs to the left of a_j in the same string, or (c) there is a terminal occurrence a_h such that a_i occurs to the left of a_h in the same string and $a_h \preceq a_j$.

Theorem 5. *Let* $\mathcal{T} \subseteq (\Sigma \cup \mathcal{N})^+$ *be a subset and* $\{S\} \Rightarrow^* \mathcal{T}$ *be a derivation. Then the binary relation* $\preceq$ *on the set of all terminal occurrences in* $\mathcal{T}$ *is a total order.*

Proof. (Sketch) By induction on m, the number of terminal occurrences in $\mathcal{T}$, where $\{S\} \Rightarrow \mathcal{T}$.

Basis: $m = k + 1$. $\mathcal{T}$ can contain only one string and the last rule used must be of type (D). Therefore, all the terminal occurrences are next to each other on the only string in $\mathcal{T}$, thus forming the total order.

Assumption: for m terminal occurrences in $\mathcal{T}$, the claim is true.

Induction: we assume that there are $m + 1$ terminal occurrences in $\mathcal{T}$. Let $\mathcal{T}_1$ be such that $\{S\} \Rightarrow^* \mathcal{T}_1$ and $\mathcal{T}_1 \Rightarrow^* \mathcal{T}$ for which rule $\alpha \to \{\beta, \gamma\}$ of type (B) and $\beta \to \theta$ of type (C) are used to introduce a new terminal occurrence a. Let L be the set of m terminal occurrences in $\mathcal{T}_1$. By the assumption, the binary relation $\preceq$ on L is a total order. Note that terminal a co-occurs with other k terminals in the same string θ. Without loss of generality, we assume a occurs to the right of terminal occurrence b and to the left of terminal occurrence c. Then $b \preceq a$ and $a \preceq c$, and for any other terminal occurrence $d \in L$, either $d \preceq b$ or $c \preceq d$, thus either $d \preceq a$ or $a \preceq d$ by the definition of $\preceq$. So the binary relationship $\preceq$ on set $L \cup \{a\}$ is also a total order. $\square$

Definition 6. Let $\Gamma = (\Sigma, \mathcal{N}, \mathcal{R}, M, S)$ be a k-tree grammar and $\mathcal{T} \subseteq (\Sigma \cup \{M\})^+$ such that $\{S\} \Rightarrow^* \mathcal{T}$. A string $a_1 a_2 \cdots a_n \in \Sigma^!$, $n \geq 3$, is *the underlying string* of $\mathcal{T}$, if for every $1 \leq i < n$, substring $a_i a_{i+1}$ occurs in some string in $\mathcal{T}$. In addition, the language defined by the grammar Γ is

$$L(\Gamma) = \{s \in \Sigma^+ : \mathcal{T} \subseteq (\Sigma \cup \{M\})^+, \{S\} \Rightarrow^* \mathcal{T}, \text{ and } uls(\mathcal{T}, s)\}$$

where predicate $uls(\mathcal{T}, s)$ asserts that s is the underlying string of $\mathcal{T}$.

For example, Figure 2(b) shows a derivation of $\mathcal{T}$ that contains four symbolic strings, for which the string `abcdefg` of 7 terminals is the underlying string.

2.2   Structure Space for Individual Strings

The introduced k-tree grammars are context-sensitive that can define crossing relationships among terminals. Let subset $\mathcal{T} \subseteq (\Sigma \cup \mathcal{N})^+$. We call two terminal occurrences *syntactically related* if they appear in the same RHS of some rule used in some derivation $\{S\} \Rightarrow^* \mathcal{T}$. We characterize such relationships of terminal occurrences in $\mathcal{T}$ with notions of graphs.

Definition 7. Let Γ be a k-tree grammar. Let $\mathcal{T} \subseteq (\Sigma \cup \mathcal{N})^+$ be such that $\{S\} \Rightarrow^* \mathcal{T}$. The *induced graph of* $\mathcal{T}$ is a labeled graph $G_{\mathcal{T}} = (V, E)$, in which vertices have one-to-one correspondence (i.e., labeled) with the terminal occurrences in $\mathcal{T}$ and edges connect vertices corresponding to syntactically related terminal occurrences. The *structure space* $\mathcal{E}(s)$ of any given string $s \in L(\Gamma)$ is defined as

$$\mathcal{E}(s) = \{G_{\mathcal{T}} : \mathcal{T} \subseteq (\Sigma \cup \{M\})^+ \text{ and } uls(\mathcal{T}, s)\}$$

For example, Figure 2(a) is the induced graph for the final set of four symbolic strings in the derivation shown in Figure 2(b), for which `abcdefg` is the underlying string.

Definition 8. Let $s = s_1 \cdots s_n \in \Sigma^+$ be a string of length n. A (labeled) graph $G = (V, E)$, where $V \subseteq \{1, 2, \cdots, n\}$, is *faithful* to s if

(a) $\forall i \in V$, vertex i is labeled with s_i; and
(b) $\forall i, j \in V$, if $i < j$ and $\neg \exists l \in V \ i < l < j$, then $(i, j) \in E$.

Lemma 9. *Let* $\{S\} \Rightarrow^* \mathcal{T}'$ *with the underlying string* $s = s_1 s_2 \cdots s_n \in \Sigma^+$. *Then for any* $\mathcal{T}$ *such that* $\{S\} \Rightarrow^+ \mathcal{T} \Rightarrow^* \mathcal{T}'$, *the induced graph of* $\mathcal{T}$ *is a faithful k-tree to string s.*

Proof. (Sketch) We prove by induction on l, the number of grammar rule applications in the derivation $\{S\} \Rightarrow^+ \mathcal{T}$ to show the induced graph $G_\mathcal{T}$ of $\mathcal{T}$ is both a k-tree and faithful to s.

$l = 1$. This is the case that rule $\{S\} \to \{X_0 a_1 X_1 \cdots a_{k+1} X_{k+1}\}$ is first used. Thus $G_\mathcal{T}$, where $\mathcal{T} = \{X_0 a_1 X_1 \cdots a_{k+1} X_{k+1}\}$, consists of $k + 1$ vertices $\{i_1, i_2, \cdots, i_{k+1}\}$ labeled with terminal co-occurrences $\{a_1, a_2, \cdots, a_{k+1}\}$. $G_\mathcal{T}$ is a $(k + 1)$-clique, thus a k-tree. It also is faithful to s since it satisfies condition (b) as no vertices other than $\{i_1, i_2, \cdots, i_{k+1}\}$ are present.

We assume the lemma to be true for the case that fewer than l rules are applied. We now prove it is also true for the case that l rules applied, $l \geq 2$. Let $\mathcal{T}_1$ be such that $\{S\} \Rightarrow^* \mathcal{T}_1 \Rightarrow^* \mathcal{T}$ and $\mathcal{T}_1 \Rightarrow^* \mathcal{T}$ be realized by either a rule of type (D) or a rule of type (B) and then a rule of type (C).

In the case of a rule of type (D) used to realize $\mathcal{T}_1 \Rightarrow^* \mathcal{T}$, no new terminal occurrences are introduced to $\mathcal{T}$. Thus $G_{\mathcal{T}_1} = G_\mathcal{T}$, proving the lemma by the assumption.

In the case of a combination of rules of types (B) and (C), one new vertex h, labeled with the new terminal occurrence b in the RHS of the rule of type C, is introduced to $G_\mathcal{T}$. New vertex h, along with the vertices labeled with $a_1, \cdots, a_t, a_{t+2}, \cdots, a_{k+1}$, forms a $(k+1)$-clique, thus $G_\mathcal{T}$ is a k-tree. In addition, let i and j be two vertices in $G_\mathcal{T}$ such that $i < j$ and there is no vertex between them. If neither is labeled with the terminal occurrence b, they should belong to $G_{\mathcal{T}_1}$ as well. By the assumption they satisfy condition (b) of Definition 8. If i (resp. j) is labeled with b, the rule of type (C) ensures that (h, i) (resp. (h, j)) is included in the new $(k+1)$-clique, thus in $G_\mathcal{T}$. Therefore, $G_\mathcal{T}$ is a faithful k-tree to s. $\qquad\square$

Let $\{S\} \Rightarrow^* \mathcal{T}$ for which s, $|s| = n$, is the underlying string. Then by Lemma 9, $G_\mathcal{T}$ is a k-tree of n vertices faithful to s. According to Definition 7, edge $(i, i+1)$ is in $G_\mathcal{T}$, for all $1 \leq i \leq n - 1$. Hence, $G_\mathcal{T}$ contains the annotated Hamiltonian path $\{(i, i+1) : 1 \leq i \leq n - 1\}$. We thus have the following.

Theorem 10. *Let Γ be a k-tree grammar and string $s \in L(\Gamma)$. The structure space $\mathcal{E}(s)$ is a set of k-trees, each containing the Hamiltonian path $\{(i, i + 1) : 1 \leq i \leq n - 1\}$, where $n = |s|$.*

On the other hand, we are interested in such k-tree grammars that for every string s in the defined language, the structure space $\mathcal{E}(s)$ contains all possible

k-trees (of size $n = |s|$) constrained by the annotated Hamiltonian path. In the following, we show that such k-grammars do exist.

Recall Definition 1 for creating all possible k-trees. Let $\kappa = \{i_1, i_2, \cdots, i_{k+1}\}$ be an existing $(k + 1)$-clique, with $i_1 < i_2 \cdots < i_{k+1}$. We call any new $(k + 1)$-clique a *child of* κ if it is formed by a newly introduced vertex along with exactly k vertices already in κ.

Lemma 11. *Let $\kappa = \{i_1, i_2, \cdots, i_{k+1}\}$ be an existing $(k + 1)$-clique. Then with the Hamiltonian path constraint, κ can have at most $k + 2$ children.*

Proof. (Sketch) A new $(k+1)$-clique can be created by introducing a new vertex in one of the $k + 2$ intervals $(1, i_1), (i_1, i_2), \cdots, (i_{k+1}, n)$ to connect to exactly k vertices in the clique κ. Therefore, it suffices to show that, for each of the $(k+2)$ intervals, at most one new $(k + 1)$-clique can be created.

Without loss of generality, assume two different new $(k+1)$-cliques κ_1 and κ_2 are created with two new vertices h and l drawn from the same interval (i_j, i_{j+1}), respectively, where $i_j < h < l < i_{j+1}$. Apparently (h, l) is not an edge. Nor can there be a path $\{(h, h+1), (h+1, h+2), \cdots, (h+m, l)\}$, where $h + m = l - 1$, for any $m \geq 1$. This is because a new vertex between h and l will only be introduced as a part of descendant of either κ_1 or κ_2 but not both. Therefore, there must be r, $0 \leq r \leq m$, such that edge $(h + r, h + r + 1)$ is not accounted for as a part of the Hamiltonian path. □

Theorem 12. *Let $k \geq 2$ be a fixed integer. There exists a k-tree grammar Γ such that $L(\Gamma) = \Sigma^*$ and, for any given string $s \in L(\Gamma)$ of length n, the structure space $\mathcal{E}(s)$ contains all k-trees constrained by the Hamiltonian path $\{(i, i + 1) : 1 \leq i < n\}$.*

Proof. (Sketch) It suffices to show that such a desired k-tree grammar has a finite number of rules.

Recall the four types of grammar rules given in Definition 3. Each rule $\{S\} \to \{\beta\}$ of type (A) induces a $(k + 1)$-clique corresponding to the co-occurrence of $k + 1$ terminals in β. Such rules can be at most $O(|\Sigma|^{k+1}|\mathcal{N}|^{k+2})$ in number. Each rule $\alpha \to \{\beta, \gamma\}$ of type (B) and each rule $\beta \to \rho$ of type (C) work together to induce an additional $(k + 1)$-clique from the $(k + 1)$-clique whose vertices are labeled with the $k + 1$ terminals that co-occurr in α. As a result of the rule applications two symbolic strings are derived. One symbolic string contains k existing terminals selected from those in α to co-occur with a new terminal occurrence b, while the other symbolic string retains the co-occurrences of $k + 1$ terminals in α but "masks off" the segment that introduces b. The latter symbolic string allows rules of types (B) and (C) to be repeatedly applied to induce more $(k + 1)$-cliques from the same terminal occurrences in α. By Lemma 11, rules of types (B) and (C) are bounded by $O(|\Sigma|^{k+2}|\mathcal{N}|^{k+4}k^2)$ in number. Finally, type (D) rules are used to terminate recursion without deriving new terminal occurrence. They are bounded by $O(|\Sigma|^{k+1}|\mathcal{N}|^{k+2})$ in number as well. □

3   Probability Computation with k-Tree Grammars

3.1   Stochastic k-Tree Grammars

Definition 13. A *stochastic k-tree grammar* (SkTG) is a pair (Γ, θ), where $\Gamma = (\Sigma, \mathcal{N}, \mathcal{R}, M, I, S)$ is a k-tree grammar and θ is a function: $\mathcal{R} \to [0, 1]$ such that for every $\alpha \in (\Sigma \cup \mathcal{N})^+$,

$$\sum_{\alpha \to \mathcal{A} \in \mathcal{R}} \theta(\alpha \to \mathcal{A}) = 1$$

We interpret the probability model θ associated with grammar rules as follows. $\theta(S \to \{\beta\})$ is the probability for co-occurrence of the $k + 1$ terminals in β. θ associated with all such type (A) rules gives a probability distribution over all co-occurrences of $k + 1$ terminals. In addition, θ distributes probabilities between rules of type (B) and of type (D) to account for the expected number of co-occurrences of $k + 1$ terminals that share the same set of at least $k - 1$ terminal occurrences. $\theta(\alpha \to \beta)$ of a type (C) rule is probability for co-occurrence of the $k + 1$ terminals in β conditional on co-occurrence of the $k + 1$ terminals in α.

Definition 14. Let $\mathcal{T} \subseteq (\Sigma \cup \mathcal{N})^+$ be such that $\{S\} \Rightarrow^* \mathcal{T}$. Then the *probability of derivation* $\{S\} \Rightarrow^* \mathcal{T}$ with (Γ, θ) is defined recursively as

$$Prob(\mathcal{T}|\Gamma, \theta) = \sum_{r \in \mathcal{R}, \mathcal{T}' \Rightarrow_r \mathcal{T}} Prob(\mathcal{T}'|\Gamma, \theta) \times \theta(r)$$

with the base case $Prob(\{S\}|\Gamma, \theta) = 1$.

Definition 15. Let (Γ, θ) be a SkTG. Then for any given string $s \in L(\Gamma)$, its probability with (Γ, θ) is defined as

$$Prob(s|\Gamma, \theta) = \sum_{\{S\} \Rightarrow^* \mathcal{T}, \, uls(\mathcal{T}, s)} Prob(\mathcal{T}|\Gamma, \theta)$$

Therefore, the probability of s under the model (Γ, θ) is computed as the sum of probabilities of all derivations of s by the grammar. In other word, $Prob(s|\Gamma, \theta)$ is the likelihood for the string s to possess at least one k-tree structure. We observe that

Proposition 16. *Let (Γ, θ) be a SkTG, the strings in the language $L(\Gamma)$ form a probabilistic space, i.e.,*

$$\sum_{s \in L(\Gamma)} Prob(s|\Gamma, \theta) = 1$$

Alternatively, it is of interest to know the most likely structure possessed by a given string s. This then is to compute the maximum probability of a derivation $\{S\} \Rightarrow^* \mathcal{T}$ for which s is the underlying string. Similar to the total probability computation, we can define maximum probability recursively,

Let $\mathcal{T} \subseteq (\Sigma \cup \mathcal{N})^+$ be such that $\{S\} \Rightarrow^* \mathcal{T}$. Then the *maximum probability of derivation* $\{S\} \Rightarrow^* \mathcal{T}$ is defined recursively as

$$Maxp(\mathcal{T}|\Gamma, \theta) = \max_{r \in \mathcal{R},\, \mathcal{T}' \Rightarrow_r \mathcal{T}} Maxp(\mathcal{T}'|\Gamma, \theta) \times \theta(r)$$

with the base case $Maxp(\{S\}|\Gamma, \theta) = 1$.

Definition 17. Let (Γ, θ) be a stochastic k-tree grammar. Then for every given string $s \in L(\Gamma)$, the maximum probability of a derivation for s is defined as

$$Maxp(s|\Gamma, \theta) = \max_{\{S\} \Rightarrow^* \mathcal{T},\, uls(\mathcal{T}, s)} Maxp(\mathcal{T}|\Gamma, \theta)$$

And *the most likely structure* for s with (Γ, θ) is the induced graph $G_{\mathcal{T}^\diamond}$ of the subset $\mathcal{T}^\diamond \subseteq (\Sigma \cup \{M\})^+$ decoded from $Maxp(s|\Gamma, \theta)$, where

$$\mathcal{T}^\diamond = arg \max_{\{S\} \Rightarrow^* \mathcal{T},\, uls(\mathcal{T}, s)} Maxp(\mathcal{T}|\Gamma, \theta)$$

3.2   Dynamic Programming Algorithms

We now show probability computations with SkTG can be done efficiently. We outline a dynamic programming strategy for computing the maximum probability function $Maxp$. The computation for the total probability function is similar. Let $s = s_1 \cdots s_n$, where $s_i \in \Sigma$, for $1 \le i \le n$, be a given terminal string.

Definition 18. Let $\alpha = X_0 a_1 X_1 \cdots a_{k+1} X_{k+1} \in (\Sigma \cup \mathcal{N})^+$ be a symbolic string and $\kappa = (l_1, l_2, \cdots, l_{k+1})$ be $k+1$ ordered integers where $1 \le l_1 < l_2, \cdots, l_{k+1} \le n$. (α, κ) is a *consistent pair* if
  (1) $a_i = s_{l_i}$, $1 \le i \le k+1$, and
  (2) For $i = 0, 1, \cdots, k+1$, $X_i = \epsilon$ iff $l_i = l_{i+1} - 1$ ($l_0 =_{df} 1$ and $l_{k+2} =_{df} n$).

Now given a pair (α, κ), we define function $f(\alpha, \kappa)$ to be the maximum probability for a derivation $\{\alpha\} \Rightarrow^* \mathcal{T}$, where $\mathcal{T} \subseteq (\Sigma \cup \{M\})^+$ for which s is the underlying string. Then function f can be recursively defined according to types of α and the types of rules α is involved with in $\mathcal{R}$.

1. $\alpha \in (\Sigma \cup \{M\})^+$:

$$f(\alpha, \kappa) = \begin{cases} 1 & (\alpha, \kappa) \text{ is a consistent pair} \\ 0 & \text{otherwise} \end{cases}$$

2. $\alpha \in (\Sigma \cup \mathcal{N})^+$ but $\alpha \ne S$:

$$f(\alpha, \kappa) = \max_{r \in \mathcal{R}} \begin{cases} f(\beta, \kappa) f(\gamma, \kappa) \theta(r) & r = \alpha \to \{\beta, \gamma\}, \text{ type (B)} \\ \max_{l_s < h < l_{s+1}, \kappa' = \kappa|_h^{l_{t+1}}} f(\beta, \kappa') \theta(r) & r = \alpha \to \{\beta\}, \text{ type (C)} \\ f(\beta, \kappa) \theta(r) & r = \alpha \to \{\beta\}, \text{ type (D)} \end{cases}$$

where for the case of r being a type (C) rule, s and t are known values given in $\beta = \alpha|_{YbZ}^{X_s}|_M^{X_t a_{t+1} X_{t+1}}$, satisfying $(s-t) > 1$ or $(t-s) \ge 1$, and $\kappa' = \kappa|_h^{l_{t+1}}$ represents the ordered set modified from κ by replacing l_{t+1} with h.

3. $\alpha = S$:

$$f(S, \kappa) = \max_{S \to \{\beta\} \in \mathcal{R}} f(\beta, \kappa)\theta(S \to \{\beta\})$$

Theorem 19. $Maxp(s|\Gamma, \theta) = \max_{\kappa \in [n]^{k+1}} f(S, \kappa)$, where $[n]^{k+1}$ is the set of all combinations of $k + 1$ integers in $[n] = \{1, 2, \cdots, n\}$.

Proof. (Sketch) We prove by induction on the number m of rule applications in a process to generate the string s with the maximum probability, where $m \geq 2$. The base case $m = 2$ is obvious. The proof of inductive step examines all possible types of rules used in the last step.   □

A dynamic programming algorithm can be implemented to compute function $f(\alpha, \kappa)$. This is to establish a table to store computed values of function f through the use of the formulae provided above (the cases 1 through 3). The table has $k+2$ dimensions, one for all α's in the grammar and the other $k+1$ are for all κ's, resulting in the $O(n^{k+2}|\Gamma|)$-time and $O(n^{k+1}|\Gamma|)$-space complexities, respectively, for every fixed k.

4   Applications and Discussions

We have introduced the stochastic k-tree grammar (SkTG) for the purpose of modeling context-sensitive yet tamable crossing co-occurrences of terminals. The recursive rules of the new grammar permit association of probability distributions in a natural way. The resulting dynamic programming algorithms for probability computation with SkTG are efficient enough, with potential for statistical analysis of real-world structures. This work is in progress in both application and further theoretical investigation.

4.1   Application in Biomolecular Structure Prediction

This work was initially motivated by the need in the analysis of biomolecules for tertiary structure prediction. A biomolecular sequence, e.g., ribonucleic acid (RNA) or protein, is a linear chain of residues interacting spatially to form a 3D structure functionally important [22,20]. One of the most desirable computational biology tasks is to predict the tertiary structure from the sequence information only [12,38]. The newly introduced SkTG offers a viable approach to this task. We briefly outline the application as follows.

Biomolecular sequences are natural strings definable over some finite alphabet Σ (e.g., $\Sigma = \{A, C, G, U\}$ for nucleic acids). A class of biomolecular sequences can be defined as a language with a SkTG in which grammar rules model statistically not only the sequential composition but also structural composition of the sequences. The task of designing SkTGs, much like that for SCFGs, is non-trivial and may often be based on experience. Equipping a designed SkTG Γ with probability parameters θ may be done through learning from known biomolecules (with or without known structures) (see next subsection for a briefly discussion).

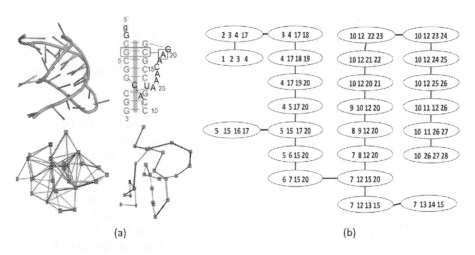

(a)   (b)

Fig. 3. Illustration of a tertiary structure prediction from BWYV (beet western yellows virus) RNA molecule sequence (PDB ID: 1L2X) that contains 28 nucleotides, coaxial helices connected with two loops, and an A-minor motif. Top of (a): Tertiary structure (drawn via pymol) and details of nucleotide interactions (http://www.biomath.nyu.edu/motifs/); (b) The induced 3-tree (containing desired interactions) corresponding to a derivation of the sequence with the maximum probability. The 3-tree is presented in terms of the tree topology connecting the created 4-cliques in the 3-tree. Bottom-left of (a): 3D representation of the 3-tree with one tetrahedron for every 4-clique; Bottom-right of (a): only backbone edges are kept from the tetrahedron representation, serving as a preliminary structure prediction from the sequence. We note that more accurate structural motif modeling of individual 4-cliques would allow more accurate prediction of the overall tertiary structure.

With an SkTG (Γ, θ), using the dynamic programming algorithm (developed in section 3) we can compute the maximum probability of an induced k-tree , e.g., $k = 3$, from a given query sequence. In such an application, every $(k+1)$-clique κ in the desired k-tree may potentially admit one of many possible configurations (i.e., all possible interaction topologies along with permissible geometry shapes) for the $k+1$ residues in κ. Therefore, the dynamic programming algorithm can be tailored to include the third argument C_κ in the probability function f defined in section 4, where C_κ is the set of all possible configurations incurred by $(k+1)$-clique κ. The information about C_κ can often be obtained from known tertiary structures of biomolecules as well. Figure 3 illustrates this approach used in a tertiary structure prediction for a small RNA molecule.

4.2   Further Theoretical Issues

SkTG is a natural extension from SCFG; in particular, k-tree grammars, for $k = 2$, can define all context-free languages. In addition, the outlined dynamic programming algorithm (in section 3.2) to compute the maximum probability can be improved. In fact, in a related work [6], the authors have developed an algorithm of time $O(n^{k+1})$, for every fixed value of k, for computing the maximum spanning k-tree that includes a designated Hamiltonian path. On the other hand, due to the long standing barrier of $O(n^3)$ for parsing context-free languages, this also suggests the time complexity upper bound $O(n^{k+1})$ has optimal order of growth in n for each $k \geq 2$.

We further note that the above efficiency issue is closely related with the parameterized complexity [7] of the following problem: computing the maximum probability of an input sentence to be produced by an input SkTG, for which k is considered a variable parameter. By the above observation, such a problem is likely parameterized intractable. Nevertheless, the interesting question remains whether an additional small parameter (e.g., significant in applications) may be associated with such problems for further improvement of computational efficiency.

Estimation of probability parameters θ for given k-tree grammars deserves more thorough investigation and it is not within the scope of this paper. However, we point out that it is highly possible to develop efficient parameter estimation algorithms for SkTG. This is because $O(n^{k+1})$-time algorithms may exist for computing the maximum and total probabilities of given language strings. Much like the analogous algorithms for SCFG, these algorithms can be used to re-estimate probability parameters θ given an initial parameter θ_0, through an EM algorithm.

Finally, we feel that future work is also needed to investigate the relationship between the k-tree grammar and other grammars that already exist (e.g., the Tree-Adjoining Grammar and its generalized versions [14]) for constrained context-sensitive languages.

Acknowledgments. This work was supported in part by NSF IIS grant (award No: 0916250).

References

1. Achawanantakun, R., Takyar, S., Sun, Y.: Grammar string: A novel ncRNA secondary structure representation. lifesciences society org, pp. 2–13 (2010)
2. Rozenknop, A.: Gibbsian context-free grammar for parsing. In: Sojka, P., Kopeček, I., Pala, K. (eds.) TSD 2002. LNCS (LNAI), vol. 2448, pp. 49–56. Springer, Heidelberg (2002)
3. Arnborg, S., Proskurowski, A.: Linear time algorithms for np-hard problems restricted to partial k-trees. Discrete Applied Mathematics 23(1), 11–24 (1989)
4. Chiang, D., Joshi, A.K., Searls, D.B.: Grammatical representations of macromolecular structure. Journal of Computational Biology 13(5), 1077–1100 (2006)
5. Dill, K.A., Lucas, A., Hockenmaier, J., Huang, L., Chiang, D., Josh, A.K.: Computational linguistics: A new tool for exploring biopolymer structures and statistical mechanics. Polymer 48, 4289–4300 (2007)
6. Ding, L., Samad, A., Li, G., Robinson, R., Xue, X., Malmberg, R., Cai, L.: Finding maximum spanning k-trees on backbone graphs in polynomial time (2013) (manuscript)

7. Downey, R.G., Fellows, M.R.: Parameterized Complexity. Springer (1999)
8. Durbin, R., Eddy, S., Krogh, A., Mitchison, G.: Biological Sequence Analysis: Probabilistic Models of Proteins and Nucleic Acids. Cambridge University Press (1998)
9. Hopcroft, J.E., Motwani, R., Ullman, J.D.: Introduction to Automata Theory, Languages, and Computation. Addison-Wesley (2007)
10. Huang, Z., Mohebbi, M., Malmberg, R., Cai, L.: RNAv: Non-coding RNA secondary structure variation search via graph homomorphism. In: Proceedings of Computational Systems Bioinformatics Conference (CSB 2010), vol. 9, pp. 56–69 (2010)
11. Huang, Z., Wu, Y., Robertson, J., Feng, L., Malmberg, R., Cai, L.: Fast and accurate search for non-coding RNA pseudoknot structures in genomes. Bioinforamtics 24(20), 2281–2287 (2008)
12. Thiim, J.F.I.M., Mardia, M., Ferkinghoff-Borg, K., Hamelryck, J.,, T.: A probabilistic model of RNA conformational space. PLoS Comput. Biol. 5(6) (2009)
13. Joshi, A.: How much context-sensitivity is necessary for characterizing structural descriptions. In: Dowty, D., Karttunen, L., Zwicky, A. (eds.) Natural Language Processing: Theoretical, Computational, and Psychological Perspectives, pp. 206–250. Cambridge University Press, NY (1985)
14. Joshi, A., Vijay-Shanker, K., Weir, D.: The convergence of mildly context-sensitive grammar formalisms. Issues in Natural Language Processing, pp. 31–81. MIT Press, Cambridge (1991)
15. Jurafsky, D., Wooters, C., Segal, J., Stolcke, A., Fosler, E., Tajchaman, G., Morgan, N.: Using a stochastic context-free grammar as a language model for speech recognition. In: Proceedings of International Conference on Acoustics, Speech and Signal Processing, pp. 189–192 (1995)
16. Klein, D., Manning, C.: Accurate unlexicalized parsing. In: Proceedings of the 41st Meeting of the Association for Computational Linguistics, pp. 423–430 (2003)
17. Knudsen, B., Hein, J.: Pfold: RNA secondary structure prediction using stochastic context-free grammars. Nucleic Acids Res. 31, 3423–3428 (2003)
18. Lari, K., Young, S.J.: The estimation of stochastic context-free grammars using the inside-outside algorithm. Computer Speech and Language 4, 35–56 (1990)
19. Martin, D., Sigal, R., Weyuker, E.J.: Computability, complexity, and languages: Fundamentals of theoretical computer science, 2nd edn. Morgan Kaufmann (1994)
20. Murzin, A.G., Brenner, S., Hubbard, T., Chothia, C.: Scop: A structural classification of proteins database for the investigation of sequences and structures. Journal of Molecular Biology 247(4), 536–540 (1995)
21. Nawrocki, E.P., Kolbe, D.L., Eddy, S.R.: Infernal 1.0: Inference of RNA alignments. Bioinformatics 25, 1335–1337 (2009)
22. Noller, H.F.: Structure of ribosomal RNA. Annual Review of Biochemistry 53, 119–162 (1984)
23. Patil, H.P.: On the structure of k-trees. Journal of Combinatorics, Information and System Sciences 11(2-4), 57–64 (1986)
24. Rivas, E., Lang, R., Eddy, S.R.: A range of complex probabilistic models for RNA secondary structure prediction that include the nearest neighbor model and more. RNA 18, 193–212 (2012)
25. Sakakibara, Y., Brown, M., Hughey, R., Mian, I.S., Sjolander, K., Underwood, R.C., Haussler, D.: Stochastic context-free grammars for tRNA modeling. Nucleic Acids Research 22, 5112–5120 (1994)
26. Salomaa, A.: Jewels of Formal Language Theory. Computer Science Press (1981)

27. Sánchez, I.A., Benedi, J.M., Linares, D.: Performance of a scfg-based language model with training data sets of increasing size. In: Proceedings of Conference on Pattern Recognition and Image Analysis, pp. 586–594 (2005)
28. Searls, D.B.: The computational linguistics of biological sequences. Artificial Intelligence and Molecular Biology, pp. 47–120 (1993)
29. Searls, D.B.: Molecules, languages and automata. In: Sempere, J.M., García, P. (eds.) ICGI 2010. LNCS, vol. 6339, pp. 5–10. Springer, Heidelberg (2010)
30. Sergio Caracciolo, S., Masbaum, G., Sokal, A., Sportiello, A.: A randomized polynomial-time algorithm for the spanning hypertree problem on 3-uniform hypergraphs. CoRR abs/0812.3593 (2008)
31. Song, Y., Liu, C., Huang, X., Malmberg, R., Xu, Y., Cai, L.: Efficient parameterized algorithms for biopolymer structure-sequence alignment. IEEE/ACM Transactions on Computational Biology and Bioinformatics 3(4), 423–431 (2006)
32. Srebro, N.: Maximum likelihood bounded tree-width Markov networks. Artificial Intelligence 143(2003), 123–138 (2003)
33. Uemura, Y., Hasegawa, A., Kobayashi, S., Yokomori, T.: Tree adjoining grammars for RNA structure prediction. Theoretical Computer Science 210, 277–303 (1999)
34. Vijay-Shanker, K., Weir, D.: The equivalence of four extensions of context-free grammars. Mathematical Systems Theory 27(6), 511–546 (1994)
35. Waters, C.J., MacDonald, B.A.: Efficient word-graph parsing and search with a stochastic context-free grammar. In: Proceedings of IEEE Workshop on Automatic Speech Recognition and Understanding, pp. 311–318 (1997)
36. Xu, J., Berger, B.: Fast and accurate algorithms for protein side-chain packing. Journal of the ACM 53(4), 533–557 (2006)
37. Xu, Y., Liu, Z., Cai, L., Xu, D.: Protein structure prediction by protein threading. In: Computational Methods for Protein Structure Prediction and Modeling, pp. 389–430. Springer I&II (2006)
38. Progress, Y.Z.: challenges in protein structure prediction. Current Opinions in Structural Biology 18(3), 342–348 (2008)
39. Weinberg, Z., Ruzzo, L.: Faster genome annotation of non-coding RNA families without loss of accuracy. In: Proceedings of Conference on Research in Computational Molecular Biology (RECOMB 2004), pp. 243–251 (2004)
40. Zimand, M.: The complexity of the optimal spanning hypertree problem. Technical Report, University of Rochester. Computer Science Department (2004)

Weighted Automata and Logics
for Infinite Nested Words

Manfred Droste and Stefan Dück[*]

Institut für Informatik, University Leipzig, 04109 Leipzig, Germany
{droste,dueck}@informatik.uni-leipzig.de

Abstract. Nested words introduced by Alur and Madhusudan are used
to capture structures with both linear and hierarchical order, e.g. XML
documents, without losing valuable closure properties. Furthermore, Alur
and Madhusudan introduced automata and equivalent logics for both finite and infinite nested words, thus extending Büchi's theorem to nested
words. Recently, average and discounted computations of weights in
quantitative systems found much interest. Here, we will introduce and
investigate weighted automata models and weighted MSO logics for infinite nested words. As weight structures we consider valuation monoids
which incorporate average and discounted computations of weights as
well as the classical semirings. We show that under suitable assumptions,
two resp. three fragments of our weighted logics can be transformed into
each other. Moreover, we show that the logic fragments have the same
expressive power as weighted nested word automata.

Keywords: nested words, weighted automata, weighted logics, quantitative automata, valuation monoids.

1 Introduction

Nested words, introduced by Alur and Madhusudan [2], capture models with
both a natural sequence of positions and an hierarchical nesting of these positions. Prominent examples include XML documents and executions of recursively
structured programs. Automata on nested words, logical specifications, and corresponding languages of nested words have been intensively studied, see [1],
[2], [17]. Recently, there has been much interest in quantitative features for the
specification and analysis of systems. Quantitative automata modeling the long-
time average or discounted behavior of systems were investigated by Chatterjee,
Doyen, and Henzinger [6], [7]. It is the goal of this paper to present quantitative
logics for such quantitative automata on nested words.

The connection between MSO logic and automata due to Büchi, Elgot, and
Trakhenbrot [5], [15], [21] has proven most fruitful. Weighted automata over
semirings (like $(\mathbb{N}, +, \cdot, 0, 1)$) were already investigated by Schützenberger [20]

[*] Supported by Deutsche Forschungsgemeinschaft (DFG), project DR 202/11-1 and
Graduiertenkolleg 1763 (QuantLA).

A.-H. Dediu et al. (Eds.): LATA 2014, LNCS 8370, pp. 323–334, 2014.

and soon developed a flourishing theory, cf. the books [3], [14], [16], [19] and the recent handbook [8]. However, an expressively equivalent weighted MSO logic was developed only recently [9]. This was extended to semiring-weighted automata and logics over finite nested words in [18], and further to strong bi-monoids as weight structures in [12]. For quantitative automata and logics, incorporating average and discounting computations of weights over words, such an equivalence was given in [11].

In this paper, we will investigate quantitative nested word automata and suitable quantitative MSO logics. We will concentrate on infinite nested words, although our results also hold for finite nested words. We employ the stair Muller nested word automata of [2], [17], since these can be determinized without losing expressive power. As weight structures we take the valuation monoids of [11]. These include infinite products as in totally complete semirings [13], but also computations of long-time averages or discountings of weights. As example for such a setting we give the calculation of the long-time ratio of bracket-free positions in prefixes of an infinite nested word. As our first main result, we show that under suitable assumptions on the valuation monoid D, two resp. three versions of our weighted MSO logic have the same expressive power. In particular, if D is commutative, then any weighted MSO-formula is equivalent to one in which conjunctions occur only between 'classical' boolean formulas and constants. In contrast to [11], our proof uses direct conversions of the formulas and thus has much better complexity than using the automata-theoretic constructions of [11]. These conversions are new even for the case of weighted logics on words.

In our second main result, we show under suitable assumptions on the valuation monoid that our weighted MSO logics have the same expressive power as weighted nested automata. These assumptions on the valuation monoid are satisfied by long-time average resp. discounted computations of weights; therefore our results apply to these settings. All our constructions of automata from formulas or conversely are effective.

2   Automata and Logics for Nested ω-words

In this section we describe basic background for classical (unweighted) automata and logics on nested-ω-words. We denote by Σ an alphabet and by Σ^ω the set of all ω-words over Σ. $\mathbb{N}$ is the set of all natural numbers without zero. For a binary relation R, we denote with $R(x,y)$ that $(x,y) \in R$.

Definition 1. *A* matching relation ν *over* $\mathbb{N}$ *is a subset of* $(\{-\infty\} \cup \mathbb{N}) \times (\mathbb{N} \cup \{\infty\})$ *such that:*

  (i) $\nu(i,j) \Rightarrow i < j$,
  (ii) $\forall i \in \mathbb{N} : |\{j : \nu(i,j)\}| \le 1 \wedge |\{j : \nu(j,i)\}| \le 1$,
  (iii) $\nu(i,j) \wedge \nu(i',j') \wedge i < i' \Rightarrow j < i' \vee j > j'$,
  (iv) $(-\infty, \infty) \notin \nu$.

A nested ω-word nw *over* Σ *is a pair* $(w,\nu) = (a_1a_2..., \nu)$ *where* $w = a_1a_2...$ *is an ω-word over* Σ *and* ν *is a matching relation over* $\mathbb{N}$. *We denote by* $NW^\omega(\Sigma)$

the set of all nested ω-words over Σ and we call every subset of $NW^\omega(\Sigma)$ a language of nested ω-words.

If $\nu(i, j)$ holds, we call i a *call position* and j a *return position*. In case of $j = \infty$, i is a *pending call* otherwise a *matched call*. In case of $i = -\infty$, j is a *pending return* otherwise a *matched return*. If i is neither call nor return, then we say i is an *internal*.

Definition 2. *A deterministic stair Muller nested word automaton (sMNWA) over Σ is a quadruple $\mathcal{A} = (Q, q_0, \delta, \mathfrak{F})$, where $\delta = (\delta_{\text{call}}, \delta_{\text{int}}, \delta_{\text{ret}})$, consisting of:*

- *a finite set of states Q,*
- *an initial state $q_0 \in Q$,*
- *a set $\mathfrak{F} \subseteq 2^Q$ of accepting sets of states,*
- *the transition functions $\delta_{\text{call}}, \delta_{\text{int}} : Q \times \Sigma \to Q$,*
- *the transition function $\delta_{\text{ret}} : Q \times Q \times \Sigma \to Q$.*

A *run* r of the sMNWA $\mathcal{A}$ on the nested ω-word $nw = (a_1 a_2 ..., \nu)$ is an infinite sequence of states $r = (q_0, q_1, ...)$ where $q_i \in Q$ for each $i \in \mathbb{N}$ and q_0 is the inital state of $\mathcal{A}$ such that for each $i \in \mathbb{N}$ the following holds:

$$\begin{cases} \delta_{\text{call}}(q_{i-1}, a_i) = q_i & \text{, if } \nu(i, j) \text{ for some } j > i \text{ (or } j = \infty) \\ \delta_{\text{int}}(q_{i-1}, a_i) = q_i & \text{, if } i \text{ is an internal} \\ \delta_{\text{ret}}(q_{i-1}, q_{j-1}, a_i) = q_i & \text{, if } \nu(j, i) \text{ for some } 1 \le j < i \\ \delta_{\text{ret}}(q_{i-1}, q_0, a_i) = q_i & \text{, if } \nu(-\infty, i) \text{ .} \end{cases}$$

We call $i \in \mathbb{N}$ a *top-level position* if there exist no positions $j, k \in \mathbb{N}$ with $j < i < k$ and $\nu(j, k)$. We define

$$Q^t_\infty(r) = \{q \in Q \mid q = q_i \text{ for infinitely many top-level positions } i\} \ .$$

A run r of an sMNWA is *accepted* if $Q^t_\infty(r) \in \mathfrak{F}$. An sMNWA $\mathcal{A}$ *accepts* the nested ω-word nw if there is an accepted run of $\mathcal{A}$ on nw. We denote with $L(\mathcal{A})$ the set of all accepted nested ω-words of $\mathcal{A}$. We call a language L of nested ω-words *regular* if there is an sMNWA $\mathcal{A}$ with $L(\mathcal{A}) = L$.

Alur and Madhusudan [2] considered nondeterministic Büchi NWA and nondeterministic Muller NWA. They showed that the deterministic versions of these automata have strictly less expressive power than the nondeterministic automata. However, refering to Löding, Madhusudan and Serre [17], Alur and Madhusudan stated that deterministic stair Muller NWA have the same expressive power as their nondeterministic versions as well as nondeterministic Büchi NWA. Moreover, the class of regular languages of nested-ω-words is closed under union, intersection and complement ([2]).

Definition 3. *The monadic second order logic for nested words $MSO(NW(\Sigma))$ contains exactly all formulas φ which are given by the following syntax:*

$$\varphi ::= \text{Lab}_a(x) \mid \text{call}(x) \mid \text{ret}(x) \mid x \le y \mid \nu(x, y) \mid x \in X \mid \neg \varphi \mid \varphi \vee \varphi \mid \exists x . \varphi \mid \exists X . \varphi$$

where $a \in \Sigma$ and x, y are first order variables and X is a second order variable.

The semantics of these formulas is given in a natural way, cf. [2]. Later we give a full definition of the semantics of *weighted* MSO-formulas. We call φ a *sentence* if φ contains no free variables. If φ is a sentence, then $L(\varphi) = \{nw \in NW^\omega(\Sigma) \mid nw \vDash \varphi\}$ is *the language defined by* φ.

Theorem 4 (Alur, Madhusudan [2]). *Let L be a language of nested ω-words over Σ. Then L is regular if and only if L is definable by some $MSO(NW(\Sigma))$-sentence φ.*

3   Weighted Stair Muller Nested Word Automata

In this section, we introduce weighted versions of stair Muller nested word automata. As weight structures, we will employ ω-valuation monoids introduced in [11]. We recall the definitions.

A monoid $(D, +, 0)$ is *complete* if it has infinitary sum operations $\sum_I : D^I \to D$ for any index set I such that

- $\sum_{i \in \varnothing} d_i = 0$, $\sum_{i \in \{k\}} d_i = d_k$, $\sum_{i \in \{j,k\}} d_i = d_j + d_k$ for $j \neq k$,
- $\sum_{j \in J}(\sum_{i \in I_j} d_i) = \sum_{i \in I} d_i$ if $\bigcup_{j \in J} I_j = I$ and $I_j \cap I_k = \varnothing$ for $j \neq k$.

Note that in every complete monoid the operation $+$ is commutative. We let D^ω comprise all infinite sequences of elements of D.

Definition 5 (Droste, Meinecke [11]). *An ω-valuation monoid $(D, +, \mathrm{Val}^\omega, 0)$ is a complete monoid $(D, +, 0)$ equipped with an ω-valuation function $\mathrm{Val}^\omega : D^\omega \to D$ with $\mathrm{Val}^\omega((d_i)_{i \in \mathbb{N}}) = 0$ if $d_i = 0$ for some $i \in \mathbb{N}$.*

A product ω-valuation monoid $(D, +, \mathrm{Val}^\omega, \diamond, 0, 1)$ (short ω-pv-monoid) is an ω-valuation monoid $(D, +, \mathrm{Val}^\omega, 0)$ with a constant $1 \in D$ and an operation $\diamond : D^2 \to D$ satisfying $\mathrm{Val}^\omega(1^\omega) = 1$, $0 \diamond d = d \diamond 0 = 0$ and $1 \diamond d = d \diamond 1 = d$ for all $d \in D$.

Let $(D, +, \mathrm{Val}^\omega, \diamond, 0, 1)$ be an ω-pv-monoid. D is called *associative* resp. *commutative* if $\diamond$ is associative resp. commutative. D is *left-+-distributive* if for all $d \in D$, for any index set I and $(d_i)_{i \in I} \in D^I$:

$$d \diamond \sum_{i \in I} d_i = \sum_{i \in I}(d \diamond d_i) \ .$$

Right-+-distributivity is defined analogously. We call D *+-distributive* if D is left- and right-+-distributive. D is *left-Val^ω-distributive* if for all $d \in D$ and $(d_i)_{i \in \mathbb{N}} \in D^\omega$:

$$d \diamond \mathrm{Val}^\omega((d_i)_{i \in \mathbb{N}}) = \mathrm{Val}^\omega((d \diamond d_i)_{i \in \mathbb{N}}) \ .$$

D is *left-multiplicative* if for all $d \in D$ and $(d_i)_{i \in \mathbb{N}} \in D^\omega$:

$$d \diamond \mathrm{Val}^\omega((d_i)_{i \in \mathbb{N}}) = \mathrm{Val}^\omega(d \diamond d_1, (d_i)_{i \geq 2}) \ .$$

D is called *conditionally commutative*, if for all $(d_i)_{i \in \mathbb{N}}, (d'_i)_{i \in \mathbb{N}} \in D^\omega$ with $d_i \diamond d'_j = d'_j \diamond d_i$ for all $j < i$, the following holds:

$$\mathrm{Val}^\omega((d_i)_{i \in \mathbb{N}}) \diamond \mathrm{Val}^\omega((d'_i)_{i \in \mathbb{N}}) = \mathrm{Val}^\omega((d_i \diamond d'_i)_{i \in \mathbb{N}}) \ .$$

We call D *left-distributive* if D is left-+-distributive and, additionally, left-Val^ω-distributive or left-multiplicative. If D is +-distributive and associative, then $(D, +, \diamond, 0, 1)$ is a complete semiring and we call $(D, +, \mathrm{Val}^\omega, \diamond, 0, 1)$ an ω-*valuation semiring*. A *cc-ω-valuation semiring* is an ω-valuation semiring D which is conditionally commutative and left-distributive.

Example 1 ([11]). We set $\bar{\mathbb{R}} = \mathbb{R} \cup \{-\infty, \infty\}$ and $-\infty + \infty = -\infty$. We let

$$(D_1, +, \mathrm{Val}^\omega, \diamond, 0, 1) = (\bar{\mathbb{R}}, \sup, \lim \mathrm{avg}, +, -\infty, 0),$$

where
$$\lim \mathrm{avg}((d_i)_{i \in \mathbb{N}}) = \liminf_{n \to \infty} \frac{1}{n} \sum_{i=1}^{n} d_i \ .$$

Let $0 < \lambda < 1$ and $\bar{\mathbb{R}}_+ = \{x \in \bar{\mathbb{R}} \mid x \geq 0\} \cup \{-\infty\}$. We put

$$(D_2, +, \mathrm{Val}^\omega, \diamond, 0, 1) = (\bar{\mathbb{R}}_+, \sup, \mathrm{disc}_\lambda, +, -\infty, 0),$$

where
$$\mathrm{disc}_\lambda((d_i)_{i \in \mathbb{N}}) = \lim_{n \to \infty} \sum_{i=1}^{n} \lambda^{i-1} d_i \ .$$

Then D_1 is a left-+-distributive and left-Val^ω-distributive ω-valuation monoid but not conditionally commutative. Furthermore, D_2 is a left-multiplicative cc-ω-valuation semiring.

Definition 6. *A weighted stair Muller nested word automaton (wsMNWA) $\mathcal{A} = (Q, I, \delta, \mathfrak{F})$, where $\delta = (\delta_{\mathrm{call}}, \delta_{\mathrm{int}}, \delta_{\mathrm{ret}})$, over the alphabet Σ and the ω-valuation monoid $(D, +, \mathrm{Val}^\omega, 0)$ consists of:*

- *a finite set of states Q,*
- *a set $I \subseteq Q$ of initial states,*
- *a set $\mathfrak{F} \subseteq 2^Q$ of accepting sets of states,*
- *the weight functions $\delta_{\mathrm{call}}, \delta_{\mathrm{int}} : Q \times \Sigma \times Q \to D$,*
- *the weight function $\delta_{\mathrm{ret}} : Q \times Q \times \Sigma \times Q \to D$.*

A *run* r of the wsMNWA $\mathcal{A}$ on the nested ω-word $nw = (a_1 a_2 ..., \nu)$ is an infinite sequence of states $r = (q_0, q_1, ...)$. We denote with $wt_\mathcal{A}(r, nw, i)$ the weight of the transition of r used at position $i \in \mathbb{N}$, defined as follows

$$wt_\mathcal{A}(r, nw, i) = \begin{cases} \delta_{\mathrm{call}}(q_{i-1}, a_i, q_i) & \text{, if } \nu(i, j) \text{ for some } j > i \\ \delta_{\mathrm{int}}(q_{i-1}, a_i, q_i) & \text{, if } i \text{ is an internal} \\ \delta_{\mathrm{ret}}(q_{i-1}, q_{j-1}, a_i, q_i) & \text{, if } \nu(j, i) \text{ for some } 1 \leq j < i \\ \delta_{\mathrm{ret}}(q_{i-1}, q_I, a_i, q_i) & \text{, if } \nu(-\infty, i) \text{ for some } q_I \in I \ . \end{cases} \tag{1}$$

Then we define the *weight* $wt_\mathcal{A}(r, nw)$ of r on nw by letting

$$wt_\mathcal{A}(r, nw) = \mathrm{Val}^\omega((wt_\mathcal{A}(r, nw, i))_{i \in \mathbb{N}}) \ .$$

We define top-level positions and the set $Q_\infty^t(r)$ as before. A run r is *accepted* if $q_0 \in I$ and $Q_\infty^t(r) \in \mathfrak{F}$. We denote with $acc(\mathcal{A})$ the set of all accepted runs in $\mathcal{A}$.

We define the *behavior of the automaton* $\mathcal{A}$ as the function $\|\mathcal{A}\| : NW^\omega(\Sigma) \to D$ given by (where as usual, empty sums are defined to be 0)

$$\|\mathcal{A}\|(nw) = \sum_{r \in acc(\mathcal{A})} wt_\mathcal{A}(r, nw)$$

$$= \sum_{r \in acc(\mathcal{A})} Val^\omega((wt_\mathcal{A}(r, nw, i))_{i \in \mathbb{N}}) .$$

We call every function $S : NW^\omega(\Sigma) \to D$ a *nested ω-word series* (short: *series*). We call a series S *regular* if there exists an automaton $\mathcal{A}$ with $\|\mathcal{A}\| = S$.

Example 2. Within the following example we call a position i of a nested ω-word $nw = (w, \nu)$ *bracketfree* if there are no positions $j, k \in (\mathbb{N} \cup \{-\infty, \infty\})$ with $j < i < k$ and $\nu(j, k)$. This requirement is stronger than i being a top-level position because it contains $-\infty$ and ∞ thus also banning i being in the scope of pending calls and pending returns. Only for well-matched nested ω-words, i.e. nested ω-words without pending edges, the two properties coincide.

We consider the series S assigning to every nested ω-word nw the greatest accumulation point of the ratio of bracketfree positions in finite prefixes of nw.

To model S we use the ω-valuation monoid $D = (\bar{\mathbb{R}}, \sup, \lim avg, -\infty)$. If we want to analyze this property for well-matched nested ω-words only, then automaton $\mathcal{A}_1$ given below recognizes S. In the general case including pending edges, automaton $\mathcal{A}_2$ recognizes S. Note that we denote the call transitions with $\langle \Sigma$ and the return transitions with $\Sigma \rangle / q$ where q has to be the state where the last open call was encountered. The weights 1 resp. 0 are given in brackets.

Automaton 1: wsMNWA $\mathcal{A}_1$ with $\mathfrak{F}_1 = \{\{q_0\}\}$

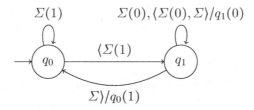

Automaton 2: wsMNWA $\mathcal{A}_2$ with $\mathfrak{F}_1 = \{\{q_2\}, \{q_p\}, \{q_2, q_p\}, \{q_0, q_1\}, \{q_0\}, \{q_1\}\}$

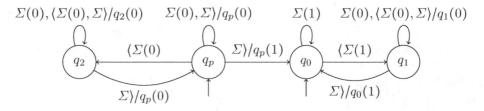

As usual, we extend the operation $+$ and $\diamond$ to series $S, T : NW^\omega(\Sigma) \to D$ by means of pointwise definitions as follows:

$$(S \star T)(nw) = S(nw) \star T(nw) \text{ for each } nw \in NW^\omega(\Sigma), \star \in \{+, \diamond\} \ .$$

We let $d \in D$ also denote the constant series with value d, i.e. $\|d\|(nw) = d$ for each $nw \in NW^\omega(\Sigma)$. For $L \subseteq NW^\omega(\Sigma)$, we define the *characteristic series* $\mathbb{1}_L : NW^\omega(\Sigma) \to D$ by letting $\mathbb{1}_L(nw) = 1$ if $nw \in L$, and $\mathbb{1}_L(nw) = 0$ otherwise. We call a series S a *regular step function* if

$$S = \sum_{i=1}^{k} d_i \diamond \mathbb{1}_{L_i} \ , \tag{2}$$

where L_i are regular languages of nested-ω-words forming a partition of $NW^\omega(\Sigma)$ and $d_i \in D$ for each $i \in \{1, ..., k\}$; so $S(nw) = d_i$ iff $nw \in L_i$ for each $i \in \{1, ..., k\}$.

An ω-pv-monoid D is *regular* if for any alphabet Σ we have: For each $d \in D$ there exists a wsMNWA $\mathcal{A}_d$ with $\|\mathcal{A}_d\| = d$. Analogously to Droste and Meinecke [11] we can show that every left-distributive ω-pv-monoid is regular.

Proposition 7. *Let D be a regular ω-pv-monoid. Then each regular step function $S : NW^\omega(\Sigma) \to D$ is regular. Furthermore, the set of all regular step functions is closed under $+$ and $\diamond$.*

Next we show that regular series are closed under projections. Consider a mapping $h : \Sigma \to \Gamma$ between two alphabets. Then h extends uniquely to an homomorphism between Σ^ω and Γ^ω, also denoted by h. Hence h is length-preserving and we can extend h to a function $h : NW^\omega(\Sigma) \to NW^\omega(\Gamma)$ by defining $h(nw) = h(w, \nu) = (h(w), \nu)$ for each $nw \in NW^\omega(\Sigma)$. Let $S : NW^\omega(\Sigma) \to D$ be a series. Then we define $h(S) : NW^\omega(\Gamma) \to D$ for each $nv \in NW^\omega(\Gamma)$ by

$$h(S)(nv) = \sum(S(nw) \mid nw \in NW^\omega(\Sigma), h(nw) = nv) \ .$$

Proposition 8. *Let D be an ω-valuation monoid, $S : NW^\omega(\Sigma) \to D$ regular and $h : \Sigma \to \Gamma$. Then $h(S) : NW^\omega(\Gamma) \to D$ is regular.*

4   Weighted MSO-Logic for Nested ω-words

In this section, we will present different fragments of our weighted MSO logic, and we give our first main result on the equivalence of these fragments. In the following D is always an ω-pv-monoid. We combine ideas of Alur and Madhusu-dan [2], Droste and Gastin [9], and Bollig and Gastin [4], and divide the syntax of the weighted logic into a boolean part and a weighted part.

Definition 9 (Syntax). *The weighted monadic second order logic for nested words $MSO(D, NW(\Sigma))$ is given by the following syntax*

$$\beta ::= \mathrm{Lab}_a(x) \mid \mathrm{call}(x) \mid \mathrm{ret}(x) \mid x \leq y \mid \nu(x, y) \mid x \in X \mid \neg\beta \mid \beta \wedge \beta \mid \forall x.\beta \mid \forall X.\beta$$

$$\varphi ::= d \mid \beta \mid \varphi \vee \varphi \mid \varphi \wedge \varphi \mid \forall x.\varphi \mid \exists x.\varphi \mid \exists X.\varphi$$

where $d \in D$, $a \in \Sigma$ and x, y, X are first resp. second order variables. We call all formulas β boolean formulas.

The set of all positions of $nw \in NW^\omega(\Sigma)$ is $\mathbb{N}$. Let $\varphi \in MSO(D, NW(\Sigma))$. We denote the set of free variables of φ by free(φ). Let $\mathcal{V}$ be a finite set of variables containing free(φ). As usual, we define a $(\mathcal{V}, nw)$-*assignment* γ as function assigning to every first order variable of $\mathcal{V}$ a position of nw and to every second order variable a subset of positions of nw. We let $\gamma[x \to i]$ (resp. $\gamma[X \to I]$) be the $(\mathcal{V} \cup \{x\}, nw)$-assignment (resp. $(\mathcal{V} \cup \{X\}, nw)$-assignment) mapping x to i (resp. X to I) and equaling γ anywhere else.

We encode a pair (nw, γ) as nested ω-word as usual over the extended alphabet $\Sigma_\mathcal{V} = \Sigma \times \{0, 1\}^\mathcal{V}$ with the same matching relation ν (cf. [9], [12]). We call $(nw, \sigma) \in NW^\omega(\Sigma_\mathcal{V})$ *valid* if σ emerges from a $(\mathcal{V}, nw)$-assignment. Clearly the language $N_\mathcal{V}$ of all valid words is regular.

Definition 10 (Semantics). *The semantics of φ is a series $[\![\varphi]\!]_\mathcal{V} : NW^\omega(\Sigma_\mathcal{V}) \to D$. If (nw, σ) is not valid, we set $[\![\varphi]\!]_\mathcal{V}(nw, \sigma) = 0$. Otherwise we define $[\![\varphi]\!]_\mathcal{V}(nw, \sigma)$ for $(nw, \sigma) = ((a_1 a_2 ..., \nu), \sigma)$ inductively as follows:*

$$[\![\text{Lab}_a(x)]\!]_\mathcal{V}(nw, \sigma) = \begin{cases} 1 & , \text{if } a_{\sigma(x)} = a \\ 0 & , \text{otherwise,} \end{cases} \qquad [\![\text{call}(x)]\!]_\mathcal{V}(nw, \sigma) = \begin{cases} 1 & , \text{if } \sigma(x) \text{ is a call} \\ 0 & , \text{otherwise,} \end{cases}$$

$$[\![\text{ret}(x)]\!]_\mathcal{V}(nw, \sigma) = \begin{cases} 1 & , \text{if } \sigma(x) \text{ is a return} \\ 0 & , \text{otherwise,} \end{cases} \qquad [\![x \le y]\!]_\mathcal{V}(nw, \sigma) = \begin{cases} 1 & , \text{if } \sigma(x) \le \sigma(y) \\ 0 & , \text{otherwise,} \end{cases}$$

$$[\![\nu(x, y)]\!]_\mathcal{V}(nw, \sigma) = \begin{cases} 1 & , \text{if } \nu(\sigma(x), \sigma(y)) \\ 0 & , \text{otherwise,} \end{cases} \qquad [\![x \in X]\!]_\mathcal{V}(nw, \sigma) = \begin{cases} 1 & , \text{if } \sigma(x) \in \sigma(X) \\ 0 & , \text{otherwise,} \end{cases}$$

$$[\![\neg\beta]\!]_\mathcal{V}(nw, \sigma) = \begin{cases} 1 & , \text{if } [\![\beta]\!]_\mathcal{V}(nw, \sigma) = 0 \\ 0 & , \text{otherwise,} \end{cases} \qquad [\![d]\!]_\mathcal{V}(nw, \sigma) = d \quad \text{for all } d \in D,$$

$$[\![\varphi \vee \psi]\!]_\mathcal{V}(nw, \sigma) = [\![\varphi]\!]_\mathcal{V}(nw, \sigma) + [\![\psi]\!]_\mathcal{V}(nw, \sigma),$$

$$[\![\varphi \wedge \psi]\!]_\mathcal{V}(nw, \sigma) = [\![\varphi]\!]_\mathcal{V}(nw, \sigma) \diamond [\![\psi]\!]_\mathcal{V}(nw, \sigma),$$

$$[\![\exists x.\varphi]\!]_\mathcal{V}(nw, \sigma) = \sum_{i \in \mathbb{N}} ([\![\varphi]\!]_{\mathcal{V} \cup \{x\}}(nw, \sigma[x \to i])),$$

$$[\![\exists X.\varphi]\!]_\mathcal{V}(nw, \sigma) = \sum_{I \subseteq \mathbb{N}} ([\![\varphi]\!]_{\mathcal{V} \cup \{X\}}(nw, \sigma[X \to I])),$$

$$[\![\forall x.\varphi]\!]_\mathcal{V}(nw, \sigma) = \text{Val}^\omega(([\![\varphi]\!]_{\mathcal{V} \cup \{x\}}(nw, \sigma[x \to i]))_{i \in \mathbb{N}}),$$

$$[\![\forall X.\beta]\!]_\mathcal{V}(nw, \sigma) = \begin{cases} 1 & , \text{if } [\![\beta]\!]_{\mathcal{V} \cup \{X\}}(nw, \sigma[X \to I]) = 1 \text{ for all } I \subseteq \mathbb{N} \\ 0 & , \text{otherwise} . \end{cases}$$

We write $[\![\varphi]\!]$ for $[\![\varphi]\!]_{\text{free}(\varphi)}$, so $[\![\varphi]\!] : NW^\omega(\Sigma_{\text{free}(\varphi)}) \to D$. If φ contains no free variables, φ is a *sentence* and $[\![\varphi]\!] : NW^\omega(\Sigma) \to D$.

Example 3. Continuing Example 2 with $D = (\bar{\mathbb{R}}, \sup, \lim \text{avg}, +, -\infty, 0)$ we define

$$\text{pcall}(x) = \text{call}(x) \wedge \forall w.\neg\nu(x, w), \quad \text{pret}(z) = \text{ret}(z) \wedge \forall u.\neg\nu(u, z),$$

$$\text{bfr}(y) = \forall x \forall z.(\neg(x < y < z \wedge \nu(x, z)) \wedge \neg(x < y \wedge \text{pcall}(x)) \wedge \neg(y < z \wedge \text{pret}(z))),$$

where $x < y < z = \neg(y \le x) \wedge \neg(z \le y)$. Then $[\![\forall y.((\text{bfr}(y) \wedge 1) \vee 0)]\!] = S = \|\mathcal{A}_2\|$.

Analogously to [9] and [12] we can show:

Proposition 11. *Let $\varphi \in MSO(D, NW(\Sigma))$ and let $\mathcal{V}$ be a finite set of variables with* $\mathrm{free}(\varphi) \subseteq \mathcal{V}$. *Then* $[\![\varphi]\!]_{\mathcal{V}}(nw, \sigma) = [\![\varphi]\!](nw, \sigma \upharpoonright \mathrm{free}(\varphi))$ *for each valid* $(nw, \sigma) \in NW^{\omega}(\Sigma_{\mathcal{V}})$. *Furthermore,* $[\![\varphi]\!]$ *is regular iff* $[\![\varphi]\!]_{\mathcal{V}}$ *is regular.*

Clearly, every boolean formula $\beta \in MSO(D, NW(\Sigma))$ can be interpreted as an unweighted MSO-formula $\psi \in MSO(NW(\Sigma))$ with $[\![\beta]\!] = \mathbb{1}_{L(\psi)}$, since $[\![\beta]\!]$ only yields the values 0 and 1. Conversely, for every formula $\psi \in MSO(NW(\Sigma))$ there exists a boolean MSO-formula $\beta \in MSO(D, NW(\Sigma))$ with $[\![\beta]\!] = \mathbb{1}_{L(\psi)}$, since we can replace disjunctions by conjunctions and negations and we can replace existential quantifiers by universal quantifiers and negations.

In order to obtain a Büchi-like theorem (as Theorem 17 below) for weighted automata on finite words, it is necessary to restrict the weighted MSO logic (cf. [9]). Therefore we introduce and study suitable fragments of $MSO(D, NW(\Sigma))$ as in the following.

Definition 12. *The set of almost boolean formulas is the smallest set of all formulas of $MSO(D, NW(\Sigma))$ containing all constants $d \in D$ and all boolean formulas, which is closed under disjunction and conjunction.*

Proposition 13. *(a) If $\varphi \in MSO(D, NW(\Sigma))$ is an almost boolean formula, then $[\![\varphi]\!]$ is a regular step function.*
(b) For every regular step function $S : NW^{\omega}(\Sigma) \to D$, there exists an almost boolean sentence φ with $S = [\![\varphi]\!]$.

Definition 14. *Let $\varphi \in MSO(D, NW(\Sigma))$. We denote by $\mathrm{const}(\varphi)$ the set of all elements of D occurring in φ. We call φ*

1. strongly-$\wedge$-restricted *if for all subformulas $\psi \wedge \theta$ of φ:*
   Either ψ and θ are almost boolean or ψ is boolean or θ is boolean.
2. $\wedge$-restricted *if for all subformulas $\psi \wedge \theta$ of φ:*
   Either ψ is almost boolean or θ is boolean.
3. commutatively-$\wedge$-restricted *if for all subformulas $\psi \wedge \theta$ of φ:*
   Either $\mathrm{const}(\psi)$ and $\mathrm{const}(\theta)$ commute or ψ is almost boolean.
4. $\forall$-restricted *if for all subformulas $\forall x.\psi$ of φ: ψ is almost boolean.*

We call a formula of $MSO(D, NW(\Sigma))$ *syntactically restricted* if it is both $\forall$-restricted and strongly-$\wedge$-restricted. Note that every subformula of a syntactically restricted formula is syntactically restricted itself.

Now we show that under suitable assumptions on the ω-pv-monoid D, particular classes of $MSO(D, NW(\Sigma))$-formulas have the same expressive power. In [11] these equivalences (for unnested words) followed from the main result and thus needed constructions of automata. Here we show the equivalence of the logic fragments directly.

Theorem 15.   *(a) Let D be left-distributive and $\varphi \in MSO(D, NW(\Sigma))$ be $\wedge$-restricted. Then there exists a strongly-$\wedge$-restricted formula*
   $\varphi' \in MSO(D, NW(\Sigma))$ *with* $[\![\varphi]\!] = [\![\varphi']\!]$. *Moreover, if φ is also $\forall$-restricted, then φ' can also be chosen to be $\forall$-restricted.*

(b) *Let D be a cc-ω-valuation semiring and let $\varphi \in MSO(D, NW(\Sigma))$ be commutatively-$\wedge$-restricted. Then there exists a strongly-$\wedge$-restricted formula $\varphi' \in MSO(D, NW(\Sigma))$ with $[\![\varphi]\!] = [\![\varphi']\!]$. Moreover, if φ is also $\forall$-restricted, then φ' can also be chosen to be $\forall$-restricted.*

Proof (sketch). We use an induction on the structure of φ. The interesting case is $\varphi = \psi \wedge \theta$ and ψ is almost boolean. By induction we can assume that ψ and θ are strongly-$\wedge$-restricted (and resp. $\forall$-restricted). As an example, we consider the case of the universal quantification in (a) as follows. Assume $\theta = \forall x.\theta_1$ and ψ does not contain x. By the induction hypothesis, we obtain a strongly-$\wedge$-restricted formula φ_1 such that $[\![\varphi_1]\!] = [\![\psi \wedge \theta_1]\!]$.

First let D be left-Val^ω-distributive. Using this assumption at equation *, we get for $\mathcal{V} = \text{free}(\psi) \cup \text{free}(\forall x.\theta_1)$ and each $(nw, \sigma) \in NW^\omega(\Sigma_{\mathcal{V}})$:

$$
\begin{aligned}
[\![\varphi]\!](nw, \sigma) &= [\![\psi \wedge \forall x.\theta_1]\!]_{\mathcal{V}}(nw, \sigma) \\
&= [\![\psi]\!]_{\mathcal{V}}(nw, \sigma) \diamond Val^\omega(([\![\theta_1]\!]_{\mathcal{V} \cup \{x\}}(nw, \sigma[x \to i]))_{i \in \mathbb{N}}) \\
&\overset{*}{=} Val^\omega(([\![\psi]\!]_{\mathcal{V}}(nw, \sigma) \diamond [\![\theta_1]\!]_{\mathcal{V} \cup \{x\}}(nw, \sigma[x \to i]))_{i \in \mathbb{N}}) \\
&= Val^\omega(([\![\psi]\!]_{\mathcal{V} \cup \{x\}}(nw, \sigma[x \to i]) \diamond [\![\theta_1]\!]_{\mathcal{V} \cup \{x\}}(nw, \sigma[x \to i]))_{i \in \mathbb{N}}) \\
&= Val^\omega(([\![\psi \wedge \theta_1]\!]_{\mathcal{V} \cup \{x\}}(nw, \sigma[x \to i]))_{i \in \mathbb{N}}) \\
&= [\![\forall x.(\psi \wedge \theta_1)]\!]_{\mathcal{V}}(nw, \sigma) \ .
\end{aligned}
$$

So $\varphi' = \forall x.\varphi_1$ is strongly-$\wedge$-restricted and $[\![\varphi]\!] = [\![\varphi']\!]$. If φ is $\forall$-restricted, θ_1 is almost boolean. In this case we can put directly $\varphi' = \forall x.(\psi \wedge \theta_1)$. Then φ' is strongly-$\wedge$-restricted and $\forall$-restricted because ψ and θ_1 are almost boolean formulas.

Now let D be left-multiplicative. Using the formulas $min(x) = \forall y.(x \leq y)$ and $min(x) \to \psi = \neg min(x) \vee (min(x) \wedge \psi)$ it can be shown that

$$
\begin{aligned}
[\![\varphi]\!] &= [\![\psi \wedge \forall x.\theta_1]\!] \\
&= [\![\forall x.((min(x) \to \psi) \wedge \theta_1)]\!] \\
&= [\![\forall x.((\neg min(x) \wedge \theta_1) \vee (min(x) \wedge \psi \wedge \theta_1))]\!] \ .
\end{aligned}
$$

Then $\varphi' = \forall x.((\neg min(x) \wedge \theta_1) \vee (min(x) \wedge \varphi_1))$ is strongly-$\wedge$-restricted since $min(x)$ is boolean. Furthermore, $[\![\varphi]\!] = [\![\varphi']\!]$. If φ is $\forall$-restricted, we can put directly $\varphi' = \forall x.((min(x) \to \psi) \wedge \theta_1)$. Then φ' is strongly-$\wedge$-restricted and $\forall$-restricted because $min(x) \to \psi$ and θ_1 are almost boolean formulas. $\qquad \square$

If D is a cc-ω-valuation semiring, clearly almost boolean formulas can be written as disjunctions of conjunctions of boolean formulas or constants from D. Our proof of Theorem 15 (b) shows the following corollary.

Corollary 16. *Let D be a commutative cc-ω-valuation semiring. Then for any formula $\varphi \in MSO(D, NW(\Sigma))$ there exists a formula $\varphi' \in MSO(D, NW(\Sigma))$ in which conjunctions occur only between boolean formulas and constants such that $[\![\varphi]\!] = [\![\varphi']\!]$.*

This follows also from a slightly modified proof of Theorem 17, but the present proof gives direct and efficient conversions of the formulas.

5   Characterization of Regular Series

In this section, we give our second main result on the expressive equivalence of weighted stair Muller nested word automata and our different fragments of weighted MSO logic.

Theorem 17. *Let D be a regular ω-pv-monoid and $S : NW^{\omega}(\Sigma) \to D$ a series.*

1. *The following are equivalent:*
   (a) S is regular.
   (b) $S = [\![\varphi]\!]$ for some syntactically restricted sentence φ of $MSO(D, NW(\Sigma))$.
2. *Let D be left-distributive. Then the following are equivalent:*
   (a) S is regular.
   (b) $S = [\![\varphi]\!]$ for some $\forall$-restricted and $\wedge$-restricted sentence φ of $MSO(D, NW(\Sigma))$.
3. *Let D be cc-ω-valuation semiring. Then the following are equivalent:*
   (a) S is regular.
   (b) $S = [\![\varphi]\!]$ for some $\forall$-restricted and commutatively-$\wedge$-restricted sentence φ of $MSO(D, NW(\Sigma))$.

Proof. '$(i) \Rightarrow (ii)$': We construct a syntactically restricted MSO-sentence simulating the given wsMNWA, thus showing all three statements.

'$(ii) \to (i)$': By Theorem 15 we may assume φ to be syntactically restricted. We prove the regularity of $[\![\varphi]\!]$ by induction on the structure of φ as follows. If φ is almost boolean, by Propositions 13(a) and 7, $[\![\varphi]\!]$ is regular. Next we have to prove that the regularity is preserved under the non-boolean operations. We only sketch the ideas. Closure under disjunction follows from Proposition 8 and a union construction of automata. If φ is a conjunction, the regularity of $[\![\varphi]\!]$ follows from a product construction of automata. The regularity of $[\![\exists x.\varphi]\!]$ and $[\![\exists X.\varphi]\!]$ follows from Proposition 8. For $\forall x.\varphi$, φ is almost boolean. Then $[\![\forall x.\varphi]\!]$ can also be shown to be regular. □

6   Conclusion

We have introduced a weighted automaton model for infinite nested words and weighted MSO logics. We could show that under suitable assumptions on the valuation monoids, two resp. three fragments of the weighted logics have the same expressive power with efficient conversions into the smallest fragment. Moreover, the weighted automata and our logic fragments have the same expressive power. The valuation monoids form very general weight structures; they model long-time average and discounted computations of weights as well as the classical complete semirings [9]. As in [2], we considered nested words possibly containing pending edges. We remark that our results also hold similarly for finite nested words, and our conversions of the weighted logic formulas also work, similarly, for other discrete structures like trees, cf. [10].

It would be interesting to investigate decision problems for weighted nested word automata, e.g., like done in [6], [7] for automata on words and with average or discounted computations of weights.

References

1. Alur, R., Arenas, M., Barceló, P., Etessami, K., Immerman, N., Libkin, L.: First-order and temporal logics for nested words. Logical Methods in Computer Science 4(4), 1–44 (2008)
2. Alur, R., Madhusudan, P.: Adding nesting structure to words. Journal of the ACM 56(3), 16:1–16:43 (2009)
3. Berstel, J., Reutenauer, C.: Rational Series and Their Languages. EATCS Monographs in Theoretical Computer Science, vol. 12. Springer (1988)
4. Bollig, B., Gastin, P.: Weighted versus probabilistic logics. In: Diekert, V., Nowotka, D. (eds.) DLT 2009. LNCS, vol. 5583, pp. 18–38. Springer, Heidelberg (2009)
5. Büchi, J.R.: Weak second-order arithmetic and finite automata. Z. Math. Logik und Grundlagen Math. 6, 66–92 (1960)
6. Chatterjee, K., Doyen, L., Henzinger, T.A.: Quantitative languages. In: Kaminski, M., Martini, S. (eds.) CSL 2008. LNCS, vol. 5213, pp. 385–400. Springer, Heidelberg (2008)
7. Chatterjee, K., Doyen, L., Henzinger, T.A.: Expressiveness and closure properties for quantitative languages. In: LICS, pp. 199–208. IEEE Computer Society (2009)
8. Droste, M., Kuich, W., Vogler, H. (eds.): Handbook of Weighted Automata. EATCS Monographs in Theoretical Computer Science. Springer (2009)
9. Droste, M., Gastin, P.: Weighted automata and weighted logics. Theor. Comput. Sci. 380(1-2), 69–86 (2007)
10. Droste, M., Götze, D., Märcker, S., Meinecke, I.: Weighted tree automata over valuation monoids and their characterization by weighted logics. In: Kuich, W., Rahonis, G. (eds.) Algebraic Foundations in Computer Science. LNCS, vol. 7020, pp. 30–55. Springer, Heidelberg (2011)
11. Droste, M., Meinecke, I.: Weighted automata and weighted MSO logics for average and long-time behaviors. Inf. Comput. 220, 44–59 (2012)
12. Droste, M., Pibaljommee, B.: Weighted nested word automata and logics over strong bimonoids. In: Moreira, N., Reis, R. (eds.) CIAA 2012. LNCS, vol. 7381, pp. 138–148. Springer, Heidelberg (2012)
13. Droste, M., Rahonis, G.: Weighted automata and weighted logics on infinite words. In: Ibarra, O.H., Dang, Z. (eds.) DLT 2006. LNCS, vol. 4036, pp. 49–58. Springer, Heidelberg (2006)
14. Eilenberg, S.: Automata, Languages, and Machines, Volume A, Pure and Applied Mathematics, vol. 59. Academic Press (1974)
15. Elgot, C.C.: Decision problems of finite automata design and related arithmetics. Transactions of the American Mathematical Society 98(1), 21–52 (1961)
16. Kuich, W., Salomaa, A.: Semirings, Automata, Languages. EATCS Monographs in Theoretical Computer Science, vol. 6. Springer (1986)
17. Löding, C., Madhusudan, P., Serre, O.: Visibly pushdown games. In: Lodaya, K., Mahajan, M. (eds.) FSTTCS 2004. LNCS, vol. 3328, pp. 408–420. Springer, Heidelberg (2004)
18. Mathissen, C.: Weighted logics for nested words and algebraic formal power series. In: Aceto, L., Damgård, I., Goldberg, L.A., Halldórsson, M.M., Ingólfsdóttir, A., Walukiewicz, I. (eds.) ICALP 2008, Part II. LNCS, vol. 5126, pp. 221–232. Springer, Heidelberg (2008)
19. Salomaa, A., Soittola, M.: Automata-Theoretic Aspects of Formal Power Series. Texts and Monographs in Computer Science. Springer (1978)
20. Schützenberger, M.P.: On the definition of a family of automata. Information and Control 4(2-3), 245–270 (1961)
21. Trakhtenbrot, B.A.: Finite automata and logic of monadic predicates. Doklady Akademii Nauk SSR 140, 326–329 (1961) (in Russian)

Algebraic Tools
for the Overlapping Tile Product

Etienne Dubourg* and David Janin**

Université de Bordeaux, LaBRI UMR 5800
351, cours de la Libération, 33405 Talence, France
{dubourg,janin}@labri.fr

Abstract. Overlapping tile automata and the associated notion of recognizability by means of (adequate) premorphisms in finite ordered monoids have recently been defined for coping with the collapse of classical recognizability in inverse monoids. In this paper, we investigate more in depth the associated algebraic tools that allow for a better understanding of the underlying mathematical theory. In particular, addressing the surprisingly difficult problem of language product, we eventually found some deep links with classical notions of inverse semigroup theory such as the notion of restricted product.

Keywords: Overlapping structures, premorphisms, Ehresmann monoids, birooted trees.

Introduction

Overlapping structures, be they linear shaped as in McAlister monoids [18], tree shaped as in free inverse monoids [23,20] or more generally higher-dimensional (overlapping) strings as in Kellendonk's tiling monoids [14,15], are promising high level models of system behaviors as already illustrated in musical application modeling [13] and the associated programming language proposal [12], or in distributed behavior modeling [3]. Be it for modeling/typing purposes or system analysis, there is incentive to develop the language theory of overlapping tiles.

Since Kellendonk's tiling monoids are inverse semigroups, such a language theory lies at the intersection between inverse semigroup theory [21,17] and formal language theory. A number of studies, such as [19,24] to mention but a few, already show deep connections between these fields. However, with Monadic Second Order logic (MSO) in the background as yardstick of expressive power (see e.g. [25]), classical language theoretic concepts and tools fail to be expressive enough [24,10]. Adaptation of the classical theory have thus recently been proposed in order to cope with such a collapse in expressive power. The resulting concepts: tile automata and quasi-recognizability have been proved to *essentially* capture MSO both for linear tiles [11] or tree-shaped tiles [9].

---

* Partially funded by the project CONTINT 2012 - ANR 12 CORD 009 02 - INEDIT.
** CNRS temporary researcher fellow (2013-2014).

A.-H. Dediu et al. (Eds.): LATA 2014, LNCS 8370, pp. 335–346, 2014.
© Springer International Publishing Switzerland 2014

Although, the resulting theory is somewhat robust – word or tree shaped tile automata are essentially non deterministic word or tree automata with adapted semantics [11,9] – the resulting language theory remains mysteriously tricky. For instance, the product of two quasi-recognizable languages is not necessarily quasi-recognizable.

In this paper, as an echo of [22] in classical algebraic language theory, continuing the newly developed theory, we study the case of positive word tiles, that is birooted words where the input root never occurs after the output root – positive tiles are the word counterpart of the positive birooted trees that form the elements of free ample monoids [5]. Our interest in studying the associated algebraic language theory is that, restricted to positive birooted words or trees, the class of quasi-recognizable languages turned out to be closed under product and star.

Although such a closure property was expected from our automata characterization od quasi-recognizable languages of word tiles [11], its proof is surprisingly technical unless, as we propose here, we first consider the restricted product: a fundamental notion in inverse semigroup theory [17] that was so far unused in our language theoretic investigation.

As a result, the proposed study not only sheds a new light on the adequate ordered monoids that are used as recognizers in quasi-recognizability, but also strengthens quite in depth the underlying theoretical framework. The fact is that our proposed recognizer definition can also be seen as a follow-up to the research track initiated by Fountain et al. [4,16,6,1] on certain semigroups with local units.

Worth being mentioned, though we restrict our study to positive birooted words as studied in [10,8,11], it is quite clear that our constructions can be extended to the case of positive birooted trees as studied in [9]. It follows that the algebraic tools proposed here are particularly well-suited for a language theory of the free ample monoid [5] whose elements are, precisely, positive birooted trees.

Organization of the Paper

In the first part, we give a formal definition of positive tiles as triplets of words, i.e. as words equipped with an input and output point, of the product of tiles and of the set of all tiles, including or not a 0 tile to make the product complete.

The following part presents our automatic tools, then the algebraic ones, mainly Ehresmann monoids and premorphisms. The definition of adequate premorphisms will bring that of the restricted product, which is an added condition to the definition of the product.

The third part presents our algebraic construction: we explore all the decompositions of each element in a restricted product. For tiles, this is equivalent to going through all the possible cutting points between input and output, which simulates a first order existential quantifier. Since we only consider positive tiles, this amounts to only going forward, "between input and output".

In the last part, we use this construction to prove the closure property under (restricted) products. For such a purpose, we use the fact that if a tile belongs to the restricted product $L_1 \bullet L_2$ of two languages L_1 and L_2, then any tile with a similar set of decompositions will also belong to the same product.

1   Overlapping Tiles

Let A be an alphabet and let A^* be the free monoid generated by A, that is, the set of finite words equipped with the concatenation operation. The empty word is denoted by 1 and, for every two words u and $v \in A^*$, we write $u \cdot v$ or simply uv for the concatenation of the words u and v.

The set A^* is ordered by the prefix order $\leq_p$ (resp. the suffix order $\leq_s$) defined, for every word u and $v \in A^*$, by $u \leq_p v$ (resp. $u \leq_s v$) when there exists $w \in A^*$ such that $uw = v$ (resp. $wu = v$).

A defined positive overlapping unidimensional tile (or just tile in the sequel) on the alphabet A is any triple u of the form $u = (u_1, u_2, u_3) \in A^* \times A^* \times A^*$. Such a tile is depicted in Figure 1. The set of defined tiles on the alphabet A

Fig. 1. A graphical representation of tile (u_1, u_2, u_3)

is denoted by $T^1(A)$. It is ordered by the *natural order* relation $\leq$ defined, for every tile $u = (u_1, u_2, u_3)$ and $v = (v_1, v_2, v_3)$, by

$$(u_1, u_2, u_3) \leq (v_1, v_2, v_3) \text{ when } u_1 \geq_s v_1, u_2 = v_2, u_3 \geq_p v_3$$

The (partial) product $u \cdot v$ of two such tiles u and v is defined, if it exists, as the greatest tile $w = (w_1, w_2, w_3)$ in the natural order such that $w_1 \geq_s u_1$, $w_1 u_2 \geq_s v_1$, $w_2 = u_2 v_2$, $v_2 w_3 \geq_p u_3$ and $w_3 \geq_p v_3$. Such a definition is depicted in Figure 2. Completing the set $T^+(A)$ of positive tiles by an undefined tile 0, the partial product is made complete by letting $u \cdot v = 0$ when there exists no such a defined product tile w and we put $u \cdot 0 = 0 = 0 \cdot u$ for every defined or undefined tile u.

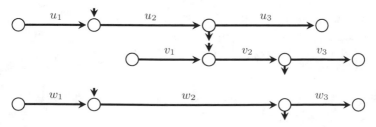

Fig. 2. A graphical representation of the product $(u_1, u_2, u_3) \cdot (v_1, v_2, v_3) = (w_1, w_2, w_3)$

It has already been shown in [10] that the set $T_0^+(A)$ of positive tiles equipped with such a product is actually a submonoid of the (inverse) monoid of McAlister [18] with unit $1 = (1, 1, 1)$. Moreover, extending the natural order to the undefined tile by letting $0 \leq u$ for every $u \in T_0^+(A)$, the natural order is indeed a partial order relation over $T_0^+(A)$ that is also stable under product, i.e. for every tiles u, v and $w \in T_0^+(A)$, if $u \leq v$ then $w \cdot u \leq w \cdot v$ and $u \cdot w \leq v \cdot w$.

For every non zero tile $u = (u_1, u_2, u_3) \in T^+(A)$, we define the *left projection* $u^L = (u_1 u_2, 1, u_3)$ and the *right projection* $u^R = (u_1, 1, u_2 u_3)$ of the tile u. These projections are extended to zero by taking $0^L = 0 = 0^R$. Then it can be shown [10] that for every u and $v \in T_0^+(A)$ we have $u \leq v$ if and only $u = u^R \cdot v$ if and only if $u = v \cdot u^L$. Moreover, a tile u is idempotent, that is $u \cdot u = u$ if and only if $u^L = u$ if and only if $u^R = u$ if and only if $u \leq 1$.

Let $U(T_0^+(A)) = \{u \in T_0^+(A) : u \leq 1\}$ be the set of *subunits* of the monoid of positive tiles. We have just seen that, in the monoid $T_0^+(A)$ idempotents and subunits coincide. It can also be shown that $U(T_0^+(A))$ ordered by the natural order is a complete lattice with product as meet.

2   Automata and Algebra for Overlapping Tiles

We review here the notion of quasi-recognizable languages of overlapping tiles. The notion of Ehresmann ordered monoids, previously left unused, that extends to adequately ordered monoids the congruence property of left and right projections defined for U-semiadequate monoids [16], turns out to play a crucial role in the analysis of the product of quasi-recognizable languages, somehow much in the same way it is of crucial importance for extending quasi-recognizability to infinite tiles [2].

Definition 1 (Tile automata [11]). *A (finite) tile automaton on the alphabet A is a triple $\mathcal{A} = \langle Q, \delta, K \rangle$ such that Q is a (finite) set of states, $\delta : A \rightarrow \mathcal{P}(Q \times Q)$ is the transition function, and $K \subseteq Q \times Q$ is the set of accepting pairs of states.*

Given $\delta^* : A^* \rightarrow \mathcal{P}(Q \times Q)$ the closure of the transition function inductively defined by $\delta(1) = \{(q, q) \in Q \times Q : q \in Q\}$ and $\delta(wa) = \{(p, q) \in Q \times Q : \exists r \in Q, (p, r) \in \delta^*(w), (r, q) \in \delta(a)\}$ for every $a \in A$ and $w \in A^*$, a run of the tile automaton $\mathcal{A}$ on a defined tile $u = (u_1, u_2, u_3)$ is defined as a pair of states $(p, q) \in Q \times Q$, such that there is a start state $s \in Q$ and an end state $e \in Q$ such that $(s, p) \in \delta^*(u_1)$, $(p, q) \in \delta^*(u_2)$ and $(q, e) \in \delta^*(u_3)$. By convention, there exists no run of the automaton $\mathcal{A}$ on the undefined tile 0.

For every tile $u \in T_0^+(A)$, we write $\varphi_{\mathcal{A}}(u) \subseteq Q \times Q$ for the set of runs of the automaton $\mathcal{A}$ on the tile u. The language $L(\mathcal{A})$ of tiles recognized by the automaton $\mathcal{A}$ is defined by $L(\mathcal{A}) = \{u \in T_0^+(A) : \varphi_{\mathcal{A}}(u) \cap K \neq \emptyset\}$.

By definition, every such recognized language is upward closed in the natural order. We have already shown that:

Theorem 2 (Logical characterization [11]). *A language of tiles $L \subseteq T_0^+(A)$ is recognizable by a finite state tile automaton if and only if it does not contain zero, it is upward closed and definable in Monadic Second Order (MSO) logic.*

And, as a corollary:

Corollary 3 (Closure property [11]). *The union, intersection, product and star of languages recognizable by finite state tile automata are recognizable by finite state tile automata.*

Remark. For every language $L \subseteq T_0^+(A)$ recognizable by a (finite) tile automaton $\mathcal{A}$, we have $L = \varphi_{\mathcal{A}}^{-1}(\varphi_{\mathcal{A}}(L))$, i.e. the language L is recognized by the mapping $\varphi_{\mathcal{A}}$. However, although the set $\mathcal{P}(Q \times Q)$ can be seen as a monoid with product $X \cdot Y = \{(p, q) \in Q \times Q : \exists r \in Q, (p, r) \in X, (r, q) \in Y\}$ for every X and $Y \subseteq Q \times Q$, the mapping $\varphi_{\mathcal{A}} : T_0^+(A) \to \mathcal{P}(Q \times Q)$ is not a monoid morphism. Indeed, we only have $\varphi_{\mathcal{A}}(u \cdot v) \subseteq \varphi_{\mathcal{A}}(u) \cdot \varphi_{\mathcal{A}}(v)$, that is, the mapping $\varphi_{\mathcal{A}}$ is a $\vee$-premorphism (see [7]). This observation leads us in [8] and [11] to make the properties of both the monoid $\mathcal{P}(Q \times Q)$ and the premorphism $\varphi_{\mathcal{A}}$ explicit in order to define an effective notion of algebraic recognizability called here quasi-recognizability.

The notion of quasi-recognizability itself is the main object of study in the present paper and is a refined version of the one proposed in [11] and [9].

The recognizers we use are called E-preordered monoid in reference to Ehresmann monoids defined in [16].

Definition 4 (Ehresmann preordered monoid). A *preordered monoid* is a monoid S equipped with a preorder relation $\preccurlyeq$, i.e. a reflexive and transitive relation, that is stable under product, i.e. for every $x, y, z \in S$, if $x \preccurlyeq y$ then $zx \preccurlyeq zy$ and $xz \preccurlyeq xy$.

Such a preordered monoid is said to be an *Ehresmann preordered monoid*, or just *E-monoid*, when it satisfies the following properties:

(A0) S possesses a minimum 0, i.e. for any $x \in S$, $0 \preccurlyeq x$ and if $x \preccurlyeq 0$ then $x = 0$,

(A1) relation $\preccurlyeq$ restricted to the set $U(S)$ is an order and the set $U(S) = \{x \in S \mid x \preccurlyeq 1\}$ of subunits of S ordered by $\preccurlyeq$ is a $\wedge$-semilattice with $\wedge$ as product,

(A2) the left projection $x^L = min\{y \in U(S) \mid xy = x\}$ and the right projection $x^R = min\{y \in U(S) \mid yx = x\}$ are defined for every $x \in S$,

(A3) left and right projections are monotonic, i.e. if $x \preccurlyeq y$ then $x^L \preccurlyeq y^L$ and $x^R \preccurlyeq y^R$ for every x and $y \in S$,

(A4) left and right projections induce right and left semi-congruence, i.e. we have $(xy)^L = (x^L y)^L$ and $(xy)^R = (xy^R)^R$ for every x and $y \in S$.

Remark. One easily checks that Property (A1) implies that all subunits are idempotents and commute since the product is a meet for the order induced on the subunits by the preorder on S.

One can also check that in the case where S is finite then Property (A1) implies Property (A2) since we have $x^R = \prod\{z \leq 1 : zx = x\}$ and $x^L = \prod\{z \leq$

$1 : xz = x\}$. In all cases, whenever $x \leq 1$ then we have $x^L = x = x^R$ hence the mappings $x \mapsto x^L$ and $x \mapsto x^R$ are indeed projections.

Surprisingly, the link between the monotonicity hypothesis (Property (A3)) and Properties (A0) to (A2) is far from being clear. Intuition says that, in the finite case at least, the Property (A3) may well be implied by the previous ones.

Property (A4) indeed equivalently says that the equivalence induced by the left (resp. right) projection is a right (resp. left) congruence. Indeed, assume that $x^L = y^L$ then, for every $z \in S$, we have $x^L z = y^L z$, and, by applying Property (A4), we have $(xz)^L = (x^L z)^L$ and $(yz)^L = (x^L z)^L$ and thus $(xz)^L = (yz)^L$. A symmetrical argument proves the right case.

Last, left and right projections are related with Green left and right preorders as follows. Recall that for a semigroup S, the left and right Green's preorders are defined, for every $x, y \in S$ by $x \leq_{\mathcal{R}} y$ when $x = yz$ for some $z \in S$ and $x \leq_{\mathcal{L}} y$ when $x = zy$ for some $z \in S$. If we assume that S is an E-monoid then one can easily check that for every x and $y \in S$ we have that if $x \leq_{\mathcal{R}} y$ then $x^R \leq y^R$ and if $x \leq_{\mathcal{L}} y$ then $x^L \leq y^L$, i.e. left and right projections are refinements of the left and right Green's classes.

Examples. Examples of E-monoids are numerous. First, every semigroup extended with a zero and trivially ordered by the relation $x \leq y$ when $x = 0$ or $x = y$ is an E-monoid. Every inverse semigroup, possibly extended with a zero, and ordered by the natural order (see [17]) is also an E-monoid with projection $x^L = x^{-1}x$ and $x^R = xx^{-1}$ for every element x. Every submonoid with a zero of an inverse monoid naturally ordered and closed under left and right projection as above is also an E-monoid. In particular, the monoid $\mathcal{T}^+(A)$ of positive tiles and naturally ordered is an E-monoid.

Less obviously, one can check that the monoid $\mathcal{P}(Q \times Q)$ of relations over the set Q, ordered by inclusion, is also an E-monoid with projection $X^L = \{(q,q) \in Q \times Q : \exists p \in Q, (p,q) \in X\}$ and $X^R = \{(p,p) \in Q \times Q : \exists q \in Q, (p,q) \in X\}$ for every $X \subseteq Q \times Q$. Every submonoid of $\mathcal{P}(Q \times Q)$ that contains $\emptyset$ (the zero for the relation product) and that is closed under the above left and right projections is an E-monoid.

Examples of E-monoids with preorders that are not partial order relations arise later in the text when defining the E-monoid of decompositions.

The following definition is an extension of the well-known notion of restricted product in inverse semigroup theory.

Definition 5 (Restricted product). Let S be an E-monoid. For every $x, y \in S$, the restricted product $x \bullet y$ of x and y is defined when $x^L = y^R$ and, in that case, it equals xy. In the sequel, we shall write $\exists x \bullet y$ to denote both the fact that the restricted product $x \bullet y$ is defined and, if needed, its value.

The restricted product is extended to subsets of S by taking $X \bullet Y = \{xy \in S : x \in X, y \in Y, \exists x \bullet y\}$.

Remark. The restricted product is associative in the sense that, for every $x, y, z \in S$, we have $\exists x \bullet (\exists y \bullet z) = \exists (\exists x \bullet y) \bullet z$. Indeed, this is a direct consequence of

Property (A4) hence the fact that for every $x, y, z \in S$, we have $(x \bullet y)^R = (xy^R)^R = (xx^L)^R = x^R$ and $(y \bullet z)^L = (y^L \bullet z)^L = (z^R \bullet z)^L = z^L$. In that case, we simply write $\exists x \bullet y \bullet z$.

We can now define adequate premorphisms and quasi-recognizability .

Definition 6. Let S be an E-monoid. A premorphism $\varphi : \mathcal{T}^+(A) \to S$ is a monotonic mapping such that $\varphi(uv) \preceq \varphi(u)\varphi(v)$. It is called adequate when, moreover, it satisfies the following properties:

▷ it preserves left and right projections, i.e. for every $u \in \mathcal{T}^+(A)$, we have $\varphi(u^R) = \varphi(u)^R$ and $\varphi(u^L) = \varphi(u)^L$,
▷ it preserves disjoint products, i.e. for every tile $u = (u_1, u_2, u_2)$ and $v = (v_1, v_2, v_3)$, if $u_3 = 1 = v_1$ then $\varphi(uv) = \varphi(u)\varphi(v)$,
▷ it also preserves restricted products, i.e. for every $u, v \in \mathcal{T}^+(A)$ such that $\exists u \bullet v$, we have $\varphi(u \bullet v) = \varphi(u) \bullet \varphi(v)$.

Remark. We have already seen that for every set Q, the monoid $\mathcal{P}(Q \times Q)$ ordered by inclusion is an E-monoid. One can check that the mapping $\varphi_{\mathcal{A}} : \mathcal{T}^+(A) \to \mathcal{P}(Q \times Q)$ defined out a tile automaton $\mathcal{A}$ as above in an adequate premorphism.

It must be mentioned that in [11], the adequate premorphisms are not required to preserve restricted product. The fact is that, when arbitrary tiles are involved as in [11], it may be the case that the premorphism $\varphi_{\mathcal{A}}$ *does not* preserve the restricted product. So the definition of adequacy given here is really suited to the case of positive tiles.

Definition 7 (Quasi-recognizability). A language $L \subseteq \mathcal{T}^+(A)$ of non zero positive tiles is quasi-recognizable (QR) when there exists a finite E-monoid S and an adequate premorphism $\varphi : \mathcal{T}^+(A) \to S$ such that $L = \varphi^{-1}(\varphi(L))$.

One can easily derive from the analogous statement proved in [9] for birooted trees and with an automata theoretic proof:

Theorem 8 (Logical characterization [9]). *A language $L \subseteq T(A)$ of non zero positive tiles is quasi-recognizable (QR) if and only if it is a finite boolean combination of upward closed MSO definable languages of non zero positive tiles.*

Remark. Observe that such a result does not imply that the class of quasi-recognizable languages in closed under product. This is especially clear considering that the same statement holds for languages of arbitrary tiles which closure under product is provably false [9].

3   Restricted Decompositions Monoid

We aim at providing algebraic tools for the product of two languages of positive tiles. Beware that by the product of two languages X and $Y \subseteq T_0^+(A)$ we mean the product $X \cdot Y = \{xy \in T_0^+(A) : x \in X, y \in Y, xy \neq 0\}$, i.e. the undefined tile is systematically omitted from the resulting point-wise product.

Applying Theorem 8, one can directly prove, by means of automata theoretic techniques, that the class of quasi-recognizable languages of positive tiles is closed under product. But the need to account for all the different configurations that may arise makes such a construction lengthy and tedious.

The algebraic tools developed here are thus defined for the restricted product and, fortunately, the arbitrary product can still be expressed quite simply in terms of the restricted one.

We define in this section out of any E-monoid, the monoid of its decompositions and show how this construction preserves in some sense quasi-recognizability on positive tiles.

Then, in the next section, such a restricted decomposition monoid can be used for achieving an algebraic proof that the restricted product (and henceforth the product) of two quasi-recognizable languages is indeed quasi-recognizable.

From now on, let S be an E-monoid preordered by the relation $\preccurlyeq$. The relation $\preccurlyeq$ is extended to pairs over $S \times S$ by taking the product preorder defined by $(x, x') \preccurlyeq (y, y')$ when $x \preccurlyeq x'$ and $y \preccurlyeq y'$ for every $(x, x'), (y, y') \in S \times S$. It is then extended to $\mathcal{P}(S \times S)$ by taking $X \preccurlyeq Y$ when for every $x \in X$, there exists $y \in Y$ so that $x \preccurlyeq y$, for every $X, Y \in \mathcal{P}(S \times S)$. Similar constructions can be found in the context of ordered semigroups in [22].

Definition 9 (Restricted decompositions monoid). We define the set $\mathcal{D}^r(S) \subseteq \mathcal{P}(S \times S)$, preordered by $\preccurlyeq$, by

$$\mathcal{D}^r(S) = \{X \in \mathcal{P}(S \times S) \mid \exists c \in S, \ (c, c^L) \in X,$$
$$(c^R, c) \in X, \forall (x, y) \in X, \ x \bullet y = c\}.$$

The product $*$ is defined for every $(x, x'), (y, y') \in S \times S$ by:

$$(x, x') * (y, y') = \{((x(x'yy')^R, x^L x'yy'), (xx'yy'^R, (xx'y)^L y')\}.$$

and extended to $\mathcal{D}^r(S)$ in a point-wise manner, that is, for every $X, Y \in \mathcal{D}^r(S)$, by

$$X * Y = \bigcup_{\substack{(x,x') \in X \\ (y,y') \in Y}} (x, x') * (y, y').$$

Lemma 10. *The set $\mathcal{D}^r(S)$ equipped with the product $*$ and ordered by the relation $\preccurlyeq$ is an Ehresmann preordered monoid.*

Proof. The detailed proof is a little long but presents no real difficulties as soon as the appropriate definition has been found. □

Let then $L \subseteq \mathcal{T}^+(A)$ be a language recognized by adequate premorphism $\varphi : \mathcal{T}^+(A) \longrightarrow S$ into the E-monoid S. We build out of φ an adequate premorphism from $\mathcal{T}^+(A)$ to the decomposition $\mathcal{D}^r(S)$ that still recognizes L.

For that purpose, let $\psi : \mathcal{T}^+(A) \to \mathcal{D}^r(S)$ be defined, for every $u \in \mathcal{T}^+(A)$ by $\psi(u) = \{(\varphi(u_1), \varphi(u_2)) \mid u = \exists u_1 \bullet u_2\}$.

Lemma 11. *The mapping $\psi : \mathcal{T}^+(A) \to \mathcal{D}^r(S)$ is an adequate premorphism that recognizes L.*

Proof. The detailed proof is a little long but, again, presents no real difficulties as soon as the appropriate definitions have been found.   □

4 Application to the Restricted and Unrestricted Products of Languages

In the previous section, given any adequate premorphism $\varphi : \mathcal{T}^+(A) \to S$ we have defined $\psi : \mathcal{T}^+(A) \to \mathcal{D}^r(S)$ that allows for computing φ on the two components of any restricted decomposition of a positive tile. In some sense, for every positive tile u, when u is seen as a FO-structure with edges labeled over the alphabet A, this construction allows for simulating any existential first order quantification over the vertices between and including the input root and the output root.

This intuition is used here to prove our main theorem:

Theorem 12. *Let $L_1, L_2 \subseteq \mathcal{T}^+(A)$ quasi-recognizable languages, the language $L_1 \bullet L_2$ is quasi-recognizable.*

Proof. Let S_1, S_2 E-monoids and $L_1, L_2 \subseteq \mathcal{T}^+(A)$ respectively recognized by adequate premorphisms $\varphi_1 : \mathcal{T}^+(A) \longrightarrow S_1$ and $\varphi_2 : \mathcal{T}^+(A) \longrightarrow S_2$. First, we define

$$\varphi : \mathcal{T}^+(A) \longrightarrow S_1 \times S_2$$
$$u \longrightarrow (\varphi_1(u), \varphi_2(u))$$

Remark that $S_1 \times S_2$ is an E-monoid and φ is an adequate premorphism recognizing both L_1 and L_2.

We now consider premorphism $\psi : \mathcal{T}^+(A) \to \mathcal{D}^r(S_1 \times S_2)$ as defined in the previous section from the adequate premorphism φ. By Lemma 10, the monoid $\mathcal{D}^r(S_1 \times S_2)$ is an E-monoid and, by Lemma 11, the mapping ψ is an adequate premorphism. We will now prove that ψ recognizes $L_1 \bullet L_2$.

Let $u_1 \in L_1$ and $u_2 \in L_2$ so that $\exists u_1 \bullet u_2$, and let $v \in \mathcal{T}^+(A)$ so that $\psi(v) = p(u_1 \bullet u_2)$. So we have $(\varphi(u_1), \varphi(u_2)) \in \psi(v)$, therefore there exists $v_1, v_2 \in \mathcal{T}^+(A)$ so that $v_1 \bullet v_2 = v$ and

$$(\varphi(u_1), \varphi(u_2)) = (\varphi(v_1), \varphi(v_2)).$$

Since φ recognizes L_1 and L_2, we have $v_1 \in L_1$ and $v_2 \in L_2$. Consequently, $v = v_1 \bullet v_2 \in L_1 \bullet L_2$.   □

We will now apply the restricted product case to the general product case.

Lemma 13. *Let $L_1, L_2 \subseteq \mathcal{T}^+(A)$ be quasi-recognizable languages. We have*

$$L_1 L_2 = \left((A^*)^L L_1 (A^*)^R \bullet L_2 \right) \cup \left(L_1 \bullet (A^*)^L L_2 (A^*)^R \right)$$
$$\cup \left((A^*)^L L_1 \bullet L_2 (A^*)^R \right) \cup \left(L_1 (A^*)^R \bullet (A^*)^L L_2 \right)$$

Proof. We first show that $L_1 L_2 \subseteq \left((A^*)^L L_1 (A^*)^R \bullet L_2\right) \cup \left(L_1 \bullet (A^*)^L L_2 (A^*)^R\right)$ $\cup \left((A^*)^L L_1 \bullet L_2 (A^*)^R\right) \cup \left(L_1 (A^*)^R \bullet (A^*)^L L_2\right)$. Let $u = (u_1, u_2, u_3) \in L_1$, $v = (v_1, v_2, v_3) \in L_2$, so that $uv \neq 0$. By definition of the product, we have four possibilities :

> ▷ if v_1 is a suffix of $u_1 u_2$ and u_3 is a prefix of $v_2 v_3$, then $wv_1 = u_1 u_2$ for a $w \in A^*$ and $u_3 w' = v_2 v_3$ for a $w' \in A^*$, hence $uv = (u_1, u_2, u_3 w') \bullet (wv_1, v_2, v_3)$ that thus belongs to $L_1 (A^*)^R \bullet (A^*)^L L_2$,
> ▷ if $u_1 u_2$ is a suffix of v_1 and u_3 is a prefix of $v_2 v_3$, then $wu_1 u_2 = v_1$ for a $w \in A^*$ and $u_3 w' = v_2 v_3$ for a $w' \in A^*$, hence $uv = (wu_1, u_2, u_3 w') \bullet (v_1, v_2, v_3)$ that thus belongs to $(A^*)^L L_1 (A^*)^R \bullet L_2$,
> ▷ if v_1 is a suffix of $u_1 u_2$ and $v_2 v_3$ is a prefix of u_3, then $wv_1 = u_1 u_2$ for a $w \in A^*$ and $v_2 v_3 w' = u_3$ for a $w' \in A^*$, hence $uv = (u_1, u_2, u_3) \bullet (wv_1, v_2, v_3 w')$ that thus belongs to $L_1 \bullet (A^*)^L L_2 (A^*)^R$,
> ▷ if $u_1 u_2$ is a suffix of v_1 and $v_2 v_3$ is a prefix of u_3, then $wu_1 u_2 = v_1$ for a $w \in A^*$ and $v_2 v_3 w' = u_3$ for a $w' \in A^*$, hence $uv = (wu_1, u_2, u_3) \bullet (v_1, v_2, v_3 w')$ that thus belongs to $L_1 \bullet (A^*)^L L_2 (A^*)^R$.

Conversely, let $u = (u_1, u_2, u_3) \in L_1$, $v = (v_1, v_2, v_3) \in L_2$, and $w, w' \in A^*$. If $\exists (wu_1, u_2, u_3 w') \bullet (v_1, v_2, v_3) = t$, or if $\exists (u_1, u_2, u_3 w) \bullet (w' v_1, v_2, v_3) = t$, or if $\exists (wu_1, u_2, u_3) \bullet (v_1, v_2, v_3 w') = t$, or if $\exists (u_1, u_2, u_3) \bullet (wv_1, v_2, v_3 w') = t$, then $t = uv$. □

We then have to show that these "completions" on the right or left (the product with $(A^*)^L$ or $(A^*)^R$) preserves quasi-recognizability. First, we prove that it preserves the recognizability by automaton with simple constructions.

Lemma 14. *Let $L \subseteq \mathcal{T}^+(A)$ be a language recognized by an automaton $\mathcal{A}$, therefore there exist automata $\mathcal{A}_r$ and $\mathcal{A}_l$ that recognize respectively $L(A^*)^R$ and $(A^*)^L L$.*

Proof. Let $\mathcal{A} = \langle Q, \delta, K \rangle$ be an automaton recognizing a language $L \subseteq \mathcal{T}^+(A)$, we define $\mathcal{A}_l = \langle Q \cup *, \delta_l, K \rangle$ and $\mathcal{A}_r = \langle Q \cup *, \delta_r, K \rangle$, with for any a, $\delta_l(a) = \delta(a) \cup \{(*, *), (*, q) \mid q \in Q\}$ and $\delta_r(a) = \delta(a) \cup \{(*, *), (q, *) \mid q \in Q\}$.

We see that any tile of the type $(a_1 a_2 \ldots a_k u, v, w)$, with $(u, v, w) \in L$, is recognized by $\mathcal{A}_l$, by a run of the form $*a_1 * a_2 * \ldots * a_k R$, R being a run of $\mathcal{A}$ on (u, v, w).

Reciprocally, any run we have over tile (u, v, w) by $\mathcal{A}_l$ is of the form $*a_1 * a_2 * \ldots * a_k R$, where $a_1 a_2 \ldots a_k$ is a prefix of u, and R a *-less run on (u', v, w) where $u = a_1 a_2 \ldots a_k u'$, i.e. a run of $\mathcal{A}$ on (u', v, w). Therefore, if (u, v, w) is recognized by $\mathcal{A}_l$, then (u', v, w) is recognized by $\mathcal{A}$, so $(u, v, w) \in (A^*)^L L$.

We demonstrate symmetrically that $\mathcal{A}_r$ recognizes $L(A^*)^R$. □

We can now show that these "completions" on the right or left preserve quasi-recognizability. This is accomplished by noting that quasi-recognizable languages are combinations of upward-closed languages, that can be recognized by automata.

Lemma 15. *Let $L \subseteq \mathcal{T}^+(A)$ be a quasi-recognizable language, therefore $L(A^*)^R$ and $(A^*)^L L$ are quasi-recognizable.*

Proof. Let $L \subseteq \mathcal{T}^+(A)$ be a quasi-recognizable language, then L is a linear combination of languages recognized by automata hence a finite union of the form $L = \bigcup_{i \in I} D_i \cap U_i$ with, for every $i \in I$, quasi-recognizable language U_i upward-closed and quasi-recognizable language D_i downward-closed. It follows that $L(A^*)^R = \bigcup_{i \in I} D_i \cap U_i(A^*)^R$ since $D_i(A^*)^R = D_i$ for every downward closed language D_i. We conclude by applying Lemma 14 that shows that $U_i(A^*)^R$ is quasi-recognizable for every upward closed language U_i.   □

Corollary 16. *Let $L \subseteq \mathcal{T}^+(A)$ be a quasi-recognizable language, then the language $(A^*)^L L(A^*)^R$ is quasi-recognizable.*

Theorem 17. *Let $L_1, L_2 \subseteq \mathcal{T}^+(A)$ quasi-recognizable languages, then $L_1 L_2$ is quasi-recognizable.*

Proof. This follows directly from Lemma 13, Lemma 15 and corollary 16, and theorem 12.   □

5   Conclusion

We have thus shown, by means of algebraic tools, that both the restricted product (Theorem 12) and the arbitrary product (Theorem 17) of two quasi-recognizable languages of positive tiles are quasi-recognizable. By using the notion of restricted decomposition monoid (Definition 9) we have eventually extended to positive tiles classical algebraic techniques that, over words, are used to simulate existential FO-quantification when letters are modeled by labeled graph edges. We do believe that a similar technique can be used to prove the closure under iterated product (Kleene star).

References

1. Cornock, C., Gould, V.: Proper two-sided restriction semigroups and partial actions. Journal of Pure and Applied Algebra 216, 935–949 (2012)
2. Dicky, A., Janin, D.: Embedding finite and infinite words into overlapping tiles. Tech. Rep. RR-1475-13, LaBRI, Université de Bordeaux, Bordeaux (2013)
3. Dicky, A., Janin, D.: Modélisation algébrique du diner des philosophes. Modélisation des Systèmes Réactifs (MSR). Journal Européen des Systèmes Automatisés (JESA) 47(1-2-3/2013) (November 2013)
4. Fountain, J.: Right PP monoids with central idempotents. Semigroup Forum 13, 229–237 (1977)
5. Fountain, J., Gomes, G., Gould, V.: The free ample monoid. Int. Jour. of Algebra and Computation 19, 527–554 (2009)
6. Hollings, C.D.: From right PP monoids to restriction semigroups: a survey. European Journal of Pure and Applied Mathematics 2(1), 21–57 (2009)

7. Hollings, C.D.: The Ehresmann-Schein-Nambooripad Theorem and its successors. European Journal of Pure and Applied Mathematics 5(4), 414–450 (2012)

8. Janin, D.: Quasi-recognizable vs MSO definable languages of one-dimensional overlapping tiles. In: Rovan, B., Sassone, V., Widmayer, P. (eds.) MFCS 2012. LNCS, vol. 7464, pp. 516–528. Springer, Heidelberg (2012)

9. Janin, D.: Algebras, automata and logic for languages of labeled birooted trees. In: Fomin, F.V., Freivalds, R., Kwiatkowska, M., Peleg, D. (eds.) ICALP 2013, Part II. LNCS, vol. 7966, pp. 312–323. Springer, Heidelberg (2013)

10. Janin, D.: On languages of one-dimensional overlapping tiles. In: van Emde Boas, P., Groen, F.C.A., Italiano, G.F., Nawrocki, J., Sack, H. (eds.) SOFSEM 2013. LNCS, vol. 7741, pp. 244–256. Springer, Heidelberg (2013)

11. Janin, D.: Overlapping tile automata. In: Bulatov, A.A., Shur, A.M. (eds.) CSR 2013. LNCS, vol. 7913, pp. 431–443. Springer, Heidelberg (2013)

12. Janin, D., Berthaut, F., DeSainte-Catherine, M., Orlarey, Y., Salvati, S.: The T-calculus : towards a structured programming of (musical) time and space. In: Workshop on Functional Art, Music, Modeling and Design (FARM). ACM Press (2013)

13. Janin, D., Berthaut, F., DeSainteCatherine, M.: Multi-scale design of interactive music systems: the libTuiles experiment. In: Sound and Music Computing (SMC) (2013)

14. Kellendonk, J.: The local structure of tilings and their integer group of coinvariants. Comm. Math. Phys. 187, 115–157 (1997)

15. Kellendonk, J., Lawson, M.V.: Tiling semigroups. Journal of Algebra 224(1), 140–150 (2000)

16. Lawson, M.V.: Semigroups and ordered categories. I. the reduced case. Journal of Algebra 141(2), 422–462 (1991)

17. Lawson, M.V.: Inverse Semigroups: The theory of partial symmetries. World Scientific (1998)

18. Lawson, M.V.: McAlister semigroups. Journal of Algebra 202(1), 276–294 (1998)

19. Margolis, S.W., Meakin, J.C.: Inverse monoids, trees and context-free languages. Trans. Amer. Math. Soc. 335, 259–276 (1993)

20. Munn, W.D.: Free inverse semigroups. Proceeedings of the London Mathematical Society 29(3), 385–404 (1974)

21. Pietrich, M.: Inverse semigroups. Wiley (1984)

22. Pin, J.E.: Algebraic tools for the concatenation product. Theoretical Comp. Science 292(1), 317–342 (2003)

23. Scheiblich, H.E.: Free inverse semigroups. Semigroup Forum 4, 351–359 (1972)

24. Silva, P.V.: On free inverse monoid languages. ITA 30(4), 349–378 (1996)

25. Thomas, W.: Handbook of Formal Languages. In: Languages, Automata, and Logic, ch. 7, vol. III, pp. 389–455. Springer, Heidelberg (1997)

Reachability Analysis with State-Compatible Automata*

Bertram Felgenhauer and René Thiemann

Institute of Computer Science, University of Innsbruck, Innsbruck, Austria
{bertram.felgenhauer,rene.thiemann}@uibk.ac.at

Abstract. Regular tree languages are a popular device for reachability analysis over term rewrite systems, with many applications like analysis of cryptographic protocols, or confluence and termination analysis. At the heart of this approach lies tree automata completion, first introduced by Genet for left-linear rewrite systems. Korp and Middeldorp introduced so-called quasi-deterministic automata to extend the technique to non-left-linear systems. In this paper, we introduce the simpler notion of state-compatible automata, which are slightly more general than quasi-deterministic, compatible automata. This notion also allows us to decide whether a regular tree language is closed under rewriting, a problem which was not known to be decidable before.

Several of our results have been formalized in the theorem prover Isabelle/HOL. This allows to certify automatically generated non-confluence and termination proofs that are using tree automata techniques.

1   Introduction

In this paper we are largely concerned with over-approximations of the terms reachable from a regular tree language L_0 by rewriting using a term rewrite system $\mathcal{R}$, that is, we are interested in regular tree languages L such that $\mathcal{R}^*(L_0) \subseteq L$. Such over-approximations have been used, among other things, in the analysis of cryptographic protocols [6], for termination analysis [7,10] and for establishing non-confluence of term rewrite systems [15].

Unfortunately, the question whether $\mathcal{R}^*(L_0) \subset L$ is undecidable in general. Tree automata completion, conceived by Genet et al. [4,5], is based on the stronger requirements that $L_0 \subseteq L$ and L is itself closed under rewriting, i.e., $\mathcal{R}(L) \subseteq L$. This is accomplished by constructing L as the language accepted by a bottom-up tree automaton $\mathcal{A}$ that is *compatible* with $\mathcal{R}$: Whenever $l\sigma$ is accepted in state q by $\mathcal{A}$, where $l \to r \in \mathcal{R}$ and σ maps variables to states of $\mathcal{A}$, we demand that $r\sigma$ is also accepted in q. If $\mathcal{A}$ is deterministic or if $\mathcal{R}$ is a left-linear term rewrite system, then compatibility ensures that $\mathcal{L}(\mathcal{A})$ is closed under rewriting by $\mathcal{R}$.

---

* This research was supported by FWF projects P22467 and P22767.

A.-H. Dediu et al. (Eds.): LATA 2014, LNCS 8370, pp. 347–359, 2014.
© Springer International Publishing Switzerland 2014

Example 1. Let $\mathcal{R} = \{f(x,x) \to x\}$ and $\mathcal{A}$ be the automaton with states $1, 2, 3$, final state 3, and transitions

$$a \to 1 \qquad\qquad a \to 2 \qquad\qquad f(1,2) \to 3$$

So $\mathcal{A}$ is non-deterministic and $\mathcal{R}$ is non-left-linear. Even though $\mathcal{A}$ is compatible with $\mathcal{R}$, $\mathcal{L}(\mathcal{A}) = \{f(a,a)\}$ is not closed under rewriting by $\mathcal{R}$, because $f(a,a)$ can be rewritten to a which is not in $\mathcal{L}(\mathcal{A})$.

However, demanding $\mathcal{A}$ to be deterministic if $\mathcal{R}$ is not left-linear may result in bad approximations.

Example 2. Let $\mathcal{R} = \{f(x,x) \to b, b \to a\}$ and $L_0 = \{f(a,a)\}$. The set of terms reachable from L_0, namely $\mathcal{R}^*(L_0) = \{f(a,a), b, a\}$, is not accepted by any deterministic, compatible tree automaton. To see why, assume that such an automaton $\mathcal{A}$ exists, and let q be the state accepting $f(a,a)$. There must be transitions $a \to q'$ (q' is unique because $\mathcal{A}$ is deterministic) and $f(q',q') \to q$ in $\mathcal{A}$. By compatibility with the rules $f(x,x) \to b$ and $b \to a$, we must have transitions $b \to q$, and $a \to q$. Since we already have the transition $a \to q'$, determinism implies $q' = q$. With the three transitions $a \to q$, $b \to q$, and $f(q,q) \to q$, $\mathcal{A}$ accepts every term over the signature $\{f, a, b\}$, which is not a very useful approximation of $\mathcal{R}^*(L_0)$.

To overcome this problem, Korp and Middeldorp introduced quasi-deterministic automata [10]. Indeed it is easy to find a quasi-deterministic automaton accepting $\mathcal{R}^*(L_0) = \{f(a,a), b, a\}$ that is compatible with $\mathcal{R}$ from the previous example.

Example 3. Let $\mathcal{A}$ be an automaton with states $1, 2$, final state 2 and transitions

$$a \to 1^* \qquad\quad a \to 2 \qquad\quad b \to 2^* \qquad\quad f(1,1) \to 2^*$$

where the stars indicate the so-called designated states for each left-hand side. Then $\mathcal{A}$ is quasi-deterministic, compatible with $\mathcal{R}$ and $\mathcal{L}(\mathcal{A}) = \{f(a,a), b, a\}$.

In this paper, we concentrate on the compatibility requirement that ensures $\mathcal{R}(L) \subseteq L$. Since there may be bugs in the implementation of tree automata completion, it is important to independently certify whether $\mathcal{R}(L) \subseteq L$ is really satisfied. Such a certifier has already been developed in [2], but it is restricted to left-linear systems and does not support the stronger quasi-deterministic automata. We extend this work by introducing *state-compatible* automata, which are deterministic but accomplish the effect of quasi-deterministic automata by relaxing the compatibility requirement instead. It turns out that as long as $\mathcal{R}$ has only non-collapsing rules, state-compatible automata and quasi-deterministic automata are equivalent. In the presence of collapsing rules, state-compatible automata can capture more approximations than quasi-deterministic ones.

We will further show that state-compatibility does not only ensure $\mathcal{R}(L) \subseteq L$, but it can also be utilized to obtain a decision procedure for the question whether a regular tree language is closed under rewriting—a problem whose decidability was hitherto unknown, as far as we know. These results have also been formalized within the theorem prover Isabelle/HOL [13], resulting in a formalized decision procedure for the question $\mathcal{R}(L) \subseteq L$. It is used to certify non-confluence proofs and termination proofs that are using the techniques of [9,10,15].

This paper is structured as follows. In Section 2 we recall basic definitions and notation. The main part of our paper is Section 3, where we introduce the notions of state-coherence and state-compatibility, and present the decision procedure. Section 4 is devoted to a comparison to quasi-deterministic automata. Details on the formalization are provided in Section 5. Finally, we conclude in Section 6.

2   Preliminaries

We assume that the reader is familiar with first order term rewriting and tree automata. For introductions to these topics see [1] and [3].

Terms over a signature $\mathcal{F}$ and a set of variables $\mathcal{V}$, denoted $\mathcal{T}(\mathcal{F}, \mathcal{V})$ (or $\mathcal{T}(\mathcal{F})$ if $\mathcal{V}$ is empty) are inductively defined as either variables $v \in \mathcal{V}$ or of the form $f(t_1, \ldots, t_n)$, where $t_1, \ldots, t_n$ are terms and $f \in \mathcal{F}$ is a function symbol of arity n. We write $\mathsf{Var}(t)$ for the set of variables in t. A term t is linear if each variable occurs in t at most once. Contexts are terms over $\mathcal{F} \cup \{\Box\}$ that contain exactly one occurrence of $\Box$. If C is a context and t a term, then $C[t]$ denotes the term obtained by replacing the $\Box$ in C by t. A substitution $\sigma : \mathcal{V} \to \mathcal{T}(\mathcal{F}, \mathcal{V})$ maps variables to terms. We write $t\sigma$ for the result of replacing each variable x in t by $\sigma(x)$.

A term rewrite system (TRS) $\mathcal{R}$ is a set of rewrite rules $l \to r$, where each rule's left-hand side l and right-hand side r are terms such that $l \notin \mathcal{V}$ and $\mathsf{Var}(r) \subseteq \mathsf{Var}(l)$. A TRS $\mathcal{R}$ defines a rewrite relation $\to_{\mathcal{R}}$, namely $s \to_{\mathcal{R}} t$ whenever there are a context C, a rule $l \to r \in \mathcal{R}$, and a substitution σ such that $s = C[l\sigma]$ and $t = C[r\sigma]$. We denote by $\mathrm{lhs}(\mathcal{R})$ the set of all left-hand sides of rules in $\mathcal{R}$. A TRS is left-linear if all its left-hand sides are linear terms. A rule $l \to r$ is called collapsing if r is a variable. The inverse, the reflexive closure, transitive closure, and the reflexive, transitive closure of a binary relation $\to$ are denoted by $\leftarrow$, $\to^=$, $\to^+$, and $\to^*$, respectively. Given a set of terms L, $\mathcal{R}(L)$ ($\mathcal{R}^*(L)$) is the set of one-step (many-step) descendants of L: $t' \in \mathcal{R}(L)$ ($t' \in \mathcal{R}^*(L)$) iff $t \to_{\mathcal{R}} t'$ ($t \to_{\mathcal{R}}^* t'$) for some $t \in L$. A language L is closed under rewriting by $\mathcal{R}$, if $\mathcal{R}(L) \subseteq L$.

A (bottom-up) tree automaton $\mathcal{A} = (\mathcal{F}, Q, Q_f, \Delta)$ over a signature $\mathcal{F}$ consists of a set of states Q disjoint from $\mathcal{F}$, a set of final states $Q_f \subseteq Q$, and a set of transitions Δ of shape $f(q_1, \ldots, q_n) \to q$ where the root $f \in \mathcal{F}$ has arity n and $q, q_1, \ldots, q_n \in Q$. (We forbid ε-transitions for the sake of simplicity.) We regard Δ as a TRS over the signature $\mathcal{F} \cup Q$, with the states as constants. A substitution σ is a state substitution if $\sigma(x) \in Q$ for all $x \in \mathcal{V}$. A term t is accepted in state q if $t \to_{\Delta}^* q$; t is accepted by $\mathcal{A}$ if it is accepted in a final state. The language accepted by $\mathcal{A}$ is $\mathcal{L}(\mathcal{A}) = \{t \mid t \to_{\Delta}^* q \text{ for some } q \in Q_f\}$. We call $\mathcal{A}$ deterministic if no two rules in Δ have the same left-hand side. For convenience,

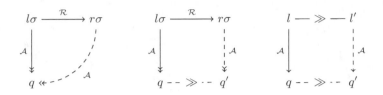

Fig. 1. Compatibility, state-compatibility, and state-coherence

we often write $\to_{\mathcal{A}}$ for $\to_{\Delta}$. Following [10], we formulate Genet's result from [5] as follows:

Definition 4. *A tree automaton $\mathcal{A}$ is* compatible *with a TRS $\mathcal{R}$ if for all state substitutions σ, rules $l \to r \in \mathcal{R}$ and states $q \in Q$, $l\sigma \to_{\mathcal{A}}^* q$ implies $r\sigma \to_{\mathcal{A}}^* q$.*

Theorem 5. *Let the tree automaton $\mathcal{A}$ be compatible with the TRS $\mathcal{R}$. Then*

1. *if $\mathcal{R}$ is left-linear, then $\mathcal{L}(\mathcal{A})$ is closed under rewriting by $\mathcal{R}$, and*
2. *if $\mathcal{A}$ is deterministic, then $\mathcal{L}(\mathcal{A})$ is closed under rewriting by $\mathcal{R}$.*

Finally, we recall that every tree automaton can be reduced to an equivalent automaton where all states are useful.

Definition 6. *Let $\mathcal{A} = (\mathcal{F}, Q, Q_f, \Delta)$ be a tree automaton. We say that a state $q \in Q$ is* reachable *if $t \to_{\mathcal{A}}^* q$ for some term $t \in \mathcal{T}(\mathcal{F})$; $q \in Q$ is* productive *if $C[q] \to_{\mathcal{A}}^* q'$ for some context C and state $q' \in Q_f$. Finally, an automaton $\mathcal{A}$ is* trim *if all its states are both reachable and productive.*

Proposition 7. *For any tree automaton $\mathcal{A}$ there is an equivalent tree automaton $\mathcal{A}'$ that is trim. If $\mathcal{A}$ is deterministic, then $\mathcal{A}'$ is also deterministic.*

3   State-Compatible Automata

3.1   Definitions

Before we get down to definitions, let us briefly analyze the failure in Example 2. What happens there is that, by the compatibility requirement, all three terms in the rewrite sequence $\mathsf{f}(\mathsf{a}, \mathsf{a}) \to_{\mathcal{R}} \mathsf{b} \to_{\mathcal{R}} \mathsf{a}$ have to be accepted in the same state. In conjunction with the determinism requirement, this is fatal. Consequently, because our goal is to obtain a deterministic automaton, we must allow a and b to be accepted in separate states, q_{a} and q_{b}. To track their connection by rewriting, we introduce a relation $\gg$ on states, such that $q_{\mathsf{b}} \gg q_{\mathsf{a}}$. In general, we require $\gg$ to be state-compatible and state-coherent, which are defined as follows (see also Figure 1).

Definition 8. *Let $\mathcal{A} = (\mathcal{F}, Q, Q_f, \Delta)$ be a tree automaton, and $\gg \subseteq Q \times Q$ be a relation on the states of $\mathcal{A}$. We say that $(\mathcal{A}, \gg)$ is* state-compatible *with a TRS*

$\mathcal{R}$ *if for all state substitutions* σ, *rules* $l \to r \in \mathcal{R}$ *and states* $q \in Q$, *if* $l\sigma \to_{\mathcal{A}}^* q$ *then* $r\sigma \to_{\mathcal{A}}^* q'$ *for some* $q' \in Q$ *with* $q \gg q'$. *We say that* $(\mathcal{A}, \gg)$ *is state-coherent if* $\{q' \mid q \in Q_f, q \gg q'\} \subseteq Q_f$, *and if for all* $f(q_1, \ldots, q_i, \ldots, q_n) \to q \in \Delta$ *and* $q_i \gg q_i'$ *there is some* $q' \in Q$ *with* $f(q_1, \ldots, q_i', \ldots, q_n) \to q' \in \Delta$ *and* $q \gg q'$.

The purpose of state-coherence is to deal with contexts in rewrite steps, as we will see in the proof of Theorem 11 below.

Example 9. Let $\mathcal{A}$ be an automaton with states $1, 2$ (both final), and transitions

$$\mathsf{a} \to 1 \qquad\qquad \mathsf{b} \to 2 \qquad\qquad \mathsf{f}(1,1) \to 2$$

Furthermore, let $2 \gg 2$ and $2 \gg 1$. Then $(\mathcal{A}, \gg)$ is state-coherent and state-compatible with $\mathcal{R} = \{\mathsf{f}(x, x) \to \mathsf{b}, \mathsf{b} \to \mathsf{a}\}$ and $\mathcal{L}(\mathcal{A}) = \{\mathsf{f}(\mathsf{a}, \mathsf{a}), \mathsf{b}, \mathsf{a}\}$. Note that this automaton was obtained from the quasi-deterministic automaton from Example 3 by keeping only the transitions to designated states. We will see in Section 4 that this construction works in general.

Remark 10. If $(\mathcal{A}, \gg)$ is state-coherent, then $(\mathcal{A}, \gg^=)$ and $(\mathcal{A}, \gg^*)$ are also state-coherent. The same holds for state-compatibility with $\mathcal{R}$.

3.2   Soundness and Completeness

Next we prove the analogue of Theorem 5 for state-coherent, state-compatible automata.

Theorem 11. *Let* $\mathcal{A}$ *be a tree automaton such that* $(\mathcal{A}, \gg)$ *is state-coherent and state-compatible with the TRS* $\mathcal{R}$ *for some relation* $\gg$. *Then*

  1. *if* $\mathcal{R}$ *is left-linear, then* $\mathcal{L}(\mathcal{A})$ *is closed under rewriting by* $\mathcal{R}$, *and*
  2. *if* $\mathcal{A}$ *is deterministic, then* $\mathcal{L}(\mathcal{A})$ *is closed under rewriting by* $\mathcal{R}$.

Proof. Let $\mathcal{A} = (\mathcal{F}, Q, Q_f, \Delta)$. First we show that whenever $l\tau \to_{\mathcal{A}}^* q$ for some substitution τ and rule $l \to r \in \mathcal{R}$, then there is a state $q' \in Q$ with $q \gg q'$ and $r\tau \to_{\mathcal{A}}^* q'$. By the assumptions, we can extract from $l\tau \to_{\mathcal{A}}^* q$ a state substitution σ such that $l\tau \to_{\mathcal{A}}^* l\sigma \to_{\mathcal{A}}^* q$: For each $x \in \mathcal{V}ar(l)$, we map x to the state reached from $\tau(x)$ in the given sequence. The state is unique either by left-linearity, or because the given automaton is deterministic. By state-compatibility, we obtain a state q' such that $q \gg q'$ and $r\tau \to_{\mathcal{A}}^* r\sigma \to_{\mathcal{A}}^* q'$.

Using state-coherence we can show by structural induction on C that whenever $C[q] \to_{\mathcal{A}}^* q_\bullet$ and $q \gg q'$, then $C[q'] \to_{\mathcal{A}}^* q_\bullet'$ for some state $q_\bullet'$ with $q_\bullet \gg q_\bullet'$.

Finally, assume that $t \in \mathcal{L}(\mathcal{A})$ and $t \to_{\mathcal{R}} t'$. Then there exist a rule $l \to r \in \mathcal{R}$, a context C and a substitution τ such that $t = C[l\tau]$ and $t' = C[r\tau]$. We have a derivation $t = C[l\tau] \to_{\mathcal{A}}^* C[q] \to_{\mathcal{A}}^* q_\bullet \in Q_f$. By the preceding observations we can find states $q \gg q'$ and $q_\bullet \gg q_\bullet'$ such that $t' = C[r\tau] \to_{\mathcal{A}}^* C[q'] \to_{\mathcal{A}}^* q_\bullet'$. Note that by state-coherence, $q_\bullet \in Q_f$ implies $q_\bullet' \in Q_f$, so that $t' \in \mathcal{L}(\mathcal{A})$.   □

Note that Theorem 11 generalizes Theorem 5 (choose $\gg$ to be the identity relation on states, which is always state-coherent). Moreover, the converse of Theorem 11 holds for trim, deterministic automata. We will prove this in Theorem 13 below, which allows us to derive our main decidability result in Corollary 14. But first let us show by example that the converse fails for some trim, non-deterministic automaton and ground TRS $\mathcal{R}$.

Example 12. Consider the TRS $\mathcal{R} = \{\mathsf{a} \to \mathsf{b}\}$ and the automaton $\mathcal{A}$ with states $0, 1, 2, 3$, final state 0, and transitions

$$
\begin{array}{llll}
\mathsf{a} \to 1 & \mathsf{f}(1) \to 0 & \mathsf{g}(1) \to 0 & \\
\mathsf{b} \to 2 & \mathsf{f}(2) \to 0 & \mathsf{b} \to 3 & \mathsf{g}(3) \to 0
\end{array}
$$

This automaton accepts $\mathcal{L}(\mathcal{A}) = \{\mathsf{f}(\mathsf{a}), \mathsf{f}(\mathsf{b}), \mathsf{g}(\mathsf{a}), \mathsf{g}(\mathsf{b})\}$, which is closed under rewriting by $\mathcal{R}$. Assume that $(\mathcal{A}, \gg)$ is state-coherent and state-compatible with $\mathcal{R}$. By state-compatibility, $\mathsf{a} \to \mathsf{b}$ begets $1 \gg 2$ or $1 \gg 3$. If $1 \gg 2$, then state-coherence, considering the transition $\mathsf{g}(1) \to 0$, requires a transition with left-hand side $\mathsf{g}(2)$, which does not exist. Similarly, if $1 \gg 3$, then $\mathsf{f}(1) \to 0$ requires a transition with left-hand side $\mathsf{f}(3)$, which does not exist.

Theorem 13. *Let $\mathcal{A}$ be a trim, deterministic tree automaton such that $\mathcal{L}(\mathcal{A})$ is closed under rewriting by the TRS $\mathcal{R}$. Then there is a relation $\gg$ such that $(\mathcal{A}, \gg)$ is state-coherent and state-compatible with $\mathcal{R}$.*

Proof. Let $\mathcal{A} = (\mathcal{F}, Q, Q_f, \Delta)$. We define $\gg$ as follows: $q \gg q'$ iff for some terms $t, t' \in \mathcal{T}(\mathcal{F})$, we have

$$
q \overset{*}{\underset{\mathcal{A}}{\leftarrow}} t \underset{\mathcal{R}}{\rightarrow} t' \overset{*}{\underset{\mathcal{A}}{\rightarrow}} q' \tag{1}
$$

Note that by virtue of $\mathcal{A}$ being deterministic, t and t' determine q and q' uniquely. We show that $(\mathcal{A}, \gg)$ is state-coherent and state-compatible.

1. (state-coherence) If $q \in Q_f$ and $q \gg q'$, then there exist terms t, t' satisfying (1). In particular, $q \in Q_f$ implies $t \in \mathcal{L}(\mathcal{A})$, and $t \to_{\mathcal{R}} t'$ implies $t' \in \mathcal{L}(\mathcal{A})$, because $\mathcal{L}(\mathcal{A})$ is closed under rewriting by $\mathcal{R}$. Because $\mathcal{A}$ is deterministic, t' determines q' uniquely, and $q' \in Q_f$ follows.
2. (state-coherence) Assume that $f(q_1, \ldots, q_n) \to q \in \Delta$ and $q_i \gg q_i'$ for some index i and state q_i'. By (1) there are t_i, t_i' such that $q_i \overset{*}{\underset{\mathcal{A}}{\leftarrow}} t_i \to_{\mathcal{R}} t_i' \overset{*}{\underset{\mathcal{A}}{\rightarrow}} q_i'$. Because all q_j are reachable, we can fix terms t_j with $t_j \overset{*}{\underset{\mathcal{A}}{\rightarrow}} q_j$ for $j \neq i$. The state q is productive, so there is a context C such that $C[q] \overset{*}{\underset{\mathcal{A}}{\rightarrow}} q_\bullet \in Q_f$. Let $t = f(t_1, \ldots, t_n)$ and $t' = f(t_1, \ldots, t_i', \ldots, t_n)$. Then $C[t] \in \mathcal{L}(\mathcal{A})$ and $C[t] \to_{\mathcal{R}} C[t']$, hence $C[t'] \in \mathcal{L}(\mathcal{A})$ as well. Consequently, there are states $q', q_\bullet'$ such that

$$
C[q] \overset{*}{\underset{\mathcal{A}}{\leftarrow}} C[t] \underset{\mathcal{R}}{\rightarrow} C[t'] \overset{*}{\underset{\mathcal{A}}{\rightarrow}} C[f(q_1, \ldots, q_i', \ldots, q_n)] \underset{\mathcal{A}}{\rightarrow} C[q'] \overset{*}{\underset{\mathcal{A}}{\rightarrow}} q_\bullet' \in Q_f
$$

In particular, we have a transition $f(q_1, \ldots, q_i', \ldots, q_n) \to q' \in \Delta$, and $q \gg q'$.

3. (state-compatibility) Assume that $l\sigma \to_{\mathcal{A}}^* q$ for a state substitution σ. All states of $\mathcal{A}$ are reachable, so there is a substitution $\tau : \mathcal{V} \to \mathcal{T}(\mathcal{F})$ with $\tau(x) \to_{\mathcal{A}}^* \sigma(x)$ for all $x \in \mathcal{V}$. Furthermore, q is productive, so that for some context C, $C[q] \to_{\mathcal{A}}^* q_\bullet \in Q_f$. We have $C[l\tau] \in \mathcal{L}(\mathcal{A})$ and $C[l\tau] \to_{\mathcal{R}} C[r\tau]$. Consequently, $C[r\tau] \in \mathcal{L}(\mathcal{A})$ and for some states $q', q'_\bullet$,

$$C[q] \xleftarrow[\mathcal{A}]{*} C[l\sigma] \xleftarrow[\mathcal{A}]{*} C[l\tau] \xrightarrow[\mathcal{R}]{} C[r\tau] \xrightarrow[\mathcal{A}]{*} C[q'] \xrightarrow[\mathcal{A}]{*} q'_\bullet \in Q_f$$

In particular, $r\tau \to_{\mathcal{A}}^* q'$. Recall that $\mathcal{A}$ is deterministic. Hence we can decompose this rewrite sequence as follows: $r\tau \to_{\mathcal{A}}^* r\sigma \to_{\mathcal{A}}^* q'$. We conclude by noting that $q \gg q'$ by the definition of $\gg$.   □

Corollary 14. *The problem $\mathcal{R}(\mathcal{L}(\mathcal{A})) \subseteq \mathcal{L}(\mathcal{A})$ is decidable.*

Proof. W.l.o.g. we may assume that $\mathcal{A}$ is deterministic. Using Proposition 7 we may also assume that $\mathcal{A}$ is trim. By Theorems 11 and 13 the problem reduces to whether there is some relation $\gg$ such that $(\mathcal{A}, \gg)$ is both state-compatible with $\mathcal{R}$ and state-coherent. But since there are only finitely many relations $\gg$ we can just test state-compatibility and state-coherence for each $\gg$.

Remark 15. As a consequence of Theorem 13, regular languages accepted by state-coherent automata that are state-compatible with a fixed TRS $\mathcal{R}$ are closed under intersection and union. This can also be shown directly by a product construction.

3.3   Deciding $\mathcal{R}(\mathcal{L}(\mathcal{A})) \subseteq \mathcal{L}(\mathcal{A})$

In the remainder of this section we show that instead of testing all possible relations $\gg$, it suffices to construct a minimal one. We proceed as follows:

1. We assume that $\mathcal{A} = (\mathcal{F}, Q, Q_f, \Delta)$ is trim and deterministic. Note that given a non-deterministic automaton, we can compute an equivalent deterministic one in exponential time. Once we have a deterministic automaton, we can compute an equivalent trim one in polynomial time.
2. For each state $q \in Q$ and rule $l \to r \subset \mathcal{R}$, check whether there is a state substitution σ such that $l\sigma \to_{\Delta}^* q$, but there is no q' with $r\sigma \to_{\Delta}^* q'$. If such a σ exists, then $\mathcal{L}(\mathcal{A})$ is not closed under rewriting by $\mathcal{R}$, and the procedure terminates.
3. In the following steps we will find the smallest relation $\gg$ that makes $(\mathcal{A}, \gg)$ both state-compatible with $\mathcal{R}$ and state-coherent, if such a relation exists.
4. For each pair of states $q, q' \in Q$ and rule $l \to r \in \mathcal{R}$, check whether there is a state substitution σ such that $l\sigma \to_{\Delta}^* q$ and $r\sigma \to_{\Delta}^* q'$. If so, *assert* $q \gg q'$. This ensures that $(\mathcal{A}, \gg)$ will be state-compatible with $\mathcal{R}$.
5. Whenever $q \gg q'$ is *asserted* for the first time for states q and q', we fail if q is final but q' is not, violating the state-coherence. Otherwise, we check Δ for transitions with q on the left-hand side. If $f(q_1, \ldots, q_i = q, \ldots, q_n) \to q_\bullet \in \Delta$, then we look for a transition with left-hand side $f(q_1, \ldots, q'_i = q', \ldots, q_n)$

in Δ. If no such transition exists, state-coherence fails, and the algorithm terminates. Otherwise, let $q'_\bullet \in Q$ be the corresponding right-hand side. We have $f(q_1, \ldots, q'_i = q', \ldots, q_n) \to q'_\bullet \in \Delta$. *Assert* that $q_\bullet \gg q'_\bullet$.

Note that step 5 is really a subroutine, and invokes itself recursively. Steps 2 and 4, which identify the applicable instances of the state-compatibility constraint, consist of a polynomial number of NP queries, and step 5 can be performed in polynomial time. The whole procedure is, therefore, in the Δ_2^P (or P^{NP}) complexity class for deterministic automata as input.

Remark 16. Using [3, Exercise 1.12.2], which shows that it is NP-hard to decide whether an instance of a term l is accepted by a tree automaton $\mathcal{A}$, we can show that deciding whether the language accepted by a deterministic automaton is closed under rewriting by a given TRS is co-NP-hard. To wit, given a term l, a tree automaton $\mathcal{A}$, a fresh unary function $\star$ and a fresh constant $\diamond$, then $\star(\mathcal{L}(\mathcal{A})) = \{\star(x) \mid x \in \mathcal{L}(\mathcal{A})\}$ is closed under rewriting by $\star(l) \to \diamond$ if and only if no instance of l is accepted by $\mathcal{A}$.

4   Relation to Quasi-deterministic Automata

We recall the definitions of compatibility and quasi-determinism from [10], and show that given a compatible, quasi-deterministic automaton, we can extract a state-compatible, deterministic automaton accepting the same language, while the opposite direction fails in the presence of collapsing rules.

Definition 17 (Definition 18 of [10]). *Let $\mathcal{A} = (\mathcal{F}, Q, Q_f, \Delta)$ be a tree automaton. For a left-hand side $l \in \mathrm{lhs}(\Delta)$ of a transition, we denote the set $\{q \mid l \to q \in \Delta\}$ of possible right-hand sides by $Q(l)$. We call $\mathcal{A}$ quasi-deterministic if for every $l \in \mathrm{lhs}(\Delta)$ there exists a designated state $p \in Q(l)$ such that for all transitions $f(q_1, \ldots, q_n) \to q \in \Delta$ and $i \in \{1, \ldots, n\}$ with $q_i \in Q(l)$, the transition $f(q_1, \ldots, q_{i-1}, p, q_{i+1}, \ldots, q_n) \to q$ belongs to Δ. Moreover, we require that $p \in Q_f$ whenever $Q(l)$ contains a final state.*

For each $l \in \mathrm{lhs}(\Delta)$ we pick a state p_l satisfying the constraints of Definition 17. We denote the set of designated states by Q_d and the set $\{l \to p_l \mid l \in \mathrm{lhs}(\Delta)\}$ by Δ_d. The notion of compatibility used for quasi-deterministic tree automata is refined slightly from the standard one, Definition 4.

Definition 18 (Definition 23 of [10]). *Let $\mathcal{R}$ be a TRS and L a language. Let $\mathcal{A} = (\mathcal{F}, Q, Q_f, \Delta)$ be a quasi-deterministic tree automaton. We say that $\mathcal{A}$ is compatible with $\mathcal{R}$ and L if $L \subseteq \mathcal{L}(\mathcal{A})$ and for each rewrite rule $l \to r \in \mathcal{R}$ and state substitution $\sigma \colon \mathcal{V}ar(l) \to Q_d$ such that $l\sigma \to^*_{\Delta_d} q$ it holds that $r\sigma \to^*_{\Delta} q$.*

Example 3 exhibits a quasi-deterministic, quasi-compatible automaton.

   We will show that for each quasi-deterministic automaton that is compatible with a TRS $\mathcal{R}$, there is a deterministic, state-coherent automaton that is state-compatible with $\mathcal{R}$ and accepts the same language. To this end, we need the following key lemma, a slight generalization of [10, Lemma 20], which shows that a quasi-deterministic automaton $\mathcal{A}$ is almost deterministic: all but the last step in a reduction can be performed using the deterministic Δ_d transitions.

Lemma 19. *Let* $\mathcal{A} = (\mathcal{F}, Q, Q_f, \Delta)$ *be a quasi-deterministic automaton. If* $t \to_\Delta^+ q$ *then* $t \to_{\Delta_d}^* \cdot \to_\Delta q$ *for all terms* $t \in \mathcal{T}(\mathcal{F} \cup Q)$ *and states* $q \in Q$.

Proof. Identical to the proof of [10, Lemma 20], except when t_i in $t = f(t_1, \dots, t_n)$ is a state. In that case, we let $p_{l_i} = q_i = t_i$. $\square$

Theorem 20. *Let* $\mathcal{A} = (\mathcal{F}, Q, Q_f, \Delta)$ *be a quasi-deterministic tree automaton that is compatible with* $\mathcal{R}$. *Then* $\mathcal{A}' = (\mathcal{F}, Q_d, Q_f \cap Q_d, \Delta_d)$ *makes* $(\mathcal{A}', \gg)$ *state-coherent and state-compatible with* $\mathcal{R}$, *where* $q \gg q'$ *if* $q = q'$ *or, for some left-hand side* $l \in \text{lhs}(\Delta)$, $q \in Q(l)$ *and* $q' = p_l$. *Furthermore,* $\mathcal{L}(\mathcal{A}') = \mathcal{L}(\mathcal{A})$.

Proof. Note that $\to_\mathcal{A} = \to_\Delta$ and $\to_{\mathcal{A}'} = \to_{\Delta_d}$.

1. (state-coherence) Assume that q is final in $\mathcal{A}'$, and $q \gg q'$. If $q = q'$ then q' is final, too. Otherwise, there is a left-hand side l such that $q \in Q(l)$ and $q' = p_l$ is the designated state of l. Since $Q(l)$ contains a final state (namely, q), q' must be final as well by Definition 17.
2. (state-coherence) Let $l = f(q_1, \dots, q_i, \dots, q_n)$ and $l' = f(q_1, \dots, q_i', \dots, q_n)$, where $q_i \gg q_i'$. Furthermore, let $l \to q \in \Delta_d$. If $q_i = q_i'$ then $l' \to q \in \Delta_d$ and $q \gg q$. Otherwise, there is a left-hand side $l^\bullet$ such that $q_i \in Q(l^\bullet)$ and $q_i' = p_{l^\bullet}$ is the designated state of $l^\bullet$. By Definition 17, there is a transition $l' \to q$ in Δ. Thus, l' is a left-hand side and $q \in Q(l')$. Furthermore, $l' \to p_{l'} \in \Delta_d$, and $q \gg p_{l'}$ follows.
3. (state-compatibility) Let σ be a state substitution and $l\sigma \to_{\Delta_d}^* q$. By compatibility, we have $r\sigma \to_\Delta^* q$. If r is a variable, we are done, noting that $q \gg q$. Otherwise, using Lemma 19, there is a left-hand side $l' \in \text{lhs}(\mathcal{A})$ such that $r\sigma \to_{\Delta_d}^* l' \to_\Delta q$. Consequently, $r\sigma \to_{\Delta_d}^* \cdot \to_{\Delta_d} p_{l'}$, and since $q \in Q(l')$, we have $q \gg p_{l'}$.
4. (accepted language) $\mathcal{L}(\mathcal{A}') \subseteq \mathcal{L}(\mathcal{A})$ is obvious. To show $\mathcal{L}(\mathcal{A}) \subseteq \mathcal{L}(\mathcal{A}')$, assume that $t \in \mathcal{L}(\mathcal{A})$, i.e., $t \to_\Delta^* q \in Q_f$. By Lemma 19, there is a left-hand side $l \in \text{lhs}(\mathcal{A})$ such that $t \to_{\Delta_d}^* l \to_\Delta q$. As in the previous item we conclude that $t \to_{\Delta_d}^* p_l$, and $q \gg p_l$. The state p_l is final by state-coherence, so $t \in \mathcal{L}(\mathcal{A}')$ follows. $\square$

In the opposite direction, we have a positive result for non-collapsing TRSs.

Theorem 21. *Let* $\mathcal{A} = (\mathcal{F}, Q, Q_f, \Delta)$ *be a deterministic automaton and the relation* $\gg \subseteq Q \times Q$ *be such that* $(\mathcal{A}, \gg)$ *is state-coherent and state-compatible with* $\mathcal{R}$. *Furthermore, assume that* $\mathcal{R}$ *contains no collapsing rules. Then the automaton* $\mathcal{A}' = (\mathcal{F}, Q, Q_f, \Delta')$ *with* $\Delta' = \{l \to q' \mid l \to q \in \Delta, q \gg^= q'\}$ *is a quasi-deterministic automaton with designated states* $p_l = q$ *for* $l \to q \in \Delta$, *such that* $\mathcal{A}'$ *is compatible with* $\mathcal{R}$ *and accepts the same language as* $\mathcal{A}$.

Proof. Verifying that the construction results in a quasi-deterministic automaton that is compatible with $\mathcal{R}$ is straight-forward. Note that applying Theorem 20 to $\mathcal{A}'$ results in some $(\mathcal{A}'', \gg'')$ with $\mathcal{L}(\mathcal{A}'') = \mathcal{L}(\mathcal{A}')$, where $\mathcal{A}''$ is $\mathcal{A}$ with states restricted to Q'_d, the right-hand sides of Δ'. This restriction preserves the accepted language. Therefore, $\mathcal{L}(\mathcal{A}) = \mathcal{L}(\mathcal{A}')$. □

If $\mathcal{R}$ contains collapsing rules, quasi-deterministic, compatible automata may be weaker than state-coherent, state-compatible ones, as the following example demonstrates.

Example 22. Let $\mathcal{R} = \{\mathsf{f}(x,x) \to x\}$. The automaton $\mathcal{A}'$ over $\{\mathsf{f}, \mathsf{a}\}$ with states $1, 2$, both final, and transitions

$$\mathsf{a} \to 1 \qquad\qquad\qquad \mathsf{f}(1,1) \to 2$$

accepts $L = \{\mathsf{f}(\mathsf{a}, \mathsf{a}), \mathsf{a}\}$. Furthermore, $(\mathcal{A}', \gg)$ is state-coherent and state-compatible with $\mathcal{R}$ if we let $2 \gg 1$.

Now assume that $\mathcal{A} = (\{\mathsf{f}, \mathsf{a}\}, Q, Q_f, \Delta)$ is a quasi-deterministic automaton and compatible with $\mathcal{R}$, and that $\mathsf{f}(\mathsf{a}, \mathsf{a}) \in \mathcal{L}(\mathcal{A})$. We will show that $\mathcal{A}$ accepts all terms over $\{\mathsf{f}, \mathsf{a}\}$. Note that since $\mathsf{f}(\mathsf{a}, \mathsf{a})$ is accepted, a must be a left-hand side of $\mathcal{A}$. Let q be the designated state of a. By Lemma 19, we have a run $\mathsf{f}(\mathsf{a}, \mathsf{a}) \to^*_{\Delta_d} \mathsf{f}(q,q) \to_\Delta q' \in Q_f$. Let $q^\bullet$ be the designated state of the left-hand side $\mathsf{f}(q,q)$. By quasi-determinism, $q^\bullet$ is a final state. Compatibility requires that $\mathsf{f}(q,q) \to_{\Delta_d} q^\bullet \ {}^*_\Delta\!\!\leftarrow q$, i.e., $q^\bullet = q$. So we have a final state q and two transitions $\mathsf{a} \to q$, $\mathsf{f}(q,q) \to q$, and $\mathcal{A}$ accepts all of $\mathcal{T}(\{\mathsf{f}, \mathsf{a}\})$.

Remark 23. In his thesis [9], Korp generalizes Definition 17 (cf. [9, Definition 3.10]) by incorporating an auxiliary relation $\succeq_{\phi_\mathcal{A}}$ that may be viewed as a precursor to our relation $\gg$. The modified definition permits smaller automata, which benefits implementations, but is more complicated than Definition 17. The modification also does not add expressive power. Indeed if $\mathcal{A} = (\mathcal{F}, Q, Q_f, \Delta)$ satisfies [9, Definition 3.10] using $\succeq_{\phi_\mathcal{A}}$, then taking $\Delta' = \{l \to q \mid l \in \mathrm{lhs}(\Delta), \phi_\mathcal{A}(l) \succeq q\}$, the automaton $\mathcal{A}' = (\mathcal{F}, Q, Q_f, \Delta')$ satisfies Definition 17, noting that $\phi_\mathcal{A}(l)$ is just another notation for the designated state p_l of l. Furthermore, $\mathcal{L}(\mathcal{A}') = \mathcal{L}(\mathcal{A})$.

5   Formalization

We have formalized all results from Section 3 as part of IsaFoR, our *Isa*belle *Fo*rmalization of *R*ewriting, in combination with executable algorithms which check state-compatibility and state-coherence. These are used in CeTA [14], a certifier for several properties related to term rewriting.

Here, `Tree_Automata.thy` starts with basic definitions on tree automata where there are two major differences to this paper: the formalization allows ε-transitions, and the set of *reachable* states $t \to^*_\Delta q$ is formalized directly as a function ta_res mapping terms to sets of states. Using a function instead of a relation has both positive and negative effects. For example, it eases proofs which are naturally performed by induction on terms, since in $f(t_1, \ldots, t_n)$ one does not have to reduce all arguments t_1 to t_n sequentially in a relation, but this is done in one step in ta_res.

On the other hand, one cannot trace derivations $t \to_{\Delta}^* q$ explicitly as there is no notion of derivation. Hence, some obvious results have to be proven explicitly by induction, e.g., that removing transitions results in a smaller accepted language.

The file continues with proofs of Proposition 7 (obtain_trimmed_ta), Theorems 11 and 13 (state_compatible_lang and ta_trim_det_closed), and Corollary 14 (closed_iff_compatible_and_coherent), where the corollary states only that $\mathcal{L}(\mathcal{A})$ is closed under $\mathcal{R}$ iff for the determinized and trimmed automaton there exists a suitable relation $\gg$. Instead of formally proving decidability by an algorithm which enumerates all possible relations, we directly formalized the algorithm of Section 3.3 to compute the least such relation. Here, we described the algorithm on an abstract level via some inference system, and its soundness and completeness manifests in theorem decide_coherent_compatible. It is later on refined to a fully executable one.

In addition to the decision procedure, we also provide Theorem 11 to demonstrate closure under rewriting when $\gg$ is supplied. The advantage of the latter is its improved runtime and its broader applicability: one does not have to iteratively construct the relation, and for left-linear TRSs, also non-deterministic automata with ε-transitions are supported, cf. state_compatible_lang. Here, for checking state-compatibility, we use a tree automaton matching algorithm (ta_match), that restricts the set of state substitutions σ that have to be considered for compatibility w.r.t. Definition 8.

Whereas Tree_Automata.thy formalizes most algorithms on an abstract level, in Tree_Automata_Impl.thy we refined those to fully executable ones. In fact, for some algorithms we just relied on the automatic refinement provided in [12] which turns set operations into operations on trees. However, for some algorithms like the matching algorithm we performed manual refinement to increase the efficiency. For example, we group the transitions of an automaton by their root symbols and store these groups in ordered trees using Isabelle's collection framework [11]. Moreover, for each $f(q_1, \dots, q_n) \to q$, we precompute the closure of q under ε-transitions. This speeds up the computation of ta_res while checking state-compatibility. In the end, we provide an executable algorithm which for given $\mathcal{A}$ and $\mathcal{R}$ checks whether $\mathcal{A}$ is closed under $\mathcal{R}$, cf. tree_aut_trs_closed.

We have extended the termination tool T_TT_2 [8] and the confluence tool CSI [15] to produce state-coherent, state-compatible automata. Since both tools use quasi-deterministic automata in their completion process, we applied the construction of Theorem 20 as a post-processing step, resulting in a state-coherent, state-compatible automaton. CeTA can then be used to certify this output. Whereas for non-confluence proofs the input can be arbitrary, for match-bounds we currently require left-linearity. The reason is that without left-linearity, the match-bounds technique requires further conditions besides closure under rewriting, which have not been formalized yet and which remain as future work.

All tools and the formalization are available at http://cl-informatik.uibk.ac.at/research/software/ (CeTA + IsaFoR version 2.12, T_TT_2 version 1.14, CSI version 0.4.)

6   Conclusion

We have introduced the class of deterministic, state-coherent automata that are state-compatible with a TRS $\mathcal{R}$. We have shown that these automata capture precisely those regular tree languages that are closed under rewriting by $\mathcal{R}$, leading to a decision procedure for checking whether a regular language is closed under rewriting. Their simple definition allowed us to formalize most of our results on state-coherent, state-compatible automata.

Even though state-coherent, state-compatible tree automata are strictly more general, we still rely on quasi-deterministic tree automata for the actual completion process in the CSI and T_TT_2 tools. Thus, they cannot exploit the full power of state-coherent and state-compatible tree automata, and will fail when analyzing TRSs like Example 22. As future work, we plan to investigate whether working directly on state-coherent, state-compatible automata can improve tree automata completion.

Acknowledgments. We would like to thank Aart Middeldorp for fruitful discussions on the topic of tree automata and helpful feedback. We are also grateful to the anonymous reviewers for their constructive feedback.

References

1. Baader, F., Nipkow, T.: Term Rewriting and All That. Cambridge University Press (1998)
2. Boyer, B., Genet, T., Jensen, T.P.: Certifying a tree automata completion checker. In: Armando, A., Baumgartner, P., Dowek, G. (eds.) IJCAR 2008. LNCS (LNAI), vol. 5195, pp. 523–538. Springer, Heidelberg (2008)
3. Comon, H., Dauchet, M., Gilleron, R., Jacquemard, F., Lugiez, D., Löding, C., Tison, S., Tommasi, M.: Tree automata techniques and applications (2007), http://tata.gforge.inria.fr
4. Feuillade, G., Genet, T., Tong, V.V.T.: Reachability analysis over term rewriting systems. Journal of Automated Reasoning 33, 341–383 (2004)
5. Genet, T.: Decidable approximations of sets of descendants and sets of normal forms. In: Nipkow, T. (ed.) RTA 1998. LNCS, vol. 1379, pp. 151–165. Springer, Heidelberg (1998)
6. Genet, T., Tang-Talpin, Y.M., Tong, V.V.T.: Verification of copy-protection cryptographic protocol using approximations of term rewriting systems. In: Proc. WITS 2003 (Workshop on Issues in the Theory of Security) (2003)
7. Geser, A., Hofbauer, D., Waldmann, J., Zantema, H.: On tree automata that certify termination of left-linear term rewriting systems. Information and Computation 205(4), 512–534 (2007)
8. Hirokawa, N., Middeldorp, A.: Tyrolean termination tool. In: Giesl, J. (ed.) RTA 2005. LNCS, vol. 3467, pp. 175–184. Springer, Heidelberg (2005)
9. Korp, M.: Termination Analysis by Tree Automata Completion. Ph.D. thesis, University of Innsbruck (2010)
10. Korp, M., Middeldorp, A.: Match-bounds revisited. Information and Computation 207(11), 1259–1283 (2009)

11. Lammich, P., Lochbihler, A.: The Isabelle collections framework. In: Kaufmann, M., Paulson, L.C. (eds.) ITP 2010. LNCS, vol. 6172, pp. 339–354. Springer, Heidelberg (2010)
12. Lochbihler, A.: Light-weight containers for Isabelle: Efficient, extensible, nestable. In: Blazy, S., Paulin-Mohring, C., Pichardie, D. (eds.) ITP 2013. LNCS, vol. 7998, pp. 116–132. Springer, Heidelberg (2013)
13. Nipkow, T., Paulson, L.C., Wenzel, M.T. (eds.): Isabelle/HOL – A Proof Assistant for Higher-Order Logic. LNCS, vol. 2283. Springer, Heidelberg (2002)
14. Thiemann, R., Sternagel, C.: Certification of termination proofs using CeTA. In: Berghofer, S., Nipkow, T., Urban, C., Wenzel, M. (eds.) TPHOLs 2009. LNCS, vol. 5674, pp. 452–468. Springer, Heidelberg (2009)
15. Zankl, H., Felgenhauer, B., Middeldorp, A.: CSI – A confluence tool. In: Bjørner, N., Sofronie-Stokkermans, V. (eds.) CADE 2011. LNCS, vol. 6803, pp. 499–505. Springer, Heidelberg (2011)

Counting Models of Linear-Time Temporal Logic

Bernd Finkbeiner and Hazem Torfah

Reactive Systems Group, Saarland University
66123 Saarbrücken, Germany
{finkbeiner,torfah}@cs.uni-saarland.de

Abstract. We investigate the model counting problem for safety specifications expressed in linear-time temporal logic (LTL). Model counting has previously been studied for propositional logic; in planning, for example, propositional model counting is used to compute the plan's robustness in an incomplete domain. Counting the models of an LTL formula opens up new applications in verification and synthesis. We distinguish word and tree models of an LTL formula. Word models are labeled sequences that satisfy the formula. Counting the number of word models can be used in model checking to determine the number of errors in a system. Tree models are labeled trees where every branch satisfies the formula. Counting the number of tree models can be used in synthesis to determine the number of implementations that satisfy a given formula. We present algorithms for the word and tree model counting problems, and compare these direct constructions to an indirect approach based on encodings into propositional logic.

Keywords: Model counting, temporal logic, model checking, synthesis, tree automata.

1   Introduction

Model counting, the problem of computing the *number of solutions* of a given logical formula, is a useful generalization of satisfiability. Many probabilistic inference problems, such as Bayesian net reasoning [13], and planning problems, such as computing the robustness of plans in incomplete domains [14], can be formulated as model counting problems of propositional logic. State-of-the-art tools for propositional model counting include Relsat [1] and c2d [6].

In this paper, we study the model counting problem for safety specifications expressed in *linear-time temporal logic* (LTL). LTL is the most commonly used specification logic for reactive systems [15] and the standard input language for model checking [2,5] and synthesis tools [4,3,7]. Just like propositional model counting generalizes SAT, LTL model counting introduces "quantitative" extensions of model checking and synthesis. In *model checking*, model counting can be used to determine not only the existence of computations that violate the specification, but also the *number* of such *violations*. For example, in a communication system, where messages are lost (with some probability) on the channel,

A.-H. Dediu et al. (Eds.): LATA 2014, LNCS 8370, pp. 360–371, 2014.
© Springer International Publishing Switzerland 2014

it is typically not necessary (or even possible) to guarantee a 100% correct trans-
mission. Instead, the number of executions that lead to a message loss is a good
indication for the quality of the implementation. In *synthesis*, model counting
can be used to determine not only the existence of an implementation that satisfies the specification, but also the *number* of such *implementations*. The number
of implementations of a specification is a helpful metric to understand how much
room for implementation choices is left by a given specification, and to estimate
the impact of new requirements on the remaining design space.

Formally, we distinguish two types of models of an LTL formula. A *word
model* of an LTL formula φ over a set of atomic propositions AP is a sequence
of valuations of AP such that the sequence satisfies φ. A *tree model* of an LTL
formula φ over a set of atomic propositions $AP = I \cup O$, partitioned into *inputs* I
and *outputs* O, is a tree that branches according to the valuations of I and that is
labeled with valuations of O, such that every path of the tree satisfies φ. In order
to guarantee that the number of models is finite, we consider *bounded models*,
i.e., words of bounded length and trees of bounded depth. This is motivated by
applications like bounded model checking [2] and bounded synthesis [8], where we
look for small error paths and small implementations, respectively, by iteratively
increasing a bound on the size of the model.

Since both bounded model checking and bounded synthesis are based on satisfiability checking, a natural idea to solve the model counting problem of LTL
is to reduce it to the propositional counting problem: for word models, this can
be done by introducing a copy of the atomic propositions for each position of
the word, for tree models, by introducing a copy of the atomic propositions for
each node in the tree. Unfortunately, however, this reduction quickly results in
intractable propositional problems. For word models, we need a linear number of
propositional variables in the bound, for tree models even an exponential number
of variables. This is critical, since propositional counting is #P-complete. Current state-of-the-art model counters cannot handle more than approximately
1000-10000 propositional variables [9]. This limit is exceeded easily, for example,
by a tree of depth 5. (Assuming 3 bits of input and a 3-bit encoding of the LTL
formula, we need approximately 100000 variables.)

In this paper, we present a model counting algorithm with much better performance. For both word and tree models, the complexity of our algorithm is
linear in the bound. This improvement is obtained by dynamic programming:
we compute the number of models *backwards*, i.e., from the last position to the
first in the case of word models, and form the leaves to the root in the case
of tree models. We show that LTL formulas can be translated to word and tree
automata that have exactly one run for every model. The number of runs is then
computed by incrementally considering larger models and computing, for each
bound, the number of models that are accepted by runs starting in a specific
state.

Analyzing the complexity of this construction, it turns out that the dramatic
improvement in the complexity with respect to the bound does not come for free,
as our constructions are more expensive in the size of the formula, compared to

the solution based on a reduction to propositional counting. In practice, however, this is not a problem, because costs in relation to the size of the formula are much more benign than costs in relation to the bound: typically, we are interested in systems with large implementations, but small specifications.

Overview. After reviewing the necessary preliminaries in Section 2, we formally define the model counting problem in Section 3. Counting algorithms for word models and tree models are presented in Sections 4 and 5, respectively.

2   Preliminaries

Transition Systems. We represent models as *labeled transition systems*. For a given finite set Υ of directions and a finite set Σ of labels, a Σ-labeled Υ-transition system is a tuple $\mathcal{S} = (S, s_0, \tau, o)$, consisting of a finite set of states S, an initial state $s_0 \in S$, a transition function $\tau : S \times \Upsilon \to S$, and a labeling function $o : S \to \Sigma$.

A *path* in a labeled transition system is a sequence $\pi : \mathbb{N} \to S \times \Upsilon$ of states and directions that follows the transition relation, i.e., for all $i \in \mathbb{N}$ if $\pi(i) = (t_i, e_i)$ then $\pi(i + 1) = (t_{i+1}, e_{i+1})$ where $t_{i+1} \in \tau(t_i, e_i)$. We call the path initial if it starts with the initial state: $\pi(0) = (t_0, e)$ from some $e \in \Upsilon$. We define the set $paths(\mathcal{S})$ as the set of all initial paths of $\mathcal{S}$.

Specifications. We use linear-time temporal logic (LTL) [15], with the usual temporal operators Next X, Until U, and the derived operators Eventually $\Diamond$ and Globally $\Box$. LTL formulas are defined over a set of atomic propositions $AP = I \cup O$, which is partitioned into a set I of input variables and a set O of output variables. We denote the satisfaction of an LTL formula φ by an infinite sequence $\sigma : \mathbb{N} \to 2^{AP}$ of valuations of the atomic propositions by $\sigma \models \varphi$. A 2^O-labeled 2^I-transition system $\mathcal{S} = (S, s_0, \tau, o)$ satisfies φ, if for all $\pi \in paths(\mathcal{S})$ the sequence $\sigma_\pi : i \mapsto o(\pi(i))$, where $o(s, e) = (o(s) \cup e)$, satisfies φ. In the remainder of the paper, we assume that all considered LTL specifications express safety properties. An infinite sequence $\sigma : \mathbb{N} \to 2^{AP}$ violates a safety property iff there is a prefix $\sigma' : [0, i] \to 2^{AP}$ of σ such that for all extensions $\hat{\sigma} : \mathbb{N} \to 2^{AP}$, $\sigma' \hat{\sigma} \not\models \varphi$. We call σ' a *bad prefix* for φ.

Universal Safety Automata. A *universal safety automaton* is a tuple $\mathcal{U} = (Q, q_0, \delta, \Sigma, \Upsilon)$, where Q denotes a finite set of states, $q_0 \in Q$ denotes the initial state, δ denotes a transition function, Σ a finite set of labels, and Υ a finite set of directions. The transition function $\delta : Q \times \Sigma \times \Upsilon \to 2^Q$ maps a state to the set of successor states reachable via a label $\sigma \in \Sigma$ and a direction $v \in \Upsilon$. A run graph of the automaton on a Σ-labeled Υ-transition system $\mathcal{S} = (S, s_0, \tau, o)$ is a directed graph $\mathcal{G} = (G, E)$ such that: The vertices $G \subseteq Q \times S$, the pair $(q_0, s_0) \in G$, and for each pair $(q, s) \in G$ there is an edge to $(q', \tau(s, v))$ for $v \in \Upsilon$ and for every $q' \in \delta(q, o(s), v)$. A transition system is accepted by the automaton if it has a run graph in the automaton.

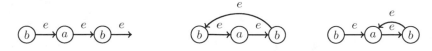

Fig. 1. A base and two word models

For each safety property expressed as an LTL formula φ, we can construct a universal safety automaton that accepts exactly the sequences that satisfy φ. If φ has length n, then the number of states of this universal safety automaton is in $2^{\mathcal{O}(n)}$. (This can be done by translating φ into an automaton that recognizes its bad prefixes, called *fine bad prefix automaton* in [12], and dualizing this automaton.)

Bottom-Up Tree Automata. Σ-labeled Υ-trees are trees where each node is labeled with a label $\alpha \in \Sigma$ and has exactly one child for every direction $\upsilon \in \Upsilon$. A *bottom-up tree automaton* is a tuple $\mathcal{T} = (T, T_F, \Delta_0, \Delta, \Sigma, \Upsilon)$ defined over Σ-labeled Υ-trees, where T is a finite set of states, $T_F \subseteq T$ denotes the set of accepting states, an initial transition relation $\Delta_0 \subseteq \Sigma \times T$ that associates a leaf node of the tree to a state of the automaton, according to the label $\alpha \in \Sigma$ of the leaf node, and the transition relation $\Delta \subseteq T^{|\Upsilon|} \times \Sigma \times T$ that determines the state labeling of a node according to the label of the node and the state labelings of the children nodes. A run of the automaton over a Σ-labeled Υ-tree is a T-labeled Υ-tree. We say that a tree is accepted by the automaton if the root of its run tree is in T_F.

3   The Model Counting Problem

A model of an LTL formula is a finite transition system. Counting the number of transition systems that satisfy a given LTL formula would not, however, be very informative, because this number is either 0 or ∞: if the formula is satisfiable, it is satisfied by some ultimately periodic model, and each unrolling of the periodic part results in a new transition system that satisfies the formula. We therefore consider *bounded models*.

We distinguish two types of bounded models, *word* and *tree* models. A *k-word model* of an LTL formula φ over $AP = I \cup O$ is a lasso sequence $\pi(0) \ldots \pi(i-1)(\pi(i), \ldots \pi(k))^{\omega} \in (2^O \times 2^I)^{\omega}$ for some $i \in \{0, ..., k\}$. We call $\pi_{\perp} = \pi(0) \ldots \pi(k) \in (2^O \times 2^I)^{k+1}$ the *base* of the model. Figure 1 shows two word models and their base.

A *k-tree model* of an LTL formula φ is a 2^O-labeled-2^I-transition system that forms a tree of depth k with additional loop-back transitions from the leaves (for every leaf and every direction, there is an edge to some state on the branch leading to the leaf). The tree without the loop-back transitions is the *base* of the model. Figure 2 shows two tree models and their base.

For an LTL formula φ and a bound k, the *k-word* (*k-tree*) *counting problem* is to compute the number of *k-word* (*k-tree*) models of φ.

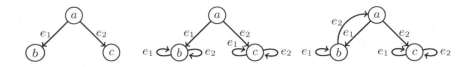

Fig. 2. A base and two tree models

4 Counting Word Models

We start by introducing an algorithm for counting word models of safety LTL formulas. In the next section we show how we can adapt the ideas of this algorithm in order to count tree models. For a given bound k and a safety specification φ, we construct a word automaton that accepts a finite sequence of size k if it is a base for a word model of φ. We introduce an algorithm based on the automaton that delivers the number of word models of φ.

4.1 An Automaton for Word Models

The following theorem shows that for each safety property expressed as an LTL specification φ and a bound k, we can construct a word automaton that accepts a word of maximum length k if it is a base of a word model of φ. In theorem 2 we show that the word automaton can be used to count the number of k-word models for the specification φ. Our starting point is the representation of the specification φ as a universal safety automaton. When a word model π satisfies φ, then there is a run graph of the universal safety automaton on π. In the run graph, every state s in π is mapped to a set of states in the universal automaton. This set is the set of universal states visited by π in the state s. We refer to this set as an *annotation* of s. Intuitively, our word automaton tries to reconstruct a possible annotation for each state for a given base of a word model. The loop in the word model corresponds to a suffix of the base. The annotation of this suffix is a repeating annotation in the run graph of the word model. The word automaton guesses the annotation of the loop-back state (the first state of the suffix), and checks, traversing the base *backwards*, whether (1) a repetition of the guessed annotation along the base is observed, and (2) an initial annotation is reached after having traversed the whole base (an annotation containing the initial state of the universal safety automaton). Since one base may correspond to several word models, the automaton also keeps track of the number of repetitions of the guessed annotation.

It remains to ensure that the automaton is unambiguous with respect to a word model, i.e, that every base has at most a single accepted annotation for a word model. So far, a single base might have multiple annotations. Such a situation is depicted in Figure 3. To prevent multiple annotations of the same base, the automaton only allows *maximal* annotations: in addition to the "positive" annotation, the automaton builds a "negative" annotation consisting of states

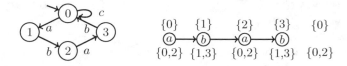

Fig. 3. On the left: a universal safety automaton; on the right: a base with two different annotations. The annotation shown above the base corresponds to a run graph of the universal automaton. The annotation shown below the base is the maximal annotation.

of the universal automaton that must not occur in the positive annotation. All states that do not occur in the positive annotation must occur in the negative annotation. Due to the determinism of the universal safety automaton, there is only one maximal annotation for each word model over its base. Note that the maximal annotation in Figure 3 (shown below the base) includes the alternative annotation shown above the base. Also note that the maximal annotation fits two different word models, the word model with the loop labeled *abab* and the word model labeled *ab*.

Theorem 1. *Given a universal safety automaton* $\mathcal{U} = (Q, q_0, \delta, \Sigma, \Upsilon)$ *with n states, and a bound k, we can construct a finite word automaton* $\mathcal{A}_{\#} = (Q_{\#}, Q_{0\#},$ $Q_{F\#}, \Delta, \Sigma, \Upsilon)$ *that accepts a sequence* $\sigma \in (\Sigma \times \Upsilon)^k$ *if σ^{-1} is the sequence of labels of a base of a word model that is accepted by* $\mathcal{U}$. *The number of states of the automaton* $\mathcal{A}_{\#}$ *is in* $2^{\mathcal{O}(n)}$.

Construction: We choose $(2^Q \times \{0, ..., k\} \times (2^Q)^{n-1}) \times 2^Q \times (2^Q)^n$ to be the state space of the word automaton. A state $((\mathcal{C}, c, \mathcal{C}_1, \ldots, \mathcal{C}_{n-1}), \mathcal{P}, \mathcal{N}_0, \ldots, \mathcal{N}_{n-1})$ is split into conjecture sets $\mathcal{C}, \mathcal{C}_1, \ldots, \mathcal{C}_{n-1}$, which once chosen cannot be manipulated by the transition relation, and tracking sets $\mathcal{P}, \mathcal{N}_0, \ldots, \mathcal{N}_{n-1}$, which keep track of the possible state annotations for a given sequence π, starting from the conjecture annotations. The conjecture $\mathcal{C}$ denotes a loop back annotation reached in some state s on π when looping back to s. Given $\mathcal{C}$, the idea is to traverse π backwards and check whether this annotation is repeated in some state s on π. If this is the case, we point to a possible word model with a loop back to s. The counter c denotes the number of valid repetitions of $\mathcal{C}$ along π. The conjectures $\mathcal{C}_1, \ldots, \mathcal{C}_{n-1}$ are used for the maximality check. As discussed earlier, when we check a finite sequence in an inverse fashion, we may find more than one valid annotation of the universal automaton for its states. To check whether the annotation is maximal, we also compute for each state a set of negative universal states that are not allowed to be in a sequence state's annotation. For a state s in the sequence, a set $\mathcal{C}_j$ involves states that lead to a dead end in the automaton $\mathcal{U}$ in the j-th loop to s. Starting with $\mathcal{N}_0 = \emptyset$ we need to loop at most $n - 1$ times to reach the largest set $\mathcal{C}_{n-1}$ of all negative universal states (the set is at most as large the set of universal states. In each loop this set either increases or we will have reached a fix-point in which all negative states are already included). An annotation is maximal if it contains all states that are not in the negative set. An initial state is a conjecture state of the form $((\mathcal{C}, 0, \mathcal{C}_1, \ldots, \mathcal{C}_{n-1}), \mathcal{C}, \emptyset, \mathcal{C}_1, \ldots, \mathcal{C}_{n-1})$ where $\mathcal{C}, \mathcal{C}_1, \ldots, \mathcal{C}_{n-1} \subseteq 2^Q$.

The sets $\mathcal{P}, \mathcal{N}_0, \ldots, \mathcal{N}_{n-1}$ are computed via the transition relation Δ. Once the automaton made a choice for an initial conjecture state $((\mathcal{C}, 0, \mathcal{C}_1, \ldots, \mathcal{C}_{n-1}), \mathcal{C}, \emptyset, \mathcal{C}_1, \ldots, \mathcal{C}_{n-1})$, Δ becomes deterministic. For a symbol $\alpha \in \Sigma \times \Upsilon$ and a state $\Lambda_\# = ((\mathcal{C}, c, \mathcal{C}_1, \ldots, \mathcal{C}_{n-1}), \mathcal{P}, \mathcal{N}_0, \ldots, \mathcal{N}_{n-1})$ the transition relation computes a state $\Lambda'_\# = \Delta(\Lambda_\#, \alpha) = ((\mathcal{C}, c', \mathcal{C}_1, \ldots, \mathcal{C}_{n-1}), \mathcal{P}', \mathcal{N}'_0, \ldots, \mathcal{N}'_{n-1})$ as follows. The set $\mathcal{P}'$ contains all the states of the universal automaton that lead to exactly the set $\mathcal{P}$ via the transition with α, i.e., $\max\{\mathcal{P}' \mid \bigcup_{q \in \mathcal{P}'} \delta(q, \alpha) = \mathcal{P}\}$. If such a set does not exist then there is no transition with α from this state. For the sets of maximality check Δ computes $\mathcal{N}'_i$ such that it contains all universal states that may lead via α to a state in $\mathcal{N}_i$ or have no transition with σ, i.e., $\max\{\mathcal{N}' \mid \forall q' \in \mathcal{N}'_i. \ \delta(q', \sigma) = \emptyset \ \vee \ \exists q \in \mathcal{N}_i. \ q \in \delta(q', \sigma)\}$.

A loop is found if the initial conjecture $\mathcal{C}$ is repeated i.e. $\mathcal{C} \subseteq \mathcal{P}$, and the maximality check holds. The latter is true when all positive states are in $\mathcal{P}$, i.e., $\overline{\mathcal{P}} = \mathcal{N}_{n-1}$, and for all $j < n-1, \mathcal{N}_j = \mathcal{C}_{j+1}$, and a fix-point for the set of negative states is reached, i.e., $\mathcal{C}_{n-1} = \mathcal{N}_{n-1}$. In this case the counter c is then incremented by one. A sequence is accepted if a state $\Lambda_\# = ((\mathcal{C}, c, \mathcal{C}_1, \ldots, \mathcal{C}_{n-1}), \mathcal{P}, \mathcal{N}_0, \ldots, \mathcal{N}_{n-1})$ is reached after reading the last symbol σ_0, s.t., $q_0 \in \mathcal{P}$ and $c > 0$.

The unambiguity of the automaton with respect to a word model follows immediately from the maximality check and the determinism of the transition relation after having chosen the initial state.   □

4.2   An Algorithm for Counting Word Models

Theorem 2. *There is a procedure that counts the number of k-word models of a safety specification expressed as an LTL formula φ in time linear in the bound k and double-exponential in the length of φ.*

Algorithm 1 describes a procedure for computing the number of word models of bases accepted by the automaton. The algorithm computes for each state of the automaton the number of bases of length i that are accepted in this state in the i-th iteration (when this state is visited in the i-th step). Ω maps each accepting state in the k-th iteration to the number of bases of length k that are accepted by the automaton. For an accepting state $q = ((\mathcal{C}, c, \mathcal{C}_1, \ldots, \mathcal{C}_{n-1}), \mathcal{P}, \mathcal{N}_0, \ldots, \mathcal{N}_{n-1})$, a base accepted in this state has a loop annotation $\mathcal{C}$ and it is repeated c times. Thus, each base accepted in q has c word models. The number of word models is computed by summing up the number of word models in each accepting state. The algorithm traverses the automaton k times, resulting in a complexity of $\mathcal{O}(k) . 2^{2^{\mathcal{O}(|\varphi|)}}$.

5   Counting Tree Models

In this section, we introduce the counting algorithm for tree models. Our starting point is again the universal safety automaton. Similar to the case of word model counting, we guess a loop annotation and check whether the annotation

$\Omega = \{(q,1)|\ q \in Q_{0\#}\}$
for $(i := 0, i \le k, i{+}{+})$ **do**
  for all $q \in \Omega$ **do**
    for all $\sigma \in \Sigma \times \Upsilon$ **do**
      $\Omega(\Delta(q,\sigma)) + := \Omega(q)$

return $\displaystyle\sum_{q=((\mathcal{C},c,\mathcal{C}_1,\dots,\mathcal{C}_{n-1}),\mathcal{P},\mathcal{N}_0,\dots,\mathcal{N}_{n-1})\in Q_{F\#}} \Omega(q) \cdot c$

Algorithm 1. Counting with $\mathcal{A}_{\#}$

is repeated when exploring the tree from its leaves upwards. However, because tree models are a composition of word models, we need to guess an annotation for each branch of the tree (for each leaf). Furthermore, as described in Section 2, a tree model must preserve the input structure of a transition system, i.e., a tree model has loop-backs from each leaf for each direction. Therefore, we have a conjecture annotation for each direction in each leaf. Traversing the tree upwards we apply then the procedure of the word case with an additional merging procedure that merges all the information received from the children in their parent tree state.

The following theorem shows that for each safety property expressed as an LTL specification φ and a bound k we can construct a bottom-up tree automaton, that accepts a tree if it is a base for a tree model of φ. Theorem 4 shows that this automaton can be used to count the number of k-tree models for the specification φ.

Theorem 3. *Given a universal safety automaton* $\mathcal{U} = (Q, q_0, \delta, \Sigma, \Upsilon)$ *with* n *states, and a bound* k, *we can construct a bottom-up tree automaton* $\mathcal{T}_{\#}$ — $(Q_{\#}, Q_{F\#}, \Delta_0, \Delta, \Sigma, \Upsilon)$ *that accepts a* Σ*-labeled* Υ*-tree of depth* k *if it is a tree base of a tree model that is accepted by* $\mathcal{U}$. *The number of states of* $\mathcal{T}_{\#}$ *is double exponential in* n.

Construction: We choose $((2^Q \to \{\bot\} \cup \{0,\dots,k\}) \times (2^Q)^{n-1}) \times 2^Q \times (2^Q)^n$ to be the state space of $\mathcal{T}_{\#}$. The universal safety automaton $\mathcal{U}$ has a unique annotation for every tree model. A state $((f, \mathcal{C}_1, \dots, \mathcal{C}_{n-1}), \mathcal{P}, \mathcal{N}_0, \dots, \mathcal{N}_{n-1})$ is again split into a conjecture part $f, \mathcal{C}_1, \dots, \mathcal{C}_{n-1}$ and a tracking part $\mathcal{P}, \mathcal{N}_0, \dots, \mathcal{N}_{n-1}$. The tracking sets assign a node of a tree with the set of its positive and negative universal states. These annotations are reached from the conjecture sets of all leaves that lead upwards to this node. The conjecture part differs from the word case in the conjecture function f. The conjecture function is a partial function that maps an annotation to the number of expected repetitions along a branch of the input tree. At leaf level, the function maps the guessed annotation $\mathcal{C}$ to some number $\mu \in \{0, \dots, k\}$. Moving upwards in the tree the transition relation counts down the number of repetitions of $\mathcal{C}$. In each node of the input tree the function f is a bookkeeping process for the repetitions of all the conjectures at leaf level up to this node. A node annotation is maximal if $\overline{\mathcal{P}} = \mathcal{N}_{n-1}$.

The sets $\mathcal{P}, \mathcal{N}_0, \ldots, \mathcal{N}_{n-1}$ are computed via the transition relations as follows. For each leaf state s labeled with $\alpha \in \Sigma$ the initial transition relation Δ_0 guesses an annotation $\mathcal{C}_{v_i}$ and sets $\mathcal{C}_{v_i}^1, \ldots, \mathcal{C}_{v_i}^{n-1}$ for each direction $v_i \in \Upsilon$. It then uses the transition relation Δ to compute the state labeling of leaf state s by computing $\Delta(\Lambda_{\#}^{v_1}, \ldots, \Lambda_{\#}^{v_{|\Upsilon|}}, \alpha)$ with $\Lambda_{\#}^{v_i} = ((f_{v_i}, \mathcal{C}_{v_i}^1, \ldots, \mathcal{C}_{v_i}^{n-1}), \mathcal{C}_{v_i}, \emptyset, \mathcal{C}_{v_i}^1, \ldots, \mathcal{C}_{v_i}^{n-1})$, where f_{v_i} is a singleton function that maps $\mathcal{C}_{v_i}$ to some number $\mu \in \{0, \ldots, k\}$.

The transition relation Δ is deterministic. For states $\Lambda_{\#}^{v_1}, \ldots, \Lambda_{\#}^{v_{|\Upsilon|}}$ with $\Lambda_{\#}^{v_i} = ((f_{v_i}, \mathcal{C}_{v_i}^1, \ldots, \mathcal{C}_{v_i}^{n-1}), \mathcal{P}_{v_i}, \mathcal{N}_{v_i}^0, \mathcal{N}_{v_i}^1, \ldots, \mathcal{N}_{v_i}^{n-1})$, and a label $\alpha \in \Sigma$, the transition relation computes a state $\Delta(\Lambda_{\#}^{v_1}, \ldots, \Lambda_{\#}^{v_{|\Upsilon|}}, \alpha) = \Lambda_{\#} = ((f, \mathcal{C}_1, \ldots, \mathcal{C}_{n-1}), \mathcal{P}, \mathcal{N}_0, \mathcal{N}_1, \ldots, \mathcal{N}_{n-1})$ such that, $\mathcal{C}_i = \bigcup_{v_i \in \Upsilon} \mathcal{C}_{v_i}^i$. $\mathcal{P}$ is the largest set that leads via the label α and direction v_i to exactly the set $\mathcal{P}_{v_i}$, i.e., $\max\{\mathcal{P} \mid \bigcup_{q \in \mathcal{P}} \delta(q, \alpha, v_i) = \mathcal{P}_{v_i}\}$. If such set does not exist then there is no transition for α from this state. To compute such a set we compute for each $\mathcal{P}_{v_i}$ a set $\widetilde{\mathcal{P}}_{v_i}$ in the same fashion as in the word case. We compute then the intersection $\bigcap_i \widetilde{\mathcal{P}}_{v_i}$ and check whether the latter condition holds. Each set $\mathcal{N}_i$ must contain all universal states that may lead via α and direction v_i to a state in $\mathcal{N}_{v_i}^i$ or have no transition with α and v_i, i.e., $\max\{\mathcal{N}_i \mid \forall q' \in \mathcal{N}_i. \; \delta(q', \alpha, v_i) = \emptyset \; \vee \; \exists q \in \mathcal{N}_{v_i}^i. \; q \in \delta(q', \alpha, v_i)\}$. Thus, it is the union of all sets $\widetilde{\mathcal{N}}_{v_i}$ that may lead to $\mathcal{N}_{v_i}^i$ and the set $\mathcal{N}_{\not\to}$ of states that reach no state via α and any v_i.

A new mapping f is also computed. The domain of f is the union of the domains of all f_{v_i}. If some $\mathcal{C}$ is shared between two domains of functions f_{v_i} and f_{v_j}, then we require that $(f_{v_i}(\mathcal{C})) = (f_{v_j}(\mathcal{C}))$. If this condition is violated then there is no transition for α from this state. For all $\mathcal{C}$ with $f_{v_i}(\mathcal{C}) = c$, if $\mathcal{C} \subseteq \mathcal{P}$, for all $j < n - 1$, $\mathcal{N}_j = \mathcal{C}_{j+1}$, $\mathcal{C}_{n-1} = \mathcal{N}_{n-1}$, and $\overline{\mathcal{P}} = \mathcal{N}_{n-1}$, then a loop with $\mathcal{C}$ is found and we assign $f(\mathcal{C}) = c - 1$. Otherwise $f(\mathcal{C}) = f_{v_i}(\mathcal{C})$. If $c \leq 0$ then $c - 1 = \bot$. A state $((f, \mathcal{C}_1, \ldots, \mathcal{C}_{n-1}), \mathcal{P}, \mathcal{N}_0, \mathcal{N}_1, \ldots, \mathcal{N}_{n-1})$ is accepting if $q_0 \in \mathcal{P}$ and for all $\mathcal{C}$ in the domain of f, $f(\mathcal{C}) = 0$ (This means the guess of the number of repetitions was correct). The progress of the transition relation is depicted in Figure 4. The unambiguity of the tree automaton with respect to tree models follows from the fact that for each tree there is exactly one maximal annotation and from the determinism of the transition relation Δ. $\qquad \square$

5.1   An Algorithm for Counting Tree Models

Theorem 4. *There is a procedure that counts the number of k-tree models of a safety specification expressed as an LTL formula φ in time linear in the bound k and triple-exponential in the length of φ.*

Algorithm 2 describes a procedure for computing the number of tree models of tree bases accepted by the automaton. The algorithm starts at the initial states computed by Δ_0. These states involve the initial conjectures for the number of

$$\mathcal{P} \overset{?}{\supseteq} \mathcal{C} \in \mathrm{Dom}(f)$$

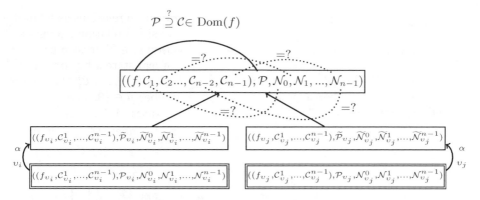

Fig. 4. A transition of the tree automaton over trees with directions v_i and v_j. The transition reads in this step a node labeled with α. The double lined states are the state labelings of the children nodes in directions v_i and v_j.

expected repetitions of an initial guessed annotation. Each initial state is mapped via the function Θ to the number of expected repetitions of the initial annotation (at this level the conjecture function is defined only over one annotation). We track in each step every possible transition in Δ. A transition involves states $q_0, ..., q_{v-1}$ and a parent state q. Recall that a transition exists only if for all the shared annotations in the domains of the conjecture functions $f_0, ...f_{r-1}$ of $q_0, ..., q_{v-1}$, the number of expected remaining repetitions is identical (this means that the initial guess was correct). For an annotation $\mathcal{C}$ shared among the domains of conjecture functions (not necessary all of them), let $c_0, ...c_h$ be initial guesses for $\mathcal{C}$ (the number of loop backs of each leaf annotated with $\mathcal{C}$). The number of possible loop combinations from those leaves is $\prod_i c_i$. In the k-th iteration each accepting state q' of $\mathcal{T}_{\#}$ is mapped to a function $\Theta(q)$ that defines for each annotation $\mathcal{C}$ in the domain of $f_{q'}$ the number of possible loop combinations for $\mathcal{C}$ in a tree accepted in q'. By multiplying all possible loop combination for each defined annotation we get the number of tree models of the tree accepted in q'. Finally, we sum up the results for all accepting states. The automaton is traversed k times before obtaining the final result.

6   Discussion

We have studied the model counting problem for safety specifications expressed in LTL. Counting word and tree models of LTL formulas opens up new "quantitative" versions of the classic model checking and synthesis problems for reactive systems: instead of just *checking* correctness and realizability, respectively, we can now judge the severity of the error by *counting* the number of error paths, and judge the specificity of the specification by *counting* the number of implementations.

$\Theta : Q_\# \times 2^Q \to \mathbb{N}$
$Q_{0\#}$: initial states guessed by Δ_0
Let $q = (f_q, \mathcal{C}_q^1, \mathcal{C}_q^{n-1}, \mathcal{P}_q, \mathcal{N}_q^0, ..., \mathcal{N}_q^{n-1})$

for all $\sigma \in \Sigma$ **do**
   for all $q \in Q_{0\#}$ **do**
      for all $\mathcal{C} \in \mathrm{Dom}(f_q)$ **do**
         $\Theta(q, \mathcal{C}) := f_q(\mathcal{C})$

for $(i := 0, i \leq k, i+1)$ **do**
   for all $(q_0, q_1, \ldots, q_{r-1}, \sigma, q) \in \Delta$ **do**
      for all $\mathcal{C} \in \mathrm{Dom}(f_q)$ **do**
         $\Theta(q, \mathcal{C}) + := \prod_i \Theta(q_i, \mathcal{C})$ /*only if $\mathcal{C}$ is defined for q_i*/

return $\displaystyle\sum_{q \in Q_{F\#}} \prod_{\mathcal{C} \in 2^Q} \Theta(q, \mathcal{C})$

Algorithm 2. Counting with $\mathcal{T}_\#$

While our algorithms are the first to specifically solve the model counting problem for safety specifications expressed in LTL, obvious competitors are the reduction to propositional model counting, as well as a direct enumeration of the models. As discussed in the introduction, the reduction to propositional counting is not a viable solution, because the reduction quickly leads to propositional constraints with far more than the 1000-10000 variables that can be handled by currently available model counters [9].

For k-word models, the complexity of our counting algorithm is double-exponential in the length of the LTL formula and linear in k. If the complexity in the formula were our main concern, we could do better better than this by exhaustively enumerating all words of length k: checking whether a *specific* sequence satisfies an LTL formula can be done in polynomial time (or even in NC [10]). However, the enumeration of all words takes exponential time in k, which is, for reasonable values of k, impractical. For k-tree models, the situation is similar: enumerating all trees would allow us to exploit inexpensive model checking algorithms for finite trees [11], but would result in double-exponential complexity in k, while our algorithm maintains linear complexity in k at the price of triple-exponential complexity in the length of the formula.

In future work, we plan to extend the model counting algorithms to full LTL, and to investigate the complexity for other fragments of LTL besides safety. The high complexity in the length of the formula results from the necessity to memorize information about each leaf of the tree. Fragments where this information is not needed, such as reachability properties, should therefore result in less expensive model counting algorithms.

References

1. Bayardo, R.J., Schrag, R.: Using csp look-back techniques to solve real-world sat instances. In: AAAI/IAAI, pp. 203–208 (1997)
2. Biere, A.: Bounded model checking. In: Handbook of Satisfiability, pp. 457–481. IOS Press (2009)
3. Bloem, R.P., Gamauf, H.-J., Hofferek, G., Könighofer, B., Könighofer, R.: Synthesizing robust systems with RATSY. In: Association, O.P. (ed.) Proceedings First Workshop on Synthesis (SYNT 2012), vol. 84, pp. 47–53. Electronic Proceedings in Theoretical Computer Science (2012)
4. Bohy, A., Bruyère, V., Filiot, E., Jin, N., Raskin, J.-F.: Acacia+, a tool for LTL synthesis. In: Madhusudan, P., Seshia, S.A. (eds.) CAV 2012. LNCS, vol. 7358, pp. 652–657. Springer, Heidelberg (2012)
5. Burch, J.R., Clarke, E.M., McMillan, K.L., Dill, D.L., Hwang, L.J.: Symbolic model checking: 10^{20} states and beyond (1992)
6. Darwiche, A.: New advances in compiling cnf into decomposable negation normal form. In: ECAI, pp. 328–332 (2004)
7. Ehlers, R.: Unbeast: Symbolic bounded synthesis. In: Abdulla, P.A., Leino, K.R.M. (eds.) TACAS 2011. LNCS, vol. 6605, pp. 272–275. Springer, Heidelberg (2011)
8. Finkbeiner, B., Schewe, S.: Bounded synthesis. International Journal on Software Tools for Technology Transfer 15(5-6), 519–539 (2013)
9. Gomes, C.P., Sabharwal, A., Selman, B.: Model counting. In: Biere, A., Heule, M., van Maaren, H., Walsh, T. (eds.) Handbook of Satisfiability, Frontiers in Artificial Intelligence and Applications, vol. 185, pp. 633–654. IOS Press, Amsterdam (2009), http://dblp.uni-trier.de/db/series/faia/faia185.html#GomesSS09
10. Kuhtz, L., Finkbeiner, B.: LTL path checking is efficiently parallelizable. In: Albers, S., Marchetti-Spaccamela, A., Matias, Y., Nikoletseas, S., Thomas, W. (eds.) ICALP 2009, Part II. LNCS, vol. 5556, pp. 235–246. Springer, Heidelberg (2009)
11. Kuhtz, L., Finkbeiner, B.: Weak kripke structures and LTL. In: Katoen, J.-P., König, B. (eds.) CONCUR 2011. LNCS, vol. 6901, pp. 419–433. Springer, Heidelberg (2011)
12. Kupferman, O., Lampert, R.: On the construction of fine automata for safety properties. In: Graf, S., Zhang, W. (eds.) ATVA 2006. LNCS, vol. 4218, pp. 110–124. Springer, Heidelberg (2006)
13. Littman, M.L., Majercik, S.M., Pitassi, T.: Stochastic boolean satisfiability. Journal of Automated Reasoning 27, 2001 (2000)
14. Morwood, D., Bryce, D.: Evaluating temporal plans in incomplete domains. In: AAAI (2012)
15. Pnueli, A.: The temporal logic of programs. In: Proceedings of the 18th Annual Symposium on Foundations of Computer Science, SFCS 1977, pp. 46–57. IEEE Computer Society, Washington, DC (1977), http://dx.doi.org/10.1109/SFCS.1977.32

ω-rational Languages:
High Complexity Classes vs. Borel Hierarchy[*]

Enrico Formenti[1], Markus Holzer[2], Martin Kutrib[2], and Julien Provillard[1]

[1] Université Nice Sophia Antipolis
CNRS, I3S, UMR 7271, 06900 Sophia Antipolis, France
{enrico.formenti,julien.provillard}@unice.fr
[2] Justus-Liebig Universität Gießen Institut für Informatik
Arndtstraße 2, 35392 Gießen, Germany
{markus.holzer,martin.kutrib}@informatik.uni-giessen.de

Abstract. The paper investigates classes of languages of infinite words with respect to the acceptance conditions of the finite automata recognizing them. Some new natural classes are compared with the Borel hierachy. In particular, it is proved that $(\mathtt{fin}, =)$ is as high as $\mathsf{F}_\sigma^\mathsf{R}$ and $\mathsf{G}_\delta^\mathsf{R}$. As a side effect, it is also proved that in this last case, considering or not considering the initial state of the FA makes a substantial difference.

Keywords: ω-rational languages, Borel hierarchy, acceptance conditions.

1   Introduction

Languages over infinite words have been used since the very introduction of symbolic dynamics. Afterwards, they have spread in a multitude of scientific fields. Computer science is more directly concerned for example by their application in formal specification and verification, game theory, logics, *etc.*.

ω-rational languages have been introduced as a natural extension of languages of finite words recognized by finite automata. Indeed, a finite automaton accepts some input u if at the end of the reading of u, the automaton reaches a final state. Clearly, when generalizing to infinite words, this accepting condition has to be changed. For this reason, new accepting conditions have been introduced in literature. For example, an infinite word w is accepted by a finite automaton $\mathcal{A}$ under the Büchi acceptance condition if and only if there exists a run of $\mathcal{A}$ which passes infinitely often through a set of accepting states while reading w. Indeed, this was introduced by Richard Büchi in the seminal work [1] in 1960.

Later on, David Muller characterized runs that pass through all elements of a given set of accepting states and visit them infinitely often [8]. Afterwards, more acceptance conditions appeared in a series of papers [4,5,11,7,6]. Each of these works was trying to capture a particular semantic on the runs or to fill some conceptual gap. Acceptance conditions are selectors for runs of the automaton under

---

[*] This work has been partially supported by the French National Research Agency project EMC (ANR-09-BLAN-0164.)

A.-H. Dediu et al. (Eds.): LATA 2014, LNCS 8370, pp. 372–383, 2014.
© Springer International Publishing Switzerland 2014

consideration. Of course, the set of selected runs is also deeply influenced by the structural properties of the FA: deterministic vs. non-deterministic, complete vs. non complete (see for instance [6]).

Each acceptance condition characterizes a class of languages. In [2], it is proved that if the acceptance condition is definable in MSO (monadic second order) logic then the class of languages it induces is ω-rational. However, more work was necessary to find which was the overall picture *i.e.* which are the relations between classes of languages induced by the acceptance conditions appeared in literature so far. The well-known Borel hierarchy constitute the backbone of such a picture. Classes in the hierarchy are ordered by set inclusion.

This paper continues the classification work closing some open questions concerning the positioning of the class of languages induced by CDFA(fin, =) (*i.e.* languages characterized by runs that pass finitely many times through all the elements of a given set of final states, recognized by Complete Deterministic Finite Automata). The motivation for a further study of the condition (fin, =) is twofold. From one hand, this class is, in a sense, surprising. Indeed, it is as high as the highest classes of the Borel hierachy but it is distinct from them. The interest of such a result is to have examples of languages that have high complexity but in which the complexity is not just determined by the topology one defines over the words (the Cantor topology here) but the complexity is determined by the intrinsic combinatorial complexity of the words themselves.

From the other hand, it is another step in the understanding of the theory of formal specification and verification of daemon processes (non-terminating processes). In this case, a run of the process is accepted only if it passes through a finite number of exceptions.

The paper also highlights an interesting phenomenon: the complexity class can be greatly influenced by the fact that one considers the very first elements of the paths (initial node) or not. In the sequel given an acceptance condition (c, R), the version in which the initial node is considered is denoted (c', R).

For example run is the set of states visited by the finite automaton while reading the input word, excluding the initial state; run' is the same as run but includes the initial state. By Proposition 21, one finds that CDFA(fin, =) $\subsetneq$ CDFA(fin', =) (CDFA stands for complete deterministic finite automata). As a consequence CDFA(fin', =) is even higher than CDFA(fin, =). The rest of the paper is devoted in proving (or disproving) the inclusion relations *wrt.* all previously known classes. The resulting hierarchy is illustrated in Figure 5.

Most of the proofs have been omitted due to a lack of space. They will appear in the long version of this article.

2   Languages and Automata

Let $\mathbb{N}$ denote the set of non-negative integers. For all $i, j \in \mathbb{N}$, $[i, j]$ is the set $\{i, i+1, \ldots, j\}$. For a set A, $|A|$ denotes the cardinality of A and $\mathcal{P}(A)$ the powerset of A. An *alphabet* is a finite set and a *letter* is an element of an alphabet. Given an alphabet Σ, a *word* over Σ is a sequence of letters from Σ. Let Σ^*

and Σ^ω denote the set of all finite words and the set of all infinite words over Σ, respectively. Let Σ^∞ denote $\Sigma^* \cup \Sigma^\omega$. For a word u, $|u|$ denotes the length of u and $|u|_a$ denotes the number of occurrences of the letter a in u. The empty word ε is the only word of length zero. For all words $u \in \Sigma^*$ and $v \in \Sigma^\infty$, uv denotes the concatenation of u with v. For all word $u \in \Sigma^\infty$, for all $0 \le i \le j < |u|$, the word $u_i u_{i+1} \ldots u_j$ is denoted by $u_{[i,j]}$.

A *language* is a subset of Σ^*, similarly an ω-*language* is a subset of Σ^ω. For a language $\mathcal{L}_1$ and for $\mathcal{L}_2 \in \Sigma^\infty$, $\mathcal{L}_1\mathcal{L}_2 = \{uv \in \Sigma^* : u \in \mathcal{L}_1, v \in \mathcal{L}_2\}$ denotes the concatenation of $\mathcal{L}_1$ with $\mathcal{L}_2$. For a language $\mathcal{L} \subseteq \Sigma^*$, let $\mathcal{L}^0 = \{\varepsilon\}$, $\mathcal{L}^{n+1} = \mathcal{L}^n\mathcal{L}$ and $\mathcal{L}^* = \bigcup_{n \in \mathbb{N}} \mathcal{L}^n$ the Kleene star of $\mathcal{L}$. For a language $\mathcal{L}$, the infinite iteration of $\mathcal{L}$ is the ω-language $\mathcal{L}^\omega = \{u_0 u_1 u_2 \cdots : \forall i \in \mathbb{N}, u_i \in \mathcal{L} \smallsetminus \{\epsilon\}\}$.

The class of *rational languages* is the smallest class of languages containing $\emptyset$, all sets $\{a\}$ (for $a \in \Sigma$) and which is closed under union, concatenation and Kleene star operations. An ω-language $\mathcal{L}$ is ω-*rational* if there exist $n \in \mathbb{N}$ and two families $\{\mathcal{L}_i\}$ and $\{\mathcal{L}'_i\}$ of n rational languages such that $\mathcal{L} = \bigcup_{i=0}^{n-1} \mathcal{L}_i \mathcal{L}'^\omega_i$. Let RAT denote the set of all ω-rational languages.

Rational languages and ω-rational languages are denoted by rational expressions. For instance, for the alphabet $\Sigma = \{0,1\}$, Σ^*1 denotes the language of words ending with a 1 while $(\Sigma^*1)^\omega$ and $\Sigma^*(0^\omega + 1^\omega)$ denote the ω-languages of words containing an infinite number of 1's, and a finite number of 0's or a finite number of 1's, respectively.

A *finite automaton* (FA) is a tuple $(\Sigma, Q, T, I, \mathcal{F})$ where Σ is an alphabet, Q a finite set of states, $T \subseteq Q \times \Sigma \times Q$ is the set of *transitions*, $I \subseteq Q$ is the set of *initial states* and $\mathcal{F} \subseteq \mathcal{P}(Q)$ is the *acceptance table*. A FA is a *deterministic finite automaton* (DFA) if $|I| = 1$ and $|\{q \in Q : (p, a, q) \in T\}| \le 1$ for all $p \in Q$, $a \in \Sigma$. It is a *complete* finite automaton (CFA) if $|\{q \in Q : (p, a, q) \in T\}| \ge 1$ for all $p \in Q$, $a \in \Sigma$. A CDFA is a FA which is both deterministic and complete.

A CDFA induces a transition function $\delta : Q \times \Sigma \to Q$ such that for all $p \in Q$ and $a \in \Sigma$, $\delta(p, a)$ is the only state such that $(p, a, \delta(p, a)) \in T$. The transition function can be extended to a function $\delta' : Q \times \Sigma^* \to Q$ by defining for all $p \in Q$, $\delta'(p, \varepsilon) = p$ and for all $p \in Q$, $a \in \Sigma$ and $u \in \Sigma^*$, $\delta'(p, au) = \delta'(\delta(p, a), u)$. We usually make no distinction between δ and δ'.

If $I = \{q_0\}$ for some state $q_0 \in Q$, we shall write $(\Sigma, Q, T, q_0, \mathcal{F})$ instead of $(\Sigma, Q, T, I, \mathcal{F})$. Similarly, if $\mathcal{F} = \{F\}$ or $\mathcal{F} = \{\{f\}\}$, we shall write (Σ, Q, T, I, F) or (Σ, Q, T, I, f) instead of $(\Sigma, Q, T, I, \mathcal{F})$, respectively.

An *infinite path* in a FA $\mathcal{A} = (\Sigma, Q, T, I, \mathcal{F})$ is a sequence $(p_i, x_i)_{i \in \mathbb{N}}$ such that $(p_i, x_i, p_{i+1}) \in T$ for all $i \in \mathbb{N}$. The (infinite) word x is the *label* of the path. A *finite path* from p to q is a sequence $(p_i, u_i)_{i \in [0,n]}$ for some n such that $p_0 = p$, for all $i \in [0, n-1]$, $(p_i, u_i, p_{i+1}) \in T$ and $(p_n, u_n, q) \in T$. The (finite) word u is the label of the path. A path is *initial* if $p_0 \in I$. A state q is *accessible* if there exists an initial path to q and $\mathcal{A}$ is *accessible* if all its states are. A *loop* is a path from a state to the same state. The FA $\mathcal{A}$ is *normalized* if it is accessible, $I = \{q_0\}$ for some $q_0 \in Q$ and q_0 does not belong to a loop.

3   Acceptance Conditions, Classes of Languages and Topology

Definition 1. *Let $\mathcal{A} = (\Sigma, Q, T, I, \mathcal{F})$ be a FA and $p = (p_i, x_i)_{i \in \mathbb{N}}$ a path in $\mathcal{A}$. Define the sets*

- $\mathrm{run}_{\mathcal{A}}(p) = \{q \in Q : \exists i > 0, p_i = q\}$,
- $\mathrm{run}'_{\mathcal{A}}(p) = \{q \in Q : \exists i \geq 1, p_i = q\}$,
- $\inf_{\mathcal{A}}(p) = \{q \in Q : \forall i > 0, \exists j \geq i, p_j = q\}$,
- $\mathrm{fin}_{\mathcal{A}}(p) = \mathrm{run}_{\mathcal{A}}(p) \smallsetminus \inf_{\mathcal{A}}(p)$,
- $\mathrm{fin}'_{\mathcal{A}}(p) = \mathrm{run}'_{\mathcal{A}}(p) \smallsetminus \inf_{\mathcal{A}}(p)$,
- $\mathrm{ninf}_{\mathcal{A}}(p) = Q \smallsetminus \inf_{\mathcal{A}}(p)$

as the sets of states appearing at least one time *(counting or not the first state of the path)*, infinitely many times, finitely many times but at least once *(counting or not the first state of the path), and either* finitely many times including never *in p, respectively.*

An *acceptance condition* for $\mathcal{A}$ is a subset of all the initial infinite paths of $\mathcal{A}$. The paths inside such a subset are called *accepting paths*. Let $\mathcal{A}$ be a FA and *cond* be an acceptance condition for $\mathcal{A}$, a word x is *accepted* by $\mathcal{A}$ (under condition *cond*) if and only if it is the label of some accepting path.

   Let $\sqcap$ be the binary relation over sets such that for all sets A and B, $A \sqcap B$ if and only if $A \cap B \neq \emptyset$.

   In this paper, we consider acceptance conditions induced by pairs $(c, \mathbf{R}) \in \{\mathrm{run}, \mathrm{run}', \inf, \mathrm{fin}, \mathrm{fin}', \mathrm{ninf}\} \times \{\sqcap, \subseteq, =\}$. A pair *cond* $= (c, \mathbf{R})$ defines an acceptance condition *cond*$_{\mathcal{A}}$ on an automaton $\mathcal{A} = (\Sigma, Q, T, I, \mathcal{F})$ as follows: an initial infinite path $p = (p_i, x_i)_{i \in \mathbb{N}}$ is accepting if and only if there exists a set $F \in \mathcal{F}$ such that $c_{\mathcal{A}}(p) \mathbf{R} F$. We denote by $\mathcal{L}_{\mathcal{A}}^{cond}$ the *language accepted by $\mathcal{A}$ under the acceptance condition cond*$_{\mathcal{A}}$, i.e., the set of all words accepted by $\mathcal{A}$ under *cond*$_{\mathcal{A}}$.

Remark 2. For acceptance conditions which use the relation $\sqcap$, we can assume that the acceptance table is reduced to one set of states, taking, if necessary, the union of all sets in the acceptance table.

Definition 3. *For all pairs cond $\in \{\mathrm{run}, \mathrm{run}', \inf, \mathrm{fin}, \mathrm{fin}', \mathrm{ninf}\} \times \{\sqcap, \subseteq, =\}$ and for all finite alphabets Σ, define the following sets*

- $\mathrm{FA}^{(\Sigma)}(cond) = \{\mathcal{L}_{\mathcal{A}}^{cond}, \ \mathcal{A} \text{ is a FA on } \Sigma\}$,
- $\mathrm{DFA}^{(\Sigma)}(cond) = \{\mathcal{L}_{\mathcal{A}}^{cond}, \ \mathcal{A} \text{ is a DFA on } \Sigma\}$,
- $\mathrm{CFA}^{(\Sigma)}(cond) = \{\mathcal{L}_{\mathcal{A}}^{cond}, \ \mathcal{A} \text{ is a CFA on } \Sigma\}$,
- $\mathrm{CDFA}^{(\Sigma)}(cond) = \{\mathcal{L}_{\mathcal{A}}^{cond}, \ \mathcal{A} \text{ is a CDFA on } \Sigma\}$

as the classes of ω-languages on Σ accepted by FA, DFA, CFA, and CDFA, respectively, under the acceptance condition derived by cond. When it is not confusing, we omit to precise the alphabet in these notations.

When Σ is endowed with discrete topology and Σ^ω with the induced product topology, let F, G, F_σ and G_δ be the collections of all closed sets, open sets, countable unions of closed set and countable intersections of open sets, respectively. For any pair A, B of collections of sets, denote by $\mathcal{B}(A)$, $A \Delta B$, and A^R the Boolean closure of A, the set $\{U \cap V : U \in A, V \in B\}$ and the set $A \cap \mathsf{RAT}$, respectively. These, indeed, are the lower classes of the Borel hierarchy. For more on this subject we refer the reader to [12] or [9], for instance.

Some of the acceptance conditions derived by pairs $(c, \boldsymbol{R})$ have been studied in the literature (see [1,8,4,5,11,7,6,10,3]). It is known that all the classes of languages induced are subclasses of RAT because the acceptance conditions are MSO-definable, see [1,2]. The known inclusions are depicted in Figure 5.

In the sequel, we deal with languages sharing the same structure. For an alphabet Σ, $a \in \Sigma$, $k \geq 0$ and $n > 0$, we denote the language

$$\{x \in \Sigma^\omega : |x|_a = k \pmod{n}\}$$

by $\mathcal{L}_{k,n}^{\Sigma,a}$ and $\tilde{\mathcal{L}}_{k,n}^{\Sigma,a}$ denotes the language $\mathcal{L}_{k,n}^{\Sigma,a} + (\Sigma^* a)^\omega$.

4   Some Relations between run and run′, and fin and fin′

The following lemma is immediate.

Lemma 4. *Let* $cond \in \{\mathtt{run}, \mathtt{inf}, \mathtt{fin}, \mathtt{ninf}\} \times \{\sqcap, \subseteq, =\}$. *If a language $\mathcal{L}$ is recognized by an automaton under condition cond, then it is recognized by a normalized automaton which is complete (resp. deterministic) if the initial one is complete (resp. deterministic) under condition cond.*

Corollary 5. *Let* $(c, \boldsymbol{R}) \in \{\mathtt{run}, \mathtt{fin}\} \times \{\sqcap, \subseteq, =\}$. *The class of languages induced by $(c, \boldsymbol{R})$ is included in the respective class of languages induced by $(c', \boldsymbol{R})$.*

Lemma 6. *Let* $\boldsymbol{R} \in \{\sqcap, \subseteq, =\}$ *and* $cond = (\mathtt{run'}, \boldsymbol{R})$. *If a language $\mathcal{L}$ is recognized by an automaton under condition cond, then it is recognized by a normalized automaton which is complete (resp. deterministic) if the initial one is complete (resp. deterministic) under condition cond.*

Proposition 7. *Let* $\boldsymbol{R} \in \{\sqcap, \subseteq, =\}$. *The conditions* $(\mathtt{run}, \boldsymbol{R})$ *and* $(\mathtt{run'}, \boldsymbol{R})$ *induce the same classes of languages.*

We will see later that Proposition 7 has no equivalence for condition based on fin. In general, the inclusion of classes induced by fin in the respective class induced by fin′ is strict.

From now on, without loss of generality, we assume that Σ is an alphabet containing $\{0, 1\}$ and we denote the set $\Sigma \smallsetminus \{1\}$ by Σ_0 and the set $\Sigma \smallsetminus \{0\}$ by Σ_1.

5   The Acceptance Conditions (fin, $\sqcap$) and (fin′, $\sqcap$)

The acceptance condition (fin, $\sqcap$) has already been studied in [6]. In this paper, we prove that the condition (fin′, $\sqcap$) defines new classes for deterministic or complete automata.

Proposition 8. *The class* $\mathrm{FA}(\mathtt{fin}', \sqcap)$ *is included in the class* $\mathrm{FA}(\mathtt{fin}, \sqcap)$.

Proposition 9. *The language* $\mathcal{L}_{0,2}^{\Sigma,1}$ *is in* $\mathrm{CDFA}(\mathtt{fin}', \sqcap)$ *but not in* $\mathrm{CFA}(\mathtt{fin}, \sqcap)$ *or in* $\mathrm{DFA}(\mathtt{fin}, \sqcap)$.

Proof. Remark that $\mathcal{L}_{0,2}^{\Sigma,1} = \mathcal{L}_{\mathcal{A}}^{(\mathtt{fin}', \sqcap)}$ for the CDFA $\mathcal{A} = (\Sigma, \{q_0, q_1\}, T, q_0, q_1)$ where $(p, a, q) \in T$ if and only if $a = 1$ and $p \neq q$ or $a \neq 1$ and $p = q$.

For the sake of argument, assume that $\mathcal{L}_{0,2}^{\Sigma,1} = \mathcal{L}_{\mathcal{A}}^{(\mathtt{fin}, \sqcap)}$ for a CFA $\mathcal{A}$. The word $x = 0^\omega$ is in $\mathcal{L}_{0,2}^{\Sigma,1}$ so there exists an accepting path $p = (p_i, x_i)_{i \in \mathbb{N}}$ in $\mathcal{A}$ under $(\mathtt{fin}, \sqcap)$. Let $k > 0$ such that $p_k \in F$ is visited finitely often in p. Let $y = 0^k 10^\omega$, y is not in $\mathcal{L}$, then all paths starting from p_k and labeled by 10^ω visit p_k infinitely often. Therefore, there exists a loop on p_k labeled by $10^{k'}$ for some $k' \in \mathbb{N}$. Inserting this loop one time in the first path, we find an accepting path labeled by y, this is a contradiction.

For the sake of argument, assume that $\mathcal{L}_{0,2}^{\Sigma,1} = \mathcal{L}_{\mathcal{A}}^{(\mathtt{fin}, \sqcap)}$ for a DFA $\mathcal{A}$. Without loss of generality, we can assume that $\mathcal{A}$ is accessible. As for all $u \in \Sigma^*$, $u0^\omega$ or $u10^\omega$ is in $\mathcal{L}_{0,2}^{\Sigma,1}$, there exists a finite initial path labeled by u and $\mathcal{A}$ is complete. We have just shown that this is not possible. $\qquad\square$

Theorem 10. *The following relations hold for the classes induced by* $(\mathtt{fin}', \sqcap)$:

1. $\mathrm{CDFA}(\mathtt{fin}, \sqcap) \subsetneq \mathrm{CDFA}(\mathtt{fin}', \sqcap)$, $\mathrm{DFA}(\mathtt{fin}, \sqcap) \subsetneq \mathrm{DFA}(\mathtt{fin}', \sqcap)$, $\mathrm{CFA}(\mathtt{fin}, \sqcap) \subsetneq \mathrm{CFA}(\mathtt{fin}', \sqcap)$,
2. $\mathrm{FA}(\mathtt{fin}, \sqcap) = \mathrm{FA}(\mathtt{fin}', \sqcap)$,
3. $\mathrm{CDFA}(\mathtt{fin}', \sqcap) \subsetneq \mathrm{CFA}(\mathtt{fin}', \sqcap) \subsetneq \mathrm{FA}(\mathtt{fin}', \sqcap)$,
4. $\mathrm{CDFA}(\mathtt{fin}', \sqcap) \subsetneq \mathrm{DFA}(\mathtt{fin}', \sqcap) \subsetneq \mathrm{FA}(\mathtt{fin}', \sqcap)$.

There are no other relations for the classes induced by $(\mathtt{fin}', \sqcap)$ *except those obtained by transitivity with previously known classes.*

Proof. The first point follows from Corollary 5 and Proposition 9. The equality $\mathrm{FA}(\mathtt{fin}, \sqcap) = \mathrm{FA}(\mathtt{fin}', \sqcap)$ holds from Corollary 5 and Proposition 8. The incomparability of $\mathrm{DFA}(\mathtt{fin}', \sqcap)$ with $\mathrm{CFA}(\mathtt{fin}', \sqcap)$ and the fact there is no other inclusions come from results of [6]. Indeed, at the one hand, $\mathsf{F}^\mathsf{R} \subseteq \mathrm{DFA}(\mathtt{fin}, \sqcap)$ but $\mathsf{F}^\mathsf{R} \not\subseteq \mathrm{CFA}(\mathtt{fin}', \sqcap)$. And, at the other hand, the language $\Sigma^* 10 \Sigma^\omega + \Sigma^* 0^\omega$ is in $(\mathrm{CDFA}(\mathtt{ninf}, \sqcap) \cap \mathrm{CFA}(\mathtt{fin}, \sqcap)) \setminus \mathrm{DFA}(\mathtt{fin}', \sqcap)$. Finally, the language $\Sigma^* 0^\omega$ is in $\mathrm{CDFA}(\mathtt{fin}, \sqcap) \cap (\mathsf{F}_\sigma^\mathsf{R} \setminus \mathsf{G}_\delta^\mathsf{R})$. $\qquad\square$

6   The Acceptance Conditions $(\mathtt{fin}, \subseteq)$ and $(\mathtt{fin}', \subseteq)$

In [3], it is proved that an automaton using the acceptance condition $(\mathtt{fin}, \subseteq)$ and $(\mathtt{fin}, =)$ can be completed without changing the recognized language. It follows that the completeness does not matter for classes induced by those conditions. The same holds for $(\mathtt{fin}', \subseteq)$ and $(\mathtt{fin}', =)$.

Proposition 11. *The class* F *is included in* CDFA($\texttt{fin}, \subseteq$) *and the class* $\mathsf{F}_\sigma \cap \mathsf{G}_\delta$ *is included in* CDFA($\texttt{fin}, =$).

Proposition 12 ([2]). *The class* CDFA($\texttt{fin}', \subseteq$) *is included in* G_δ.

Proposition 13. *The language* $(\Sigma^* 1)^\omega$ *is in* CDFA($\texttt{fin}, \subseteq$) $\smallsetminus \mathsf{F}_\sigma^{\mathsf{R}}$.

Lemma 14. *Let* $\mathcal{L}$ *be a language in* FA($\texttt{fin}, \subseteq$) *(resp. in* FA($\texttt{fin}', \subseteq$)*) such that there exists* $a, b \in \Sigma$, $u \in \Sigma^*$ *and for all* $k \in \mathbb{N}$, $ba^k u a^\omega \in \mathcal{L}$ *(resp.* $a^k u a^\omega \in \mathcal{L}$*). Then* ba^ω *(resp.* a^ω*) is in* $\mathcal{L}$.

Proof. Let $\mathcal{A} = (\Sigma, Q, T, I, \mathcal{F})$ such that $\mathcal{L} = \mathcal{L}_{\mathcal{A}}^{(\texttt{fin}, \subseteq)}$ (resp. $\mathcal{L} = \mathcal{L}_{\mathcal{A}}^{(\texttt{fin}', \subseteq)}$). Let $n = |Q|$, as $x = ba^n u a^\omega$ (resp. $x = a^n u a^\omega$) is in $\mathcal{L}$, there exists an accepting path $p = (p_i, x_i)_{i \in \mathbb{N}}$ in $\mathcal{A}$. There exists k, k' such that $1 \leq k < k' \leq n + 1$ (resp. $0 \leq k < k' \leq n$) and $p_k = p_{k'}$. Choose k minimal. We define a path $p' = (p_i', y_i)_{i \in \mathbb{N}}$ in $\mathcal{A}$ where $y = ba^\omega$ (resp. $y = a^\omega$), for all $i \in [0, k], p_i' = p_i$ and for all $i \in \mathbb{N}, p_{k+i}' = p_{k+(i \bmod k'-k)}$. If p' is accepting, we can conclude. If not, then, by minimality of k, $\texttt{fin}(p') = \{p_i : i \in [1, k-1]\}$ (resp. $\texttt{fin}'(p') = \{p_i : i \in [0, k-1]\}$) is not included in any $F \in \mathcal{F}$. But as p is accepting, there exists $F \in \mathcal{F}$ such that $\texttt{fin}(p) \subseteq F$ (resp. $\texttt{fin}'(p) \subseteq F$). That means there exists $q \in \texttt{fin}(p')$ (resp. $q \in \texttt{fin}'(p')$) such that $q \in \texttt{inf}(p)$. Let $k_0 \in [1, k-1]$ (resp. $k_0 \in [0, k-1]$) be minimal such that $p_{k_0} \in \texttt{inf}(p)$. Then by definition of $\texttt{inf}(p)$, we can find an index k_0' such that $p_{k_0'} = p_{k_0}$, $k_0' \geq |u| + n + 1$ and for all $i \geq k_0', p_i \in \texttt{inf}(p)$. We define a path $p'' = (p_i'', y_i)_{i \in \mathbb{N}}$ in $\mathcal{A}$ where for all $i \in [0, k_0], p_i'' = p_i$ and for all $i \in \mathbb{N}, p_{k_0'+i}' = p_{k_0'+i}$. By minimality of k_0 and by definition of k_0', $\texttt{fin}(p'') = \{p_i : i \in [1, k_0-1]\} \subseteq \texttt{fin}(p) \subseteq F$ (resp. $\texttt{fin}'(p'') = \{p_i : i \in [0, k_0-1]\} \subseteq \texttt{fin}'(p) \subseteq F$) and p'' is an accepting path labeled by y. $\qquad\square$

Proposition 15. *The language* $\Sigma^* 1 \Sigma^\omega$ *is in* CDFA($\texttt{ninf}, \sqcap$) $\cap \mathsf{G}^{\mathsf{R}} \smallsetminus$ FA($\texttt{fin}', \subseteq$).

Proposition 16. *The language* $\tilde{\mathcal{L}}_{0,2}^{\Sigma, 1}$ *is in* CDFA($\texttt{fin}', \subseteq$) $\smallsetminus$ FA($\texttt{fin}, \subseteq$).

Proposition 17. *The language* $\mathcal{L} = \Sigma_0(\tilde{\mathcal{L}}_{0,2}^{\Sigma, 1} + \tilde{\mathcal{L}}_{0,3}^{\Sigma, 1})$ *is in* FA($\texttt{fin}, \subseteq$) *but not in* CDFA($\texttt{fin}', \subseteq$).

Proof. We have $\mathcal{L} = \mathcal{L}_{\mathcal{A}}^{(\texttt{fin}, \subseteq)}$ for the FA $\mathcal{A} = (\Sigma, \{q_0, q_1, q_2, q_3, q_4, q_5\}, T, q_0, \{\{q_2\}, \{q_4, q_5\}\})$ where T is depicted on Figure 1. For the sake of argument, assume that $\mathcal{L} = \mathcal{L}_{\mathcal{A}}^{(\texttt{fin}', \subseteq)}$ for a CDFA $\mathcal{A} = (\Sigma, Q, T, q_0, \mathcal{F})$. Let $\delta : Q \to Q$ be the transition function of $\mathcal{A}$.

We first show that if u and v are two words such that u is a prefix of v starting by a 0 and $\delta(q_0, u) = \delta(q_0, v)$ then $|u|_1 = |v|_1 \pmod 6$. Let us denote $k = |u|_1 \pmod 6$ and $k' = |v|_1 \pmod 6$. If x is an ω-word, then the set of states visited finitely often by the path labeled by ux is included in the set of states visited finitely often by the path labeled by vx. Then, whenever ux is rejected for some x, vx is rejected. We take $x = 1^{(5-k)} 0^\omega$ (resp. $x = 1^{(7-k)} 0^\omega$), as ux is not in the language, it is rejected and vx is also rejected. We deduce that

$|vx|_1 = k' + 5 - k$ (resp. $|vx|_1 = k' + 7 - k$) is congruent to 1 or 5 modulo 6. This implies that $k = k'$. Let $n = |Q|$ and $x = 010^n 10^\omega$. As x is in $\mathcal{L}$, there exists $F \in \mathcal{F}$ such that $\mathtt{fin}'(p) \subseteq F$ where p is the path labeled by x. Let $S = \{q_0\} \cup \{\delta(q_0, x_{[0,k]}) : k \in [0, n+1]\}$, according to the above lemma, $S \subseteq \mathtt{fin}'(p)$. Moreover, we can find two integers $i < j$ such that $\delta(q_0, 010^i) = \delta(q_0, 010^j)$, then the path p' labeled by $y = 010^\omega$ is such that $\mathtt{run}'(p') = S$. Finally, $\mathtt{fin}'(p') \subseteq \mathtt{run}'(p') = S \subseteq \mathtt{fin}'(p) \subseteq F$ and y is recognized by $\mathcal{A}$ but $y \notin \mathcal{L}$. We get a contradiction. □

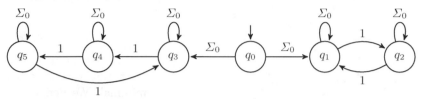

Fig. 1. A FA recognizing $\Sigma_0(\tilde{\mathcal{L}}_{0,2}^{\Sigma,1} + \tilde{\mathcal{L}}_{0,3}^{\Sigma,1})$ under the condition $(\mathtt{fin}, \subseteq)$

Proposition 18. *The language* $\mathcal{L} = \Sigma(11\Sigma^* + 0)^\omega$ *is in* $\mathrm{FA}(\mathtt{fin}, \subseteq) \smallsetminus G_\delta$.

Proof. We have $\mathcal{L} = \mathcal{L}_{\mathcal{A}}^{(\mathtt{fin}, \subseteq)}$ for the FA $\mathcal{A} = (\Sigma, \{q_0, q_1, q_2, q_3\}, T, q_0, q_1)$ where T is depicted on Figure 2. It is straightforward to prove that $\mathcal{L}$ is not in G_δ. □

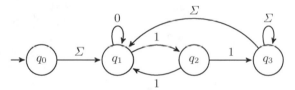

Fig. 2. A FA recognizing $\Sigma(11\Sigma^* + 0)^\omega$ under the condition $(\mathtt{fin}, \subseteq)$

Theorem 19. *The classes induced by* $(\mathtt{fin}, \subseteq)$ *and* $(\mathtt{fin}', \subseteq)$ *satisfy the following relations:*

1. $\mathsf{F} \subsetneq \mathrm{CDFA}(\mathtt{fin}, \subseteq) \subsetneq \mathrm{CDFA}(\mathtt{fin}', \subseteq) \subsetneq G_\delta$,
2. $\mathrm{CDFA}(\mathtt{fin}, \subseteq) \subsetneq \mathrm{FA}(\mathtt{fin}, \subseteq)$ *and* $\mathrm{CDFA}(\mathtt{fin}', \subseteq) \subsetneq \mathrm{FA}(\mathtt{fin}', \subseteq)$,
3. $\mathrm{FA}(\mathtt{fin}, \subseteq) \subsetneq \mathrm{FA}(\mathtt{fin}', \subseteq)$.

There is no other relation for the classes induced by $(\mathtt{fin}, \subseteq)$ *and* $(\mathtt{fin}', \subseteq)$ *except those obtained by transitivity with previously known classes.*

Proof. The inclusions of the first point comes from the Proposition 11, Corollary 5 and Proposition 12, respectively. By Propositions 13, 16 and 15, respectively, the inclusions are strict. The inclusions of the second point are clear and by Proposition 17 it is strict. The inclusions of the third point are a consequence of the Corollary 5 and by Proposition 16 they are strict.

The incomparability with the other known classes comes from Proposition 15 which proves that G and $\mathrm{CDFA}(\mathtt{ninf}, \sqcap)$ are not subclasses of $\mathrm{FA}(\mathtt{fin}', \subseteq)$ and from Propositions 13 and 18 which prove that $\mathrm{CDFA}(\mathtt{fin}, \subseteq)$ is not a subclass of F_σ and $\mathrm{FA}(\mathtt{fin}, \subseteq)$ is not a subclass of G_δ, respectively. □

7   The Acceptance Condition (fin, $=$) and (fin$'$, $=$)

In the previous section we have proved that the class CFA(fin, $\subseteq$) is pretty high in the hierarchy. However, it is incomparable with $\mathsf{F}_\sigma^R \cap \mathsf{G}_\delta^R$ and it does not contain any open language. In this section, we are going to show two more classes which have nicer properties.

Lemma 20. *Let $a, b \in \Sigma$ be two distinct letters and $\mathcal{L}$ a language such that $\mathcal{L} \cap \{a, b\}^* b^\omega = \mathcal{L}_{0,2}^{\{a,b\},a}$. If $\mathcal{L} = \mathcal{L}_{\mathcal{A}}^{(\mathtt{fin},=)}$ or $\mathcal{L} = \mathcal{L}_{\mathcal{A}}^{(\mathtt{fin}',=)}$ for a CDFA $\mathcal{A}$ then $\mathcal{A}$ has a loop on its initial state labeled by b^k for some $k > 0$.*

Proof. Let $\mathcal{A} = (\Sigma, Q, T, q_0, \mathcal{F})$ be a DFA such that $\mathcal{L} = \mathcal{L}_{\mathcal{A}}^{(\mathtt{fin},=)}$ or $\mathcal{L} = \mathcal{L}_{\mathcal{A}}^{(\mathtt{fin}',=)}$. For the sake of argument, assume that q_0 does not belong to a loop labeled by b's. Let δ be the transition function of $\mathcal{A}$. For all word x, denote by p_x the path in $\mathcal{A}$ labeled by the word x.

Define a sequence of integers $(k_i)_{i \in \mathbb{N}}$ such that, denoting the finite word $b^{k_0} a b^{k_1} a \ldots a b^{k_i}$ by u_i, for all $i \in \mathbb{N}$, $\delta(q_0, u_i)$ does not belong to a loop labeled by b's but $\delta(q_0, u_i 0)$ does. As q_0 is not on a loop labeled by b's, we define k_0 as $\max\left\{ j \in \mathbb{N} : \forall j' > j, \delta(q_0, b^{j'}) \neq \delta(q_0, b^j) \right\}$. Assume that k_i is defined for some $i \in \mathbb{N}$. Then, the state $\delta(q_0, u_i a)$ does not belong to a loop labeled by b's. Indeed, otherwise the words $x = u_i b^\omega$ and $y = u_i a b^\omega$ verify $\mathtt{fin}_{\mathcal{A}}(p_x) = \mathtt{fin}_{\mathcal{A}}(p_y)$ and $\mathtt{fin}'_{\mathcal{A}}(p_x) = \mathtt{fin}'_{\mathcal{A}}(p_y)$ (in both cases, the states which appear in those sets are states reached by reading u_i in $\mathcal{A}$ counting or not the first state). This is not possible because only one of this words is accepted by $\mathcal{A}$. We define k_{i+1} as $\max\left\{ j \in \mathbb{N} : \forall j' > j, \delta(q_0, u_i 10^{j'}) \neq \delta(q_0, u_i 10^j) \right\}$.

Since Q is finite, there exists $i < j$ such that $\delta(q_0, u_i) = \delta(q_0, u_j)$. The words $x = u_j b^\omega$ and $y = u_j a b^\omega$ verify $\mathtt{fin}_{\mathcal{A}}(p_x) = \mathtt{fin}_{\mathcal{A}}(p_y)$ and $\mathtt{fin}'_{\mathcal{A}}(p_x) = \mathtt{fin}'_{\mathcal{A}}(p_y)$ (see Figure 3) but as above only one of these words is accepted by $\mathcal{A}$. We get a contradiction. $\qquad\square$

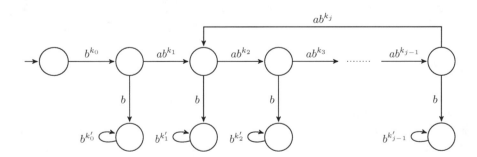

Fig. 3. A figure illustrating the construction in Lemma 20 with $i = 1$

Proposition 21. *The language $\mathcal{L}_{0,2}^{\Sigma,1}$ is in* CDFA($\mathtt{fin'},=$) $\smallsetminus$ CDFA($\mathtt{fin},=$).

Proof. We have $\mathcal{L}_{0,2}^{\Sigma,1} = \mathcal{L}_{\mathcal{A}}^{(\mathtt{fin'},=)}$ for the CDFA $\mathcal{A} = (\Sigma, \{q_0, q_1, q_2\}, T, q_0, \{\emptyset, \{q_0, q_1\}\})$ where $(p, a, q) \in T$ if and only if $a \neq 1$ and $p = q$ or $a = 1$ and $(p, q) \in \{(q_0, q_1), (q_1, q_2), (q_2, q_1)\}$. If $\mathcal{L}_{0,2}^{\Sigma,1}$ would be recognized by a CDFA $\mathcal{B}$ under condition ($\mathtt{fin},=$), $\mathcal{B}$ could be assumed normalized by Lemma 4. But as $\mathcal{L}_{0,2}^{\Sigma,1} \cap \{0,1\}^* 0^\omega = \mathcal{L}_{0,2}^{\{0,1\},1}$, by Lemma 20, this automaton should have a loop on its initial state. This is not possible and $\mathcal{L}_{0,2}^{\Sigma,1}$ is not in CDFA($\mathtt{fin},=$). $\square$

Proposition 22. *The language $\mathcal{L} = \mathcal{L}_{0,2}^{\Sigma,0} + \mathcal{L}_{0,2}^{\Sigma,1}$ is not in* CDFA($\mathtt{fin'},=$).

Proof. For the sake of argument, assume that $\mathcal{L} = \mathcal{L}_{\mathcal{A}}^{(\mathtt{fin'},=)}$ for a CDFA $\mathcal{A} = (\Sigma, Q, T, q_0, F)$. As $\mathcal{L} \cap \{0,1\}^* 0^\omega = \mathcal{L}_{0,2}^{\Sigma,1}$, by Lemma 20, there exists k such that there exists a loop on q_0 labeled by 0^k. Symmetrically, there exists k' such that there exists a loop on q_0 labeled by $1^{k'}$. As $0^\omega \in \mathcal{L}$, $\emptyset \in F$. The path p labeled by $x = (0^k 1^{k'})^\omega$ verifies $\mathtt{fin'}(p) = \emptyset \in F$. Then x is recognized but x is not in $\mathcal{L}$. We have a contradiction. $\square$

Remark 23. Using similar methods as in the proof of Lemma 20 and Proposition 22, we can prove that the language $\Sigma(\mathcal{L}_{1,2}^{\Sigma,0} + \mathcal{L}_{1,2}^{\Sigma,1})$ is not in CDFA($\mathtt{fin'},=$). Since CDFA($\mathtt{fin'},=$) is clearly closed under complementation, $\Sigma(\tilde{\mathcal{L}}_{0,2}^{\Sigma,0} \cap \tilde{\mathcal{L}}_{0,2}^{\Sigma,1})$ is not in CDFA($\mathtt{fin'},=$).

Proposition 24. *The language $\mathcal{L} = \mathcal{L}_{1,2}^{\Sigma,0} + \mathcal{L}_{1,2}^{\Sigma,1}$ is in* CFA($\mathtt{fin},\sqcap$) *but not in* CDFA($\mathtt{fin'},=$).

Proof. By Proposition 9 and using the non-determinism, it is clear that $\mathcal{L}$ is in CFA($\mathtt{fin},\sqcap$). By Remark 23, $\mathcal{L} \notin$ CDFA($\mathtt{fin'},=$). $\square$

Proposition 25. *The language $\mathcal{L} = \Sigma(\tilde{\mathcal{L}}_{0,2}^{\Sigma,0} \cap \tilde{\mathcal{L}}_{0,2}^{\Sigma,1})$ is in* FA($\mathtt{fin},\subseteq$) *but not in* CDFA($\mathtt{fin'},=$).

Proof. We have $\mathcal{L} = \mathcal{L}_{\mathcal{A}}^{(\mathtt{fin},\subseteq)}$ for the CFA $\mathcal{A} = (\Sigma, \{q_0, q_1, q_2, q_3, q_4, q_5, q_6, q_7\}, T, q_0, \{q_2, q_3, q_4, q_6\})$ where T is depicted in Figure 4 (here $\bar{\Sigma}$ means $\Sigma \smallsetminus \{0,1\}$). This automaton is split in two disjoint parts. A path which visits the state q_5 is successful if and only if q_5 (and then q_7) is visited an infinite number of times, if and only if its label contains an infinite number of occurrences of the pattern 01, if and only if its label contains in infinite number of a's and b's.

A path visiting q_1 is successful if and only if q_1 is visited an infinite number of times. Let p be a successful path visiting q_1, let ax be its label where $a \in \Sigma$ and $x \in \Sigma^\omega$. If $|x|_0$ (resp. $|x|_1$) is finite, the set $\mathtt{inf}_{\mathcal{A}}(p)$ is included in $\{q_1, q_2\}$ or in $\{q_3, q_4\}$ (resp., in $\{q_1, q_3\}$ or in $\{q_2, q_4\}$). Since p is successful, q_1 is in $\mathtt{inf}_{\mathcal{A}}(p)$, therefore $\mathtt{inf}_{\mathcal{A}}(p)$ is included in $\{q_1, q_2\}$ (resp., in $\{q_1, q_3\}$) and $|x|_0$ (resp., $|x|_1$) is even. The converse is clear. By Remark 23, $\mathcal{L} \notin$ CDFA($\mathtt{fin'},=$). $\square$

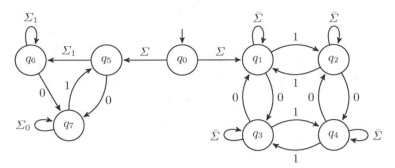

Fig. 4. A FA recognizing $\Sigma(\tilde{\mathcal{L}}_{0,2}^{\Sigma,0} \cap \tilde{\mathcal{L}}_{0,2}^{\Sigma,1})$ under the condition $(\mathtt{fin}, \subseteq)$

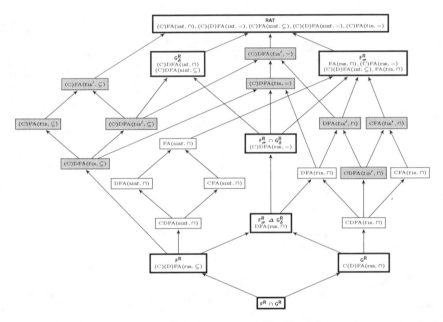

Fig. 5. The hierarchy of classes of ω-languages. The Borel hierarchy appears in bold boxes. Grayed boxes show the new classes studied in this paper. Arrows represent inclusions between classes. Classes in the same box are equal. The possibly missing arrows are from classes induced by $(\mathtt{ninf}, \sqcap)$ to $\mathrm{CDFA}(\mathtt{fin}', =)$, the question is open in this case.

8 Conclusions

This paper is a step further in the study of the hierarchy of ω-languages induced by accepting conditions found in the literature. Figure 5 illustrates the hierarchy and highlights the contribution of this paper.

This research can be continued along several directions. First of all, some inclusions of classes induced by $(\mathtt{ninf}, \sqcap)$ into $\mathrm{CDFA}(\mathtt{fin}', =)$ are still open.

Secondly, in [2], the authors proved that a slight generalization of classical Büchi result: all second order definable accepting conditions induce ω-rational languages. It would be very interesting to study what is the impact of weaker fragments of logic over the classification provided here.

Another promising research direction considers the closure properties of the newly found classes of ω-languages.

Finally, the decidability of the new classes is certainly a promising research direction.

References

1. Büchi, J.R.: Symposium on decision problems: On a decision method in restricted second order arithmetic. In: Ernest Nagel, P.S., Tarski, A. (eds.) Logic, Methodology and Philosophy of Science Proceeding of the 1960 International Congress. Studies in Logic and the Foundations of Mathematics, vol. 44, pp. 1–11. Elsevier (1960)
2. Cervelle, J., Dennunzio, A., Formenti, E., Provillard, J.: Acceptance conditions for ω-languages and the Borel hierarchy (2013) (submitted)
3. Dennunzio, A., Formenti, E., Provillard, J.: Acceptance conditions for ω-languages. In: Yen, H.-C., Ibarra, O.H. (eds.) DLT 2012. LNCS, vol. 7410, pp. 320–331. Springer, Heidelberg (2012)
4. Hartmanis, J., Stearns, R.E.: Sets of numbers defined by finite automata. American Mathematical Monthly 74, 539–542 (1967)
5. Landweber, L.H.: Decision problems for omega-automata. Mathematical Systems Theory 3(4), 376–384 (1969)
6. Litovsky, I., Staiger, L.: Finite acceptance of infinite words. Theoretical Computer Science 174(1-2), 1–21 (1997)
7. Moriya, T., Yamasaki, H.: Accepting conditions for automata on ω-languages. Theoretical Computer Science 61, 137–147 (1988)
8. Muller, D.E.: Infinite sequences and finite machines. In: Proceedings of the 1963 Proceedings of the Fourth Annual Symposium on Switching Circuit Theory and Logical Design, SWCT 19, pp. 3–16. IEEE Computer Society (1963)
9. Perrin, D., Pin, J.E.: Infinite words, automata, semigroups, logic and games. In: Pure and Applied Mathematics, vol. 141. Elsevier (2004)
10. Staiger, L.: ω-languages. In: Handbook of formal languages, vol. 3, pp. 339–387. Springer (1997)
11. Staiger, L., Wagner, K.W.: Automatentheoretische und automatenfreie charakterisierungen topologischer klassen regulärer folgenmengen. Elektronische Informationsverarbeitung und Kybernetik 10(7), 379–392 (1974)
12. Wagner, K.W.: On ω-regular sets. Information and Control 43(2), 123–177 (1979)

On Context-Diverse Repeats
and Their Incremental Computation

Matthias Gallé and Matías Tealdi

Xerox Research Centre Europe
Grenoble, France

Abstract. The context in which a substring appears is an important
notion to identify – for example – its semantic meaning. However, existing
classes of repeats fail to take this into account directly. We present here
xkcd-repeats, a new family of repeats characterized by the number of
different symbols at the left and right of their occurrences. These repeats
include as special extreme cases maximal and super-maximal repeats.

We give sufficient and necessary condition to bound their number
linearly in the size of the sequence, and show an optimal algorithm that
computes them in linear time – given a suffix array –, independent on
the size of the alphabet, as well as two other algorithms that are faster
in practice.

Additionally, we provide an independent and general framework that
allows to compute these (and other) repeats incrementally; extending the
application space of repeats in a streaming framework.

1 Introduction

Inferring *constituents* is a basic step for many applications involving textual documents. These are the semantic blocks that define the meaning of a document,
and may be single words, multi-words expressions or even partial (or several) sentences. They can be used to represent the document, and an accurate description
is crucial to tasks such as classification, clustering, topic detection or knowledge
extraction. It has long been known that the importance of a constituent does
not rely only on its own properties (like frequency, lengths or composition) but
also on the *context* it appears in. J.R. Frith famously said "You shall know a
word by the company it keeps" and the "distributional" approach has been used
successfully in natural-language applications [16].

These constituents are also crucial in a more fundamental task, that is to infer
the structure of a document. In grammatical inference for instance – where it is
supposed that the document samples were generated by a grammar – prior to
detecting how different rules are related to each other, one has to find which of
the substrings of the document may or not correspond to the same constituent.
A crucial step for this is the notion of the context in which a substring appears
in, which in the most basic setting is just the character to the left and to the
right of the occurrence of a given word. ADIOS for instance [17] uses as fundamental signal to decide on the set of constituents the fraction of different context

A.-H. Dediu et al. (Eds.): LATA 2014, LNCS 8370, pp. 384–395, 2014.

a substring appears in. Zellig Harris substitutability theory – strongly related to the idea of context of a constituent – got implemented in distinct way in the literature, like in ABL [19] or through the mutual information criterion of Clark [5,6]. Selecting which are the right substrings to consider, prior to deciding how they associate to each other hierarchically, is also important in the related Smallest Grammar Problem [4].

Unfortunately, existing notions from stringology offers only limited links to these theories. Existing classes of words only tangentially takes into account the notion of context; and applied algorithms, like those cited above, rely on trivial straightforward and brute force algorithms to detect them.

We present here a new class of repeats, named *xkcd*-repeats, that is explicitly defined by the number of different context a substring appears in. These repeats define a family of classes where maximal and super-maximal repeats are extreme cases. We give bounds on their number with respect to the size of the sequence and compare three algorithms to compute them, relying on different ideas.

This document is structured as follows: first we give some basic definitions on sequences and data structures in Sect. 2. We then define our new class of repeats (Sect. 3) and give three algorithms to compute them, running in $\mathcal{O}(|\Sigma|n)$, $\mathcal{O}(n \log n / \log \log n)$ and $\mathcal{O}(n)$ time respectively (Sect. 4). In Sect. 5 we give a general framework to compute those and other repeats incrementally, which we believe is of independent interest.

Due to lack of space we have omitted most of the proofs, advancing only the main arguments.

2 Definitions

A *sequence* s is a concatenation of symbols $s[1], \ldots, s[n]$, with $s[i] \in \Sigma$, the alphabet. The *length* of s, $|s|$ is the numbers of symbols, which we will generally denote by n. When necessary we will suppose that s starts and ends with different unique symbols ($s[0] = \$_1, s[n+1] = \$_2, \$_1 \neq \$_2$ and $\$_1, \$_2 \notin \Sigma$). Another sequence $\omega \in \Sigma^*$ is said to *occur* in s at position k if $\omega[i] = s[k+i]$ for $i = 1 \ldots |\omega|$. The set of occurrences of ω in s is denoted by $occ_s(\omega)$ (or just $occ(\omega)$ if s is clear from the context) and $|occ_s(\omega)|$ is the *support* of ω. If $|occ_s(\omega)| \geq 2$, ω is called a *repeat* of s and $\mathcal{R}(s)$ is the set of all repeats of the sequence s.

The size of the left (right) context of a word ω in s is defined as the number of different symbols appearing to the left (right) of all occurrences of ω: $lc_s(\omega) = |\{s[i-1] : i \in occ(\omega)\}|$ ($rc_s(\omega) = |\{s[i+|\omega|] : i \in occ(\omega)\}|$).

Due to lack of space we will assume that the reader is familiar with the suffix tree and array data structure. We will denote by sa the suffix array itself (the positions of the lexigraphically ordered permutation of suffixes), and by lcp the array holding the lengths of the longest common prefix of successive suffixes of sa. Suffix arrays and lcp array can be constructed in linear time [14].

3   Context-Diverse Repeats

Indexing the set $\mathcal{R}(s)$ permits to analyze potential constituents of the sequence or to perform other indexing or counting operations. However, the total set of repeats is highly redundant and can grow quadratically with the size of the sequence ($|\mathcal{R}(s)| \in \mathcal{O}(|s|^2)$). These "maps bigger than the empire" (as Apostolico referred to this [2], borrowing an expression from Borges), lead to the popular use of maximal classes of repeats. Such a class should ideally focus only on the "interesting" repeats. We detail below three of these classes:

Maximal Repeat: a repeat is said to be maximal, if it cannot be extended without losing support. Formally: ω is a maximal repeat iff there is no other repeat ω' such that ω occurs in ω' and $|occ(\omega)| = |occ(\omega')|$. The number of maximal repeats grows linearly with the size of the sequence, although their total number of occurrences can still be quadratic [8].

Largest-Maximal Repeats (or near-supermaximal repeat [10]): a repeat is said to be largest-maximal if it has at least one occurrence that is not strictly included in another occurrence. Formally, ω is largest-maximal iff there do not exist other repeats $\omega_1, \ldots, \omega_k$ such that ω occurs in all of them and $pos(\omega) \subseteq \bigcup_{i=1}^{k} pos(\omega_i)$, where $pos(\omega) = \bigcup_{i \in occ(\omega)}\{i, \ldots, i + |\omega|\}$, the set of positions that ω covers over the sequence. Largest-maximal repeats are a subset of maximal repeats and therefore linear. The total number of occurrences grows in the worst case at least as $\Omega(n^{\frac{3}{2}})$, although a tight bound is unknown [8].

Super-Maximal Repeat: A repeat is said to be super-maximal if it does not occur in any other repeat. The total number of occurrences of super-maximal repeat is linear.

As an example, consider the sequence $dab\,Wab\,Xac\,Yac\,Zdab$. The only super-maximal repeats here are ac and dab. In addition to these ab is also a largest-maximal repeat because it appears once without any other repeat covering it. Finally, a is also a maximal repeat as there is no other repeat that contains it and has the same support.

The following two lemmas provide the fundamental motivation of this paper, as they show how to characterize maximal and super-maximal repeats by the number of their left and right context.

Lemma 1. *A repeat ω is maximal in s iff $lc(\omega), rc(\omega) \geq 2$.*

Lemma 2. *A repeat ω is super-maximal in s iff $lc(\omega) = rc(\omega) = |occ(\omega)|$.*

Context-diverse ($xkcd$) repeats fill the whole range of these two extremes, by permitting to vary the values of these contexts:

Definition 3. *$\omega \in \mathcal{R}(s)$ is said to be x–right-context-diverse ($xrcd$) in s if $rc_s(\omega) \geq x$. It is said to be k–left-context-diverse ($klcd$) in s if $lc_s(\omega) \geq k$.*
Finally, ω is a $\langle x, k \rangle$–context-diverse ($xkcd$) in s if it is $xrcd$ and $klcd$.

The following theorem follows from Lemma 1 and 2:

Theorem 4.

1. ω is a maximal repeat in s iff it is $\langle 2, 2 \rangle$–context-diverse.
2. ω is a super-maximal repeat in s iff it is $\langle |occ(\omega)|, |occ(\omega)| \rangle$–context-diverse.

And the following Corollary follows from here, and from the linearity of right- and left-maximal repeats [10]:

Corollary 5. *The number of xkcd-repeats is $\mathcal{O}(n)$ iff $\max(x, k) \geq 2$. It is $\Theta(n^2)$ otherwise.*

A notable exception of a class of repeats that cannot be captured by this notion of xkcd-repeats are *largest-maximal repeats*. For a repeat to be largest-maximal, it has to have at least one occurrence with a right-(left-)context different from all right-(left-) contexts of remaining occurrences. Such *context-uniqueness* cannot be captured with the rc and lc functions.

4 Computation

We will compare three algorithms to compute xkcd-repeats, all using the suffix array. We will analyze their running time and compare them empirically.

First note that a straightforward way of computing all xkcd-repeat would be the following two-stage approach: first, compute all repeats $\mathcal{R}(s)$. Then, for each repeat ω inspect all occurrences and store two sets of symbols: those occurring to the left and to the right (this is, $\{s[i-1]\}$ and $\{s[i+|\omega|]\}, \forall i \in occ_s(\omega)$). xkcd-repeats are then those where the size of these sets are greater than x and k, respectively. Unfortunately, such an approach is $\Omega(n^2)$, as there may be this number of repeats in s. If we are only interested in xkcd-repeats such that $\max(x, k) \geq 2$, we can precompute only the left (or right) maximal repeats, whose number is linear (see Corollary 5). However, the total number of *occurrences* of such repeats is still $\Theta(n^2)$.

4.1 Simple Algorithm

Repeats can easily be computed through the enhanced suffix array because all occurrences of a given repeat are adjacent in the suffix array. Moreover, information about the right context can also be easily obtained through the same way, by analyzing how the lcp values evolve.

We keep a stack of the current analyzed repeat and traverse the suffix array in order. If the value of the lcp value remains equal it indicates just another occurrence of the current repeat (which is at the top of the stack), but with a different right context. An increase in the *lcp* value indicates not only the presence of another repeat, but also that the current one is not adding an additional right-context until the newly found repeat is popped out of the stack. Finally, a decrease in the *lcp* value indicates the last occurrence of a repeat, and triggers the eventual output of the current analysed repeat.

Unfortunately, information of the left-context is much harder to get, as it is spread out over the suffix array. An easy way of collecting this (which we will improve later on) is to store all the symbols appearing as left context. When a repeat is popped out, these are then inherited by the topmost repeat in the stack. Recording all the left-contexts adds an extra $|\Sigma|$ factor to the space and time complexity. In our implementation we actually used a set implementation which adds an additional $\log(|\Sigma|)$, but which allows a better trade-off with the memory requirements than using a bit array for example.

The exact algorithm is depicted in Alg. 1, which presupposes the existence of the *lcp* array and *sa* array. Each repeat ω is represented by a tuple $\langle p, \ell \rangle$, where p is the leftmost occurrence of repeat ω in s and $\ell = |\omega|$. While having the leftmost occurrence is not necessary at this stage, we will use this later on. In line 23 a new repeat is added to the stack. Keeping track of where this repeat started (variable *st*) is an important detail [15], which we extend with tracking also the set of left context seen so far (variable *stlc*). Note that, thanks to the way the suffix array is build, the repeat added here only has two different right contexts.

4.2   Using Dynamic Range Computation

What we need in order to determine if a repeat has enough left contexts is to compute the number of different elements in the virtual array $lc[i : p]$, where $lc = [s[sa[1] - 1], \ldots, s[sa[n] - 1]]$. The way Alg. 1 achieves this is by storing explicitly these different elements in a set, but there are more efficient ways (the problem is called *color counting* in the stringology community, see [12, Problem 12]). The best know solution [3] requires $\mathcal{O}(\log n / \log \log n)$ time for each query. In our implementation we used a simpler and very easy to implement solution based upon Fenwick trees (also called BIT trees) [7]. Given a sequence of integers x a Fenwick tree permits to compute prefix sums $psum_x(k) = \sum_{i=1}^{k} x[i]$ in time $\log(|x|)$, and to modify $x[i]$ also in time $\log(|x|)$.

We traverse the suffix array as in Alg. 1, updating a Fenwick tree over a binary array *islast* which contains a 1 at position i if the last occurrence of $lc[i]$ – *so far* – is i, and 0 otherwise. This update is done at the beginning of the outermost loop in Alg. 1. Therefore, when a repeat is popped out the number of its different left contexts is given by the value $psum_{islast}(i) - psum_{islast}(p - 1)$[1]. To perform the update over *islast* we need an additional array *last* of size $|\Sigma|$ that for each symbol σ keeps its right-most position so far in lc. The update then becomes simply to set $islast[last[lc[i]]] = 0$ (except if this is the first occurrence of $lc[i]$) and $islast[i] = 1$; and of course $last[lc[i]] = i$. Remember that all updates of *islast* cost $\log n$ as its Fenwick tree has also to be updated. This algorithm then runs in time $\mathcal{O}(n \log n)$.

---

[1] $psum_{islast}(i)$ is the number of different symbols encountered so far as left context, and can therefore be kept as global variable.

Algorithm 1. Computation of $xkcd$-repeats in $\mathcal{O}(|\Sigma|n)$

$xkcd$ (s, sa, lcp, x, k)

Input: sequence s, suffix array sa, lcp-array, minimal value of right and left context diversity x, k

Output: $xkcd$-repeats in the form $\langle p, \ell \rangle$

```
 1: T = empty stack
 2: ⟨p, ℓ, r, lc⟩ := ⟨0, 0, 1, {$}⟩
 3: T.push(⟨p, ℓ, r, lc⟩) // ensures that the stack never becomes empty
 4: for all  i ∈ [2..n + 1] do
 5:     st := sa[i − 1]
 6:     stlc := {s[st − 1]}
 7:     while T.top().ℓ > lcp[i] do // last occurrence of a repeat
 8:         ⟨p, ℓ, r, lc⟩ := T.pop() // repeat of length ℓ with leftmost occurrence p and
                r, |lc| different right/left context
 9:         st := p
10:         s.top().p := min(s.top().p, p)
11:         stlc := lc
12:         if r ≥ x ∧ |lc| ≥ k then
13:             output ⟨p, ℓ⟩ // has i − p occurrences
14:         end if
15:         T.top().lc := T.top().lc ∪ lc
16:     end while
17:     if T.top().ℓ = lcp[i] then // new occurrence of same repeat
18:         T.top().r := T.top().r + 1
19:         T.top().lc := T.top().lc ∪ {s[sa[i] − 1]}
20:         s.top().p := min(s.top().p, sa[i])
21:     else // new repeat, which already has i − st occurrences
22:         stlc := stlc ∪ {s[sa[i] − 1]}
23:         T.push(⟨min(sa[i − 1], sa[i], st), lcp[i], 2, stlc⟩)
24:     end if
25: end for
```

4.3   A Truly Linear Algorithm

Supposing a fixed alphabet is acceptable for many applications, including genetic sequences analysis on the original alphabet (of size 4 or 20). However, as soon as new symbols are added [4] or the analyzed sequences are natural language (with one identifier for each word) having the time complexity depending on the size of Σ can become a problem. We present an algorithm that computes all $xkcd$ repeats in optimal linear time, even for an integer alphabet.

Like in many algorithms based on suffix data structures, reasoning with the right context in Alg. 1 is straightforward. It is the left context which adds complexity. The algorithm is basically traversing the lcp-interval tree [1], so that any statistics on the left context that can be summarized in constant time and that can be computed based on the statistics of the children does not present problems. This is the case for maximal repeats (where we are interested in context diversity), super-maximal repeats (context uniqueness of all occurrences) [15] and largest-maximal repeats (context uniqueness of at least one occurrence). However, for $xkcd$-repeats we are interested in the number of different context, and to compute them based on the information of the children we need to know exactly which symbols appeared as left contexts.

Our optimal solution diverges therefore from other algorithms, and we divide it in three stages: compute (i) $xrcd$-repeats, (ii) $klcd$-repeats and (iii) merge them. In order to keep linearity, we consider therefore the case of $x, k \geq 2$. This is, it will not compute $xkcd$-repeats that are not maximal. The reason for this is that doing otherwise would risk either phase (i) or (ii) to be potentially quadratic.

Computation of $xrcd$-Repeats. In its most basic form, this is just a simplified version of Alg. 1, where we ignore everything related to the left context. We are, however, interested in having a constant representation for each repeat that permits to compare it afterwards (in step (iii)). It becomes important now to have a canonical definition of a repeat, which we take here to be its leftmost occurrences, together with its length. As shown in Alg. 1, the left-most position for a node in the lcp-interval tree is one of those statistics that can easily be computed from the left-most occurrences of its children.

The final output of this phase is an array of lists denoted by q_{xrcd}. For each position $1 \leq p \leq n$, the list q_{xrcd} contains the length of all $xrcd$-repeats whose first occurrence in s is at position p. An important fact is that each list should be sorted in strict decreasing order.

The following proposition shows that this order is obtained for free.

Proposition 6. *If Alg. 1 outputs $\langle p_1, \ell_1 \rangle$ before $\langle p_2, \ell_2 \rangle$ and $\ell_1 < \ell_2$ then $p_1 \neq p_2$*

Computation of $klcd$-Repeats. Computing $klcd$ is equivalent to compute $xrcd$ repeats, but on a prefix array, (the array of lexicographically ordered prefixes instead of suffixes). Alternatively, this can be achieved by using the suffix array of the reversed string ($\overleftarrow{s}$) and running the same algorithm described to retrieve $xrcd$-repeats. However, in order to compare these repeats to the $xrcd$-pairs, we need to compute the maximal (right-most) occurrence (replace min by max in Alg. 1). An $\langle p, \ell \rangle$ $xrcd$-repeat on $\overleftarrow{s}$ corresponds therefore to the $xlcd$-repeat $\langle n - p + \ell, \ell \rangle$ of s. We define $inv(p, \ell) = n - (p + \ell)$, which gives the index over s corresponding to position p over $\overleftarrow{s}$. Moreover, if p is the right-most occurrence of a repeat ω in $\overleftarrow{s}$, then $inv(p, |\omega|)$ is the left-most occurrence of $\overleftarrow{\omega}$ in s.

Equivalently to step (i), the output here is an array of lists $\overleftarrow{q}_{krcd}$, where $\overleftarrow{q}_{krcd}[p]$ contains the length ℓ of all $krcd$-repeats whose last occurrence in $\overleftarrow{s}$ is at position p.

In order to do the merging in step (iii) in linear time we need to transform $\overleftarrow{q}_{krcd}$ to q_{klcd} in linear time, such that the lists at each position are sorted. Prop. 6 ensures this for q_{xrcd} already. The $klcd$ however are computed as $krcd$ on the reversed string, and it is not trivial which should be the right order to traverse them to keep order. The following proposition resolves this:

Proposition 7. *Let $\langle p, \ell_1 \rangle$ and $\langle p, \ell_2 \rangle$ be two klcd-repeats (with p being the left-most occurrence), such that $\ell_1 < \ell_2$, and let p'_1, p'_2 be such that $p = inv(p'_1, \ell_1) = inv(p'_2, \ell_2)$. Then, $p'_1 > p'_2$.*

So, to keep the order in the lists of q_{klcd} of the original string, it is enough to traverse the $\langle p, \ell \rangle$ krcd-repeats of $\overleftarrow{s}$ in decreasing order of p and to add them in this order into $q_{klcd}[inv(p, \ell)]$.

Merging. Prop 6 and 7 ensures that the lists $q_{xrcd}[i]$ and $q_{klcd}[i]$ are sorted in decreasing and increasing order, respectively. Finding the intersection between them can then be done in linear time

Theorem 8. *The xkcd-repeats of sequence s can be computed in time $\mathcal{O}(|s|)$, independent of the size of the alphabet.*

4.4   Comparison

We compared the execution time of all three algorithms for sequences of different alphabet size, over two kind of sequences: randomly generated (uniform and independent distribution of symbols) over different alphabet size, and an English wikipedia dump (of Sept 2012), where each tokenized word[2] was assigned an integer identifier. There were 9 million different symbols, distributed as expected by a power law. Herdan's law says that the number of words type in a sequence of size n is expected to be $k \times n^{\beta}$, with typical values of $30 \leq k \leq 100$ and $\beta \approx 0.5$ [11]. However, in our case we did not perform any cleaning and included all the XML meta-data of the wikipedia dump. For the random sequences, we report the average over 5 runs for each point, and for the wikipedia sequence we took increasing prefixes.

Because the linear algorithm requires to construct two suffix arrays, the choice of which algorithm to use is crucial. We used Yuta Mori's implementation[3] of the SAIS algorithm [20], the fastest of the variants we tried, and which also runs in linear time. While we measured only user time, we did not have major problems with swapping in the size of sequences considered here (the machine had 32GB of RAM).

As can been seen in Fig. 1, only with very large alphabets the linear algorithm effectively outperforms the $\mathcal{O}(|\Sigma|n \log |\Sigma|)$ one. In all cases, the BIT implementation outperforms all other. However, it should be noted that most of the time in the linear algorithm is used in the construction of the two suffix arrays and lcp (85% of the total time, compared with 68% for the BIT version). Any improvement in this, or creating both suffix arrays at the same time [13] would directly impact these plots.

---

[2] We used the NLTK library of python for tokenization.
[3] https://sites.google.com/site/yuta256/sais

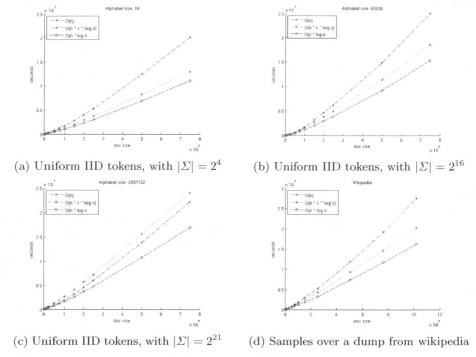

(a) Uniform IID tokens, with $|\Sigma| = 2^4$ (b) Uniform IID tokens, with $|\Sigma| = 2^{16}$

(c) Uniform IID tokens, with $|\Sigma| = 2^{21}$ (d) Samples over a dump from wikipedia

Fig. 1. (User) Time usage of all three algorithms, over different document lengths. All datapoints are the average over 5 runs.

5   Incremental Computation

Existing algorithms to compute repeats, including those presented so far, work in an off-line mode, where the whole sequence is supposed to be available. This assumption may be true in bioinformatics applications, but many use-cases around natural language documents work in a streaming setting (ex: online news analysis, real-time document classification, etc). In [9] we advocated the use of repeats to model this type of documents, showing the advantages over other approaches. However, this can not be adapted directly to such streaming settings without paying an extremely high efficiency toll. We present therefore here a general method to compute – iteratively – a set of characteristics on context-diverse repeats. For this we will rely on the incremental suffix tree creation algorithm of Ukkonen [18]. Formally, the problem we try to solve is[4]:

Problem 9. For a class of repeats $\mathbf{X}$, and given documents $D = \{d_1 \ldots d_k\}$, document d_{k+1} and a suffix tree on $S = d_1.d_2.\ldots.d_k$ where all nodes corresponding to $\mathbf{X}(S)$ are marked and augmented with their number of occurrences; return a suffix tree on $S.d_{k+1}$ where all nodes corresponding to $\mathbf{X}(S.d_{k+1})$ are marked and augmented with their number of occurrences.

---

[4] We define the concatenation of s_1 and s_2 as $s_1.s_2 = s_1\$s_2$, with $\$$ a new symbol.

Our approach can easily be extended to keep the list of occurrences or other information on the selected repeats.

An algorithm resolving this problem is said to be optimal if it runs in time $\mathcal{O}(\max(n, |\mathbf{X}_{update}(S,d)|)$, where $\mathbf{X}_{update}(S,d)$ are the updates that have to be performed: $\mathbf{X}_{update}(S,d) = \{w \in d : w \in \mathbf{X}(S.d)\} \cup (\mathbf{X}(S) \setminus \mathbf{X}(S.d))$. Note that the size of this set can be much larger than the length of the newly added document:

Proposition 10. $|\mathbf{MR}_{update}(S,d)| \in \Theta(|d|^2)$

5.1   Cover of a New Document

When a new document d_{k+1} is added to S, the new suffix tree will have $|d_{k+1}|+1$ leaves corresponding to each of the suffixes of d_{k+1}. Any ancestor of these leaves can potentially change its maximal class because it is repeated for the first time or because of a change in its context sets. It is therefore this set of nodes we will be traversing after each update:

Definition 11. *We define $cover(S, d_{k+1})$ as the set of substrings of d_{k+1} which are right-maximal repeat in $S.d_{k+1}$. This corresponds therefore to those internal nodes of the suffix tree which are ancestor of leaves added for d_{k+1}.*

However, the cover set should not be traversed in any arbitrary order: each node should of course be visited only once, and only after having visited all its children. Any information on the occurrence of an internal node v can be obtained by aggregating correctly this same information of the children of v. The way we achieve this is by ordering the nodes with respect to the lengths of their represented substring. This defines therefore a partial order on the nodes of the suffix tree, where $v < w$ iff v is a prefix of w and $|v| < |w|$. The nodes can then be correctly traversed by using this order with a priority queue, which is initialized with all new leaves. The order in which non-ordered pairs are selected is not important: the priority property ensures that when a node v is visited, all his descendants in the cover were visited before.

This choice makes this algorithm run in time $\mathcal{O}((|cover| + n)\log(n))$, where the additional n factor is due to the leaves of d_{k+1} and the logarithm factor is due to the complexity of insertion and deletion in the queue. The queue could be replaced by an array p of lists of node of maximal size n, where $p[i]$ is the list of node whose depth is equal to i and an array *added* of boolean value where $added[v]$ asserts if the node v was added to p. These two arrays can be updated in constant time and the additional cost in memory trades off with speed, making the cover traversal $\mathcal{O}(|cover| + n)$.

Updating those nodes belonging to a given class is then straightforward, given that each node is enriched with additional information: we keep for each node v two sets of symbols: the first lc_{unique} will contain those symbols c such that c is left context of a leaf-child of v and there is not another leaf of v with the same left-context. The second set, lc will be disjoint to this one and contain all other characters which are left-context of a leaf of v but are not in lc_{unique}. With these definitions, each node can easily be updated with the following rules:

- v is a maximal repeat iff $|v.lc| + |v.lc_{unique}| > 1 \wedge |children(v)| > 1$
- v is $\langle x, k \rangle$-cd iff $|v.lc| + |v.lc_{unique}| \geq x \wedge |children(v)| \geq k$
- v is a supermaximal repeat iff $|v.lc_{unique}| = |children(v)|$
- v is a largest maximal repeat iff $|v.lc_{unique}| > 0$

Of course, both sets have also to be updated. This can be done in a straight-forward manner by updating both sets of the parent of each visited node, adding a additional $|\Sigma|$ factor to the final complexity.

Updating the total number of occurrences can be done similarly. The number of occurrences of a repeat is the number of leaves in the subtree rooted at the node representing it, a value hold in an additional variable per node. In addition, each node will keep an auxiliary variable with the number of new leaves it has due to incorporation of d_{k+1} to the suffix tree. In a first traversal of the cover each node updates this auxiliary variable of his parent, and in a second traversal each node updates its count of subtrees adding the number of new leaves.

6   Conclusions

In this paper we investigated a new class of repeat that includes explicitely the context in which it appears. This *xkcd* family is a generic class that includes maximal and supermaximal repeats as special cases. We also gave three different algorithms to compute these repeats, allowing for a rich choice depending on the size of the alphabet. In addition we also studied the problem of computing these repeats incrementally, to support the addition of new documents to the analyzed collection. This resulted in a general framework relying on Ukkonen's suffix tree creation algorithm.

We believe that these advances will allow the use of these notions from stringology to model natural language documents, an approach which has not sufficiently be studied in our opinion. The notion of the *cover* for the incremental computation of repeats is of independent interest, and we plan to study its characteristic further. Note for example that if the goal is only to mark maximal repeats on the final suffix tree, then the cover permits to do so in optimal time (this is, in time proportional to the number of nodes that change state).

References

1. Abouelhoda, M.I., Kurtz, S., Ohlebusch, E.: Replacing suffix trees with enhanced suffix arrays. Journal of Discrete Algorithms 2, 53–86 (2004)
2. Apostolico, A.: Of maps bigger than the empire (invited paper). In: SPIRE, pp. 2–9 (2001)
3. Bose, P., He, M., Maheshwari, A., Morin, P.: Succinct Orthogonal Range Search Structures on a Grid with Applications to Text Indexing. In: Dehne, F., Gavrilova, M., Sack, J.-R., Tóth, C.D. (eds.) WADS 2009. LNCS, vol. 5664, pp. 98–109. Springer, Heidelberg (2009)

4. Carrascosa, R., Coste, F., Gallé, M., Infante-Lopez, G.: The smallest grammar problem as constituents choice and minimal grammar parsing. MDPI Algorithms 4(4), 262–284 (2011)
5. Clark, A.: Learning deterministic context free grammars: The Omphalos competition. Machine Learning, 93–110 (January 2007)
6. Clark, A., Eyraud, R., Habrard, A.: A polynomial algorithm for the inference of context free languages. In: ICGI, pp. 29–42 (July 2008)
7. Fenwick, P.M.: A New Data Structure for Cumulative Frequency Tables. Softw. Pract. Exper. 24, 327–336 (1994)
8. Gallé, M.: Searching for Compact Hierarchical Structures in DNA by means of the Smallest Grammar Problem. Université de Rennes 1 (February 2011)
9. Gallé, M.: The bag-of-repeats representation of documents. In: SIGIR, pp. 1053–1056 (2013)
10. Gusfield, D.: Algorithms on Strings, Trees, and Sequences: Computer Science and Computational Biology. Cambridge University Press (January 1997)
11. Manning, C., Raghavan, P., Schütze, H.: Introduction to Inf Retrieval. Cambridge UP (2009)
12. Navarro, G.: Spaces, Trees and Colors: The Algorithmic Landscape of Document Retrieval on Sequences. arXiv (2013)
13. Ohlebusch, E., Beller, T., Abouelhoda, M.I.: Computing the Burrows Wheeler transform of a string and its reverse in parallel. Journal of Discrete Algorithms 1, 1 13 (2013), http://linkinghub.elsevier.com/retrieve/pii/S1570866713000397
14. Puglisi, S., Smyth, W.F., Turpin, A.: A taxonomy of suffix array construction algorithms. ACM Computing Surveys 39(2) (July 2007), http://portal.acm.org/citation.cfm?id=1242471.1242472
15. Puglisi, S.J., Smyth, W.F., Yusufu, M.: Fast optimal algorithms for computing all the repeats in a string. In: Prague Stringology Conference, pp. 161–169 (2008)
16. Schütze, H.: Automatic Word Sense Discrimination. Comput. Ling. 24(1) (1998)
17. Solan, Z., Horn, D., Ruppin, E., Edelman, S.: Unsupervised learning of natural languages. PNAS, 11629 11634 (January 2005)
18. Ukkonen, E.: Online construction of suffix trees. Algorithmica 14, 249–260 (1995)
19. van Zaanen, M.: ABL: Alignment-based learning. In: International Conference on Computational Linguistics, pp. 961–967 (2000)
20. Zhang, S., Nong, G., Chan, W.H.: Fast and space efficient linear suffix array construction. In: DCC. IEEE Computer Society, Washington, DC (2008)

Ordered Counter-Abstraction
Refinable Subword Relations for Parameterized Verification

Pierre Ganty[1] and Ahmed Rezine[2],[*]

[1] IMDEA Software Institute, Spain
pierre.ganty@imdea.org
[2] Linköping University, Sweden
ahmed.rezine@liu.se

Abstract. We present an original refinable subword based symbolic representation for the verification of linearly ordered parameterized systems. Such a system consists of arbitrary many finite processes placed in an array. Processes communicate using global transitions constrained by their relative positions (i.e., priorities). The model can include binary communication, broadcast, shared variables or dynamic creation and deletion of processes. Configurations are finite words of arbitrary lengths. The successful monotonic abstraction approach uses the subword relation to define upward closed sets as symbolic representations for such systems. Natural and automatic refinements remained missing for such symbolic representations. For instance, subword based relations are simply too coarse for automatic forward verification of systems involving priorities. We remedy to this situation and introduce a symbolic representation based on an original combination of counter abstraction with subword based relations. This allows us to define an infinite family of relaxation operators that guarantee termination by a new well quasi ordering argument. The proposed automatic analysis is at least as precise and efficient as monotonic abstraction when performed backwards. It can also be successfully used in forward, something monotonic abstraction is incapable of. We implemented a prototype to illustrate the approach.

Keywords: counter abstraction, well quasi ordering, reachability, parameterized verification.

1 Introduction

We introduce in this paper an original adaptation of counter abstraction and use it for the verification of safety properties for linearly ordered parameterized systems. Typically, such a system consists of an arbitrary number of identical processes placed in a linear array. Each process is assumed to have a finite number of states (for example obtained by predicate abstraction). The arbitrary size of these systems results in an infinite number of possible configurations. Examples of

---
[*] Supported in part by the CENIIT research organization (project 12.04).

A.-H. Dediu et al. (Eds.): LATA 2014, LNCS 8370, pp. 396–408, 2014.

linearly ordered parameterized systems include mutual exclusion algorithms, bus protocols, telecommunication protocols, and cache coherence protocols. The goal is to check correctness (here safety) regardless of the number of processes.

Configurations of such a system are finite words of arbitrary lengths over the finite set Q of process states. Processes change state using transitions that might involve universal or existential conditions. Transition t below is constrained by a universal condition. Here, a process (with array index) i may fire t if all processes with indices $j > i$ (i.e., to the right, hence $\forall_R$) are in states $\{q_1, q_2, q_3\} \subseteq Q$.

$$t : q_5 \rightarrow q_6 : \forall_R \{q_1, q_2, q_3\} \tag{1}$$

An existential condition requires that some (instead of all) processes are in certain states. Regular model checking [13,3] is an important technique for the uniform verification of infinite state systems in general, and of linearly ordered parameterized systems in particular. It uses finite state automata to represent sets of configurations, and transducers (i.e., finite state automata over pairs of letters) to capture transitions of the system. Verification boils down to the repeated calculation of several automata-based constructions among which is the application of the transducers to (typically) heavier and heavier automata representing more and more complex sets of reachable configurations. Acceleration [3], widening [6,17] and abstraction [7] methods are used to ease termination.

In order to combat this complexity, the framework of monotonic abstraction [2,1] uses upward closed sets (wrt. a predefined pre-order) as symbolic representations. This introduces an over-approximation, as sets of states generated during the analysis are not necessarily upward closed. The advantage is to use minimal constraints (instead of arbitrary automata) to succinctly represent infinite sets of configurations. The approach typically adopts the subword relation as the pre-order for the kind of systems we consider in this work[1]. The analysis starts with upward closed sets representing the bad configurations and repeatedly approximates sets of predecessors by closing them upwards. Termination is guaranteed by well quasi ordering [12]. The scheme proved quite successful [2,1] but did not propose refinements for eliminating false positives in ordered systems like the ones we consider here.

In this work, we describe an original integration of upward closed based symbolic representation and of threshold based counter abstraction. The resulting symbolic representation allows for the introduction of original relaxation operators that can be used in classical over-approximate-check-refine reachability schemes. The idea of counter abstraction [16,10] is to keep track of the number of processes that satisfy a certain property. A typical property for a process is to be in some state in Q. A simple approach to ensure termination is then to count up to a prefixed threshold. After the threshold, any number of processes satisfying the property is assumed possible. This results in a finite state system that can be exhaustively explored. If the approximation is too coarse, the threshold can be augmented. For systems like those we consider in this paper, automatically

---

[1] As a concrete example, if $q_5 \in Q$, then the word $q_5 q_5$ would represent all configurations in $(Q^* q_5 Q^* q_5 Q^*)$ since $q_5 q_5$ is subword of each one of them.

finding the right properties and thresholds can get very challenging. Consider for instance the transition t above (1). It is part of Burns mutual exclusion algorithm, where q_6 models access to the critical section [2]. Suppose we want to compute the t-successors of configurations only containing processes in state q_5. These are in fact reachable in Burns algorithm. Plain counter abstraction would capture that all processes are at state q_5. After one step it would capture that there is one process at state q_6 and all other processes are at state q_5 (loosing that q_6 is at the right of all q_5). After the second step it would conclude that configurations with at least two q_6 are reachable (mutual exclusion violation). Observe that increasing the threshold will not help as it will not preserve the relative positions of the processes. Upward closure based representations will also result in a mutual exclusion violation if used in forward. Suppose we use $q_5 q_5$ as a minimal constraint. Upward closure wrt. to the subword relation would result in the set $(Q^* q_5 Q^* q_5 Q^*)$ which already allows two processes at state q_6 to coexist. Even when using the refined ordering of [1], upward closure would result in $(\{q_5\}^* q_5 \{q_5\}^* q_5 \{q_5\}^*)$. After one step, the obtained $(\{q_5\}^* q_5 \{q_5\}^* q_6)$ will be approximated with $(\{q_5, q_6\}^* q_5 \{q_5, q_6\}^* q_6 \{q_5, q_6\}^*)$, again violating mutual exclusion. Approximations are needed to ensure termination (the problem is undecidable in general [4]). Indeed, without approximation, one would differentiate among infinite numbers of sets, like in the following sequence:

$$(\{q_5\}^* q_6), \ (\{q_5\}^* q_6 \{q_5\}^* q_6), \cdots (\{q_5\}^* q_6 \{q_5\}^* \cdots \{q_5\}^* q_6) \tag{2}$$

The idea of this work is to combine threshold-based counter abstraction with subword-based upward closure techniques in order to propose an infinite number of infinite abstract domains allowing increasing precision of the analysis while still ensuring termination. To achieve this, we introduce the notion of a *counted word*. A counted word has a *base* and a number of formulae (called *counters*). Like in monotonic abstraction, a base (a word in Q^*) is used as a minimal element and denotes all larger words wrt. the subword relation. In addition, the counters are used to constrain the denotation of the base. We associate two counters to each position in the base: a *left* and a *right counter*. For each state q in Q, the left counter of a position constrains how many of the processes to the left of the position can be in state q (i.e. constrains Parikh images of allowed prefixes). Right counters constrain allowed suffixes to the right of the positions. For example $(\{q_5\}^* q_6)$ is captured by the counted word φ_1 defined below:

$$
\begin{aligned}
\varphi_1 &= \left(\begin{bmatrix} v_{q_5} \geq 0 \\ \wedge v_{q_6} = 0 \end{bmatrix}, q_6, \begin{bmatrix} v_{q_5} = 0 \\ \wedge v_{q_6} = 0 \end{bmatrix} \right), \\
\varphi_2 &= \left(\begin{bmatrix} v_{q_5} \geq 0 \\ \wedge v_{q_6} = 0 \end{bmatrix}, q_6, \begin{bmatrix} v_{q_5} \geq 0 \\ \wedge v_{q_6} = 1 \end{bmatrix} \right) \left(\begin{bmatrix} v_{q_5} \geq 0 \\ \wedge v_{q_6} = 1 \end{bmatrix}, q_6, \begin{bmatrix} v_{q_5} = 0 \\ \wedge v_{q_6} = 0 \end{bmatrix} \right), \cdots \\
\varphi_k &= \left(\begin{bmatrix} v_{q_5} \geq 0 \\ \wedge v_{q_6} = 0 \end{bmatrix}, q_6, \begin{bmatrix} v_{q_5} \geq 0 \\ \wedge v_{q_6} = (k-1) \end{bmatrix} \right) \cdots \left(\begin{bmatrix} v_{q_5} \geq 0 \\ \wedge v_{q_6} = (k-1) \end{bmatrix}, q_6, \begin{bmatrix} v_{q_5} = 0 \\ \wedge v_{q_6} = 0 \end{bmatrix} \right)
\end{aligned}
$$

In φ_1, the base q_6 denotes $(Q^* q_6 Q^*)$. This is constrained to $(\{q_5\}^* q_6 Q^*)$ by the right counter $\begin{bmatrix} v_{q_5} \geq 0 \\ \wedge v_{q_6} = 0 \end{bmatrix}$ and to $(\{q_5\}^* q_6)$ by the left counter $\begin{bmatrix} v_{q_5} = 0 \\ \wedge v_{q_6} = 0 \end{bmatrix}$. Sequence (2) can then be captured by the counted words $\varphi_1, \varphi_2, \ldots \varphi_k$. This gain in precision comes at the price of loosing termination. We therefore propose *relaxation* operators. Each operator comes with a cut-off, i.e., thresholds associated to each state in Q. If a counter requires $(v_q = k)$ with k larger than the threshold imposed by the cut-off, we weaken $(v_q = k)$ into $(v_q \geq k)$. Using a well quasi ordering argument, we show that this is enough to ensure termination of the analysis that relaxes all generated representations. If a spurious trace is generated, we increase the thresholds in order to obtain a more precise relaxation that eliminates the spurious trace. We implemented a prototype that allows, for the first time, to use in a forward exploration scheme upward closure based representations to verify classical linearly ordered parameterized systems.

Related work. Other verification efforts with a termination guarantee typically consider decidable subclasses [10,9], or use approximations to obtain systems on which the analysis is decidable [16,8,15]. For example, the authors in [9] propose a forward framework with systematic refinement to decide safety properties for a decidable class. The problem we consider here is undecidable. The authors in [15] use heuristics to deduce cut-offs in order to check invariants on finite instances. In [16] the authors use counter abstraction and truncate the counters in order to obtain a finite state system. This might require manual insertion of auxiliary variables to capture the relative order of processes in the array. Environment abstraction [8] combines predicate and counter abstraction. It results in what is essentially a finite state approximated system. Hence, it can require considerable interaction and human ingenuity to find the right predicates. Our approach handles linearly ordered systems in a uniform manner. It automatically adds precision based on the spurious traces it might generate.

Outline. Section (2) gives preliminaries and formalizes the notion of counters. Section (3) uses these counters to define counted words and to state some of their properties that will be useful to build a symbolic representation for the verification of parameterized systems. Section (4) formally describes the considered class of parameterized systems and reports on using counted words as a symbolic representation to solve their reachability problem. We conclude in Section (5).

2   Preliminaries

Preliminaries. Fix a finite alphabet Σ and let Σ^* be the set of finite words over Σ. Let $w \cdot w'$ be the concatenation of the words w and w', ϵ be the empty word, $w \sqcup w'$ be the shuffle set $\{w_1 \cdot w_1' \cdot w_2 \cdots w_m' \mid w = w_1 \cdots w_n \text{ and } w' = w_1' \cdots w_m'\}$. We use $\mathbb{N}$ for the set of natural numbers and $\bar{n}$, with $n \in \mathbb{N}$, to mean $\{1, \ldots, n\}$. Assume a word $w = \sigma_1 \cdots \sigma_n$ where $\sigma_i \in \Sigma$ for $i \in \bar{n}$. We write $|w|$ for the size n, $w_{[i,j]}$ to mean the word $\sigma_i \cdot \sigma_{i+1} \cdots \sigma_j$, $w_{[i]}$ for the letter σ_i, $hd(w)$ for the letter σ_1, $tl(w)$ for the suffix $w_{[2,n]}$, and $w^{\bullet}$ for the set $\{\sigma_1, \ldots, \sigma_n\}$. A multiset m is a mapping $\Sigma \to \mathbb{N}$. We write $m \preceq m'$ to mean that $m(\sigma) \leq m'(\sigma)$ for each $\sigma \in \Sigma$. We write $m \oplus m'$ to mean the multiset satisfying $(m \oplus m')(\sigma) = m(\sigma) + m'(\sigma)$ for each $\sigma \in \Sigma$. If $m' \preceq m$, then the multiset $m \ominus m'$ is defined and verifies $(m \ominus m')(\sigma) = m(\sigma) - m'(\sigma)$ for each σ in Σ. The Parikh image $w^{\#}$ of a word w is the multiset that gives the number of occurrences of each letter σ in w. Given a set S and a pre-order (i.e., a reflexive and transitive binary relation) $\sqsubseteq$ on S, the pair $(S, \sqsubseteq)$ is said to be a well quasi ordering (wqo for short) if there is no infinite sequence $e_1, e_2, \ldots$ of elements of S with $e_i \not\sqsubseteq e_j$ for all $1 \leq i < j$.

A *counter* over an alphabet Σ is a conjunction of simple constraints that denotes a set of multisets. We fix a set of integer variables V_Σ that is in a one to one correspondence with Σ. Each variable v is associated to a letter σ in Σ. We write v_σ to make the association clear. Intuitively, v_σ is used to count the number of occurrences of the associated letter σ in a word in Σ^*. A counter basically captures multisets over Σ by separately imposing a constraint on each letter in Σ. Indeed, we define a counter cr to be either [`false`] or a conjunction $[\wedge_{\sigma \in \Sigma}(v_\sigma \sim k)]$ where $\sim$ is in $\{=, \geq\}$, each v_σ is a variable ranging over $\mathbb{N}$ and each k is a constant in $\mathbb{N}$. Assume in the following a counter cr. For a letter σ in Σ, we write $cr(\sigma)$ to mean the strongest predicate of the form $(v_\sigma \sim k)$ implied by the counter cr. We write $\mathbf{1}_\sigma$ (resp. $\mathbf{0}$) to mean the counter $[\wedge_{\sigma_i \in \Sigma}(v_{\sigma_i} = b_{\sigma_i})]$ with $b_{\sigma_i} = 1$ for $\sigma_i = \sigma$ and $b_{\sigma_i} = 0$ otherwise (resp. $b_{\sigma_i} = 0$ for all $\sigma_i \in \Sigma$). A substitution is a set $\{v_1 \leftarrow u_1, \ldots\}$ of pairs (s.t. $v_i \neq v_j$ if $i \neq j$) where $v_1, \ldots$ are variables, and $u_1, \ldots$ are either all variables or all natural numbers. Given a substitution S, we write $cr[S]$ to mean the formula obtained by replacing, for each pair $v_i \leftarrow u_i$, each occurrence of v_i in cr by u_i. We sometimes regard a multiset m as the substitution $\{v_\sigma \leftarrow m(\sigma) | \sigma \text{ in } \Sigma\}$. For a multiset m, the formula $cr[m]$ takes a Boolean value. In the case where it evaluates to true (resp. false), we say that m satisfies (resp. doesn't satisfy) the counter cr and that the counter cr accepts (resp. does not accept) the multiset m. Given a word w in Σ^*, we abuse notation and write $cr[w]$ to mean that $(w^\#)$ satisfies cr. We write $[\![cr]\!]$ to mean the set $\{m | cr[m] \text{ and } m \text{ is a multiset over } \Sigma\}$. We define the cut-off of cr, written $\kappa(cr)$, to be the multiset that associates to each letter σ in Σ the value $k + 1$ if $cr(\sigma) = (v_\sigma = k)$ and 0 otherwise. Observe that if $\kappa(cr)(\sigma) \neq 0$ for all $\sigma \in \Sigma$, then cr accepts a single multiset, while if $\kappa(cr)(\sigma) = 0$ for all $\sigma \in \Sigma$, then cr accepts an upward closed set of multisets wrt. $\preceq^2$. We write $\mathbb{C}$ for the set of counters over Σ. Given a natural k, we write $\mathbb{C}_k$ to mean $\{cr | \kappa(cr)(\sigma) \leq k \text{ for each } \sigma \in \Sigma\}$. Observe that for any counter $cr \in \mathbb{C}_k$, $((cr(\sigma) = (v_\sigma = k')) \implies k' < k)$.

Example 1. For the counter $cr = [v_a = 0 \wedge v_b = 2 \wedge v_c \geq 1]$ over $\Sigma = \{a, b, c\}$, we have that: $\kappa(cr)(a) = 1$, $\kappa(cr)(b) = 3$, and $\kappa(cr)(c) = 0$. In addition, cr is in $\mathbb{C}_3$.

Operations on counters. Assume two counters cr and cr'. The predicate $(cr \sqsubseteq_\mathbb{C} cr')$ is defined as the conjunction $\wedge_{\sigma \in \Sigma}(cr \sqsubseteq_\mathbb{C} cr')(\sigma)$, where $(cr \sqsubseteq_\mathbb{C} cr')(\sigma)$ is defined in Table (1). In addition, let cr'' be any of the counters $(cr \sqcap_\mathbb{C} cr')$, $(cr \ominus_\mathbb{C} cr')$, or $(cr \oplus_\mathbb{C} cr')$. The counter cr'' is defined as the conjunction $\wedge_{\sigma \in \Sigma} cr''(\sigma)$, where $cr''(\sigma)$ is stated in Table (1). Observe that $(cr \sqsubseteq_\mathbb{C} cr') \implies [\![cr']\!] \subseteq [\![cr]\!]$, that $[\![cr \sqcap_\mathbb{C} cr']\!] = [\![cr]\!] \cap [\![cr']\!]$, that $[\![cr \oplus_\mathbb{C} cr']\!] = \{m_1 \oplus m_2 | cr[m_1] \text{ and } cr'[m_2]\}$, and that $[\![cr \ominus_\mathbb{C} cr']\!] = \{m_1 \ominus m_2 | cr[m_1] \text{ and } cr'[m_2]\}$.

---

[2] A set M of multisets is upward closed wrt $\preceq$ if $m \preceq m'$ and $m \in M$ imlpy $m' \in M$.

Table 1. Contribution of each $\sigma \in \Sigma$ to the predicate $cr \sqsubseteq_C cr'$ and to the counters $cr \sqcap_C cr'$, $cr \oplus_C cr'$ and $cr \ominus_C cr'$

$cr(\sigma)$	$cr'(\sigma)$	$(cr \sqsubseteq_C cr')(\sigma)$	$(cr \sqcap_C cr')(\sigma)$	$(cr \oplus_C cr')(\sigma)$	$(cr \ominus_C cr')(\sigma)$
$v_\sigma = b$	$v_\sigma = b'$	$b = b'$	$(b = b')?v_\sigma = b : \text{false}$	$v_\sigma = b + b'$	$(b \geq b')?v_\sigma = b - b' : \text{false}$
	$v_\sigma \geq b'$	false	$(b \geq b')?v_\sigma = b : \text{false}$		$v_\sigma \geq max(0, b - b')$
$v_\sigma \geq b$	$v_\sigma = b'$	$b' \geq b$	$(b' \geq b)?v_\sigma = b' : \text{false}$		
	$v_\sigma \geq b'$		$v_\sigma \geq max(b, b')$		

Lemma 2. *For $k \in \mathbb{N}$, $(\mathbb{C}_k, \sqsubseteq_C)$ is a well quasi ordering. From every infinite sequence $cr_1, cr_2, \ldots$ we can extract an infinite sequence $cr_{i_1} \sqsubseteq_C cr_{i_2} \sqsubseteq_C \cdots$*

Proof. Let $cr_1, cr_2, \ldots$ be an infinite sequence. Fix a letter σ. If the number of counters for which $cr_m(\sigma)$ is not an equality is infinite, then remove all the counters for which $cr_m(\sigma)$ is an equality. Otherwise, by definition of $\mathbb{C}_k$, there is a $b_0 < k$ such that the number of counters for which $cr_m(\sigma) = (v_\sigma = b_0)$ is infinite. Keep those counters and remove all others from the resulting sequence. By repeating this procedure for each letter σ in Σ, we obtain a new infinite sequence of counters $cr_{m_1}, cr_{m_2}, \ldots$ for which, for each m_i, m_j, $cr_{m_i}(\sigma) = (v_\sigma = b)$ iff $cr_{m_j}(\sigma) = (v_\sigma = b)$. Fix a letter σ for which $cr_{m_1}(\sigma) = (v_\sigma \geq b)$. It is possible to extract from the resulting sequence another infinite sequence $cr_{n_1}, cr_{n_2}, \ldots$ such that if $cr_{n_i}(\sigma) = (v_\sigma \geq b_{n_i})$ and $cr_{n_j}(\sigma) = (v_\sigma \geq b_{n_j})$ with $n_i < n_j$, then $b_{n_i} \leq b_{n_j}$. By repeating this for each letter σ, we obtain an infinite sequence in which $cr_{i_1} \sqsubseteq_C cr_{i_2} \sqsubseteq_C \cdots$. $\square$

3   Counted Words

A *counted word* φ is a finite sequence $(l_1, \sigma_1, r_1) \cdots (l_n, \sigma_n, r_n)$ in $(\mathbb{C} \times \Sigma \times \mathbb{C})^*$. The *base* of φ (written $\overline{\varphi}$) is the word $\sigma_1 \cdots \sigma_n$ in Σ^*. We write $\overleftarrow{\varphi}$ (resp. $\overrightarrow{\varphi}$) to mean the counter $[\wedge_{\sigma \in \Sigma}(v_\sigma \geq 0)]$ (resp. $[\wedge_{\sigma \in \Sigma}(v_\sigma \geq 0)]$) if $\varphi = \epsilon$, and l_1 (resp. r_n) otherwise. We refer to $l_1, \ldots l_n$ (resp. $r_1, \ldots r_n$) as the left (resp. right) counters of φ. The counted word φ is *well formed* if $l_i[(\overline{\varphi})_{[1, i-1]}]$ and $r_i[(\overline{\varphi})_{[i+1, n]}]$ evaluate to true for each $i \in \overline{n}$. We assume ϵ is *well formed*. The following lemma constrains the possible predicates in a well formed counted word.

Lemma 3 (Well formedness). *Let $\varphi = (l_1, \sigma_1, r_1) \cdots (l_n, \sigma_n, r_n)$ be a well formed word. For each $i \in \overline{n}$, the counter $l_i(\sigma)$ (resp. $r_i(\sigma)$) either equals: $(v_\sigma = (\overline{\varphi}_{[1, i-1]})^\#(\sigma))$ (resp. $(v_q = (\overline{\varphi}_{[i+1, n]})^\#(\sigma)))$, or $(v_\sigma \geq k)$ for some k in $\{0, \ldots (\overline{\varphi}_{[1, i-1]})^\#(\sigma)\}$ (resp. in $\{0, \ldots (\overline{\varphi}_{[i+1, n]})^\#(\sigma)\}$).*

Denotation. If $w = \sigma_1 \cdots \sigma_m$, $\varphi = (l_1, \sigma_1, r_1) \cdots (l_n, \sigma_n, r_n)$, and $h : \overline{n} \to \overline{m}$ is an increasing injection, we write $w \models^h \varphi$ to mean that all following three conditions hold for each $i \in \overline{n}$ i) $\overline{\varphi}_{[i]} = w_{[h(i)]}$, and ii) $l_i(w_{[1, h(i) - 1]})$, and iii) $r_i(w_{[h(i) + 1, n]})$. Intuitively, there is an injection h that ensures $\overline{\varphi}$ is subword of w, and s.t. words to the left and right of each image of h respectively respect corresponding left and right counters in φ. We write $w \models \varphi$ if $w \models^h \varphi$ for some injection h, and $[\![\varphi]\!]$ to mean $\{w | w \models \varphi\}$. We let $[\![\epsilon]\!] = \Sigma^*$. Observe that every well formed word has a non-empty denotation since $\overline{\varphi} \models \varphi$. We use $\mathbb{CW}$ to mean the set of well formed counted words.

Example 4. $\varphi = \left(\left[\begin{smallmatrix} v_a = 0 \\ \wedge v_b \geq 0 \end{smallmatrix} \right], a, \left[\begin{smallmatrix} v_a \geq 0 \\ \wedge v_b \geq 0 \end{smallmatrix} \right] \right) \left(\left[\begin{smallmatrix} v_a = 1 \\ \wedge v_b = 0 \end{smallmatrix} \right], a, \left[\begin{smallmatrix} v_a = 0 \\ \wedge v_b \geq 0 \end{smallmatrix} \right] \right)$ and $\llbracket \varphi \rrbracket = aab^*$.

Normalization of well formed words. Counters in a counted word are not independent. Consider for instance $\varphi = (l_1, a, r_1)(l_2, a, r_2)$ in Example (4). We can change $l_1(b)$ to $(v_b = 0)$ without affecting the denotation of φ. The reason is that any prefix accepted by l_1 will have to be allowed by l_2. It is therefore vacuous for l_1 to accept words containing b, and more generally to accept more than $l_2 \ominus_\mathbb{C} \mathbf{1}_a$ (defined by well formedness). Also, observe that l_2 and r_2 imply we can change $r_1(a)$ from $(v_a \geq 0)$ to $(v_a = 1)$. We strengthen a well formed word using the normalization rules depicted in Table (2).

Lemma 5 (Normalization rules). *Applying any of the rules of Table (2) on a well formed word φ does preserve its denotation, and hence its well formedness.*

Table 2. Normalization rules. For example, the rule $l_{i<j}$ states we can replace counter l_i in word $(\varphi_p \cdot (l_i, \sigma_i, r_i) \cdot \varphi_m \cdot (l_j, \sigma_j, r_j) \cdot \varphi_s)$ by $(l_i \sqcap_\mathbb{C} l_{i,j})$. The introduced counters are $l_{i,j} = (l_j \ominus_\mathbb{C} (\mathbf{1}_{\sigma_i} \oplus_\mathbb{C} \dots \oplus_\mathbb{C} \mathbf{1}_{\sigma_{j-1}}))$, $r_{j,i} = (r_i \ominus_\mathbb{C} (\mathbf{1}_{\sigma_{i+1}} \oplus_\mathbb{C} \dots \oplus_\mathbb{C} \mathbf{1}_{\sigma_j}))$, $l'_{j,i} = (l_i \oplus_\mathbb{C} \mathbf{1}_{\sigma_i} \oplus_\mathbb{C} r_i) \ominus_\mathbb{C} (r_j \oplus_\mathbb{C} \mathbf{1}_{\sigma_j})$, and $r'_{i,j} = (r_j \oplus_\mathbb{C} \mathbf{1}_{\sigma_j} \oplus_\mathbb{C} l_j) \ominus_\mathbb{C} (l_i \oplus_\mathbb{C} \mathbf{1}_{\sigma_i})$.

$$\frac{\varphi_p \cdot (l_i, \sigma_i, r_i) \cdot \varphi_m \cdot (l_j, \sigma_j, r_j) \cdot \varphi_s}{\varphi_p \cdot (l_i \sqcap_\mathbb{C} l_{i,j}, \sigma_i, r_i) \cdot \varphi_m \cdot (l_j, \sigma_j, r_j) \cdot \varphi_s} \; l_{i<j} \qquad \frac{\varphi_p \cdot (l_i, \sigma_i, r_i) \cdot \varphi_m \cdot (l_j, \sigma_j, r_j) \cdot \varphi_s}{\varphi_p \cdot (l_i, \sigma_i, r_i) \cdot \varphi_m \cdot (l_j, \sigma_j, r_j \sqcap_\mathbb{C} r_{j,i}) \cdot \varphi_s} \; r_{i<j}$$

$$\frac{\varphi_p \cdot (l_i, \sigma_i, r_i) \cdot \varphi_m \cdot (l_j, \sigma_j, r_j) \cdot \varphi_s}{\varphi_p \cdot (l_i, \sigma_i, r_i) \cdot \varphi_m \cdot ((l_j \sqcap_\mathbb{C} l'_{j,i}), \sigma_j, r_j) \cdot \varphi_s} \; l'_{i \neq j} \qquad \frac{\varphi_p \cdot (l_i, \sigma_i, r_i) \cdot \varphi_m \cdot (l_j, \sigma_j, r_j) \cdot \varphi_s}{\varphi_p \cdot (l_i, \sigma_i, r_i \sqcap_\mathbb{C} r'_{i,j}) \cdot \varphi_m \cdot (l_j, \sigma_j, r_j) \cdot \varphi_s} \; r'_{i \neq j}$$

Proof. Sketch. Let φ' be the word obtained from φ by applying one of the above rules. Such a rule only strengthens the counters. Hence, $\llbracket \varphi \rrbracket \supseteq \llbracket \varphi' \rrbracket$. Assume w in Σ^* with $w \models^h \varphi$. We show $w \models^h \varphi'$ holds. We describe the cases $l_{i<j}$ and $r'_{i \neq j}$. We start with $l_{i<j}$ and show that $l_{i,j}(w_{[1,h(i)-1]})$. We know $l_j(w_{[1,h(j)-1]})$ from $w \models^h \varphi$. We also know $l_j(w_{[h(1)]} \cdot w_{[h(2)]} \cdots w_{[h(j-1)]})$ by well formedness of φ and $w \models^h \varphi$. Observe that due to the allowed predicates in the counters, if $cr[m]$ and $cr[m'']$ for some multisets $m \preceq m''$, then $cr[m']$ for any multiset $m \preceq m' \preceq m''$. Also, observe that: $(w_{[h(1)]} \cdot w_{[h(2)]} \cdots w_{[h(j-1)]})^\# \preceq (w_{[1,h(i)-1]} \cdot w_{[h(i)]} \cdot w_{[h(i+1)]} \cdots w_{[h(j-1)]})^\# \preceq (w_{[1,h(j)-1]})^\#$. We get $(l_j \ominus_\mathbb{C} (\mathbf{1}_{\sigma_i} \oplus_\mathbb{C} \dots \oplus_\mathbb{C} \mathbf{1}_{\sigma_{j-1}}))(w_{[1,h(i)-1]})$ and hence the result. For $r'_{i \neq j}$, we show $r'_{i,j}(w_{[h(i)+1,|w|]})$. Observe that $w \models^h \varphi$ ensures $l_i(w_{[1,h(i)-1]})$, $\mathbf{1}_{\sigma_i}(w_{[h(i)]})$, $l_j(w_{[1,h(j)-1]})$, $\mathbf{1}_{\sigma_j}(w_{[h(j)]})$ and $r_j(w_{[h(j)+1,|w|]})$. Hence, $(l_j \oplus_\mathbb{C} \mathbf{1}_{\sigma_j} \oplus_\mathbb{C} r_j)[w]$ and $(l_i \oplus_\mathbb{C} \mathbf{1}_{\sigma_i})[w_{[1,h(i)]}]$. The result follows from $w^\# = (w_{[1,h(i)]})^\# \oplus (w_{[h(i)+1,|w|]})^\#$. $\square$

Lemma 6 (Normalization). *Procedure Normalize repeatedly applies the rules of Table (2). The resulting counted word is independent of the application order.*

Proof. First termination, At each rule, manipulated and obtained counted words are well formed. Using Lemma (3), we deduce all counters belong to a finite lattice where rules are monotonic functions that only strengthen one counter.

Unicity can be obtained by contradiction. Suppose two different counted words are obtained as normalizations of the same well formed word. The words can only differ in their counters. Pick different corresponding counters. Given the allowed forms for the predicates (Lemmata (3) and (5)), we deduce that at least one predicate associated to some letter is strictly stronger in one of the counters. If we apply to the word with a weaker predicate, the sequence of rules that were applied to the word with a stronger predicate, we would get a strictly stronger predicate. This contradicts having reached a fixpoint. $\qquad\square$

Procedure Normalize($(l_1, \sigma_1, r_1) \cdots (l_n, \sigma_n, r_n)$)

---

1 **repeat**
2 $\quad (l'_1, \sigma'_1, r'_1) \cdots (l'_n, \sigma'_n, r'_n) \leftarrow (l_1, \sigma_1, r_1) \cdots (l_n, \sigma_n, r_n)$;
3 $\quad$ **for** $i \leftarrow 1$ **to** n **do**
4 $\qquad$ **for** $j \leftarrow 1$ **to** $i - 1$ **do**
5 $\qquad\quad r_i \leftarrow r_i \sqcap_{\mathbb{C}} \left(r_j \ominus_{\mathbb{C}} \left(\mathbf{1}_{\sigma_{j+1}} \oplus_{\mathbb{C}} \ldots \mathbf{1}_{\sigma_i} \right) \right)$;
6 $\qquad\quad l_i \leftarrow l_i \sqcap_{\mathbb{C}} \left((l_j \oplus_{\mathbb{C}} \mathbf{1}_{\sigma_j} \oplus_{\mathbb{C}} r_j) \ominus_{\mathbb{C}} (r_i \oplus_{\mathbb{C}} \mathbf{1}_{\sigma_i}) \right)$;
7 $\qquad\quad r_i \leftarrow r_i \sqcap_{\mathbb{C}} \left((l_j \oslash_{\mathbb{C}} \mathbf{1}_{o_j} \oplus_{\mathbb{C}} r_j) \ominus_{\mathbb{C}} (l_i \oplus_{\mathbb{C}} \mathbf{1}_{\sigma_i}) \right)$;
8 $\qquad$ **for** $j \leftarrow i + 1$ **to** n **do**
9 $\qquad\quad l_i \leftarrow l_i \sqcap_{\mathbb{C}} \left(l_j \ominus_{\mathbb{C}} \left(\mathbf{1}_{\sigma_{j+1}} \oplus_{\mathbb{C}} \ldots \mathbf{1}_{\sigma_i} \right) \right)$;
10 $\qquad\quad l_i \leftarrow l_i \sqcap_{\mathbb{C}} \left((l_j \oplus_{\mathbb{C}} \mathbf{1}_{\sigma_j} \oplus_{\mathbb{C}} r_j) \ominus_{\mathbb{C}} (r_i \oplus_{\mathbb{C}} \mathbf{1}_{\sigma_i}) \right)$;
11 $\qquad\quad r_i \leftarrow r_i \sqcap_{\mathbb{C}} \left((l_j \oplus_{\mathbb{C}} \mathbf{1}_{\sigma_j} \oplus_{\mathbb{C}} r_j) \ominus_{\mathbb{C}} (l_i \oplus_{\mathbb{C}} \mathbf{1}_{\sigma_i}) \right)$;
12 **until** $(l_1, \sigma_1, r_1) \cdots (l_n, \sigma_n, r_n) \sqsubseteq_{\mathrm{NCW}} (l'_1, \sigma'_1, r'_1) \cdots (l'_n, \sigma'_n, r'_n)$;
13 **return** $(l_1, \sigma_1, r_1) \cdots (l_n, \sigma_n, r_n)$;

Normalized words and entailment. We write $\mathbb{NCW}$ to mean the set of normalized words in $\mathbb{CW}$. Assume two normalized words $\varphi = (l_1, \sigma_1, r_1) \cdots (l_n, \sigma_n, r_n)$ and $\varphi' = (l'_1, \sigma'_1, r'_1) \cdots (l'_m, \sigma'_m, r'_m)$. We say that φ is h-entailed by φ' for some increasing injection $h : \overline{n} \to \overline{m}$, and write $\varphi \sqsubseteq^h_{\mathrm{NCW}} \varphi'$, to mean that for each $i \in \overline{n}$, $\overline{\varphi}_{[i]} = \overline{\varphi'}_{[h(i)]}$, $l_i \sqsubseteq_{\mathbb{C}} l'_{h(i)}$, and $r_i \sqsubseteq_{\mathbb{C}} r'_{h(i)}$. We write $\varphi \sqsubseteq_{\mathrm{NCW}} \varphi'$ if $\varphi \sqsubseteq^h_{\mathrm{NCW}} \varphi'$ for some h. Observe that $([v_a \geq 0], a, [v_a = 0]) \not\sqsubseteq_{\mathrm{NCW}} ([v_a = 0], a, [v_a \geq 0])$, but $[\![([v_a \geq 0], a, [v_a = 0])]\!] = [\![([v_a = 0], a, [v_a \geq 0])]\!] = a^+$.

Lemma 7 (Entailment). $\sqsubseteq_{\mathrm{NCW}}$ *is reflexive and transitive. It can be checked in linear time in the length of the counted words and* $\varphi \sqsubseteq_{\mathrm{NCW}} \varphi'$ *implies* $[\![\varphi']\!] \subseteq [\![\varphi]\!]$.

Word cut-offs. Similarly to the cut-offs defined in Section (2) for counters, the cut-off of a well formed word φ is a multiset $\kappa(\varphi)$. It associates to each letter σ the natural number $max \{\kappa(cr)(\sigma) | cr$ is a counter in $\varphi\}$. In Example (4), $\kappa(\varphi)(a) = 2$ and $\kappa(\varphi)(b) = 1$. We say that a counted word φ has a k-cut-off if all its counters are in $\mathbb{C}_k$. For example, counted words with a 0-cut-off only have inequalities in their counters (i.e. denote upward closed sets). We write $\mathbb{CW}_k$ ($\mathbb{NCW}_k$) to mean the set of (normalized) well formed counted words that have a k-cut-off.

Theorem 8 (WQO). *For* $k \in \mathbb{N}$, $(\mathbb{NCW}_k, \sqsubseteq_{\mathrm{NCW}})$ *is a well quasi ordering.*

Proof. Higman's Lemma [12] states that if $(\Sigma, \preceq)$ is a wqo, then the pair $(\Sigma^*, \preceq^*)$ is also a wqo[3]. We let $\Gamma = \mathbb{C}_k \times \Sigma \times \mathbb{C}_k$ and $(l, \sigma, r) \preceq (l', \sigma', r')$ if $l \sqsubseteq_{\mathbb{C}} l'$ and $\sigma =$

---

[3] $\sigma_1 \cdots \sigma_n \preceq^* \sigma'_1 \cdots \sigma'_m$ iff there is a strictly increasing $h : \overline{n} \to \overline{m}$ with $\sigma_i \preceq \sigma'_{h(i)}$.

σ' and $r \sqsubseteq_C r'$. Observe that $\mathrm{NCW}_k \subseteq \Gamma^*$, and that $\preceq^*$ coincides with $\sqsubseteq_{\mathrm{NCW}}$. Hence, showing $(\Gamma, \preceq)$ is a wqo establishes the result. Given an infinite sequence we can extract an infinite subsequence $(l_{m_1}, \sigma_{m_1}, r_{m_1}), (l_{m_2}, \sigma_{m_2}, r_{m_2}), \ldots$ in which $\sigma_{m_i} = \sigma_{m_j}$ for all $i \neq j$ and use Lemma (2). □

Procedure $\mathrm{zip}(z, (p{:}s), (p'{:}s'))$

---

1 collect $:= \emptyset$;
2 **if** $(s \neq \epsilon)$ **then**
3 $\quad$ | **if** $\kappa(\overrightarrow{p'})(\overline{hd(s)}) = 0$ and $\kappa(\overleftarrow{s'})(\overline{hd(s)}) = 0$ **then**
4 $\quad$ | $\quad$ | collect $\cup := \mathrm{zip}(z \cdot hd(s), (p \cdot hd(s) : tl(s)), (p' : s'))$
5 **if** $(s \neq \epsilon$ and $s' \neq \epsilon)$ **then**
6 $\quad$ | **if** $(\overleftarrow{hd(s)} \sqcap_C hd(s')) \neq \mathtt{false})$ and $(\overline{hd(s)} = \overline{hd(s')})$ and
   $\quad$ $(hd(s) \sqcap_C \overrightarrow{hd(s')}) \neq \mathtt{false})$ **then**
7 $\quad$ | $\quad$ | $e := (\overleftarrow{hd(s)} \sqcap_C hd(s'), \overrightarrow{hd(s)}, hd(s) \sqcap_C \overrightarrow{hd(s')})$;
8 $\quad$ | $\quad$ | collect $\cup := \mathrm{zip}(z \cdot e, (p \cdot hd(s) : tl(s)), (p' \cdot hd(s') : tl(s')))$
9 **if** $(s' \neq \epsilon)$ **then**
10 $\quad$ | **if** $\kappa(\overrightarrow{p})(\overline{hd(s')}) = 0$ and $\kappa(\overleftarrow{s})(\overline{hd(s')}) = 0$ **then**
11 $\quad$ | $\quad$ | collect $\cup := \mathrm{zip}(z \cdot hd(s'), (p : s), (p' \cdot hd(s') : tl(s')))$
12 **if** $(s = \epsilon$ and $s' = \epsilon)$ **then**
13 $\quad$ | collect $:= \{\mathrm{Normalize}(z)\}$
14 **return** collect;

Meet of counted words. Given φ, φ' in NCW, the result of Procedure (zip) is a set $(\varphi \sqcap_{\mathrm{NCW}} \varphi')$ of normalized counted words whose denotation coincides with $\llbracket \varphi \rrbracket \cap \llbracket \varphi' \rrbracket$. This recursive procedure builds a constrained shuffle of φ and φ'. It takes as arguments five counted words z, p, s, p', s', with $\varphi = (p \cdot s)$ and $\varphi' = (p' \cdot s')$. We write $(z, (p : s), (p' : s'))$ for clarity. Intuitively, each call tries to complete the first argument z to obtain a counted word that entails both $(p \cdot s)$ and $(p' \cdot s')$. The procedure starts with $(\epsilon, (\epsilon : \varphi), (\epsilon : \varphi'))$ and collects all such counted words z. At each call, it considers contributions to z from $hd(s)$ (lines (2-4)), $hd(s')$ (lines (9-11)), or both $hd(s)$ and $hd(s')$ (lines (5-8)). The contributions are completed by further recursive calls. Results are collected in the local variable collect. Lines (2-4) capture the situation where a state in z is mapped to $hd(s)$ and tolerated by the counters of φ' (test at line (3)). Lines (5-8) correspond to a state in z simultaneously mapped to $hd(s)$ and $hd(s')$. The words s and s' contain states that are still not treated. Termination is obtained with the ranking function $|s| + |s'|$. The following lemma establishes correctness.

Lemma 9 (intersection). *For $\varphi, \varphi' \in \mathrm{NCW}$, $\mathrm{zip}(\epsilon, (\epsilon : \varphi), (\epsilon : \varphi'))$ returns a finite set $\{\varphi_1, \ldots \varphi_n\}$ such that $\cup_{i \in \bar{n}} \llbracket \varphi_i \rrbracket = \llbracket \varphi \rrbracket \cap \llbracket \varphi' \rrbracket$.*

Relaxation. Relaxing a counter $cr = [\wedge_{\sigma \text{ in } \Sigma}(v_\sigma \sim k)]$, wrt. a multiset ρ, written $\nabla_\rho(cr)$, results in the counter $[\wedge_{\sigma \text{ in } \Sigma}(v_\sigma \sim' k)]$ s.t. $(v_\sigma \sim' k)$ equals $(v_\sigma \geq k)$ if $(v_\sigma \sim k)$ was $(v_\sigma = k)$ in cr with $k \geq \rho(\sigma)$, and $(v_\sigma \sim k)$ otherwise. In other words, relaxation wrt. ρ replaces by inequalities those equalities that involve constants larger or equal to what is allowed by ρ. Relaxation of a counted word φ wrt. a multiset ρ is simply the word $\nabla_\rho(\varphi)$ obtained by normalizing the result of relaxing all counters in φ wrt. ρ. We let ∇_{NCW} be the set $\{\nabla_\rho|\ \rho$ is a multiset over $\Sigma\}$.

Lemma 10 (Relaxation). $\nabla_\rho(\varphi) \sqsubseteq_{\mathrm{NCW}} \varphi$ *for any* $\varphi \in \mathrm{NCW}$ *and multiset* ρ. *In addition,* $\kappa(\nabla_\rho(\varphi))(\sigma) \leq max(0, 2\rho(\sigma) - 1)$ *for each* $\sigma \in \Sigma$.

Proof. Sketch. Suppose $\nabla_\rho(\varphi) \not\sqsubseteq_{\mathrm{NCW}} \varphi$, then there is a counter cr_φ in φ that does not entail a corresponding counter cr_∇ in $\nabla_\rho(\varphi)$. This is not possible. Indeed, before normalization, $\nabla_\rho(\varphi)$ and φ are both well formed with the same base and normalization in $\nabla_\rho(\varphi)$ starts with weaker counters than those in φ . By applying to φ the sequence of normalization rules used to normalize $\nabla_\rho(\varphi)$, we obtain (by monotonicity) that the counters in $\nabla_\rho(\varphi)$ are weaker than those in φ. The strongest cut-off $(2\rho(\sigma) - 1)$ is obtained when both left and right counters in some tuple (l, σ, r) associate the predicate $v_\sigma = (\rho(\sigma) - 1)$ to the letter σ. One can show by induction on the number of applications of the normalization rules, that for any letter σ', $\kappa(l \oplus_C \mathbf{1}_\sigma \oplus_C r)(\sigma') < max(0, 2\rho(\sigma') - 1)$. □

4   Reachability for Linear Parameterized Systems

Linear Parameterized Systems with Global Conditions. Such a system consists of arbitrary many finite processes placed in an array. Formally, a *linear parameterized system* is a pair $\mathcal{P} = (Q, T)$, where Q is a finite set of *local states* and T is a finite set of *transitions*. A transition is either *local* or *global*. A local transition is of the form $q \to q'$. It allows a process to change its local state from q to q' independently of the local states of the other processes. A global transition is of the form $q \to q' : \mathbb{Q}P$, where $\mathbb{Q} \in \{\exists_L, \exists_R, \exists_{LR}, \forall_L, \forall_R, \forall_{LR}\}$ and $P \subset Q$. For instance, the condition $\forall_L P$ means that "all processes to the left should be in local states that belong to P". This work is well suited for extensions involving binary or broadcast communication, shared variables or dynamic creation and deletion of processes. We omit them for clarity. A parameterized system (Q, T) induces an infinite-state transition system where $C = Q^*$ is the set of *configurations* and $\longrightarrow$ is a transition relation on C. For configurations $c = c_1qc_2$, $c' = c_1q'c_2$, and a transition $t \in T$, we write $c \longrightarrow_t c'$ to mean:

- t is a local transition of the form $q \to q'$, or
- t is a global transition $q \to q' : \mathbb{Q}P$, and one of the following holds:
  - either $\mathbb{Q}P = \exists_L P$ and $c_1{}^\bullet \cap P \neq \emptyset$, or $\mathbb{Q}P = \exists_R P$ and $c_2{}^\bullet \cap P \neq \emptyset$, or $\mathbb{Q}P = \exists_{LR}P$ and $(c_1{}^\bullet \cup c_2{}^\bullet) \cap P \neq \emptyset$.
  - or $\mathbb{Q}P = \forall_L P$ and $c_1{}^\bullet \subseteq P$, or $\mathbb{Q}P = \forall_R P$ and $c_2{}^\bullet \subseteq P$, or $\mathbb{Q}P = \forall_{LR}P$ and $(c_1{}^\bullet \cup c_2{}^\bullet) \subseteq P$.

We write $\longrightarrow$ to mean $\cup_{t \in T} \longrightarrow_t$ and use $\stackrel{*}{\longrightarrow}$ to denote its reflexive transitive closure. We assume all processes have the same *initial* state. We use *Init* to denote the set of *initial* configurations. *Init* is infinite. Using standard techniques (see e.g. [18]), checking safety properties (expressed as regular languages) can be translated into instances of the following reachability problem: given $\mathcal{P} = (Q, T)$ and a possibly infnite set C_F of configurations, check whether $Init \stackrel{*}{\longrightarrow} C_F$.

The reachability scheme. We use NCW over Q as a symbolic representation for configurations of (Q, T). We require *Init* and C_F to be captured using a (finite set of) counted words. We repeatedly compute (lemma (11)) in forward (resp. backward) the set of successor (resp. predecessor) configurations starting from *Init* (resp. C_F). We use lemma (9) to check intersection with C_F (resp. *Init*). We use $\sqsubseteq_{\mathrm{NCW}}$ (lemma (7)) to maintain a set of pairwise unrelated elements capturing configurations that are forward (resp. backward) reachable from *Init* (resp. C_F). For termination, we systematically apply some relaxation ∇_ρ that imposes bounded cut-offs (lemma (10) and theorem (8)). We start with $\rho = \mathbf{0}$ and increase it in case the over-approximation induced by ∇_ρ results in a spurious trace. Strengthening the cut-off ρ results in a more precise (and hence more expensive) analysis. We use the following heuristic to eliminate encountered spurious traces without making the analysis unecessary expensive. We follow without relaxation, the trace obtained using ∇_ρ and identify the letters in Q for which the relaxation in ρ is responsible for generating the supious trace. We then only increase the cut-offs for those letters.

Lemma 11 (Post and Pre). *Given $\varphi \in$ NCW and a transition t, we can compute two sets of counted words $post_t(\varphi)$ and $pre_t(\varphi)$ s.t. $\cup_{\varphi' \in post_t(\varphi)} \llbracket \varphi' \rrbracket$ equals $\{c' \mid c \longrightarrow_t c'$ and $c \in \llbracket \varphi \rrbracket\}$ and $\cup_{\varphi' \in pre_t(\varphi)} \llbracket \varphi' \rrbracket$ equals $\{c' \mid c' \longrightarrow_t c$ and $c \in \llbracket \varphi \rrbracket\}$.*

Example 12. $post_t(\varphi)$ for $t = (a \to b : \exists_R \{a\})$ and $\varphi = \left(\left[\begin{smallmatrix} v_a \geq 0 \\ \wedge v_b = 0 \end{smallmatrix} \right], a, \left[\begin{smallmatrix} v_a \geq 0 \\ \wedge v_b = 0 \end{smallmatrix} \right] \right)$ is
$$\left\{ \left(\left[\begin{smallmatrix} v_a \geq 0 \\ \wedge v_b = 0 \end{smallmatrix} \right], b, \left[\begin{smallmatrix} v_a \geq 1 \\ \wedge v_b = 0 \end{smallmatrix} \right] \right) \left(\left[\begin{smallmatrix} v_a \geq 0 \\ \wedge v_b = 1 \end{smallmatrix} \right], a, \left[\begin{smallmatrix} v_a \geq 0 \\ \wedge v_b = 0 \end{smallmatrix} \right] \right) \right\}$$

Table 3. NCW based forward analysis of mutex algorithms

	Forward expoloration					Backwards exploration				
	refinements	time	steps	words	safe	refinements	time	steps	words	safe
I	3	0.11	17	875	$\checkmark$	1	0.02	2	151	$\checkmark$
II	7	5.85	171	5143	$\checkmark$	1	0.18	19	3026	$\checkmark$
III	10	>1200	>2000	>68000	$\times$	3	158.3	1567	194425	$\checkmark$
IV	11	>1200	>2800	>120000	$\times$	2	138.1	932	233604	$\checkmark$

Experimental Results. We have implemented the introduced scheme in Ocaml and run experiments on an Intel Core 2 Duo 2.26 GHz laptop with 4GB of memory. We have considered four classical mutex algorithms: Burns [2], compact [5] and refined [14] versions of Szymanski's algorithm, and the related Gribomont-Zenner mutex [11] (respectively rows I,II,III and IV in Table (3)). Our prototype takes as input descriptions for the systems introduced at the beginning of this section. We give running times (seconds), number of refinements, maximum numbers (per refinement) of steps and of generated counted words. We write "$\checkmark$" to mean unreachability is established. We allocate a budget of 20 minutes per refinement and write $\times$ if the budget is exhausted. Unlike forward analysis, backwards analysis profits from the fact that C_F is typically upward closed.

5 Conclusions

We have introduced a new symbolic representation for the verification of parameterized systems where processes are organized in a linear array. The new representation combines counter abstraction together with upward closure based techniques. It allows for an approximated analysis with a threshold-based precision (or relaxation) that can be uniformly tuned. Based on the representation, we implemented a counter example based refinement scheme that illustrated the applicability and the relevance of the approach, both for forward and for backward analysis. Possible futur work can investigate more general representations to apply to heap or graph manipulating programs.

References

1. Abdulla, P.A., Delzanno, G., Rezine, A.: Approximated context-sensitive analysis for parameterized verification. In: Lee, D., Lopes, A., Poetzsch-Heffter, A. (eds.) FMOODS 2009. LNCS, vol. 5522, pp. 41–56. Springer, Heidelberg (2009)
2. Abdulla, P.A., Delzanno, G., Ben Henda, N., Rezine, A.: Regular model checking without transducers (On efficient verification of parameterized systems). In: Grumberg, O., Huth, M. (eds.) TACAS 2007. LNCS, vol. 4424, pp. 721–736. Springer, Heidelberg (2007)
3. Abdulla, P.A., Jonsson, B., Nilsson, M., d'Orso, J.: Regular model checking made simple and efficient. In: Brim, L., Jančar, P., Křetínský, M., Kučera, A. (eds.) CONCUR 2002. LNCS, vol. 2421, pp. 116–130. Springer, Heidelberg (2002)
4. Apt, K., Kozen, D.: Limits for automatic verification of finite-state concurrent systems. Information Processing Letters 22, 307–309 (1986)
5. Arons, T., Pnueli, A., Ruah, S., Xu, J., Zuck, L.: Parameterized verification with automatically computed inductive assertions. In: Berry, G., Comon, H., Finkel, A. (eds.) CAV 2001. LNCS, vol. 2102, pp. 221–234. Springer, Heidelberg (2001)
6. Boigelot, B., Legay, A., Wolper, P.: Iterating transducers in the large. In: Hunt Jr., W.A., Somenzi, F. (eds.) CAV 2003. LNCS, vol. 2725, pp. 223–235. Springer, Heidelberg (2003)
7. Bouajjani, A., Habermehl, P., Vojnar, T.: Abstract regular model checking. In: Alur, R., Peled, D.A. (eds.) CAV 2004. LNCS, vol. 3114, pp. 372–386. Springer, Heidelberg (2004)
8. Clarke, E., Talupur, M., Veith, H.: Environment abstraction for parameterized verification. In: Emerson, E.A., Namjoshi, K.S. (eds.) VMCAI 2006. LNCS, vol. 3855, pp. 126–141. Springer, Heidelberg (2006)
9. Geeraerts, G., Raskin, J.F., Van Begin, L.: Expand, Enlarge and Check: new algorithms for the coverability problem of WSTS. Journal of Computer and System Sciences 72(1), 180–203 (2006)
10. German, S.M., Sistla, A.P.: Reasoning about systems with many processes. Journal of the ACM 39(3), 675–735 (1992)
11. Gribomont, E.P., Zenner, G.: Automated verification of szymanski's algorithm. In: Steffen, B. (ed.) TACAS 1998. LNCS, vol. 1384, pp. 424–438. Springer, Heidelberg (1998)
12. Higman, G.: Ordering by divisibility in abstract algebras. Proc. London Mathematical Society (3) 2(7), 326–336 (1952)

13. Kesten, Y., Maler, O., Marcus, M., Pnueli, A., Shahar, E.: Symbolic model checking with rich assertional languages. In: Grumberg, O. (ed.) CAV 1997. LNCS, vol. 1254, pp. 424–435. Springer, Heidelberg (1997)
14. Manna, Z., Pnueli, A.: An exercise in the verification of multi – process programs. In: Feijen, W., van Gasteren, A., Gries, D., Misra, J. (eds.) Beauty is Our Business, pp. 289–301. Springer (1990)
15. Pnueli, A., Ruah, S., Zuck, L.: Automatic deductive verification with invisible invariants. In: Margaria, T., Yi, W. (eds.) TACAS 2001. LNCS, vol. 2031, pp. 82–97. Springer, Heidelberg (2001)
16. Pnueli, A., Xu, J., Zuck, L.D.: Liveness with $(0, 1, \infty)$-counter abstraction. In: Brinksma, E., Larsen, K.G. (eds.) CAV 2002. LNCS, vol. 2404, p. 107. Springer, Heidelberg (2002)
17. Touili, T.: Regular Model Checking using Widening Techniques. Electronic Notes in Theoretical Computer Science 50(4) (2001), Proc. Workshop on Verification of Parametrized Systems (VEPAS 2001), Crete (July 2001)
18. Vardi, M.Y., Wolper, P.: An automata-theoretic approach to automatic program verification. In: Proc. LICS 1986, pp. 332–344 (June 1986)

On SAT Representations of XOR Constraints

Matthew Gwynne and Oliver Kullmann

Computer Science Department, Swansea University, UK
http://cs.swan.ac.uk/~csmg/
http://cs.swan.ac.uk/~csoliver/

Abstract. We consider the problem of finding good representations, via boolean conjunctive normal forms F (clause-sets), of systems S of XOR-constraints $x_1 \oplus \cdots \oplus x_n = \varepsilon$, $\varepsilon \in \{0, 1\}$ (also called parity constraints), i.e., systems of linear equations over the two-element field. These representations are to be used as parts of SAT problems $F^* \supset F$, such that F^* has "good" properties for SAT solving. The basic quality criterion is "arc consistency", that is, for every partial assignment φ to the variables of S, all assignments $x_i = \varepsilon$ forced by φ are determined by unit-clause propagation on the result $\varphi * F$ of the application. We show there is no arc-consistent representation of polynomial size for arbitrary S. The proof combines the basic method by Bessiere et al. 2009 ([2]) on the relation between monotone circuits and "consistency checkers", adapted and simplified in the underlying report Gwynne et al. [10], with the lower bound on monotone circuits for monotone span programs in Babai et al. 1999 [1]. On the other side, our basic positive result is that computing an arc-consistent representation is fixed-parameter tractable in the number m of equations of S. To obtain stronger representations, instead of mere arc-consistency we consider the class $\mathcal{PC}$ of propagation-complete clause-sets, as introduced in Bordeaux et al. 2012 [4]. The stronger criterion is now $F \in \mathcal{PC}$, which requires for *all* partial assignments, possibly involving also the auxiliary (new) variables in F, that forced assignments can be determined by unit-clause propagation. We analyse the basic translation, which for $m = 1$ lies in $\mathcal{PC}$, but fails badly so already for $m = 2$, and we show how to repair this.

Keywords: arc consistency, parity constraints, monotone circuits, monotone span programs, unit-propagation complete, acyclic incidence graph.

1 Introduction

Recall that the two-element field $\mathbb{Z}_2$ has elements $0, 1$, where addition is XOR, which we write as $\oplus$, while multiplication is AND, written $\cdot$. A linear system S of equations over $\mathbb{Z}_2$, in matrix form $A \cdot x = b$, where A is an $m \times n$ matrix over $\{0, 1\}$, with m the number of equations, n the number of variables, while $b \in \{0, 1\}^m$, yields a boolean function f_S, which assigns 1 to a total assignment of the n variables of S iff that assignment is a solution of S. The task of finding "good" representations of f_S by conjunctive normal forms F (clause-sets, to be

A.-H. Dediu et al. (Eds.): LATA 2014, LNCS 8370, pp. 409–420, 2014.

precise), for the purpose of SAT solving, shows up in many applications, for example cryptanalysing the Data Encryption Standard and the MD5 hashing algorithm in [5], translating Pseudo-Boolean constraints to SAT in [6], and in roughly 1 in 6 benchmarks from SAT 2005 to 2011 according to [19].

The basic criterion for a good F is "arc-consistency". See Chapter 3 of [24] for an overview of "arc-consistency" at the constraint level, see [7] for discussion of the support encoding, a SAT translation of explicitly-given constraints which maintains arc-consistency, and see [6] for an overview of maintaining arc-consistency when translating Pseudo-boolean constraints to SAT. To define arc-consistency for SAT, we use r_1 for unit-clause propagation, and we write $\varphi * F$ for the application of a partial assignment φ to a clause-set F.[1] For a boolean function f, a CNF-representation F of f is arc-consistent iff for every partial assignment φ to the variables of f the reduced instantiation $F' := r_1(\varphi * F)$ has no forced assignments anymore, that is, for every remaining variable v and $\varepsilon \in \{0,1\}$ the result $\langle v \to \varepsilon \rangle * F'$ of assigning ε to v in F' is satisfiable.

We show that there is no polynomial-size arc-consistent representation of arbitrary S (Theorem 7). The proof combines the translation of arc-consistent CNF-representations of f into monotone circuits computing a monotonisation $\widehat{f}$, motivated by [2] and proven in the underlying report [10], with the lower bound on monotone circuit sizes for monotone span programs (msp's) from [1]. Besides this fundamental negative result, we provide various forms of good representations of systems S with bounded number of equations. Theorem 12 shows that there is an arc-consistent representation with $O(n \cdot 2^m)$ many clauses. The remaining results use a stronger criterion for a "good" representation, namely they demand that $F \in \mathcal{PC}$, where $\mathcal{PC}$ is the class of "unit-propagation complete clause-sets" as introduced in [4] — while for arc-consistency only partial assignments to the variables of f are considered, now partial assignments for all variables in F (which contains the variables of f, and possibly further auxiliary variables) are to be considered. For $m = 1$ the obvious translation X_1, by subdividing the constraints into small constraints, is in $\mathcal{PC}$ (Lemma 10). For $m = 2$ we have an intelligent representation X_2 in $\mathcal{PC}$ (Theorem 13), while the use of X_1 (piecewise) is still feasible for full (dag-)resolution, but not for tree-resolution. We conjecture (Conjecture 14) that Theorem 12 and Theorem 13 can be combined, which would yield a fixed-parameter tractable algorithm for computing a representation $F \in \mathcal{PC}$ for arbitrary S with the parameter m.

It is well-known that translating each XOR to its prime implicates can result in hard (unsatisfiable) instances for resolution. This goes back to the "Tseitin formulas" introduced in [26], which were proven hard for full resolution in [27], and generalised to (empirically) hard satisfiable instances in [12]. Thus, to tackle XOR-constraints, some solvers integrate XOR reasoning. EqSatz ([23]) extracts XOR clauses from its input and applies DP-resolution plus incomplete XOR reasoning. CryptoMiniSAT ([25]) integrates Gaußian elimination during

---

[1] r_1 has been generalised to r_k for $k \in \mathbb{N}_0$ in [14,15]. In the underlying report [10] we discuss this form of generalised unit-clause propagation, where for example r_2 is failed-literal elimination, but in this paper we concentrate on r_1.

search, allowing both explicitly specified XOR clauses and also XOR clauses extracted from CNF input, however in the newest version 3.3 the XOR handling during search is removed, since it is deemed too expensive. [17] integrates XOR reasoning into MiniSat in a similar manner to SMT, while [18] expands on this by reasoning about equivalence classes of literals created by binary XORs. [20] learns conflicts in terms of "parity (XOR) explanations". [21] extends the reasoning from "Gaußian elimination" to "Gauß-Jordan elimination", which also detects forced literals, not just inconsistencies. Still, for leading current SAT solvers usage of SAT translations is important. Considering such translations of XORs to CNF, [19] identifies the subsets of "tree-like" systems of XOR constraints, where one obtains an arc-consistent CNF representation; our results on acyclic systems strengthens this. Additionally they consider equivalence reasoning, where for "cycle-partitionable" systems of XOR constraints this reasoning is sufficient to derive all conclusions. They also show how to eliminate the need for such special reasoning by another arc-consistent CNF representation. In general, the idea is to only use Gaußian elimination for such parts of XOR systems which the SAT solver is otherwise incapable of propagating on. Existing propagation mechanisms, especially unit-clause propagation, and to a lesser degree equivalence reasoning, are very fast, while Gaußian elimination is much slower. Experimental evaluation on SAT 2005 benchmarks instances showed that such CNF translations can outperform dedicated XOR reasoning modules.

Viewing a linear system S as a constraint on $\mathrm{var}(S)$, one can encode evaluation via Tscitin's translation, obtaining a CNF-representation F with the property that for every *total* assignment φ, i.e., $\mathrm{var}(\varphi) = \mathrm{var}(S)$, we have that $r_1(\varphi * F)$ either contains the empty clause or is empty.[2] However, as Theorem 7 shows, there is no polysize representation which treats all *partial* assignments. Gaußian elimination handles all partial assignments in polynomial time (detects unsatisfiability of $\varphi * F$ for all partial assignments φ), but this can not be integrated into the CNF formalism (by using auxiliary variables and clauses), since algorithms always need total assignments, and so partial assignments φ would need to be encoded — the information "variable v not assigned" (i.e., $v \notin \mathrm{var}(\varphi)$) needs to be represented by *setting* some auxiliary variable, and this must happen by a mechanism outside of the CNF formalism. It is an essential strength of the CNF formalism to allow partial instantiation; if we want these partial instantiations also to be easily understandable by a SAT solver, then the results of [2] and our results show that there are restrictions. Yet there is little understanding of these restrictions. There are many examples where arc-consistent and stronger representations are possible, while the current non-representability results, one in [2], one in this article and a variation on [2] in [16], rely on non-trivial lower bounds on monotone circuit complexity; in fact, as we show in [10], there is a polysize arc-consistent representation of a boolean function f iff the monotonisation $\widehat{f}$, encoding partial assignments to f, has polysize monotone circuits.

---

[2] In Subsection 9.4.1 of [11] this class of representations is called $\exists \mathcal{UP}$; up to linear-time transformation it is the same as representations by boolean circuits.

2   Preliminaries

We follow the general notations and definitions as outlined in [13]; for full details
see the underlying report [10] (which contains additional and generalised results).
We use $\mathbb{N} = \{1, 2, \ldots\}$ and $\mathbb{N}_0 = \mathbb{N} \cup \{0\}$. Let $\mathcal{VA}$ be the infinite set of variables,
and let $\mathcal{LIT} = \mathcal{VA} \cup \{\overline{v} : v \in \mathcal{VA}\}$ be the set of literals, the disjoint union
of variables as positive literals and complemented variables as negative literals.
We use $\overline{L} := \{\overline{x} : x \in L\}$ to complement a set L of literals. A clause is a
finite subset $C \subset \mathcal{LIT}$ which is complement-free, i.e., $C \cap \overline{C} = \emptyset$; the set of
all clauses is denoted by $\mathcal{CL}$. A clause-set is a finite set of clauses, the set of all
clause-sets is $\mathcal{CLS}$. By $\mathrm{var}(x) \in \mathcal{VA}$ we denote the underlying variable of a literal
$x \in \mathcal{LIT}$, and we extend this via $\mathrm{var}(C) := \{\mathrm{var}(x) : x \in C\} \subset \mathcal{VA}$ for clauses
C, and via $\mathrm{var}(F) := \bigcup_{C \in F} \mathrm{var}(C)$ for clause-sets F. The possible literals in a
clause-set F are denoted by $\mathrm{lit}(F) := \mathrm{var}(F) \cup \overline{\mathrm{var}(F)}$. Measuring clause-sets
happens by $n(F) := |\mathrm{var}(F)|$ for the number of variables, $c(F) := |F|$ for the
number of clauses, and $\ell(F) := \sum_{C \in F} |C|$ for the number of literal occurrences.
A special clause-set is $\top := \emptyset \in \mathcal{CLS}$, the empty clause-set, and a special clause is
$\bot := \emptyset \in \mathcal{CL}$, the empty clause. A partial assignment is a map $\varphi : V \to \{0, 1\}$ for
some finite $V \subset \mathcal{VA}$, where we set $\mathrm{var}(\varphi) := V$, and where the set of all partial
assignments is $\mathcal{PASS}$. For $v \in \mathrm{var}(\varphi)$ let $\varphi(\overline{v}) := \overline{\varphi(v)}$ (with $\overline{0} = 1$ and $\overline{1} = 0$).
We construct partial assignments by terms $\langle x_1 \to \varepsilon_1, \ldots, x_n \to \varepsilon_n \rangle \in \mathcal{PASS}$
for literals $x_1, \ldots, x_n$ with different underlying variables and $\varepsilon_i \in \{0, 1\}$. For
$\varphi \in \mathcal{PASS}$ and $F \in \mathcal{CLS}$ we denote the result of applying φ to F by $\varphi * F$,
removing clauses $C \in F$ containing $x \in C$ with $\varphi(x) = 1$, and removing literals
x with $\varphi(x) = 0$ from the remaining clauses. By $\mathcal{SAT} := \{F \in \mathcal{CLS} \mid \exists \varphi \in$
$\mathcal{PASS} : \varphi * F = \top\}$ the set of satisfiable clause-sets is denoted, and by $\mathcal{USAT} :=$
$\mathcal{CLS} \setminus \mathcal{SAT}$ the set of unsatisfiable clause-sets. By $r_1 : \mathcal{CLS} \to \mathcal{CLS}$ we denote
unit-clause propagation, that is $r_1(F) := \{\bot\}$ if $\bot \in F$, $r_1(F) := F$ if F contains
only clauses of length at least 2, while otherwise a unit-clause $\{x\} \in F$ is chosen,
and recursively we define $r_1(F) := r_1(\langle x \to 1 \rangle * F)$; it is easy to see that the
final result $r_1(F)$ does not depend on the choice of unit-clauses. Reduction by r_1
applies certain **forced assignments** to the (current) F, which are assignments
$\langle x \to 1 \rangle$ such that the opposite assignment yields an unsatisfiable clause-set,
that is, where $\langle x \to 0 \rangle * F \in \mathcal{USAT}$; the literal x here is also called a **forced
literal**. Two clauses $C, D \in \mathcal{CL}$ are resolvable iff they clash in exactly one
literal x, that is, $C \cap \overline{D} = \{x\}$, in which case their resolvent is $(C \cup D) \setminus \{x, \overline{x}\}$
(with resolution literal x). A resolution tree is a full binary tree formed by the
resolution operation. We write $T : F \vdash C$ if T is a resolution tree with axioms
(the clauses at the leaves) all in F and with derived clause (at the root) C. A
prime implicate of $F \in \mathcal{CLS}$ is a clause C such that a resolution tree T with
$T : F \vdash C$ exists, but no T' exists for some $C' \subset C$ with $T' : F \vdash C'$; the
set of all prime implicates of F is denoted by $\mathbf{prc}_0(F) \in \mathcal{CLS}$. Two clause-sets
$F, F' \in \mathcal{CLS}$ are equivalent iff $\mathrm{prc}_0(F) = \mathrm{prc}_0(F')$. A clause-set F is unsatisfiable
iff $\mathrm{prc}_0(F) = \{\bot\}$. If F is unsatisfiable, then every literal $x \in \mathcal{LIT}$ is a forced
literal for F, while otherwise x is forced for F iff $\{x\} \in \mathrm{prc}_0(F)$.

3 Propagation-Hardness and $\mathcal{PC}$

A clause-set F is a "CNF-representation" of a boolean function f, if the satisfying assignments of F projected to the variables of f are precisely the satisfying assignments of f. Stronger, F is an "arc-consistent" representation of f, if F is a CNF-representation of f and $\mathrm{phd}^{\mathrm{var}(f)}(F) \leq 1$ holds, which is defined as follows.

Definition 1. *For $F \in \mathcal{CLS}$ and $V \subseteq \mathcal{VA}$ the relation $\mathbf{phd}^V(F) \leq 1$ holds (F has p(ropagation)-hardness at most 1 relative to V) if for all partial assignments $\varphi \in \mathcal{PASS}$ with $\mathrm{var}(\varphi) \subseteq V$ the clause-set $F' := \mathrm{r}_1(\varphi * F)$ has no forced literals $x \in \mathrm{lit}(F')$, that is, for all $x \in \mathrm{lit}(F')$ the clause-set $\langle x \to 0 \rangle * F'$ is satisfiable. We write "$\mathrm{phd}(F)$" for "$\mathrm{phd}^{\mathrm{var}(F)}(F)$". The class $\mathcal{PC} \subset \mathcal{CLS}$ is the set of all F with $\mathrm{phd}(F) \leq 1$ (the class of **unit-propagation-complete clause-sets**).*

See [8,9] and the underlying report [10] for the general picture, where the measure $\mathrm{phd}^V(F) \in \mathbb{N}_0$ is defined in general. We now present the basic graph-theoretic criterion for $\bigcup_{i \in I} F_i \in \mathcal{PC}$ for clause-sets $F_i \in \mathcal{PC}$.

Definition 2. *For a finite family $(F_i)_{i \in I}$ of clause-sets $F_i \subset \mathcal{CLS}$ the **incidence graph** $B((F_i)_{i \in I})$ is the bipartite graph, where the two parts are given by $\bigcup_{i \in I} \mathrm{var}(F_i)$ and I, while there is an edge between v and i if $v \in \mathrm{var}(F_i)$. We say that $(F_i)_{i \in I}$ **is acyclic** if $B((F_i)_{i \in I})$ is acyclic (i.e., has no cycle as an (undirected) graph, or, equivalently, is a forest). A single clause-set $F \in \mathcal{CLS}$ is **acyclic** if $(\{C\})_{C \in F}$ is acyclic.*

The following central lemma is kind of folklore in the CSP literature; for a complete proof see the underlying report [10].

Lemma 3. *Consider an acyclic family $(F_i)_{i \in I}$ of clause-sets. If no F_i has forced assignments, then also $\bigcup_{i \in I} F_i$ has no forced assignments.*

We obtain a sufficient criterion for the union of unit-propagation complete clause-sets to be itself unit-propagation complete:

Theorem 4. *Consider an acyclic family $(F_i)_{i \in I}$ of clause-sets. If for all $i \in I$ we have $F_i \subset \mathcal{PC}$, then also $\bigcup_{i \in I} F_i \in \mathcal{PC}$.*

Proof. Let $F := \bigcup_{i \in I} F_i$, and consider a partial assignment φ with $F' \neq \{\bot\}$ for $F' := \mathrm{r}_1(\varphi * F)$. We have to show that F' has no forced assignments. For all $i \in I$ we have $\mathrm{r}_1(\varphi * F_i) \neq \{\bot\}$, and thus $\mathrm{r}_1(\varphi * F_i)$ has no forced assignments (since $F_i \in \mathcal{PC}$). So $\bigcup_{i \in I} \mathrm{r}_1(\varphi * F_i)$ has no forced assignments by Lemma 3. Using that for $A, B \in \mathcal{CLS}$ holds $\mathrm{r}_1(A \cup \mathrm{r}_1(B)) = \mathrm{r}_1(A \cup B)$, we get $F' = \mathrm{r}_1(\bigcup_{i \in I} \varphi * F_i) = \mathrm{r}_1(\bigcup_{i \in I} \mathrm{r}_1(\varphi * F_i)) = \bigcup_{i \in I} \mathrm{r}_1(\varphi * F_i)$, whence F' has no forced assignments. $\square$

Two special cases of acyclic $(F_i)_{i \in I}$ are of special importance to us, and are spelled out in the following corollary (see [10] for full details).

Corollary 5. *Consider a family $(F_i)_{i \in I}$ of clause-sets with $F_i \in \mathcal{PC}$ for all $i \in I$. Then each of the following conditions implies $\bigcup_{i \in I} F_i \in \mathcal{PC}$:*

1. *Any two different clause-sets have at most one variable in common, and the variable-interaction graph is acyclic. (The variable-interaction graph has vertex-set I, while there is an edge between $i, j \in I$ with $i \neq j$ if $\mathrm{var}(F_i) \cap \mathrm{var}(F_j) \neq \emptyset$.)*
2. *There is a variable v with $\mathrm{var}(F_i) \cap \mathrm{var}(F_j) \subseteq \{v\}$ for all $i, j \in I$, $i \neq j$.*

4   XOR-Clause-Sets

As usual, an **XOR-constraint** (also known as "parity constraint") is a (boolean) constraint of the form $x_1 \oplus \cdots \oplus x_n = \varepsilon$ for literals $x_1, \ldots, x_n$ and $\varepsilon \in \{0, 1\}$, where $\oplus$ is the addition in the 2-element field $\mathbb{Z}_2 = \{0, 1\}$. Note that $x_1 \oplus \cdots \oplus x_n = y$ is equivalent to $x_1 \oplus \cdots \oplus x_n \oplus y = 0$, while $x \oplus x = 0$ and $x \oplus \overline{x} = 1$, and $0 \oplus x = x$ and $1 \oplus x = \overline{x}$. Two XOR-constraints are *equivalent*, if they have exactly the same set of solutions. We represent XOR-constraints by **XOR-clauses**, which are just ordinary clauses $C \in \mathcal{CL}$, but now under a different interpretation, namely implicitly interpreting C as the XOR-constraints $\bigoplus_{x \in C} x = 0$. And instead of systems of XOR-constraints we just handle **XOR-clause-sets** F, which are sets of XOR-clauses, that is, ordinary clause-sets $F \in \mathcal{CLS}$ with a different interpretation. So two XOR-clauses C, D are equivalent iff $\mathrm{var}(C) = \mathrm{var}(D)$ and the number of complements in C has the same parity as the number of complements in D. That clauses are sets is justified by the commutativity of XOR, while repetition of literals is not needed due to $x \oplus x = 0$. Clashing literal pairs can be removed by $x \oplus \overline{x} = 1$ and $1 \oplus y = \overline{y}$, as long as there is still a literal left. So every XOR-constraint can be represented by an XOR-clause except of inconsistent XOR-constraints, where the simplest form is $0 = 1$; we can represent this by two XOR-clauses $\{v\}, \{\overline{v}\}$. In our theoretical study we might even assume that the case of an inconsistent XOR-clause-set is filtered out by preprocessing.

The appropriate theoretical background for XOR-constraints is the theory of systems of linear equations over a field (here the two-element field). To an XOR-clause-set F corresponds a system $A(F) \cdot v = b(F)$, using ordinary matrix notation. To make this correspondence explicit we use $n := n(F)$, $m := c(F)$, $\mathrm{var}(F) = \{v_1, \ldots, v_n\}$, and $F = \{C_1, \ldots, C_m\}$. Now F yields an $m \times n$ matrix $A(F)$ over $\mathbb{Z}_2$ together with a vector $b(F) \in \{0, 1\}^m$, where the rows $A(F)_{i,-}$ of $A(F)$ correspond to the clauses $C_i \in F$, where a coefficient $A(F)_{i,j}$ of v_j is 0 iff $v_j \notin \mathrm{var}(C_i)$, and $b_i = 0$ iff the number of complementations in C_i is even.

A partial assignment $\varphi \in \mathcal{PASS}$ satisfies an XOR-clause-set F iff $\mathrm{var}(\varphi) \supseteq \mathrm{var}(F)$ and for every $C \in F$ the number of $x \in C$ with $\varphi(x) = 1$ is even. An XOR-clause-set F implies an XOR-clause C if every satisfying partial assignment φ for F is also a satisfying assignment for $\{C\}$. The satisfying total assignments for an XOR-clause-set F correspond one-to-one to the solutions of $A(F) \cdot v = b$ (as elements of $\{0, 1\}^n$), while implication of XOR-clauses C by F correspond to single equations $c \cdot v = d$, which follow from the system, where c is an $1 \times n$-matrix over $\mathbb{Z}_2$, and $d \in \mathbb{Z}_2$. A **CNF-representation** of an XOR-clause-set $F \in \mathcal{CLS}$ is a clause-set $F' \in \mathcal{CLS}$ with $\mathrm{var}(F) \subseteq \mathrm{var}(F')$, such that the projections of the satisfying total assignments for F' (as CNF-clause-set) to $\mathrm{var}(F)$ are precisely the satisfying (total) assignments for F (as XOR-clause-set).

The resolution operation for CNF-clauses is the basic semantic operation, and analogically for XOR-clauses we have the addition of clauses, which corresponds to set-union, that is, from two XOR-clauses C, D follows $C \cup D$. Since we do not allow clashing literals, some rule is supposed here to translate $C \cup D$ into an equivalent $E \in \mathcal{CL}$ in case the two clauses are not inconsistent together. More generally, for an arbitrary XOR-clause-set F we can consider the *sum*, written as $\oplus F \in \mathcal{CL}$, which is defined as the reduction of $\bigcup F$ to some clause $\oplus F := E \in \mathcal{CL}$, assuming that the reduction does not end up in the situation $\{v, \overline{v}\}$ for some variable v — in this case we say that $\oplus F$ is inconsistent (which is only possible for $c(F) \geq 2$). The following fundamental lemma translates witnessing of unsatisfiable systems of linear equations and derivation of implied equations into the language of XOR-clause-sets; it is basically a result of linear algebra, and the proof is provided in the underlying report [10].

Lemma 6. *Consider an XOR-clause-set $F \in \mathcal{CLS}$.*

1. *F is unsatisfiable if and only if there is $F' \subseteq F$ such that $\oplus F'$ is inconsistent.*
2. *Assume that F is satisfiable. Then for all $F' \subseteq F$ the sum $\oplus F'$ is defined, and the set of all these clauses is modulo equivalence precisely the set of all XOR-clauses which follow from F.*

5   No Arc-Consistent Representations in General

We now show, if there were polynomial size arc-consistent representations of all XOR-clause-sets, then all "monotone span programs" (msp's) could be computed by monotone boolean circuits, which is not possible by [1]. The first step here is to translate msp's into linear systems S. An msp computes a boolean function $f(x_1, \ldots, x_n) \in \{0, 1\}$ (with $x_i \in \{0, 1\}$), by using auxiliary boolean variables $y_1, \ldots, y_m$, and for each $i \in \{1, \ldots, n\}$ a linear system $A_i \cdot y = b_i$, where A_i is an $m_i \times m$ matrix over $\mathbb{Z}_2$. For the computation of $f(x_1, \ldots, x_n)$, a value $x_i = 0$ means the system $A_i \cdot y = b_i$ is active, while otherwise it is inactive; the value of f is 0 if all the active systems together are unsatisfiable, and 1 otherwise. Obviously f is monotonically increasing. The task is now to put that machinery into a single system S of equations. The main idea is to "dope" each equation of every $A_i \cdot y = b_i$ with a dedicated new boolean variable added to the equation, making that equation trivially satisfiable, independently of everything else; all these new variables together are called $z_1, \ldots, z_N$, where $N = \sum_{i=1}^{n} m_i$ is the number of equations in S. If all the doping variable used for a system $A_i \cdot y = b_i$ are set to 0, then they disappear and the system is active, while if they are not set, then the system is trivially satisfiable, and thus is deactivated. Now consider an arc-consistent representation F of S. Note that the x_i are not part of F, but the variables of F are $y_1, \ldots, y_m$ together with $z_1, \ldots, z_N$, where the latter represent in a sense the $x_1, \ldots, x_n$. Using F we can compute f by setting the z_j accordingly (if $x_i = 0$, then all z_j belonging to $A_i \cdot y = b_i$ are set to 0, if $x_i = 1$, then these variables stay unassigned), running r_1 on the system, and output 0 iff the empty clause was produced by r_1. By Theorem 6.1 in [10], based on [2], finally from F we obtain a monotone circuit $\mathcal{C}$ computing f, whose size is polynomial in $\ell(F)$, where by [1] the size of $\mathcal{C}$ is $N^{O(\log N)}$.

Theorem 7. *There is no polynomial p s.t. for all XOR-clause-sets $F \in \mathcal{CLS}$ there is a representation $F' \in \mathcal{CLS}$ with $\ell(F') \leq p(\ell(F))$ and $\mathrm{phd}^{\mathrm{var}(F)}(F') \leq 1$.*

Proof. We consider representations of monotone boolean functions $f : \{0,1\}^n \to \{0,1\}$ (that is, $\boldsymbol{x} \leq \boldsymbol{y} \Rightarrow f(\boldsymbol{x}) \leq f(\boldsymbol{y})$) by monotone span programs (msp's). The input variables are given by $x_1, \ldots, x_n$. Additionally $m \in \mathbb{N}_0$ boolean variables $y_1, \ldots, y_m$ can be used, where m is the dimension, which we can also be taken as the size of the span program. For each $i \in \{1, \ldots, n\}$ there is a linear system $A_i \cdot y = b_i$ over $\mathbb{Z}_2$, where A_i is an $m_i \times m$ matrix with $m_i \leq m$. For a total assignment φ, i.e., $\varphi \in \mathcal{PASS}$ with $\mathrm{var}(\varphi) = \{x_1, \ldots, x_n\}$, the value $f(\varphi)$ is 0 if and only if the linear systems given by $\varphi(x_i) = 0$ together are unsatisfiable, that is, $\{y \in \{0,1\}^m \mid \forall i \in \{1, \ldots, n\} : \varphi(x_i) = 0 \Rightarrow A_i \cdot y = b_i\} = \emptyset$. W.l.o.g. we assume that each system $A_i \cdot y = b_i$ is satisfiable.

Consider for each $i \in \{1, \ldots, n\}$ an XOR-clause-set $A_i' \in \mathcal{CLS}$ representing $A_i \cdot y = b_i$ (so $\mathrm{var}(A_i') \supseteq \{y_1, \ldots, y_m\}$), where, as always, new variables for different A_i' are used, that is, for $i \neq j$ we have $(\mathrm{var}(A_i') \cap \mathrm{var}(A_j')) \setminus \{y_1, \ldots, y_m\} = \emptyset$. We use the process $\mathrm{D} : \mathcal{CLS} \to \mathcal{CLS}$ of "doping", as introduced in [11], where $\mathrm{D}(F)$ is obtained from F by adding to each clause a new variable. Let $A_i'' := \mathrm{D}(A_i')$, where the doping variables for different i do not clash; we denote them (altogether) by $z_1, \ldots, z_N$. Let $F := \bigcup_{i=1}^n A_i''$. Consider a CNF-representation F' of the XOR-clause-set F.

We have $f(\varphi) = 0$ iff $\varphi' * F' \in \mathcal{USAT}$, where φ' is a partial assignment with φ' assigning only doping variables z_j, namely if $\varphi(x_i) = 0$, then all the doping variables used in $\mathrm{D}(A_i')$ are set to 0, while if $\varphi(x_i) = 1$, then nothing is assigned here. The reason is that by setting the doping variables to 0 we obtain the original system $A_i \cdot y = b_i$, while by leaving them in, this system becomes satisfiable whatever the assignments to the y-variables are.

Now assume that we have $\mathrm{phd}^{\{z_1, \ldots, z_N\}}(F') \leq 1$. By Theorem 6.1 in [10] we obtain from F' a monotone circuit $\mathcal{C}$ (using only and's and or's) of size polynomial in $\ell(F')$ with input variables $z_1', z_1'', \ldots, z_N', z_N''$, where

- $z_j' = z_j'' = 1$ means that z_j has not been assigned,
- $z_j' = 1, z_j'' = 0$ means $z_j = 0$,
- $z_j' = 0, z_j'' = 1$ means $z_j = 1$,
- while $z_j' = 0, z_j'' = 0$ means "contradiction" (forcing the output of $\mathcal{C}$ to 0).

The value of $\mathcal{C}$ is 0 iff the corresponding partial assignment applied to F' yields an unsatisfiable clause-set. In $\mathcal{C}$ we now replace the inputs z_j by inputs x_i, which in case of $x_i = 0$ sets $z_j' = 1, z_j'' = 0$ for all related j, while in case of $x_i = 1$ all related z_j', z_j'' are set to 1.[3] This is now a monotone circuit computing f. By [1], Theorem 1.1, thus it is not possible that F' is of polynomial size in F. $\qquad\square$

---

[3] In other words, for the j related to i always all z_j' are set to 1, while $z_j'' = x_i$.

6 The Translations X_0, X_1

After having shown that there is no "small" arc-consistent representation of arbitrary XOR-clause-sets F, the task is to find "good" CNF-representations for special F. First we consider $c(F) = 1$, that is, a single XOR-clause C, to which we often refer as "$x_1 \oplus \cdots \oplus x_n = 0$". There is precisely one equivalent clause-set, i.e., there is exactly one representation without new variables, namely $\boldsymbol{X_0(C)} := \mathrm{prc}_0(x_1 \oplus \cdots \oplus x_n = 0)$, the set of prime implicates of the underlying boolean function, which is unique since these prime implicates are not resolvable. $X_0(C)$ has 2^{n-1} clauses for $n \geq 1$ (while for $n = 0$ we have $X_0(C) = \top$), namely precisely those full clauses (containing all variables) over $\{\mathrm{var}(x_1), \ldots, \mathrm{var}(x_n)\}$ where the parity of the number of complementations is different from the parity of the number of complementations in C. Note that for two XOR-clauses C, D we have $X_0(C) = X_0(D)$ iff C, D are equivalent. More generally, we define $X_0 : \mathcal{CLS} \to \mathcal{CLS}$, where the input is interpreted as XOR-clause-set and the output as CNF-clause-set, by $\boldsymbol{X_0(F)} := \bigcup_{C \in F} X_0(C)$. By Theorem 4:

Lemma 8. *If $F \in \mathcal{CLS}$ is acyclic, then $X_0(F) \in \mathcal{PC}$.*

An early and influential example of unsatisfiable clause-sets are the "Tseitin formulas" introduced in [26], which are obtained as applications of X_0 to XOR-clause-sets derived from graphs; see the underlying report [10] for various discussions. In [26] an exponential lower bound for regular resolution refutations of (special) Tseitin clause-sets was shown, and thus unsatisfiable Tseitin clause-sets in general have high hardness. This was extended in [27] to full resolution, and thus unsatisfiable Tseitin clause-sets in general have also high "w-hardness" (see [10]; as hardness captures tree-resolution, w-hardness captures dag-resolution).

In the following we refine $X_0 : \mathcal{CLS} \to \mathcal{CLS}$ in various ways, by first transforming an XOR-clause-set F into another XOR-clause-set F' representing F, and then using $X_0(F')$. If the XOR-clause-set F contains large clauses, then $X_0(F)$ is not feasible, and the XOR-clauses of F have to be broken up into short clauses, which we consider now. As we have defined how a CNF-clause-set can represent an XOR-clause-set, we can define that an XOR-clause-set F' represents an XOR-clause-set F, namely if the satisfying assignments of F' projected to the variables of F are precisely the satisfying assignments of F.

Definition 9. *Consider an XOR-clause $C = \{x_1, \ldots, x_n\} \in \mathcal{CL}$. The **natural splitting** of C is the XOR-clause-set F' obtained as follows, using $n := |C|$:*

- *If $n \leq 2$, then $F' := \{C\}$.*
- *Otherwise choose pairwise different new variables $y_2, \ldots, y_{n-1}$, and let $F' := \{x_1 \oplus x_2 = y_2\} \cup \{y_{i-1} \oplus x_i = y_i\}_{i \in \{3, \ldots, n-1\}} \cup \{y_{n-1} \oplus x_n = 0\}$, (i.e., $F' = \{\{x_1, x_2, y_2\}\} \cup \{\{y_{i-1}, x_i, y_i\}\}_{i \in \{3, \ldots, n-1\}} \cup \{\{y_{n-1}, x_n\}\}$).*

Then F' is, as XOR-clause-set, a representation of $\{C\}$. Let $\boldsymbol{X_1(C)} := X_0(F')$.

We have for $F := X_1(C)$: If $n \leq 2$, then $n(F) = c(F) = n$, and $\ell(F) = 2^{n-1} \cdot n$. Otherwise $n(F) = 2n - 2$, $c(F) = 4n - 6$ and $\ell(F) = 12n - 20$. Corollary 5, Part 2, applies to F' from Definition 9, and thus:

Lemma 10. *For $C \in \mathcal{CL}$ we have $X_1(C) \in \mathcal{PC}$.*

We define $X_1 : \mathcal{CLS} \to \mathcal{CLS}$, where the input is interpreted as XOR-clause-set and the output as CNF-clause-set, by $X_1(F) := \bigcup_{C \in F} X_1(C)$ for $F \in \mathcal{CLS}$, where some choice for the new variables is used, so that the new variables for different XOR-clauses do not overlap. By Theorem 4 and Lemma 10:

Theorem 11. *If $F \in \mathcal{CLS}$ is acyclic, then $X_1(F) \in \mathcal{PC}$.*

A precursor to Theorem 11 is found in Theorem 1 of [19], where it is stated that tree-like XOR clause-sets are "UP-deducible", which is precisely the assertion that for acyclic $F \in \mathcal{CLS}$ the representation $X_1(F)$ is arc-consistent. We now show that the problem of computing an arc-consistent CNF-representation for an XOR-clause-set F is fixed-parameter tractable in the parameter $c(F)$.

Theorem 12. *Consider a satisfiable XOR-clause-set $F \in \mathcal{CLS}$. Let $F^* := \{\oplus F' : F' \subseteq F\} \in \mathcal{CLS}$ (recall Lemma 6); F^* is computable in time $O(\ell(F) \cdot 2^{c(F)})$. Then $X_1(F^*)$ is a CNF-representation of F with $\mathrm{phd}^{\mathrm{var}(F)}(X_1(F^*)) \leq 1$.*

Proof. Consider some partial assignment φ with $\mathrm{var}(\varphi) \subseteq \mathrm{var}(F)$, let $F' := \mathrm{r}_1(\varphi * F^*)$, and assume there is a forced literal $x \in \mathrm{lit}(F')$ for F'. Then the XOR-clause $C := \{y \in \mathcal{LIT} : \varphi(y) = 0\} \cup \{\overline{x}\}$ follows from F. By Lemma 6 there is $F' \subseteq F$ with $\oplus F' = C$ modulo equivalence of XOR-clauses. So we have (modulo equivalence) $X_1(C) \subseteq F^*$, where due to $X_1(C) \in \mathcal{PC}$ (Lemma 10) the forced literal x for $\varphi * X_1(C)$ is set by r_1, contradicting the assumption.   □

Theorem 4 in [22] yields the weaker bound $O(4^{n(F)})$ for the number of clauses in an arc-consistent representation of F (w.l.o.g. $c(F) \leq n(F)$). In Conjecture 14 we state our belief that we can strengthen Theorem 12 by establishing $\mathrm{phd}(F') \leq 1$ for an appropriate, more intelligent representation F' of F. We now turn to the problem of understanding and refining the basic translation X_1 for two clauses.

7   Translating Two XOR-Clauses

The analysis of the translation $X_1(\{C, D\})$ for two XOR-clauses C, D in the underlying report [10] shows, that this representation is very hard for tree-resolution, but easy for full and for width-restricted resolution. So it might be usable for (conflict-driven) SAT solvers. But indeed we can provide a representation in $\mathcal{PC}$ as follows; note that an XOR-clause-set $\{C, D\}$ is unsatisfiable iff $|C \cap \overline{D}|$ is odd and $\mathrm{var}(C) = \mathrm{var}(D)$.

Theorem 13. *Consider two XOR-clauses $C, D \in \mathcal{CL}$. To simplify the presentation, using $V := \mathrm{var}(C) \cap \mathrm{var}(D)$, we assume $|V| \geq 2$, and $|C| > |V|$ as well as $|D| > |V|$. Thus w.l.o.g. $|C \cap D| = |V|$. Let $I := C \cap D$. Choose $s \in \mathcal{VA} \setminus \mathrm{var}(\{C, D\})$, and let $I' := I \cup \{s\}$. Let $C' := (C \setminus I) \cup \{s\}$ and $D' := (D \setminus I) \cup \{s\}$. Now $\{I', C', D'\}$ is an XOR-clause-set which represents the XOR-clause-set $\{C, D\}$. Let $\mathbf{X_2(C, D)} := X_1(\{I', C', D'\})$. Then $X_2(C, D) \in \mathcal{PC}$ is a CNF-representation of the XOR-clause-set $\{C, D\}$.*

Proof. That $\{I', C', D'\}$ represents $\{C, D\}$ is obvious, since s is the sum of the common part. $\{I', C', D'\}$ is acyclic (besides the common variable s the three variable-sets are disjoint), and thus by Theorem 11 we get $X_2(C, D) \in \mathcal{PC}$. □

Conjecture 14. We can combine a generalisation of Theorem 13 with Theorem 12 and obtain $X_* : \mathcal{CLS} \to \mathcal{PC}$, which computes for an XOR-clause-set $F \in \mathcal{CLS}$ a CNF-representation $X_*(F)$ such that $\ell(X_*(F)) = 2^{O(c(F))} \cdot \ell(F)^{O(1)}$.

8   Open Problems and Future Research Directions

Regarding lower bounds, the main question for Theorem 7 is to obtain sharp bounds on the size of shortest representations F' with $\mathrm{phd}^{\mathrm{var}(F)}(F') \leq 1$. Turning to upper bounds, in Lemma 8, Theorem 11, and Theorem 13 we have established methods to obtain representations in $\mathcal{PC}$, while Conjecture 14 says, that computing a representation in $\mathcal{PC}$ should be fixed-parameter tractable in the number of XOR-clauses. See [10] for more open problems and conjectures.

References

1. Babai, L., Gál, A., Wigderson, A.: Superpolynomial lower bounds for monotone span programs. Combinatorica 19(3), 301–319 (1999)
2. Bessiere, C., Katsirelos, G., Narodytska, N., Walsh, T.: Circuit complexity and decompositions of global constraints. In: Twenty-First International Joint Conference on Artificial Intelligence (IJCAI 2009), pp. 412–418 (2009)
3. Biere, A., Heule, M.J., van Maaren, H., Walsh, T.: Handbook of Satisfiability, Frontiers in Artificial Intelligence and Applications, vol. 185. IOS Press (February 2009)
4. Bordeaux, L., Marques-Silva, J.: Knowledge compilation with empowerment. In: Bieliková, M., Friedrich, G., Gottlob, G., Katzenbeisser, S., Turán, G. (eds.) SOFSEM 2012. LNCS, vol. 7147, pp. 612–624. Springer, Heidelberg (2012)
5. Courtois, N.T., Bard, G.V.: Algebraic cryptanalysis of the Data Encryption Standard. In: Galbraith, S.D. (ed.) Cryptography and Coding 2007. LNCS, vol. 4887, pp. 152–169. Springer, Heidelberg (2007)
6. Eén, N., Sörensson, N.: Translating pseudo-boolean constraints into SAT. Journal on Satisfiability, Boolean Modeling and Computation 2, 1–26 (2006)
7. Gent, I.P.: Arc consistency in SAT. In: van Harmelen, F. (ed.) 15th European Conference on Artificial Intelligence (ECAI 2002), pp. 121–125. IOS Press (2002)
8. Gwynne, M., Kullmann, O.: Generalising and unifying SLUR and unit-refutation completeness. In: van Emde Boas, P., Groen, F.C.A., Italiano, G.F., Nawrocki, J., Sack, H. (eds.) SOFSEM 2013. LNCS, vol. 7741, pp. 220–232. Springer, Heidelberg (2013)
9. Gwynne, M., Kullmann, O.: Generalising Unit-Refutation Completeness and SLUR via Nested Input Resolution. Journal of Automated Reasoning 52(1), 31–65 (2014), doi:10.1007/s10817-013-9275-8
10. Gwynne, M., Kullmann, O.: On SAT representations of XOR constraints. Tech. Rep. arXiv:1309.3060v4 [cs.CC], arXiv (December 2013)
11. Gwynne, M., Kullmann, O.: Trading inference effort versus size in CNF knowledge compilation. Tech. Rep. arXiv:1310.5746v2 [cs.CC], arXiv (November 2013)

12. Haanpää, H., Järvisalo, M., Kaski, P., Niemelä, I.: Hard satisfiable clause sets for benchmarking equivalence reasoning techniques. Journal of Satisfiability, Boolean Modeling and Computation 2, 27–46 (2006)
13. Kleine Büning, H., Kullmann, O.: Minimal unsatisfiability and autarkies. In: Biere, et al. (eds.) [3], ch. 11, pp. 339–401
14. Kullmann, O.: Investigating a general hierarchy of polynomially decidable classes of CNF's based on short tree-like resolution proofs. Tech. Rep. TR99-041, Electronic Colloquium on Computational Complexity (ECCC) (October 1999)
15. Kullmann, O.: Upper and lower bounds on the complexity of generalised resolution and generalised constraint satisfaction problems. Annals of Mathematics and Artificial Intelligence 40(3-4), 303–352 (2004)
16. Kullmann, O.: Hardness measures and resolution lower bounds. Tech. Rep. arXiv:1310.7627v1 [cs.CC], arXiv (October 2013)
17. Laitinen, T., Junttila, T., Niemelä, I.: Extending clause learning DPLL with parity reasoning. In: Coelho, H., Studer, R., Wooldridge, M. (eds.) ECAI 2010 – 19th European Conference on Artificial Intelligence, pp. 21–26. IOS Press (2010)
18. Laitinen, T., Junttila, T., Niemelä, I.: Equivalence class based parity reasoning with DPLL(XOR). In: ICTAI 2011 – 23rd International Conference on Tools with Artificial Intelligence, pp. 649–658 (2011)
19. Laitinen, T., Junttila, T., Niemelä, I.: Classifying and propagating parity constraints. In: Milano, M. (ed.) CP 2012. LNCS, vol. 7514, pp. 357–372. Springer, Heidelberg (2012)
20. Laitinen, T., Junttila, T., Niemelä, I.: Conflict-driven XOR-clause learning. In: Cimatti, A., Sebastiani, R. (eds.) SAT 2012. LNCS, vol. 7317, pp. 383–396. Springer, Heidelberg (2012)
21. Laitinen, T., Junttila, T., Niemelä, I.: Extending clause learning SAT solvers with complete parity reasoning. In: ICTAI 2012 – 24th International Conference on Tools with Artificial Intelligence, pp. 65–72 (2012)
22. Laitinen, T., Junttila, T., Niemelä, I.: Simulating parity reasoning. In: McMillan, K., Middeldorp, A., Voronkov, A. (eds.) LPAR-19 2013. LNCS, vol. 8312, pp. 568–583. Springer, Heidelberg (2013)
23. Li, C.M.: Equivalency reasoning to solve a class of hard SAT problems. Information Processing Letters 76, 75–81 (2000)
24. Rossi, F., van Beek, P., Walsh, T. (eds.): Handbook of Constraint Programming. Foundations of Artificial Intelligence. Elsevier (2006)
25. Soos, M., Nohl, K., Castelluccia, C.: Extending SAT solvers to cryptographic problems. In: Kullmann, O. (ed.) SAT 2009. LNCS, vol. 5584, pp. 244–257. Springer, Heidelberg (2009)
26. Tseitin, G.: On the complexity of derivation in propositional calculus. In: Seminars in Mathematics, vol. 8. V.A. Steklov Mathematical Institute, Leningrad (1968); English translation: Slisenko, A.O.(ed.) Studies in mathematics and mathematical logic, Part II, pp. 115–125 (1970)
27. Urquhart, A.: Hard examples for resolution. Journal of the ACM 34, 209–219 (1987)

Minimal Triangulation Algorithms for Perfect Phylogeny Problems

Rob Gysel

Department of Computer Science
University of California, Davis
1 Shields Avenue, Davis CA 95616, USA
rsgysel@ucdavis.edu

Abstract. In this paper, we show that minimal triangulation techniques similar to those proposed by Bouchitté and Todinca can be applied to a variety of perfect phylogeny (or character compatibility) problems. These problems arise in the context of supertree construction, a critical step in estimating the Tree of Life.

Keywords: perfect phylogeny, minimal triangulation.

1 Introduction

The perfect phylogeny problem, also known as the character compatibility problem, is a classic NP-hard [6,30] problem in phylogenetics [13,29] related to supertree construction. The supertree problem takes as input a collection of phylogenies whose species set partially overlap, and asks for a phylogeny on the entire species set. Given a collection of partially labeled unrooted phylogenies, one can construct two-state partial characters that have a perfect phylogeny precisely when the collection has a compatible supertree [29]. Supertree construction is required for estimating the Tree of Life [3]. Rodrigo and Ross [27] called for the development of compatiblity-based algorithms for supertree construction, because its criterion is more intuitive than *parsimony*[1].

Solutions to the perfect phylogeny problem are characterized by the existence of restricted triangulations of the partition intersection graph [12,24,30], and minimal triangulations of the partition intersection graph also play an important role in two variants of this problem. The first, the maximum compatibility problem, asks to find the largest subset of a set of given characters that has a perfect phylogeny [8,18], and the second, asks if a set of characters has a unique perfect phylogeny [28,15]. To our knowledge, advances in the field of minimal triangulations have not been extended to these problems, although the use of such methods to solve at least the perfect phylogeny problem may have been alluded to in [14]. In this paper, we extend the potential maxclique approach developed by Bouchitté and Todinca [10], which was improved in [14], to solve

---

[1] Matrix representation with parsimony (MRP) is one of the most popular supertree methods in use [11].

A.-H. Dediu et al. (Eds.): LATA 2014, LNCS 8370, pp. 421–432, 2014.

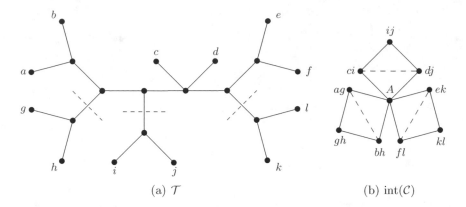

(a) $\mathcal{T}$ (b) int($\mathcal{C}$)

Fig. 1. A phylogeny $\mathcal{T}$ displaying $\mathcal{C} = \{abcdef|gh|ij|kl, ag|dj|fl, bh|ci|ek\}$ and the corresponding partition intersection graph int($\mathcal{C}$). A denotes $abcdef$. The dashed edges of $\mathcal{T}$ are distinguished by $abcdef|gh|ij|kl$. Removing these edges results in the four subtrees defined by $\mathcal{T}(abcdef)$, $\mathcal{T}(gh)$, $\mathcal{T}(ij)$, and $\mathcal{T}(kl)$. The dashed edges of int($\mathcal{C}$) define a proper triangulation, and the solid edges are obtained from pairs of states that are shared by at least one species. Note $\mathcal{T}$ does not display $ag|bh$, $ci|dj$, or $ek|fl$.

the perfect phylogeny problem and its variants. This approach is motivated by the following: first, the algorithms in [9,14] run in time polynomial in the number of minimal separators of the graph, and second, that data generated by the coalescent-based program *ms* [21] often results in a partition intersection graph with a reasonable number of minimal separators [16], despite there being an exponential number of minimal separators in general. In order to unify our approach, we use a weighted variant of the minimum-fill problem. Minimum-fill is NP-hard [31] and is an active area of research [5,14].

Given full characters (i.e. defined on every species), the perfect phylogeny problem is solvable in polynomial time when the number of characters is fixed [23] or when the number of states is bounded [1], but is NP-complete for two-state partial characters [29]. The maximum compatibility problem is NP-hard for two-state full characters [29]. The unique perfect phylogeny problem is CoNP-complete even when a perfect phylogeny for the characters is given [7,19]. Our results apply to partial characters, and some of our results apply to unbounded characters. See [20] for a survey on minimal triangulations, [13,29] for further reading on the perfect phylogeny / character compatibility problem, and [15] for further reading on unique perfect phylogeny. Due to space considerations, some proofs have been sketched or removed. Details may be found in a preliminary version of this paper [17].

2   Definitions and Results

Let X be a set of *species* or *taxa*. A *phylogeny* is a pair $\mathcal{T} = (T, \phi)$ where T is an undirected tree and ϕ is a bijective map from X to the leaves of T. A *character on* X is a partition $\chi = A_1|A_2|\ldots|A_r$ of a subset of X, and it is a *two-state character* if $r = 2$. For $i = 1, 2, \ldots, r$ the set A_i is a *state* of χ. Given a state A of a character, the minimal subtree of T that connects $\phi(A)$ is denoted $\mathcal{T}(A)$. A phylogeny $\mathcal{T}$ *displays* a character χ if, for each pair of distinct states A and A' of χ, the trees $\mathcal{T}(A)$ and $\mathcal{T}(A')$ have no nodes in common. Given a set $\mathcal{C}$ of characters on X, the *perfect phylogeny problem* is to determine if there is a phylogeny $\mathcal{T}$ that displays every character in $\mathcal{C}$. In this case, we call $\mathcal{T}$ a *perfect phylogeny* for $\mathcal{C}$, and say that $\mathcal{C}$ is *compatible*.

The perfect phylogeny problem reduces to a graph theoretic problem that we detail now. A graph is *chordal* if any cycle it has on four or more vertices has a *chord*, that is, an edge between two non-consecutive vertices in the cycle. For a non-chordal graph $G = (V, E)$, any chordal supergraph $H = (V, E \cup F)$ is a *triangulation of* G with *fill edges* F. If there is no $F' \subset F$ such that $H' = (V, E \cup F')$ is a triangulation, then H is a *minimal triangulation of* G.

Given a set of characters $\mathcal{C}$, the *partition intersection graph* $\mathrm{int}(\mathcal{C})$ is the graph with vertex set $\{(A, \chi) \mid \chi \in \mathcal{C} \text{ and } A \text{ is a state of } \chi\}$, and two vertices (A, χ) and (A', χ') are adjacent in $\mathrm{int}(\mathcal{C})$ if and only if A and A' have non-empty intersection. If A_1 and A_2 are states of a character χ, then A_1 and A_2 are disjoint because χ is a partition of a subset of X, so (A_1, χ) and (A_2, χ) are not adjacent in $\mathrm{int}(\mathcal{C})$. The vertex (A, χ) has *state* A and *character* χ. Let H be a (minimal) triangulation of $\mathrm{int}(\mathcal{C})$ with fill edges F. If, for each fill edge in F, the vertices incident to the fill edge have different characters, then H is a *proper (minimal) triangulation of* $\mathrm{int}(\mathcal{C})$. This may be viewed as coloring each vertex (A, χ) of $\mathrm{int}(\mathcal{C})$ by its character χ, resulting in a properly colored graph whose coloring is fixed by $\mathcal{C}$, and proper triangulations are those whose fill edges preserve the proper coloring. If u and v are vertices of $\mathrm{int}(\mathcal{C})$ that have the same character/color, we say that u and v are *monochromatic*, and if a triangulation of $\mathrm{int}(\mathcal{C})$ has uv as an edge, we say that uv is a *monochromatic fill edge* of the triangulation. See Figure 1 for an example of these concepts.

For the remainder of this section, we characterize solutions to perfect phylogeny problems as constrained minimal triangulations of the partition intersection graph, and state our algorithmic results. These problems will then be discussed in terms of minimum-weight minimal triangulations in Section 3, and we prove our computational results in Section 4, all of which rely on Algorithm 1. The connection between triangulations and perfect phylogeny stems from the following result.

Theorem 1. *[12,24,30] Let $\mathcal{C}$ be a set of characters on X. Then $\mathcal{C}$ is compatible if and only if $\mathrm{int}(\mathcal{C})$ has a proper minimal triangulation.*

The set of minimal separators of $\mathrm{int}(\mathcal{C})$ are denoted $\Delta_{\mathrm{int}(\mathcal{C})}$, and in general, $|\Delta_{\mathrm{int}(\mathcal{C})}| = O(2^{r|\mathcal{C}|})$. For exposition, we defer their definition to Section 3. Our first algorithmic result is the following.

Theorem 2. *Let $\mathcal{C}$ be a set of characters on X with at most r states per character. There is an $O(|X||\mathcal{C}|^2 + (r|\mathcal{C}|)^4|\Delta_{\mathrm{int}(\mathcal{C})}|^2)$ time algorithm that solves the perfect phylogeny problem.*

If C is not compatible, then the *maximum compatibility problem* is to determine the largest subset C^* of C that is compatible. We call C^* a *maximum compatible subset of C*. A character χ is *broken* by a fill edge $(A, \chi)(A', \chi')$ if $\chi = \chi'$. For a triangulation H of int(C), the *displayed characters* of H are those in C that are not broken by any fill edge of H.

Theorem 3. *[8,18] Let C be a set of characters on X. Then C^* is a maximum compatible subset of C if and only if there is a minimal triangulation H^* of int(C) that has C^* as its displayed characters, and for every other minimal triangulation H' of int(C) with displayed characters C', $|C'| \leq |C^*|$.*

Let C be a set of characters on X and $w : C \to \mathbb{R}^{>0}$ a weight on C. For a subset C' of C, define $w(C') = \sum_{\chi \in C'} w(\chi)$. The $w-$*maximum compatibility problem* is to find $C^* = \arg\max w(C')$, where the maximum is taken over all compatible subsets C' of C. Theorem 3 is generalized as follows, whose proof we omit due to space considerations.

Theorem 4. *Let C be a set of characters on X weighted by w. Then C^* is a $w-$maximum compatible subset of C if and only if there is a minimal triangulation H^* of int(C) that has C^* as its displayed characters, and for any other minimal triangulation H' of int(C) with displayed characters C', $w(C') \leq w(C^*)$.*

Our second algorithmic result is as follows.

Theorem 5. *Let C be a set of ($w-$weighted) two-state characters on X. There is an $O(|X||C|^2 + |C|^4|\Delta_{\text{int}(C)}|^2)$ time algorithm that solves the (w-)maximum compatibility problem.*

The *unique perfect phylogeny problem* is to determine if a perfect phylogeny $\mathcal{T} = (T, \phi)$ for C is unique. An edge uv of T is *distinguished* by a character χ if contracting uv results in a phylogeny that does not display χ, and $\mathcal{T}$ is *distinguished* by C if each edge of $\mathcal{T}$ is distinguished by a character of C. A phylogeny $\mathcal{T} = (T, \phi)$ is *ternary* if every internal node of T has degree three.

Theorem 6. *[28] Let C be a set of characters on X. Then C has a unique perfect phylogeny $\mathcal{T} = (T, \phi)$ if and only if the following conditions hold:*

1. *there is a ternary perfect phylogeny $\mathcal{T}$ for C and $\mathcal{T}$ is distinguished by C;*
2. *int(C) has a unique proper minimal triangulation.*

A perfect phylogeny $\mathcal{T}$ can be constructed from a proper minimal triangulation in polynomial time (see [29] and [4]). Checking if $\mathcal{T}$ is ternary and distinguished by C is also easy to do: an edge uv is distinguished by χ if and only if u is a node of $\mathcal{T}(A)$ and v is a node of $\mathcal{T}(A)$ for distinct states A, A' of χ. So if it is known that int(C) has a unique proper minimal triangulation, it is possible to determine if C has a unique perfect phylogeny in polynomial time. On the other hand, it has recently been shown [7,19] that if a perfect phylogeny is given for a set of characters, it is still CoNP-complete to determine if it is the unique perfect phylogeny for those characters. That is, determining if $\text{int}(\mathcal{C})$ has a unique proper minimal triangulation is CoNP-complete [19]. From this point of view, the intractibility in utilizing Theorem 6 lies in determining whether or not $\text{int}(\mathcal{C})$ has a unique proper minimal triangulation. This makes our last algorithmic result of interest.

Theorem 7. *Let $\mathcal{C}$ be a set of characters on X with at most r states. There is an $O(|X||\mathcal{C}|^2 + (r|\mathcal{C}|)^4|\Delta_{\text{int}(\mathcal{C})}|^2)$ time algorithm that determines if $\text{int}(\mathcal{C})$ has a unique proper minimal triangulation, i.e. it solves the unique perfect phylogeny problem.*

3   Characterizations via Weighted Minimum-Fill

In this section, we characterize solutions to the perfect phylogeny problem and its variants with respect to a weighted-variant of the *minimum-fill problem*, which asks for the fewest number of fill edges required to triangulate a graph. In order for our results to be useful in the next section, each result will be given with respect to minimal triangulations.

Suppose G is a non-complete graph. If U is a subset of G's vertices, then the *potential fill edges* $\text{pf}(U)$ of U are pairs of vertices of U that are not edges of G. A *fill weight* on $G = (V, E)$ is a function $F_w : \text{pf}(V) \to \mathbb{R}^{\geq 0}$. For a triangulation H of G with fill weight F_w, the *weight* of H is $F_w(H) = \sum F_w(f)$ where the sum occurs over all fill edges of H. We will call H a F_w*-minimum triangulation* of G if, for every other triangulation H' of G, $F_w(H) \leq F_w(H')$. In this case we write $\text{mfi}_{F_w}(G) = F_w(H)$. If $F_w(H) = 0$, then H is a F_w*-zero triangulation* of G. If H is a F_w-minimum or F_w-zero triangulation that is also a minimal triangulation of G, then H is a F_w*-minimum minimal triangulation* or F_w*-zero minimal triangulation*, respectively.

Let $\mathcal{C}$ be a set of characters on X. A natural fill weight on $\text{int}(\mathcal{C})$ is $F_{w(\mathcal{C})}$, defined as $F_{w(\mathcal{C})}(uv) = 1$ if u and v are monochromatic, and $F_{w(\mathcal{C})}(uv) = 0$ otherwise. This fill weight indicates when a fill edge breaks a character.

Lemma 8. *A collection $\mathcal{C}$ of characters on X are compatible if and only if $\text{int}(\mathcal{C})$ has a $F_{w(\mathcal{C})}$-zero minimal triangulation.*

Proof. A triangulation H of $\text{int}(\mathcal{C})$ is proper if and only if $F_{w(\mathcal{C})}(H) = 0$. The result now follows from Theorem 1.   □

The following two lemmas, which follow from results in [8,18], will be helpful for proving Theorem 4.

Lemma 9. *Suppose $\mathcal{C}$ is a set of characters and $\mathcal{C}' \subseteq \mathcal{C}$ is compatible. Then there is a minimal triangulation of $\text{int}(\mathcal{C})$ and $\mathcal{C}'$ is a subset of its displayed characters.*

Lemma 10. *Suppose $\mathcal{C}$ is a set of characters and H is a triangulation of $\text{int}(\mathcal{C})$. Then the displayed characters of H are a compatible subset of $\mathcal{C}$.*

Suppose $\mathcal{C}$ is a set of characters on X weighted by w. Then w induces a fill weight F_w of int($\mathcal{C}$) defined by $F_w(uv) = w(\chi)$ if u and v are monochromatic and colored by χ, and $F_w(uv) = 0$ otherwise.

Lemma 11. *Let $\mathcal{C}$ be a collection of two-state characters weighted by w, and suppose H is a triangulation of* int($\mathcal{C}$) *with displayed characters $\mathcal{C}(H)$. Then $w(\mathcal{C}) = F_w(H) + w(\mathcal{C}(H))$.*

Proof. (Sketch) For each χ in $\mathcal{C}$ there is exactly one potential fill edge uv of int($\mathcal{C}$) such that u and v are monochromatic because χ has two states, and both vertices are colored by χ. Either uv is a fill edge of H, so χ is broken by uv and uv contributes to the second term of the RHS, or $\chi \in \mathcal{C}(H)$ and uv contributes to the first term of the RHS. □

Theorem 12. *Let $\mathcal{C}$ be a collection of two-state characters weighted by w. Then $\mathcal{C}^*$ is a w−maximum compatible subset of $\mathcal{C}$ if and only if there is a F_w-minimum minimal triangulation H^* of* int($\mathcal{C}$) *that has $\mathcal{C}^*$ as its displayed characters.*

Proof. Suppose that $\mathcal{C}^*$ is a w−maximum compatible subset of $\mathcal{C}$. By Theorem 4, there is a minimal triangulation H^* of int($\mathcal{C}$) that has $\mathcal{C}^*$ as its displayed characters. For the sake of contradiction suppose H^* is not a F_w−minimum minimal triangulation, so there is a triangulation H of int($\mathcal{C}$) such that $F_w(H) < F_w(H^*)$. Letting $\mathcal{C}(H)$ be the displayed characters of H, by Lemma 11 we have $w(\mathcal{C}) - w(\mathcal{C}(H)) < w(\mathcal{C}) - w(\mathcal{C}^*)$ and therefore $w(\mathcal{C}^*) < w(\mathcal{C}(H))$. By Lemma 10 $\mathcal{C}(H)$ is compatible, so $\mathcal{C}^*$ can not be a w−maximum compatible subset of $\mathcal{C}$, a contradiction. Therefore H^* must be a F_w−minimum minimal triangulation.

Now let H' be a F_w−minimum minimal triangulation of int($\mathcal{C}$) with displayed characters $\mathcal{C}(H')$. Then $F_w(H') = F_w(H^*)$ by F_w−minimization, and $w(\mathcal{C}) - w(\mathcal{C}(H')) = w(\mathcal{C}) - w(\mathcal{C}^*)$ by Lemma 11 so $w(\mathcal{C}(H')) = w(\mathcal{C}^*)$. The set $\mathcal{C}(H)$ is compatible by Lemma 10, so $\mathcal{C}(H)$ is an optimal solution. □

Maximum compatibility reduces to weighted maximum compatibility by assigning each character a weight of one. This character weighting induces the fill weight $F_{w(\mathcal{C})}$, giving the following corollary.

Corollary 13. *Let $\mathcal{C}$ be a collection of two-state characters. Then $\mathcal{C}^*$ is a maximum compatible subset of $\mathcal{C}$ if and only if there is a $F_{w(\mathcal{C})}$-minimum minimal triangulation H^* of* int($\mathcal{C}$) *that has $\mathcal{C}^*$ as its displayed characters.*

We conclude this section by characterizing solutions to unique perfect phylogeny. Let $G = (V, E)$ be an undirected graph and $S \subseteq V$. We will use $G - S$ to denote the graph obtained from G by removing the vertices S and edges that are incident to a vertex in S. If x, y are connected vertices in G but disconnected in $G - S$, then S is an xy−*separator*. When no proper subset of S is also an xy−separator, then S is a *minimal xy−separator*[2]. If there is at least one pair of vertices x and y such that S is a minimal xy−separator, then S is a *minimal separator* of G.

---

[2] Note that a minimal xy−separator S is defined with respect to x and y. There may be a different pair of vertices u,v of G such that S is a non-minimal uv−separator.

The set of minimal separators of G is denoted by Δ_G. Two minimal separators S and S' are *parallel* if there is a connected component C of $G - S$ such that $S' \subseteq S \cup C$ (this relationship happens to be symmetric [25]). Suppose Φ is a subset of G's minimal separators. The graph G_Φ is obtained from G by adding the fill edge uv whenever $uv \in \mathrm{pf}(S)$ for some S in Φ, and we say G is obtained by *saturating* each minimal separator in Φ. We say Φ is *pairwise-parallel* if every two minimal separators S and S' in Φ are parallel. The following result characterizes the minimal triangulations of a graph in terms of its minimal separators.

Theorem 14. *[25,26] see also [22] Let G a graph and Δ_G its minimal separators. If H is a minimal triangulation of G, then Δ_H is a maximal pairwise-parallel set of minimal separators of G and $H = G_{\Delta_H}$. Conversely, if Φ is any maximal pairwise-parallel set of minimal separators of G, then G_Φ is a minimal triangulation of G and $\Delta_{G_\Phi} = \Phi$.*

An important observation from this theorem is that if H is a minimal triangulation of G, then $\Delta_H \subseteq \Delta_G$. Let $\mathcal{C}$ be a set of characters on X and F_w be a fill weight on $\mathrm{int}(\mathcal{C})$. We will use $\Delta_{F_w}^{\min}$ to denote the set of minimal separators S of $\mathrm{int}(\mathcal{C})$ such that there is a F_w–minimum minimal triangulation H of $\mathrm{int}(\mathcal{C})$ with $S \in \Delta_H$.

Theorem 15. *Suppose $\mathcal{C}$ is a collection of characters on X. Then $\mathrm{int}(\mathcal{C})$ has a unique proper minimal triangulation if and only if*

1. *$\mathrm{int}(\mathcal{C})$ has a $F_{w(\mathcal{C})}$–zero minimal triangulation; and*
2. *$\Delta_{F_{w(\mathcal{C})}}^{\min}$ is a maximal set of pairwise-parallel minimal separators of $\mathrm{int}(\mathcal{C})$.*

Proof. Suppose $\mathrm{int}(\mathcal{C})$ has a unique proper minimal triangulation H^*. Then H^* is a $F_{w(\mathcal{C})}$–zero minimal triangulation of $\mathrm{int}(\mathcal{C})$, and further, every $F_{w(\mathcal{C})}$–minimum minimal triangulation of $\mathrm{int}(\mathcal{C})$ is a $F_{w(\mathcal{C})}$–zero minimal triangulation of $\mathrm{int}(\mathcal{C})$. Each minimal separator of H^* is a minimal separator of $\mathrm{int}(\mathcal{C})$ by Theorem 14, so $\Delta_{H^*} \subseteq \Delta_{F_{w(\mathcal{C})}}^{\min}$. Alternatively, if $S \in \Delta_{F_{w(\mathcal{C})}}^{\min}$, then S is a minimal separator of a $F_{w(\mathcal{C})}$–zero minimal triangulation of $\mathrm{int}(\mathcal{C})$. This minimal triangulation is proper, so $S \in \Delta_{H^*}$ by uniqueness. Therefore $\Delta_{F_{w(\mathcal{C})}}^{\min} = \Delta_{H^*}$, and $\Delta_{F_{w(\mathcal{C})}}^{\min}$ is a maximal pairwise-parallel set of minimal separators of $\mathrm{int}(\mathcal{C})$ by Theorem 14.

To prove the converse, suppose that $\mathrm{int}(\mathcal{C})$ has a $F_{w(\mathcal{C})}$–zero minimal triangulation, and $\Delta_{F_{w(\mathcal{C})}}^{\min}$ is a maximal set of pairwise-parallel minimal separators of $\mathrm{int}(\mathcal{C})$. By Theorem 14, the graph H obtained from $\mathrm{int}(\mathcal{C})$ by saturating each minimal separator in $\Delta_{F_{w(\mathcal{C})}}^{\min}$ is a minimal triangulation of $\mathrm{int}(\mathcal{C})$, and further, for each fill edge uv of H, there is a $S' \in \Delta_{F_{w(\mathcal{C})}}^{\min}$ such that $u, v \in S'$. By definition there is some $F_{w(\mathcal{C})}$–zero minimal triangulation that has S' as a minimal separator. This triangulation has uv as a fill edge by Theorem 14 so $F_{w(\mathcal{C})}(uv) = 0$. Therefore H is an $F_{w(\mathcal{C})}$–zero minimal triangulation of $\mathrm{int}(\mathcal{C})$, so H is a proper minimal triangulation of $\mathrm{int}(\mathcal{C})$.

Now let H' be any proper minimal triangulation of $\mathrm{int}(\mathcal{C})$. Then H' is an $F_{w(\mathcal{C})}$–zero minimal triangulation of $\mathrm{int}(\mathcal{C})$, so $\Delta_{H'} \subseteq \Delta_{F_{w(\mathcal{C})}}^{\min}$. We assumed

$\Delta^{\min}_{F_w(\mathcal{C})}$ is pairwise-parallel, and $\Delta_{H'}$ is maximal with respect to being pairwise-parallel by Theorem 14, so $\Delta_{H'} = \Delta^{\min}_{F_w(\mathcal{C})}$. Thus both H and H' are obtained from int$(\mathcal{C})$ by saturating each minimal separator of $\Delta_{H'} = \Delta^{\min}_{F_w(\mathcal{C})}$, so $H' = H$. Therefore H is the unique proper minimal triangulation of int$(\mathcal{C})$. $\square$

4 Finding Weighted Minimum Triangulations

To prove our algorithmic results, now we show that both $\text{mfi}_{F_w}(G)$ and $\Delta^{\min}_{F_w}$ can be computed in $O(|X||\mathcal{C}|^2 + (r|\mathcal{C}|)^4 |\Delta_{\text{int}(\mathcal{C})}|^2)$ time.

A *block* of a graph G is a pair (S, C) where $S \in \Delta_G$ and C is a connected component of $G - S$, and it is *full* or *full with respect to* S if every vertex of S has at least one neighboring vertex that is in C (we write $N(C) = S$). The *realization* of a block (S, C) is the graph $R(S, C)$ with vertex set $S \cup C$, and for any u and v in $S \cup C$, uv is an edge of $R(S, C)$ if either uv is an edge of G or $uv \in \text{pf}(S)$.

Kloks, Kratsch, and Spinrad [22] related minimal triangulations to their minimal separators and block structure to show that minimum fill can be computed recursively. The following lemma follows with a slight modification of the proof of Theorem 3.4 in [22], so we omit it.

Lemma 16. *Let G be a non-complete graph and F_w be a fill weight on G. Then*

$$\text{mfi}_{F_w}(G) = \min_{S \in \Delta_G} \left(\text{fill}_{F_w}(S) + \sum_C \text{mfi}_{F_w}(R(S, C)) \right)$$

where the sum occurs over the connected components C of $G - S$ and $\text{fill}_{F_w}(S) = \sum_{uv \in \text{pf}(S)} F_w(uv)$.

It turns out that non-full blocks with respect to $S \in \Delta_G$ are full blocks with respect to a different minimal separator of G. They also allow us to compute $\text{mfi}_{F_w}(R(S, C))$, which is a useful fact for later when we restrict our attention to full blocks of G.

Lemma 17. *[9] Let G be a graph, $S \in \Delta_G$, and C be a connected component of $G - S$. If $N(C) = S' \subset S$, then (S', C) is a full block of G (i.e. $S' \in \Delta_G$). Further, if $E' \subseteq \text{pf}(C)$, then the graph obtained from $R(S, C)$ by adding the fill edges in E' is a minimal triangulation of $R(S, C)$ if and only if the graph obtained from $R(S', C)$ by adding the fill edges in E' is a minimal triangulation of $R(S, C)$.*

This gives us the following, an extension of Corollary 4.5 in [9].

Lemma 18. *Let G be a graph, $S \in \Delta_G$, and C be a connected component of $G - S$. If $N(C) = S' \subset S$, then $\text{mfi}_{F_w}(R(S, C)) = \text{mfi}_{F_w}(R(S', C))$ for any fill weight F_w.*

Proof. (Sketch) Let H be a triangulation of $R(S, C)$ such that $\text{mfi}_{F_w}(R(S, C)) = F_w(H)$. Restricting H to $C \cup S'$ yields a triangulation of $R(S', C)$, because any cycles in this graph must appear in H as well. Therefore $\text{mfi}_{F_w}(R(S, C)) \geq \text{mfi}_{F_w}(R(S', C))$.

To obtain $\text{mfi}_{F_w}(R(S, C)) \leq \text{mfi}_{F_w}(R(S', C))$, note that any triangulation H' of $R(S', C)$ induces a triangulation of $R(S, C)$ via its fill edges. The only cycles that appear in $R(S, C)$ after adding the fill edges of H' but are not also cycles of $R(S, C)$ must contain a vertex of $S - S'$. This vertex must be adjacent to two vertices of S', and S' induces a clique in $R(S, C)$ because S has been saturated. These vertices form a chord for this cycle. □

In order to compute $\text{mfi}_{F_w}(R(S, C))$, we need the notion of a potential maximal clique. Let G be a graph and K be a subset of its vertices. Then K is a *potential maximal clique* of G if there is a minimal triangulation H of G and K is a maximal clique of H. That is, every pair of vertices in K are adjacent in H, and no proper superset of K has this property. The set of potential maximal cliques of G is denoted by Π_G. The next two lemmas describe the interplay between potential maximal cliques, minimal separators, and blocks.

Lemma 19. *[9] Let G be a graph and K be a potential maximal clique of G. Then $S \in \Delta_G$ and $S \subseteq K$ if and only if $N(C) - S$ for some connected component C of $G - K$.*

Therefore if $K \in \Pi_G$ and $C_1, C_2, \ldots, C_k$ are the connected components of $G - K$, each (S_i, C_i) where $N(C_i) = S_i$ is a full block of G (i.e. $S_i \in \Delta_G$). These blocks are called the blocks *associated* to K.

Lemma 20. *[9] Suppose G is a graph, $S \in \Delta_G$, and (S, C) is a full block. Then $H(S, C)$ is a minimal triangulation of $R(S, C)$ if and only if*

1. *there is a potential maximal clique K of G such that $S \subset K \subseteq (S, C)$; and*
2. *letting (S_i, C_i) for $1 \leq i \leq p$ be the blocks associated to K such that $S_i \cup C_i \subset S \cup C$, we have $E(H) = \bigcup_{i=1}^{p} E(H_i) \cup \text{pf}(K)$ where H_i is a minimal triangulation of $R(S_i, C_i)$ for each $1 \leq i \leq p$.*

The following lemma is an extension of Corollary 4.8 in [9].

Lemma 21. *Let (S, C) be a full block of G and F_w be a fill weight on G. Then*

$$\text{mfi}_{F_w}(R(S, C)) = \min_{S \subset K \subseteq (S, C)} \left(\text{fill}_{F_w}(K) - \text{fill}_{F_w}(S) + \sum \text{mfi}_{F_w}(R(S_i, C_i)) \right) \quad (1)$$

where the minimum is taken over all $K \in \Pi_G$ such that $S \subset K \subseteq (S, C)$, and (S_i, C_i) are the blocks associated to K in G such that $S_i \cup C_i \subset S \cup C$.

Proof. Omitted due to space constraints. See [17] for a proof. □

Theorem 22. *Let $\mathcal{C}$ be a set of partial characters on X with at most r states, and F_w be a fill weight on $\text{int}(\mathcal{C})$. There is an $O(|X||\mathcal{C}|^2 + (r|\mathcal{C}|)^4 |\Delta_{\text{int}(\mathcal{C})}|^2)$ algorithm that computes $\text{mfi}_{F_w}(\text{int}(\mathcal{C}))$ and $\Delta_{F_w}^{\min}$.*

Data: Partial characters $\mathcal{C}$ on X with at most r states
Result: $\mathrm{mfi}_{F_w}(\mathrm{int}(\mathcal{C}))$ and $\Delta_{F_w}^{\min}$
compute $\mathrm{int}(\mathcal{C})$;
compute $\Delta_{\mathrm{int}(\mathcal{C})}$ and $\Pi_{\mathrm{int}(\mathcal{C})}$;
// Find the F_w−minimum fill value for each full block
compute all the full blocks (S, C) and sort them by the number of vertices;
for *each full block (S, C) taken in increasing order* **do**

> $\mathrm{mfi}_{F_w}(R(S, C)) \leftarrow \mathrm{fill}_{F_w}(S \cup C)$ if (S, C) is inclusion-minimal;
> and $\mathrm{mfi}_{F_w}(R(S, C)) \leftarrow \infty$ otherwise;
> **for** *each potential maximal clique K s.t. $S \subset K \subseteq S \cup C$* **do**
>
> > compute the blocks (S_i, C_i) associated to K s.t. $S_i \cup C_i \subset S \cup C$;
> > newfill $\leftarrow \mathrm{fill}_{F_w}(K) - \mathrm{fill}_{F_w}(S) + \sum_i \mathrm{mfi}_{F_w}(R(S_i, C_i))$;
> > $\mathrm{mfi}_{F_w}(R(S, C)) \leftarrow \min(\mathrm{mfi}_{F_w}(R(S, C)), \text{newfill})$;
>
> **end**

end
$\mathrm{mfi}_{F_w}(\mathrm{int}(\mathcal{C})) \leftarrow \infty$;
// Find the F_w−minimum fill value for minimal triangulations
    containing S
for *each minimal separator S of $\mathrm{int}(\mathcal{C})$* **do**

> compute the blocks (S_i, C_i) associated to S where $N(C_i) = S_i$;
> $\mathrm{mfi}_{F_w}(S) \leftarrow \mathrm{fill}_{F_w}(S) + \sum_i \mathrm{mfi}_{F_w}(R(S_i, C_i))$;
> $\mathrm{mfi}_{F_w}(\mathrm{int}(\mathcal{C})) \leftarrow \min(\mathrm{mfi}_{F_w}(\mathrm{int}(\mathcal{C})), \mathrm{mfi}_{F_w}(S))$;

end
$\Delta_{F_{w(\mathcal{C})}}^{\min} \leftarrow \{S \in \Delta_{F_{w(\mathcal{C})}} \text{ s.t. } \mathrm{mfi}_{F_w}(S) = \mathrm{mfi}_{F_w}(\mathrm{int}(\mathcal{C}))\}$;
return $\mathrm{mfi}_{F_w}(\mathrm{int}(\mathcal{C}))$ and $\Delta_{F_w}^{\min}$;

Algorithm 1. Computing F_w−minimum fill

Proof. Our approach is described in Algorithm 1. Constructing $\mathrm{int}(\mathcal{C})$ can be done in $O((|X| + r^2)|\mathcal{C}|^2)$ time as follows. There are at most $r|\mathcal{C}|$ vertices of $\mathrm{int}(\mathcal{C})$, one per part of each character. Recall that a pair of vertices (A, χ) and (A', χ') of $\mathrm{int}(\mathcal{C})$ form an edge if and only if there is some $a \in A \cap A'$. For each $a \in X$, let $\mathcal{C}(a)$ be the vertices of $\mathrm{int}(\mathcal{C})$ whose state contains a. These sets are computed in $O(|X||\mathcal{C}|)$ amortized time by scanning each state of each character. The edges of $\mathrm{int}(\mathcal{C})$ are now found by examining each pair $(A_1, \chi_1)(A_2, \chi_2)$ in $\mathcal{C}(a)$ for all $a \in X$. To address redundancy, order the characters and states of each characters, then construct a table to check if (A_1, χ_1) and (A_2, χ_2) has already been found as an edge. Examining $\mathcal{C}(a)$ to find these edges takes $O(|\mathcal{C}|^2)$ time (because a is in at most one state per character), and constructing the redundancy table takes $O((r|\mathcal{C}|)^2)$ time, for a total of $O((|X| + r^2)|\mathcal{C}|^2)$ time.

For a general graph G with $|V|$ vertices and $|E|$ edges, it is possible to compute Δ_G in $O(|V|^3|\Delta_G|)$ time [2] and Π_G in $O(|V|^2|E||\Delta_G|^2)$ time [10]. Let n be the number of vertices of $\mathrm{int}(\mathcal{C})$. The full block computation and nested for loop can be implemented in $O(n^3|\Pi_{\mathrm{int}(\mathcal{C})}|)$ time, which follows from the proof of Theorem 3.4 in [14]. It is known that $|\Pi_G| \leq |V||\Delta_G|^2 + |V||\Delta_G| + 1$ [10], so the nested for loop takes $O(n^4|\Delta_{\mathrm{int}(\mathcal{C})}|^2)$ time.

Consider the second for loop and let $S \in \Delta_{\text{int}(\mathcal{C})}$. The blocks associated to S are found in $O(n^2)$ time by searching the graph to find the connected components, and then computing $N(C)$ for each connected component C of $G - S$ (in the second computation, each edge of the graph is examined at most once). By Lemma 18, each (S_i, C_i) is a full block of G, so we have calculated $\text{mfi}_{F_w}(R(S_i, C_i))$ during the first for loop. The calculation on line 17 matches the one in Lemma 16 because $\text{mfi}_{F_w}(R(S, C_i)) = \text{mfi}_{F_w}(R(S_i, C_i))$ by Lemma 18. It takes $O(n^2)$ time to compute $\text{fill}_{F_w}(S)$, so the second for loop takes $O(n^2|\Delta_{\text{int}(\mathcal{C})}|)$ time. The last line of the algorithm takes $O(|\Delta_{\text{int}(\mathcal{C})}|)$ time. Aside from the $O(|X||\mathcal{C}|^2)$ term, the bottleneck of the algorithm is the first nested for loop and calculating $\Pi_{\text{int}(\mathcal{C})}$, so the entire algorithm runs in $O(|X||\mathcal{C}|^2 + (r|\mathcal{C}|)^4|\Delta_{\text{int}(\mathcal{C})}|^2)$ time. □

Theorems 2, 5, and 7 follow from Theorem 22.

Acknowledgements. This research was partially supported by NSF grants IIS-0803564 and CCF-1017580. We thank the anonymous referees for their helpful comments that led to improvements in the exposition.

References

1. Agarwala, R., Fernández-Baca, D.: A polynomial-time algorithm for the perfect phylogeny problem when the number of character states is fixed. SIAM Journal on Computing 23, 1216–1224 (1994)
2. Berry, A., Bordat, J., Cogis, O.: Generating all the minimal separators of a graph. International Journal of Foundations of Computer Science 11(3), 397–403 (2000)
3. Bininda-Emonds, O.R.: The evolution of supertrees. Trends in Ecology and Evolution 19(6), 315–322 (2004)
4. Blair, J., Peyton, B.: An introduction to chordal graphs and clique trees. In: George, J., Gilbert, J., Liu, J.H. (eds.) Graph Theory and Sparse Matrix Computations, IMA Volumes in Mathematics and its Applications, vol. 56, pp. 1–27. Springer (1993)
5. Bodlaender, H., Heggernes, P., Villanger, Y.: Faster parameterized algorithms for minimum fill-in. Algorithmica 61, 817–838 (2011)
6. Bodlaender, H., Fellows, M., Warnow, T.: Two strikes against perfect phylogeny. In: Kuich, W. (ed.) ICALP 1992. LNCS, vol. 623, pp. 273–283. Springer, Heidelberg (1992)
7. Bonet, M., Linz, S., John, K.S.: The complexity of finding multiple solutions to betweenness and quartet compatibility. IEEE/ACM Transactions on Computational Biology and Bioinformatics 9(1), 273–285 (2012)
8. Bordewich, M., Huber, K., Semple, C.: Identifying phylogenetic trees. Discrete Mathematics 300(1-3), 30–43 (2005)
9. Bouchitté, V., Todinca, I.: Treewidth and minimum fill-in: Grouping the minimal separators. SIAM Journal on Computing 31(1), 212–232 (2001)
10. Bouchitté, V., Todinca, I.: Listing all potential maximal cliques of a graph. Theoretical Computer Science 276(1-2), 17–32 (2002)
11. Brinkmeyer, M., Griebel, T., Böcker, S.: Polynomial supertree methods revisited. Advances in Bioinformatics (2011)

12. Buneman, P.: A characterisation of rigid circuit graphs. Discrete Mathematics 9(3), 205–212 (1974)
13. Fernández-Baca, D.: The perfect phylogeny problem. In: Cheng, X., Du, D.Z. (eds.) Steiner Trees in Industry, pp. 203–234. Kluwer (2001)
14. Fomin, F., Kratsch, D., Todinca, I., Villanger, Y.: Exact algorithms for treewidth and minimum fill-in. SIAM Journal on Computing 38(3), 1058–1079 (2008)
15. Grünewald, S., Huber, K.: Identifying and defining trees. In: Gascuel, O., Steel, M. (eds.) Reconstructing Evolution: New Mathematical and Computational Advances, pp. 217–246. Oxford University Press (2007)
16. Gusfield, D.: The multi–state perfect phylogeny problem with missing and removable data: solutions via integer–programming and chordal graph theory. Journal of Computational Biology 17(3), 383–399 (2010)
17. Gysel, R.: Potential maximal clique algorithms for perfect phylogeny problems. Pre-print: arXiv 1303.3931 (2013)
18. Gysel, R., Gusfield, D.: Extensions and improvements to the chordal graph approach to the multistate perfect phylogeny problem. IEEE/ACM Transactions on Computational Biology and Bioinformatics 8(4), 912–917 (2011)
19. Habib, M., Stacho, J.: Unique perfect phylogeny is intractable. Theoretical Computer Science 476, 47 – 66 (2013)
20. Heggernes, P.: Minimal triangulations of graphs: a survey. Discrete Mathematics 306(3), 297–317 (2006)
21. Hudson, R.: Generating samples under a wright-fisher neutral model of genetic variation. Bioinformatics 18(2), 337–338 (2002)
22. Kloks, T., Kratsch, D., Spinrad, J.: On treewidth and minimum fill-in of asteroidal triple-free graphs. Theoretical Computer Science 175(2), 309–335 (1997)
23. McMorris, F., Warnow, T., Wimer, T.: Triangulating vertex–colored graphs. SIAM Journal of Discrete Mathematics 7, 296–306 (1994)
24. Meacham, C.: Theoretical and computational considerations of the compatibility of qualitative taxonomic characters. In: Felsenstein, J. (ed.) Numerical Taxonomy. NATO ASI Series G, vol. 1, pp. 304–314. Springer (1983)
25. Parra, A., Scheffler, P.: How to use the minimal separators of a graph for its chordal triangulation. In: Fülöp, Z. (ed.) ICALP 1995. LNCS, vol. 944, pp. 123–134. Springer, Heidelberg (1995)
26. Parra, A., Scheffler, P.: Characterizations and algorithmic applications of chordal graph embeddings. Discrete Applied Mathematics 79(1-3), 171–188 (1997)
27. Ross, H., Rodrigo, A.: An assessment of matrix representation with compatibility in supertree construction. In: Bininda-Emonds, O. (ed.) Phylogenetic supertrees: Combining information to reveal the Tree of Life, pp. 35–63. Kluwer Academic Publishers (2004)
28. Semple, C., Steel, M.: A characterization for a set of partial partitions to define an X-tree. Discrete Mathematics 247(1-3), 169–186 (2002)
29. Semple, C., Steel, M.: Phylogenetics. In: Oxford Lecture Series in Mathematics and Its Applications. Oxford University Press (2003)
30. Steel, M.: The complexity of reconstructing trees from qualitative characters and subtrees. Journal of Classification 9(1), 91–116 (1992)
31. Yannakakis, M.: Computing the minimum fill-in is NP-complete. SIAM Journal on Algebraic Discrete Methods 2, 77–79 (1981)

On Computability and Learnability
of the Pumping Lemma Function

Dariusz Kalociński*

Institute of Philosophy
University of Warsaw
ul. Krakowskie Przedmieście 3
00-927 Warsaw, Poland
dariusz.kalocinski@gmail.com

Abstract. On the basis of the well known pumping lemma for regular languages we define such a partial function $f : \mathbb{N} \to \mathbb{N}$ that for every e it yields the least pumping constant for the language W_e. We ask whether f is computable. Not surprisingly f turns out to be non-computable. Then we check whether f is algorithmically learnable. This is also proved not to be the case. Further we investigate how powerful oracle is necessary to actually learn f. We prove that f is learnable in $0'$. We also prove some facts relating f to arithmetical hierarchy.

Keywords: pumping lemma, computability, algorithmic learning, arithmetical hierarchy, reducibility.

1 The Pumping Lemma Function

From automata theory one knows the pumping lemma for regular languages [4]. Before we formulate the lemma, let us fix some terminology. Let W_e denote the domain of the partial function computed by the Turing machine with Gödel number e. By $R(e,c)$ we understand the following statement: *for each word $\omega \in W_e$, if $|\omega| > c$, then there are words α,β,γ such that $\omega = \alpha\beta\gamma$, $\beta \neq \varepsilon$, $|\alpha\beta| \leq c$ and for all $i \in \mathbb{N}$ $\alpha\beta^i\gamma \in W_e$* . We may now formulate the pumping lemma for regular languages easily: for each e such that W_e is a regular language there is a positive integer c such that $R(e,c)$. On the basis of this familiar result, one can define a partial function $f : \mathbb{N} \to \mathbb{N}$ (here called the pumping lemma function) that for each $e \in \mathbb{N}$ yields the least c such that $R(e,c)$, if any such c exists, and otherwise is undefined. A natural question arises: is f computable? The solution to this question is negative. Another problem which arises is as follows: is the complement of the graph of f recursively enumerable? This is also proved not to be the case. Further we investigate algorithmic learnability of f. We prove that f is not algorithmically learnable. Then we ask how powerful oracle is necessary to actually learn f. We prove that f is algorithmically learnable with the halting problem in oracle. We finish the article by relating f to arithmetical hierarchy.

---

* The author is a PhD student in the Department of Logic, Institute of Philosophy, University of Warsaw.

A.-H. Dediu et al. (Eds.): LATA 2014, LNCS 8370, pp. 433–440, 2014.

2   Terminology and Notation

In this section we provide terminology and notation used throughout the article. Notions used locally are defined when they are needed. For further details on computability consult [1,6].

In Sect. 1 we have already introduced some notation, namely W_e and $R(e, c)$. Sometimes we use R for the binary relation expressed by the formula $R(e, c)$. The appropriate meaning of R shall be clear from the context. By G_h we denote the graph of the (possibly partial) function h. The operation of taking the complement of a relation $S \subseteq \mathbb{N}^k$ is defined as follows: $S' = \mathbb{N}^k - S$. Inputs of algorithms are words or numbers. We can assume that algorithms are Turing machines that work with words over binary alphabet. Binary words are easily coded as numbers. We do not make any explicit distinction between numeric inputs and string inputs - a particular usage will be clear from context. The length of the word x is denoted by $lh(x)$ or $|x|$. By $\leq_{bl}$ we denote bounded lexicographical order on strings. Let x, y be words. We say x is less or equal to y with respect to bounded lexicographical order (in symbols $x \leq_{bl} y$), if $|x| < |y|$ or both $|x| = |y|$ and x is lexicographically less or equal to y. The characteristic function of the set $A \subseteq \mathbb{N}$ is denoted by c_A. By coding of pairs we mean some fixed reasonable coding, for example Cantor pairing function. By π_i, $i = 1, 2$, we mean the canonical projection of a pair on the i-th coordinate. The symbol $\leq_m$ refers to the relation of many-one reducibility. We write $h : A \leq_m B$ to express the fact that h is total recursive and $x \in A \Leftrightarrow h(x) \in B$, for all $x \in \mathbb{N}$. We use $T(e, x, c)$ for the Kleene predicate, where e is a Gödel number of a Turing machine, x stands for an input and c for a computation. $U(c)$ refers to the output of a computation c. By EMPTY we denote the emptiness problem, i.e the set $\{e \in \mathbb{N} : W_e = \emptyset\}$. NOTEMPTY stands for the non-emptiness problem, i.e. the set $\mathbb{N} - \text{EMPTY}$. TOT denotes the totality problem, i.e. $\{e : \forall x \exists c\, T(e, x, c)\}$. The halting problem $\{(e, x) \in \mathbb{N}^2 : \exists c\, T(e, x, c)\}$ is denoted by HALT. We use the standard notation Σ_k^0, Π_k^0, Δ_k^0 for the classes of sets in arithmetical hierarchy.

3   Non-computability Results

Lemma 1. EMPTY $\leq_m R$.

Proof. We define the function $r(e) = (\sigma(e), 1)$, where σ is the total computable function obtained through *smn* theorem from $g(e, x)$ which is computed as follows: We examine whether W_e is empty. If it is empty, the computation goes on forever; otherwise emptiness checking procedure stops. In that case we measure the length of x and if it is even we return 1. Otherwise we loop forever.

Therefore if $W_e = \emptyset$, then $W_{\sigma(e)} = \emptyset$ and $(\sigma(e), 1) \in R$. If $W_e \neq \emptyset$ then $W_{\sigma(e)}$ contains all words of even length. In this case clearly $(\sigma(e), 1) \notin R$ (otherwise W_e would have contained words of odd length). $\qquad \square$

Lemma 2. *If $R(e, c)$ then $(\forall d > c)\, R(e, d)$.*

Proof. Directly from the definition of $R(e, c)$. $\qquad\qquad\qquad\qquad\square$

Theorem 3. *f is not computable.*

Proof. Suppose for the sake of contradiction that f is computable. Then of course R is recursively enumerable, that is Σ_1^0 (use the fact that the graph of recursive function is r.e. and, bearing in mind the Lemma 2, devise an algorithm for enumerating R). Taking into account that EMPTY is Π_1^0-complete, we have $A \leq_m$ EMPTY $\leq_m R \in \Sigma_1^0$ for all $A \in \Pi_1^0$. It follows that $\Pi_1^0 \subseteq \Sigma_1^0$, which is impossible, because it is well known that $\Pi_1^0 - \Sigma_1^0 \neq \emptyset$. $\qquad\square$

Lemma 4. NOTEMPTY $\leq_m R$.

Proof. We define the function $r(e) = (\sigma(e), 1)$, where σ is the total computable function obtained through *smn* theorem from $g(e, x)$ which is computed as follows. If the length of x is even, then stop. Otherwise start checking, whether $W_e \neq \emptyset$. If $W_e \neq \emptyset$ then - when a word α is found such that $\alpha \in W_e$ - stop. Otherwise (if $W_e = \emptyset$) we loop forever.

Therefore if $W_e = \emptyset$ then $W_{\sigma(e)}$ contains all words of even length, so $(\sigma(e), 1) \notin R$. If $W_e \neq \emptyset$, then $W_{\sigma(e)}$ contains all words and $(\sigma(e), 1) \in R$. $\qquad\square$

Lemma 5. *If G'_f is r.e., then R' is r.e.*

Proof. Let $p : \mathbb{N}^2 \to \mathbb{N}$ be the recursive partial characteristic function of G'_f, i.e.

$$p(x, y) = \begin{cases} 1 & \text{if } (x, y) \notin G_f \\ \text{undefined} & \text{otherwise} \end{cases} . \tag{1}$$

We define $h : \mathbb{N}^2 \to \mathbb{N}$ - the recursive partial characteristic function of R':

$$h(x, y) = \Pi_{i=0}^{y} p(x, i) . \tag{2}$$

Assume $(x, y) \in R'$. Then it must be the case that $(\forall i \leq y) (x, i) \in R'$. Therefore $(\forall i \leq y) (x, i) \notin G_f$ and thus $(\forall i \leq y) p(x, i) = 1$.

Now assume $(x, y) \notin R'$. Then $(x, y) \in R$. So it must be the case that $f(x) \leq y$. Therefore $\Pi_{i=0}^{y} p(x, i)$ is undefined, since $p(x, f(x))$ is undefined and $0 \leq f(x) \leq y$. $\qquad\square$

Theorem 6. *G'_f is not r.e.*

Proof. Suppose on the contrary that G'_f is r.e. Then by Lemma 5 R' is r.e. We use the following fact from recursion theory: $A \leq_m B \Leftrightarrow A' \leq_m B'$. We apply it to NOTEMPTY $\leq_m R$ and obtain EMPTY $\leq_m R'$. Then the reasoning is analogous to the proof of the Theorem 3. $\qquad\square$

4   Non-learnability Result

We established some lower bounds on the complexity of the pumping lemma function: we know that f is not recursive and that G'_f is not r.e. In this section we show that f is not algorithmically learnable which means that G_f is not Σ_2^0.

Definition 7. *Let* $f : \mathbb{N}^k \to \mathbb{N}$ *be a (possibly partial) function. We say that* f *is algorithmically learnable (shortly: learnable) if there is a total computable function* $g_t(\overline{x})$[1] *such that for all* $\overline{x} \in \mathbb{N}^k$:

$$lim_{t \to \infty} g_t(\overline{x}) = f(\overline{x}) , \tag{3}$$

which means that neither $f(\overline{x})$ *nor* $lim_{t \to \infty} g_t(\overline{x})$ *exist or - alternatively - both* $f(\overline{x})$ *and* $lim_{t \to \infty} g_t(\overline{x})$ *exist and are equal.*

The following lemma is a familiar result from the algorithmic learning theory. Classical papers related to the subject are [2,3,5].

Lemma 8. *Let* $f : \mathbb{N}^k \to \mathbb{N}$. f *is learnable if and only if* G_f *is* Σ_2^0.

Proof. ($\Rightarrow$) Let f be learnable and $g_t(\overline{x})$ be total computable function satisfying the Equation (3). We observe that $G_f(\overline{x}, y) \Leftrightarrow (\exists t \in \mathbb{N})(\forall k > t) y = g_k(\overline{x})$. The formula $y = g_k(\overline{x})$ defines a recursive ternary relation, because g is total computable.

($\Leftarrow$) Let f be such that G_f is Σ_2^0. Choose a recursive relation $A \subseteq \mathbb{N}^{k+3}$ such that $G_f(\overline{x}, y) \Leftrightarrow \exists z \forall w A(\overline{x}, y, z, w)$. We define an infinite procedure $G(\overline{x})$ (Alg. 1) that is easily convertible to the appropriate definition of total computable learning function $g_t(\overline{x})$ satisfying Equation 3.

Data: $\overline{x}$
Result: y contains hypothesized value $f(\overline{x})$
$p, w, y, z \leftarrow 0$;
while *true* **do**
  $y \leftarrow \pi_1(p)$;
  $z \leftarrow \pi_2(p)$;
  if $c_A(\overline{x}, y, z, w) = 1$ **then**
    $w \leftarrow w + 1$;
  else
    $p \leftarrow p + 1$;
    $w \leftarrow 0$;
  end
end

Algorithm 1. The infinite procedure $G(\overline{x})$

The procedure simply searches for y satisfying $\exists z \forall w A(\overline{x}, y, z, w)$ by enumerating all possible pairs (y, z). If there is such y, the procedure will finally spot it together with the relevant witness z. Therefore $f(\overline{x})$ will be finally stored in the variable y and from that point the contents of y will never change. On the other hand, if there is no such y, the contents of the variable y will continue to change ad infinitum. □

---

[1] Expression $g_t(\overline{x})$ shall be read as $g(t, \overline{x})$. In precise terms, $g_t(\overline{x})$ stands for a sequence of functions. We use indexed t to distinguish the discrete time parameter from the input.

Lemma 9. $\mathsf{TOT} \leq_m R$.

Proof. Let $H(x, m, t)$ mean $(\exists c < t)T(x, m, c)$. Define a relation $S(x, y)$:

$$S(x, y) \Leftrightarrow_{\mathrm{df}} 2 \mid lh(y) \vee (\exists t)(\forall m \leq_{\mathrm{bl}} y)\, H(x, m, t) \ . \tag{4}$$

Of course, S is r.e. By $p_S(x, y)$ we denote recursive partial characteristic function of S. By smn theorem, there is a total computable function σ such that $\{\sigma(x)\}(y) \simeq p_S(x, y)$, where $\{\cdot\}$ refers to the function computed by Turing machine having Gödel number $\cdot$. Define $r(x) =_{\mathrm{df}} (\sigma(x), 1)$. We prove that $r : \mathsf{TOT} \leq_m R$.

$(\Rightarrow)$ Let $x' \in \mathsf{TOT}$. We begin by showing that $S(x', y)$ holds for all y. Fix y'. If $2 \mid lh(y')$, then obviously $S(x', y')$. Assume that $2 \nmid lh(y')$. Because $x' \in \mathsf{TOT}$, we choose a finite sequence of numbers $(t_\omega)_{\omega \leq_{\mathrm{bl}} y'}$ such that $H(x', \omega, t_\omega)$ holds for $\omega \leq_{\mathrm{bl}} y'$. Let $T = max\{t_\omega : \omega \leq_{\mathrm{bl}} y'\}$. Then we have $(\forall \omega \leq_{\mathrm{bl}} y')\, H(x', \omega, T)$, and - what follows - $(\exists t)(\forall \omega \leq_{\mathrm{bl}} y')\, H(x', \omega, t)$. Therefore $S(x', y')$.

The following are equivalent:

$(\forall y)\, S(x', y)$,
$(\forall y)\, p_S(x', y) = 1$,
$(\forall y)\, \{\sigma(x')\}(y) = 1$,
$\sigma(x') \in \mathsf{TOT}$.

It remains to show that $r(x') - (\sigma(x'), 1) \in R$, which is trivially true.

$(\Leftarrow)$ Let $x' \notin \mathsf{TOT}$. Choose y_0 to be the smallest word with respect to $\leq_{\mathrm{bl}}$ such that $\neg(\exists c)T(x', y_0, c)$. Observe that $(\forall y)\, [y_0 \leq_{\mathrm{bl}} y \wedge 2 \nmid lh(y) \Rightarrow \neg S(x', y)]$. For let $y_0 \leq_{\mathrm{bl}} y$, $2 \nmid lh(y)$ and suppose $S(x', y)$. It follows, that $(\exists t)H(x', y_0, t)$ which is a contradiction.

Let k be any number satisfying $lh(y_0) < 2k$. Observe that $1^{2k} \in W_{\sigma(x')}$, because $2 \mid lh(1^{2k})$. The only possible division of 1^{2k} into $\alpha\beta\gamma$ satisfying conditions $lh(\alpha\beta) \leq 1$, $\beta \neq \varepsilon$ is $\alpha = \varepsilon$, $\beta = 1$, $\gamma = 1^{2k-1}$. Consider $\alpha\beta^0\gamma = 1^{2k-1}$. Clearly $y_0 <_{\mathrm{bl}} 1^{2k-1}$. Therefore $\neg S(x', 1^{2k-1})$ and thus $1^{2k-1} \notin W_{\sigma(x')}$. We may then conclude that $r(x') = (\sigma(x'), 1) \notin R$. $\qquad \square$

We are ready to prove the main non-learnability theorem.

Theorem 10. *f is not learnable.*

Proof. Suppose for the sake of contradiction that f is learnable. By the Lemma 8, G_f is Σ_2^0. R can be defined as follows: $R(x, y) \Leftrightarrow \exists c(G_f(x, c) \wedge c \leq y)$. The right side of the equivalence is easily convertible to a Σ_2^0-formula. Thus R is Σ_2^0. Due to the Lemma 9, TOT is Σ_2^0. However, TOT is not Σ_2^0. Contradiction. $\quad \square$

5   Learnability Result

So far we have proved only negative results concerning computability or learnability of f. The question now rises how much we would need to strengthen our computational capabilities to turn this task into something learnable. In terms of recursion theory: how complex oracle we need to make f learnable? In this section we prove that f is learnable in $0'$.

Theorem 11. f *is learnable in* $0'$.

Proof. We choose $\mathsf{HALT} = \{(e, x) \in \mathbb{N}^2 : (\exists c)T(e, x, c)\}$ for the oracle. By $\phi(e, x)$ we denote the formula provided below. The formula $\phi(e, x)$ expresses the fact that $R(e, x)$:

$$x > 0 \wedge (\forall \omega) \{[\overbrace{(e, \omega) \in \mathsf{HALT} \wedge lh(\omega) \geq x}^{\sigma(x, e, \omega)}] \Rightarrow (\exists \alpha, \beta, \gamma \leq_{\mathrm{bl}} \omega)$$
$$\underbrace{[\alpha\beta\gamma = \omega \wedge lh(\alpha\beta) \leq x \wedge \beta \neq \varepsilon}_{\theta_1(x, \omega, \alpha, \beta, \gamma)} \wedge (\forall i) \underbrace{(e, \alpha\beta^i\gamma) \in \mathsf{HALT}}_{\theta_2(e, \alpha, \beta, \gamma, i)})]\} \ . \tag{5}$$

Observe that the relation $\{(x, \omega, e, \alpha, \beta, \gamma, i) : \theta_1 \wedge \theta_2\}$ is recursive in HALT. There is a relation η, recursive in HALT, such that $(\exists \alpha, \beta, \gamma \leq_{\mathrm{bl}} \omega)(\forall i)(\theta_1 \wedge \theta_2) \Leftrightarrow (\forall j)\,\eta(x, e, \omega, j)$.

The following are equivalent:

$\phi(e, x)$,

$x > 0 \wedge (\forall \omega) \{\sigma \Rightarrow (\exists \alpha, \beta, \gamma \leq_{\mathrm{bl}} \omega)[\theta_1 \wedge (\forall i)\,\theta_2]\}$,

$(\forall \omega) \{x > 0 \wedge [\sigma \Rightarrow (\exists \alpha, \beta, \gamma \leq_{\mathrm{bl}} \omega)(\forall i)(\theta_1 \wedge \theta_2)]\}$,

$(\forall \omega) \{x > 0 \wedge [\sigma \Rightarrow (\forall j)\,\eta]\}$,

$(\forall \omega)(\forall j) \{\underbrace{(x > 0 \wedge (\neg\sigma \vee \eta))}_{\xi(x, e, \omega, j)}\}$.

Thus, the relation defined by $\phi(e, x)$ is Π_1^0 in HALT. Now we express the fact that x is the least number such that $\phi(e, x)$:

$$\phi_{\mathrm{inf}}(e, x) := \phi(e, x) \wedge (\forall y < x)\neg(\forall \omega)(\forall j)\,\xi(e, x, \omega, j) \ . \tag{6}$$

The right conjunct of ϕ_{inf} is equivalent to a formula of the form $(\exists z)\zeta(e, x, z)$, where $\zeta(e, x, z)$ is recursive in HALT. Therefore we have

$$\phi_{\mathrm{inf}}(e, x) \Leftrightarrow (\forall \omega, j)\,\xi(e, x, \omega, j) \wedge (\exists z)\zeta(e, x, z) \ . \tag{7}$$

This easily leads us - by familiar first-order transformations - to the Σ_2^0 definition of the relation expressed by $\phi_{\mathrm{inf}}(e, x)$, in terms of relations recursive in HALT. Observe that the relation defined by $\phi_{\mathrm{inf}}(e, x)$ is G_f. Thus, G_f is Σ_2^0 in HALT. By the relativized version of the Lemma 8 f is learnable in HALT. $\square$

6   Supplement

We end the article by providing supplementary results relating f to arithmetical hierarchy.

Theorem 12. R *is* Π_2^0-*complete*.

Proof. We have already proved that $\mathsf{TOT} \leq_{\mathrm{m}} R$ (Lemma 9). It remains to show that R is Π_2^0. Consider the following definition of R:

$$x > 0 \wedge (\forall \omega) \{[(\exists c)T(e, \omega, c) \wedge \overbrace{lh(\omega) \geq x}^{\sigma(x, \omega)}] \Rightarrow (\exists \alpha, \beta, \gamma \leq_{\mathrm{bl}} \omega)$$
$$\underbrace{[\alpha\beta\gamma = \omega \wedge lh(\alpha\beta) \leq x \wedge \beta \neq \varepsilon}_{\theta(x, \omega, \alpha, \beta, \gamma)} \wedge (\forall i)(\exists c)T(e, \alpha\beta^i\gamma, c)]\} \tag{8}$$

Consider the implication enclosed in curly brackets:

$$((\exists c)T(e,\omega,c) \wedge \sigma) \Rightarrow (\exists \alpha, \beta, \gamma \leq_{\mathrm{bl}} \omega)(\theta \wedge (\forall i)(\exists c)T(e, \alpha\beta^{i}\gamma, c)) \qquad (9)$$

We proceed by equivalent reformulations of (9):

$$(\forall c)(\neg T(e,\omega,c) \vee \neg\sigma) \vee (\exists \alpha, \beta, \gamma \leq_{\mathrm{bl}} \omega)(\theta \wedge (\forall i)(\exists c)T(e, \alpha\beta^{i}\gamma, c)) ,$$

$$(\forall c)(\neg T(e,\omega,c) \vee \neg\sigma) \vee \underbrace{(\exists \alpha, \beta, \gamma \leq_{\mathrm{bl}} \omega)(\forall i)(\exists d)(\theta \wedge T(e, \alpha\beta^{i}\gamma, d))}_{\varphi} ,$$

$$(\forall c)(\neg T(e,\omega,c) \vee \neg\sigma) \vee \underbrace{(\forall i)(\exists d)\psi(x,e,\omega,i,d)}_{\varphi'} ,$$

$$(\forall c)(\forall i)(\exists d)(\neg T(e,\omega,c) \vee \neg\sigma \vee \psi(x,e,\omega,i,d)) .$$

Thus, (8) is equivalent to:

$$(\forall \omega)(\forall c)(\forall i)(\exists d)[x > 0 \wedge (\neg T(e,\omega,c) \vee \neg\sigma \vee \psi(x,e,\omega,i,d))] . \qquad (10)$$

We have used familiar first order tautologies and the fact that bounded quantifier prefix in φ can be somehow shifted inside, resulting in an equivalent Π_2^0-formula φ' such that its subformula ψ expresses a recursive relation. The above argument clearly shows that R is Π_2^0. $\qquad \square$

Theorem 13. G_f is Δ_3^0.

Proof. Let us denote by $\phi(e,x)$ the Π_2^0-formula (10) defining R. As in the proof of the Theorem 11 define G_f in the following way:

$$\phi(e,x) \wedge (\forall y < x)\neg\phi(e,y) \qquad (11)$$

The formula (11) is equivalent to a formula of the form $(\forall\exists \ldots \wedge \exists\forall \ldots)$, where $\ldots$ stand for some formulae expressing recursive relations. Proper shifts of quantifiers lead us to Π_3^0- and Σ_3^0-formula defining G_f. This clearly shows that G_f is Δ_3^0. $\qquad \square$

7   Remarks about Practical Significance

We addressed the problem of determining the least constant from the pumping lemma with a view to applying results to formal language learning framework.

One of the directions for further investigation can be as follows. Consider the machine placed in an unknown environment. The environment presents positive and negative examples of an unknown language L from an unknown class. Note that the environment that exhaustively presents both positive and negative examples can be viewed as an oracle for the input language L. Thus the Theorem 11 may be applied, since in such an environment the machine is equipped with an analogue of the halting problem and the only queries to the halting problem that are important for determining the least constant for L are of the form ,,$\alpha \in L$?". Suppose the machine is equipped with the learning procedure for f as described above. The pumping lemma for regular languages gives a necessary condition for a language to be regular. Let the input language $L = W_e$, for some e. If $\neg \exists c R(e, c)$, then W_e is not regular and the learning procedure for f diverges. This fact can be used as a heuristic and the machine can hypothesize that the input language is not regular. On the other hand, if $\exists c R(e, c)$, then the learning procedure for f converges and the machine can use this fact as a heuristic and conjecture that input language is regular or at least exclude certain languages from consideration.

We can put further constraints on input languages to make the applications more practical. Since for every regular language L there is a Turing machine e that computes the characteristic function c_L in linear time, we can restrain the working time of input machine e by a suitable quadratic polynomial p, with no worry of omitting any regular language. If the computation of e on the input α does not stop after $p(|\alpha|)$ steps, the answer to ,,$\alpha \in W_e$?" is set to no. By including such time constraints, we in fact obtain the learning algorithm of the pumping lemma function for languages decidable in quadratic polynomial time. In this setting, the learning algorithm may be used as a supplementary heuristic for hypothesizing whether input language is regular.

Acknowledgments. I would like to thank M.T. Godziszewski (University of Warsaw) for fruitful discussions. Furthermore, I would like to thank W. Homenda (Warsaw University of Technology) who taught me formal language theory. He proposed to address the problem of learning the pumping lemma function. Eventually, a team of computer science students, including me, M. Figat and B.V. Nam, literally wrote the learning algorithm for the context-free languages. Last but not least I would like to thank M. Mostowski (University of Warsaw) who taught me logic, computability theory and algorithmic learning theory.

References

1. Cutland, N.: Computability, an Introduction to Recursive Function Theory. Cambridge University Press (1980)
2. Gold, E.M.: Limiting recursion. J. Symbolic Logic 30, 28–48 (1965)
3. Gold, E.M.: Language identification in the limit. Information and Control 10, 447–474 (1967)
4. Hopcroft, J.E., Ullman, J.D.: Introduction to Automata Theory, Languages and Computation. Addison-Wesley, Cambridge (1979)
5. Putnam, H.: Trial and error predicates and the solution to a problem of Mostowski. J. Symbolic Logic 30, 49–57 (1965)
6. Shoenfield, J.R.: Recursion Theory. Lecture Notes in Logic, vol. 1. Springer, Berlin (1993)

Interval Temporal Logic Semantics
of Box Algebra

Hanna Klaudel[1], Maciej Koutny[2], and Zhenhua Duan[3]

[1] IBISC, Université d'Évry-Val-d'Essonne
23 boulevard de France, 91037 Évry Cedex, France
hanna.klaudel@ibisc.univ-evry.fr
[2] School of Computing Science, Newcastle University
Claremont Tower, Claremont Road, Newcastle upon Tyne, NE1 7RU, UK
maciej.koutny@newcastle.ac.uk
[3] Institute of Computing Theory and Technology, Xidian University
P.O. Box 177, No.2 South Tai Bai Road, Xi'an 710071, P.R. China
zhhduan@mail.xidian.edu.cn

Abstract. By focusing on two specific formalisms, viz. Box Algebra and
Interval Temporal Logic, we extend the recently introduced translation
of Petri nets into behaviourally equivalent logic formulas. We remove restrictions concerning the way in which the control flow of a concurrent
system is modelled, and allow for a fully general synchronisation operator. Crucially, we strengthen the notion of equivalence between a Petri
net and the corresponding logic formula, by proving such an equivalence
at the level of transition based executions of Petri nets, rather than just
by considering their labels.

Keywords: Interval Temporal Logic, Box Algebra, Petri net, composition, semantics, general synchronisation, step sequence, equivalence.

1 Introduction

In general, temporal logics [1] and Petri nets [4] are regarded as fundamentally
different approaches to the specification and analysis of concurrent systems.
Temporal logics allow one to specify both the system designs and correctness
requirements within the same framework, and the verification of correctness can
be done by checking the satisfaction of logic formulas. Petri nets, on the other
hand, are an automata inspired model with semantics based on actions and local
states which allows one to capture causal relationships in systems' behaviour.
As a result, verification of behavioural properties can then be carried out using
invariant techniques [14] based on the graph structure of nets, or model checking
techniques based on partial order reductions [15].

To establish a semantical link between logics and Petri nets, we focused in [5]
on two specific formalisms, Interval Temporal Logic (ITL) [11,12] and Box Algebra (BA) [2], which are closely related by the their compositional approaches to
constructing systems' descriptions. In particular, in both ITL and BA the control flow of a system is specified using commonly used programming operators,

A.-H. Dediu et al. (Eds.): LATA 2014, LNCS 8370, pp. 441–452, 2014.

such as sequence, choice, parallelism and iteration. The synchronisation between concurrently executed subsystems is, however, achieved in different ways and therefore needs to be handled carefully when relating the two models. In [5] we proposed a translation from a submodel of BA to semantically equivalent ITL formulas.

In this paper, we extend the results of [5] which provided a compositional translation from a sub-model of BA [2]. In particular, we drop constraints forbidding the use of the parallel composition outside the topmost level, and consider a fully general (multi-way) synchronisation operator. Crucially, we strengthen the notion of equivalence between nets and the corresponding formulas, by proving such an equivalence at the level of transition based executions of Petri nets, rather than just by considering their labels. Full details, examples and proofs are provided in: `http://www.cs.ncl.ac.uk/publications/trs/papers/1373.pdf`

Throughout the paper, $\mathbb{N}$ denotes the set of all positive integers, $\mathbb{N}_0 = \mathbb{N} \cup \{0\}$ and $\mathbb{N}_\omega = \mathbb{N}_0 \cup \{\omega\}$, where ω denotes the first transfinite ordinal. We extend to $\mathbb{N}_\omega$ the standard arithmetic comparison operators, assuming that $n < \omega$, for all $n \in \mathbb{N}_0$. Moreover, we define $\preceq$ as $\leq$ without the pair (ω, ω). The concatenation operator for sequences of sets will be denoted by "$\circ$", and for sequences of symbols by "$\cdot$". For a symbol s and a set of sequences S, we will write $s \cdot S$ to denote the set $\{s \cdot s' \mid s' \in S\}$. We also denote $\varnothing^\infty = \{\varnothing\varnothing \dots\}$ and $\varnothing^* = \{\epsilon, \varnothing, \varnothing\varnothing, \dots\}$, i.e., $\varnothing^\infty$ comprises a single infinite sequence. The k-th element of a sequence θ is denoted by $\theta_{(k)}$, and its length by $|\theta|$.

In what follows, $\mathcal{A}$ is a set of actions, and a synchronisation relation ρ is any set of tuples of actions $(a_1, \dots, a_n, a)$ with $n \geq 1$. Intuitively, the a_i's represent n concurrent actions which can be synchronised to yield a single action with the label a. To reflect this intuition we will denote $(a_1, \dots, a_n, a)$ by $a_1 \dots a_n \mapsto a$.

2   Box Algebra

The Box Algebra is based on box expressions using which one can capture common programming constructs. The standard semantics of box expressions is given through a mapping into Petri nets called boxes.

Box Expressions. The syntax given below defines two kinds of box expressions, namely non-synchronised expressions E capturing the control flow in a concurrent system, and synchronised expressions F (below a is an action and ρ a synchronisation relation):

$$E ::= \mathsf{stop} \mid a \mid E\,;E \mid E \,\square\, E \mid E \,\|\, E \mid [\![E \circledast E \circledast E]\!]$$
$$F ::= E \,\mathsf{sco}\, \rho$$

The intuition behind the above syntax is that: (i) stop denotes a blocked process; (ii) a denotes a process which can execute an action $a \in \mathcal{A}$ and terminate; (iii) $E\,;E'$ denotes sequential composition; (iv) $E \,\square\, E'$ denotes choice composition; (v) $E \,\|\, E'$ denotes parallel composition; (vi) $[\![E \circledast E' \circledast E'']\!]$ denotes a loop with an initial part E, iterated part E', and terminal part E''; and (vii) E sco ρ denotes scoping which enforces all the synchronisations specified by ρ and then blocks all the original actions present in E.

Box Nets. A box is a tuple $N = (P, T, F, \ell)$ such that: (i) P and T are disjoint finite sets of respectively places (representing local states) and transitions (representing activities); (ii) $F \subseteq (P \times T) \cup (T \times P)$ is a flow relation; and (iii) ℓ is a labelling associating an entry or internal or exit status $\ell(p) \in \{\mathsf{e}, \mathsf{i}, \mathsf{x}\}$ with every place p, and an action $\ell(t) \in \mathcal{A}$ with every transition t. The entry, internal and exit places of N are given respectively by $N^e = \ell^{-1}(\mathsf{e})$, $N^i = \ell^{-1}(\mathsf{i})$ and $N^x = \ell^{-1}(\mathsf{x})$. Moreover, $N^{ei} = N^e \cup N^i$, $N^{ix} = N^i \cup N^x$, $N^{pl} = P$ and $N^{tr} = T$.

Markings are the global states of a box N, each marking being a mapping $M : N^{pl} \to \mathbb{N}_0$. The initial (or entry) and final (or exit) markings of N, denoted respectively by M_N^{init} and M_N^{fin}, are defined, for every $p \in P$, by: (i) $M_N^{init}(p)$ is 1 if $p \in N^e$, and otherwise 0; and (ii) $M_N^{fin}(p)$ is 1 if $p \in N^x$, and otherwise 0.

The change of a marking of N results from a simultaneous execution of a step of transitions. Formally, a step of N is any set of transitions $U \subseteq N^{tr}$. It is enabled at a marking M if, for every $p \in N^{pl}$, $M(p) \geq |\{t \in U \mid (p,t) \in F\}|$. An enabled step U can be executed leading to a marking M' given by $M'(p) = (M(p) - |\{t \in U \mid (p,t) \in F\}|) + |\{t \in U \mid (t,p) \in F\}|$, for every $p \in N^{pl}$. We denote this by $M[U\rangle M'$.

The semantics of N is given through sequences of executed steps starting from the default initial marking M_N^{init}. We will assume that each such step sequence is infinite which is a harmless requirement as any finite step sequence can be extended by an infinite sequence of empty steps (note that $M[\varnothing\rangle M$ for every marking M). In addition, we single out finite step sequences which lead from the default initial marking M_N^{init} to the default final marking M_N^{fin}. Each such step sequence can be seen as a successfully terminated execution of N.

A step sequence of N is an infinite sequence $\theta = U_1 U_2 \ldots$ of steps such that there exist markings $M_1, M_2, \ldots$ satisfying $M_N^{init}[U_1\rangle M_1[U_2\rangle M_2 \ldots$. Moreover, we define a terminated step sequence of N as a finite sequence $\theta = U_1 \ldots U_m$ $(m \geq 0)$ of steps such that there exist markings $M_1, \ldots, M_{m-1}$ satisfying $M_N^{init}[U_1\rangle M_1[U_2\rangle M_2 \ldots M_{m-1}[U_m\rangle M_N^{fin}$. The sets of step sequences and terminated sequences of N are respectively denoted by $\mathsf{sts}(N)$ and $\mathsf{tsts}(N)$, and the set of all (executable) steps occurring in $\mathsf{sts}(N)$ by N^{steps}.

Composite Boxes. BA derives boxes compositionally from box expressions. The labelling of places provides the necessary device for composing boxes along the entry and exit interfaces, i.e., the entry places N^e and exit places N^x. The derived boxes have a very specific form for their places and transitions, making it easier to establish connections with ITL formulas. Intuitively, we use the syntax of a non-synchronised box expression E to construct concrete places and transitions of the corresponding box N, by embedding paths from the root of the parse tree of E in the identities of places and transitions. In what follows, finite sequences in the set $\Pi = \{\,;_L, ;_R, \Box_L, \Box_R, \|_L, \|_R, \circledast_L, \circledast_M, \circledast_R\}^*$ will be called syntax paths (with ϵ denoting the empty syntax path). Note that symbols appearing in syntax paths correspond to the arguments of operators used in box expressions (with 'L' indicating the left argument, etc). For two syntax paths, π_1 and π_2, we denote $\pi_1|\pi_2$ if $\{\pi_1, \pi_2\} = \{\pi \cdot \|_L \cdot \pi', \pi \cdot \|_R \cdot \pi''\}$, for some π, π' and π''. Intuitively, two actions of a non-synchronised box expression are concurrent iff their positions, π_1 and π_2, in the parse tree are such that $\pi_1|\pi_2$.

The form of each place in compositionally defined boxes will be p_Z, where $p \in \{\mathsf{e},\mathsf{i},\mathsf{x}\}$ and $Z \subset \Pi\cdot\{\mathsf{e},\mathsf{x}\}$, while each transition will be of the form a_W, where $a \in \mathcal{A}$ and $W \subset \Pi$. Moreover, for brevity, the sets Z and W will be written as comma separated lists without brackets.

For a syntax path $\pi \in \Pi$ and transition a_W, we denote $\pi\cdot a_W = a_{\pi\cdot W}$. This prefix notation extends in the usual way to sets of transitions and sequences of sets of transitions as well as (sets of) places. The specific form of places and transitions, together with the systematic way in which boxes are manipulated below, will mean that for a compositional box $N = (P,T,F,\ell)$ it will be the case that, for all $p_Z \in P$ and $a_W \in T$, $\ell(p_Z) = p$ and $\ell(a_W) = a$, as well as $(p_Z, a_W) \in F \Leftrightarrow \exists \pi \in W \colon \pi\cdot\mathsf{e} \in Z$ and $(a_W, p_Z) \in F \Leftrightarrow \exists \pi \in W \colon \pi\cdot\mathsf{x} \in Z$. As a result, we will represent such a box simply by $N = (P,T)$. Below we present a compositional way of constructing composite boxes. For examples, see Figure 1.

Constants: With the blocking expression stop and a single-action expression $a \in \mathcal{A}$, we respectively associate two boxes, $\mathsf{N_{stop}} = (\{\mathsf{e_e}, \mathsf{x_x}\}, \varnothing)$ and $\mathsf{N}_a = (\{\mathsf{e_e}, \mathsf{x_x}\}, \{a_\epsilon\})$, depicted in Figure 1.

Sequence: $N\,;K = (P_L \cup P_R \cup \mathcal{X}, T_L \cup T_R)$, where $T_L = ;_L\cdot N^{tr}$, $T_R = ;_R\cdot K^{tr}$, $P_L = ;_L\cdot N^{ei}$, $P_R = ;_R\cdot K^{ix}$ and $\mathcal{X} = \{\mathsf{i}_{;_L\cdot Z \,\cup\, ;_R\cdot W} \mid \mathsf{x}_Z \in N^x \wedge \mathsf{e}_W \in K^e\}$, combines the exit interface of N with the entry interface of K. The entry interface of the resulting box is that of N, and the exit interface is that of K.

Choice: $N\,\Box\,K = (P_L \cup P_R \cup \mathcal{X} \cup \mathcal{Y}, T_L \cup T_R)$, where $T_L = \Box_L\cdot N^{tr}$, $T_R = \Box_R\cdot K^{tr}$, $P_L = \Box_L\cdot N^i$, $P_R = \Box_R\cdot K^i$, $\mathcal{X} = \{\mathsf{e}_{\Box_L\cdot Z \,\cup\, \Box_R\cdot W} \mid \mathsf{e}_Z \in N^e \wedge \mathsf{e}_W \in K^e\}$ and $\mathcal{Y} = \{\mathsf{x}_{\Box_L\cdot Z \,\cup\, \Box_R\cdot W} \mid \mathsf{x}_Z \in N^x \wedge \mathsf{x}_W \in K^x\}$, combines together the entry interfaces of the two boxes creating a new entry interface, as well as their exit interfaces creating a new exit interface.

Parallelism: $N \parallel K = (P_L \cup P_R, T_L \cup T_R)$, where $T_L = \|_L\cdot N^{tr}$, $T_R = \|_R\cdot K^{tr}$, $P_L = \|_L\cdot N^{pl}$ and $P_R = \|_R\cdot K^{pl}$, puts next to each other the boxes N and K. The new entry (resp. exit) interface is simply the union of the entry (resp. exit) interfaces of the composed boxes.

Iteration: $[\![N \circledast K \circledast J]\!] = (P_L \cup P_M \cup P_R \cup \mathcal{X}, T_L \cup T_M \cup T_R)$, where $T_L = \circledast_L\cdot N^{tr}$, $T_M = \circledast_M\cdot K^{tr}$, $T_R = \circledast_R\cdot J^{tr}$, $P_L = \circledast_L\cdot N^{ei}$, $P_M = \circledast_M\cdot K^i$, $P_R = \circledast_R\cdot J^{ix}$ and $\mathcal{X} = \{\mathsf{i}_{\circledast_L\cdot Z \,\cup\, \circledast_M\cdot W \,\cup\, \circledast_M\cdot V \,\cup\, \circledast_R\cdot Y} \mid \mathsf{x}_Z \in N^x \wedge \mathsf{e}_W \in K^e \wedge \mathsf{x}_V \in K^x \wedge \mathsf{e}_Y \in J^e\}$, combines the exit interfaces of N and K with the entry interfaces of K and J. The new entry interface is that of N, and the exit interface is that of J.

One can provide a very useful static characterisation of all the executable steps of a box N constructed using the above rules. In what follows, $N^{psteps} = \{\{a^1_{\pi_1}, \ldots, a^n_{\pi_n}\} \subseteq N^{tr} \mid \forall i < j \colon \pi_i|\pi_j\}$ is the set of all potential steps of N.

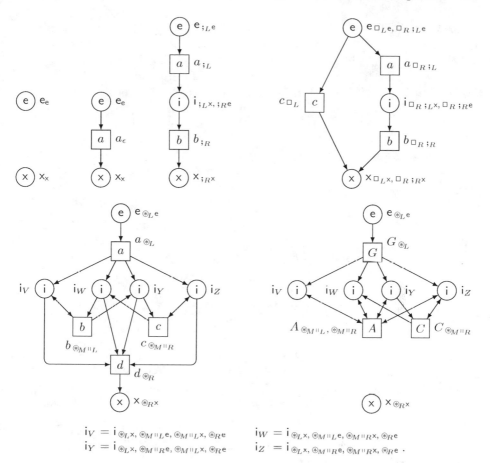

Fig. 1. Diagrams of composite boxes with place and transition labels shown inside the nodes. Top row (left to right): the diagrams of N_{stop}, N_a, N_a ; N_b (corresponding to a ; b) and $N_c \square (N_a ; N_b)$ (corresponding to $c \square (a ; b)$). Bottom row (left to right): the diagrams of $[\![N_a \circledast (N_b \| N_c) \circledast N_d]\!]$ and $([\![N_a \circledast (N_b \| N_c) \circledast N_d]\!])\,\mathsf{sco}\,\{a \mapsto G, bc \mapsto A, c \mapsto C\}$ (the four internal places are defined below the diagrams).

Fact 1. *Let N be any net constructed from the N_a's as well as the operators of sequence, choice, parallelism and iteration. Then $N^{steps} = N^{psteps}$. Moreover, if N_{stop} can also be used in the construction, then $N^{steps} \subseteq N^{psteps}$.*

For a synchronisation relation ρ and N constructed as above, we define $\rho_N = \{(U, a_{\{\pi_1,...,\pi_n\}}) \mid U = \{a^1_{\pi_1}, ..., a^n_{\pi_n}\} \in N^{psteps} \wedge (a^1, ..., a^n, a) \in \rho\}$. Taking the left N in the bottom row of Figure 1 and $\rho = \{a \mapsto G, bc \mapsto A, c \mapsto C\}$, we obtain $N^{psteps} = \{\varnothing, \{a_{\circledast L}\}, \{b_{\circledast M \| L}\}, \{c_{\circledast M \| R}\}, \{d_{\circledast R}\}, \{b_{\circledast M \| L}, c_{\circledast M \| R}\}\}$ and then

$$\rho_N = \{(\{a_{\circledast L}\}, \{G_{\circledast L}\}), (\{c_{\circledast M \| R}\}, \{C_{\circledast M \| R}\}), (\{b_{\circledast M \| L}, c_{\circledast M \| R}\}, \{A_{\circledast M \| L, \circledast M \| R}\})\}$$

Scoping: $N \text{ sco } \rho = (N^{pl}, \mathcal{V})$, where $\mathcal{V} = \{t \mid (U, t) \in \rho_N\}$, has the same places as N. Moreover, for each potential step of N, one creates a new transition (representing a synchronisation of two or more actions of N if the potential step is not a singleton). After that all the transitions of N are removed.

By defining a suitable synchronisation relation ρ, one can capture all practical forms of synchronisation of two or more actions (e.g., $bc \mapsto A$ above), as well as action relabelling (e.g., $a \mapsto G$ and $c \mapsto C$) and action restriction (as for d).

From Expressions to Boxes. The semantics of box expressions is obtained by transforming them compositionally into the corresponding boxes, and then adopting the execution semantics of the latter. Formally, we define a mapping box(.) from box expressions to boxes, in the following way (below $op \in \{\,;\,, \Box, \|\,\}$ and $const \in \{\text{stop}\} \cup \mathcal{A}$) :

$$
\begin{aligned}
\text{box}(const) &= \mathsf{N}_{const} \\
\text{box}(E \text{ } op \text{ } E') &= \text{box}(E) \text{ } op \text{ } \text{box}(E') \\
\text{box}(\llbracket E \circledast E' \circledast E'' \rrbracket) &= \llbracket \text{box}(E) \circledast \text{box}(E') \circledast \text{box}(E'') \rrbracket \\
\text{box}(E \text{ sco } \rho) &= \text{box}(E) \text{ sco } \rho
\end{aligned}
\tag{1}
$$

From now on, by a box we will mean a composite box constructed using (1).

Behavioural Properties of Composite Boxes. Both finite and infinite step sequences of composite boxes exhibit compositional properties, i.e., one can (easily) derive the semantics of a composite box from the semantics of the composed boxes. This is demonstrated by a series of results which follow from the general properties of boxes [2].

Fact 2. *Let $a \in \mathcal{A}$. Then* $\text{sts}(\mathsf{N}_{\text{stop}}) = \varnothing^{\infty}$ *and* $\text{sts}(\mathsf{N}_a) = \varnothing^{\infty} \cup \varnothing^* \circ \{\{a_\epsilon\}\} \circ \varnothing^{\infty}$ *as well as* $\text{tsts}(\mathsf{N}_{\text{stop}}) = \varnothing$ *and* $\text{tsts}(\mathsf{N}_a) = \varnothing^* \circ \{\{a_\epsilon\}\} \circ \varnothing^*$.

For choice and parallelism, the semantics of a composite box can easily be expressed in terms of the semantics of the composed boxes, using an auxiliary notion of parallel composition of step sequences. For two step sequences, θ and τ, of equal length, $\theta \| \delta$ is a step sequence of the same length such that $(\theta \| \delta)_{(k)} = \theta_{(k)} \cup \delta_{(k)}$, for all k. Moreover, for two sets of step sequences, Θ and Δ, the set $\Theta \| \Delta$ comprises all step sequences $\theta \| \delta$, where $\theta \in \Theta$ and $\delta \in \Delta$ are of equal length.

Fact 3. *Let* $\mathsf{f} \in \{\text{sts}, \text{tsts}\}$. *Then*

$$
\begin{aligned}
\mathsf{f}(N \Box K) &= \Box_L \cdot \mathsf{f}(N) \cup \Box_R \cdot \mathsf{f}(N) \\
\mathsf{f}(N \| K) &= {}_{\|L} \cdot \mathsf{f}(N) \| {}_{\|R} \cdot \mathsf{f}(N) \\
\text{sts}(N \,;\, K) &= {}_{;L} \cdot \text{sts}(N) \cup {}_{;L} \cdot \text{tsts}(N) \circ {}_{;R} \cdot \text{sts}(K) \\
\text{tsts}(N \,;\, K) &= {}_{;L} \cdot \text{tsts}(N) \circ {}_{;R} \cdot \text{tsts}(K) \\
\text{sts}(\llbracket N \circledast K \circledast J \rrbracket) &= \circledast_L \cdot \text{sts}(N) \cup \circledast_L \cdot \text{tsts}(N) \circ (\circledast_M \cdot \text{tsts}(K))^* \circ \circledast_M \cdot \text{sts}(K) \cup \\
&\qquad \circledast_L \cdot \text{tsts}(N) \circ (\circledast_M \cdot \text{tsts}(K))^* \circ \circledast_R \cdot \text{sts}(J) \\
\text{tsts}(\llbracket N \circledast K \circledast J \rrbracket) &= \circledast_L \cdot \text{tsts}(N) \circ (\circledast_M \cdot \text{tsts}(K))^* \circ \circledast_R \cdot \text{tsts}(J) \ .
\end{aligned}
$$

Finally, we consider a box $N \operatorname{sco} \rho$, where N is constructed from some non-synchronised box expression. In this case, relating step sequences of $N \operatorname{sco} \rho$ and N is more involved. First, we define a relation $\tilde{\rho}_N$ comprising all pairs $(U, \{t_1, \ldots, t_k\})$ $(k \geq 0)$, where $U \in N^{psteps}$ and $\{t_1, \ldots, t_k\} \subseteq \operatorname{box}(N \operatorname{sco} \rho)^{tr}$ are such that there is a partition $U_1, \ldots, U_k$ of U satisfying $(U_j, t_j) \in \rho_N$, for each $j \leq k$. Moreover, for two equal length sequences of sets of transitions, τ and θ, we denote $(\tau, \theta) \in \tilde{\rho}_N$ if $(\tau_{(j)}, \theta_{(j)}) \in \tilde{\rho}_N$, for all j.

Fact 4. *Let* $\mathsf{f} \in \{\mathsf{sts}, \mathsf{tsts}\}$. *Then* $\mathsf{f}(N \operatorname{sco} \rho) = \{\theta \mid \exists \tau \in \mathsf{f}(N) : (\tau, \theta) \in \tilde{\rho}_N\}$.

Streamlined Box Expressions. The subsequent translation from BA to ITL is much simpler for the class of streamlined synchronised expressions. We call a box expression $E \operatorname{sco} \rho$ streamlined if, for each transition $a_\pi \in \operatorname{box}(E)^{tr}$ there is exactly one transition $b_W \in \operatorname{box}(E \operatorname{sco} \rho)^{tr}$ such that $\pi \in W$. In other words, if each transition in E 'contributes' to exactly one transition in $\operatorname{box}(E \operatorname{sco} \rho)$.

It turns out that each synchronised box expression $F = E \operatorname{sco} \rho$ can be transformed into a semantically equivalent streamlined expression $\mathsf{stl}(F) = F' = E' \operatorname{sco} \rho'$, in the following way. First, for every $a_\pi \in \operatorname{box}(E)^{tr}$, let

$$\mathsf{trans}(a_\pi) = \{b_W \in \operatorname{box}(F)^{tr} \mid \pi \in W\} = \{t \mid \exists (U, t) \in \rho_{\operatorname{box}(E)} : a_\pi \in U\} \,.$$

For example, if we take $[\![a \circledast (b \,\|\, c) \circledast d]\!] \operatorname{sco} \{a \mapsto G, bc \mapsto A, c \mapsto C\}$ with the corresponding box depicted in Figure 1, then we have:

$$\mathsf{trans}(a_{\circledast_L}) = \{G_{\circledast_L}\} \qquad \mathsf{trans}(b_{\circledast_{M^{\|}L}}) = \{A_{\circledast_{M^{\|}L}, \circledast_{M^{\|}R}}\}$$
$$\mathsf{trans}(d_{\circledast_R}) = \varnothing \qquad \mathsf{trans}(c_{\circledast_{M^{\|}R}}) = \{A_{\circledast_{M^{\|}L}, \circledast_{M^{\|}R}}, C_{\circledast_{M^{\|}R}}\}$$

Then a suitable E' is obtained by replacing each occurrence of an action $a \in \mathcal{A}$ in E corresponding to transition a_π in $\operatorname{box}(E)$ (i.e., identified by the path in the syntax tree of E which corresponds to π), by:

- stop if $\mathsf{trans}(a_\pi) = \varnothing$, and b_W if $\mathsf{trans}(a_\pi) = \{b_W\}$;
- $b^1_{W_1} \square (\ldots \square (b^{m-1}_{W_{m-1}} \square b^m_{W_m}) \ldots)$ if $\mathsf{trans}(a_\pi) = \{b^1_{W_1}, \ldots, b^m_{W_m}\}$ and $m \geq 2$ (we assume a fixed total ordering on the transitions of $\operatorname{box}(F)$ so that the enumeration of $\mathsf{trans}(a_\pi)$ can be made unique).

Furthermore, $\rho' = \{b^{|W|}_W \mapsto b_W \mid b_W \in \operatorname{box}(E \operatorname{sco} \rho)^{tr}\}$ defines a suitable synchronisation relation. For the above example, we obtain $F' = E' \operatorname{sco} \rho'$, with $E' = ([\![\gamma \circledast (\alpha \,\|\, (\alpha \square \zeta)) \circledast \mathsf{stop}]\!])$ and $\rho' = \{\gamma \mapsto \gamma, \alpha\alpha \mapsto \alpha, \zeta \mapsto \zeta\}$, where $\gamma = G_{\circledast_L}$, $\alpha = A_{\circledast_{M^{\|}L}, \circledast_{M^{\|}R}}$ and $\zeta = C_{\circledast_{M^{\|}R}}$. The box corresponding to F' is shown on the right in the bottom row of Figure 1.

We now observe that each transition of $\operatorname{box}(F')$ is of the form $(b_W)_Y$, where $b_W \in \operatorname{box}(E \operatorname{sco} \rho)^{tr}$. As a result, we can define a bijection

$$\lambda : \operatorname{box}(F')^{tr} \to \operatorname{box}(F)^{tr} \tag{2}$$

by setting $\lambda((b_W)_Y) = b_W$, for each transition $(b_W)_Y$ in $\operatorname{box}(F')$. Crucially, we then obtain

Fact 5. *The nets* $\operatorname{box}(F')$ *and* $\operatorname{box}(F)$ *are isomorphic after replacing each transition label* b_W *by* b. *Moreover,* $\lambda(\mathsf{f}(\operatorname{box}(F'))) = \mathsf{f}(\operatorname{box}(F))$, *for* $\mathsf{f} \in \{\mathsf{sts}, \mathsf{tsts}\}$.

3   Interval Temporal Logic

We now provide the syntax and semantics of a fragment of ITL, including only those constructs (basic and derived) which are used in the subsequent translation of box expressions. In particular, we assume that V is a countable set of boolean variables, all such variables being (the identities of) transitions of boxes created using the box mapping.[1] The fragment of the ITL logic we need is defined by:

$$\phi ::= \text{ true } \mid \text{ flip}(v) \mid \text{ skipstable}(v) \mid \phi \wedge \phi' \mid \phi \vee \phi' \mid \phi \,; \phi' \mid \phi^* \mid \text{ inf}$$

where $v \in V$. The intuition behind the syntax is as follows: $\text{flip}(v)$ inverts the value of a boolean variable v over a unit interval while $\text{skipstable}(v)$ keeps the value of v over a unit interval; "$;$" is a sequential composition operator (called chop); "$*$" is an iterative version of chop; and inf indicates an infinite interval.

A state is a mapping which assigns values to the (boolean) variables V, and an interval σ is a possibly infinite non-empty sequence of states. Its length, $|\sigma|$, is ω if σ is infinite, and otherwise its number of states minus 1. To simplify definitions, we will denote σ as $\langle \sigma_0, \sigma_1, \ldots, \sigma_{|\sigma|} \rangle$, where $\sigma_{|\sigma|}$ is undefined if σ is infinite. With such a notation, we denote $hd_\sigma = \sigma_0$ and $tl_\sigma = \sigma_{|\sigma|}$ (whenever σ is finite). Moreover, for $0 \leq j \preceq k \leq |\sigma|$, we denote $\sigma_{j..k} = \langle \sigma_j, \ldots, \sigma_k \rangle$ and $\sigma^j = \langle \sigma_0, \ldots, \sigma_j \rangle$ and $\sigma^{(j)} = \langle \sigma_j, \ldots, \sigma_{|\sigma|} \rangle$. The meaning of formulas is given by the satisfaction relation $\models$ defined as follows:

- $\sigma \models \text{true}$
- $\sigma \models \text{flip}(v)$ iff $|\sigma| = 1$ and $hd_\sigma(v) = \neg tl_\sigma(v)$.
- $\sigma \models \text{skipstable}(v)$ iff $|\sigma| = 1$ and $hd_\sigma(v) = tl_\sigma(v)$.
- $\sigma \models \phi \vee \phi'$ iff $\sigma \models \phi$ or $\sigma \models \phi'$.
- $\sigma \models \phi \wedge \phi'$ iff $\sigma \models \phi$ and $\sigma \models \phi'$.
- $\sigma \models \phi \,; \phi'$ iff one of the following holds: (i) $|\sigma| = \omega$ and $\sigma \models \phi$, or (ii) there is $r \preceq |\sigma|$ such that $\sigma^r \models \phi$ and $\sigma^{(r)} \models \phi'$.
- $\sigma \models \phi^*$ iff one of the following holds: (i) $|\sigma| = 0$, or (ii) there are $0 = r_0 \leq r_1 \leq \cdots \leq r_{n-1} \preceq r_n = |\sigma|$ such that, for all $1 \leq l \leq n$, $\sigma_{r_{l-1}..r_l} \models \phi$, or (iii) $|\sigma| = \omega$ and there are infinitely many integers $0 = r_0 \leq r_1 \leq \ldots$ such that $\lim_{j \to \omega} r_j = \omega$ and for all $l \geq 1$, $\sigma_{r_{l-1}..r_l} \models \phi$.
- $\sigma \models \text{inf}$ iff $|\sigma| = \omega$.

The set of variables occurring in a formula ϕ will be denoted by $\text{var}(\phi)$, and we will denote, for a finite set of logic variables $V' \subseteq V$:

$$\text{skipstable}(V') = \begin{cases} \bigwedge_{v \in V'} \text{skipstable}(v) & \text{if } V' \neq \varnothing \\ \text{true} & \text{otherwise} \end{cases}$$

$$\text{flip}(V') = \begin{cases} \bigvee_{v \in V'} \text{flip}(v) \wedge \text{skipstable}(V' \setminus \{v\}) & \text{if } V' \neq \varnothing \\ \text{inf} & \text{otherwise .} \end{cases}$$

To capture the semantical link between a box expression and the corresponding formula, with each ITL formula ϕ and interval σ satisfying $\sigma \models \phi$,

---

[1] The ITL syntax is as in [5] except for the names of logic variables in V.

we associate a sequence of sets $\gamma_\sigma = \Gamma_1 \ldots \Gamma_{|\sigma|}$, where each Γ_j is given by $\Gamma_j = \{v \in \mathsf{var}(\phi) \mid \sigma_{j-1}(v) \neq \sigma_j(v)\}$. That is, each Γ_j records all the variables which flipped their values at the point of entering the state σ_j. As a result, γ_σ can provide a direct interpretation of σ in terms of sequences of steps of transitions of box nets. Formally, for any ITL formula ϕ, $\mathsf{sts}(\phi) = \{\gamma_\sigma \mid \sigma \models \phi \wedge |\sigma| = \omega\}$ and $\mathsf{tsts}(\phi) = \{\gamma_\sigma \mid \sigma \models \phi \wedge |\sigma| < \omega\}$.

The following results are used in the proofs of behavioural equivalence between box expressions and the corresponding logic formulas (c.f. Fact 3).

Proposition 1. *Let ϕ and ϕ' be formulas with disjoint sets of variables, $\psi = \mathsf{skipstable}(\mathsf{var}(\phi))^*$, $\psi' = \mathsf{skipstable}(\mathsf{var}(\phi'))^*$ and $\mathsf{f} \in \{\mathsf{sts}, \mathsf{tsts}\}$. Then*

$$\mathsf{f}(\phi \wedge \psi' \ \vee \ \phi' \wedge \psi) = \mathsf{f}(\phi) \cup \mathsf{f}(\phi')$$
$$\mathsf{f}(\phi \wedge \phi') = \mathsf{f}(\phi) \parallel \mathsf{f}(\phi')$$
$$\mathsf{sts}((\phi \wedge \psi')\,;(\phi' \wedge \psi)) = \mathsf{sts}(\phi) \ \cup \ \mathsf{tsts}(\phi) \circ \mathsf{sts}(\phi')$$
$$\mathsf{tsts}((\phi \wedge \psi')\,;(\phi' \wedge \psi)) = \mathsf{tsts}(\phi) \circ \mathsf{tsts}(\phi')$$

Proposition 2. *Let ϕ_1, ϕ_2 and ϕ_3 be formulas with mutually disjoint sets of variables and $\psi_{i,j} = \mathsf{skipstable}(\mathsf{var}(\phi_i) \cup \mathsf{var}(\phi_j))^*$. Then*

$$\mathsf{sts}((\phi_1 \wedge \psi_{2,3})\,;((\phi_2 \wedge \psi_{1,3})^*\,;(\phi_3 \wedge \psi_{1,2})) = \mathsf{sts}(\phi_1) \ \cup$$
$$\mathsf{tsts}(\phi_1) \circ \mathsf{tsts}(\phi_2)^* \cap \mathsf{sts}(\phi_2) \ \cup \ \mathsf{tsts}(\phi_1) \circ \mathsf{tsts}(\phi_2)^* \circ \mathsf{sts}(\phi_3)$$
$$\mathsf{tsts}((\phi_1 \wedge \psi_{2,3})\,;((\phi_2 \wedge \psi_{1,3})^*\,;(\phi_3 \wedge \psi_{1,2})) = \mathsf{tsts}(\phi_1) \circ \mathsf{tsts}(\phi_2)^* \circ \mathsf{tsts}(\phi_3) \ .$$

Proposition 3. *Let ϕ' be a formula obtained from ϕ by a consistent renaming of variables given by a bijection λ and $\mathsf{f} \in \{\mathsf{sts}, \mathsf{tsts}\}$. Then $\mathsf{f}(\phi') - \lambda(\mathsf{f}(\phi))$.*

Proposition 4. *Let $\pi{\cdot}\phi$, where $\pi \in \Pi$, be obtained from ϕ by replacing each variable $a_{\pi'}$ with $a_{\pi{\cdot}\pi'}$ and $\mathsf{f} \in \{\mathsf{sts}, \mathsf{tsts}\}$. Then $\mathsf{f}(\pi{\cdot}\phi) = \pi{\cdot}\mathsf{f}(\phi)$.*

4   From Box Expressions to Logic Formulas

To make the presentation more accessible, we will first show how to translate non-synchronised box expressions, after that we will extend the translation to the streamlined synchronised expressions, and finally we will deal with the case of general synchronised expressions.

The translation for non-synchronised expressions is as follows (note that except the formula for $\mathsf{itl}(a)$, skipstable is applied to sets of variables):

$$\mathsf{itl}(\mathsf{stop}) = \mathsf{inf}$$
$$\mathsf{itl}(a) = \mathsf{skipstable}(a_\epsilon)^*\,;\mathsf{flip}(a_\epsilon)\,;\mathsf{skipstable}(a_\epsilon)^*$$
$$\mathsf{itl}(E\,;F) = \,;_L{\cdot}\mathsf{itl}(E) \wedge \mathsf{skipstable}(\,;_R{\cdot}\mathsf{box}(F)^{tr})^*\ ;$$
$$\,;_R{\cdot}\mathsf{itl}(F) \wedge \mathsf{skipstable}(\,;_L{\cdot}\mathsf{box}(E)^{tr})^*$$
$$\mathsf{itl}(E\,\square\,F) = \square_L{\cdot}\mathsf{itl}(E) \wedge \mathsf{skipstable}(\square_R{\cdot}\mathsf{box}(F)^{tr})^* \ \vee$$
$$\square_R{\cdot}\mathsf{itl}(F) \wedge \mathsf{skipstable}(\square_L{\cdot}\mathsf{box}(E)^{tr})^*$$
$$\mathsf{itl}(E\,\|\,F) = \|_L{\cdot}\mathsf{itl}(E) \ \wedge \ \|_R{\cdot}\mathsf{itl}(F)$$
$$\mathsf{itl}([\![E \ \circledast \ F \ \circledast \ G]\!]) = \circledast_L{\cdot}\mathsf{itl}(E) \wedge \mathsf{skipstable}(\circledast_M{\cdot}\mathsf{box}(F)^{tr} \cup \circledast_R{\cdot}\mathsf{box}(G)^{tr})^* \ ;$$
$$(\circledast_M{\cdot}\mathsf{itl}(F) \wedge \mathsf{skipstable}(\circledast_L{\cdot}\mathsf{box}(E)^{tr} \cup \circledast_R{\cdot}\mathsf{box}(G)^{tr})^*)^* \ ;$$
$$\circledast_R{\cdot}\mathsf{itl}(G) \wedge \mathsf{skipstable}(\circledast_L{\cdot}\mathsf{box}(E)^{tr} \cup \circledast_M{\cdot}\mathsf{box}(F)^{tr})^* \ .$$

For example, if we take the box expressions $E = a\,;b$ and $F = c\,\square\,(a\,;b)$ generating two of the boxes in Figure 1, then we obtain:

$$\mathsf{itl}(E) = \big(\mathsf{sfs}(a_{\,;_L}) \wedge \mathsf{skipstable}(b_{\,;_R})^*\big) \;;\; \big(\mathsf{sfs}(b_{\,;_R}) \wedge \mathsf{skipstable}(a_{\,;_L})^*\big)$$

$$\mathsf{itl}(F) = \big(\mathsf{sfs}(c_{\square_L}) \wedge \mathsf{skipstable}(\{a_{\square_R\,;_L}, b_{\square_R\,;_R}\})^*\big) \vee$$
$$\big(\big((\mathsf{sfs}(a_{\square_R\,;_L}) \wedge \mathsf{skipstable}(b_{\square_R\,;_R})^*\big) \;;$$
$$\big(\mathsf{sfs}(b_{\square_R\,;_R}) \wedge \mathsf{skipstable}(a_{\square_R\,;_L})^*\big)\big) \wedge \mathsf{skipstable}(c_{\square_L})^*\big)$$

where, for every $v \in V$, $\mathsf{sfs}(v) = \mathsf{skipstable}(v)^*\,;\mathsf{flip}(v)\,;\mathsf{skipstable}(v)^*$.

It can easily be checked that the variables occurring in $\mathsf{itl}(E)$ are precisely the transitions of $\mathsf{box}(E)$. Crucially, however, the step semantics of a non-synchronised box expression and the corresponding ITL formula coincide, i.e., we have $\mathsf{sts}(\mathsf{itl}(E)) = \mathsf{sts}(\mathsf{box}(E))$ and $\mathsf{tsts}(\mathsf{itl}(E)) = \mathsf{tsts}(\mathsf{box}(E))$. Such a result is very strong as it basically states that the behavioural properties of non-synchronised box expressions related to the sequencing of executed actions can be reinterpreted as properties of the translated formulas, assuming that an execution of a transition is 'simulated' by a flipping of the corresponding boolean variable. Extending such a result to streamlined expressions highlights the way in which the box expression synchronisation mechanism (through merging transitions), and the ITL synchronisation mechanism (through flipping variables in different parts of a formula) match each other.

Having dealt with non-synchronised expression, we can turn our attention to a streamlined box expression $F = E \operatorname{sco} \rho$. In such a case, $\mathsf{itl}(F)$ is obtained from $\mathsf{itl}(E)$ by replacing each occurrence of each variable v by the unique variable in $\mathsf{trans}(v)$. We then obtain that

$$\mathsf{sts}(\mathsf{itl}(F)) = \mathsf{sts}(\mathsf{box}(F)) \quad \text{and} \quad \mathsf{tsts}(\mathsf{itl}(F)) = \mathsf{tsts}(\mathsf{box}(F)) . \tag{3}$$

Finally, suppose that $F = E \operatorname{sco} \rho$ is an arbitrary synchronised expression. Given Fact 5, we could now simply take the corresponding streamlined expression $\mathsf{stl}(F)$ and, after consistently renaming variables according to the bijection λ defined in (2), obtain the desired result. Having said that, we can proceed without pre-processing and conservatively extend the previous translation. More precisely, for any synchronised expression $F = E \operatorname{sco} \rho$ we construct $\mathsf{itl}(F)$ from $\mathsf{itl}(E)$ by replacing:

- each sub-formula $\mathsf{flip}(t)$ by $\mathsf{flip}(\mathsf{trans}(t))$;
- each sub-formula $\mathsf{skipstable}(t)$ by $\mathsf{skipstable}(\mathsf{trans}(t))$; and
- each sub-formula $\mathsf{skipstable}(V')$ by $\mathsf{skipstable}(\bigcup \mathsf{trans}(V'))$.

We then obtain our main result (the proof of which relies on (3) and therefore it additionally justifies our interest in streamlined expressions).

Theorem 5. *Let $F = E \operatorname{sco} \rho$ be a synchronised box expression. Moreover, let $\mathsf{f} \in \{\mathsf{sts}, \mathsf{tsts}\}$. Then $\mathsf{f}(\mathsf{itl}(F)) = \mathsf{f}(\mathsf{box}(F))$.*

Consider a non-streamlined expression $F = E \operatorname{sco} \{a \mapsto G, bc \mapsto A, c \mapsto C\}$, where $E = [\![a \circledast (b \,\|\, c) \circledast d]\!]$. Note that boxes corresponding to E and F are shown in the bottom row of Figure 1. We first derive

$$\begin{aligned}
\mathsf{itl}(E) = {}& \mathsf{sfs}(a_{\circledast L}) \wedge \mathsf{skipstable}(\{b_{\circledast M^{||}L}, c_{\circledast M^{||}R}, d_{\circledast R}\})^* \;; \\
& (\mathsf{sfs}(b_{\circledast M^{||}L}) \wedge \mathsf{sfs}(c_{\circledast M^{||}R}) \wedge \mathsf{skipstable}(\{a_{\circledast L}, d_{\circledast R}\})^*)^* \;; \\
& \mathsf{sfs}(d_{\circledast R}) \wedge \mathsf{skipstable}(\{a_{\circledast L}, b_{\circledast M^{||}L}, c_{\circledast M^{||}R}\})^* \;.
\end{aligned}$$

To prepare for applying scoping we derive

$$\begin{aligned}
\mathsf{trans}(a_{\circledast L}) &= \{G_{\circledast L}\} = \{\gamma\} & \mathsf{trans}(b_{\circledast M^{||}L}) &= \{A_{\circledast M^{||}L, \circledast M^{||}R}\} & &= \{\alpha\} \\
\mathsf{trans}(d_{\circledast R}) &= \varnothing & \mathsf{trans}(c_{\circledast M^{||}R}) &= \{A_{\circledast M^{||}L, \circledast M^{||}R}, C_{\circledast M^{||}R}\} & &= \{\alpha, \zeta\}
\end{aligned}$$

which leads to

$$\begin{aligned}
\mathsf{itl}(F) = {}& \mathsf{sfs}(\gamma) \wedge \mathsf{skipstable}(\{\alpha, \zeta\})^* \;; \\
& (\mathsf{sfs}(\alpha) \wedge \mathsf{sfs}(\{\alpha, \zeta\}) \wedge \mathsf{skipstable}(\gamma)^*)^* \;; \\
& \mathsf{inf} \wedge \mathsf{skipstable}(\{\gamma, \alpha, \zeta\})^* \;.
\end{aligned}$$

Hence, in an equivalent form, we obtain:

$$\begin{aligned}
\mathsf{itl}(F) = {}& \mathsf{sfs}(\gamma) \wedge \mathsf{skipstable}(\{\alpha, \zeta\})^* \;; \\
& (\mathsf{sfs}(\alpha) \wedge \\
& \quad (\mathsf{skipstable}(\{\alpha, \zeta\})^* \;; \mathsf{flip}(\alpha) \wedge \mathsf{skipstable}(\zeta) \;; \mathsf{skipstable}(\{\alpha, \zeta\})^* \\
& \quad \vee \\
& \quad \mathsf{skipstable}(\{\alpha, \zeta\})^* \;; \mathsf{flip}(\zeta) \wedge \mathsf{skipstable}(\alpha) \;; \mathsf{skipstable}(\{\alpha, \zeta\})^*) \\
& \quad \wedge \mathsf{skipstable}(\gamma)^*)^* \;; \\
& \mathsf{inf} \wedge \mathsf{skipstable}(\{\gamma, \alpha, \zeta\})^* \;.
\end{aligned}$$

Note that $\mathsf{tsts}(F) = \varnothing$ and

$$\begin{aligned}
\mathsf{sts}(F) = {}& \varnothing^\infty \;\cup \\
& \varnothing^* \circ \{\gamma\} \circ \varnothing^\infty \;\cup \\
& \varnothing^* \circ \{\gamma\} \circ (\varnothing^* \circ \{\alpha\})^* \varnothing^\infty \;\cup \\
& \varnothing^* \circ \{\gamma\} \circ (\varnothing^* \circ \{\alpha\})^* \circ \varnothing^* \circ \{\zeta\} \circ \varnothing^\infty \;.
\end{aligned}$$

5 Conclusions

In the past, different kinds of logics have been used as formalism for expressing correctness properties of systems specified using Petri nets. When it comes to the relationship between logics and Petri nets, we feel that the work on the connections between linear logic [7] and Place Transition nets was the closest one. However, the main concern there was the handling of multiple token occurrences in net places whereas here nets can hold at most two tokens in a single place ever. Another way in which logics and Petri nets were discussed was reported in [13] which provided a characterisation of Petri net languages in terms of second-order logical formulas.

The results presented in this paper provide a very close structural connection between BA and ITL. In our future work we plan to investigate what is the subset of ITL which can be modelled by BA. A longer time goal is the development of a hybrid verification methodology combining ITL and BA techniques. For example, sequential algorithms and infinite data structures could be treated by ITL techniques [3,8,10], while intensive parallel or communicating aspects of systems could be treated, e.g., by net unfoldings [6,9].

Acknowledgement. We would like to thank the referees for their comments and useful suggestions. This research was supported by the 973 Program Grant 2010CB328102, NSFC Grant 61133001, ANR SYNBIOTIC and EPSRC GAELS and UNCOVER projects.

References

1. Ben-Ari, M., Manna, Z., Pnueli, A.: The temporal logic of branching time. In: White, J., Lipton, R.J., Goldberg, P.C. (eds.), pp. 164–176. ACM Press (1981)
2. Best, E., Devillers, R., Koutny, M.: Petri Net Algebra. Springer (2001)
3. Cau, A., Zedan, H.: Refining interval temporal logic specifications. In: Rus, T., Bertrán, M. (eds.) AMAST-ARTS 1997, ARTS 1997, and AMAST-WS 1997. LNCS, vol. 1231, pp. 79–94. Springer, Heidelberg (1997)
4. Desel, J., Juhás, G.: What is a petri net? In: Ehrig, H., Juhás, G., Padberg, J., Rozenberg, G. (eds.) Unifying Petri Nets. LNCS, vol. 2128, pp. 1–25. Springer, Heidelberg (2001)
5. Duan, Z., Klaudel, H., Koutny, M.: ITL semantics of composite Petri nets. J. Log. Algebr. Program 82(2), 95–110 (2013)
6. Esparza, J.: Model checking using net unfoldings. Sci. Comput. Program. 23(2-3), 151–195 (1994)
7. Girard, J.Y.: Linear logic. Theor. Comput. Sci. 50, 1–102 (1987)
8. Janicke, H., Cau, A., Siewe, F., Zedan, H.: Dynamic access control policies: Specification and verification. Comput. J. 56(4), 440–463 (2013)
9. Khomenko, V., Koutny, M.: Towards an efficient algorithm for unfolding petri nets. In: Larsen, K.G., Nielsen, M. (eds.) CONCUR 2001. LNCS, vol. 2154, pp. 366–380. Springer, Heidelberg (2001)
10. Moszkowski, B.C.: Executing Temporal Logic Programs. Cambridge University Press (1986)
11. Moszkowski, B.C.: A complete axiom system for propositional interval temporal logic with infinite time. Logical Methods in Computer Science 8(3) (2012)
12. Moszkowski, B.C., Manna, Z.: Reasoning in interval temporal logic. In: Clarke, E., Kozen, D. (eds.) Logic of Programs 1983. LNCS, vol. 164, pp. 371–382. Springer, Heidelberg (1984)
13. Parigot, M., Pelz, E.: A logical approach of Petri net languages. Theor. Comput. Sci. 39, 155–169 (1985)
14. Silva, M., Teruel, E., Colom, J.M.: Linear algebraic and linear programming techniques for the analysis of place/transition net systems. In: Reisig, W., Rozenberg, G. (eds.) APN 1998. LNCS, vol. 1491, pp. 309–373. Springer, Heidelberg (1998)
15. Valmari, A., Hansen, H.: Can stubborn sets be optimal? Fundam. Inform. 113(3-4), 377–397 (2011)

Are Good-for-Games Automata Good for Probabilistic Model Checking?[*]

Joachim Klein, David Müller, Christel Baier, and Sascha Klüppelholz

Institute of Theoretical Computer Science
Technische Universität Dresden, 01062 Dresden, Germany
{klein,david.mueller,baier,klueppel}@tcs.inf.tu-dresden.de

Abstract. The potential double exponential blow-up for the generation of deterministic ω-automata for linear temporal logic formulas motivates research on weaker forms of determinism. One of these notions is
the *good-for-games* property that has been introduced by Henzinger and
Piterman together with an algorithm for generating good-for-games automata from nondeterministic Büchi automata. The contribution of our
paper is twofold. First, we report on an implementation of this algorithms
and exhaustive experiments. Second, we show how good-for-games automata can be used for the quantitative analysis of systems modeled by
Markov decision processes against ω-regular specifications and evaluate
this new method by a series of experiments.

1 Introduction

The automata-theoretic approach to formal verification relies on the effective
translation of specifications, e.g., formulas of some temporal logic such as linear
temporal logic (LTL) into automata over infinite words (ω-automata) [34,6,12].
The verification problem for finite-state system models is then solvable by analyzing the product of the system model and the automaton for the formula. In
the classical setting where the system model can be seen as a nondeterministic
automaton, *nondeterministic ω-automata* suffice. For some applications, such as
game-based synthesis and probabilistic model-checking problems, the nondeterminism of the ω-automaton poses a problem. Used as a monitor to determine
the winning strategies of turn-based two-player games, the lack of look-ahead
beyond the players' choices in general precludes the use of nondeterministic
automata. Similarly, in probabilistic model checking, the lack of look-ahead beyond the probabilistic choices renders nondeterministic automata unsuitable in
general. In these settings, the use of *deterministic ω-automata* resolves these
problems at the cost of a further worst-case exponential determinization construction [26,31,25]. Thus, there is considerable interest in methods that tackle
the worst-case double exponential time-complexity of algorithms caused by the

---

[*] Funded by the DFG through the Graduiertenkolleg 1763 (QuantLA), the CRC 912
HAEC, the cluster of excellence cfAED and the project QuaOS and by the ESF young
researcher group IMData 100098198 and the EU-FP-7 grant 295261 (MEALS).

A.-H. Dediu et al. (Eds.): LATA 2014, LNCS 8370, pp. 453–465, 2014.

construction of deterministic ω-automata for LTL formulas. This includes for example variants of the determinization construction for nondeterministic Büchi automata (NBA) [28,32], heuristics [14,15] and the direct translation from fragments of LTL to deterministic automata [24,17,1]. Instead of reducing the number of states, [27] provides a translation from non-confluent NBA that aims to generate a compact symbolic representation of the generated deterministic automata based on binary decision diagrams (BDDs).

There are also several attempts to avoid determinization in certain scenarios [21,18] and provide better theoretical complexity and performance in practice. Henzinger and Piterman [13] introduce a special property for nondeterministic automata, being *good-for-games* (GFG), that is fulfilled by all deterministic automata but still permits nondeterministic choices. [13] proposes an algorithm, called the HP-algorithm here, for the construction of a nondeterministic GFG automaton with parity acceptance from an NBA that is amenable to a symbolic representation. The number of states in the constructed GFG automaton is still exponential in the number of states of the given NBA, but a smaller worst-case bound on the number of states can be provided than for Safra's determinization algorithm [31]. Among others, [4] introduced the notion *determinizable-by-pruning* for automata that have an embedded deterministic automaton for the same language. [4] states the existence of GFG automata that are not determinizable-by-pruning, but we are not aware of any result stating the existence of languages where GFG automata are more succinct than deterministic ones. To the best of our knowledge, the HP-algorithm is the sole published algorithm for the construction of GFG automata and it has not been implemented or experimentally evaluated yet.

In the context of probabilistic model checking for finite-state Markov chains, [8,3] propose the use of unambiguous automata that can be generated from LTL formulas with a single exponential blow-up in the worst case. Alternative approaches that also lead to single exponential-time model-checking algorithms for Markov chains and LTL specifications have been presented in [5] using weak alternating automata and in [7] using an iterative approach to integrate the effect of the temporal modalities of a given LTL formula φ in the Markov chain. Given that the analogous problem is 2EXPTIME-complete for models where nondeterministic and probabilistic choices alternate [7], there is no hope to generalize these results for Markov decision processes (MDPs). Only for the qualitative analysis of MDPs where the task is to show that an ω-regular path property holds with probability 1, no matter how the nondeterminism is resolved, Büchi automata that are deterministic-in-limit are shown to be sufficient [34,7].

Contribution. The purpose of our paper is to study whether GFG automata are adequate in the context of probabilistic model checking, both at the theoretical and the practical level. At the theoretical level, we answer in the affirmative and provide algorithms for the computation of maximal or minimal probabilities for path properties specified by GFG automata in finite-state Markov decision processes (MDPs). The time complexity of our algorithm is polynomial in the size of the given MDP and GFG automaton. To evaluate the GFG-based approach empirically, we have implemented the HP-algorithm (and various variants) symbolically using binary decision diagrams (BDDs). In a series of experiments, we study the performance of the HP-algorithm – from LTL formula via NBA to GFG automaton – compared to the determinization implementation of LTL2DSTAR [14,15] based on Safra's construction. We have furthermore implemented the GFG-based approach for the analysis of MDPs in the popular probabilistic model checker PRISM [22] and evaluated its performance in practice.

Outline. Section 2 briefly introduces our notations for ω-automata and MDPs. The applicability of GFG automata for the quantitative analysis of MDPs is shown in Section 3. In Section 4, we study the HP-algorithm in detail and present a few heuristics that have been integrated in our implementation. Section 5 reports on our experiments, Section 6 contains some concluding remarks. Omitted proofs and other additional material can be found in the technical report [16].

2 Preliminaries

Throughout the paper, the reader is supposed to be familiar with the basic principles of ω-automata, games and temporal logics. For details we refer to [6,12]. We briefly summarize our notations for ω-automata, present the definition of good-for-games automata [13] and provide a condensed survey of the relevant principles of Markov decision processes (MDPs). Further details on MDPs and their use in the context of model checking can be found e.g. in [30,2].

Automata over Infinite Words. An ω-automaton $\mathcal{A} = (Q, \Sigma, \delta, q_0, Acc)$ is a tuple, where Q is a finite set of states, Σ is a finite alphabet, $\delta : Q \times \Sigma \to 2^Q$ is the (nondeterministic) transition function and $q_0 \in Q$ is the initial state. The last component Acc is the acceptance condition (see below). The size of $|\mathcal{A}|$ denotes the number of states in $\mathcal{A}$. $\mathcal{A}$ is said to be *complete*, if $\delta(q, \sigma) \neq \varnothing$ for all states $q \in Q$ and all symbols $\sigma \in \Sigma$. $\mathcal{A}$ is called *deterministic*, if $|\delta(q, \sigma)| \leq 1$ for all $q \in Q$ and $\sigma \in \Sigma$. A *run* in $\mathcal{A}$ for an infinite word $w = \sigma_0 \sigma_1 \sigma_2 \ldots \in \Sigma^\omega$ is a sequence $\rho = q_0 q_1 q_2 \ldots \in Q^\omega$ starting in the initial state q_0 such that $q_{i+1} \in \delta(q_i, a)$ for all $i \in \mathbb{N}$. We write $\inf(\rho)$ to denote the set of all states occurring infinitely often in ρ. A run ρ is called accepting, if it meets the acceptance condition Acc, denoted $\rho \models Acc$. We consider here the following three types of acceptance conditions and describe their constraints for infinite runs:

- *Büchi:* $Acc = F$ is a set of states, i.e., $F \subseteq Q$, with the meaning $\Box\Diamond F$
- *parity:* Acc is a function $col : Q \to \mathbb{N}$ assigning to each state q a parity color and requiring that the least parity color appearing infinitely often is even
- *Rabin:* Acc is a set consisting of pairs (E, F) with $E, F \subseteq Q$, imposing the constraint $\bigvee_{(E,F) \in Acc} (\Diamond\Box\neg E \wedge \Box\Diamond F)$

Büchi acceptance can be seen as a special case of parity acceptance which again can be seen as a special case of Rabin acceptance. We use the standard notations NBA (NRA, NPA) for nondeterministic Büchi (Rabin, parity) automata and

DBA, DRA, DPA for their deterministic versions. The language of $\mathcal{A}$, denoted $\mathcal{L}(\mathcal{A})$, consists of all infinite words $w \in \Sigma^\omega$ that have at least one accepting run in $\mathcal{A}$, i.e., $w \in \mathcal{L}(\mathcal{A})$ iff there exists a run ρ for w with $\rho \models Acc$.

It is well-known that the classes of languages recognizable by NBA, NRA, NPA, DRA or DPA are the same (the so-called ω-regular languages), while DBA are less powerful. For each LTL formula φ with atomic propositions in some finite set AP, the semantics of φ can be described as an ω-regular language $\mathcal{L}(\varphi)$ over the alphabet $\Sigma = 2^{AP}$ and there is an NBA $\mathcal{A}$ for φ (i.e., $\mathcal{L}(\varphi) = \mathcal{L}(\mathcal{A})$) whose size is at most exponential in the formula length $|\varphi|$.

Good-for-Games (GFG) Automata. The formal definition of GFG automata [13] relies on a game-based view of ω-automata. Given an ω-automaton $\mathcal{A}$ as before, we consider $\mathcal{A}$ as the game arena of a turn-based two-player game, called *monitor game*: if the current state is q then player 1 chooses a symbol $\sigma \in \Sigma$ whereas the other player (player 0) has to answer by a successor state $q' \in \delta(q, \sigma)$, i.e., resolve the nondeterminism. In the next round q' becomes the current state. A *play* is an alternating sequence $\varsigma = q_0 \sigma_0 q_1 \sigma_1 q_2 \sigma_2 \ldots$ of states and (action) symbols in the alphabet Σ starting with the initial state q_0. Intuitively, the σ_i's are the symbols chosen by player 1 and the q_i's are the states chosen by player 0 in round i. Player 0 wins the play ς if ς is infinite and if $\varsigma|_\Sigma = \sigma_0 \sigma_1 \sigma_2 \ldots \in \mathcal{L}(\mathcal{A})$ then $\varsigma|_Q = q_0 q_1 q_2 \ldots$ is an accepting run. A strategy for player 0 is a function $\mathfrak{f} : (Q \times \Sigma)^+ \to Q$ with $\mathfrak{f}(\ldots q \sigma) \in \delta(q, \sigma)$. A play $\varsigma = q_0 \sigma_0 q_1 \sigma_1 q_2 \ldots$ is said to be $\mathfrak{f}$-*conform* or a $\mathfrak{f}$-play if $q_i = \mathfrak{f}(\varsigma \downarrow i)$ for all $i \geq 1$ where $\varsigma \downarrow i = q_0 \sigma_0 \ldots \sigma_{i-2} \ldots q_{i-1} \sigma_i$ is the prefix of ρ that ends with the chosen symbol in round i. An automaton $\mathcal{A}$ is called *good-for-games* if there is a strategy $\mathfrak{f}$ such that player 0 wins each $\mathfrak{f}$-play. Such strategies will be called *GFG-strategies* for $\mathcal{A}$. Obviously, each deterministic automaton enjoys the GFG property. GFG automata with Rabin or parity condition cover the full class of ω-regular languages, while GFG automata with Büchi acceptance do not [4]. For illustrating examples of GFG automata see [16].

Markov Decision Processes (MDP). MDPs are an operational model for systems that exhibit nondeterministic and probabilistic choices. For the purposes of this paper, we formalize an MDP by a tuple $\mathcal{M} = (S, Act, P, s_0, AP, \ell)$ where S is a finite set of states, $s_0 \in S$ is the initial state, Act a finite set of actions and $P : S \times Act \times S \to [0, 1]$ is the transition probability function satisfying:

$$\sum_{s' \in S} P(s, \alpha, s') \in \{0, 1\} \qquad \text{for all } s \in S, \alpha \in Act.$$

We write $Act(s)$ for the set of actions α that are enabled in s, i.e., $P(s, \alpha, s') > 0$ for some $s' \in S$, in which case $s' \mapsto P(s, \alpha, s')$ is a distribution formalizing the probabilistic effect of taking action α in state s. We refer to the triples (s, α, s') with $P(s, \alpha, s') > 0$ as a step. The choice between the enabled actions is viewed to be nondeterministic. For technical reasons, we require $Act(s) \neq \varnothing$ for all states s. The last two components AP and ℓ serve to formalize properties of paths in $\mathcal{M}$. Formally, AP is a finite set of atomic propositions and $\ell : S \to 2^{AP}$ assigns to each state s the set $\ell(s)$ of atomic propositions that hold in s. Paths in $\mathcal{M}$ are finite or infinite sequences $\pi = s_0\,\alpha_0\,s_1\,\alpha_2\,s_2\,\alpha_3\ldots$ starting in the initial state s_0 that are built by consecutive steps, i.e., $P(s_i, \alpha_i, s_{i+1}) > 0$ for all i. The trace of π is the word over the alphabet $\Sigma = 2^{AP}$ that arises by taking the projections to the state labels, i.e., $trace(\pi) = \ell(s_0)\,\ell(s_1)\,\ell(s_2)\ldots$. For an LTL formula φ over AP we write $\pi \models \varphi$ if $trace(\pi) \in \mathcal{L}(\varphi)$.

As the monitor game in nondeterministic automata, MDPs can be seen as stochastic games, also called a $1\frac{1}{2}$-player games. The first (full) player resolves the nondeterministic choice by selecting an enabled action α of the current state s. The second (half) player behaves probabilistically and selects a successor state s' with $P(s, \alpha, s') > 0$. Strategies for the full player are called *schedulers*. Since the behavior of $\mathcal{M}$ is purely probabilistic if some scheduler $\mathfrak{s}$ is fixed, one can reason about the probability of path events. If L is an ω-regular language then $\mathrm{Pr}^{\mathfrak{s}}_{\mathcal{M}}(L)$ denotes the probability under $\mathfrak{s}$ for the set of infinite paths π with $trace(\pi) \in L$. In notations like $\mathrm{Pr}^{\mathfrak{s}}_{\mathcal{M}}(\varphi)$ or $\mathrm{Pr}^{\mathfrak{s}}_{\mathcal{M}}(\mathcal{A})$ we identify LTL formulas φ and ω-automata $\mathcal{A}$ with their languages. For the mathematical details of the underlying sigma-algebra and probability measure, we refer to [30,2].

For a worst-case analysis of a system modeled by an MDP $\mathcal{M}$, one ranges over all initial states and all schedulers (i.e., all possible resolutions of the nondeterminism) and considers the maximal or minimal probabilities for some ω-regular language L. Depending on whether L represents a desired or undesired path property, the quantitative worst-case analysis amounts to computing $\mathrm{Pr}^{\min}_{\mathcal{M}}(\varphi) = \min_{\mathfrak{s}} \mathrm{Pr}^{\mathfrak{s}}_{\mathcal{M}}(L)$ or $\mathrm{Pr}^{\max}_{\mathcal{M}}(L) = \max_{\mathfrak{s}} \mathrm{Pr}^{\mathfrak{s}}_{\mathcal{M}}(L)$.

3   Automata-Based Analysis of Markov Decision Processes

We address the task to compute the maximal or minimal probability in an MDP $\mathcal{M}$ for the path property imposed by a nondeterministic ω-automaton $\mathcal{A}$. The standard approach, see e.g. [2], assumes $\mathcal{A}$ to be deterministic and relies on a product construction where the transitions of $\mathcal{M}$ are simply annotated with the unique corresponding transition in $\mathcal{A}$. Thus, $\mathcal{M} \otimes \mathcal{A}$ can be seen as a refinement of $\mathcal{M}$ since $\mathcal{A}$ does not not affect $\mathcal{M}$'s behaviors, but attaches information on $\mathcal{A}$'s current state for the prefixes of the traces induced by the paths of $\mathcal{M}$.

We now modify the standard definition of the product for nondeterministic ω-automaton. The crucial difference is that the actions are now pairs $\langle \alpha, p \rangle$ consisting of an action in $\mathcal{M}$ and a state in $\mathcal{A}$, representing the nondeterministic alternatives in both the MDP $\mathcal{M}$ and the automaton $\mathcal{A}$. Formally, let $\mathcal{M} = (S, Act, P, s_0, AP, \ell)$ be an MDP and $\mathcal{A} = (Q, \Sigma, \delta, q_0, Acc)$ a complete nondeterministic ω-automaton with $\Sigma = 2^{AP}$. The product MDP is

$$\mathcal{M} \otimes \mathcal{A} = (S \times Q, Act \times Q, P', \langle s_0, q_0 \rangle, AP, \ell')$$

where the transition probability function P' is given by $P'(\langle s, q \rangle, \langle \alpha, p \rangle, \langle s', q' \rangle) = P(s, \alpha, s')$ if $p = q' \in \delta(q, \ell(s))$. In all other cases $P'(\langle s, q \rangle, \langle \alpha, p \rangle, \langle s', q' \rangle) = 0$. The assumption that $\mathcal{A}$ is complete yields that for each $\alpha \in Act(s)$ there is some action $\langle \alpha, q' \rangle \in Act(\langle s, q \rangle)$ for all states s in $\mathcal{M}$ and q in $\mathcal{A}$. The labeling function is given by $\ell'(\langle s, q \rangle) = \{q\}$. Thus, the traces in $\mathcal{M} \otimes \mathcal{A}$ are words over the alphabet Q. Likewise, $\mathcal{A}$'s acceptance condition Acc can be seen as a language over Q, which permits to treat Acc as a property that the paths in $\mathcal{M} \otimes \mathcal{A}$ might or might not have. We prove in [16]:

Theorem 1. *For each MDP $\mathcal{M}$ and nondeterministic ω-automaton $\mathcal{A}$ as above:*

(a) $\mathrm{Pr}^{\max}_{\mathcal{M} \otimes \mathcal{A}}(Acc) \leq \mathrm{Pr}^{\max}_{\mathcal{M}}(\mathcal{A})$

(b)   *If $\mathcal{A}$ is good-for-games then:* $\mathrm{Pr}^{\max}_{\mathcal{M} \otimes \mathcal{A}}(Acc) = \mathrm{Pr}^{\max}_{\mathcal{M}}(\mathcal{A})$

Theorem 1 (b) shows that with a slightly modified definition of the product, the techniques that are known for the quantitative analysis of MDPs against deterministic ω-automata specifications are also applicable for GFG automata. The computation of maximal probabilities for properties given by an ω-regular acceptance condition Acc (e.g., Büchi, Rabin or parity) can be carried out by a graph analysis that replaces Acc with a reachability condition and linear programming techniques for computing maximal reachability probabilities. See e.g. [2]. The time complexity is polynomial in the size of the $\mathcal{M}$ and $\mathcal{A}$. Thus, if the specification is given in terms of an LTL formula φ then the costs of our GFG-based approach are dominated by the generation of a GFG automaton for φ. Minimal probabilities can be handled by using $\mathrm{Pr}^{\min}_{\mathcal{M}}(\varphi) = 1 - \mathrm{Pr}^{\max}_{\mathcal{M}}(\neg\varphi)$.

[20,19] proves that a double exponential blow-up for translating LTL to deterministic ω-automata (of any type) is unavoidable. We adapted the proof in [19] for GFG automata (see [16]). Thus, the double exponential time complexity of the GFG-based approach is in accordance with the known 2EXPTIME-completeness for the analysis of MDPs against LTL specifications [7].

Theorem 2. *There exists a family of LTL formulas $(\varphi)_{n \in \mathbb{N}}$ such that $|\varphi_n| = \mathcal{O}(n)$, while every GFG automaton $\mathcal{A}_n$ for φ_n has at least $2^{2^{\Omega(n)}}$ states.*

4   From LTL to GFG Automata

We have previously shown [14,15] that it is possible in practice, using the tool LTL2DSTAR, to obtain deterministic ω-automata for a wide range of LTL formula φ via the translation to an NBA and Safra's determinization construction [31] refined by various heuristics. Here, we are interested in replacing Safra's determinization algorithm with the HP-algorithm [13] to generate a GFG automaton instead of a deterministic automaton. We first provide an outline of the HP-algorithm and then explain a few new heuristics.

The HP-algorithm transforms an NBA $\mathcal{B} = (Q, \Sigma, \delta, q_0, F)$ with $|Q| = n$ states into a GFG automaton $\mathcal{A}$ with parity acceptance and at most $2^n \cdot n^{2n}$ states and $2n$ parity colors (or an NRA with n Rabin pairs), which improves on the upper bound given for Safra's determinization algorithm. We recall here the main concepts, for a formal description we refer to [13]. Like Safra's construction, the HP-algorithm relies on the simultaneous tracking of multiple subset constructions to determine acceptance or rejection in the NBA. However, while the states of Safra's DRA organize the subsets in trees, the HP-algorithm uses a simpler, linear arrangement of the subsets. The state space $P = (2^Q \times 2^Q)^n$ of the GFG automaton $\mathcal{A}$ consists of n pairs of subsets of NBA states Q, i.e., states of the form $p = \langle (A_1, B_1), \ldots, (A_n, B_n) \rangle$ where $B_i \subseteq A_i \subseteq Q$, plus some additional constraints on the state space. Each set B_i serves to mark those states in A_i that were reached via some accepting state in F of the NBA. The successor state in $\mathcal{A}$ for symbol σ is obtained by applying the transition function δ to each of the subsets and adding states in F to the B_i subsets. In crucial difference to Safra's construction, the HP-algorithm however then introduces significant nondeterminism by allowing $\mathcal{A}$ to discard an arbitrary number of states in any of the subsets. For $p = \langle \ldots (A_i, B_i) \ldots \rangle$, the set A_i' in a σ-successor p' of $\mathcal{A}$ thus does not correspond to $A_i' = \delta(A_i, \sigma)$ but there is a nondeterministic choice between any A_i' satisfying $A_i' \subseteq \delta(A_i, \sigma)$, including the empty set. Whenever some A_i is empty, $\mathcal{A}$ can "reset" A_i by setting A_i to some subset of the first set A_1. Such resets are reflected in the acceptance condition of $\mathcal{A}$ as "bad" events for the pair i, as they signify that the previously tracked runs terminated. The "good" events in the acceptance condition occur whenever all states in an A_i are marked as having recently visited F, i.e., whenever $A_i = B_i \neq \varnothing$. In the next step, B_i' is then cleared and the tracking of visits to F starts anew. Infinitely many "good" events without "bad" events then correspond to the existence of an accepting run in the NBA $\mathcal{B}$. The HP-algorithm relies on the GFG-strategy to resolve the nondeterminism in the constructed automaton $\mathcal{A}$, i.e., which states in the subsets are kept, which are dropped and when to reset. There is a large amount of nondeterminism and a lot of combinatorial possibilities in the reachable state space of $\mathcal{A}$. This is confirmed by our experiments, e.g., applying the construction to the two-state NBA for $\Diamond\Box a$ already yields a GFG automaton with 16 states, where LTL2DSTAR generates a two-state DRA. As stated in [13], the HP-algorithm is thus not well-suited for an explicit representation for $\mathcal{A}$, but is intended for a symbolic implementation. In this context, [13] briefly discusses the possibility of variants of the transition function in the GFG automaton that either apply more or less strict constraints on the relationship enforced between the (A_i, B_i) pairs in each state. In particular, [13] posits that introducing even further nondeterminism (and increasing the number of possible states) by loosening a disjunctness requirement on the A_i may lead to a smaller symbolic representation. In our experiments, we will refer to this as the *loose variant*.

Iterative Approach. In the context of games, [13] proposes an iterative approach to the HP-algorithm by successively constructing the automata $\mathcal{A}^m$ obtained by using only the first m of the n pairs, i.e., by setting $A_i = B_i = \varnothing$ for all $m < i \leq n$. In the acceptance condition this reduces the number of required parity colors to $2m$ and Rabin pairs to m as well. For these automata, $\mathcal{L}(\mathcal{A}^m) = \mathcal{L}(\mathcal{A}) = \mathcal{L}(\mathcal{B})$, but there is no guarantee that $\mathcal{A}^m$ for $m < n$ is good-for-games by construction. We start with $m = 1$ and increase m until early success or reaching $m = n$. Our experimental results indeed show that early termination appears rather often.

We now explain how the iterative approach of [13] can be integrated in the GFG-based quantitative analysis of MDPs against LTL specifications. Suppose, e.g., that the task is to show that $\mathrm{Pr}_{\mathcal{M}}^{\max}(\varphi) \geq \theta$ for some LTL formula φ and threshold $\theta \in {]}0,1]$. Let $\mathcal{B}$ be an n-state NBA with $\mathcal{L}(\mathcal{B}) = \mathcal{L}(\varphi)$ and $\mathcal{A}^m$ the automaton obtained using only the first $m \leq n$ pairs in the HP-algorithm applied to $\mathcal{B}$. Let Acc^m denote the acceptance condition of $\mathcal{A}^m$. By Theorem 1 (a):

$$\text{If } \mathrm{Pr}_{\mathcal{M} \otimes \mathcal{A}^m}^{\max}(Acc^m) \geq \theta \text{ for some } m \leq n \text{ then } \mathrm{Pr}_{\mathcal{M}}^{\max}(\varphi) \geq \theta.$$

Moreover, $\mathrm{Pr}_{\mathcal{M} \otimes \mathcal{A}^m}^{\max}(Acc^m) \leq \mathrm{Pr}_{\mathcal{M} \otimes \mathcal{A}^{m+1}}^{\max}(Acc^{m+1})$ for $m < n$. These observations suggest an approach that resembles the classical *abstraction-refinement* schema: starting with $m = 1$, we carry out the quantitative analysis of $\mathcal{M} \otimes \mathcal{A}^m$ against Acc^m and successively increase m until $\mathrm{Pr}_{\mathcal{M} \otimes \mathcal{A}^m}^{\max}(Acc^m) \geq \theta$ or $\mathcal{A}^m$ is GFG (which is the case at the latest when $m = n$). As an additional heuristic to increase the performance of the linear programming techniques that are applied for the quantitative analysis of $\mathcal{M} \otimes \mathcal{A}^m$ against Acc^m, one can reuse the results computed for $\mathcal{M} \otimes \mathcal{A}^{m-1}$ and Acc^{m-1} as initial values.

It remains to explain how to check whether $\mathcal{A}^m$ has the GFG property. For details we refer to [16]. In this aspect, our prototype implementation departs from [13] and checks whether $\mathcal{A}^m$ is GFG by solving a Rabin game (itself an NP-complete problem) constructed from $\mathcal{A}^m$ and a DRA for $\neg\varphi$ constructed with LTL2DSTAR while [13] proposes an algorithm based on checking fair simulation. To study the impact of the iterative approach in terms of the number of required iterations and the size of the resulting GFG automata, the choice of the GFG test is irrelevant.

Union Operator for Disjunctive Formulas. For generating a deterministic automaton from an LTL formula, we have shown in [14] that optionally handling disjunctive LTL formulas of the form $\varphi = \varphi_1 \vee \varphi_2$ by constructing DRA $\mathcal{A}_1$ and $\mathcal{A}_2$ for the subformulas φ_1 and φ_2 and then obtaining the DRA $\mathcal{A}_1 \cup \mathcal{A}_2$ for the language $\mathcal{L}(\mathcal{A}_1) \cup \mathcal{L}(\mathcal{A}_2)$ via a product construction can be very beneficial in practice. The definition of $\mathcal{A}_1 \cup \mathcal{A}_2$ used in [14] can easily be extended to NRA. The GFG property is preserved by the union construction. See [16].

5   Implementation and Experiments

We have implemented the HP-algorithm in a tool we refer to as LTL2GFG. Based on LTL2GFG, we have additionally implemented the GFG-based quantitative analysis of MDPs in PRISM. After a brief overview of LTL2GFG, we report on our experiments and comparison with the determinization approach of LTL2DSTAR.

LTL2GFG. Given an LTL formula φ, our implementation LTL2GFG constructs a symbolic, BDD-based representation of a GFG-NPA for φ. It first converts φ into an (explicitly represented) NBA $\mathcal{B}$. In our experiments, we use LTL2BA v1.1 [11] for this task. To facilitate an efficient symbolic representation of the various subsets used in the HP-algorithm, $\mathcal{B}$ is then converted to a symbolic representation, using a unary encoding of the $|Q| = n$ states of $\mathcal{B}$, i.e., using one boolean variable q_i per state. The state space of the GFG-automaton $\mathcal{A}$, i.e., the n pairs (A_i, B_i) is likewise encoded by n^2 boolean variables $a_{i,j}$ and $b_{i,j}$, i.e.,

Table 1. Statistics for the automata $\mathcal{A}_\varphi$ constructed for the 94 benchmark formulas. Number of $\mathcal{A}_\varphi$ constructed within a given timeframe and a given range of BDD sizes.

	aborted	$\mathcal{A}_\varphi$ with constr. time				$\mathcal{A}_\varphi$ with BDD size					
		$<1s$	$<10s$	$<1m$	$<30m$	<10	$<10^2$	$<10^3$	$<10^4$	$<10^5$	$\geq 10^5$
LTL2DSTAR std.	0	90	91	92	94	4	65	87	90	91	3
no opt.	0	90	90	92	94	3	48	78	89	90	4
LTL2GFG std.	39	40	47	48	55	3	6	19	26	36	19
std., dynamic	45	34	36	48	49	5	8	19	36	39	10
loose, dynamic	34	43	49	56	60	5	14	31	47	56	4
lo., union, dyn.	29	52	59	61	65	4	13	35	54	60	5
lo., iterative	20	74	74	74	74	3	19	39	60	74	0
lo., it., un., dyn.	18	70	72	74	76	4	32	63	70	76	0

$a_{i,j}$ is true iff NBA state $q_j \in A_i$ and $b_{i,j}$ is true iff $q_j \in B_i$ for $1 \leq i, j \leq n$. To allow the encoding of the transition relations of $\mathcal{A}$ and $\mathcal{B}$, each state variable has a primed copy, i.e., q_i', $a_{i,j}'$ and $b_{i,j}'$ and each of the k atomic proposition in φ is represented by a boolean variable l_i. For a BDD-based symbolic representation, the order of the variables is crucial. The state variables and their copies are always kept adjacent. The standard variable ordering used by LTL2GFG is then an interleaving of the $a_{i,j}$ and $b_{i,j}$ variables with the q_j variables, i.e.,
$$l_1 < \ldots < l_k < q_1 < \ldots < q_j < a_{1,j} < b_{1,j} < a_{2,j} < b_{2,j} < \ldots < q_{j+1} < \ldots.$$
LTL2GFG uses the JINC C++ BDD library for the symbolic representation.

Experimental Results for the HP-algorithm. We report here on a number of experiments with LTL2GFG using the benchmark formulas used in the evaluation of LTL2DSTAR in [14,15], i.e., 39 LTL formulas from the literature [10,33] and 55 pattern formulas [9] that represent common specification patterns. All our experiments were carried out on a computer with 2 Intel E5-2680 8-core CPUs at 2.70 GHz with 384GB of RAM running Linux and with a memory limit of 10 GB and a time-out of 30 minutes for each formula.

For the automata $\mathcal{A}_\varphi$, we report on the number of BDD nodes in the encoding of the transition function, as this the most crucial aspect. To allow a fair comparison with the explicit determinization in LTL2DSTAR, we consider symbolic encodings of the DRA $\mathcal{A}_\rho$ obtained from LTL2DSTAR 0.5.1. This encoding uses $\lceil \log_2 n \rceil$ boolean variables to straightforwardly encode the n state indices in $\mathcal{A}_\varphi$, which is the same encoding employed in PRISM for its DRA-based approach to LTL model checking.

Table 1 presents statistics for the construction of DRA with LTL2DSTAR and GFG-NPA with LTL2GFG for the benchmark formulas. The LTL2DSTAR results are given once with standard settings and for a variant where all optimizations are disabled, i.e., with purely Safra's construction. For LTL2GFG, we start with the pure HP-algorithm and consider variants with the "loose" transition definition, the union construction, and with dynamic reordering of the variable order. We also give statistics for the iterative approach, where LTL2GFG constructs the partial automata $\mathcal{A}^m$ until it can be shown (via solving a Rabin game [29]) that the automaton is GFG.

Table 2. Results of the iterative approach in LTL2GFG, for the loose variant. M is the minimal value $m \leq n$ for which the partial NPA $\mathcal{A}^m$ could be shown to be GFG.

	with n NBA states											
	2	3	4	5	6	7	8	9	10	11	12	>12
number of φ	13	17	13	9	8	3	3	1	4	2	4	11
number of φ, $M < n$	11	17	13	8	8	2	2	1	0	0	1	3
number of φ, $M = 1$	11	8	5	4	2	1	1	0	0	0	1	3
number of φ, $M = 2$	2	9	8	4	6	1	1	1	0	0	0	0
number of φ, GFG check aborted	0	0	0	1	0	1	1	0	4	2	3	8

LTL2DSTAR constructed most of the automata in a few seconds, the most difficult was constructed in 95s and had 1.2 million BDD nodes. Apart from the most difficult automata, the BDD sizes range in the hundreds and thousands. For all the LTL2GFG variants, a significant fraction of automata could not be constructed in the time and memory limits, around 40% for the standard HP-algorithm, and dropping to around 20% for the best variant. The loose variant by itself had a mixed effect, but in conjunction with dynamic reordering was generally beneficial. The union construction was very beneficial for the disjunctive formulas. For example, the automata for $\Box \Diamond a \to \Box \Diamond b$ could not be constructed in the time limits with the standard HP-algorithm but could be handled using the union construction. The iterative approach was successful as well in obtaining smaller automata, which is explained by the fact that for a large number of formulas it could be shown that the partial automata $\mathcal{A}^1$ or $\mathcal{A}^2$ were already GFG, as detailed in Table 2. For the iterative approach we were mostly focused on experimental data for the minimal value m for which $\mathcal{A}^m$ becomes GFG and the effect on the BDD size. Different algorithms or implementations for the GFG check than the one used in LTL2GFG lead to the same final GFG automata, but could improve the performance. At the end, despite the various approaches implemented in LTL2GFG, there were only 6 formulas with relatively small automata where the BDD size of the smallest GFG automaton was smaller than that of the DRA obtained from LTL2DSTAR (172 nodes instead of 229 nodes, 219 instead of 347, and the other 4 automata differing by 1 or 2 at a size of less than 20 nodes). We do not report here in detail on the number of reachable states in the automata, as none of the GFG automata had a smaller number of states than the DRA generated by LTL2DSTAR. In particular, the automata obtained without the iterative approach often had millions and more states.

Implementation in PRISM. We have extended the MTBDD-based, symbolic engines of PRISM 4.1 with an implementation of our algorithm for computing $\text{Pr}_{\mathcal{M}}^{\max}(\varphi)$ using GFG automata for φ (and $\text{Pr}_{\mathcal{M}}^{\min}(\varphi)$ using a GFG automaton for $\neg \varphi$). We import the BDD of $\mathcal{A}$ generated with LTL2GFG into PRISM and perform the product with $\mathcal{M}$ and analysis in $\mathcal{M} \otimes \mathcal{A}$ symbolically. In its standard approach, PRISM constructs an explicit DRA with an integrated version of LTL2DSTAR, which is then symbolically encoded as described before. The analysis is then carried out symbolically as well.

Experiments in PRISM. As a benchmark, we used a PRISM model [23] for parts of the WLAN carrier-sense protocol of IEEE 802.11. As was to be expected given our results on the automata construction, the GFG-based analysis did not improve on the standard approach. Even using the optimal variant of LTL2GFG for each formula, ignoring the automata construction times, and for cases where the product $\mathcal{M} \otimes \mathcal{A}$ had a comparable BDD size for the GFG- and DRA-based approach, the model checking using the GFG automata took significantly longer. For further details, we refer to [16].

6   Conclusion

We have shown that GFG automata can replace deterministic automata for the quantitative analysis of MDPs against ω-regular specifications without increasing the asymptotic worst-case time complexity. To evaluate the GFG-based approach from the practical side, we implemented the HP-algorithm, integrated several heuristics, and performed exhaustive experiments for the LTL-to-GFG construction and for probabilistic model checking. Our experimental results are a bit disappointing, as the generated GFG automata were often larger than DRA generated by the implementation of Safra's algorithms in LTL2DSTAR, both in the number of states and in the symbolic BDD-based representations. Thus, our empirical results are in contrast to the expectation that the HP-algorithm yields GFG automata that are better suited for symbolic approaches rather than DRA generated by Safra's algorithm. Also in the context of probabilistic model checking, the GFG-based approach turned out to be more time- and memory-consuming than the traditional approach with deterministic automata. However, it is still too early to discard the concept of GFG automata for practical purposes. Our negative empirical results might be an artefact of the HP-algorithm, which is – to the best of our knowledge – the only known algorithm for the generation of GFG automata that are not deterministic. Future directions are the design of other algorithms for the construction of succinct GFG automata. Alternatively, one might seek for automata types that are still adequate for probabilistic model checking and other areas, but rely on weaker conditions than the GFG property.

References

1. Babiak, T., Blahoudek, F., Křetínský, M., Strejček, J.: Effective translation of LTL to deterministic rabin automata: Beyond the (F,G)-fragment. In: Van Hung, D., Ogawa, M. (eds.) ATVA 2013. LNCS, vol. 8172, pp. 24–39. Springer, Heidelberg (2013)
2. Baier, C., Katoen, J.P.: Principles of Model Checking. MIT Press (2008)
3. Benedikt, M., Lenhardt, R., Worrell, J.: Two variable vs. linear temporal logic in model checking and games. Logical Methods in Computer Science 9(2) (2013)
4. Boker, U., Kuperberg, D., Kupferman, O., Skrzypczak, M.: Nondeterminism in the presence of a diverse or unknown future. In: Fomin, F.V., Freivalds, R., Kwiatkowska, M., Peleg, D. (eds.) ICALP 2013, Part II. LNCS, vol. 7966, pp. 89–100. Springer, Heidelberg (2013)

5. Bustan, D., Rubin, S., Vardi, M.Y.: Verifying ω-regular properties of Markov chains. In: Alur, R., Peled, D.A. (eds.) CAV 2004. LNCS, vol. 3114, pp. 189–201. Springer, Heidelberg (2004)
6. Clarke, E., Grumberg, O., Peled, D.: Model Checking. MIT Press (2000)
7. Courcoubetis, C., Yannakakis, M.: The complexity of probabilistic verification. Journal of ACM 42(4), 857–907 (1995)
8. Couvreur, J.M., Saheb, N., Sutre, G.: An optimal automata approach to LTL model checking of probabilistic systems. In: Vardi, M.Y., Voronkov, A. (eds.) LPAR 2003. LNCS, vol. 2850, pp. 361–375. Springer, Heidelberg (2003)
9. Dwyer, M., Avrunin, G., Corbett, J.: Patterns in property specifications for finite-state verification. In: ICSE 1999, pp. 411–420. ACM (1999)
10. Etessami, K., Holzmann, G.J.: Optimizing Büchi automata. In: Palamidessi, C. (ed.) CONCUR 2000. LNCS, vol. 1877, pp. 153–167. Springer, Heidelberg (2000)
11. Gastin, P., Oddoux, D.: Fast LTL to Büchi automata translation. In: Berry, G., Comon, H., Finkel, A. (eds.) CAV 2001. LNCS, vol. 2102, pp. 53–65. Springer, Heidelberg (2001)
12. Grädel, E., Thomas, W., Wilke, T. (eds.): Automata, Logics, and Infinite Games. LNCS, vol. 2500. Springer, Heidelberg (2002)
13. Henzinger, T., Piterman, N.: Solving games without determinization. In: Ésik, Z. (ed.) CSL 2006. LNCS, vol. 4207, pp. 395–410. Springer, Heidelberg (2006)
14. Klein, J., Baier, C.: Experiments with deterministic ω-automata for formulas of linear temporal logic. Theoretical Computer Science 363(2), 182–195 (2006)
15. Klein, J., Baier, C.: On-the-fly stuttering in the construction of deterministic ω-automata. In: Holub, J., Žďárek, J. (eds.) CIAA 2007. LNCS, vol. 4783, pp. 51–61. Springer, Heidelberg (2007)
16. Klein, J., Müller, D., Baier, C., Klüppelholz, S.: Are good-for-games automata good for probabilistic model checking (extended version). Tech. rep., Technische Universität Dresden (2013), http://wwwtcs.inf.tu-dresden.de/ALGI/PUB/LATA14/
17. Křetínský, J., Garza, R.L.: Rabinizer 2: Small deterministic automata for LTL\GU. In: Van Hung, D., Ogawa, M. (eds.) ATVA 2013. LNCS, vol. 8172, pp. 446–450. Springer, Heidelberg (2013)
18. Kupferman, O., Piterman, N., Vardi, M.: Safraless compositional synthesis. In: Ball, T., Jones, R.B. (eds.) CAV 2006. LNCS, vol. 4144, pp. 31–44. Springer, Heidelberg (2006)
19. Kupferman, O., Rosenberg, A.: The blow-up in translating LTL to deterministic automata. In: van der Meyden, R., Smaus, J.-G. (eds.) MoChArt 2010. LNCS, vol. 6572, pp. 85–94. Springer, Heidelberg (2011)
20. Kupferman, O., Vardi, M.: From linear time to branching time. ACM Transactions on Computational Logic 6(2), 273–294 (2005)
21. Kupferman, O., Vardi, M.: Safraless decision procedures. In: FOCS 2005, pp. 531–542. IEEE Computer Society (2005)
22. Kwiatkowska, M., Norman, G., Parker, D.: Probabilistic symbolic model checking with PRISM: A hybrid approach. International Journal on Software Tools for Technology Transfer (STTT) 6(2), 128–142 (2004)
23. Kwiatkowska, M., Norman, G., Sproston, J.: Probabilistic model checking of the IEEE 802.11 wireless local area network protocol. In: Hermanns, H., Segala, R. (eds.) PROBMIV 2002, PAPM-PROBMIV 2002, and PAPM 2002. LNCS, vol. 2399, pp. 169–187. Springer, Heidelberg (2002)
24. Latvala, T.: Efficient model checking of safety properties. In: Ball, T., Rajamani, S.K. (eds.) SPIN 2003. LNCS, vol. 2648, pp. 74–88. Springer, Heidelberg (2003)

25. Löding, C.: Optimal bounds for transformations of ω-automata. In: Pandu Rangan, C., Raman, V., Sarukkai, S. (eds.) FST TCS 1999. LNCS, vol. 1738, pp. 97–109. Springer, Heidelberg (1999)

26. Michel, M.: Complementation is more difficult with automata on infinite words. CNET, Paris (1988)

27. Morgenstern, A., Schneider, K.: From LTL to symbolically represented deterministic automata. In: Logozzo, F., Peled, D.A., Zuck, L.D. (eds.) VMCAI 2008. LNCS, vol. 4905, pp. 279–293. Springer, Heidelberg (2008)

28. Piterman, N.: From nondeterministic Büchi and Streett automata to deterministic parity automata. Logical Methods in Computer Science 3(3:5), 1–21 (2007)

29. Piterman, N., Pnueli, A.: Faster solutions of Rabin and Streett games. In: LICS 2006, pp. 275–284. IEEE (2006)

30. Puterman, M.: Markov Decision Processes: Discrete Stochastic Dynamic Programming. John Wiley & Sons, Inc., New York (1994)

31. Safra, S.: On the complexity of ω-automata. In: FOCS, pp. 319–327. IEEE (1988)

32. Schewe, S.: Tighter bounds for the determinisation of Büchi automata. In: de Alfaro, L. (ed.) FOSSACS 2009. LNCS, vol. 5504, pp. 167–181. Springer, Heidelberg (2009)

33. Somenzi, F., Bloem, R.: Efficient Büchi automata from LTL formulae. In: Emerson, E.A., Sistla, A.P. (eds.) CAV 2000. LNCS, vol. 1855, pp. 248–263. Springer, Heidelberg (2000)

34. Vardi, M., Wolper, P.: An automata-theoretic approach to automatic program verification. In: LICS 1986, pp. 332–344. IEEE Computer Society (1986)

Top-Down Tree Edit-Distance
of Regular Tree Languages*

Sang-Ki Ko[1], Yo-Sub Han[1], and Kai Salomaa[2]

[1] Department of Computer Science, Yonsei University
50, Yonsei-Ro, Seodaemun-Gu, Seoul 120-749, Republic of Korea
{narame7,emmous}@cs.yonsei.ac.kr
[2] School of Computing, Queen's University
Kingston, Ontario K7L 3N6, Canada
ksalomaa@cs.queensu.ca

Abstract. We study the edit-distance of regular tree languages. The edit-distance is a metric for measuring the similarity or dissimilarity between two objects, and a regular tree language is a set of trees accepted by a finite-state tree automaton or described by a regular tree grammar. Given two regular tree languages L and R, we define the edit-distance $d(L, R)$ between L and R to be the minimum edit-distance between a tree $t_1 \in L$ and $t_2 \in R$, respectively. Based on tree automata for L and R, we present a polynomial algorithm that computes $d(L, R)$. We also suggest how to use the edit-distance between two tree languages for identifying a special common string between two context-free grammars.

Keywords: tree edit-distance, regular tree languages, tree automata, dynamic programming.

1 Introduction

It is an important problem to measure the similarity or dissimilarity between data in many applications [14,22,24]. For example, there are several similarity measures between two strings [7,11,21] and one of the most well-known measures is is the Levenshtein distance [11], which is often called the *edit-distance* in the literature. The edit-distance problem is, then, to compute the shortest distance between inputs. Researchers extended the edit-distance problem between strings into the edit-distance problem between a string and a language, or between two languages [3,8,9,12,13,19,20]. Another extension of the string edit-distance problem is the tree edit-distance problem [5,10,16,17,23]. The *tree edit-distance* problem is to find the minimum number of edit operations required to transform one tree into the other. The tree edit-distance plays an important role for calculating the similarity between structural data such as XML documents [18].

* Ko and Han were supported by the Basic Science Research Program through NRF funded by MEST (2012R1A1A2044562), and Salomaa was supported by the Natural Sciences and Engineering Research Council of Canada Grant OGP0147224.

A.-H. Dediu et al. (Eds.): LATA 2014, LNCS 8370, pp. 466–477, 2014.

Consider a tree t of size m (namely, there are m nodes in t) and let m_h and m_l be the height and the number of leaves of t. Tai [17] considered the problem of computing the tree edit-distance as a generalization of the string edit-distance problem and designed an $\mathcal{O}(m_l^2 n_l^2 mn)$ algorithm, where m and n are the sizes of input trees. Later, Shasha and Zhang [23] improved Tai's algorithm and presented an $\mathcal{O}(mn \cdot \min\{m_h, m_l\} \cdot \min\{n_h, n_l\})$ algorithm, which runs in $\mathcal{O}(m^2 n^2)$ time in the worst-case. Klein [10] further improved this algorithm and obtained an $\mathcal{O}(m^2 n \log n)$ algorithm, where $n \geq m$. Recently, Demaine et al. [5] suggested an $\mathcal{O}(m^2 n(1 + \log \frac{n}{m}))$ time algorithm, for $n \geq m$, using an optimal decomposition strategy. Note that all these algorithms allows both insertion or deletion of internal nodes in a tree.

Selkow [16] considered the tree edit-distance model that requires insertion and deletion to be allowed only at leaf nodes and called this problem the *top-down tree edit-distance problem*. Then, he designed an $\mathcal{O}(mn)$ algorithm for the problem. Researchers successfully applied the top-down tree edit-distance to several applications [2,14,15]. For instance, Nierman and Jagadish [14] considered several tree edit-distance definitions for clustering XML documents and demonstrated that top-down tree edit-distance guarantees less mis-clusterings than the general tree edit-distance and, thus, is a better clustering scheme.

We examine the top-down tree edit-distance of two regular tree languages accepted by tree automata. There are many results on the problem of computing the distance between languages [1,3,8,9,12]. A regular tree language is a set of trees, and is specified by either a regular tree grammar or a finite-state tree automaton. We propose an $\mathcal{O}(m^2 n^2)$ algorithm for computing the edit-distance between two regular tree languages of k-bounded trees and an $\mathcal{O}(m^2 n^2 \log mn)$ algorithm for the edit-distance between two regular tree languages of unbounded trees, where m and n are sizes of two input tree automata.

In Section 2, we give basic definitions and notations. Then, we introduce the tree edit-distance problem in Section 3. We propose an algorithm for the edit-distance between two regular tree languages of k-bounded trees in Section 4. We also consider the unranked case in Section 5. Then, we show that our result can be applied to the problem of measuring the similarity between two context-free string languages in Section 6 and conclude the paper in Section 7.

2   Preliminaries

A ranked alphabet Σ is a pair of a finite set of characters and a function $r : \sigma \to \mathbb{N} \cup \{0\}$. We denote the set of elements of rank m by $\Sigma_m \subseteq \Sigma$, where $m \geq 0$. The set F_Σ consists of Σ-labeled trees, where a node labeled by $\sigma \in \Sigma_m$ for $m \geq 0$ has exactly m children. We denote the set of trees over Σ by F_Σ, which is the smallest set S satisfying the following condition: if $m \geq 0, \sigma \in \Sigma_m$ and $t_1, \ldots, t_m \in S$, then $\sigma(t_1, \ldots, t_m) \in S$. In an unranked tree each node has a finite number of children but the label of a node does not determine the number of children and there is no apriori upper bound on the number of children. Unranked trees can be defined as above by replacing the condition "$\sigma \in \Sigma_m$" by "$\sigma \in \Sigma$".

A *finite-state automaton* (FA) A is specified by a tuple $A = (\Sigma, Q, F, \delta)$, where Σ is an alphabet, Q is a finite set of states, $F \subseteq Q$ is a set of final states, and δ is a transition function. Given an FA $A = (Q, \Sigma, F, \delta)$, we define the size $|A|$ of A to be $|Q| + |\text{dom}(\delta)|$. Note that an FA accepts a regular language.

A *nondeterministic ranked tree automaton* (NTA) A is specified by a tuple (Q, Σ, F, δ), where Q is a finite set of states, Σ is a ranked alphabet, $F \subseteq Q$ is a set of final states, and δ associates to each $\sigma \in \Sigma_m$ a mapping $\sigma_\delta : Q^m \to 2^Q, m \geq 0$. For each tree $t = \sigma(t_1, \ldots, t_m) \in F_\Sigma$, we define inductively the set $t_\delta \subseteq Q$ by setting $q \in t_g$ if and only if there exist $q_i \in (t_i)_\delta$, for $1 \leq i \leq m$, such that $q \in \sigma_\delta(q_1, \ldots, q_m)$. Intuitively, t_δ consists of the states of Q that A may reach by reading the tree t. Thus, the tree language accepted by A is defined as follows: $L(A) = \{t \in F_\Sigma \mid t_\delta \cap Q_f \neq \emptyset\}$. We define the size $|A|$ of a ranked TA A to be $|Q| + \sum_{\sigma_\delta(q_1, \ldots, q_m) \to q}(r(\sigma) + 1)$. The automaton A is a *deterministic ranked tree automaton* (DTA) if, for each $\sigma \in \Sigma_m$, where $m \geq 0$, σ_δ is a partial function $Q^m \to Q$. The nondeterministic (bottom-up or top-down) and deterministic bottom-up tree automata accept the family of *regular tree languages of ranked trees*.

A *nondeterministic unranked tree automaton* is specified by a tuple $A = (\Sigma, Q, F, \delta)$, where Σ is an alphabet, Q is a finite set of states, $F \subseteq Q$ is a set of final states, and δ is a transition relation. For each $q \in Q$ and $\sigma \in \Sigma$, we define $\delta(q, \sigma)$ to be the horizontal language associated with q and σ. We denote an FA for the horizontal language $\delta(q, a)$ of A by $H_{q,\sigma}^A$. Then, according to the transition relation δ, each $\sigma \in \Sigma$ defines a partial function $\sigma_\delta : Q^* \to Q$, where, for $w \in Q^*$, $q \in Q$, $q \in \sigma_\delta(w)$ if $w \in H_{q,\sigma}^A$. The tree language accepted by A is defined as follows: $L(A) = \{t \in T_\Sigma \mid t \xrightarrow{*} q_f \in F\}$. We define the size $|A|$ of an unranked TA A to be $|Q| + \sum_{q \in Q, \sigma \in \Sigma}(|H_{q,\sigma}^A| + 1)$. Naturally a ranked tree automaton is a special case of an unranked tree automaton, where for $\sigma \in \Sigma_m$ and $q \in Q$ we have always $H_{q,\sigma}^A \subseteq Q^m$.

For a tree t, we assume that all nodes are ordered in postorder and, thus, t is an *ordered tree*. Let $t[i]$ be the ith node of t and $\text{des}(t[i])$ be the set of all descendants of $t[i]$ including $t[i]$ itself. Thus, $t[l(i) \ldots i]$ is the subtree rooted at $t[i]$, that is the subtree consisting of node i and all its descendants. Similarly, we define $\text{anc}(t[i])$ to be the set of all ancestors of $t[i]$ including $t[i]$. The size $|t|$ of t is the number of nodes in t. We denote the character labeling a node $t[i]$ by $\sigma(i)$. Let θ be the empty tree. We say that $\text{yield}(t)$ is a sequence of leaves in t. A forest is a sequence of trees and ordered when it has a left-to-right order among the trees. We only consider the ordered trees and the ordered forests in this paper. We refer the reader to the literature [4,6] for more details on tree automata.

3   Tree Edit-Distance

Given an alphabet Σ, let $\Omega = \{(a \to b) \mid a, b \in \Sigma \cup \{\lambda\}\} \setminus \{(\lambda, \lambda)\}$ be a set of edit operations. There are three edit operations: *deletions* $(a \to \lambda)$, *insertions* $(\lambda \to a)$ and *substitutions* $(a \to b)$ for $a \neq b$. We associate a non-negative edit cost to each edit operation $\omega_i \in \Omega$ as a function $\mathsf{C} : \Omega \to \mathbb{R}_+$. We assume that C is a distance metric satisfying the following conditions:

(i) $\mathsf{C}(a \to b) \geq 0, \mathsf{C}(a \to a) = 0$,
(ii) $\mathsf{C}(a \to b) = \mathsf{C}(b \to a)$ and
(iii) $\mathsf{C}(a \to c) \leq \mathsf{C}(a \to b) + \mathsf{C}(b \to c)$, where $a, b, c \in \Sigma \cup \{\lambda\}$.

An *edit script* $S \in \Omega^*$ between two trees t_1 and t_2 is a sequence of edit operations transforming t_1 into t_2. The cost $\mathsf{C}(S)$ of $S = s_1 s_2 \cdots s_n$ is $\mathsf{C}(S) = \sum_{i=1}^n \mathsf{C}(s_i)$. An *optimal edit script* between t_1 and t_2 is an edit script of minimum cost and the minimum cost is the *tree edit-distance* between t_1 and t_2.

Definition 1. *The tree edit-distance $d(t_1, t_2)$ of two trees t_1 and t_2 is*

$$d(t_1, t_2) = \min\{\mathsf{C}(S) \mid S \text{ is an edit script transforming } t_1 \text{ into } t_2\}.$$

That is, if S is an optimal edit script that transforms t_1 into t_2, then $\mathsf{C}(S) = d(t_1, t_2)$. We, in particular, consider the *top-down tree edit-distance*, which allows deletions and insertions of nodes only at leaves; namely, a node can be inserted or deleted only at leaf level. Thus, when an edit script S consists of edit operations where insertions or deletions occur only at leaf level, we say that S is a *top-down edit script*. We define the top-down tree edit-distance as follows:

Definition 2. *The top-down tree edit-distance $d(t_1, t_2)$ of two trees t_1 and t_2 is*

$$d(t_1, t_2) - \min\{\mathsf{C}(S) \mid S \text{ is a top-down edit script transforming } t_1 \text{ into } t_2\}.$$

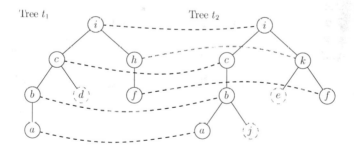

Fig. 1. A (top-down) mapping example between two trees t_1 and t_2

The example mapping in Fig. 1 depicts an edit script S that transforms t_1 into t_2 by deleting a node d, inserting two nodes j and e, and substituting a node h with k. Thus, the corresponding edit script S is

$$S = (a \to a)(\lambda \to j)(b \to b)(d \to \lambda)(c \to c)(\lambda \to e)(f \to f)(h \to k)(i \to i).$$

Let T_1 and T_2 be the sets of nodes in t_1 and t_2, respectively. We define a triple (M, t_1, t_2) to be a mapping from t_1 to t_2, where $M \subseteq T_1 \times T_2$ is a set of pair of nodes (i, j) for $1 \leq i \leq |t_1|$ and $1 \leq j \leq |t_2|$. We use M instead of (M, t_1, t_2) for simplicity when there is no confusion. We assume that trees are ordered in postorder. For any pair of (i_1, j_1) and (i_2, j_2) in M, the mapping M has the following restrictions:

  (i) $i_1 = i_2$ if and only if $j_1 = j_2$ (one-to-one)
 (ii) $i_1 < i_2$ if and only if $j_1 < j_2$ (sibling order preserved)
(iii) $t_1[i_1] \in \mathtt{anc}(t_1[i_2])$ if and only if $t_2[j_1] \in \mathtt{anc}(t_2[j_2])$ (ancestor order preserved)

We say that a node $t_1[i]$ is *touched by a line* if there exists a pair $(i, j) \in M$. Let I and J be the sets of nodes in t_1 and t_2, respectively, that are not touched by any line in M. Then, we define the cost $\mathtt{C}(M)$ of M to be

$$\mathtt{C}(M) = \sum_{(i,j) \in M} \mathtt{C}(\sigma(i) \to \sigma(j)) + \sum_{i \in I} \mathtt{C}(\sigma(i) \to \lambda) + \sum_{j \in J} \mathtt{C}(\lambda \to \sigma(j)).$$

Next we extend the concept of the edit-distance as a distance metric between a tree and a tree language.

Definition 3. *We define the edit-distance $d(t, L)$ between a tree t and a tree language L over Σ to be*

$$d(t, L) = \inf\{d(t, t') \mid t' \in L\}.$$

Then, we define the edit-distance between two tree languages as follows:

Definition 4. *We define the edit-distance $d(L, R)$ between two tree languages L and R over Σ to be*

$$d(L, R) = \inf\{d(t, t') \mid t \in L \text{ and } t' \in R\}.$$

In other words, the edit-distance between L and R is the minimum edit-distance between the most similar pair of trees from two tree languages. Note that these distance measures are symmetric. Thus, $d(t, t') = d(t', t)$, $d(t, L) = d(L, t)$, and $d(L, R) = d(R, L)$.

4   Edit-Distance of Regular Tree Languages

Before we tackle the main problem, we introduce *k-bounded tree automata*, which have more expressive power than ranked TAs. Note that if we insert or delete a node in a ranked tree, then it does not preserve its rank anymore. Therefore, instead of considering ranked TAs, we use TAs that allow only a finite number of children.

Definition 5. *A k-bounded tree automaton is specified by a tuple*

$$A = (\Sigma, Q, F, \delta),$$

where Σ is an alphabet, Q is a finite set of states, $F \subseteq Q$ is a set of final states, and δ associates to each $\sigma \in \Sigma$ a mapping $\sigma_g : Q^{\leq k} \to 2^Q$.

We define the size $|A|$ of a k-bounded TA A to be $|Q| + \sum_{\sigma_\delta(q_1,\ldots,q_l)\to q}(l+1)$. A k-bounded TA is, thus, an unranked TA where there exists a constant k such that any node can have at most k children. Note that a ranked TA is a restricted variant of a k-bounded TA. If Σ is a ranked alphabet and $k = \max\{r(\sigma) \mid \sigma \in \Sigma\}$, then any ranked TA over Σ is k-bounded.

We introduce a polynomial algorithm for computing the tree edit-distance between two regular tree languages L and R described by k-bounded TAs. By Definition 4, the edit-distance between L and R is the edit-distance between two closest pair of trees $t \in L$ and $t' \in R$.

Let $A = (\Sigma, Q, F, \delta)$ and $B = (\Sigma', Q', F', \delta')$ be two k-bounded TAs accepting regular tree languages L and R, respectively. From A, let A_q be a new TA that has a unique final state $q \in Q$. For simplicity, we denote $d(L(A_q), L(B_{q'}))$ by $d(q, q')$. Then, we have the following statement.

Proposition 6. *Let $A = (\Sigma, Q, F, \delta)$ and $B = (\Sigma', Q', F', \delta')$ be two k-bounded TAs accepting two tree languages L and R. Then,*

$$d(L, R) = \min\{d(q, q') \mid q \in F, q' \in F'\}.$$

Proposition 6 states that we can compute the edit-distance between two regular tree languages by computing the minimum edit-distance between a pair (q, q') of states q, q', where q is a final state of A and q' is a final state of B. We show that it is possible to compute such distances in polynomial time by recursively computing the distance between all pairs of states from A and B. Before we give the main algorithm, we first define the edit distance between two sequences of states. We denote the minimum edit-distance between two forests from two sequences of states $q_1 q_2 \ldots q_i$ and $q_1' q_2' \ldots q_j'$ by $d(q_1 q_2 \ldots q_i, q_1' q_2' \ldots q_j')$. From now, we denote the sequence $q_1 q_2 \ldots q_i$ of states by $\mathbb{S}_{1,i}$ and the sequence $q_1' q_2' \ldots q_j'$ by $\mathbb{S}_{1,j}'$. In other words, $\mathbb{S}_{m,n} \in Q^{\leq k}$ and $\mathbb{S}_{m,n}' \in (Q')^{\leq k}$ for $m \leq n$.

Given two k-bounded TAs $A = (\Sigma, Q, F, \delta)$ and $B = (\Sigma', Q', F', \delta')$, we define the following sets for the edit-distance $d(\mathbb{S}_{1,i}, \mathbb{S}_{1,j}')$ between two sequences of states.

(i) $I(\mathbb{S}_{1,i}, \mathbb{S}_{1,j}') = \{d(\mathbb{S}_{1,i}, \mathbb{S}_{1,j-1}') + d(\lambda, \mathbb{T}') + \mathsf{C}(\lambda, \sigma') \mid \sigma_{\delta'}'(\mathbb{T}') = q_j'\}$,

(ii) $D(\mathbb{S}_{1,i}, \mathbb{S}_{1,j}') = \{d(\mathbb{S}_{1,i-1}, \mathbb{S}_{1,j}') + d(\mathbb{T}, \lambda) + \mathsf{C}(\sigma, \lambda) \mid \sigma_\delta(\mathbb{T}) = q_i\}$, and

(iii) $S(\mathbb{S}_{1,i}, \mathbb{S}_{1,j}') = \{d(\mathbb{S}_{1,i-1}, \mathbb{S}_{1,j-1}') + d(\mathbb{T}, \mathbb{T}') + \mathsf{C}(\sigma, \sigma') \mid \sigma_\delta(\mathbb{T}) = q_i, \sigma_{\delta'}'(\mathbb{T}') = q_j'\}$,

where $\mathbb{T} \in Q^*$ and $\mathbb{T}' \in (Q')^*$. Let I, D, and S denote insertion, deletion and substitution, respectively. Based on the three sets, we present a recursive equation for computing the edit-distance between two forests accepted by two sequences of states from A and B as follows:

Lemma 7. *Given two k-bounded TAs $A = (\Sigma, Q, F, \delta)$ and $B = (\Sigma', Q', F', \delta')$, the edit-distance $d(\mathbb{S}_{1,i}, \mathbb{S}_{1,j}')$ is defined as*

$$\min[I(\mathbb{S}_{1,i}, \mathbb{S}_{1,j}') \cup D(\mathbb{S}_{1,i}, \mathbb{S}_{1,j}') \cup S(\mathbb{S}_{1,i}, \mathbb{S}_{1,j}')],$$

where $\mathbb{S}_{1,i} \in Q^{\leq k}$ and $\mathbb{S}_{1,j}' \in (Q')^{\leq k}$.

We observe that the computation of $d(\mathbb{S}_{1,i}, \mathbb{S}'_{1,j})$ may have a self-dependency problem in the computation. For example, consider the set $S(\mathbb{S}_{1,i}, \mathbb{S}'_{1,j})$, which is the 3rd case in the proof of Lemma 7.

$$S(\mathbb{S}_{1,i}, \mathbb{S}'_{1,j}) = \{d(\mathbb{S}_{1,i-1}, \mathbb{S}'_{1,j-1}) + d(\mathbb{T}, \mathbb{T}') + \mathsf{C}(\sigma, \sigma') \mid \sigma_\delta(\mathbb{T}) = q_i, \sigma'_{\delta'}(\mathbb{T}') = q'_j\}.$$

Then, the computation of $S(\mathbb{S}_{1,i}, \mathbb{S}'_{1,j})$ requires the value of $d(\mathbb{S}_{1,i}, \mathbb{S}'_{1,j})$ when $\mathbb{T} = \mathbb{S}_{1,i}$ and $\mathbb{T}' = \mathbb{S}'_{1,j}$. This implies that we need the value of $d(\mathbb{S}_{1,i}, \mathbb{S}'_{1,j})$ for computing the value of $d(\mathbb{S}_{1,i}, \mathbb{S}'_{1,j})$. We solve this dependency problem by using induction on the height of the optimal mapping. Assume that we have two trees t_1, t_2 and an optimal mapping $M \subseteq t_1 \times t_2$. We construct a new mapping M' from M by removing all insertions and deletions. Then, the resulting mapping M' consists of the pairs $(i, j) \in t_1 \times t_2$ such that $i \neq \lambda$ and $j \neq \lambda$. We call M' the *trimmed mapping*. Now we define the *height n edit-distance* to be the edit-distance between two trees, where the height of the corresponding optimal trimmed mapping for the edit-distance is at most n. Let $d_n(q, q')$ be the *height n edit-distance* between two states q and q'. Note that the similar notation can be used for the height n edit-distance between two sequences of states such as $d_n(\mathbb{S}_{1,i}, \mathbb{S}'_{1,j})$.

Then we define the following sets for recurrence of the height n edit-distance between two sequences of states.

(i) $I_n(\mathbb{S}_{1,i}, \mathbb{S}'_{1,j}) = \{d_n(\mathbb{S}_{1,i}, \mathbb{S}'_{1,j-1}) + d_0(\lambda, \mathbb{T}') + \mathsf{C}(\lambda, \sigma') \mid \sigma'_{\delta'}(\mathbb{T}') = q'_j\}$,

(ii) $D_n(\mathbb{S}_{1,i}, \mathbb{S}'_{1,j}) = \{d_n(\mathbb{S}_{1,i-1}, \mathbb{S}'_{1,j}) + d_0(\mathbb{T}, \lambda) + \mathsf{C}(\sigma, \lambda) \mid \sigma_\delta(\mathbb{T}) = q_i\}$, and

(iii) $S_n(\mathbb{S}_{1,i}, \mathbb{S}'_{1,j}) = \{d_n(\mathbb{S}_{1,i-1}, \mathbb{S}'_{1,j-1}) + d_{n-1}(\mathbb{T}, \mathbb{T}') + \mathsf{C}(\sigma, \sigma') \mid \sigma_\delta(\mathbb{T}) = q_i, \sigma'_{\delta'}(\mathbb{T}') = q'_j\}$.

Then, $d_n(q, q')$ should be the minimum of the following sets:

(i) $I_n(q, q') = \{d_0(q, \lambda) + d_0(\lambda, \mathbb{T}') + \mathsf{C}(\lambda, \sigma') \mid \sigma'_{\delta'}(\mathbb{T}') = q'\}$,

(ii) $D_n(q, q') = \{d_0(\lambda, q') + d_0(\mathbb{T}, \lambda) + \mathsf{C}(\sigma, \lambda) \mid \sigma_\delta(\mathbb{T}) = q\}$, and

(iii) $S_n(q, q') = \{d_{n-1}(\mathbb{T}, \mathbb{T}') + \mathsf{C}(\sigma, \sigma') \mid \sigma_\delta(\mathbb{T}) = q, \sigma'_{\delta'}(\mathbb{T}') = q'\}$.

Note that the following equations hold:

- $d_n(q, q') = \infty$ if $n < 0$,
- $d_0(\lambda, \lambda) = 0$,
- $d_0(q, \lambda) = \min\{|t| \mid t \in L(A_q)\}$, and
- $d_0(q, q') = \min\{|t| \mid t \in L(A_q)\} + \min\{|t'| \mid t' \in L(B_{q'})\}$.

Now we establish the following lemma.

Lemma 8. *Given two k-bounded TAs $A = (\Sigma, Q, F, \delta)$ and $B = (\Sigma', Q', F', \delta')$, the height n edit-distance $d_n(q, q')$ is defined as*

$$\min[d_{n-1}(q, q') \cup I_n(q, q') \cup D_n(q, q') \cup S_n(q, q')],$$

where $q \in Q$, $q' \in Q'$ and $n \geq 0$.

Proof. We prove by induction on n. First, we start with the case when $n = 0$. Consider a mapping M between two trees $t \in L(A_q)$ and $t' \in L(B_{q'})$, where $|t| = i$ and $|t'| = j$. We construct the mapping M in postorder, therefore, the mapping between two root nodes of t and t' should be the last to consider. By the definition, the height of the trimmed mapping of M should be 0. This implies that there is no mapping for substitutions between two trees. Since $d_{-1}(q, q')$ is ∞, we compute the case when $n = 0$ by only considering insertions and deletions.

Assume that the case $n = l$ holds. Then, we prove that the case $n = l + 1$ also holds. By the assumption, we know that the case when the height of the trimmed mapping is lower than $l + 1$ is considered by the first term $d_l(q, q')$. Now we should prove for the case when the height of the trimmed mapping is $l + 1$. There are three cases to consider:

(i) $t[i]$ is not touched by a line in M. Then, we have $(i, \lambda) \in M$. Note that insertions and deletions do not change the height of the trimmed mapping. Therefore,

$$D_n(q, q') = \min\{d_0(\lambda, q') + d_0(\mathbb{T}, \lambda) + \mathsf{C}(\sigma, \lambda) \mid \sigma(\mathbb{T}) = q \in \delta\}.$$

(ii) $t'[j]$ is not touched by a line in M. Then, we have $(\lambda, j) \in M$. Symmetrically,

$$I_n(q, q') = \min\{d_0(q, \lambda) + d_0(\lambda, \mathbb{T}') + \mathsf{C}(\lambda, \sigma') \mid \sigma'(\mathbb{T}') = q' \in \delta'\}.$$

(iii) $t[i]$ and $t'[j]$ are touched by lines in M. Thus, $(i, j) \in M$ by the mapping restrictions. That means we need an optimal mapping between two forests $t[1 \ldots i - 1]$ and $t'[1 \ldots j - 1]$. The height of optimal trimmed mapping between these two forests is l. Therefore,

$$S_n(q, q') = \min\{d_{n-1}(\mathbb{T}, \mathbb{T}') + \mathsf{C}(\sigma, \sigma') \mid \sigma(\mathbb{T}) = q \in \delta, \sigma'(\mathbb{T}') = q' \in \delta'\}.$$

Since all possible optimal mappings between two trees can be computed by the definition, we prove the lemma.   □

One remaining issue is how many times we should iterate the computation of recurrence for computing the correct edit-distance between two regular tree languages. We can show that $|Q||Q'|$ iterations are enough for computing the edit-distance between two regular languages as the height of the optimal trimmed mapping is at most $|Q||Q'|$ to avoid the repetition of the same state pair. We establish the following result.

Lemma 9. *Given two k-bounded TAs $A = (\Sigma, Q, F, \delta)$ and $B = (\Sigma', Q', F', \delta')$, $d_{mn}(q, q') = d(q, q')$, where $q \in Q$, $q' \in Q'$, $m = |Q|$ and $n = |Q'|$.*

We analyze the time complexity of Algorithm 1 for computing the top-down tree edit-distance between two regular languages given by two k-bounded TAs and establish the following result.

Theorem 10. *Let* $A = (\Sigma, Q, F, \delta)$ *and* $B = (\Sigma', Q', F', \delta')$ *be two* k-*bounded TAs. Then, we can compute the edit-distance* $d(L(A), L(B))$ *in* $\mathcal{O}(m^2 n^2)$ *time, where* $m = |A|$ *and* $n = |B|$.

Algorithm 1. Computing $d(L(A), L(B))$

---

    input : Two k-bounded TAs $A = (\Sigma, Q, F, \delta)$ and $B = (\Sigma, Q', F', \delta')$
    output: $d(L(A), L(B))$

1 $d_0(\lambda, \lambda) \leftarrow 0$;
2 **for** $q \in Q$ **do**
3 | $d_0(q, \lambda) \leftarrow \min\{|t| \mid t \in L(A_q)\}$;
4 **end**
5 **for** $q' \in Q'$ **do**
6 | $d_0(\lambda, q') \leftarrow \min\{|t'| \mid t' \in L(B_{q'})\}$;
7 **end**
8 **for** $i \leftarrow 0$ **to** $|Q||Q'|$ **do**
9 | **for** $q \in Q$ **do**
10 | | **for** $q' \in Q'$ **do**
11 | | | $d_i(q, q') \leftarrow \min[d_{i-1}(q, q') \cup I_i(q, q') \cup D_i(q, q') \cup S_i(q, q')]$;
12 | | **end**
13 | **end**
14 **end**
15 **return** $\min\{d_{|Q||Q'|}(q, q') \mid q \in F, q' \in F'\}$;

---

5   Unbounded Case

It is well known that unranked tree automata describe regular tree languages of unranked and unbounded trees, which are the generalizations of regular tree languages of ranked and bounded trees [4]. We generalize the edit-distance computation between two regular tree languages accepted by k-bounded TAs to the unbounded case in this section.

Unlike in a k-bounded or ranked TA, we have a regular language over the state set Q called a horizontal language instead of a sequence of states in an unranked TA. Therefore, we compute the minimum edit-distance between two forests accepted by two sequences of states from two horizontal languages. Let $A = (\Sigma, Q, F, \delta)$ and $B = (\Sigma', Q', F', \delta')$ be two unranked TAs. Then, the edit-distance between two states $q \in Q$ and $q' \in Q'$ is defined as the minimum of the following three sets.

(i) $I(\mathbb{S}_{1,i}, \mathbb{S}'_{1,j}) = \{d(\mathbb{S}_{1,i}, \mathbb{S}'_{1,j-1}) + d(\lambda, \mathbb{T}') + \mathsf{C}(\lambda, \sigma') \mid \sigma'_{\delta'}(\mathbb{T}') = q'_j, \mathbb{T}' \in L (H^B_{q',\sigma'})\}$,

(ii) $D(\mathbb{S}_{1,i}, \mathbb{S}'_{1,j}) = \{d(\mathbb{S}_{1,i-1}, \mathbb{S}'_{1,j}) + d(\mathbb{T}, \lambda) + \mathsf{C}(\sigma, \lambda) \mid \sigma_{\delta}(\mathbb{T}) = q_k, \mathbb{T} \in L(H^A_{q,\sigma})\}$, and

(iii) $S(\mathbb{S}_{1,i}, \mathbb{S}'_{1,j}) = \{d(\mathbb{S}_{1,i-1}, \mathbb{S}'_{1,j-1}) + d(\mathbb{T}, \mathbb{T}') + \mathsf{C}(\sigma, \sigma') \mid \sigma_{\delta}(\mathbb{T}) = q, \sigma'_{\delta'}(\mathbb{T}') = q', \mathbb{T} \in L(H^A_{q,\sigma}), \mathbb{T}' \in L(H^B_{q',\sigma'})\}$.

Similarly to the bounded case, we define the recurrence for the edit-distance between two forests accepted by two sequences of states from A and B as follows:

Lemma 11. *Given two unranked TAs $A = (\Sigma, Q, F, \delta)$ and $B = (\Sigma', Q', F', \delta')$, the height n edit-distance $d(\mathbb{S}_{1,i}, \mathbb{S}'_{1,j})$ is defined as*

$$\min[I(\mathbb{S}_{1,i}, \mathbb{S}'_{1,j}) \cup D(\mathbb{S}_{1,i}, \mathbb{S}'_{1,j}) \cup S(\mathbb{S}_{1,i}, \mathbb{S}'_{1,j})],$$

where $\mathbb{S}_{1,i} \in Q^*$ and $\mathbb{S}'_{1,j} \in (Q')^*$.

Now we consider the runtime for computing the edit-distance between two unranked TAs. The main difference compared with the bounded case is that we need to compute the edit-distance between two forests accepted by two regular horizontal languages instead of two fixed sequences of states.

Corollary 12 (Mohri [12]). *Given two FAs A and B, we can compute the edit-distance $d(L(A), L(B))$ in $\mathcal{O}(mn \cdot \log mn)$ time, where $m = |A|$ and $n = |B|$.*

Theorem 13. *Let $A = (\Sigma, Q, F, \delta)$ and $B = (\Sigma', Q', F', \delta')$ be two unranked TAs. Then, we can compute the edit-distance $d(L(A), L(B))$ in $\mathcal{O}(m^2 n^2 \cdot \log mn)$ time, where $m = |A|$ and $n = |B|$.*

6   An Application of the Tree Edit-Distance Problem

It is known that the edit-distance between two context-free languages is not computable [12]. Moreover, the emptiness of the intersection of two context-free languages is also undecidable. Now we show that it is possible to check whether or not two CFGs have a common string whose derivation trees are structurally equivalent by relying on the edit-distance computation between two regular tree languages.

Proposition 14 (Comon et al. [4]). *The following statements hold.*

- *Given a context-free grammar G, the set of derivation trees of $L(G)$ is a regular tree language.*
- *Given a regular tree language L, yield(L) is a context-free language.*
- *There exists a regular tree language that is not a set of derivation trees of a context-free language.*

Based on Proposition 14, we establish the following result.

Lemma 15. *Given two CFGs G and G', we can determine whether or not there exists a common string $w \in L(G) \cap L(G')$ whose derivation trees from G and G' are structurally equivalent.*

7   Conclusions

We have studied the top-down tree edit-distance between two regular tree languages. The tree edit-distance between two tree languages is the minimum tree edit-distance between two trees from two languages. We, in particular, have considered the restricted version of the general tree edit-distance problem called the top-down tree edit-distance. We have proposed an $\mathcal{O}(m^2n^2)$ algorithm for computing the edit-distance between two regular tree languages given by two k-bounded TAs of sizes m and n. For the edit-distance between two unranked TAs of sizes m and n, we have designed an $\mathcal{O}(m^2n^2\log mn)$ algorithm.

Given two CFGs G and G', we have also shown that it is decidable to determine whether or not there exists a common string whose derivation trees from G and G' are structurally equivalent using the proposed algorithm.

References

1. Bunke, H.: Edit distance of regular languages. In: Proceedings of the 5th Annual Symposium on Document Analysis and Information Retrieval, pp. 113–124 (1996)
2. Chawathe, S.S.: Comparing hierarchical data in external memory. In: Proceedings of the 25th International Conference on Very Large Data Bases, pp. 90–101 (1999)
3. Choffrut, C., Pighizzini, G.: Distances between languages and reflexivity of relations. Theoretical Compututer Science 286(1), 117–138 (2002)
4. Comon, H., Dauchet, M., Gilleron, R., Löding, C., Jacquemard, F., Lugiez, D., Tison, S., Tommasi, M.: Tree automata techniques and applications (2007), http://www.grappa.univ-lille3.fr/tata (release October 12, 2007)
5. Demaine, E.D., Mozes, S., Rossman, B., Weimann, O.: An optimal decomposition algorithm for tree edit distance. ACM Transactions on Algorithms 6(1), 2:1–2:19 (2009)
6. Gécseg, F., Steinby, M.: Tree languages. In: Handbook of Formal Languages, Vol. 3: Beyond Words, pp. 1–68. Springer-Verlag New York, Inc. (1997)
7. Hamming, R.W.: Error Detecting and Error Correcting Codes. Bell System Technical Journal 26(2), 147–160 (1950)
8. Han, Y.-S., Ko, S.-K., Salomaa, K.: Computing the edit-distance between a regular language and a context-free language. In: Yen, H.-C., Ibarra, O.H. (eds.) DLT 2012. LNCS, vol. 7410, pp. 85–96. Springer, Heidelberg (2012)
9. Han, Y.-S., Ko, S.-K., Salomaa, K.: Approximate matching between a context-free grammar and a finite-state automaton. In: Konstantinidis, S. (ed.) CIAA 2013. LNCS, vol. 7982, pp. 146–157. Springer, Heidelberg (2013)
10. Klein, P.N.: Computing the edit-distance between unrooted ordered trees. In: Proceedings of the 6th Annual European Symposium on Algorithms, pp. 91–102 (1998)
11. Levenshtein, V.I.: Binary codes capable of correcting deletions, insertions, and reversals. Soviet Physics Doklady 10(8), 707–710 (1966)
12. Mohri, M.: Edit-distance of weighted automata: General definitions and algorithms. International Journal of Foundations of Computer Science 14(6), 957–982 (2003)
13. Myers, G.: Approximately matching context-free languages. Information Processing Letters 54, 85–92 (1995)
14. Nierman, A., Jagadish, H.V.: Evaluating structural similarity in XML documents. In: Proceedings of the 5th International Workshop on the Web and Databases, pp. 61–66 (2002)

15. Reis, D.C., Golgher, P.B., Silva, A.S., Laender, A.F.: Automatic web news extraction using tree edit distance. In: Proceedings of the 13th International Conference on World Wide Web, pp. 502–511 (2004)
16. Selkow, S.: The tree-to-tree editing problem. Information Processing Letters 6(6), 184–186 (1977)
17. Tai, K.C.: The tree-to-tree correction problem. Journal of the ACM 26(3), 422–433 (1979)
18. Tekli, J., Chbeir, R., Yetongnon, K.: Survey: An overview on XML similarity: Background, current trends and future directions. Computer Science Review 3(3), 151–173 (2009)
19. Wagner, R.A.: Order-n correction for regular languages. Communications of the ACM 17, 265–268 (1974)
20. Wagner, R.A., Fischer, M.J.: The string-to-string correction problem. Journal of the ACM 21, 168–173 (1974)
21. Winkler, W.E.: String comparator metrics and enhanced decision rules in the fellegi-sunter model of record linkage. In: Proceedings of the Section on Survey Research, pp. 354–359 (1990)
22. Yang, R., Kalnis, P., Tung, A.K.H.: Similarity evaluation on tree-structured data. In: Proceedings of the 2005 ACM SIGMOD International Conference on Management of Data, pp. 754–765 (2005)
23. Zhang, K., Shasha, D.: Simple fast algorithms for the editing distance between trees and related problems. SIAM Journal on Computing 18(6), 1245–1262 (1989)
24. Zhang, Z., Cao, R.L.S., Zhu, Y.: Similarity metric for XML documents. In: Proceedings of Workshop on Knowledge and Experience Management (2003)

DFA with a Bounded Activity Level

Marius Konitzer and Hans Ulrich Simon

Fakultät für Mathematik, Ruhr-Universität Bochum, 44780 Bochum, Germany
{marius.konitzer,hans.simon}@rub.de

Abstract. Lookahead DFA are used during parsing for sake of resolving conflicts (as described in more detail in the introduction). The parsing of an input string w may require many DFA-explorations starting from different letter positions. This raises the question how many of these explorations can be active at the same time. If there is a bound on this number depending on the given DFA M only (i.e., the bound is valid for all input strings w), we say that M has a *bounded activity level*. The main results in this paper are as follows. We define an easy-to-check property of DFA named *prefix-cyclicity* and show that precisely the non prefix-cyclic DFA have a bounded activity level. Moreover, the largest possible number ℓ_M of mutually overlapping explorations of a given non prefix-cyclic DFA M with $t + 1$ states, the so-called *maximum activity level of* M, is bounded from above by $2^t - 1$, and this bound is tight. We show furthermore that the maximum activity levels of equivalent DFA coincide so as to form an invariant of the underlying regular language, which leads us to a characterization of prefix-cyclicity in terms of the Nerode relation. We finally establish some complexity results. For instance, the problem of computing ℓ_M for a given non prefix-cyclic DFA M is shown to be PSPACE-complete.

Keywords: parsing, lookahead DFA, computational complexity.

1 Introduction

LR-regular (LRR) parsing [12] is one of the few parsing techniques utilizing unbounded lookahead. LRR languages properly include the deterministic context-free languages [7]. LRR parsers allow for a large amount of interesting grammars with practical relevance (such as the original version of the Java language [6]), which cannot be handled by any LR(k) parser. The parsers generated with the algorithm from [11] clearly have linear runtime, although they are a little cumbersome. The algorithm is rather of theoretical interest as membership in the class of LR-regular grammars is undecidable and as some implementation details remain unclear. Practical LRR parsing techniques such as [1] and [4] basically work like the well-known LR(k) shift-reduce parsers [7], yet use regular lookaheads of arbitrary length instead of the normal fixed length ones. Starting with an inconsistent LR(0) automaton, practical LRR parser generation techniques set out to build disjoint prefix-free regular envelopes for each inconsistent LR(0) state. This aims at separating the state's conflicting suffix languages from each

A.-H. Dediu et al. (Eds.): LATA 2014, LNCS 8370, pp. 478–489, 2014.

other. These regular envelopes are typically built as prefix-free deterministic finite automata (DFA), so called *lookahead DFA*, which are used for lookahead exploration during parsing whenever necessary. Different lookahead explorations operating on a common substring of the input string may overlap each other. As explained in the abstract (and formally defined in section 2), this leads to the notion of the maximum activity level ℓ_M associated with a given DFA M (set to ∞ in the unbounded case).

If the number of mutually overlapping explorations on strings of length n is bounded from above by $B \leq n$, the whole parser has time bound $O(Bn)$ on inputs of length n (as illustrated in Fig. 1). If the parser employs prefix-cyclic DFA, this leads to the time bound $O(n^2)$.[1] If, however, only non prefix-cyclic lookahead DFA are employed during parsing, then B does not depend on n (but still depends on the sizes of the lookahead DFA).[2] As for of a fixed LR-regular grammar with a fixed collection of non prefix-cyclic lookahead DFA, one may think of B as a (possibly large) constant. But once we think in terms of practical LRR parser generators, the dependence on the sizes of the employed lookahead DFA becomes an issue.

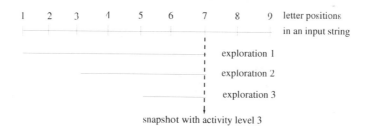

Fig. 1. The parser and all of the $B = 3$ DFA-explorations require up to n computational steps, respectively

The notion of prefix-cyclicity (among other related notions) was introduced and exploited in [8]. However, the run time analysis in [8] treats the parameter B as a constant whenever it does not depend on n. In this paper, we take care of the dependence of B on the sizes of the employed lookahead DFA and study the dependence of ℓ_M on the number t of M's non-initial states. We extend the work in [8] in various directions. Section 3 presents the (tight) upper bound $2^t - 1$ on ℓ_M. Section 4 casts ℓ_M as an invariant of the underlying language $L(M)$. Section 5 characterizes prefix-cyclicity in terms of the Nerode-relation. Section 6 is devoted to complexity issues. Specifically, it is shown that the computation of ℓ_M for a given non prefix-cyclic DFA M is PSPACE-complete.

---

[1] An LRR-grammar leading indeed to quadratic run time for parsing with (unbounded) lookahead DFA is found in [10].

[2] See [8] for several grammar constructs leading to non prefix-cyclic lookahead DFA (e.g. HTML forms [4] and Ada calls [2,10]).

2   Definitions and Notations

Let M be a Deterministic Finite Automaton (DFA) given by its finite set of states, Q, its input alphabet, Σ, its partially defined transition function δ : $Q \times \Sigma \to Q$, and its initial state $q_0 \in Q$. (For the time being, we do not need to distinguish between accepting and non-accepting states.) If M reads symbol a in state q and $\delta(q, a)$ is undefined, then we may think of M as terminating its computation. As usual, the mapping δ can be extended to a partially defined mapping $\delta^* : Q \times \Sigma^* \to Q$:

$$\delta^*(q, \varepsilon) = q$$

$$\delta^*(q, aw) = \begin{cases} \text{undefined} & \text{if } \delta(q, a) \text{ is undefined} \\ \delta^*(\delta(q, a), w) & \text{otherwise} \end{cases}$$

Here, ε denotes the empty string, $q \in Q$, $a \in \Sigma$, and $w \in \Sigma^*$. If not undefined, then $\delta^*(q, w)$ is the state reached by a computation of M that was started in state q and has processed all letters of w. We say that w is *fully processed by* M if $\delta^*(q_0, w)$ is not undefined. Suppose that $w = a_1 \cdots a_n \in \Sigma^n$. Then, for all $1 \le i < j \le n$, $w_{i,j}$ denotes the substring $a_i \cdots a_{j-1}$. We say that M *has activity level ℓ at position j of input w* if there exist $1 \le i_1 < \ldots < i_\ell < j$ such that, for all $l = 1, \ldots, \ell$, $w_{i_l, j}$ is fully processed by M. We say that M has an *unbounded activity level* if for any $\ell \ge 1$ there is a string w and a letter position j such that M has activity level ℓ at position j of input w. We define $\ell_M = \infty$ if M has an unbounded activity level, and as the highest possible activity level otherwise. Note that ℓ_M represents the largest possible number of mutually overlapping explorations when M is used as a lookahed DFA as described in Section 1.

3   DFA with a Bounded Activity Level

Section 3.1 characterizes DFA with an unbounded activity level: exactly the "prefix-cyclic" DFA M are the ones with $\ell_M = \infty$. In Section 3.2, it is shown that $\ell_M \le 2^t - 1$ for any non prefix-cyclic DFA with $t+1$ states. It is furthermore shown that there exists a non prefix-cyclic DFA M with $t + 1$ states and $\ell_M = 2^t - 1$ (so that the general upper bound is tight).

3.1   Characterization of DFA with an Unbounded Activity Level

A DFA $M = (Q, \Sigma, \delta, q_0)$ with a partially defined transition function δ is said to be *prefix-cyclic* if it satisfies the following condition:

$$\exists q \in Q, \exists w \in \Sigma^+ : \delta^*(q_0, w) = q = \delta^*(q, w) \tag{1}$$

DFA with an unbounded activity level can be characterized as follows:

Theorem 1 ([8]). *A DFA M is prefix-cyclic iff it has an unbounded activity level.*

Proof. If M is prefix-cyclic as witnessed by $q \in Q$ and $w \in \Sigma^+$, then the strings $(w^\ell)_{\ell \geq 1}$ and the letter positions $i_l = 1 + (l-1)|w|$ for $l = 1, \ldots, \ell$ witness that M has an unbounded activity level.

Suppose now that M has an unbounded activity level. Let $\ell \geq 1$ be a sufficiently large number whose precise definition is given below. Pick a string w and letter positions $1 \leq i_1 < \ldots < i_\ell < j$ which witness that $\ell_M \geq \ell$. Let K_ℓ denote the complete graph with ℓ nodes. Consider the edge-coloring of K_ℓ where each edge $\{l, l'\}$ such that $l < l'$ is colored $\delta^*(q_0, w_{i_l,i_{l'}})$. Note that this coloring uses $t := |Q|$ colors. Let $r(3,t)$ denote the smallest number of nodes of a complete graph such that any t-coloring of its edges leads to at least one monochromatic triangle.[3] It is well-known [5,3,12] that

$$2^t < r(3,t) < 1 + \frac{e - e^{-1} + 3}{2} \cdot t! < 3t! \ .$$

Let now $\ell := r(3,t) < 3t!$. Then, with the coloring defined above (as for any t-coloring), K_ℓ has at least one monochromatic triangle. By construction of the coloring, this means that there exist $1 \leq l < l' < l'' < j$ such that $\delta^*(q_0, w_{i_l,i_{l'}}) = \delta^*(q_0, w_{i_l,i_{l''}}) = \delta^*(q_0, w_{i_{l'},i_{l''}})$. Setting $q := \delta^*(q_0, w_{i_l,i_{l'}})$, we obtain

$$\delta^*(q, w_{i_{l'},i_{l''}}) = \delta^*(q_0, w_{i_l,i_{l''}}) = \delta^*(q_0, w_{i_{l'},i_{l''}}) = q$$

so that (1) holds with $w_{i_{l'},i_{l''}}$ in the role of w. It follows that M is prefix-cyclic. $\qquad\square$

We obtain the following

Corollary 2 ([8]). *Suppose that the DFA* $M = (Q, \Sigma, \delta, q_0)$ *is not prefix-cyclic and has* $t + 1$ *states. Then* $\ell_M < r(3,t)$.

Proof. There can be no string $w \in \Sigma^+$ such that $\delta^*(q_0, w) = q_0$ because, otherwise, Condition (1) would be satisfied with q_0 in the role of q. Assume for sake of contradiction that $\ell_M \geq r(3,t)$. An inspection of the second part of the proof of Theorem 1 shows that this leads to a t-coloring (the color q_0 is not used!) of the complete graph with $r(3,t)$ nodes so that there is a monochromatic triangle and, consequently, M would be prefix-cyclic (in contradiction to the assumption made in Corollary 2). $\qquad\square$

3.2   Tight Bounds on the Activity Level (Arbitrary Alphabet)

Suppose $M = (Q, \Sigma, \delta, q_0)$ is not prefix-cyclic. Let $t = |Q| - 1$ denote the number of non-initial states. According to Corollary 2, $\ell_M < r(3,t)$. Typically, bounds obtained from Ramsey theory are far from being tight. We will however show in this section that the upper bound $r(3,t)$ on ℓ_M is not so far from the truth. We begin our considerations with another upper bound on ℓ_M.

Theorem 3. *For any non prefix-cyclic DFA* M *with* $t + 1$ *states:* $\ell_M \leq 2^t - 1$.

---

[3] In Ramsey Theory, $r(3,t)$ is known as the "triangular Ramsey Number with t colors".

Proof. Let $\ell = \ell_M$. Pick a string w and letter positions $1 \leq i_1 < \ldots < i_\ell < j$ such that, for all $l = 1, \ldots, \ell$, the substrings $w_{i_l, j}$ are fully processed. For convenience, set $i_{\ell+1} = j$. For $l' = 1, \ldots, \ell$, the "l'-snapshot" is defined as the set

$$Q_{l'} := \{\delta^*(q_0, w_{i_l, i_{l'+1}}) : l = 1, \ldots, l'\} \subseteq Q \setminus \{q_0\} .$$

In other words: if we consider the l' computational processes created by starting M in positions $i_1, \ldots, i_{l'}$, then $Q_{l'}$ records the set of states of these processes when they have reached position $i_{l'+1}$. Note that $Q_{l'} \neq \emptyset$ for all $l' = 1, \ldots, \ell$ so that there can be at most $2^t - 1$ distinct snapshots. All what remains to do is showing that they actually are distinct. Suppose for sake of contradiction that $Q_{l'} = Q_{l''}$ for some $1 \leq l' < l'' \leq \ell$. It follows that we can push the activity level beyond any given bound m simply by replacing the substring $u = w_{i_{l'}, i_{l''}}$ of w by u^m. As an unbounded activity level would imply that M is prefix-cyclic, we arrived at a contradiction. It follows that the snapshots are distinct and, therefore, $\ell \leq 2^t - 1$.   □

The following result shows that the bound in Theorem 3 is tight:

Theorem 4. *There exists a non prefix-cyclic DFA M with $t + 1$ states and alphabet size t such that $\ell_M \geq 2^t - 1$.*

Proof. Let $M = (Q, \Sigma, \delta, q_0)$ be given by $Q = \{q_0, q_1, \ldots, q_t\}$, $\Sigma = \{a_1, \ldots, a_t\}$, and

$$\delta(q_i, a_j) = \begin{cases} q_j & \text{if } i < j \\ q_i & \text{if } i > j \\ \text{undefined} & \text{if } i = j \end{cases} . \tag{2}$$

The following statements obviously hold for any $q_k \in Q$ and any $w \in \Sigma^+$:

$$\delta^*(q_k, w) = q_k \Leftrightarrow k \geq 2 \wedge w \in \{a_1, \ldots, a_{k-1}\}^+$$
$$\delta^*(q_0, w) = q_k \Rightarrow \text{letter } a_k \text{ occurs in } w$$

It follows that M is not prefix-cyclic because Condition (1) cannot be satisfied. With the following inductively defined strings $w(1), \ldots, w(t)$, we will be able to push the activity level up to $2^t - 1$:

$$w(1) = a_1 \quad \text{and} \quad w(k) = w(k-1)a_k w(k-1)$$

The first members of this sequence evolve as follows:

$$w(1) = a_1 , \quad w(2) = a_1 a_2 a_1 , \quad w(3) = a_1 a_2 a_1 a_3 a_1 a_2 a_1 , \quad \ldots$$

Clearly $|w(t)| = 2^t - 1$. We claim that all $2^t - 1$ suffixes of $w(t)$ are fully processed by M (which would readily imply that $\ell_M \geq 2^t - 1$). The claim is obtained from the following observations:

1. The snapshot[4] after reading $w(k)$ contains state q_i with multiplicity 2^{i-1} for $i = 1, \ldots, k$ (and no other states).

---

[4] Here, the snapshot is considered as a multiset so as to take multiplicities of states into account.

2. The snapshot after reading $w(k)a_{k+1}$ contains the state q_{k+1} with multiplicity 2^k (and no other states).

The second statement immediately follows from the first one. The first statement for $w(k)$ immediately follows inductively from the second statement for $w(k-1)a_k$.   □

4   Activity Level of DFA with a Binary Alphabet

We argue in Section 4.1 that the maximum activity level ℓ_M of a DFA M can be associated with the language $L(M) = L$ generated by M (and can therefore be written ℓ_L). In Section 4.2, we define a mapping $L \mapsto L_{bin}$ that transforms a prefix-closed language over an arbitrary alphabet into a corresponding prefix-closed language over a binary alphabet. It is analyzed how ℓ_L and $\ell_{L_{bin}}$ are related. In Section 4.3, we show that there exists a non prefix-cyclic DFA M over a binary alphabet that has $1 + 3t\lceil \log t \rceil$ states and satisfies $\ell_M \geq 2^t - 1$. A comparison with Theorem 4 shows that the restriction of having a binary input alphabet does not reduce the largest possible activity-level dramatically.

4.1   Activity Level as an Invariant of the Underlying Language

A DFA $M = (Q, \Sigma, \delta, q_0, F)$ with a partially defined transition function δ is called *prefix-closed* if $F = Q$, i.e., all states of M are accepting. Note that, in this case, the language $L(M)$ coincides with the set of input strings which are fully processed by M. A language L is called *prefix-closed* if $w \in L$ implies that every prefix of w belongs to L too. In other words, any extension of a string $w \notin L$ does not belong to L either. Obviously the following holds:

- If M is a prefix-closed DFA, then $L(M)$ is a prefix-closed regular language.
- Any prefix-closed regular language can be be recognized by a prefix-closed DFA.

We define the *maximum activity level of a language L* as follows:

$$\ell_L = \sup\{\ell : (\exists w_1, \ldots, w_\ell \in \Sigma^+, \forall l = 1, \ldots, \ell : w_l \cdots w_\ell \in L)\}$$

For ease of later reference, we say that $w_1, \ldots, w_\ell$ are *witnesses for $\ell_M \geq \ell$* if $w_l \cdots w_\ell \in L$ for $l = 1, \ldots, \ell$.

Let M be a prefix-closed DFA. It is then evident from the definition of ℓ_M and ℓ_L that $\ell_M = \ell_{L(M)}$. Thus, $\ell_M = \ell_{M'}$ for any DFA M' such that $L(M) = L(M')$.

4.2   From an Arbitrary to a Binary Alphabet

For a language L, let $\text{Pref}(L)$ denote the language of all prefixes of strings from L. So L is prefix-closed iff $\text{Pref}(L) = L$.

Let $L \subseteq \Sigma^*$ be a prefix-closed language over the alphabet $\Sigma = \{a_0, \ldots, a_{K-1}\}$, and let $k = \lceil \log K \rceil$ so that every letter a_j can be encoded as a binary string bin(j) of length precisely k (with leading zeros if necessary). Let R denote the homomorphism from Σ^* to $\{0,1\}^*$ that is induced by $a_j \mapsto \text{bin}(j)$. Then $R(L) = \{R(w) : w \in L\}$ is the image of L under mapping R. Note that the length of any string in $R(\Sigma^*)$ is a multiple of k. The language $L_{bin} = \text{Pref}(R(L))$ is called the *binary version* of L in what follows. Note that L_{bin} is prefix-closed by construction.

Lemma 5. *With these notations, the following holds for any prefix-closed language $L \subseteq \Sigma^*$:*

1. *If $w \in L$ then $R(w) \in L_{bin}$.*
2. *If $x \in L_{bin}$ and $|x|$ is a multiple of k, then there exists a string $v \in L$ such that $x = R(v)$.*

Proof. The first statement is obvious from $w \in L \Leftrightarrow R(w) \in R(L)$ and $R(L) \subseteq \text{Pref}(R(L)) = L_{bin}$. As for the second statement, $x \in L_{bin} = \text{Pref}(R(L))$ implies that there exists a suffix $y \in \{0,1\}^*$ such that $xy \in R(L)$. $xy \in R(L)$ implies that $|xy|$ is a multiple of k and, since $|x|$ is a multiple of k by assumption, $|y|$ is a multiple of k too. The definition of $R(L)$ now implies that there exist strings $v, w \in \Sigma^*$ such that $x = R(v)$, $y = R(w)$ and $vw \in L$. Since L is prefix-closed by assumption, it follows that $v \in L$. □

Theorem 6. *With the above notations, the following holds for any prefix-closed language L:*

1. $\ell_L = \infty$ *iff* $\ell_{L_{bin}} = \infty$.
2. *If $\ell_L < \infty$, then $\ell_L \leq \ell_{L_{bin}} \leq k \cdot \ell_L + 1$.*

Proof. It suffices to show that, for all $\ell \geq 1$,

$$\ell_L \geq \ell \Rightarrow \ell_{L_{bin}} \geq \ell \text{ and } \ell_{L_{bin}} \geq k\ell + 1 \Rightarrow \ell_L \geq \ell \ .$$

Let $w_1, \ldots, w_\ell \in \Sigma^+$ be witnesses for $\ell_L \geq \ell$. It readily follows that the strings $R(w_1), \ldots, R(w_\ell) \in \{0,1\}^+$ are witnesses for $\ell_{L_{bin}} \geq \ell$.
Let now $x_1, \ldots, x_{\ell'} \in \{0,1\}^+$ be witnesses for $\ell_{L_{bin}} \geq \ell'$, i.e.,

$$\forall l = 1, \ldots, \ell' : y_l := x_l \cdots x_{\ell'} \in L_{bin} \ .$$

Let us introduce for the moment the following additional assumption:

$$\forall l = 1, \ldots, \ell' : |y_l| \text{ is a multiple of } k \ , \tag{3}$$

which is equivalent to saying that, for $l = 1, \ldots, \ell'$, $|x_l|$ is a multiple of k. Then the second statement in Lemma 5 (and the fact that R is a homomorphism) let us conclude that there exist strings $w_1, \ldots, w_{\ell'} \in \Sigma^+$ with the following properties:

- For all $l = 1, \ldots, \ell'$, $R(w_l) = x_l$.
- $w_1, \ldots, w_{\ell'}$ are witnesses for $\ell_L \geq \ell' = \ell_{L_{bin}}$.

Of course our assumption of $|y_l|$ being a multiple of k is not justified. However, if $\ell' = k\ell + 1$, we can argue as follows. We put y_l in a bucket with number $|y_l| \bmod k \in \{0, \ldots, k-1\}$. By the pigeon-hole principle, there must exist a number $\kappa \in \{0, \ldots, k-1\}$ such that the bucket with number κ contains $\ell + 1$ suffixes of $x_1 \cdots x_{\ell'}$, say $y_{l_1}, \ldots, y_{l_\ell}, y_{l_{\ell+1}}$ when ordered according to decreasing length. Note that the shortest suffix, $y_{l_{\ell+1}}$, is a common suffix of all the other ones. Let us erase the suffix $y_{l_{\ell+1}}$ from $y_{l_1}, \ldots, y_{l_\ell}$, respectively, and obtain the new sequence $y'_{l_1}, \ldots, y'_{l_\ell}$. Note that $y'_{l_1}, \ldots, y'_{l_\ell}$ still belong to L_{bin} because L_{bin} is prefix-closed. From $|y_l| \bmod k = \kappa$ for $l = 1, \ldots, \ell + 1$, we can conclude that $|y'_l| \bmod k = 0$ for $l = 1, \ldots, \ell$. Thus, assumption (3) on which our previous analysis was based is now satisfied, indeed, with ℓ in the role of ℓ'. It follows that $\ell_L \geq \ell$ provided that $\ell_{L_{bin}} \geq k\ell + 1$. $\qquad\square$

4.3   A Lower Bound on the Activity Level (Binary Alphabet)

The following lower bound, valid for a DFA with a binary input alphabet, should be compared with the (only slightly superior) lower bound from Theorem 4 (which however makes use of a DFA with a very large input alphabet).

Theorem 7. *There exists a non prefix-cyclic DFA with a binary input alphabet, $1 + 3t\lceil \log t \rceil$ states and an activity level of at least $2^t - 1$.*

Proof. Let $M = (Q, \Sigma, \delta, q_0)$ be the DFA from the proof of Theorem 4 except for the following technical modification: we use alphabet $\Sigma = \{a_0, \ldots, a_{t-1}\}$ instead of $\{a_1, \ldots, a_t\}$, and state set $Q = \{q_{-1}, q_0, \ldots, q_{t-1}\}$ instead of $\{q_0, q_1, \ldots, q_t\}$. Now q_{-1} is the initial state. As before, δ is given by (2).[5] We know from Theorem 4 that $\ell_M \geq 2^t - 1$. Let $L = L(M)$. We will design a DFA $M' = (Q', \{0, 1\}, \delta', q'_0)$ for L_{bin}. According to Theorem 6, $\ell_{M'} \geq \ell_M \geq 2^t - 1$. It suffices therefore to make sure that $|Q'| - 1 + 3t\lceil \log t \rceil$ states are sufficient for the design of M'. Let $k = \lceil \log t \rceil$. Q' is now chosen as the union of $\{q'_0\}$ with the following set:

$$\{(b, \kappa, s) : \ b \in \{\text{bin}(i) : i = 0, 1, \ldots, t-1\}, \kappa \in \{0, 1, \ldots, k-1\}, s \in \{<, >, ?\}\}$$

The intuition behind this definition is as follows:

- The computation of M on input $w \in \Sigma^*$ is simulated by running the computation of M' on input $R(w)$ where R is the homomorphism induced by $a_j \mapsto \text{bin}(j)$ from Section 4.2.
- When M is in state q_i, then M' keeps $b = \text{bin}(i) \in \{0, 1\}^k$ in its finite control. The parameter κ indicates how many bits of $\text{bin}(j)$ are processed by M' already when M is currently processing symbol a_j. The flag s is set to "$<$" (resp. "$>$") if M' already knows that $i < j$ (resp. $i > j$). Before a successful comparison of i and j, the flag s is set to "$?$".

---

[5] Reason for the modification: we will map the underlying language L to L_{bin}, and we won't leave the bit pattern $\text{bin}(0)$ unused.

It is not hard to see that the transition function δ' of M' can be defined such that M' simulates a transition $\delta(q_i, a_j)$ of DFA M. The main point is that the comparison of the binary encodings of two numbers i and j can easily be done bitwise (where $\mathrm{bin}(i)$ is kept in the finite control and $\mathrm{bin}(j)$ is processed from left to right on the input tape). The details of this simulation, omitted here due to space constraints, are found in the full version of the paper.   □

5   Unbounded Activity Level and Nerode Relation

We briefly remind the reader that the Nerode relation induced by a language $L \subseteq \Sigma^*$, denoted $\overset{L}{\equiv}$, is a right-congruent equivalence relation on Σ^* that has finitely many equivalence classes iff L is regular. The equivalence classes can then be viewed as the states of the so-called Nerode DFA M_L for L.

Theorem 8. *Let L be a prefix-closed regular language. Then, $\ell_L = \infty$ iff there exists a non-empty string $w \in L$ such that $w \overset{L}{\equiv} w^2$.*

The proof of this theorem, omitted here due to space constraints, is based on the fact that the condition $w \overset{L}{\equiv} w^2$ is equivalent to the condition $q := \delta^*(q_0, w) = \delta^*(q, w)$ for the transition function δ of the (prefix-free version of the) Nerode DFA M_L.

6   Some Complexity Issues

The following theorem is immediate from Theorem 1 and from the definition of "prefix-cyclic" in (1):

Theorem 9 ([8]). *It can be decided within $O(t^2)$ steps whether a DFA M of size t has an unbounded activity level.*

Proof. Let q_0 denote the initial state of M, and let $M' = M$. Compute the product automaton of M and M' and check whether its transition graph contains a non-trivial path from (q_0, q) to (q, q) for some state q.   □

Suppose now that M is not prefix-cyclic so that it has a bounded activity level. It turns out that the computation of the maximum activity level, ℓ_M, is a hard problem:

Theorem 10. *Given a non prefix-cyclic DFA M with $t + 1$ states and given a threshold T, the problem to decide whether $\ell_M > T$ is PSPACE-complete.*

Proof. Membership in PSPACE can be seen as follows. Guess a string $w = a_1 a_2 \ldots a_n$ letter by letter, start a computation of M on each letter of w so that up to i computations could potentially be active after the first i letters $a_1, \ldots, a_i$ have been processed. For each non-initial node q, keep track of the number $j(q)$ of active computations which are in state q after having processed $a_1, \ldots, a_i$

(where the variable $j(q)$ must be updated whenever a new letter is processed). Furthermore keep track of $J := \sum_q j(q)$. As soon as $J > T$ accept. Since, at any time, only t numbers in the range from 0 to $T + 1$ are stored, this is a space-efficient non-deterministic procedure for the given decision problem. Because of Savich's theorem, it can be turned into a space-efficient deterministic procedure. In order to show PSPACE-hardness, we present a polynomial reduction from "Finite Automata Intersection (FAI)" to our problem. FAI, which is known to be PSPACE-complete [9], is the following problem: given $T \geq 2$ and a list $M_1, \ldots, M_T$ of DFA with the same input alphabet Σ and with one accepting state per DFA, does there exist an input string $w \in \Sigma^*$ that is accepted by every DFA in the list? In the sequel, the initial state of M_j is denoted q_0^j and its (unique) accepting state is denoted q_+^j. We may consider the state sets $Q_1, \ldots, Q_T$ of $M_1, \ldots, M_T$, respectively, as pairwise disjoint. We plan to build a non prefix-cyclic DFA M from $M_1, \ldots, M_T$ such that $\ell_M > T$ iff there exists a string w that is accepted by $M_1, \ldots, M_T$. To this end, let $\vdash, \dashv \notin \Sigma$ be new symbols, and let q_0, q_+ be new states. M with input alphabet $\Sigma \cup \{\vdash, \dashv\}$, state set $\{q_0, q_+\} \cup \bigcup_{i=1}^{T} Q_i$ and initial state q_0, has precisely the following transitions:

- M inherits all transitions from $M_1, \ldots, M_T$.
- When reading $\vdash$ in state q_0, M moves to state q_0^1.
- When reading $\vdash$ in state q_0^j for some $j < T$, M moves to state q_0^{j+1}.
- When reading $\dashv$ in state q_0 or in state q_+^j for some $j \in \{1, \ldots, T\}$, M moves to state q_+.

Suppose that there is a string w which is accepted by all of $M_1, \ldots, M_T$. Consider the string $\vdash^T w \dashv$ and assume that we start a computation of M on each of the T occurrences of $\vdash$. Note that the computation started on the j-th occurrence of $\vdash$ will be in state q_0^{T+1-j} when reaching the first letter of w. Thus we run a computation on w for each of the given T DFA. After having processed w, the T active computations are in states $q_+^1, \ldots, q_+^T$, respectively. When processing the final letter of $\dashv$, we again start a new computation of M, which leads to a state transition from q_0 to q_+. In addition, we have the state transitions from q_+^j to q_+ for $j = 1, \ldots, T$. Thus, we have now $T + 1 > T$ active computations running simultaneously (though all of them will be finished at the very next step).

As for the reverse direction, we have to show that an activity level exceeding T can be reached only if there exists a word over the alphabet Σ that is accepted by all of $M_1, \ldots, M_T$. The main technical observations are as follows:

- In state q_0 only the symbols $\vdash$ and $\dashv$ are processed. Thus, a new computation can be successfully started on these two letters only.
- Any occurrence of symbol $\vdash$ will terminate all computations which are not in a state from $\{q_0, q_0^1, \ldots, q_0^{T-1}\}$.
- In state q_+ no symbol is processed.
- Symbol $\dashv$ is processed only when M is in state q_0 or q_+^j for some $j \in \{1, \ldots, T\}$. Thereafter, all still active computations are in state q_+ (so that these computations are terminated in the next step).

From these observations, it easily follows that we have at most T active computations (one for each of the DFA $M_1, \ldots, M_T$) as long as $\dashv$ is not processed. If $\dashv$ is processed, the number of active computations can be at most $T + 1$ (implying that M is not prefix-cyclic). Moreover the case of $T + 1$ active computations can occur only on strings which contain the symbols $\vdash$ and $\dashv$ and which have the property that the word between the last occurrence of $\vdash$ and the first occurrence of $\dashv$ is accepted by all of $M_1, \ldots, M_T$. $\square$

The procedure for the determination of ℓ_M that was suggested in the proof of Theorem 10 shows membership in PSPACE but is not a realistic one. We briefly sketch a procedure that is realistic at least for DFA with a small number of states. It is based on the notion of a "snapshot" (similar to the notion that was used in the proof of Theorem 3). Let $M = (Q, \Sigma, \delta, q_0)$ be the given non prefix-cyclic DFA. A set $S \subseteq Q \setminus \{q_0\}$ is called a *snapshot* if $S = \emptyset$ or if there exists a string $w = a_1 \ldots a_n$ and letter positions $1 \leq i_1 < \ldots i_\ell \leq n$ such that $S = \{\delta^*(q_0, w_{i_l,n+1}) : l = 1, \ldots, \ell\}$ and such that all strings $w_{i_l,n+1}$ are fully processed. We build a directed snapshot graph $G = (V, E)$ as follows. V is defined as the set of snapshots. V and E are computed iteratively as follows:

1. Initially set $V := \{\emptyset\}$ and $E := \emptyset$.
2. For each $S \in V$ and for each $a \in \Sigma$ such that $\delta(q, a)$ is defined for all $q \in S$, add $S' := \{\delta(q, a) : q \in S\}$ and $S'' := S' \cup \{\delta(q_0, a)\}$ to V. Moreover, add the edges (S, S') and (S, S'') to E, where edge (S, S'') is declared "special".

The second step is applied to every node only once. Intuitively, an edge represents a possible next a-transition of a collection of currently active DFA-explorations (provided that no exploration terminates when processing a), where a special edge reflects the option to start a new computation on a. An example is shown in Fig. 2.

It is easy (and similar to the proof of Theorem 3) to show the following:

- The strongly connected components of $G = (V, E)$ do not contain special edges (because, otherwise, $\ell_M = \infty$).
- If we assign length 1 to special edges and length 0 to all remaining ones, then ℓ_M coincides with the total length of a longest path in G.

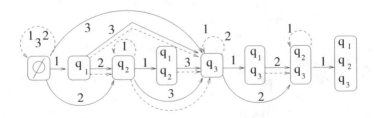

Fig. 2. The snapshot graph induced by the DFA from the proof of Theorem 4 for $t = 3$: an edge is labeled i if it represents an a_i-transition. Special edges are solid, the remaining ones are dashed.

In order to compute the total length of the longest path efficiently, an auxiliary directed acyclic "super-graph" G' is computed as follows:

1. The "super-nodes" in G' are the strongly connected components of G.
2. An edge e' is drawn from a strongly connected component K_1 to another strongly connected component K_2 iff E contains an edge e leading from a node in K_1 to a node in K_2. The edge e' is declared "special" iff the underlying edge $e \in E$ can be chosen as a special one.

Since strongly connected components do not contain special edges, the total length of the longest path in G equals the total length of the longest path in G'. But the latter quantity is easy to compute provided that the nodes of G' are processed in topological order. If everything is implemented properly then the run-time is linear in the size of snapshot graph G. Note, however, that this size can be exponential in the size t of the given DFA M.

Acknowledgements. We would like to thank four anonymous referees for their comments and suggestions. We furthermore thank Eberhard Bertsch for communicating the problem statement to us and for many helpful suggestions.

References

1. Bermudez, M.E., Schimpf, K.M.: Practical arbitrary lookahead LR parsing. Journal of Computer and System Sciences 41(2), 230–250 (1990)
2. Boullier, P.: Contribution à la construction automatique d'analyseurs lexicographiques et syntaxiques. Ph.D. thesis, Université d'Orléans (1984)
3. Chung, F.R.K., Grinstead, C.M.: A survey of bounds for classical Ramsey numbers. Journal of Graph Theory 7(1), 25–37 (1983)
4. Farré, J., Fortes Gálvez, J.: A bounded graph-connect construction for LR-regular parsers. In: Wilhelm, R. (ed.) CC 2001. LNCS, vol. 2027, pp. 244–258. Springer, Heidelberg (2001)
5. Fredricksen, H.: Schur numbers and the Ramsey numbers N(3,3,..,3;2). Journal of Combinatorial Theory, Series A 27(3), 376–377 (1979)
6. Gosling, J., Joy, B., Steele, G.: The Java™ Language Specification. Addison-Wesley (1996)
7. Knuth, D.E.: On the translation of languages from left to right. Information and Control 8(6), 607–639 (1965)
8. Konitzer, M.: Laufzeitanalyse und Optimierung von Parsern für LR-reguläre Grammatiken. Ph.D. thesis, Ruhr-University Bochum (2013)
9. Kozen, D.: Lower bounds for natural proof systems. In: Proceedings of the 18th Symposium on Foundations of Computer Science, pp. 254–266 (1977)
10. Schmitz, S.: Approximating Context-Free Grammars for Parsing and Verification. Ph.D. thesis, Université de Nice-Sophia Antipolis (2007)
11. Čulik, K., Cohen, R.: LR-regular grammars - an extension of LR(k) grammars. Journal of Computer and System Sciences 7(1), 66–96 (1973)
12. Wan, H.: Upper bounds for Ramsey numbers R(3,3,...,3) and Schur numbers. Journal of Graph Theory 26(3), 119–122 (1997)

Learning Sequential Tree-to-Word Transducers

Grégoire Laurence[1,3], Aurélien Lemay[1,3], Joachim Niehren[1,4],
Sławek Staworko[1,3], and Marc Tommasi[2,3]

[1] Links project, INRIA & LIFL (CNRS UMR8022)
[2] Magnet Project, INRIA & LIFL (CNRS UMR8022)
[3] University of Lille, France
[4] INRIA, Lille, France

Abstract. We study the problem of learning sequential top-down tree-to-word transducers (STWs). First, we present a Myhill-Nerode characterization of the corresponding class of sequential tree-to-word transformations ($\mathcal{STW}$). Next, we investigate what learning of STWs means, identify fundamental obstacles, and propose a learning model with abstain. Finally, we present a polynomial learning algorithm.

1 Introduction

The main motivation of this paper is to study learnability of a class of tree-to-word transformations. Tree-to-word transformations are ubiquitous in computer science. They are the core of many computation paradigms from the evaluation of abstract syntactic trees to modern programming languages XSLT. For these reason, they are better suited to model general XML transformations as opposed to tree-to-tree transducers [7,13,14].

Following the work of [12], we study the class of deterministic sequential top-down tree-to-word transducers (STWs). STWs are finite state machines that traverse the input tree in top-down fashion and at every node produce words obtained by the concatenation of constant words and the results from processing the child nodes. STWs capture a large subclass of deterministic nested-word to word transducers (dN2W), which have recently been the object of an enlivened interest [8,18,19]. STWs take as an input a tree in a regular tree language and output words from a context-free grammar.

Despite of some limitations mainly due to the fact they are deterministic and top-down[1], STWs remain however very powerful. They are capable of: concatenation in the output, producing arbitrary context-free languages, deleting inner nodes, and verifying that the input tree belongs to the domain even when deleting parts of it. These features are often missing in tree-to-tree transducers, and for instance, make STWs incomparable with the class of top-down tree-to-tree transducers [7,13]. The class of STWs has several interesting properties, in particular a normal form has been proposed in [12].

---

[1] Non-determinism quickly leads to fundamental limitations. For instance, equivalence of non-deterministic string transducers is known to be undecidable [11].

A.-H. Dediu et al. (Eds.): LATA 2014, LNCS 8370, pp. 490–502, 2014.

An open question raised in [12] was the existence of a Myhill-Nerode Theorem for STWs. We solve this question and this result is the first main contribution of this paper. Myhill-Nerode Theorem provides canonical representation of languages – here, transformations – based on residuals. The Myhill-Nerode Theorem also opens a way towards grammatical inference for tree-to-word transformation. Indeed, as pointed by many authors, machine learning of formal languages is essentially a matter of estimation of equivalence classes of the target language. The second contribution is then a learning algorithm for the class of STW.

This learnability result is to be placed in a tradition of learning results for other class of grammars, starting from Gold results [10] for regular languages of words. This result has served as a basis for a host of learning algorithms including inference of regular languages of word and trees [15,16] (see also [6] for a survey of the area), learning of DTDs and XML Schema [2,1], XML transformations [13], and XML queries [3,20].

The Myhill–Nerode Theorem proof starts from the identification of the class of earliest STWs (eSTWs) given in [12]. The main difficulty is to be able to characterize residual languages of a STW transformation and then define a canonical representative for eSTWs. This proof relies on an original algorithm capable to decompose a residual transformation into a form close to a rule of eSTW. In order to obtain a learning algorithm (a la RPNI [16,15]) an important step is to decide the consistency of a transducer with a finite transformation. Unfortunately, we prove that this consistency check is NP-complete. Nevertheless, we present a learning result in a slightly modified framework where the learning algorithm can abstain from answering. We prove that we can define a characteristic sample whose cardinality is within a polynomial bound of the size of the canonical transducer of the transformation to infer. Using this last result, we present here the first polynomial time learning algorithm for the class of STW.

2   Sequential Top-Down Tree-to-Word Transducers

Words and Trees. For a finite set Δ of symbols, we denote by Δ^* the free monoid on Δ , i.e. the set of finite words over Δ with the concatenation operator · and the empty word ε. For a word u, $|u|$ is its length. For $u = u_p \cdot u_f \cdot u_s$, u_p is a *prefix* of u, u_f a *factor* of u, and u_s a *suffix* of u. The *longest common prefix* of a set of words L, denoted $lcp(L)$, is the longest word u that is a prefix of every word in L. Also, $lcs(L)$ is the *longest common suffix* of L. For $w = u \cdot v$, the *left quotient* of w by u is $u^{-1} \cdot w = v$ and the *right quotient* of w by v is $w \cdot v^{-1} = u$.

A *ranked alphabet* is a finite set of ranked symbols $\Sigma = \bigcup_{k \geq 0} \Sigma^{(k)}$, where $\Sigma^{(k)}$ is the set of k-ary symbols. We sometimes write $f^{(k)}$ to indicate explicitly that $f \in \Sigma^{(k)}$. A *tree* is a ranked ordered term over Σ. By T_Σ we denote the set of all trees over Σ. A tree language is a subset of T_Σ. A *context* C is a tree over $\Sigma \cup \{x^{(0)}\}$ with only one leaf labeled x representing a hole. By $C[t]$ we denote the tree obtained by replacing x by the tree t. A *path* is a word over $\bigcup_{k>0} \Sigma^{(k)} \times \{1, \ldots, k\}$, which identifies a node in a tree by the labels of its ancestors: ε is the root node and if a node at path p is labeled with f, then

$p \cdot (f, i)$ is the i-th child of the node. By $paths(t)$ we denote the set of paths of a tree t. Similarly, for a set of trees T, $paths(T) = \bigcup_{t \in T} paths(t)$. Words, trees, paths and contexts have canonical well-founded orders that are consistent with the size of object and can be tested efficiently. Using these orders, functions $\min_{Path}$, $\min_{Tree}$ and $\min_{Ctx}$ allow to obtain the minimal element of a set of resp. paths, trees or contexts.

Transducers. A *deterministic sequential top-down tree-to-word transducer* (STW) is a tuple $M = (\Sigma, \Delta, Q, init, \delta)$, where Σ is a ranked alphabet of *input trees*, Δ is a finite alphabet of *output words*, Q is a finite set of states, $init \in \Delta^* \cdot Q \cdot \Delta^*$ is the *initial rule*, and δ is a partial *transition function* from $Q \times \Sigma$ to $(\Delta \cup Q)^*$ such that if $\delta(q, f^{(k)})$ is defined, then it has k occurrences of elements from Q. In the sequel, we call the state of the initial rule the *initial state*. We denote STWs the class of deterministic sequential top-down tree-to-word transducers and $\mathcal{STW}$ the class of transformations represented by an STW.

We often view δ as a set of *transition rules* , i.e. a subset of $Q \times \Sigma \times (\Delta \cup Q)^*$, which allows us to quantify over δ. Also, the transition function is extended to paths over Σ as follows: $\delta(q, \varepsilon) = q$ and $\delta(q, (f, i) \cdot p) = \delta(q_i, p)$, where $\delta(q, f) = u_0 \cdot q_1 \cdot u_1 \cdot \ldots \cdot q_k \cdot u_k$. The *size* of the STW M is the number of its states and the length of its rules, including the length of words used in the rules. The semantic of the STW M is the transformation $[\![M]\!]$ defined with the help of auxiliary transformations (for $q \in Q$) in a mutual recursion:

$$[\![M]\!]_q(f(t_1, \ldots, t_k)) = \begin{cases} u_0 \cdot [\![M]\!]_{q_1}(t_1) \cdot u_1 \cdot \ldots \cdot [\![M]\!]_{q_k}(t_k) \cdot u_k, \\ \qquad\qquad \text{if } \delta(q, f) = u_0 \cdot q_1 \cdot u_1 \ldots q_k \cdot u_k, \\ \text{undefined}, \quad \text{if } \delta(q, f) \text{ is undefined.} \end{cases}$$

Now, $[\![M]\!](t) = u_0 \cdot [\![M]\!]_{q_0}(t) \cdot u_1$, where $init = u_0 \cdot q_0 \cdot u_1$. Two transducers are *equivalent* iff they define the same transformation. Also, for a transformation τ, $dom(\tau)$ is the domain of τ and $ran(\tau)$ is its range.

We also use *deterministic top-down tree automata* (DTA) that define path-closed tree languages. Recall that a tree language $L \subseteq T_\Sigma$ is *path-closed* if $L = \{t \in T_\Sigma \mid paths(t) \subseteq paths(L)\}$. We refer the reader to [5] for a precise definition and point out that a DTA is essentially an STW with empty output alphabet thus defining a constant transformation mapping every element of its domain to the empty word.

Earliest Transducers. The construction of the canonical transducer, the core of the Myhill-Nerode characterisation of $\mathcal{STW}$, is inspired by the normal form for STWs. The usual choice to define normal forms of transducers is to produce the output as early as possible. This idea initially comes from normalisation of word-to-word transducers, as in [4], and is also employed in [13,9] for tree-to-tree transducers. In [12], we have proposed the following normal form for STWs. An STW $M = (\Sigma, \Delta, Q, init, \delta)$ is *earliest* (eSTW) iff:

(E$_1$) $lcp(ran([\![M]\!]_q)) = \varepsilon$ and $lcs(ran([\![M]\!]_q)) = \varepsilon$ for every state q,

(E$_2$) for $init = u_0 \cdot q_0 \cdot u_1$, $lcp(ran([\![M]\!]_{q_0}) \cdot u_1) = \varepsilon$ and for $\delta(q, f) = u_0 \cdot q_1 \cdot \ldots \cdot q_k \cdot u_k$, then $\forall 1 \leq i \leq k$, $lcp(ran([\![M]\!]_{q_i}) \cdot u_i \cdot \ldots \cdot ran([\![M]\!]_{q_k}) \cdot u_k) = \varepsilon$. $\qquad \square$

Essentially, $(\mathbf{E_1})$ ensures that the output is produced as *up* as possible during the parsing while $(\mathbf{E_2})$ ensures output is produced as *left* as possible. We also observe that in an eSTW, transformations $[\![M]\!]_q$ associated with states have the property that the *lcp* an *lcs* of their output is empty. It is know that for every STW there exists a unique minimal equivalent eSTW [12].

Example 1. Consider the transformation τ_1 that takes as an input a tree t over the signature $\Sigma = \{f^{(2)}, a^{(0)}, b^{(0)}\}$ and output a word on $\Delta = \{\#\}$ that counts the number of symbols in t (i.e. $\tau_1(f(f(a,b),a)) = \#\#\#\#\#)$. This can be done by transducer $M_1 = (\Sigma, \Delta, Q_1 = \{q\}, init_1 = q, \delta_1)$ with $\delta_1(q,a) = \delta_1(q,b) = \#$ and $\delta_1(q,f) = q \cdot \# \cdot q$. However M_1 is not earliest: the output always starts with an $\#$ which can be produced earlier ($(\mathbf{E_1})$ is not satisfied), and the symbol $\#$ in the rule $\delta_1(q,f)$ could be produced before the states ($(\mathbf{E_2})$ is not satisfied).

Consider $M_1' = (\Sigma, \Delta, Q_1' = \{q\}, init_1' = \# \cdot q, \delta_1')$ with $\delta_1'(q,a) = \delta_1'(q,b) = \varepsilon$ and $\delta_1'(q,f) = \#\# \cdot q \cdot q$. This transducer also represent τ_1 but is earliest.

3   A Myhill-Nerode Theorem for $\mathcal{STW}$

In this section we present the construction of a canonical eSTW $Can(\tau)$ that captures an arbitrary $\mathcal{STW}$ transformation τ. Because STWs process the input tree in a top-down fashion, we shall decompose τ into several transformations that capture the transformation performed by τ on the children of the input tree. The decomposition is then used to recursively define the notion of residual $p^{-1}\tau$ of τ w.r.t. a path p, essentially the transformation performed by τ at the node reached with p of its input tree. Residuals are used to define in the standard way the Myhill-Nerode equivalence relation and the canonical transducer $Can(\tau)$.

Decomposition. We fix a transformation τ and let $Left(\tau) = lcp(ran(\tau))$ and $Right(\tau) = lcs(Left(\tau)^{-1}ran(\tau))$. τ is *reduced* if $Left(\tau) = Right(\tau) = \varepsilon$. The *core* of τ is defined as $Core(\tau) = \{(t, Left(\tau)^{-1} \cdot w \cdot Right^{-1}(\tau)) \mid (t, w) \in \tau\}$. While not every transformation is reduced, its core is and it preserves the essence of the original transformation .

A *decomposition* of a reduced τ for $f \in \Sigma$ is a sequence $u_0\tau_1 u_1 \ldots u_{k-1}\tau_k u_k$, where $u_0, \ldots, u_k$ are words and $\tau_1, \ldots, \tau_k$ transformations, that satisfy the following natural conditions:

$(\mathbf{D_1})$ $dom(\tau_i) = \{t_i \mid f(t_1, \ldots, t_k) \in dom(\tau)\}$,
$(\mathbf{D_2})$ $\forall t = f(t_1, \ldots, t_k) \in dom(\tau),\ \tau(t) = u_0\tau_1(t_1)\ldots\tau_k(t_k)u_k$,

and to ensure the uniqueness of decomposition we impose two additional conditions that are obtained by reformulation of $(\mathbf{E_1})$ and $(\mathbf{E_2})$:

$(\mathbf{C_1})$ τ_i is reduced, and
$(\mathbf{C_2})$ $lcp(ran(\tau_i) \cdot u_i \cdot \ldots \cdot ran(\tau_k) \cdot u_k) = \varepsilon$.

We point out that not every transformation can be decomposed.

Example 2. Take $\tau_{\text{swap}} = \{(f(a,a), aa), (f(a,b), ba), (f(b,a), ab), (f(b,b), bb)\}$ that outputs leaves in reverse order. This transformation can not be performed by an STW and there is no decomposition for it.

Residuals. The *residual* of a transformation τ at a path p is defined recursively: $\varepsilon^{-1}\tau = Core(\tau)$ and $(p \cdot (f,i))^{-1}\tau = \tau_i$ if $u_0 \cdot \tau_1 \ldots \tau_n \cdot u_n$ is the unique decomposition of $p^{-1}\tau$ for f.

Example 3. Consider the transformation τ_1 of example 1. $Core(\tau_1) = \tau_1'$ where τ_1' gives the number of symbol in t minus one. Then $\varepsilon^{-1}\tau_1 = \tau_1'$. Also, The decomposition of τ_1' for f is $\#\# \cdot ((f,1)^{-1}\tau_1) \cdot ((f,2)^{-1}\tau_1)$. Observe that $(f,1)^{-1}\tau_1 = (f,2)^{-1}\tau_1 = \tau_1'$, so this decomposition is in fact $\#\# \cdot \tau_1' \cdot \tau_1'$. Also, $lcp(\tau_1') = lcs(\tau_1') = \varepsilon$. This decomposition is consistent with the rules of M_2.

Example 4. Consider the transformation τ_2 that takes as an input a tree $t = f(t_1, t_2)$ over $\Sigma = \{f^{(2)}, a^{(0)}, b^{(0)}\}$ and output a word over $\Delta = \{\#\}$ such that the number of $\#$ is equal to the number of f and a symbols of t_1 plus the number of f and b symbols of t_2, e.g. $\tau_2(f(f(a,b), b))) = \#^3$ (2 $\#$ for $f(a,b)$ and one for b). As τ_2 is reduced, $\varepsilon^{-1}\tau_2 = Core(\tau_2) = \tau_2$. The decomposition of τ for f is $\tau_3 \cdot \tau_4$ with $\tau_3(a) = \#$, $\tau_3(b) = \varepsilon$. The decomposition of τ_3 at f is $\# \cdot \tau_3 \cdot \tau_3$. Similarly, $\tau_4(a) = \varepsilon$, $\tau_4(b) = \#$, and its decomposition at f is $\# \cdot \tau_4 \cdot \tau_4$.

Naturally, not every transformation has well-defined residuals. However any $\mathcal{STW}$ transformation has them and there is a strict correspondence between the residuals and the states of an eSTW defining the transformation.

Lemma 5. *Let M be an eSTW with initial state q_0. For any $p \in paths(dom(\llbracket M \rrbracket))$ we have $p^{-1}\llbracket M \rrbracket = \llbracket M \rrbracket_{\delta(q_0, p)}$.*

This result suggest, and we proved it later on, that the existence of residuals of a transformation for any path of its domain is an important necessary condition for being $\mathcal{STW}$. Consequently, we say that τ is *sequential top-down* if and only if $p^{-1}\tau$ exists for every $p \in paths(dom(\tau))$.

Canonical Transducer. Having defined residuals the construction of the canonical transducer $Can(\tau)$ for a transformation τ is standard. The *Myhill-Nerode equivalence relation* $\equiv_\tau$ on paths of τ is defined in the standard manner: $p_1 \equiv_\tau p_2$ iff $p_1^{-1}\tau = p_2^{-1}\tau$ for $p_1, p_2 \in paths(dom(\tau))$. The Myhill-Nerode equivalence class of a path p w.r.t. τ is $[p]_\tau = \{p' \in paths(dom(\tau)) \mid p \equiv_\tau p'\}$. We say that τ has finite Myhill-Nerode index if $\equiv_\tau$ has a finite number of equivalence classes.

The *canonical transducer* $Can(\tau) = (\Sigma, \Delta, Q, init, \delta)$ of a sequential top-down transformation τ of finite Myhill-Nerode index follows: 1) the set of states is $Q = \{[p]_\tau \mid p \in paths(dom(\tau))\}$; 2) the initial rule is $init = Left(\tau) \cdot [\varepsilon]_\tau \cdot Right(\tau)$; 3) for every state $[p] \in Q$ and every $f \in \Sigma$ such that $p^{-1}\tau$ has a decomposition $u_0 \cdot \tau_1 \cdot u_1 \ldots \tau_k \cdot u_k$ for f, the canonical transducer $Can(\tau)$ has the transition rule $\delta([p], f) = u_0 \cdot [p \cdot (f,1)]_\tau \cdot u_1 \cdot \ldots \cdot u_{k-1} \cdot [p \cdot (f,k)]_\tau \cdot u_k$.

Theorem 6. *For any transformation τ the following conditions are equivalent: 1) τ is definable by an* STW; *2) τ is sequential top-down and has a finite Myhill-Nerode index; 3) $Can(\tau)$ is the unique minimal e*STW *defining τ.*

Direction (1) to (2) requires normalizing the STW into an eSTW and using Lemma 5. Point (3) is obtained from (2) by establishing that the minimal eSTW is in fact $Can(\tau)$ (modulo state renaming). Direction (3) to (1) is trivial.

Example 7. The canonical transducer of τ_2 (as in example 4) is $Can(\tau_2) = M_2$ defined as follow: $M_2 = (\Sigma, \Delta, Q = \{q_1, q_2, q_3\}, init = q_1, \delta)$ with $\delta(q_1, f) = \# \cdot q_2 \cdot q_3$, $\delta(q_2, f) = \# \cdot q_2 \cdot q_2$, $\delta(q_2, a) = \#$, $\delta(q_2, b) = \varepsilon$, $\delta(q_3, f) = \# \cdot q_3 \cdot q_3$, $\delta(q_3, a) = \varepsilon$ and $\delta(q_3, b) = \#$. This is consistent with decompositions observed in example 4 if one identifies q_1 with ε, q_2 with $(f, 1)$ and q_3 with $(f, 2)$.

4   Learning STWs

In this section we present a learning algorithm for $\mathcal{STW}$ transformations.

4.1   Learning Framework

First, we investigate the question of the meaning of what learning a transformation means and pursue an answer that is inspired by the Gold learning model in polynomial time and data [10]. Essentially, we are interested in a polynomial time algorithm that takes a finite *sample* $S \subseteq T_\Sigma \times \Delta^*$ and constructs an STW M transducer *consistent* with S i.e., $S \subseteq [\![M]\!]$. Unfortunately, unless P $=$ NP, the following result precludes the existence of such an algorithm.

Theorem 8. *Checking if there exists an* STW *consistent with a given sample is NP-complete.*

To overcome this difficulty, we shall allow the algorithm to *abstain* i.e., return a special *Null* for cases when an STW consistent with the input sample cannot be easily constructed. Naturally, this opens the door to a host of trivial algorithms that return *Null* for all but a finite number of hard-coded inputs. To remove such trivial algorithms from consideration we shall essentially require that the learning algorithm of interest can infer any $\mathcal{STW}$ τ from sufficiently informative samples, called *characteristic sample* of τ: the learning algorithm should be able to output an eSTW defining τ. Furthermore, we require the characteristic sample to use a number of examples bounded by a polynomial of the number of equivalence classes of $\equiv_\tau$.

Another obstacle comes from the fact that DTAs are not learnable from positive examples alone and learning DTA from a set of positive examples can be easily reduced to learning STW. To remove this obstacle, we assume that a DTA D capturing the domain of the goal transformation is given on the input. Note that this domain automaton could also be obtained by learning method, such as the RPNI algorithm for trees [15].

If a class of transformation satisfies all the above properties, we say that it is learnable with abstain from polynomial time and data. In the following, we aim to obtain the following result.

Theorem 9. $\mathcal{STW}$ *transformations represented by* e$\mathcal{STW}$ *are learnable with abstain from polynomial time and data.*

4.2   Learning Algorithm

We now present the learning algorithm for $\mathcal{STW}$. This algorithm essentially attempts to emulate the construction of the canonical transducer, using a finite sample of the transformation.

The Core Algorithm. The main procedure of the learning algorithm follows closely the construction of the canonical transducer. It takes as an input a sample S of a target transformation τ, as well as a DTA D that represents $dom(\tau)$.

Algorithm 1. learner$_D(S)$

1: $P := paths(dom(S))$; $Q := \emptyset$
2: $state := \textbf{new } hashtable\langle Path, Path\rangle()$
3: **while** $P \neq \emptyset$ **do**
4:      $p := \min_{Path}(P)$
5:      $P' := \{p' \in Q \mid p \simeq_{S,D} p'\}$
6:      **if** $P' \neq \emptyset$ **then** (*p can be merged*)
7:          $P := P \setminus \{p' \in P \mid p$ is prefix of $p'\}$
8:          $state[p] := \min_{Path}(P')$
9:      **else**
10:          $P := P \setminus \{p\}$; $Q := Q \cup \{p\}$
11:          $state[p] := p$
12: $init := Left(S) \cdot state[\varepsilon] \cdot Right(S)$
13: **for** $p \in Q$ **do**
14:      **for** $f \in \Sigma$ s.t. $\exists i$ with $p.(f,i) \in paths(dom(S))$ **do**
15:          **for** $i \in 1,...k$, Let $p_i = state[p.(f,i)]$
16:          $(u_0, \_, u_1, \ldots, u_k) := \textsf{decomp}(residual(S,p),f)$
17:          $\delta(p,f) := u_0 \cdot p_1 \cdot u_1 \cdot \ldots \cdot p_k \cdot u_k$
18: $M := (\Sigma, \Delta, Q, init, \delta)$
19: **if** $S \subseteq [\![M]\!]$ **and** $dom([\![M]\!]) \subseteq [\![D]\!]$ **then return** M **else return** *Null*

The algorithm consists of 2 parts. First, in lines 3 to 11, it attempts to identify the set of states of the canonical transducer. For this, it builds a function *state* that associates with every path the minimal path in its equivalence class that represents the corresponding residual. This is based on the predicate $\simeq_{S,D}$ which is an emulation of the Myhill-Nerode equivalence relation $\equiv_\tau$ on an finite sample of τ. Note that if $\simeq_{S,D}$ behaves exactly as $\equiv_\tau$, and assuming $paths(dom(S))$ contains all smallest paths representative of each residual, this procedure produces exactly the set Q of states of $Can(\tau)$. The exact implementation of the predicate $\simeq_{S,D}$ is explained later.

Second part, line 12, builds the other elements of the transducer. This uses the procedure decomp to compute decomposition of samples in a manner emulating decomposition of transformations and is explained in detail later.

We point out the algorithm may fail to produce an eSTW consistent with S. Therefore, in line 19 the consistence of the constructed eSTW is verified and the algorithm abstains from answer if the test fails. The following lemma is therefore trivial.

Lemma 10. *For a sample S and a* DTA *D,* learner$_D(S)$ *produces an eSTW M in time polynomial in the size of S or abstains from answer.*

This results assumes the existence of polynomial procedures for $\simeq_{S,D}$, decomp and residual, which we present next.

Decomposition. The above learning algorithm relies on the ability to decompose a sample. This is done by the following procedure. It takes as an input a sample S which is supposed to be representative of a transformation τ, and a symbol $f^{(k)}$ such that there are f rooted trees in S. From this, it outputs a sequence $u_0 \cdot S_1 \cdot u_1 \ldots S_k \cdot u_k$ which ideally is the proper decomposition of S w.r.t. to τ.

Algorithm 2. decomp$(S, f^{(k)})$

1: Let $S_f = \{(t, w) \in S \mid t$ is of the form $f(t_1, \ldots t_k)\}$
2: Let $s = f(s_1, \ldots, s_k)$ be the tree $\min_{Tree}(dom(S_f))$ and $w_s := S(s)$
3: **for** $i := 1, \ldots, k$ **do** $D_i := \{t_i \mid f(s_1, \ldots, s_{i-1}, t_i, s_{i+1}, \ldots, s_k) \in dom(S_f)\}$
4: $u_0 := lcp(\{w \mid (t, w) \in S_f\})$
5: $prefix_0 = u_0$
6: **for** $i := 1, \ldots, k$ **do**
7:     $prefix_i := lcp\{w \mid \exists t_{i+1}, \ldots, t_k. (f(s_1, \ldots, s_i, t_{i+1}, \ldots, t_k), w) \in S_f\}$
8:     $suffix_i := prefix_i^{-1} \cdot w_s$
9:     $S_i' := \emptyset$
10:    **for** $t \in D_i$ **do**
11:        $w := prefix_{i-1}^{-1} \cdot S(f(s_1, \ldots, s_{i-1}, t, s_{i+1}, \ldots, s_k)) \cdot suffix_i^{-1}$
12:        $S_i' := S_i' \cup \{(t, w)\}$
13:    $u_i := lcs(ran(S_i'))$
14:    $S_i := \{(t, w \cdot u_i^{-1}) \mid (t, w) \in S_i'\}$
15: **return** $(u_0, S_1, u_1 \ldots, S_k, u_k)$

From the minimal tree $s = f(s_1, \ldots, s_k)$ of $dom(S)$ rooted by f, the algorithm essentially tries to decompose $w_s = S(s)$ into $u_0 S_1(s_1) \ldots S_k(s_k)u_k$, as defined by the formal definition of decomp(S, f). Note this is defined only if there are some f rooted trees in $dom(S)$. The word u_0 is simply $Left(S_f)$. Then, for each i, $prefix_i$ is built such that it is equal to $u_0 S_1(s_1) \ldots S_i(s_i)u_i$ and so $suffix_i = prefix_i^{-1}w_s = S_{i+1}(s_{i+1}) \ldots u_k$. From this, residual transformations S_i and words u_i can be built simultaneously. For any tree $t_i \in (f, i)^{-1} dom(S)$, we consider the tree $t = f(s_1, \ldots, s_{i-1}, t_i, s_{i+1}, \ldots, s_k))$ (which belongs to $dom(S)$ if is path-closed or well constructed) and compute $S'_i(t_i) = S_i(t_i) \cdot u_i = prefix_{i-1}^{-1} S(t) suffix_i^{-1}$. The word u_i is obtained as $lcs(ran(S'_i))$, which allow to obtain $S_i(t_i) = S'_i(t_i) \cdot u_i^{-1}$.

If the sample is rich enough (a notion that will be made precise in the next section), the lcp and lcs of the different elements are computed correctly and the algorithm outputs exactly what it supposed to. If the sample is not rich enough, it may possibly produce a decomposition which is not necessarily sound: there may be a tree $t = f(t_1, \ldots, t_k)$ such that which $S(t) \neq u_0 \cdot S_1(t_1) \cdot u_1 \ldots S_k(t_k) \cdot u_k$. However, in any case, the algorithm answers in time polynomial in the size of S.

Residuals and Equivalence. From the decomposition procedure, it is possible to build the residual of a sample for a path p. $residual(S, p)$ is computed in a manner analogous to $p^{-1}\tau$: for $p = \varepsilon$, $residual(S, p) = reduce(S)$, and for $p = p' \cdot (f, i)$, we compute $S' = residual(S, p)$ and $residual(S, p) = S_i$, where $decomp(S', f) = u_1 \cdots S_1 \ldots S_k \cdot u_k$. Note that again, $residual(S, p)$ is a polynomial time procedure.

From this, we can define the relation $\simeq_{S,D}$ which tries to emulate $\equiv_\tau$. Recall that $p_1 \equiv_\tau p_2$ iff $p_1^{-1}\tau = p_2^{-1}\tau$ and note that two transformations are identical if they have the same domain and agree on every tree. Because the residuals $p_1^{-1}\tau$ and $p_2^{-1}\tau$ are represented with finite samples $S_1 = residual(S, p_1)$ and $S_2 = residual(S, p_2)$ and their domains need not be necessarily equal, the predicate $p_1 \simeq_{S,D} p_2$ uses the DTA D to verify that the domains of the residuals $p_1^{-1}\tau$ and $p_2^{-1}\tau$ are equal and then checks that for every tree in common both samples S_1 and S_2 produce the same results.

Again, all those procedures are polynomial. Note however that they behave correctly (i.e. $p \simeq_{S,D} p' \Leftrightarrow p \equiv_\tau p'$ for instance) only if the sample is rich enough. What it means exactly is defined in the next section.

4.3   A Characteristic Sample

In the following, we identify a characteristic sample for STW transformation τ: $CharSet(\tau)$ is a finite set of examples such that whenever learner is provided a superset of $CharSet(\tau)$ as input, it outputs $can(\tau)$.

The Characteristic Sample. We first introduce some notations and definitions. For $p \in paths(dom(\tau))$ let c_p be the minimal context with x at path p. The finite set of all minimal representatives of equivalence classes of $\equiv_\tau$ is $StatePath(\tau) = \{\min_{Path}([p]_\tau) \mid p \in paths(dom(\tau))\}$. We also define $EdgePath(\tau)$, which adds to the shortest paths their extensions with one additional step i.e., $EdgePath(\tau) = StatePath(\tau) \cup \{p \cdot (f, i) \in paths(dom(\tau)) \mid p \in StatePath(\tau)\}$.

Example 11. τ_2 has 3 distincts residuals: $\varepsilon^{-1}\tau_2$, $(f, 1)^{-1}\tau_2$ and $(f, 2)^{-1}\tau$. Therefore, $StatePath(\tau_2) = \{\varepsilon, (f, 1), (f, 2)\}$ and $EdgePath(\tau_2) = StatePath(\tau_2) \cup \{(f, 1)(f, 1), (f, 1)(f, 2), (f, 2)(f, 1), (f, 2)(f, 2)\}$.

Let us consider a path $p \in EdgePath(\tau)$, and a set of trees $T \subseteq T_\Sigma$. Then, T is *structurally representative* for τ with respect to p if

(S$_0$) the tree $\min_{Tree}(dom(p^{-1}\tau))$ belongs to T;
(S$_1$) $lcp((p^{-1}\tau)(T)) = \varepsilon$ and $lcs((p^{-1}\tau)(T)) = \varepsilon$;
(S$_2$) $lcp(ran(p^{-1}\tau) \setminus \{\varepsilon\}) = lcp((p^{-1}\tau)(T) \setminus \{\varepsilon\})$.

Additionally, we say that T is *discriminant* for τ with respect to p if

(DI) for any $p_0 \in StatePath(\tau)$, if $T_{p,p_0} = \{t \in dom(p^{-1}\tau) \cap dom(p_0^{-1}\tau) \mid p^{-1}\tau(t) \neq p_0^{-1}\tau(t)\}$ is nonempty, then $\min_{Tree}(T_{p,p_0})$ belongs to T.

For a path p, conditions **(S$_0$)**, **(S$_1$)** and **(S$_2$)** ensure that T contains all elements needed to correctly decompose the residual transformation $p^{-1}\tau$. Condition **(DI)** ensures that T contains witnesses necessary to distinguish different equivalence classes.

Example 12. Consider transformation τ_2 and take for instance $p = (f,1)$. The tree $T_{p,\varepsilon}$ is the smallest tree whose image differs in $p^{-1}\tau_2$ and $\varepsilon^{-1}\tau_2$. In fact, $T_{p,\varepsilon} = f(a,a)$ as $p^{-1}\tau_2(f(a,a)) = \#^2$ and $\varepsilon^{-1}\tau_2(f(a,a)) = \#^3$. For other $p' \in \{(f,2), (f,2)(f,1), (f,2)(f,2)\}$, $T_{p,p'} = a$.

To satisfy condition **(S$_1$)** and **(S$_2$)**, one can take $\{a, b, f(u,u)\} \in T_p$. This allows to satisfy **(S$_1$)** as $lcp(\{(p^{-1}\tau_2)(a), (p^{-1}\tau_2)(b), (p^{-1}\tau_2)(f(a,a))\}) = lcp(\{\#, \varepsilon, \#^3\}) = \varepsilon$ and the same for lcs. For **(S$_2$)**, we have $lcp(\{(p^{-1}\tau_2)(a), (p^{-1}\tau_2)(b), (p^{-1}\tau_2)(f(a,a))\} \setminus \{\varepsilon\}) = lcp(\{\#, \varepsilon, \#^3\} \setminus \{\varepsilon\}) = \#$ which is indeed equal to $lcp(ran(p^{-1}\tau) \setminus \{\varepsilon\})$.

Let τ be a transformation in $\mathcal{STW}$ and let p be a path in $EdgePath(\tau)$. A sample S is characteristic for τ at path p if (i) $S \subseteq p^{-1}\tau$ and ; (ii) for all paths p_0 such that $p \cdot p_0 \in EdgePath(\tau)$, the set of trees $c_{p_0}^{-1}dom(S)$ is discriminant and structurally representative for τ with respect to $p \cdot p_0$. A sample is *characteristic* for τ if it is characteristic for τ at path ε.

An important property is that it is possible to build a characteristic sample whose cardinality is with a polynomial bound on the number of distinct residuals of τ. Indeed, to have property **(DI)**, one need a quadratic number of trees while conditions **(S$_0$)**, **(S$_1$)**, and **(S$_2$)** all require a linear number of trees. We denote by $CharSet(\tau, p)$ the minimal characteristic sample for τ at path p and by $CharSet(\tau)$ the set $CharSet(\tau, \varepsilon)$. This yields the following lemma.

Lemma 13. *For any e*STW *M there exists a characteristic sample $CharSet(\llbracket M \rrbracket)$ of cardinality polynomial in the size of M.*

We also point out that any sample S consistent with $\llbracket M \rrbracket$ that contains $CharSet(\llbracket M \rrbracket)$ is also characteristic for $\llbracket M \rrbracket$.

Example 14. From previous example, one can build a characteristic sample for τ. In particular, the minimal context for $(f,1)$ is $f(x,a)$. In example 12, it is argued that trees $\{a, b, f(a,a)\}$ are in T_p, which means that $CharSet(\tau)$ contains $(f(a,a), \#^3)$, $(f(b,a), \#^2)$ $(f(f(a,a),a), \#^5)$. A similar approach has to be also considered for all other elements of $EdgePath(\tau)$ to obtain the full $CharSet(\tau)$.

Decomposition of Characteristic Samples. It remains to see that from the characteristic sample of a transduction, the procedures used by the learning algorithm behave as expected. We begin with the decomposition. The first lemma shows that the factors of a decomposition are identified whenever a superset of the characteristic sample is provided to the decomposition procedure.

Lemma 15. *Let $\tau \in \mathcal{STW}$ and $p \in StatePath(\tau)$. Let S be a characteristic set for τ at path p, For any $f \in \Sigma^{(k)}$ such that the decomposition of $p^{-1}\tau$ at f is $u_0 \cdot \tau_1 \ldots \tau_k \cdot u_k$, then $\mathsf{decomp}(S, f) = u_0 \cdot S_1 \ldots S_k \cdot u_k$ where each S_i is characteristic for τ at path $p \cdot (f, i)$*

This decomposition lemma relies on the idea that the properties required by the formal definition can be observed locally on a characteristic sample: for instance property $(\mathbf{D_1})$ and $(\mathbf{D_2})$ simply comes from consistency of the sample ($S \subseteq \tau$), while $(\mathbf{C_1})$ is observable on S thanks to property $(\mathbf{S_1})$. However, $(\mathbf{C_2})$ does not translate directly into a property that a characteristic sample should fulfill. This is of course the role played by property $(\mathbf{S_2})$.

The link between $(\mathbf{S_2})$ and $(\mathbf{C_2})$ is actually an indirect consequence of following property: let W and W' be two sets of words in Δ^*, if $lcp(W \setminus \{\varepsilon\}) = lcp(W' \setminus \{\varepsilon\})$, and $lcp(W) = lcp(W')$, then $lcp(\{w \cdot u \mid w \in W\}) = \varepsilon$ for a $u \in \Delta^*$ implies that $lcp(\{w' \cdot u \mid w' \in W'\}) = \varepsilon$.

Now, consider a transformation $\tau \in \mathcal{STW}$, a path $p \in StatePath(\tau)$ and a sample S characteristic for τ in a path p. If we consider $\mathsf{decomp}(p^{-1}\tau, f) = u_0\tau_1 \ldots \tau_k u_k$, then for any $i \in \{1, \ldots, k\}$ we have $lcp\{\tau_i(t_i) \cdot u_i \cdot \ldots \cdot \tau_k(t_k) \cdot u_k \mid t_i \in (f, i)^{-1} dom(S), \ldots, t_k \in (f, k)^{-1} dom(S)\} = \varepsilon$.. This is a direct consequence of above property and the fact that S satisfy $(\mathbf{S_1})$ and $(\mathbf{S_2})$, and allows us to prove Lemma 15.

As the construction of residuals $\mathsf{residual}(S, p)$ relies on the decomposition, Lemma 15 has the important consequence that those residuals can be computed properly for any $p \in EdgePath(\tau)$. This gives the following two results. First, if S is characteristic for τ, and $p \in EdgePath(\tau)$, then $\mathsf{residual}(S, p)$ is characteristic for τ w.r.t. p. Second, as a consequence and because of $(\mathbf{DI})$, if $p, p' \in EdgePath(\tau)$ then $p \simeq_{S,D} p' \Leftrightarrow p \equiv_\tau p'$. Ultimately, this indicates that from a sample S characteristic for τ, the learning algorithm builds $Can(\tau)$:

Lemma 16. *Let $\tau \in \mathcal{STW}$ and D a DTA with $[\![D]\!] = dom(\tau)$. From any sample S characteristic with τ, $\mathsf{learner}_D(S) = Can(\tau)$.*

This, along with Lemmas 10 and 13 proves Theorem 9.

5   Conclusion

We presented the first polynomial time learning algorithm for tree to string transformation. This algorithm present the particularity to abstain answering at some point. This is due to the fact that the consistency problem is NP-complete for STW, and so, it is simply not possible to provide a transducer consistant with some input sample.

Also note that the language of strings outputed by an STW are context free languages. Therefore, inference of STW is linked to inference of Context Free Grammars (CFG) and can be seen as the inference of a CFG using words and their derivative trees as input. This work may therefore bring some highlight to the problem of Context Free Grammar inference.

References

1. Bex, G.J., Gelade, W., Neven, F., Vansummeren, S.: Learning deterministic regular expressions for the inference of schemas from XML data. ACM Transactions on the Web 4(4) (2010)
2. Bex, G.J., Neven, F., Schwentick, T., Vansummeren, S.: Inference of concise regular expressions and DTDs. ACM TODS 35(2) (2010)
3. Carme, J., Gilleron, R., Lemay, A., Niehren, J.: Interactive learning of node selecting tree transducers. Machine Learning 66(1), 33–67 (2007)
4. Choffrut, C.: Minimizing subsequential transducers: A survey. TCS 292(1), 131–143 (2003)
5. Comon, H., Dauchet, M., Gilleron, R., Löding, C., Jacquemard, F., Lugiez, D., Tison, S., Tommasi, M.: Tree automata techniques and applications (October 2007), Available online since 1997: http://tata.gforge.inria.fr
6. de la Higuera, C.: A bibliographical study of grammatical inference. Pattern Recognition 38, 1332–1348 (2005)
7. Engelfriet, J., Maneth, S., Seidl, H.: Deciding equivalence of top-down XML transformations in polynomial time. Journal of Computer and System Science 75(5), 271–286 (2009)
8. Filiot, E., Raskin, J.-F., Reynier, P.-A., Servais, F., Talbot, J.-M.: Properties of visibly pushdown transducers. In: Hliněný, P., Kučera, A. (eds.) MFCS 2010. LNCS, vol. 6281, pp. 355–367. Springer, Heidelberg (2010)
9. Friese, S., Seidl, H., Maneth, S.: Minimization of deterministic Bottom-Up tree transducers. In: Gao, Y., Lu, H., Seki, S., Yu, S. (eds.) DLT 2010. LNCS, vol. 6224, pp. 185–196. Springer, Heidelberg (2010)
10. Gold, E.M.: Complexity of automaton identification from given data. Inform. Control 37, 302–320 (1978)
11. Griffiths, T.V.: The unsolvability of the equivalence problem for Lambda-Free nondeterministic generalized machines. Journal of the ACM 15(3), 409–413 (1968)
12. Laurence, G., Lemay, A., Niehren, J., Staworko, S., Tommasi, M.: Normalization of sequential Top-Down Tree-to-Word transducers. In: Dediu, A.-H., Inenaga, S., Martín-Vide, C. (eds.) LATA 2011. LNCS, vol. 6638, pp. 354–365. Springer, Heidelberg (2011)
13. Lemay, A., Maneth, S., Niehren, J.: A learning algorithm for Top-Down XML transformations. In: 29th PODS 2010, pp. 285–296. ACM Press (2010)
14. Martens, W., Neven, F., Gyssens, M.: Typechecking top-down XML transformations: Fixed input or output schemas. Inf. Comput. 206(7), 806–827 (2008)
15. Oncina, J., García, P.: Inference of recognizable tree sets. Tech. report, Dept de Sistemas Informáticos y Computación, Univ. de Alicante (1993), DSIC-II/47/93
16. Oncina, J., Gracia, P.: Identifying regular languages in polynomial time. In: Advances in Structural and Syntactic Pattern Recognition, pp. 99–108 (1992)
17. Papadimitriou, C.: Computational complexity. Addison-Wesley (1994)

18. Raskin, J.-F., Servais, F.: Visibly pushdown transducers. In: Aceto, L., Damgård, I., Goldberg, L.A., Halldórsson, M.M., Ingólfsdóttir, A., Walukiewicz, I. (eds.) ICALP 2008, Part II. LNCS, vol. 5126, pp. 386–397. Springer, Heidelberg (2008)
19. Staworko, S., Laurence, G., Lemay, A., Niehren, J.: Equivalence of deterministic nested word to word transducers. In: Kutyłowski, M., Charatonik, W., Gębala, M. (eds.) FCT 2009. LNCS, vol. 5699, pp. 310–322. Springer, Heidelberg (2009)
20. Staworko, S., Wieczorek, P.: Learning XML twig queries. CoRR, abs/1106.3 (2011)

Probabilistic Simulation
for Probabilistic Data-Aware Business Processes

Haizhou Li[1,*], François Pinet[2], and Farouk Toumani[1]

[1] LIMOS, CNRS, Blaise Pascal University, France
{li,ftoumani}@isima.fr
[2] IRSTEA, Clermont-Ferrand, France
francois.pinet@irstea.fr

Abstract. This paper studies modelling and analysis issues in the context of a probabilistic data-aware business process. It uses as formal model to describe process behaviours a labelled transitions system in which transitions are guarded by conditions defined over a probabilistic database and presents an approach for testing probabilistic simulation preorder in this context. A complexity analysis reveals that the problem is in 2-EXPTIME, and is EXPTIME-hard, w.r.t. expression complexity while it matches probabilistic query evaluation w.r.t. data-complexity.

Keywords: probabilistic database, testing simulation relaition, data-aware business process.

1 Introduction

There has been over the last few years an increasing interest around the role played by data in business processes. Indeed, in many applications the executions of processes, as specified in a control-flow, may be also governed by conditions defined over variables or over a database. This motivates the emergence of data-aware and data-centric perspectives for process modelling, approaches that promote data to first-class citizens in process models [7]. However, whereas many traditional applications manipulate precise data, there is a wide range of new applications, such as collaborative business processes, monitoring large-scale physical systems (e.g., energy efficient buildings) or privacy preservation, that need to manage imprecise and uncertain data [6,16,15]. The aforementioned applications stress the need for business process models that are able to handle imprecise data. We study in this paper the underlying modelling and analysis issues. We use as formal model to describe process semantics a Labelled Transitions System (LTS) in which transitions are guarded by conditions defined over a global database which, in spirit of [6], *contains an explicit representation of the uncertainty.* We call such a model a *probabilistic data-aware business process* (pd-process). Our choice of LTSs is motivated by the prominent role played by

---

* The author is funded by "La région Auvergne et le Feder".

A.-H. Dediu et al. (Eds.): LATA 2014, LNCS 8370, pp. 503–515, 2014.
© Springer International Publishing Switzerland 2014

this formalism for representing behaviours of systems. We rest on recent developments in the emerging field of probabilistic databases [6,11] to include imprecise data within labelled transitions systems and formally define the semantics of the obtained pd-process model.

The paper focuses on the problem of analysing pd-processes. Historically, two principal methods of equal importance have been used in the literature to analyse LTSs: temporal logic, used to verify whether a given process satisfies certain properties and equivalence or preorder relations. Regarding the second class of methods, simulation preorder is a refinement relation on processes that has been proved to be very useful in many applications. Simulation equivalence plays a crucial role in model checking since it preserves relevant properties of many temporal logics (e.g., CTL*) and hence can be exploited to minimize the state space explored by verification algorithms [5]. Simulation equivalence has also been used directly for verification of business processes [12] as well as for web service analysis and composition [3,4].

Main contributions. The paper presents a formal framework to describe probabilistic data-aware business processes and defines a probabilistic simulation preorder in this context. The provided definition of simulation is: (i) semantic, in the sense that it is based on a containment relation between the possible execution trees of pd-processes, and (ii) conservative, since it matches classical notion of simulation in non-probabilistic case. The paper describes then a refinement approach that enables to characterize simulation preorder in pd-processes. Finally, a complexity analysis is conducted w.r.t. two dimensions: (i) expression complexity, defined in terms of the size of the LTS of a pd-process and (ii) data-complexity, defined in terms of the size of the probabilistic database. The paper establishes upper and lower bounds in expression complexity and shows that there is no overhead w.r.t. probabilistic query evaluation in data-complexity. This latter result is interesting because data-complexity is the most significant factor in our context.

Indeed, the general topic of this paper is not totally new since a satisfactory verification theory for probabilistic processes has been a long-standing research problem and numerous probabilistic process models have already been proposed in the literature [14,17]. Two major classes of models are worth to distinguish: fully probabilistic models, which replace non-determinism with probabilistic choice, and non-deterministic ones, which distinguish between non-determinism and probabilistic choice [2,14]. Several types of probabilistic simulation relations have been defined in the literature depending on the used probabilistic model and target criteria, e.g., testing, composition [13,9,17]. The pd-process model falls in the category of non-deterministic models since it makes a difference between non-determinism and probabilities. Moreover, pd-processes exhibit intricated correlations between transitions since probabilities associated to transitions guards depend on the considered possible world of the probabilistic database and this makes pd-processes non-markovian, i.e., predictions for the future of the process depend on the process's full history. As a consequence, pd-process semantics does not coincide with semantics of existing probabilistic processes for which simulation has been investigated in the literature. This makes any attempt to extend existing techniques to characterize simulation preorder in the setting of pd-processes far from being straightforward (e.g., in most considered probabilistic settings, simulation is PTIME [17] while it is shown in this paper that pd-process simulation is EXPTIME-hard).

Organization. Section 2 reviews basic concepts from the theory of probabilistic databases useful for this work. Section 3 defines the pd-process model and provides its formal semantics. This section provides also a semantic definition of simulation preroder in pd-processes. Section 4 describes a refinement approach for checking simulation relations between pd-processes and provides complexity results. Section 5 draws conclusions in terms of future research directions.

2   Overview on Probabilistic Databases

A finite probability space is a pair (Ω, Pr) where Ω is the finite set of *outcomes*, and $Pr : \Omega \to [0, 1]$ s.t. $\sum_{\omega \in \Omega} Pr(\omega) = 1$. For $A \subseteq \Omega$, we take $Pr(A) = \sum_{\omega \in A} Pr(\omega)$. A set $\{t_1, ..., t_n\} \subseteq \Omega$ is independent if $Pr(t_1, ..., t_n) = Pr(t_1) \times ... \times Pr(t_n)$.

We assume readers familiar with basic database concepts (e g , see [1]). The issues underlying management of imprecision and uncertainty in data have attracted the attention of the research community since a long time. Several models have been proposed over the time to handle uncertain data. In recent years, the field of probabilistic databases gained momentum under the driving force of a wide spectrum of new applications [6,11]. In this paper, we are interested in particular by probabilistic relational databases defined over a finite domain. Informally, a probabilistic database is defined as a database that includes relations whose tuples are associated with probabilities.

Table 1. Example of a probabilistic database (D_{ins})

Relation Profit

	AgeMin	AgeMax	Educ. Lev	Lic Year	Driv. Rec	City	Profit	Pr
t_2	25	50	college	3	medium	paris	high	60%
t_3	20	30	college	3	medium	paris	medium	50%

Table 1 shows an example of a probabilistic relation[1], noted Profit, in the field of insurance risk assessment. The content of the relation Profit is indeed not certain and, hence, the relation records a confidence with each prediction or analysis result. This is materialised by the attribute Pr which associates a probability with each tuple in a (probabilistic) relation (i.e., Pr gives the marginal probability of each tuple). The standard semantics of probabilistic databases is

---

[1] In this example, and only for illustration purposes, tuples are assumed to be independent (e.g., this is why the sum of probabilities of the tuples t_2 and t_3 of relation Profit is > 1). Such an assumption is not mandatory for the proposed approach.

defined based on the notion of *possible worlds*. The intuition is that the precise content of a probabilistic database is unknown but instead the finite set of potential instances, each with some probability, can be computed. For example, one possible world of the relation Profit is a collection of tuples $W_2 = \{t_2\}$ with probability 30%. Hence, a probabilistic database can be viewed as a probabilistic distribution over a finite set of possible (complete) databases. Given a probabilistic database D, we denote by $\mathcal{W}(D)$ the finite set of its possible worlds (i.e., its possible instances). Formally, a probabilistic database D defines a finite probability space $(\mathcal{W}(D), Pr)$, whose set of outcomes $\mathcal{W}(D)$ forms all the possible instances of the probabilistic database D. Each possible world $W \in \mathcal{W}(D)$ is associated with a probability given by $Pr(W)$, with $\sum\limits_{W \in \mathcal{W}(D)} Pr(W) = 1$.

Given such a framework, a crucial question is then related to query evaluation, i.e., *the problem of calculating the probability of tuples occurring in query answers*. In this paper, we are interested in particular by *boolean queries*, i.e., queries that return as unique answers either true or false.

In the context of a probabilistic database $(\mathcal{W}(D), Pr)$, the problem of the evaluation of a boolean query q consists in computing the probability of query q to return as answer the value true. Such a probability is defined as follows $Pr(q) = \sum\limits_{\substack{W \in \mathcal{W}(D)| \\ q_1(W) = true}} Pr(W)$. In other words, $Pr(q)$ is given by the sum of the probabilities of all the possible worlds where q is evaluated to true. However, in most practical situations it is not feasible to compute the set $\mathcal{W}(D)$ and then explicitly evaluates a query q on each world in $\mathcal{W}(D)$. Indeed, $\mathcal{W}(D)$ is usually very large. To cope with this problem, existing works have developed techniques to efficiently evaluate queries on concise representations of probabilistic databases [6,11]. Not surprisingly, there is a trade-off between the expressiveness of the representation model and computational tractability of query evaluation. This is why, most existing works adopt some restricting assumptions, often expressed as a form of independence of tuples [6]. There are also a few approaches that support modelling complex correlations in probabilistic databases [11]. It is worth to mention that, while we rely on existing techniques to handle probabilistic data, our approach remains insensitive w.r.t. the assumptions underlying the representation model. We require only a system that is able to evaluate boolean queries over a probabilistic database, a requirement which is within the reach of most existing probabilistic database management systems.

3   Probabilistic Data-Aware Business Processes

We present below a formal model to describe pd-processes.

Definition 1 (probabilistic data-aware process). *A probabilistic data-aware process (pd-process) is a tuple $A = (S, s_0, D, Act, G, \Delta, F)$, where: S is a finite set of states, with $s_0 \in S$, the starting state, and $F \subseteq S$ the set of final states. D is a probabilistic database with possible worlds $\mathcal{W}(D)$. Act is a finite set of actions or activities. G is a finite set of guards defined as boolean queries over the database D. $\Delta \subseteq S \times Act \times G \times S$, the transition relations, are a set of guarded transitions.*

A pd-process is essentially a non-deterministic LTS whose transitions are guarded by boolean queries over a probabilistic database. The probabilities of transitions determine the branching choices available during a given process execution. It should be noted that probabilities of transitions are not independent. Arbitrary and complex correlations between transitions probabilities may arise depending on the considered probabilistic database and on the connections that exist between transitions guards (e.g., disjoint guards, containment, overlapping, ...). As a consequence, pd-processes are non-markovian, i.e., predictions for the future of the process depend on the process's full history. Fig 1 (a) shows an example of a pd-process, called NormCalc, which illustrates a simple business process about insurance premium calculation. The guards implement boolean queries over a probabilistic database which is not fully represented in this paper due to the space limitation (e.g., see [10] for a detailed example).

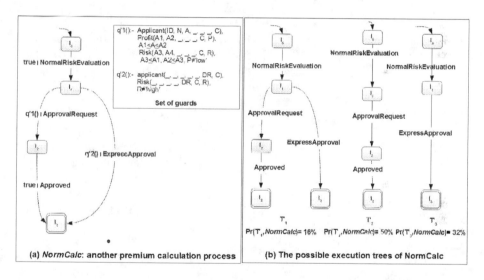

Fig. 1. Example of an insurance premium calculation business process

Semantics. Various classes of process semantics have been studied in the literature [8]. A line of demarcation between existing semantics lies in the distinction between *linear time* and *branching time* semantics. Following branching time semantics, possible executions allowed by a process are characterized in terms of trees, called *execution trees*, instead of paths. In the case of pd-processes, execution trees depend on the guard evaluation which is determined by the considered possible world of the probabilistic database.

To formally define the notions of execution trees, we use the following definition of a tree: A tree is a set $\tau \subseteq \mathbb{N}^*$ such that if $xn \in \tau$, for $x \in \mathbb{N}^*$ and $n \in \mathbb{N}$, then $x \in \tau$ and $xm \in \tau$ for all $0 \leq m < n$. The elements of τ represent nodes: the empty word ϵ is the root of τ, and for each node x, the nodes of the form xn, for $n \in \mathbb{N}$, are children of x. Given a pair of sets L and M, an $\langle L, M \rangle$-labelled tree is a triple (τ, λ, δ), where τ is a tree, $\lambda : \tau \to L$ is a node labelling function that maps each node of τ to an element in L, and $\delta : \tau \times \tau \to M$ is an edge labelling function that maps each edge (x, xn) of τ to an element in M. Then, every path $\rho = \epsilon, n_0, n_0 n_1, \ldots$ of τ generates a sequence $\Gamma(\rho) = \lambda(\epsilon).\delta(\epsilon, n_0).\lambda(n_0).\delta(n_0, n_0 n_1).\lambda(n_0 n_1).\ldots$ of alternating labels from L and M. Informally, if L and M correspond to the sets of states S and actions Act of a pd-process A, then we can use an $\langle S, Act \rangle$-labeled tree to characterize the semantics of A.

Definition 2 (Execution trees and possible execution trees). *Let $A = (S, s_0, D, Act, G, \Delta, F)$ be a pd-process.*

- *An execution tree of A in a world $W \in \mathcal{W}(D)$ is a $\langle S, Act \rangle$-labeled tree $T = (\tau, \lambda, \delta)$ such that: (i) $\lambda(\epsilon) = s_0$ and for every leaf $x \in \tau$ we have $\lambda(x) \in F$, and (ii) for each edge (x, xn) of τ, there exists a guard $g \in G$ such that $(\lambda(x), \delta(x, xn), g, \lambda(xn)) \in \Delta$ and $g(W) = True$. We denote by $Tr(A, W)$ the set of execution trees of A in the world W.*
- *A $\langle S, Act \rangle$-labelled tree $T = (\tau, \lambda, \delta)$ is a **possible execution tree** of A iff $\exists W \in \mathcal{W}(D)$ and $\exists \lambda' : \tau \to S$ such that $T' = (\tau, \lambda', \delta) \in Tr(A, W)$. The probability of a possible execution tree T of A is: $Pr(T, A) = \sum\limits_{\substack{W \in \mathcal{W}(D) \\ T' \in Tr(A, W)}} Pr(W)$.*

We denote by $Tr(A)$ the set of all possible execution trees of a process A.

The set of execution trees of a given pd-process may be infinite. A possible execution tree T is simply an execution tree augmented with the probability of occurrence of T. Note that two execution trees are considered *equal* if they differ only w.r.t. the labels of their states (e.g., T is equal to T' in Definition 2). Hence, a probability of a possible execution tree T is calculated as the sum of the probabilities of the possible worlds to which T, modulo renaming of states, belongs. Fig.1 (b) shows the execution trees of the NormCalc pd-process of Fig.1 (a) and their corresponding probabilities.

Simulation preorder. The notion of *simulation* is used in the literature to compare LTSs with respect to their branching structures [8]. Usually simulation is defined as a relation between the states of the considered processes. In the case of pd-processes, and due to the tight connection between possible execution trees and possible worlds of a probabilistic database, it is not easy to provide such a structural definition (i.e., as a general relation between states). This is because, whether or not a given state s is simulated by another state s' depends on the considered possible world. As a consequence, instead of a structural definition, we provide below a semantic definition of simulation.

Definition 3 (Simulation relation between pd-processes). *Let*

$$A = (S, s_0, D, Act, G, \Delta, F) \text{ and } A' = (S', s_0', D', Act', G', \Delta', F')$$

be two pd-processes. Then, A is simulated by A', noted $A \lesssim A'$, iff: $\forall T = (\tau, \lambda, \delta) \in Tr(A), \exists \lambda' : \tau \to S'$ such that $T' = (\tau, \lambda', \delta) \in Tr(A')$, and $Pr(T, A) \le Pr(T', A')$.

Hence, semantics of simulation is defined as a containment between the sets of possible execution trees of the considered pd-processes, i.e., if a process A is simulated by a process A', then every possible execution tree of A is also a possible execution tree of A' (modulo renaming of states) with an equal or higher probability. It is worth noting that the provided definition of simulation is conservative in the sense that when it is applied to pd-processes with non-probabilistic databases (i.e., having probability of each tuple equal to 1), it matches non-probabilistic simulation on conventional LTSs.

Unfortunately, Definition 3 is *semantic* (i.e., it defines the meaning of simulation as a relation between possible execution trees) and not structural (i.e., does not define a relation between states and transitions of the processes). Therefore, there is no direct way to derive a simulation algorithm from such a definition (since testing inclusion between potentially infinite sets of possible execution trees is not feasible). As a consequence, we propose a method in the next section to decompose a pd-process into a set of (unguarded) automata that can be analyzed separately to structurally characterize simulation relation between pd-processes.

4   Computing Probabilistic Simulation

We describe below a refinement of a pd-process structure into a set of automata, called *world-partition automata*, that can be used to *structurally* characterize simulation. We recall that for a set G, the set 2^G denotes the power set of G (i.e., the set of all subsets of G).

Let $A = (S, s_0, D, Act, G, \Delta, F)$ be a pd-process with $G = \{q_1, \ldots, q_n\}$ a set of boolean queries used as guards of transitions in A. Let P_G be a set of boolean queries obtained as follows: (i) $\forall P \in 2^G$, $q_P := (\bigwedge_{q \in P} q) \wedge (\bigwedge_{q' \notin P} \neg q')$, and (ii) $P_G = \{q_P \mid P \in 2^G\}$. Table 2 presents the partitions of guards of the NormCalc pd-process of Fig.1(a).

Table 2. Example partitions

Process	Partitions	Associated query	P. worlds	Probability
NormCalc	P_1'	$q_{P_1'} = q_1' \wedge q_2'$	$\{W_4, \ldots\}$	$Pr(q_{P_1'}) = 16\%$
	P_2'	$q_{P_2'} = q_1' \wedge \neg q_2'$	$\{W_5, \ldots\}$	$Pr(q_{P_2'}) = 34\%$
	P_3'	$q_{P_3'} = \neg q_1' \wedge q_2'$	$\{W_2, \ldots\}$	$Pr(q_{P_3'}) = 16\%$
	P_4'	$q_{P_4'} = \neg q_1 \wedge \neg q_2$	$\{W_6, \ldots\}$	$Pr(q_{P_4'}) = 34\%$

Note that, the set P_G forms a partition of the possible worlds of the database D in the sense given by the following lemma.

Lemma 4. *Let* $A = (S, s_0, D, Act, G, \Delta, F)$ *and let* P_G *be the set of guards constructed as explained above. Then,* $\forall W \in \mathcal{W}(D)$, *there exists a unique* $q_P \in P_G$ *such that* $q_P(W) = true$.

The proof of this lemma is straightforward since, by construction of P_G, $\forall W \in \mathcal{W}(D)$ we have: (i) $\bigvee_{q_P \in P_G} q_P(W) = true$, and (ii) $\forall q_{pi}, q_{pj} \in P_G$, with $i \neq j$, then $q_{pi}(W) \wedge q_{pj}(W) = false$. Hence, each boolean query $q_P \in P_G$ identifies a unique subset of $\mathcal{W}(D)$ (i.e., the set $\{W \in \mathcal{W}(D) \mid q_P(W) = true\}$). In the sequel, we use the term *partition* q_P to refer to the subset of $\mathcal{W}(D)$ identified by q_P. We introduce below the notion of *world-partition automata* as a mean to split the behavior described by a given pd-process w.r.t. the possible worlds of the underlying probabilistic database. More precisely, the goal is to split the set of possible execution trees of a pd-process A into subsets of trees each of which is described by a distinct unguarded automaton.

Definition 5 (World-partition automata). *Let $A = (S, s_0, D, Act, G, \Delta, F)$ be a pd-process and let $P_G = \{q_{P_1}, \ldots, q_{P_n}\}$ defined as previously. A world-partition automata of A using P_G is a set of automata $A_{P_G} = \{A_{P_1}, \ldots, A_{P_n}\}$, where, $\forall q_{P_i} \in P_G$, a corresponding automaton $A_{P_i} = (S, s_0, D_{P_i}, Act, G_{P_i}, \Delta_{P_i}, F)$ is constructed from A as follows: (i) the components S, s_0, Act, and F, remain unchanged, (ii) the set of guards is: $G_{P_i} = \{true\}$ and the database is $D_{P_i} = \emptyset$, (iii) the set of transitions is: $\Delta_{P_i} = \{(s, a, true, s') \mid (s, a, g, s') \in \Delta \text{ and } g \in P_i\}$.*

The probability function Pr is extended to world-partition automaton as follows: $\forall A_P \in A_{P_G}$, then $Pr(A_P) = \sum_{\substack{W \in \mathcal{W}(D) \\ q_P(W)=true}} Pr(W)$

Hence, an automaton $A_{P_i} \in A_{P_G}$ is simply a copy of the process A from which are removed the transitions having a guard g satisfying the condition $Pr(g \wedge q_{P_i}) = 0$ (or equivalently, $\forall W \in \mathcal{W}(D) \mid g \wedge q_{P_i}(W) = \mathsf{false}$). Note that such a test can be achieved easily by checking whether $g \in P_i$ (since we have: $Pr(g \wedge q_{P_i}) = 0$ iff $g \notin P_i$). From item (ii) of this definition, each automaton $A_{P_i} \in A_{P_G}$ is an unguarded automaton (i.e., all its guards are set to true). Fig.2 (a) shows the world-partition automata of the NormCalc pd-process represented at Fig.1(a).

Intuitively, a world-partition automaton $A_P \in A_{P_G}$ of an automaton A describes the behavior of A in all the possible worlds belonging to the partition q_P. The following lemma makes explicit the connection between the behavior of a pd-process A and the behaviors described by its world-partition automata.

Lemma 6. *Let $A = (S, s_0, D, Act, G, \Delta, F)$ be a pd-process and let A_{P_G} its set of world-partition automata. Then: (i) let $W \in \mathcal{W}(D)$ be a possible world of D that belongs to a partition $q_P \in A_{P_G}$. Then, T is an execution tree of A in the world W iff $T \in Tr(A_P)$. (ii) $T \in Tr(A)$ with $Pr(T, A) > 0$ iff $\exists \{A_{P_{i_1}}, \ldots, A_{P_{i_l}}\} \subseteq A_{P_G}$ such that: $T \in Tr(A_{P_{i_j}})$, $\forall i_j \in \{i_1, \ldots, i_l\}$, and $Pr(T, A) = \sum_{i_j \in \{i_1, \ldots, i_l\}} Pr(A_{P_{i_j}})$*

Proof (lemma 6). The case (i)($\Leftarrow$) is straightforward.

(i) ($\Rightarrow$) Let $W \in \mathcal{W}(D)$ and $q_P \in A_{P_G} \mid q_P(W) = \mathsf{true}$ and let $T = (\tau, \lambda, \delta) \in Tr(A, W)$. By definition 2, $\forall(x, xn) \in \tau$, $\exists g \in G$ such that

$$(\lambda(x), \delta(x, xn), g, \lambda(xn)) \in \Delta \text{ and } g(W) = True.$$

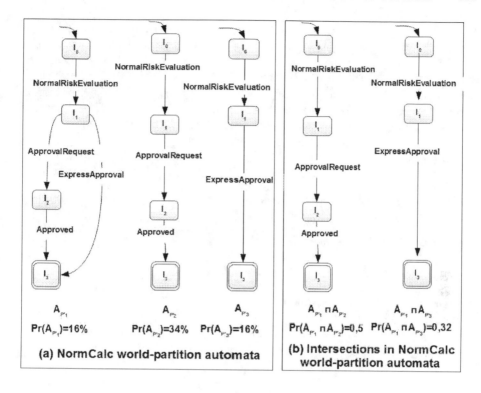

Fig. 2. Example of world-partition automata

Let G_T be the set containing such guards g. From lemma 4, we have $G_T \subseteq P$. Hence, $T \in Tr(A_P, W)$, by construction of A_P.

(ii)($\Rightarrow$) Assume $T \in Tr(A)$ with $Pr(T, A) > 0$. Hence, from definition 2, there exists $\{W_{i_1}, \ldots, W_{i_l}\} \subseteq \mathcal{W}(D) \mid T \in Tr(A, W_{i_j})$ and

$$Pr(T, A) = \sum_{i_j \in \{i_1, \ldots, i_l\}} Pr(W_{i_j}).$$

From (1), we can derive that $\exists\{A_{P_{i_1}}, \ldots, A_{P_{i_l}}\} \subseteq A_{P_G}$ such that $T \in Tr(A_{P_{i_j}})$, $\forall i_j \in \{i_1, \ldots, i_l\}$, and

$$Pr(T, A) = \sum_{i_j \in \{i_1, \ldots, i_l\}} Pr(A_{P_{i_j}}).$$

(ii)($\Leftarrow$) Assume that $\exists\{A_{P_{i_1}}, \ldots, A_{P_{i_l}}\} \subseteq A_{P_G}$ such that: $T \in Tr(A_{P_{i_j}})$, $\forall i_j \in \{i_1, \ldots, i_l\}$. Hence, $T \in Tr(A)$ and $Pr(T, A) > 0$ (from (1) and definition 2).

While a world-partition automata A_{P_G} enables to split an original process A into a set of automata that describe all the possible execution trees of A, it is still not easy to reason separately on elements of A_{P_G} to test simulation.

This is due to the fact that a probability of a possible tree T of A may be obtained from a subset of A_{P_G} (and not only from a unique element of A_{P_G}) (c.f., lemma 6). Such problematic execution trees belong to intersections of elements of A_{P_G}. Therefore, to characterize precisely the probabilities of every possible execution tree by a unique automata, there is a need to compute the closure of world-partition automata w.r.t. the intersection operation. The closure of world-partition automata, called closure-automata, is formally defined below after the introduction of some needed notation. Let $A = (S, s_0, D, Act, G, \Delta, F)$ be a pd-process and let A_{P_G} its corresponding world-partition automata. For a set $\psi \in 2^{A_{P_G}}$ (i.e., a subset of A_{P_G}), we define $A_{\sqcap_\psi} := \bigcap_{A_P \in \psi} A_P$. Therefore, $A_{\sqcap_\psi}$ is an unguarded automata which describes the behavior common to all the automata of the set ψ.

Definition 7. *(closure-automata) Let*

$$A = (S, s_0, D, Act, G, \Delta, F)$$

a pd-process and let A_{P_G} its corresponding world-partition automata. The closure of the world-partition automata of A is given by the set $C_L(A_{P_G}) = \{A_{\sqcap_\psi} \mid \psi \in 2^{A_{P_G}}\}$, where each transition system $A_{\sqcap_\psi} \in C_L(A)$ is associated with a probability distribution

$$Pr(A_{\sqcap_\psi}) = \sum_{A_P \in \psi} Pr(A_P).$$

From the definition of closure automata, we could easily find that a closure automaton is the intersection of several world-partition automata with the sum of probabilities. Fig.2 (b) presents the closure automata of the world-partition automata depicted at Fig.2(a). As a main technical result of this paper, the next theorem provides a structural characterization of the simulation relation between two pd-processes.

Theorem 8. *Let A and B be two pd-processes. Let $C_L(A_{P_G})$ be the closure-automata of A and let $B_{P_{G'}}$ be the world-partition automata of B. Then:*
$A \precsim B$ **iff** $\forall A_{\sqcap_\psi} \in C_L(A_{P_G})$, $Pr(A_{\sqcap_\psi}) \leq \sum_{\substack{B_{P'} \in B_{P_{G'}} \\ A_{\sqcap_\psi} \precsim B_{P'}}} Pr(B_{P'})$.

Proof (sketch).

($\Leftarrow$) Assume that $\forall A_{\sqcap_\psi} \in C_L(A_{P_G})$, we have $Pr(A_{\sqcap_\psi}) \leq \sum_{\substack{B_{P'} \in B_{P_{G'}} \\ A_{\sqcap_\psi} \precsim B_{P'}}} Pr(B_{P'})$ (i).

Let $T = (\tau, \lambda, \delta) \in Tr(A)$ with $Pr(T, A) > 0$.
By lemma 6, $\exists \{A_{P_{i_1}}, \ldots, A_{P_{i_l}}\} \subseteq A_{P_G}$ such that:

$$T \in Tr(A_{P_{i_j}}), \forall j \in \{1, \ldots, l\}, \text{ and } Pr(T, A) = \sum_{i_j \in \{i_1, \ldots, i_l\}} Pr(A_{P_{i_j}}).$$

Let $A_{\sqcap_\psi} := A_{P_{i_1}} \sqcap \ldots \sqcap A_{P_{i_l}}$.

From assumption (i), we derive: $\exists \{B_{P'_{i_1}}, \ldots, B_{P'_{i_k}}\} \subseteq B_{P'_G}$ such that $A_{\sqcap_\psi} \precsim B_{P'_{i_j}}$, $j \in [1, k]$, and $Pr(\sqcap_\psi) \leq \sum_{j \in [1,k]} Pr(B_{P'_{i_j}})$. Hence, $\forall B_{P'_{i_j}}$, with $j \in [1, k]$, $\exists \lambda' : \tau \to S'$ such that $T' = (\tau, \lambda', \delta) \in Tr(B_{P'_{i_j}})$. From lemma 6, $T' \in Tr(B)$ and $Pr(T', B) \geq \sum_{j \in [1,k]} Pr(B_{P'_{i_j}}) \geq Pr(T, A)$. Hence, $A \precsim B$ (by definition 3).

$(\Rightarrow)$ Assume that $A \precsim B$ and $\exists A_{\sqcap_\psi} \in C_L(A_{P_G})$ such that $Pr(A_{\sqcap_\psi}) \geq \sum_{\substack{B_{P'} \in B_{P_{G'}} \\ A_{\sqcap_\psi} \precsim B_{P'}}} Pr(B_{P'})$. In this case, the *maximal* execution tree of $A_{\sqcap_\psi}$ does not have any corresponding tree in B with enough probability. Hence, $A \nprecsim B$ (which contradict the assumption)

Complexity analysis. Let $A = (S, s_0, D_A, Act, G, \Delta, \Gamma)$ be a pd-process. We use $|X|$ to denote the cardinality of a set X. We extend this notation to pd-processes and we write $|A|$ to denote the size of the process A defined in terms of its total number of guards, transitions and states (i.e., $|A| = |S| + |\Delta| + |G|$). We use also the notation $|D_A|$ to denote the size of the probabilistic database used by A defined in terms of total number of tuples in D. We study the complexity of the problem of checking simulation between two pd-processes A and B w.r.t. two dimensions: (i) *expression complexity*, which assumes that $|D_A| + |D_B|$ is fixed while $|A| + |B|$ is variable, and (ii) *data complexity*, which assumes that $|A| + |B|$ is fixed while $|D_A| + |D_B|$ is variable.

Theorem 9. *Let A and B be two pd-processes. The problem of checking whether $A \precsim B$ is:*

(i) in $\mathcal{O}(f(|D|))$ in data complexity, where $f(|D|)$ is the data-complexity of computing the probabilities of a boolean query on a probabilistic database D,

(ii) EXPTIME-hard w.r.t. the expression complexity,

(iii) can be solved in 2-EXPTIME w.r.t. $|A| + |B|$.

The proof is omitted due to lack of space. It can be found in the extended version of this paper [10]. Therefore, checking simulation between pd-processes is intractable w.r.t. the size of the LTSs while, interestingly, it does not introduce additional overhead w.r.t. to probabilistic query evaluation in data-complexity. We refer to [6] for detailed results regarding the complexity of this latter problem (i.e., complexity of function $f(|D|)$) in the context of disjoint-independent databases. This result is encouraging because data-complexity is the most significant factor in our context. Indeed, the size of the database of a pd-process can be expected to be several order of magnitude higher than the size of its LTS.

5   Conclusion

We have proposed a framework based on probabilistic database theory to describe data-aware business processes and studied the problem of checking simulation in this context. We have shown that the proposed framework does not introduce overhead w.r.t. data-complexity of probabilistic query evaluation (the most important complexity measure in the setting of pd-processes). A prototype implementation is currently under development and will be used in experiments conducted in collaboration with the French Research Institute of Sciences and Technology for Environment and Agriculture (Irstea). Regarding future research directions, the expression complexity of the studied problem deserves a further analysis to get tighter bounds. Moreover, beyond the simulation problem investigated in this paper, our general purpose is to extend to pd-processes setting analysis and verification techniques that have already proved successful for non-probabilistic business processes (e.g., checking whether the executions of a pd-process satisfy some desirable properties expressed in some temporal logic).

References

1. Abiteboul, S., Hull, R., Vianu, V.: Foundations of Databases (1995)
2. Baier, C.: Polynomial time algorithms for testing probabilistic bisimulation and simulation. In: Alur, R., Henzinger, T.A. (eds.) CAV 1996. LNCS, vol. 1102, pp. 38–49. Springer, Heidelberg (1996)
3. Benatallah, B., Casati, F., Toumani, F.: Representing, analysing and managing web service protocols. DKE 58(3), 327–357 (2006)
4. Berardi, D., Cheikh, F., Giacomo, G.D., Patrizi, F.: Automatic service composition via simulation. IJFCS 19(2), 429–451 (2008)
5. Edmund, M.: Clarke and Bernd-Holger Schlingloff. Model checking. In: Handbook of Automated Reasoning, pp. 1635–1790. Elsevier and MIT Press (2001)
6. Dalvi, N.N., Suciu, D.: Management of probabilistic data: foundations and challenges. In: PODS, pp. 1–12 (2007)
7. Damaggio, E., Deutsch, A., Vianu, V.: Artifact systems with data dependencies and arithmetic. ACM TODS 37(3), 22:1–22:36 (2012)
8. van Glabbeek, R.J.: The Linear Time – Branching Time Spectrum (extended abstract). In: Baeten, J.C.M., Klop, J.W. (eds.) CONCUR 1990. LNCS, vol. 458, pp. 278–297. Springer, Heidelberg (1990)
9. Jonsson, B., Yi, W.: Testing preorders for probabilistic processes can be characterized by simulations. TCS 282(1), 33–51 (2002)
10. Li, H., Pinet, F., Toumani, F.: Probabilistic simulation for probabilistic data-aware business processes. Technical report, LIMOS, Ref:RR-13-14 (2013), limos.isima.fr/IMG/pdf/RR-13-14.pdf
11. Sen, P., Deshpande, A., Getoor, L.: Exploiting shared correlations in probabilistic databases. Proc. VLDB Endow. 1(1), 809–820 (2008)
12. Puhlmann, F., Weske, M.: Investigations on soundness regarding lazy activities. In: Dustdar, S., Fiadeiro, J.L., Sheth, A.P. (eds.) BPM 2006. LNCS, vol. 4102, pp. 145–160. Springer, Heidelberg (2006)
13. Segala, R., Lynch, N.A.: Probabilistic simulations for probabilistic processes. Nord. J. Comput. 2(2), 250–273 (1995)
14. Sokolova, A., De Vink, E.P.: Probabilistic automata: System types, parallel composition and comparison. In: Validation of Stochastic Systems: A Guide to Current Research, pp. 1–43 (2004)

15. Tranquillini, S., Spieß, P., Daniel, F., Karnouskos, S., Casati, F., Oertel, N., Mottola, L., Oppermann, F.J., Picco, G.P., Römer, K., Voigt, T.: Process-based design and integration of wireless sensor network applications. In: Barros, A., Gal, A., Kindler, E. (eds.) BPM 2012. LNCS, vol. 7481, pp. 134–149. Springer, Heidelberg (2012)
16. Wombacher, A.: A-posteriori detection of sensor infrastructure errors in correlated sensor data and business workflows. In: Rinderle-Ma, S., Toumani, F., Wolf, K. (eds.) BPM 2011. LNCS, vol. 6896, pp. 329–344. Springer, Heidelberg (2011)
17. Zhang, L.: Decision Algorithms for Probabilistic Simulations. Phd thesis (2008)

Expressiveness of Dynamic Networks of Timed Petri Nets

María Martos-Salgado and Fernando Rosa-Velardo [*]

Departamento de Sistemas Informáticos y Computación
Universidad Complutense de Madrid, C/Profesor José García Santesmases, Spain
mrmartos@estumail.ucm.es, fernandorosa@sip.ucm.es

Abstract. We study dynamic networks of infinite-state timed processes, where each process is a Petri net carrying a single real valued clock. We compare their expressiveness with other models within the class of Well-Structured Transition Systems, using coverability languages. We prove that unbounded places are a strict resource, meaning that extra unbounded places provides (strictly) with extra expressiveness. Also, we prove that if no unbounded places are allowed, then the obtained model is equivalent to Timed Petri nets. We conclude that dynamic networks of Timed Petri Nets are strictly more expressive than Timed Petri Nets.

1   Introduction

Petri nets, one of the best known models for distributed and concurrent systems, have been extended with time in many ways. A comparison of these extensions can be found e.g. in [6], where the authors prove that the class of Petri nets with time relative to arcs is the most expressive one. A formalism belonging to this class is that of Timed-Arc Petri Nets (*TdPN*) [4], in which tokens are endowed with a real-valued clock, that can be dynamically created and destroyed.

In [12] we extend the work in [4] by allowing dynamic process creation. We call the resulting model Timed-Arc ν-Petri nets (ν-*TdPN*).[1] Hence, our model manages infinitely-many timed processes, each of which is infinite-state (a potentially unbounded Petri net). This class can serve as the basis of parameterized verification of infinite-state timed processes. It is defined as an extension of an (untimed) model, called ν-*PN* [13], in which tokens are names, that can be created fresh and matched with other names. Names can be understood as identifiers of processes, that can be spawned and can synchronize with each other.

We consider that each process has a single real-valued clock (as in [4] under a counting abstraction). Each transition specifies which are the possible ages of the processes involved, and how this age is updated.[2]In [12] we successfully apply the theory of regions of [3]. More precisely, working with regions we prove that

---

[*] Authors supported by the Spanish projects STRONGSOFT TIN2012-39391-C04-04 and PROMETIDOS S2009/TIC-1465.
[1] This class is referred to as *locally synchronous* ν-Petri nets in [12].
[2] Actually, read-only constraints could be considered within the same setting.

A.-H. Dediu et al. (Eds.): LATA 2014, LNCS 8370, pp. 516–527, 2014.

ν-*TdPN* belong to the class of Well-Structured Transition Systems (WSTS) [8,1], for which coverability is decidable.

Several works [9,2,5,7] study the languages generated by different extensions of Petri Nets, by associating a label with each transition. A finite run of the net defines a word. Moreover, several acceptance conditions, like reachability, coverability or no condition, may be considered. These languages are commonly used to compare the expressiveness of different models.

Here, we compare the expressiveness of ν-*TdPN* with other well-structured models. In [9] coverability languages (those obtained with coverability as acceptance condition) are proposed as a measure to compare the expressiveness of WSTS. In [2,9,7,5] Petri nets (*PN*), Petri nets with transfers and resets (*AWN*), ν-*PN* and Data Nets (DN), an extension of ν-*PN* with ordered data, are compared, proving the following strict relations: $PN \prec AWN \prec \nu\text{-}PN \prec DN$. Moreover, *DN* and *TdPN* are proved to be equivalent in [5].

This paper puts ν-*TdPN* in the picture in two different ways. First, we prove that the class of ν-*TdPN* with no unbounded places is actually equivalent to *TdPN*. Second, we prove that unbounded places are strict resources, meaning that the class of ν-*TdPN* with k unbounded places is strictly less expressive that the class of ν-*TdPN* with $k + 1$ unbounded places. For that purpose we use the framework developed in [5]. In particular, we obtain that $TdPN \prec \nu\text{-}TdPN$, so that ν-*TdPN* is the most expressive model within the WSTS class, up to our knowledge, out of those whose relative expressive power has been studied.[3]

Outline. The rest of the paper is organized as follows. Section 2 gives notations and results we use throughout the paper. In Section 3 we define ν-*TdPN* and in Section 4 we study its expressiveness. In Section 5 we present our conclusions.

2   Preliminaries

Let $\mathbb{R}_{\geq 0} = [0, \infty)$ and $\mathbb{N} = \{0, 1, 2, \ldots\}$, and for $n \in \mathbb{N}$ let $n^+ = \{1, \ldots, n\}$ and $n^* = \{0, \ldots, n\}$. We denote by $\mathcal{I}$ the set of real open, closed and mixed intervals with natural endpoints (the right endpoint can be infinite). $(X, \leq)$ is a *partial order* (po) if $\leq$ is a reflexive, transitive and antisymmetric binary relation on X. A po $(X, \leq)$ is a *well partial order* (wpo) if for every infinite sequence $x_0, x_1, \ldots \in X$ there are i and j with $i < j$ st $x_i \leq x_j$.

Multisets and Words. A (finite) *multiset* m over X is a mapping $m : X \to \mathbb{N}$ st $\{x \in X \mid m(x) > 0\}$ is finite. We denote by $X^\oplus$ the set of multisets over X. We use set notation for multisets with repetitions to account for multiplicities. For $m_1, m_2 \in X^\oplus$ we define $m_1 + m_2 \in X^\oplus$ by $(m_1 + m_2)(x) = m_1(x) + m_2(x)$ and $m_1 \subseteq m_2$ if $m_1(x) \leq m_2(x)$ for every $x \in X$. When $m_1 \subseteq m_2$ we can define $m_2 - m_1 \in X^\oplus$ by $(m_2 - m_1)(x) = m_2(x) - m_1(x)$. We denote by $\emptyset$ the empty

---

[3] The expressive powers of WSTS based on trees or graphs have not been compared with others using the techniques in [9], up to our knowledge.

multiset, that is, $\emptyset(a) = 0$ for every $a \in A$. For any set X we denote by $X^{\circledast}$ the set of finite words over the alphabet X. The empty word is denoted by ϵ.

Labelled Transition Systems and WSTS. A *labelled transition system* is a tuple $\mathcal{S} = \langle X, \rightarrow, x_0, \Sigma \rangle$ where X is the set of states, $x_0 \in X$ is the initial state, $\epsilon \notin \Sigma$ and $\rightarrow \subseteq X \times (\Sigma \cup \{\epsilon\}) \times X$ is the transition relation. We write $x \xrightarrow{a} x'$ instead of $(x, a, x') \in \rightarrow$. A label ϵ denotes a silent transition. For $u \in \Sigma^{\circledast}$, we write $x \xrightarrow{u} x'$ if there is a sequence $x = x_0 \xrightarrow{a_1} x_1 \ldots x_{n-1} \xrightarrow{a_n} x_n = x'$ and $u = a_1 \ldots a_n$ (notice that some a_is may be ϵ). A *labelled well structured transition system* (WSTS) is a tuple $\mathcal{S} = \langle X, \rightarrow, x_0, x_f, \leq, \Sigma \rangle$, where $\langle X, \rightarrow, x_0, \Sigma \rangle$ is a labelled transition system, x_f is a final state, and $\leq$ is a wpo on X, st for all $x_1, x_2, x_1' \in X$ and $u \in \Sigma^{\circledast}$ st $x_1 \leq x_1'$ and $x_1 \xrightarrow{u} x_2$ there is $x_2' \in X$ such that $x_1' \xrightarrow{u} x_2'$ and $x_2 \leq x_2'$ (monotonicity). The coverability language of $\mathcal{S}$ is $L(\mathcal{S}) = \{u \in \Sigma^{\circledast} \mid x_0 \xrightarrow{u} x, \ x \geq x_f\}$.

Labelled Timed-Arc Petri Nets. A *Labelled Timed-Arc Petri Net* (*TdPN*) is a tuple $N = \langle P, T, F, H, \lambda, \Sigma \rangle$, where P and T are finite disjoint sets of places and transitions, respectively, $\lambda : T \rightarrow \Sigma \cup \{\epsilon\}$ and $F, H : P \times T \rightarrow \mathcal{I}^{\oplus}$. A marking of a *TdPN* is a finite multiset M over $P \times \mathbb{R}_{\geq 0}$. Abusing notation, we define $M(p)$ as the multiset of ages of tokens in place p at M. There are two types of transitions: timed transitions and discrete transitions. Given a marking $M = \{(p_1, d_1) \ldots, (p_n, d_n)\}$ we write $M \xrightarrow{\epsilon} M'$ if $M' = \{(p_1, d_1 + d) \ldots, (p_n, d_n + d)\}$ for some $d \geq 0$. Given $t \in T$ and a marking M we write $M \xrightarrow{\lambda(t)} M'$ if for each $p \in P$ with $F(p, t) = \{I_1, \ldots, I_n\}$ and $H(p, t) = \{J_1, \ldots, J_m\}$, there are $In = \{r_1, \ldots, r_n\}$ and $Out = \{r_1', \ldots, r_m'\}$ st: (i) $In \subseteq M'(p)$, (ii) $r_i \in I_i$ for any $i \in n^+$, (iii) $r_i' \in J_i$ for any $j \in m^+$ and (iv)$M'(p) = (M(p) - Out) + In$.

ν-Petri Nets. We fix infinite sets Id of names, Var of variables and a subset of variables $\Upsilon \subset Var$ for fresh name creation. A *ν-PN* [13] is a tuple $N = \langle P, T, F, H, \lambda, \Sigma \rangle$, where P and T are finite disjoint sets, $\lambda : T \rightarrow \Sigma \cup \{\epsilon\}$ and $F, H : T \rightarrow (P \times Var)^{\oplus}$ are the input and output functions, respectively. We say that $x \in Var(t)$ iff there is $p \in P$ with $(p, x) \in F(t) + H(t)$. A *marking* is a finite multiset over $P \times Id$. Intuitively, each name represents a different process. A *mode* is an injection $\sigma : Var(t) \rightarrow Id$. Modes are extended homomorphically to $(P \times Var(t))^{\oplus}$. A transition t is *enabled* with mode σ for a marking M if $\sigma(F(t)) \subseteq M$ and for every $\nu \in \Upsilon$, $(p, \sigma(\nu)) \notin M$ for any p. The last condition is used to create new names, not in the current marking. Then we have $M \xrightarrow{\lambda(t)} M'$, where $M' = (M - \sigma(F(t)) + \sigma(H(t))$. For an example, see Fig. 1 (for now, disregard the intervals in the arcs), in which places are represented by circles, transitions by rectangles, and F and H are represented by labelled arcs. Tokens are represented as names in places. Transition t can be fired from the marking in the second net, reaching the marking in the third one, with mode σ, with $\sigma(x) = a$, $\sigma(y) = b$ and $\sigma(\nu) = c$. In particular, the firing of t creates a new name c in p_4.

3   Dynamic Networks of Timed Petri Nets

Now we define Timed-Arc ν-*PN* (ν-*TdPN*). In ν-*TdPN* each process has a single clock. The age of the processes involved in the firing of a transition must be in the range determined by it. Also, these ages are updated according to the transition.

Definition 1 (Timed-Arc ν-PN). *A* Timed-Arc ν-*PN* (ν-*TdPN*) *is a tuple* $N = \langle P, T, F, H, \mathcal{G}, \lambda, \Sigma \rangle$, *where:*

- *P and T are finite disjoint sets,*
- $\lambda : T \to \Sigma \cup \{\epsilon\}$,
- *for $t \in T$, $F_t, H_t : Var \to P^\oplus$ are the input and output functions of t,*
- *for $t \in T$, $\mathcal{G}_t : Var \to \mathcal{I} \times \mathcal{I}$ is the time constraints function of t.*

For each $t \in T$ we define $Var(t) = \{x \in Var \mid F_t(x) + H_t(x) \neq \emptyset\}$, assumed to be finite, and we split it into $nfVar(t) = Var(t) \setminus \Upsilon$ and $fVar(t) = Var(t) \cap \Upsilon$. Focus in the first net of Fig. 1. The input and output functions are represented by labelled arcs. The figure represents a net with a transition t, with $F_t(x) = \{p_1\}$ and $H_t(x) = \{p_3\}$. Moreover, $\mathcal{G}(x) = ((0, 1], (2, 4))$ (analogously for y and ν).

Definition 2 (Markings). *A marking M of a ν-TdPN is an expression of the form* $a_1 : (m_1, r_1), \ldots, a_n : (m_n, r_n)$, *where* $Id(M) = \{a_1, \ldots, a_n\} \subset Id$ *are pairwise different names, and for each $i \in n^+$, $m_i \in P^\oplus$ and $r_i \in \mathbb{R}_{\geq 0}$.*

We treat markings of ν-*TdPN* as multisets over elements of the form $a:(m, r)$, which we call *instances* (or process instances). Hence, $a:(m, r)$ is an instance with name a, tokens according to m, and age r. For example, the marking represented in the first net of Fig. 1, consist of an instance with name a and an instance with name b, with tokens in p_1 and p_3, and p_1 and p_2, respectively. Moreover, the ages of the instances are 0 and 0.5. We use $M, M', \ldots$ to range over markings. Let us now define the semantics of ν-*TdPN*.

Definition 3 (Time delay). *Given* $M = a_1 : (m_1, r_1), \ldots, a_n : (m_n, r_n)$ *and* $d \in \mathbb{R}_{\geq 0}$, *we write M^{+d} to denote the marking* $a_1:(m_1, r_1+d), \ldots, a_n:(m_n, r_n+d)$, *in which the age of every instance has increased by d. We write $M \xrightarrow{\epsilon} M^{+d}$.*

Notice that time delays are silent transitions. Now we define the firing of transitions, for which we need the following notations. We denote by $\mathcal{G}_t^1(x)$ and $\mathcal{G}_t^2(x)$ the first and second component of $\mathcal{G}_t(x)$, respectively. Intuitively, for a transition to fire the instance corresponding to x must have an age in $\mathcal{G}_t^1(x)$ and this age is set to any value in $\mathcal{G}_t^2(x)$. We say M is the $\emptyset$-*contraction* of M' if M is obtained by removing every instance of the form $a:(\emptyset, r)$ from M'.

Definition 4 (Firing of transitions). *Let $t \in T$ with $nfVar(t) = \{x_1, \ldots, x_n\}$ and $fVar(t) = \{\nu_1, \ldots, \nu_k\}$. We say t is enabled at marking M if:*

- *There is a marking $\overline{M}$ st $M = a_1 : (m_1, r_1), \ldots, a_n : (m_n, r_n) + \overline{M}$,*
- *for each $i \in n^+$, $F_t(x_i) \subseteq m_i$ and $r_i \in \mathcal{G}_t^1(x_i)$.*

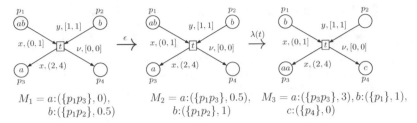

$$M_1 = a{:}(\{p_1p_3\}, 0), \qquad M_2 = a{:}(\{p_1p_3\}, 0.5), \qquad M_3 = a{:}(\{p_3p_3\}, 3), b{:}(\{p_1\}, 1),$$
$$b{:}(\{p_1p_2\}, 0.5) \qquad\qquad b{:}(\{p_1p_2\}, 1) \qquad\qquad c{:}(\{p_4\}, 0)$$

Fig. 1. Firing of a transition in a ν-$TdPN$

Then, t can be fired, *and taking*

- $\{b_1, \ldots, b_k\}$ *pairwise different names not in $Id(M)$,*
- $m_i' = (m_i - F_t(x_i)) + H_t(x_i)$ *for all $i \in n^+$,*
- $m_j'' = H_t(\nu_j)$ *for all $j \in k^+$,*
- r_i' *any value in $\mathcal{G}_t^2(x_i)$, for all $i \in n^+$,*
- r_j'' *any value in $\mathcal{G}_t^2(\nu_j)$, for all $j \in k^+$,*

we can reach M', denoted by $M \xrightarrow{\lambda(t)} M'$, where M' is the $\emptyset$-contraction of

$$a_1{:}(m_1', r_1'), \ldots, a_n{:}(m_n', r_n'), b_1{:}(m_1'', r_1''), \ldots, b_k{:}(m_k'', r_k'') + \overline{M}$$

We implicitly assume an initial marking M_0 with no empty instances, thus obtaining the transition system induced by N. Since we are taking the $\emptyset$-contraction, instances in reachable markings are not empty.

Example 5. Fig. 1 depicts a ν-$TdPN$ with three different markings. In the first marking the transition t is not fireable, because no instance with an age in $[1, 1]$ has a token in place p_2, and therefore, the second enabling condition is not fulfilled. However, after waiting 0.5 units of time, the marking M_2 is reached, and t becomes enabled. Then, we can fire t reaching, for example, the marking M_3 in the figure by assigning a to x and b to y.

The state space of ν-$TdPN$ is infinite in various dimensions. It encompasses an unbounded number of (unbounded) instances, each with a real clock. Moreover, the induced transition system is not finitely branching, since any marking has infinitely-many successors due to arbitrary time delays. In [12], we use the theory of regions [3,4] to obtain a finitary representation of each ν-$TdPN$. We fix a ν-$TdPN$ $N = \langle P, T, F, H, \mathcal{G}, \lambda, \Sigma \rangle$ and denote by $\max_N$ (or simply $\max$) the maximum integer bound appearing in $\mathcal{G}$ and $n^* \cup \{\infty\}$ as n_∞^*.

In order to obtain the region associated to a marking, we partition the instances in M into three multisets:

- The multiset M_1 of instances with an integer age of at most max,
- The multiset M_2 of instances younger than max, with a non-integer age,
- The multiset M_3 of instances older than max.

We put instances of M_1 in A_0 with their ages (and forgetting their names). We keep in $A_1 \ldots A_n$ the instances in M_2, ordered according to the fractional part of their ages, and storing only their integer part. Finally, A_∞ absorbs all instances in M_3, losing the information about their concrete age.

Hence, a *region* is an expression $A_0 * A_1 * \ldots * A_n * A_\infty$ with $n \geq 0$, and $A_i \in (P^\oplus \times I_i)^\oplus$ for every $i \in n_\infty^*$ and $I_0 = \max^*$, $I_i = (max - 1)^*$ for $i \in n^+$ and $I_\infty = \{\max + 1\}$. For instance, $A_0 \in (P^\oplus \times \max^*)^\oplus$ and $A_1 \in (P^\oplus \times (max - 1)^*)^\oplus$. We assume $A_i \neq \emptyset$ for any $i \in n^+$, and $m \neq \emptyset$ for all $(m, r) \in A_i$, for any $i \in n_\infty^*$. Each marking M of a ν-*TdPN* has a region associated to it.

Example 6. Let $M = a : (\{pq\}, 1.5)$, $b : (\{p\}, 2)$, $c : (\{ppq\}, 4.5)$, $d : (\{q\}, 2.3)$, $e : (\{qq\}, 0.3)$ be a marking of a ν-*TdPN* with max $= 3$. The region of M is $A_0 * A_1 * A_2 * A_\infty$ with $A_0 = \{(\{p\}, 2)\}$, $A_1 = \{((\{qq\}, 0), (\{q\}, 2)\}$, $A_2 = \{((\{pq\}, 1)\}$ and $A_\infty = (\{ppq\}, 4)$. In A_0 we store the instance with an integer age, in A_1 and A_2 we store instances younger than max with ages of fractional part 0.3 and 0.5 respectively, and in A_∞ we store the instance older than max.

In [12] we use these regions to obtain finitary transition systems over countable domains and we prove that the transition systems are WSTS. In particular, we can solve the control-state reachability problem (whether a given place can be marked) by reducing it to a coverability problem.

Proposition 7 ([12]). ν-*TdPN* are WSTS.

4   Expressiveness Results

In this section we prove that ν-*TdPN* are strictly more expressive than *TdPN*. We compare classes of WSTS by comparing the families of coverability languages they accept, as advocated for instance in [9,2].

For two classes of WSTS, $\mathbf{S}_1$ and $\mathbf{S}_2$, we write $\mathbf{S}_1 \preceq \mathbf{S}_2$ whenever for every $\mathcal{S}_1 \in \mathbf{S}_1$ there is $\mathcal{S}_2 \in \mathbf{S}_2$ st $L(\mathcal{S}_1) = L(\mathcal{S}_2)$. We write $\mathbf{S}_1 \simeq \mathbf{S}_2$ when $\mathbf{S}_1 \preceq \mathbf{S}_2$ and $\mathbf{S}_2 \preceq \mathbf{S}_1$, and we write $\mathbf{S}_1 \prec \mathbf{S}_2$ if $\mathbf{S}_1 \preceq \mathbf{S}_2$ and $\mathbf{S}_2 \npreceq \mathbf{S}_1$. In [2,9,5] the following relations are proved: $PN \prec AWN \prec \nu$-$PN \prec DN \simeq TdPN$.

In order to obtain a finer analysis of the expressiveness of ν-*TdPN* we partition that class into $\bigcup_{k \geq 0} \nu$-$TdPN_k$, where ν-$TdPN_k$ denotes the class of ν-*TdPN* with at most k unbounded places.[4] A place p is *bounded* if there is some $b \in \mathbb{N}$ st every instance $a : (m, r)$ satisfies $m(p) \leq b$ in every reachable marking.[5] If a ν-*TdPN* has P as set of unbounded places and m bounded places, we can represent each instance as an element of $Q \times P^\oplus$ with $Q = \{0, \ldots, b\}^m$.

Let us first see that ν-$TdPN_0$ is at least as expressive as *TdPN*.

Proposition 8. $TdPN \preceq \nu$-$TdPN_0$

---

[4] Alternatively, we could consider the class with exactly k unbounded places, though we claim these two classes are equivalent with respect to coverability languages.

[5] This is actually an undecidable problem [13], though this is not important for the study of expressiveness.

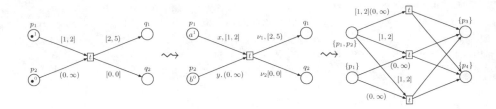

Fig. 2. Illustrating Prop. 2 and Prop. 3

Proof. Given $N \in TdPN$ we build $N' \in \nu\text{-}TdPN_0$ st $L(N) = L(N')$. The net N' has the same sets of places and transitions (with the same labels) as N, respectively. We simulate a token in p with age r by an instance with a single token in p, and with age r. Each transition is simulated by a transition (with the same label, hence accepting the same language) that (i) removes instances/tokens with clocks with the proper values and (ii) creates *fresh* instances, again with clocks with the proper values. Therefore, for each input place p_i of a transition t of N labelled by an interval I, we add to N' an arc labelled by (r_i, I), as depicted in Fig. 2 (first and second nets). Analogously, for each output place q_j of a transition t of N labelled by an interval I, we add to N' an arc labelled by (ν_j, I) (to represent the creation of new tokens, we create new instances). If the initial marking of the $TdPN$ is $\{(p_1, r_1), \ldots, (p_n, r_n)\}$ we consider $a_1 : (p_1, r_1), \ldots, a_n : (p_n, r_n)$ (for arbitrary $a_1, \ldots, a_n$) as initial marking of the $\nu\text{-}TdPN$ (analogously for the final marking). Each instance in N' has exactly one token, so that $N' \in \nu\text{-}TdPN_0$. □

The converse of the previous result is also true.

Proposition 9. $\nu\text{-}TdPN_0 \preceq TdPN$

Proof. Let $N \in \nu\text{-}TdPN_0$, so that each instance in each reachable marking is given by a control-state $q \in Q$ and the value of its clock $r \in \mathbb{R}_{\geq 0}$. We perform a standard counting abstraction (see Fig. 2, second and third nets): we build a $TdPN$ with Q as set of places, so that each token (with a clock value) in a place q represents an instance in state q (with the same clock value). Then, a transition $t \in T$ is simulated by consuming a token from q (with a legal clock value) for each instance that has to be in state q in order to fire t in N (and analogously for postconditions). □

Corollary 10. $TdPN \simeq \nu\text{-}TdPN_0$

Now we prove that $\nu\text{-}TdPN \npreceq TdPN$, ie, that the language of some $\nu\text{-}TdPN$ cannot be obtained as the language of any $TdPN$. We do it indirectly, by proving that $\nu\text{-}TdPN_k \prec \nu\text{-}TdPN_{k+1}$, which by the previous corollary in particular implies that $TdPN \prec \nu\text{-}TdPN_1$. Clearly, $\nu\text{-}TdPN_k \preceq \nu\text{-}TdPN_{k+1}$ holds. Let us prove that $\nu\text{-}TdPN_{k+1} \npreceq \nu\text{-}TdPN_k$

In [5] a framework for the strict comparison of WSTS is developed. This framework is based on two concepts: *reflections* and *witness* languages, which can be used to prove non-inclusions of families of coverability languages. Let $(X, \leq_X)$ and $(Y, \leq_Y)$ be two wpos. A mapping $\varphi : X \to Y$ is a reflection if $\varphi(x) \leq_Y \varphi(x')$ implies $x \leq_X x'$ for all $x, x' \in X$. A reflection is an *isomorphism* if it is bijective and $x \leq_X x'$ implies $\varphi(x) \leq_Y \varphi(x')$ (note that this is the symmetric property of monotonicity). We write $X \sqsubseteq_{refl} Y$ if there is a reflection from X to Y. We extend the relation $\sqsubseteq_{refl}$ to classes of wpo by $\mathbf{X} \sqsubseteq_{refl} \mathbf{X}'$ if for any $X \in \mathbf{X}$ there is $X' \in \mathbf{X}'$ st $X \sqsubseteq_{refl} X'$.

Witness languages represent the capability of a WSTS to recognize a state space. They are useful to prove strict relations between classes of WSTS because they can be proven *not* to be recognizable by some classes of WSTS.

Given an alphabet $\Sigma = \{\mathbf{a}_1, \ldots, \mathbf{a}_k\}$, we consider a disjoint copy $\overline{\Sigma} = \{\overline{\mathbf{a}_1}, \ldots, \overline{\mathbf{a}_k}\}$. This notation is extended to words and languages, as expected. A Σ-*representation* of a wpo X is any surjective partial function $\gamma : \Sigma^{\circledast} \to X$. Intuitively, every $u \in \Sigma^{\circledast}$ with $\gamma(u) = x$ is a possible encoding or representation of $x \in X$. We denote by $dom(\gamma)$ the domain of γ. For a Σ-representation γ of X, we define $L_\gamma = \{u\overline{v} \mid u, v \in dom(\gamma) \text{ and } \gamma(v) \leq \gamma(u)\}$, and we say L_γ is a witness of X. The fact that a WSTS can recognize such L_γ *witnesses* that it can represent the structure of X: it can accept all words starting with some u (representing some state $\gamma(u)$), followed by some v that represents $\gamma(v) \leq \gamma(u)$. In particular, it must be able to accept $u\overline{u}$ for any $u \in dom(\gamma)$.

Example 11. Let $X = Q \times \mathbb{N}$, with Q finite, with its standard order $\leq$ ($(q, n) \leq (p, m)$ iff $q = p$ and $n \leq m$). Taking $\Sigma = \{a\} \cup Q$, a Σ-representation of X is $\gamma : \Sigma^{\circledast} \to X$ with $\gamma(qa^n) = (q, n)$, so $L_\gamma = \{qa^n\overline{q}\overline{a}^m \mid m \leq n\}$ is a witness of X.

In order to apply the framework to two classes of WSTS we must prove that both classes are *self-witnessing*. A class of WSTS $\mathbf{S}$ is self-witnessing if it can accept encodings (over some alphabet) of their state spaces. Formally, if $\mathbf{S}$ is a class of WSTS whose state spaces are included in the class of wpos $\mathbf{X}$, $(\mathbf{X}, \mathbf{S})$ is *self-witnessing* if, for all $X \in \mathbf{X}$, there is $S \in \mathbf{S}$ that recognizes a witness of X.

Proposition 12 ([5]). *Let $(\mathbf{X}, \mathbf{S})$ and $(\mathbf{X}', \mathbf{S}')$ be self-witnessing WSTS classes. If $\mathbf{S} \preceq \mathbf{S}'$ then $\mathbf{X} \sqsubseteq_{refl} \mathbf{X}'$.*

Next we define the state space of ν-*TdPN* (using regions) in terms of standard set constructions, like products, multisets or words.

Definition 13. *We define $\mathbf{X}_k$ as the class of sets $X_{\max^*}^{\oplus} \times (X_{(\max-1)^*}^{\oplus})^{\circledast} \times X_{\{\max+1\}}^{\oplus}$ for some $\max \in \mathbb{N}$ and P, Q finite sets with $|P| = k$, where for every $I \subseteq (\max+1)^*$, $X_I = Q \times P^{\oplus} \times I$.*

As an example, consider the region in Ex. 6, assuming q is the only bounded place, bounded by 2. We take $Q = \{q_0, q_1, q_2\}$, where q_i state with i tokens in q, and $P = \{p\}$. Then, we can write this region as

$$(\{(q_0, \{p\}, 2)\}, \{(q_2, \emptyset, 0), (q_1, \emptyset, 2)\}\{(q_1, \{p\}, 1)\}, \{(q_1, \{p^2\}, 4)\})$$

that belongs to $X_{\max^*}^{\oplus} \times (X_{(\max-1)^*}^{\oplus})^{\circledast} \times X_{\{\max+1\}}^{\oplus}$, for $\max = 3$.

By lack of space we do not show in all detail the order in $\mathbf{X}_k$. It is the standard order induced by the equality over finite sets in products, multisets and words (see [5] for details). For $X \in \mathbf{X}_k$, we will write Q_X, P_X and max_X to refer to Q, P and max as above (or just Q, P and max, abusing notation). We can represent each region as an element of some $X \in \mathbf{X}_k$. If $R = A_0 * A_1 * \ldots * A_n * A_\infty$ is a region, A_0 is represented by an element of $X_{max*}^\oplus$, $A_1 * \ldots * A_n$ by a word over the alphabet $X_{(max-1)*}^\oplus$ and A_∞ by an element of $X_{\{max+1\}}^\oplus$. In particular, each $(m, r) \in A_i$ is represented by some $(q, \{p_1, \ldots, p_n\}, k) \in Q \times P^\oplus \times I$, where $q \in Q$ is a control state corresponding to the bounded part of m. In order to apply Prop. 12 to prove $\nu\text{-}TdPN_{k+1} \not\preceq \nu\text{-}TdPN_k$ we have to see that every $(\mathbf{X}_k, \nu\text{-}TdPN_k)$ is self-witnessing and that $\mathbf{X}_{k+1} \not\sqsubseteq_{refl} \mathbf{X}_k$.

Now, given $X \in \mathbf{X}_k$, we define a Σ-representation γ_X of X, and therefore, we obtain a witness L_{γ_X} of X. We need auxiliary functions γ_1^I, γ_2^I and γ_3^I.

Definition 14. *Given $X \in \mathbf{X}_k$, let $\Sigma = Q \cup P \cup (max+1)^* \cup \{*, \#, \&\}$. We define $\gamma_1^I : \Sigma^\circledast \to X_I$, $\gamma_2^I : \Sigma^\circledast \to X_I^\oplus$ and $\gamma_3^I : \Sigma^\circledast \to (X_I^\oplus)^\circledast$ as follows:*

- *$\gamma_1^I(qp_1 \ldots p_n k) = (q, \{p_1, \ldots, p_n\}, k)$, with $p_i \in P$, $q \in Q$ and $k \in I$,*
- *$\gamma_2^I(u_1 \# \ldots \# u_n) = \{\gamma_1^I(u_1), \ldots, \gamma_1^I(u_n)\}$, with $u_i \in dom(\gamma_1^I)$ for every i,*
- *$\gamma_3^I(v_1 * \ldots * v_n) = \gamma_2^I(v_1) \ldots \gamma_2^I(v_n)$, where $v_i \in dom(\gamma_2^I)$ for every i.*

Finally, we define the partial function $\gamma_X : \Sigma^\circledast \to X$ as

$$\gamma_X(u \& v \& w) = (\gamma_2^{max^*}(u), \gamma_3^{(max-1)^*}(v), \gamma_2^{\{max+1\}}(w))$$

γ_X is surjective, so that it is a Σ-representation of X and L_γ is a witness of X. Although not explicitly mentioned in the results from [5] shown above, if $X \in \mathbf{X}_{k+1}$ and $\mathbf{X}_{k+1} \not\sqsubseteq_{refl} \mathbf{X}_k$ then L_{γ_X} proves that $\nu\text{-}TdPN_{k+1} \not\preceq \nu\text{-}TdPN_k$.

Proposition 15. *$(\mathbf{X}_k, \nu\text{-}TdPN_k)$ is self-witnessing.*

Proof. Given $X \in \mathbf{X}_k$ we have to prove that there is $N \in \nu\text{-}TdPN_k$ st $L(N) = L_{\gamma_X}$. Notice that N must have at most k unbounded places. N operates in two phases: the first phase generates u with $\gamma(u) = R = A_0 * A_1 * \ldots * A_n * A_\infty$, and the second one recognizes any $\overline{v}$ with $\gamma(v) \leq R$. In turn, each of the phases has three consecutive sub-phases, dealing with A_0, $A_1 * \ldots * A_n$ and A_∞, respectively. We use control places to move from one subphase to the next (with transitions labeled by $\&$ or $\overline{\&}$). In order to differentiate between phases, we say that we generate words in the first one, but we recognize them in the second.

We explain the generation of $A_1 * \ldots * A_n$ (the other phases are simpler). Let $A_i = \{(q_1^i, m_1^i, k_1^i), \ldots, (q_{n_i}^i, m_{n_i}^i, k_{n_i}^i)\}$. We use a different name to represent each process instance. Moreover, instances in the same A_i have the same age. We use a place now that holds the name (with age 0) of the instance currently being generated. For a given i, we start by firing a transition of the form t_q (labelled by $q \in Q$) which copies the name in now to a place q. This firing is followed by the (possibly multiple) firing of transitions t_p (labelled by $p \in P$), each copying the name in now to a place p. These firings are followed by the firing of some t_k (labelled by $k \in (max-1)^*$), which copies the name in now to a place k.

Therefore, a word u with $\gamma_1^{(\max-1)^*}(u) = (q_1^i, m_1^i, k_1^i) \in X_{(\max-1)^*}$ can be produced. Next, the name in now is moved to a place all, and replaced by a fresh name, with age 0 (transition $t_\#$, labeled by $\#$). These actions can be repeated to generate (the encoding of) any element in $X_{(\max-1)^*}^\oplus$. Notice that they all demand that the instance involved has age 0. At any point, instead of firing $t_\#$ we can fire t_* (labeled by $*$), which has the same effect as $t_\#$, but it is only enabled if the instance in now has a non-null age, so that some time must elapse. Hence, we start accepting the instances with a higher fractional part, in A_{i+1}.

After this phase, there is any number of different names (each representing an instance) in all, some of which have the same age (those instances with an age with the same fractional part). Moreover, for any a in all, a belongs to some of the places in P, and exactly to one place q and one place k. The transitions in the second phase (the recognizing phase) demand that the age of the instances involved is exactly 1. Moreover, they are all labeled with symbols in $\overline{\Sigma}$.

This phase starts by taking any name in all and putting it back to now. First, a transition of the form $\overline{t}_q$ (labelled by $\overline{q}$) can remove a token from q with the same name as the one in now. Then, transitions of the form $\overline{t}_p$ (labeled by $\overline{p}$) can be fired, each consuming a name from p matching the name in now. At any point, a transition of the form $\overline{t}_k$, labeled by $\overline{k}$ (with $k \in (\max-1)^*$) can be fired, which consumes from k a name matching the one in now. Thus, if the current name represented an instance (q, m, k) then (an encoding of) any (q, m', k) with $m' \subseteq m$ can be recognized. The name in now can be replaced by a name taken from all (transition $\overline{t}_\#$ labeled by $\overline{\#}$) in order to recognize the next instance.

At any point, time can elapse, so that another instance in now reaches age 1. Then, t_* can be fired, labeled by $\overline{*}$, with the same effect as $\overline{t}_\#$. Notice that when time elapses, all the names with age greater than 1 are lost (the encodings of the instances they represent cannot be recognized). This is consistent with the fact that we must recognize (the encoding of) a state which is $less\ or\ equal$ than the one we generated. Notice also that even in the first phase, names with ages older than 1 become garbage. However, it is possible to generate all the names in the first phase with an age smaller than 1, so that the same state can be recognized.

The order between instances is not preserved within each A_i (this is not demanded by the order in $X_{(\max-1)^*}^\oplus$), but it is preserved between different A_i's, since older instances reach the age of 1 before. To conclude, we consider as final marking the one with a token in the control-state marked in the second phase (the recognizing one). Note that the built net has $max = 1$.   □

Thus, to apply Prop. 12 we only have to see that $\mathbf{X}_{k+1} \not\sqsubseteq_{refl} \mathbf{X}_k$. In order to prove it we use ordinal theory (see Prop. 16 below). Let us explain the needed concepts about ordinals (for more details see [5]). Each ordinal α is equal to the set of ordinals $\{\beta \mid \beta < \alpha\}$ below it, and the class of ordinals is totally ordered by inclusion. Every total well order $(X, \leq)$ is isomorphic to a unique ordinal $ot(X, \leq)$, called the $order\ type$ of X. In the context of ordinals, we define $0 = \emptyset$, $n = \{0, \ldots, n-1\}$ and $\omega = \mathbb{N}$, ordered by the usual order. The ordinals below ε_0 (those bounded by a tower $\omega^{\omega^{\cdot^{\cdot^{\cdot^{\omega}}}}}$) can be represented by the hierarchy of ordinals in Cantor Normal Form (CNF), recursively given by $C_0 = \{0\}$, and $C_{n+1} = \{\omega^{\alpha_1} + \ldots + \omega^{\alpha_p} \mid p \in \mathbb{N}, \alpha_1, \ldots, \alpha_p \in C_n$ and $\alpha_1 \geq \ldots \geq \alpha_p\}$ ordered by $\omega^{\alpha_1} + \cdots + \omega^{\alpha_p} \leq \omega^{\alpha'_1} + \cdots + \omega^{\alpha'_q}$ iff $(\alpha_1, \ldots, \alpha_p) \leq_{lex} (\alpha'_1, \ldots, \alpha'_q)$, where $\leq_{lex}$ is the lexicographic order.

Each ordinal below ε_0 has a unique CNF. We abbreviate $\alpha + \overset{k}{\cdots} + \alpha = \alpha * k$. A *linearization* of a po $\leq$ is a total order $\leq'$ st $x \leq y \Rightarrow x \leq' y$. A linearization of a wpo is well and total, hence isomorphic to an ordinal. The *maximal order type* of $(X, \leq)$ is $ot(X, \leq) = sup \{ot(X, \leq') \mid \leq'$ linearization of $\leq\}$. The following result states that we can prove $\mathbf{X}_{k+1} \not\sqsubseteq_{refl} \mathbf{X}_k$ by comparing their ordinal types.

Proposition 16 ([15]). *For X and Y wpos, if $X \sqsubseteq_{refl} Y$ then $ot(X) \leq ot(Y)$.*

Using [11,14,15], we can compute the order type of products, domains of finite words or finite multisets. In particular, we will need the following result.

Lemma 17. *For every Q, P and I finite, $ot((Q \times P^{\oplus} \times I)^{\oplus \circledast}) = \omega^{\omega^{\omega^{\omega^{|P|}*|Q|*|I|}}}$. In particular, $ot(P^{\oplus \oplus \circledast}) = \omega^{\omega^{\omega^{\omega^{|P|}}}}$.*

Proof. Let X be any wpo with $\omega \leq ot(X) < \varepsilon_0$ and Y a finite set. In [11,14] it is proved that $ot(X \times Y) = ot(X) * |Y|$ and $ot(X^{\circledast}) = \omega^{\omega^{ot(X)}}$, and [15] proves that $ot(X^{\oplus}) = \omega^{ot(X)}$. Moreover, $P^{\oplus}$ is isomorphic to $\mathbb{N}^{|P|}$, so that $ot(P^{\oplus}) = ot(\mathbb{N}^{|P|}) = \omega^{|P|}$. This allows us to compute the ordinals in the lemma. $\square$

Proposition 18. *$\nu\text{-}TdPN_k \prec \nu\text{-}TdPN_{k+1}$ for each $k \geq 0$.*

Proof. Trivially $\nu\text{-}TdPN_k \preceq \nu\text{-}TdPN_{k+1}$ holds. In order to prove $\nu\text{-}TdPN_{k+1} \not\preceq \nu\text{-}TdPN_k$ it is enough to apply Prop. 12, since both classes are self-witnessing (Prop. 15). Let us see that $\mathbf{X}_{k+1} \not\sqsubseteq_{refl} \mathbf{X}_k$. We consider only the part of the state spaces composed of words (of multisets), the one playing the relevant part. Let us take $X_{k+1} = P^{\oplus \oplus *} \in \mathbf{X}_{k+1}$ (i.e., with only one control-state and $max = 1$), so that $|P| = k+1$. For any $X_k \in \mathbf{X}_k$ we have that $X_k = (Q \times P'^{\oplus} \times I)^{\oplus *}$ for some $max \in \mathbb{N}$ and finite P' and Q with $|P'| = k$. By the previous lemma, $ot(X_{k+1}) = \omega^{\omega^{\omega^{\omega^{k+1}}}}$ and $ot(X_k) = \omega^{\omega^{\omega^{\omega^k * |I| * |Q|}}}$, which satisfy $ot(X_{k+1}) \not\leq ot(X_k)$. Since this is true for any $X_k \in \mathbf{X}_k$ we have that $\mathbf{X}_{k+1} \not\sqsubseteq_{refl} \mathbf{X}_k$. $\square$

Corollary 19. *$TdPN \simeq \nu\text{-}TdPN_0 \prec \nu\text{-}TdPN_1 \prec \nu\text{-}TdPN_2 \prec \ldots \prec \nu\text{-}TdPN$*

5   Final Remarks and Future Work

In this paper we have compared $\nu\text{-}TdPN$ with other classes of WSTS, proving that it is the most expressive of the studied classes. In particular, we have proved that $TdPN \prec \nu\text{-}TdPN$ by applying the framework in [5]. Interestingly, the number of unbounded places is a strict resource and the class $\nu\text{-}TdPN_0$, with no unbounded places, is equivalent to $TdPN$ with respect to coverability languages.

The state space of a ν-$TdPN$ is determined by three factors, the sizes of P and Q, and max (as denoted in the paper). In the analysis of the expressiveness, we have only used the size of P, and not the size of Q or max. Actually, the computation of the ordinal types in the proof of Prop. 18 suggests that max plays the same role as the number of control-states. Thus, either max plays no role in the expressiveness of ν-$TdPN$, or it does, but it cannot be proved by an analysis of the order types. We believe that any ν-$TdPN$ with $max = 0$ can be simulated by a ν-PN extended with broadcast primitives.

Regarding complexity, since ν-$TdPN$ are more expressive than $TdPN$, the complexity of the control-state reachability problem is non-primitive recursive. It would be interesting to obtain a finer-grained complexity analysis, as in [10].

References

1. Abdulla, P.A., Cerans, K., Jonsson, B., Tsay, Y.K.: Algorithmic analysis of programs with well quasi-ordered domains. Inf. Comput 160(1-2), 109–127 (2000)
2. Abdulla, P.A., Delzanno, G., Begin, L.V.: A classification of the expressive power of well-structured transition systems. Inf. Comput 209(3), 248–279 (2011)
3. Abdulla, P.A., Jonsson, B.: Verifying networks of timed processes (extended abstract). In: Steffen, B. (ed.) TACAS 1998. LNCS, vol. 1384, pp. 298–312. Springer, Heidelberg (1998)
4. Abdulla, P.A., Nylén, A.: Timed Petri nets and bqos. In: Colom, J.-M., Koutny, M. (eds.) ICATPN 2001. LNCS, vol. 2075, pp. 53–70. Springer, Heidelberg (2001)
5. Bonnet, R., Finkel, A., Haddad, S., Rosa-Velardo, F.: Ordinal theory for expressiveness of well structured transition systems. Inf. Comput. (2012)
6. Boyer, M., Roux, O.H.: On the compared expressiveness of arc, place and transition time Petri nets. Fundam. Inform. 88(3), 225–249 (2008)
7. Delzanno, G., Rosa-Velardo, F.: On the coverability and reachability languages of monotonic extensions of petri nets. Theor. Comput. Sci. 467, 12–29 (2013)
8. Finkel, A., Schnoebelen, P.: Well-structured transition systems everywhere! Theor. Comput. Sci. 256(1-2), 63–92 (2001)
9. Geeraerts, G., Raskin, J.-F., Begin, L.V.: Well-structured languages. Acta Inf. 44(3-4), 249–288 (2007)
10. Haddad, S., Schmitz, S., Schnoebelen, P.: The ordinal-recursive complexity of timed-arc Petri nets, data nets, and other enriched nets. IEEE LICS, 355–364 (2012)
11. de Jongh, D.H.J., Parikh, R.: Well partial orderings and hierarchies. Indagationes Mathematicae, 195–207 (1977)
12. Martos-Salgado, M., Rosa-Velardo, F.: Dynamic networks of infinite-state timed processes. Technical Report 9/13, UCM (2013), http://antares.sip.ucm.es/~frosa/
13. Rosa-Velardo, F., de Frutos-Escrig, D.: Decidability and complexity of Petri nets with unordered data. Theor. Comput. Sci. 412(34), 4439–4451 (2011)
14. Schmidt, D.: Well-partial orderings and their maximal order types. Habilitationsscrift (1979)
15. Weiermann, A.: A computation of the maximal order type of the term ordering on finite multisets. In: Ambos-Spies, K., Löwe, B., Merkle, W. (eds.) CiE 2009. LNCS, vol. 5635, pp. 488–498. Springer, Heidelberg (2009)

Distinguishing Pattern Languages with Membership Examples

Zeinab Mazadi, Ziyuan Gao, and Sandra Zilles

Department of Computer Science
University of Regina, Regina, SK, Canada S4S 0A2
{mazadi2z,gao257,zilles}@cs.uregina.ca

Abstract. This article determines two learning-theoretic combinatorial parameters, the teaching dimension and the recursive teaching dimension, for various families of pattern languages over alphabets of varying size. Our results and formal proofs are of relevance to recent studies in computational learning theory as well as in formal language theory.

1   Introduction

A pattern is a nonempty finite string of variable symbols and constant symbols, the latter being chosen from a fixed alphabet Σ. Each pattern generates a formal language, which contains all words obtained by replacing the variables in the pattern with nonempty words over Σ [1]. Since their introduction by Angluin in 1980 [1], pattern languages have continually served as interesting objects of study both in computational learning theory and in formal language theory. A large variety of very recent studies include for example the following.

(1) In computational learning theory: learning extensions of pattern languages [8,9] and novel models of learning pattern languages [7,8].

(2) In formal language theory: embedding pattern languages into the Chomsky hierarchy [11,15], the complexity of the membership problem for pattern languages [5,8,14], and decision problems on comparing pattern languages [6].

Pattern languages have also found applications in bioinformatics [2] and for text editing in automatic program synthesis [13].

The present article concerns a formal study of structural properties of pattern languages. The first question we study is the following: given a family $\mathcal{L}$ of pattern languages, how many (and which) membership examples are required in the worst case to distinguish any language $L \in \mathcal{L}$ from all other languages in $\mathcal{L}$? Here, by *membership example* we mean a word together with the information whether or not it belongs to L. This question is of interest (i) from a formal language point of view, because it reveals interesting structural properties of families of pattern languages, and (ii) from a computational learning theory point of view, because it corresponds to the complexity of learning pattern languages from helpfully chosen examples, i.e., in models of learning from teachers. The number of membership examples required in the worst case directly corresponds to the widely studied *teaching dimension (TD)* complexity parameter [10,16].

A.-H. Dediu et al. (Eds.): LATA 2014, LNCS 8370, pp. 528–540, 2014.

Our second question concerns a second complexity parameter of a formal model of teaching, namely the so-called *recursive teaching dimension (RTD)* [18]. The RTD is also based on membership examples and it lower-bounds the TD, but its definition is more involved (see Section 2). Our motivation for studying RTD is that it, unlike the TD, turns out to be related to other central notions in computational learning theory, namely to the VC-dimension and to sample compression schemes [4,3]. In general, lower values of TD and RTD mean more efficient learnability (where efficiency is measured by the number of membership examples exchanged between the teacher and the learner).

To the best of our knowledge, this paper is the first to study TD and RTD for families of formal languages, and the first to study RTD for any infinite family of objects. We compute both parameters for a variety of families of pattern languages (arbitrary pattern languages, one-variable pattern languages, and regular pattern languages) as well as for various alphabet sizes. Among the interesting observations we make are the following: (i) for some families of pattern languages, TD can be infinite where RTD is only 2; (ii) TD can depend on the size of the underlying alphabet; (iii) different interesting subfamilies of pattern languages exhibit different values of TD while the RTD is 2 in all cases where we could determine it. Not surprisingly, the proof techniques vary substantially when varying the alphabet from singleton over finite non-singleton to infinite. Similarly, the particular sets of membership examples used for distinguishing a language often obey a different structure when the alphabet is changed.

The particular sets of membership examples used for distinguishing a pattern language (in the TD model as well as in the RTD model) give insights into structural properties of pattern languages that can be useful for further studies in formal language theory. Further, the fact that RTD equals 2 in so many cases of non-trivial families of pattern languages suggests the possibility of designing application scenarios in which pattern languages can be learned very efficiently from helpful teachers.

2   Preliminaries

Throughout this document, Σ denotes a nonempty and either finite or countably infinite set, called the alphabet. A language over Σ is a subset of Σ^*, i.e., a set of finite words formed over Σ. The complement of a language L w.r.t. Σ^* is denoted by $\overline{L}$. For any two languages L, L' over Σ, the term $L \triangle L'$ denotes the symmetric difference between L and L'. For any set Γ of symbols, Γ^+ is the set of nonempty strings over Γ; for any $w \in \Gamma^+$ and any $p \in \mathbb{N}$, we denote the pth symbol of w by $w[p]$ and the length of w by $|w|$. We often omit any reference to Σ when the choice of alphabet is clear from the context.

2.1   Teaching Dimension and Recursive Teaching Dimension

Let $W \subseteq \Sigma^*$ be a set of words. Two languages L, L' are said to agree on W if $W \cap (L \triangle L') = \emptyset$, i.e., if $W \cap (L \cup L') \subseteq L \cap L'$. Further suppose that $\mathcal{L}$ is a family of languages and $L \in \mathcal{L}$. The set W is called a distinguishing set for L w.r.t. $\mathcal{L}$, if $W \cap (L \triangle L') \neq \emptyset$ for every $L' \in \mathcal{L} \setminus \{L\}$, i.e., if no other language in $\mathcal{L}$ agrees with L on W. For example, $\{a, b\}$ is the smallest distinguishing set for $L = \emptyset$ w.r.t. $\mathcal{L} = \{\emptyset, \{a\}, \{b\}\}$.

The size and structure of distinguishing sets often reveal insights into structural properties of the underlying family $\mathcal{L}$ of languages. In particular, they determine how many and which so-called *membership examples* are required to distinguish a specific language from all other languages in the family. A membership example for a language L is a word w, together with the information whether or not $w \in L$. Such distinguishing information is useful for example in the context of formal learning models in the field of computational learning theory. There, the size of a smallest possible distinguishing set is referred to as the teaching dimension of L w.r.t. $\mathcal{L}$.

Definition 1. *[10,16] Let Σ be any alphabet and $\mathcal{L}$ any family of languages over Σ. Let $L \in \mathcal{L}$. The size of a smallest distinguishing set for L w.r.t. $\mathcal{L}$ is called the teaching dimension of L w.r.t. $\mathcal{L}$, denoted by $\mathrm{td}(L, \mathcal{L})$. The teaching dimension of $\mathcal{L}$ is then defined as $\sup\{\mathrm{td}(L, \mathcal{L}) \mid L \in \mathcal{L}\}$ and denoted by $\mathrm{TD}(\mathcal{L})$.*

For example, if $\mathcal{L}_{1,\emptyset}$ consists of all singleton languages (over any alphabet) and the empty language, then every singleton has a teaching dimension of 1, while the empty set has a teaching dimension of ∞, so that $\mathrm{TD}(\mathcal{L}_{1,\emptyset}) = \infty$.

Another complexity parameter recently studied in computational learning theory is the recursive teaching dimension. It refers to the size of distinguishing sets required in a series of nested subfamilies of the language family.

Definition 2. *[18,12] Let Σ be any alphabet and $\mathcal{L}$ any family of languages over Σ. Let $\mathcal{L}_0 = \mathcal{L}$. For each $k \in \mathbb{N}$ with $\mathcal{L}_k \neq \emptyset$, let*

$$\mathcal{L}_k^{min} = \{L \in \mathcal{L}_k \mid \mathrm{td}(L, \mathcal{L}_k) = \min_{L' \in \mathcal{L}_k} (\mathrm{td}(L', \mathcal{L}_k))\}$$

be the set of languages in $\mathcal{L}_k$ whose teaching dimension w.r.t. $\mathcal{L}_k$ is smallest. Further, let $d_k = \mathrm{td}(L, \mathcal{L}_k)$ for $L \in \mathcal{L}_k^{min}$ be that teaching dimension, and let $\mathcal{L}_{k+1} = \mathcal{L}_k \setminus \mathcal{L}_k^{min}$. The recursive teaching dimension of $\mathcal{L}$ is then defined as $\sup\{d_k \mid k \in \mathbb{N} \text{ and } \mathcal{L}_k \neq \emptyset\}$ and denoted by $\mathrm{RTD}(\mathcal{L})$.

Intuitively, starting from $\mathcal{L}$, one recursively removes the languages with the smallest teaching dimension and, in the next step, considers the smallest teaching dimension over the remaining family of languages. $\mathrm{RTD}(\mathcal{L})$ is the largest smallest teaching dimension encountered in this process. One obtains the same value when recursively removing only a nonempty subset of the languages with the smallest teaching dimension. It further does not affect the value of RTD if, at any stage k, a language $L \in \mathcal{L}_k$ with $\mathrm{td}(L, \mathcal{L}_k) < \mathrm{RTD}(\mathcal{L})$ is removed immediately, even if $L \notin \mathcal{L}_k^{min}$. We will use these properties in our proofs. A recursive teaching sequence is any sequence $((\mathcal{F}_0, d_0), (\mathcal{F}_1, d_1), \ldots)$ where (i) the families $\mathcal{F}_i$ form a partition of $\mathcal{L}$, and (ii) $d_i = \mathrm{td}(L, \mathcal{L} \setminus \bigcup_{0 \leq j < i} \mathcal{F}_j)$ for all i and all

$L \in \mathcal{F}_i$. Then $\sup\{d_i \mid i \in \mathbb{N} \text{ and } \mathcal{F}_i \neq \emptyset\}$ ($\geq \mathrm{RTD}(\mathcal{L})$) is called the order of the recursive teaching sequence.

$\mathrm{RTD}(\mathcal{L}_{1,\emptyset}) = 1$: one first removes all singletons with a teaching dimension of 1, leaving only the empty set, to be removed with a teaching dimension of 0.

RTD has some interesting properties; for example, in many cases it is upper-bounded by the VC-dimension, which is a central parameter in learning theory, and it is of importance for the study of sample compression schemes, see [3,4].

2.2 Pattern Languages

Next we define the concepts of pattern and pattern language, as introduced by Angluin [1]. Let Σ be an alphabet and $X = \{x_1, x_2, \dots\}$ a countably infinite set of *variables*, disjoint from Σ. A *pattern* π is a nonempty finite string of *constant symbols* from Σ and variables from X, i.e., $\pi \in (\Sigma \cup X)^*$. Each pattern π generates a pattern language $L(\pi)$, defined as the set of words in Σ^+ obtained by substituting (nonempty) words from Σ^+ for variables. For instance, if $\Sigma = \{a, b\}$ and $\pi = x_1 a x_1 x_2 b$, then $L(\pi) = \{w_1 a w_1 w_2 b \mid w_1, w_2 \in \Sigma^+\}$.[1] For a set Π of patterns, $\mathcal{L}(\Pi)$ denotes the family of languages generated by patterns in Π.

A pattern π is (i) a constant pattern if $\pi \in \Sigma^*$, (ii) a one-variable pattern if π contains at most one variable (possibly with repetitions), and (iii) a regular pattern [17] if π has no repeated variables. Languages generated by one-variable (regular) patterns are called one-variable (regular, resp.) pattern languages.

The main focus of this paper is to determine TD and RTD for various families of pattern languages, in particular for arbitrary (Section 3), one-variable (Section 4) and regular pattern languages (Section 5). We begin by stating two straightforward properties of distinguishing sets of some such families.

Proposition 3. *Let Σ be any alphabet, Π any set of patterns containing all constant patterns, and $\pi \in \Pi$. Then every finite distinguishing set for $L(\pi)$ w.r.t. $\mathcal{L}(\Pi)$ contains at least one word in $L(\pi)$.*

Proof. For finite $W \subseteq \overline{L(\pi)}$ and any $w \notin W$, $L(w)$ agrees with $L(\pi)$ on W. □

Proposition 4. *Let Σ be any alphabet, Π any set of patterns containing x_1, and $\pi \in \Pi \setminus \{x_1\}$. Then every distinguishing set for $L(\pi)$ w.r.t. $\mathcal{L}(\Pi)$ contains at least one word in $\overline{L(\pi)}$.*

Proof. $L(x_1)$ agrees with $L(\pi)$ on every $W \subseteq L(\pi)$. □

3   Arbitrary Pattern Languages

A well-known fact is that every word is contained in only finitely many pattern languages [1]. This implies that every pattern language can be distinguished

---

[1] These languages have been called non-erasing languages to distinguish from the type of pattern language obtained when allowing the variables to be replaced with the empty string. The latter, called erasing pattern languages or extended pattern languages, were introduced by Shinohara [17]. In this article, we use the term *pattern language* only to refer to a non-erasing pattern language.

from all other pattern languages using only finitely many membership examples, independent of the underlying alphabet:

Theorem 5. *Let Σ be any alphabet and π any pattern. Then the teaching dimension of $L(\pi)$ w.r.t. the family of all pattern languages is finite.*

Proof. Let $w \in L(\pi)$. Let Π_w be the set of all patterns that generate w; it contains only patterns of length at most $|w|$ and is hence finite. Every finite set of words containing w and, for each $\pi' \in \Pi_w \setminus \{\pi\}$, one word $w' \in L(\pi) \triangle L(\pi')$, is a distinguishing set for $L(\pi)$ w.r.t. the family of all pattern languages.   □

However, it turns out that there is no finite upper bound on the teaching dimension of an arbitrary pattern language w.r.t. the family of all such languages, again independent of the underlying alphabet.

Theorem 6. *Let Σ be any alphabet. Then the teaching dimension of the family of all pattern languages equals ∞.*

Proof. Assume each pattern language has a teaching dimension of at most $m \in \mathbb{N}$ w.r.t. the family of all pattern languages. Let $k = p_1 \cdot \ldots \cdot p_m$ for m pairwise distinct prime numbers $p_1, \ldots, p_m$. Let $\pi = x^k$ for some $x \in X$. By assumption, π has a distinguishing set S of at most m words w.r.t. the family of all pattern languages. By Proposition 3, at most $m - 1$ words in S are in $\overline{L(\pi)}$.

For $i \in \{1, \ldots, m\}$, define $t_i = \frac{k}{p_i}$. Then $L(x^{t_1}), \ldots, L(x^{t_m})$ are pairwise distinct pattern languages, all of which are proper supersets of $L(\pi)$. Thus, for each $i \in \{1, \ldots, m\}$ there is a word $w_i \in L(x^{t_i}) \setminus L(\pi)$ with $w_i \in S$.

Since S contains at most $m - 1$ words in $\overline{L(\pi)}$, there must be two distinct indices $i, j \in \{1, \ldots, m\}$ such that $w_i = w_j$. In particular, $w_i \in (L(x^{t_i}) \cap L(x^{t_j})) \setminus L(\pi)$. However, one can easily verify that $L(x^{t_i}) \cap L(x^{t_j}) = L(\pi)$, which yields a contradiction. Therefore, m does not exist.   □

In contrast to the infinity of the TD, we now show that the RTD of the family of all pattern languages is 2, when the alphabet Σ is infinite. This shows how large the difference between the two complexity parameters can be.

Theorem 7. *Let $|\Sigma| = \infty$. Then the recursive teaching dimension of the family of all pattern languages equals 2.*

Proof. (Sketch.) Let $\mathcal{L}$ be the family of all pattern languages. For any pattern π, let $V_\pi = \{i \mid \pi[i] \in X\}$ and let M_π be the multi-set of numbers of occurrences of variables in π, e.g., $M_\pi = \{1, 2, 2\}$ for $\pi = x_1 x_1 a x_2 b a x_3 x_2$. For multi-sets $\{i_1, \ldots, i_s\}$ and $\{j_1, \ldots, j_t\}$ we write $\{i_1, \ldots, i_s\} \preceq \{j_1, \ldots, j_t\}$ iff there is a partition $I_1, \ldots, I_t$ of $\{i_1, \ldots, i_s\}$ such that, for all $\ell \in \{1, \ldots, t\}$, $j_\ell = \sum_{i \in I_\ell} i$.

Let $\mathcal{L}_\pi = \{L(\rho) \mid |\pi| = |\rho|$ and $M_\pi \preceq M_\rho\} \cup \{L(\rho) \mid |\pi| = |\rho|$ and $|V_\pi| \geq |V_\rho|$ and $M_\pi \not\preceq M_\rho$ and $M_\rho \not\preceq M_\pi\} \cup \{L(\rho) \mid |\rho| > |\pi|\}$. Then two facts hold:

1. If $M_\pi \neq \emptyset$, then $td(L(\pi), \mathcal{L}_\pi) = 2$. In particular, $S = \{w_1, w_2\} \subseteq L(\pi)$ is a distinguishing set for $L(\pi)$ w.r.t. $\mathcal{L}_\pi$, where $|w_1| = |w_2| = |\pi|$, the sets $\{w_1[i] \mid i \in V_\pi\}$, $\{w_2[i] \mid i \in V_\pi\}$, and $\{\pi[i] \mid i \notin V_\pi\}$ are pairwise disjoint and $w_s[i] \neq w_s[j]$ for $s \in \{1, 2\}$, $1 \leq i < j \leq |\pi|$, and $i, j \in V_\pi$ with $\pi[i] \neq \pi[j]$.

2. If $M_\pi = \emptyset$, i.e., if $\pi \in \Sigma^+$, then $\mathrm{td}(L(\pi), \mathcal{L}_\pi) = 1$. This is witnessed by the distinguishing set $S = \{\pi\}$.

$\mathrm{RTD}(\mathcal{L}) \leq 2$ is witnessed by the following recursive teaching sequence: Let $\mathcal{F}_{-1} = \emptyset$. At stage i, one removes all languages in the set $\mathcal{F}_i = \{L(\rho) \mid \rho \in \Pi$ with $[|\rho| \leq |\pi|$ and $M_\pi \nprec M_\rho]$ for all π with $L(\pi) \in \mathcal{L}\backslash\bigcup_{j\leq i-1} \mathcal{F}_j\}$. Note that, for any i and any pattern π with $L(\pi) \in \mathcal{F}_i$, the set $\mathcal{L}\backslash\bigcup_{j\leq i-1}\mathcal{F}_j$ corresponds to the set $\mathcal{L}_\pi$. Thus, by the two statements above, the distinguishing sets required at each stage in the removal process have size at most 2, which yields $\mathrm{RTD}(\mathcal{L}) \leq 2$. $\mathrm{RTD}(\mathcal{L}) \geq 2$ follows immediately from Propositions 3 and 4. □

For finite alphabets, the recursive teaching dimension of the family of all pattern languages remains open. The remainder of this paper therefore deals with comparing TD and RTD for interesting subfamilies of pattern languages.

As a first example, for finite non-singleton alphabets, there is a rather narrow family of pattern languages with an RTD of 2 and a TD of ∞.

Theorem 8. *Let $2 \leq |\Sigma| < \infty$. Let Π_{xx} be the family of all patterns of the form $x_1 \ldots x_k$ (for $k \geq 1$) or $x_1 \ldots x_{j-1} x_i x_{j+1} \ldots x_k$ (for $1 \leq i < j \leq k$). Then*

1. *the teaching dimension of $\mathcal{L}(\Pi_{xx})$ equals ∞, and*
2. *the recursive teaching dimension of $\mathcal{L}(\Pi_{xx})$ equals 2.*

Proof. Assertion 1. Assume $\mathrm{TD}(\mathcal{L}(\Pi_{xx})) \leq m$ for some $m \in \mathbb{N}$. Let $k = |\Sigma|^{m-1} + 1$. By Proposition 4, there is a distinguishing set S for $L(x_1 \ldots x_k)$ w.r.t. $\mathcal{L}(\Pi_{xx})$ that contains at most $m - 1$ words in $L(x_1 \ldots x_k)$. Let $\{w_1, \ldots, w_{m-1}\}$ be a set of $m - 1$ distinct words in $L(x_1 \ldots x_k)$ containing these words. Since S is a distinguishing set for $L(x_1 \ldots x_k)$ w.r.t. $\mathcal{L}(\Pi_{xx})$, there is no pattern $\pi \in \Pi_{xx}$ such that $\{w_1, \ldots, w_{n-1}\} \subseteq L(\pi) \subset L(x_1 \ldots x_k)$. Let $n = \min\{|w_1|, \ldots, |w_{m-1}|\}$. Note that $n \geq k = |\Sigma|^{m-1} + 1$.

Let $l_1 = |\Sigma|^{m-2} + 1$. There must be l_1 many positions $p_1^1, \ldots, p_{l_1}^1 \leq n$ and a constant $\sigma_1 \in \Sigma$ such that $w_1[p_1^1] = w_1[p_2^1] = \ldots = w_1[p_{l_1}^1] = \sigma_1$. Otherwise no symbol in Σ would occur at least l_1 times in $w_1[1], \ldots, w_1[n]$ which would imply $n \leq (l_1 - 1)|\Sigma| = |\Sigma|^{m-1} < k \leq n$. Similarly, if $l_2 = |\Sigma|^{n-3} + 1$, there must be l_2 many positions $p_1^2, \ldots, p_{l_2}^2$ among $p_1^1, \ldots, p_{l_1}^1$ and a constant $\sigma_2 \in \Sigma$ such that $w_2[p_1^2] = w_2[p_1^2] = \ldots = w_2[p_{l_2}^2] = \sigma_2$. Otherwise, $l_1 \leq (l_2 - 1)|\Sigma| = |\Sigma|^{m-2} < l_1$.

In general, for $1 \leq i \leq m - 1$, let $l_i = |\Sigma|^{m-1-i} + 1$. Then there are l_i many positions in which all words w_j for $1 \leq j \leq i$ have only repetitions of some constant $\sigma_j \in \Sigma$. Finally, there must be $l_{m-1} = 2$ repetitions of a single constant in w_{m-1}, such that all words $w_1, \ldots, w_{m-2}$ have repeated constants in the same positions. Let us call these two positions p_1 and p_2, $p_1 < p_2$. Thus, $\pi = x_1 \ldots x_{p_2-1} x_{p_1} x_{p_2+1} \ldots x_k$ generates a proper subset of $L(x_1 \ldots x_k)$ that agrees with $L(x_1 \ldots x_k)$ on all words in S. Hence, S is not a distinguishing set for $L(x_1 \ldots x_k)$ w.r.t. $\mathcal{L}(\Pi_{xx})$—a contradiction.

Assertion 2. (Sketch.) $\mathrm{RTD}(\mathcal{L}(\Pi_{xx})) \geq 2$ is easy to verify. To see that $\mathrm{RTD}(\mathcal{L}(\Pi_{xx})) \leq 2$, let $z = |\Sigma|$ and consider the following recursive teaching sequence of order 2:

- $\mathcal{L}_0^{min} = \{L(x_1)\}$ with $d_0 = 1$,
- for all $k \in \{1, \ldots, z-1\}$, $\mathcal{L}_{2k}^{min} = \{L(x_1 \ldots x_{k+1})\}$ with $d_{2k} = 1$, and $\mathcal{L}_{2k+1}^{min} = \{L(x_1 \ldots x_{j-1} x_i x_{j+1} \ldots x_{k+1}) \mid 1 \leq i < j \leq k+1\}$ with $d_{2k+1} = 1$,
- for all $k \in \mathbb{N}$, $\mathcal{L}_{2z+2k}^{min} = \{L(x_1 \ldots x_{j-1} x_i x_{j+1} \ldots x_{z+k}) \mid 1 \leq i < j \leq z+k\}$ with $d_{2z+2k} = 2$, and $\mathcal{L}_{2z+2k+1}^{min} = \{L(x_1 \ldots x_{z+k})\}$ with $d_{2z+2k+1} = 1$. □

The subsequent sections focus on two more "natural" families, namely the one-variable pattern languages and the regular pattern languages.

4   One-Variable Pattern Languages

The languages used in Theorem 8 are generated by a family of patterns with an unbounded number of variables. One might ask whether it is this property that creates the infinite gap between TD and RTD for that family. An opposite extreme is the family of one-variable pattern languages. As it turns out, even for this family, the teaching dimension is infinite, while the recursive teaching dimension is only 2—independent of the underlying alphabet.

Theorem 9. *Let Σ be any alphabet. Then the following two statements hold.*

1. *The teaching dimension of the family of one-variable pattern languages is ∞.*
2. *The recursive teaching dimension of the family of one-variable pattern languages is 2.*

Proof. Assertion 1 is an immediate consequence of the proof of Theorem 6, which uses exclusively one-variable patterns. It remains to prove Assertion 2.

Let $\mathcal{L}$ be the family of one-variable pattern languages. By Propositions 3 and 4, $\mathrm{RTD}(\mathcal{L}) \geq 2$. It thus suffices to find a recursive teaching sequence of order 2.

Claim 1. For any $n \geq 1$ and any $\pi \in \Sigma^+$ with $|\pi| = n$, the teaching dimension of $L(\pi)$ $(= \{\pi\})$ w.r.t. the family of languages generated by one-variable patterns of length greater than n or by constant patterns of any length is 1.

Proof of Claim 1. Obviously, $L(\pi)$ is the only language in the family of interest that contains π, i.e., $\{\pi\}$ is a distinguishing set for $L(\pi)$.

Claim 2. Suppose $|\Sigma| \geq 2$. Then, for any $n \geq 1$ and any non-constant one-variable pattern π with $|\pi| = n$, the teaching dimension of $L(\pi)$ w.r.t. the family of languages generated by one-variable patterns of length at least n is 2.

Proof of Claim 2. Let w_1 and w_2 be two distinct words of length n in $L(\pi)$ such that $w_1[p] \neq w_2[p]$ for any $p \in \{1, \ldots, n\}$ with $\pi[p] \in X$. It is not hard to see that $\{w_1, w_2\}$ is a distinguishing set for $L(\pi)$ w.r.t. the family of interest.

Claim 3. Suppose $|\Sigma| = 1$. Let $n \geq k > 1$, and let π be any non-constant one-variable pattern of length n with exactly k occurrences of a variable. Then the teaching dimension of $L(\pi)$ w.r.t. the family of languages generated by one-variable patterns of length greater than n or by one-variable patterns of length n that contain more than k variable positions is 2.

Proof of Claim 3. Let $\Sigma = \{a\}$. Then the set $\{a^n, a^{n+k}\} \subset L(\pi)$ is a distinguishing set for $L(\pi)$ w.r.t. the family of interest. No pattern of length greater than n generates a^n and no one-variable pattern of length n with more than k variable occurrences generates a^{n+k}. All one-variable patterns of length n with exactly k variable positions are equivalent in the sense that they all generate the same language, namely $L(\pi)$.

These claims together imply that a recursive teaching sequence of order 2 is obtained by removing languages in increasing order of the length of the underlying patterns, and for each length to proceed in increasing order of the number of variable positions in the generating patterns. □

5 Regular Pattern Languages

Note that all patterns in the family Π_{xx} in Theorem 8 contain at most one repeated variable; such a variable then has exactly two occurrences in the pattern. Thus the patterns in Π_{xx} look very similar to regular patterns. It would be interesting to know whether the single repetition of a single variable in a pattern is crucial for obtaining the infinity of the teaching dimension for the corresponding family $\mathcal{L}(\Pi_{xx})$ of languages by comparing to the teaching dimension of the family of regular pattern languages.

Unfortunately, for finite non-singleton alphabets, i.e., for the setting of Theorem 8, we have not been able to determine the teaching dimension of the family of regular pattern languages. Interestingly though, we can prove that the corresponding parameter is exactly 3 for singleton alphabets and exactly 5 for infinite alphabets. This is in contrast with all our previous results proving the teaching dimension of particular families of pattern languages to be ∞. It further shows that the size of the alphabet can have an influence on the teaching dimension of families of pattern languages. In both cases, the recursive teaching dimension equals 2, which is again smaller than the respective teaching dimension.

We begin with the case of singleton alphabets.

Theorem 10. *Let $|\Sigma| = 1$. Then the following two statements hold.*

1. *The teaching dimension of the family of regular pattern languages equals 3.*
2. *The recursive teaching dimension of the family of regular pattern languages equals 2.*

Proof. (Sketch.) Let $\mathcal{L}$ be the family of regular pattern languages over $\Sigma = \{a\}$. Every regular pattern of length $n \geq 1$ that contains exactly k variables, for $1 \leq k \leq n$, generates the same language as the regular pattern $a^{n-1}x_1$. Thus, we need to consider only constant patterns and patterns of the form $a^{n-1}x_1$.

Both assertions can then easily be verified by observing that the constant pattern a^n does not generate the word a^{n+1}, while any other regular pattern that generates a^n does generate a^{n+1}. Thus $\{a^n, a^{n+1}\}$ is a distinguishing set for $L(a^n)$ w.r.t. $\mathcal{L}$, and $\{a^n, a^{n+1}, a^{n-1}\}$ is a distinguishing set for $L(a^{n-1}x_1)$ w.r.t. $\mathcal{L}$ (it is not hard to see that no smaller distinguishing sets exist). To establish a recursive teaching dimension of 2, languages can be removed according to increasing length of their generating patterns, thus making the word a^{n-1} in the distinguishing set for $L(a^{n-1}x_1)$ obsolete. $\square$

The result on the RTD generalizes to arbitrary alphabets:

Theorem 11. *Let Σ be any alphabet. Then the recursive teaching dimension of the family of regular pattern languages equals 2.*

Proof. For singleton alphabets, this was already proven in Theorem 10. So suppose Σ contains at least 2 distinct symbols a and b. By Propositions 3 and 4, RTD is at least 2. A recursive teaching sequence of order 2 looks as follows:

$$((\mathcal{F}_1^0, 2), (\mathcal{F}_1^1, 1), (\mathcal{F}_2^0, 2), (\mathcal{F}_2^1, 2), (\mathcal{F}_2^2, 1), \ldots, (\mathcal{F}_n^0, 2), (\mathcal{F}_n^1, 2), \ldots, (\mathcal{F}_n^n, 1), \ldots,$$

where, for $n \geq 1$ and $n \geq k \geq 0$, the family $\mathcal{F}_n^k$ consists of all languages generated by regular patterns of length n that contain exactly $n - k$ variables.

For $k < n$, a distinguishing set of size 2 for any language $L(\pi)$ in $\mathcal{F}_n^k$ w.r.t. the family of languages occurring later in the sequence consists of two words in $L(\pi)$ of length n, namely one in which all $n - k$ variable symbols in π are replaced by the symbol a and one in which they are replaced by b. Languages in $\mathcal{F}_m^i$ for $m > n$ and any i do not contain these words, because their shortest words are of length $m > n$. Languages in $\mathcal{F}_n^i$ for any $i > k$ do not contain both these words at the same time, because that would require their underlying patterns to contain at least $n - k$ variables, while they contain only $n - i < n - k$ many.

For $k = n$, any language in $\mathcal{F}_n^k$ is generated by a constant pattern $\pi \in \Sigma^+$ of length n, while all languages occurring later in the sequence are generated by patterns of length greater than n. Thus π itself forms a distinguishing set of size 1 for $L(\pi)$ w.r.t. the family of languages occurring later in the sequence. $\square$

For the TD however, Theorem 10 does not generalize to larger alphabets, as summarized in the following two theorems. Unlike the case for singleton alphabets, for $|\Sigma'| \geq 2$ the positioning of variables in a regular pattern π may affect the language generated. Hence the distinguishing sets for $L(\pi)$ w.r.t. the family of all regular pattern languages over Σ' "encode" the variable positions in π.

Theorem 12. *Let $2 \leq |\Sigma| < \infty$. Then the teaching dimension of the family of regular pattern languages is at least 5.*

Theorem 13. *Let $|\Sigma| = \infty$. Then the teaching dimension of the family of regular pattern languages equals 5.*

Our proof of Theorem 13 is quite involved and has to be shortened substantially here. We sketch it by presenting the main lemmas establishing Theorems 12 and 13 and by explaining the core ideas for proving these lemmas.

Lemma 14. *Let $|\Sigma| \geq 2$ and $n \geq 3$. Let $\pi = c_1 x_1 c_2 x_2 \ldots x_{n-1} c_n$ for some $c_1, c_n \in \Sigma^+$ and $c_2, \ldots, c_{n-1} \in \Sigma^*$ such that at least one of the c_i for $1 < i < n$ is nonempty. Then the teaching dimension of $L(\pi)$ w.r.t. the family of regular pattern languages is at least 5.*

Proof. Let $\mathcal{L}$ be the family of regular pattern languages. First, note that any distinguishing set for $L(\pi)$ w.r.t. $\mathcal{L}$ must contain at least 2 words from $L(\pi)$.

Second, it is shown that any distinguishing set for $L(\pi)$ w.r.t. $\mathcal{L}$ must have at least three words from $\overline{L(\pi)}$. Let $i \in \{2, \ldots, n-1\}$ such that c_i is nonempty. For $j = 1, i, n$, let π_j be the pattern derived from π by replacing the first symbol of c_j with a variable x that does not occur in π. Then $L(\pi_j) \supset L(\pi)$: if some word w is generated from π by substituting w_t for x_t whenever $1 \leq t \leq n-1$, then w can be generated from π_j by replacing x with the first symbol of c_j, and substituting w_t for x_t whenever $1 \leq t \leq n-1$. Thus, any distinguishing set W for $L(\pi)$ w.r.t. $\mathcal{L}$ contains some $v_j \in L(\pi_j) - L(\pi)$. To show that W contains at least three words in $\overline{L(\pi)}$, it suffices to prove $L(\pi_j) \cap L(\pi_\ell) \subseteq L(\pi)$ for $j, \ell \in \{1, i, n\}$, $j \neq \ell$. Consider any $w \in L(\pi_1) \cap L(\pi_i)$. Since $w \in L(\pi_i)$, c_1 is a prefix of w. Further, $w \in L(\pi_1)$ implies that there are nonempty strings $w_1, \ldots, w_{n-2}$ such that $c_2 w_1 c_3 w_3 \ldots w_{n-2} c_n$ is a suffix of w, and this suffix must start at a position in w which is at least equal to $|c_1| + 2$. Thus w can be expressed in the form $c_1 S c_2 w_1 c_3 w_3 \ldots w_{n-2} c_n$ for some nonempty string S, and therefore $w \in L(\pi)$. Analogous proofs apply to the cases $\{j = 1, \ell = n\}$ and $\{j = i, \ell = n\}$. □

This lemma immediately yields Theorem 12. To prove Theorem 13, we proceed with the following lemmas.

Lemma 15. *Let $|\Sigma| = \infty$ and $n \geq 3$. Let $\pi = c_1 x_1 c_2 x_2 \ldots x_{n-1} c_n$ for some $c_1, c_n \in \Sigma^+$ and $c_2, \ldots, c_{n-1} \in \Sigma^*$ such that at least one of the c_i for $1 < i < n$ is nonempty. Then the teaching dimension of $L(\pi)$ w.r.t. the family of regular pattern languages equals 5.*

Proof. Let a, b, y, z be four distinct constants in Σ none of which occur in π. For any $c \in \Sigma^*$, define $\widetilde{c} = c$ if $|c| = 0$, $\widetilde{c} = y$ if $|c| = 1$, and $\widetilde{c} = c[1]c[2] \ldots c[|c| - 1]yc[2]c[3] \ldots c[|c|]$ if $|c| > 1$. Let $\pi(a)$ be the string derived from π by substituting a for all the variables in π; define $\pi(b)$ analogously. The required distinguishing set W of size 5 consists of two words in $L(\pi)$, namely $\pi(a)$ and $\pi(b)$, as well as three words σ, τ, and η from $\overline{L(\pi)}$, namely $\tau = \widetilde{c}_1 z c_2 z \ldots z c_n$, $\eta = c_1 z c_2 z \ldots z \widetilde{c}_n$, and $\sigma = c_1 z \widetilde{c}_2 w_1 \widetilde{c}_3 w_2 \ldots w_{n-3} \widetilde{c}_{n-1} z c_n$, where, for all $i \leq n - 3$, w_i is defined by

$$
w_i = \begin{cases} z c_{i+2} z & \text{if } c_{i+1} \text{ is a substring of } c_{i+2}; \\ z c_{i+2} z c_{i+1} z & \text{if } c_{i+1} \text{ is not a substring of } c_{i+2}. \end{cases}
$$

We first show that σ satisfies the crucial property that for any i with $2 \leq i \leq n - 1$, σ can be expressed as $\sigma = c_1 w_1' c_2 w_2' \ldots c_{i-1} w_{i-1}' \widetilde{c}_i w_i' c_{i+1} w_{i+1}' \ldots w_{n-1}' c_n$ for some nonempty words $w_1', w_2', \ldots, w_{n-1}'$; denote this property by $(*_i)$.

Consider the expression $c_1 z \widetilde{c}_2 w_1 \widetilde{c}_3 w_2 \ldots w_{n-3} \widetilde{c}_{n-1} z c_n$ for σ. For any i with $2 < i < n - 1$, σ contains the sequence of words $c_1, w_1, w_2, \ldots, w_{i-2}, \widetilde{c}_i, w_{i-1}, w_i, \ldots$,

w_{n-3}, z, c_n. Furthermore, for all $i < n - 2$, both c_{i+1} and c_{i+2} are substrings of w_i that do not overlap with the end points of w_i. It follows that σ can be expressed in the form $c_1 w_1' c_2 \ldots w_{i-2}' c_{i-1} w_{i-1}' \widetilde{c_i} w_i' c_{i+1} w_{i+1}' \ldots c_{n-1} w_{n-1}' c_n$ for some nonempty words $w_1', w_2', \ldots, w_{n-1}'$ whenever $2 < i < n-1$, as required. For $i = 2$, consider the sequence of words $c_1, z, \widetilde{c_2}, w_1, w_2, \ldots, w_{n-3}, z, c_n$ contained in σ. For all j with $1 \leq j \leq n - 3$, c_{j+2} is a substring of w_j that does not intersect the end points of w_j. Hence property $(*_1)$ holds for σ. Similarly, for $i = n - 1$, σ contains the sequence of words $c_1, w_1, w_2, \ldots, w_{n-3}, \widetilde{c_{n-1}}, z, c_n$, and for all $j < n - 2$, c_{j+1} is a substring of w_j disjoint with the end points of w_j. Hence property $(*_{n-1})$ holds for σ.

The claim that W is indeed a distinguishing set for $L(\pi)$ w.r.t. $\mathcal{L}$ can be deduced immediately from the following two assertions. The proofs of these assertions are non-trivial, but have to be omitted due to space constraints.

1. $\sigma \notin L(\pi)$, $\tau \notin L(\pi)$, and $\eta \notin L(\pi)$.
2. If a regular pattern ρ with $L(\rho) \neq L(\pi)$ generates both $\pi(a)$ and $\pi(b)$, then:
   - If c_1 is not a prefix of ρ, then $\tau \in L(\rho)$.
   - If c_n is not a suffix of ρ, then $\eta \in L(\rho)$.
   - If c_1 is a prefix of ρ and c_n is a suffix of ρ, then $\sigma' \in L(\rho)$ for every σ' satisfying $(*_2)$ through $(*_{n-1})$. $\qquad\square$

The next lemmas can be proven following similar ideas as for Lemma 15.

Lemma 16. *Let $|\Sigma| = \infty$ and $n \geq 2$. Let $\pi = x_1 c_1 x_2 c_2 \ldots c_{n-1} x_n$ for some $c_1, c_2, \ldots, c_{n-1} \in \Sigma^*$ such that at least one of the c_i for $1 \leq i < n$ is nonempty. Then the teaching dimension of $L(\pi)$ w.r.t. the family of regular pattern languages equals 3.*

Lemma 17. *Let $|\Sigma| = \infty$ and $n \geq 2$. Let $\pi = c x_1 c_1 x_2 \ldots c_{n-1} x_n$ and $\rho = x_1 c_1 x_2 c_2 \ldots c_{n-1} x_n c$ for some $c \in \Sigma^+$ and some $c_1, c_2, \ldots, c_{n-1} \in \Sigma^*$ such that at least one of the c_i for $1 \leq i < n$ is nonempty. Then both the teaching dimension of $L(\pi)$ w.r.t. the family of regular pattern languages and the teaching dimension of $L(\rho)$ w.r.t. the family of regular pattern languages equal 4.*

Together with the facts that constant patterns generate languages of teaching dimension 2 and constant-free patterns generate languages of teaching dimension 3, both w.r.t. the family of regular pattern languages over infinite alphabets, the previous four lemmas establish Theorem 13.

For alphabets of size 2, we determined the exact teaching dimension for a few special cases of regular pattern languages. We omit the rather lengthy proofs.

Proposition 18. *Let $\Sigma = \{a, b\}$ and $n \geq 3$. Let $\pi = a x_1 b x_2 a x_3 b \ldots a x_{2n-1} b$. Then the teaching dimension of $L(\pi)$ w.r.t. the family of regular pattern languages equals 5.*

Proposition 19. *Let $|\Sigma| = 2$. Let $\pi = x_1 c x_2$ for some $c \in \Sigma^+$. Then the teaching dimension of $L(\pi)$ w.r.t. the family of regular pattern languages equals 3.*

6   Conclusions

We determined TD and RTD for a variety of families of pattern languages and under consideration of the size of the underlying alphabet:

	$2 \leq |\Sigma| \leq \infty$	$|\Sigma| = 1$	$|\Sigma| = \infty$
arbitrary patterns	$\text{TD} = \infty$	$\text{TD} = \infty$	$\text{TD} = \infty$
	$\text{RTD} \geq 2$	$\text{RTD} \geq 2$	$\text{RTD} = 2$
one-variable patterns	$\text{TD} = \infty$	$\text{TD} = \infty$	$\text{TD} = \infty$
	$\text{RTD} = 2$	$\text{RTD} = 2$	$\text{RTD} = 2$
regular patterns	$\text{TD} \geq 5$	$\text{TD} = 3$	$\text{TD} = 5$
	$\text{RTD} = 2$	$\text{RTD} = 2$	$\text{RTD} = 2$

In all cases in which we calculated both parameters, RTD is lower than TD. Our study of regular pattern languages shows that the alphabet size can indeed affect the value of TD. Furthermore, even when a teaching complexity parameter is not affected by the alphabet size, often the proof techniques used are different for different alphabet sizes. In particular, the distinguishing sets themselves may exhibit a different structure when changing the alphabet.

For the families we studied, the value of RTD did not prove to be affected by the alphabet size; however, the RTD of the family of all pattern languages over finite alphabets remains open. Further, we could not determine the TD of the family of regular pattern languages over finite non-singleton alphabets.

References

1. Angluin, D.: Finding patterns common to a set of strings. J. Comput. Syst. Sci. 21, 46–62 (1980)
2. Arikawa, S., Miyano, S., Shinohara, A., Kuhara, S., Mukouchi, Y., Shinohara, T.: A machine discovery from amino acid sequences by decision trees over regular patterns. New Generation Comput. 11, 361–375 (1993)
3. Darnstädt, M., Doliwa, T., Simon, H.U., Zilles, S.: Order compression schemes. In: Jain, S., Munos, R., Stephan, F., Zeugmann, T. (eds.) ALT 2013. LNCS, vol. 8139, pp. 173–187. Springer, Heidelberg (2013)
4. Doliwa, T., Simon, H.U., Zilles, S.: Recursive teaching dimension, learning complexity, and maximum classes. In: Hutter, M., Stephan, F., Vovk, V., Zeugmann, T. (eds.) Algorithmic Learning Theory. LNCS, vol. 6331, pp. 209–223. Springer, Heidelberg (2010)
5. Fernau, H., Schmid, M.L.: Pattern matching with variables: A multivariate complexity analysis. In: Fischer, J., Sanders, P. (eds.) CPM 2013. LNCS, vol. 7922, pp. 83–94. Springer, Heidelberg (2013)
6. Freydenberger, D.D., Reidenbach, D.: Bad news on decision problems for patterns. Inf. Comput. 208, 83–96 (2010)
7. Freydenberger, D.D., Reidenbach, D.: Inferring descriptive generalisations of formal languages. J. Comp. Sys. Sci. 79, 622–639 (2013)
8. Geilke, M., Zilles, S.: Learning relational patterns. In: Kivinen, J., Szepesvári, C., Ukkonen, E., Zeugmann, T. (eds.) ALT 2011. LNCS, vol. 6925, pp. 84–98. Springer, Heidelberg (2011)

9. Geilke, M., Zilles, S.: Polynomial-time algorithms for learning typed pattern languages. In: Dediu, A.-H., Martín-Vide, C. (eds.) LATA 2012. LNCS, vol. 7183, pp. 277–288. Springer, Heidelberg (2012)
10. Goldman, S.A., Kearns, M.J.: On the complexity of teaching. J. Comput. Syst. Sci. 50, 20–31 (1995)
11. Jain, S., Ong, Y.S., Stephan, F.: Regular patterns, regular languages and context-free languages. Inf. Process. Lett. 110, 1114–1119 (2010)
12. Mazadi, Z.: Learning pattern languages from a small number of helpfully chosen examples, M.Sc. Thesis, University of Regina (2013)
13. Nix, R.P.: Editing by example. ACM Trans. Program. Lang. Syst. 7, 600–621 (1985)
14. Reidenbach, D., Schmid, M.L.: Patterns with bounded treewidth. In: Dediu, A.-H., Martín-Vide, C. (eds.) LATA 2012. LNCS, vol. 7183, pp. 468–479. Springer, Heidelberg (2012)
15. Reidenbach, D., Schmid, M.L.: Regular and context-free pattern languages over small alphabets. In: Yen, H.-C., Ibarra, O.H. (eds.) DLT 2012. LNCS, vol. 7410, pp. 130–141. Springer, Heidelberg (2012)
16. Shinohara, A., Miyano, S.: Teachability in computational learning. New Generation Comput. 8, 337–347 (1991)
17. Shinohara, T.: Polynomial time inference of extended regular pattern languages. In: Goto, E., Furukawa, K., Nakajima, R., Nakata, I., Yonezawa, A. (eds.) RIMS 1982. LNCS, vol. 147, pp. 115–127. Springer, Heidelberg (1983)
18. Zilles, S., Lange, S., Holte, R., Zinkevich, M.: Models of cooperative teaching and learning. J. Mach. Learn. Res. 12, 349–384 (2011)

Extended Two-Way Ordered Restarting Automata for Picture Languages

Friedrich Otto[1] and František Mráz[2],*

[1] Fachbereich Elektrotechnik/Informatik, Universität Kassel,
34109 Kassel, Germany
otto@theory.informatik.uni-kassel.de
[2] Charles University, Faculty of Mathematics and Physics,
Malostranské nám. 25, 118 25 Prague 1, Czech Republic
frantisek.mraz@mff.cuni.cz

Abstract. We introduce a two-dimensional variant of the deterministic restarting automaton for processing rectangular pictures. Our device has a window of size three-by-three, in a rewrite step it can only replace the symbol in the central position of its window by a symbol that is smaller with respect to a fixed ordering on the tape alphabet, and it can only perform (extended) move-right and move-down steps. This automaton is strictly more expressive than the deterministic Sgraffito automaton, but its word problem can still be solved in polynomial time, and when restricted to one-dimensional input, it only accepts the regular languages.

Keywords: restarting automaton, ordered rewriting, picture language.

1 Introduction

In the literature one finds many different types of grammars and automata for defining classes of picture languages (for a survey see, e.g., [5]). In particular, a lot of work has been devoted to defining and characterizing a class of picture languages that would correspond to the class of regular 'string' languages. Eventually, an agreement has been reached that the class REC of *recognizable languages* of Giammarresi and Restivo [4] is such a class. Unfortunately, this class contains some NP-complete languages [8], which means that in general the membership problem for a recognizable picture language can be quite complex.

Motivated by that observation the current authors started a research program for finding a two-dimensional automaton that is conceptually simple, that is more powerful than the class DREC of deterministic recognizable languages of [1], but that only accepts the regular languages when restricted to the one-dimensional case (that is, string languages), that has a membership problem that is decidable in polynomial time, and that has nice closure properties.

---

* F. Mráz was supported by the Grant Agency of the Czech Republic under the projects P103/10/0783 and P202/10/1333.

A.-H. Dediu et al. (Eds.): LATA 2014, LNCS 8370, pp. 541–552, 2014.

As a first such model, the *Sgraffito automaton* was introduced and studied in cooperation with D. Průša [11,12,13]. The nondeterministic Sgraffito automaton is too powerful, as it accepts all recognizable picture languages, but the deterministic Sgraffito automaton meets most of these properties, and it is more expressive than the *four-way alternating automaton* [7] and the deterministic four-way one-marker automaton [2], and it accepts the *sudoku-deterministically recognizable* picture languages [3].

In the next step, the restarting automaton, which was introduced in [6] as a formal device to model the linguistic technique of *analysis by reduction*, has been extended to models that process two-dimensional inputs. The first such model is the *restarting tiling automaton*, which is a stateless device with a two-by-two window [10]. In each cycle it scans the current picture based on a given scanning strategy until, at some place, it performs a rewrite step and restarts. If no rewrite operation can be performed, then the automaton halts after scanning the current picture completely. It is said to accept if at that point the current picture satisfies certain local conditions, similar to a tiling automaton (see, e.g., [5]).

Then in [9] the current authors introduced the *deterministic two-dimensional three-way ordered restarting automaton* (or det-2D-3W-ORWW-automaton, for short) that works more in the original spirit of restarting automata. Such an automaton has a window of size three-by-three, and it scans a given rectangular input picture starting at the top left corner. Based on the current state and the contents of its window, it can change its state and move either to the right, down, or up, but not to the left. It keeps on moving until it either halts, accepting or rejecting, or until it performs a rewrite step, in which it replaces the symbol in the middle of its window by a symbol that is strictly smaller with respect to a given ordering on its tape alphabet. After performing such a rewrite, the automaton restarts immediately. When restricted to one-dimensional inputs (that is, strings), then this device just accepts the regular languages. For two-dimensional inputs, however, it is quite powerful, as it can simulate the deterministic Sgraffito automaton. However, because it can perform up and down movements, while it can only perform move-right but no move-left steps, this automaton clearly favours vertical operations over horizontal operations. Hence, it is not surprising that it can accept picture languages consisting of one-column pictures such that the corresponding string languages are not regular.

Here we restrict these automata even further by considering a *two-way* variant that can only perform move-right and move-down steps. However, such an automaton would not be able to scan a given rectangular picture completely within a single cycle, and accordingly, it appears that it would be quite weak. Therefore, we introduce an extended variant, the *deterministic two-dimensional extended two-way ordered restarting automaton* (det-2D-x2W-ORWW-automaton) for which the move operations are somewhat more general: when a move-right step is executed, while the central position of the window is placed in row i of the last column, then the window is moved such that its central position is in row $i + 1$ of the first column. Analogously, when a move-down step is executed, while the central position of the window is placed in column j of the bottommost row, then the window is moved such that its central position is in column $j + 1$ of the topmost row. In order to avoid infinite sequences of move operations, we require that in any cycle, the automaton can either use extended move-left or extended move-down steps, but not both.

Here we show that this automaton is quite expressive, as already its stateless variant can simulate the deterministic Sgraffito automaton. However, when restricted to picture languages that only consist of one-column pictures, then the det-2D-x2W-ORWW-automaton can only accept languages that are obtained by the operation of rotation from regular (string) languages. Hence, it follows that our automata cannot accept all those picture languages that are accepted by det-2D-3W-ORWW-automata. In fact, it turns out that the class of picture languages that our automata accept is incomparable under inclusion to the class of picture languages that are accepted by det-2D-3W-ORWW-automata.

2   Picture Languages

Here we use the common notation and terms on pictures and picture languages (see, e.g., [5]). Let Σ be a finite alphabet, and let $P \in \Sigma^{*,*}$ be a *picture* over Σ, that is, a two-dimensional array of symbols from Σ. If P is of size $m \times n$, then we write $P \in \Sigma^{m,n}$, and we take $\ell_1(P)$ ($\ell_2(P)$) to denote the number of rows (columns) of P. Further, $P(i, j)$ denotes the symbol at row i and in column j for all $1 \le i \le m$ and $1 \le j \le n$. We introduce a set of five special markers (*sentinels*) $\mathcal{S} = \{\vdash, \dashv, \top, \bot, \#\}$, and we assume that $\Sigma \cap \mathcal{S} = \emptyset$ for any alphabet Σ considered. In order to enable an automaton to detect the border of P easily, we define the *boundary picture* $\widehat{P}$ over $\Sigma \cup \mathcal{S}$ of size $(m + 2) \times (n + 2)$. It is illustrated by the following schema:

#	$\top$	$\top$	$\cdots$	$\top$	$\top$	#
$\vdash$						$\dashv$
$\vdots$			P			$\vdots$
$\vdash$						$\dashv$
#	$\bot$	$\bot$	$\cdots$	$\bot$	$\bot$	#

We now restate in short the definition of the det-2D-3W-ORWW-automaton from [9]. This automaton has a read/write window of size three-by-three, which it can move across a given bordered picture $\widehat{P}$. For doing so, it uses the set $\mathcal{H} = \{R, D, U\}$ of possible *window movements*, where R denotes a step to the *right*, D a step *down*, and U a step *up*. Observe that no movement to the left is allowed.

Definition 1. *A deterministic two-dimensional three-way ordered restarting automaton, a* det-2D-3W-ORWW-*automaton for short, is given through a 7-tuple* $M = (Q, \Sigma, \Gamma, \mathcal{S}, q_0, \delta, >)$, *where Q is a finite set of states containing the initial state q_0, Σ is a finite input alphabet, Γ is a finite tape alphabet containing Σ such that $\Gamma \cap \mathcal{S} = \emptyset$, $>$ is a partial ordering on Γ, and $\delta : Q \times (\Gamma \cup \mathcal{S})^{3,3}$*

$\rightarrow (Q \times \mathcal{H}) \cup \Gamma \cup \{\mathsf{Accept}\}$ *is the transition function that satisfies the following four restrictions for all* $q \in Q$ *and all* $C \in (\Gamma \cup \mathcal{S})^{3,3}$:

1. *if* $C(1,2) = \top$, *then* $\delta(q, C) \neq (q', \mathrm{U})$ *for all* $q' \in Q$,
2. *if* $C(2,3) = \dashv$, *then* $\delta(q, C) \neq (q', \mathrm{R})$ *for all* $q' \in Q$,
3. *if* $C(3,2) = \bot$, *then* $\delta(q, C) \neq (q', \mathrm{D})$ *for all* $q' \in Q$,
4. *if* $\delta(q, C) = b \in \Gamma$, *then* $C(2,2) > b$ *with respect to the ordering* $>$.

To simplify the presentation we say that the window of M is *at position* (i, j) to mean that the field in the center of the window is at row i and column j. Given a picture $P \in \Sigma^{m,n}$ as input, M begins its computation in state q_0 with its read/write window reading the subpicture of size 3×3 of $\widehat{P}$ at the upper left corner, that is, the window is at position $(1,1)$ of P. Applying its transition function, M now moves through $\widehat{P}$ until it reaches a state q and a position with current contents C of the read/write window such that either $\delta(p, C)$ is undefined, or $\delta(p, C) = \mathsf{Accept}$, or $\delta(p, C) = b$ for some letter $b \in \Gamma$ such that $C(2,2) > b$. In the first case, M gets stuck, and so the current computation ends without accepting, in the second case, M halts and accepts, and in the third case, M replaces the symbol $C(2,2)$ by the symbol b, moves its read/write window back to the upper left corner, and reenters its initial state q_0. This latter step is therefore called a *combined rewrite/restart step*. A picture $P \in \Sigma^{*,*}$ is *accepted* by M, if the computation of M on input P ends with an Accept instruction. By $L(M)$ we denote the language consisting of all pictures over Σ that M accepts.

In principle it could happen that M does not terminate on some input picture, as it may get stuck on a column, just moving up and down. Therefore, it is *required explicitly* that M halts on all input pictures (see [9])! To avoid this cumbersome requirement, we now modify this model by allowing only move-right and move-down steps, albeit in an extended form.

Definition 2. *A deterministic two-dimensional extended two-way ordered restarting automaton, a det-2D-x2W-ORWW-automaton for short, is given through a 7-tuple* $M = (Q, \Sigma, \Gamma, \mathcal{S}, q_0, \delta, >)$, *where all components are defined as for a det-2D-3W-ORWW-automaton with the restriction that* $\mathcal{H} = \{\mathrm{R}, \mathrm{D}\}$ *is taken in the definition of the transition function* δ. *However, the move-right and move-down steps are extended as follows:*

1. *As long as the window does neither contain the right border marker nor the bottom marker, move-right and move-down steps can be used freely.*
2. *When the widow contains the right border marker, but not the bottom marker, then an extended move-right step shifts the window to the beginning of the next row, that is, if the central position of the window is on the last field of row i for some $i < \ell_1(P)$, then it is now placed on the first field of row $i + 1$.*
3. *When the widow contains the bottom marker, but not the right border marker, then an extended move-down step shifts the window to the top of the next column, that is, if the central position of the window is on the bottom-most field of column j for some $j < \ell_2(P)$, then it is now placed on the top-most field of column $j + 1$.*

4. In any cycle, as soon as M executes an extended move-right (move-down) step, then for the rest of this cycle, it cannot execute any extended move-down (move-right) step.

Finally, M is called a stateless det-2D-x2W-ORWW-*automaton (or a* stl-det-2D-x2W-ORWW-*automaton) if it has just a single state. To simplify the presentation, the components Q and q_0 are suppressed for such an automaton.*

Acceptance is defined in the same way as for det-2D-3W-ORWW-automata, and the language consisting of all pictures accepted by M is denoted as $L(M)$.

When restricted to one-row pictures $P \in \Sigma^{1,*}$, then the det-2D-x2W-ORWW-automaton coincides with the det-ORWW-automaton of [9]. Thus, we obtain the following result, where the part on stateless variants is an easy extension.

Corollary 3. *When restricted to one-dimensional input, then the* det-2D-x2W-ORWW-*automaton just accepts the regular string languages. This also holds for the stateless variant.*

We start our investigation with a simple example.

Example 4. Let $\Sigma = \{0, 1\}$, and let $L_{\text{perm}} \subseteq \Sigma^{*,*}$ be the picture language

$$L_{\text{perm}} = \{ P \in \Sigma^{*,*} \mid \ell_1(P) = \ell_2(P) \geq 1,$$
$$\text{each row and column contains exactly one symbol 1} \}.$$

We describe a stl-det-2D-x2W-ORWW-automaton M_{perm} that accepts this language. Obviously, a det-2D-x2W-ORWW-automaton (i.e. with states) could easily check whether each row (column) of the given input picture P contains exactly one occurrence of the symbol 1 by traversing P row by row (column by column) using extended move-right (move-down) steps. However, as it could not do both these traversals in a single cycle, it must use its ability to perform rewrite operations for switching from the one traversal to the other. As M_{perm} is reset to the initial state and the initial position after each rewrite step, it cannot remember which traversal it has already completed. A stateless automaton must use markings on its tape to control which phase of the above computations it is in.

Let $\Gamma = \Sigma \cup \{0', 1', 0'_1, 0'', 1'', 0''_1\}$, and let $1 > 0 > 1' > 0' > 0'_1 > 1'' > 0'' > 0''_1$ be the ordering on Γ to be used. Let us describe how a stl-det-2D-x2W-ORWW-automaton can check that a given row contains exactly one occurrence of the symbol 1.

If the first symbol in the row is from Σ, then M_{perm} moves across this row from left to right. In order to check that there is a unique occurrence of the symbol 1 in this row, the automaton starts to rewrite symbols in the row from right to left. The symbol 0 is rewritten into $0'$ only if it is followed by the right sentinel or by another $0'$. The symbol 1 is rewritten into $1'$ only if it is followed by the right sentinel or by $0'$. If 1 is followed by $1'$, then the automaton rejects, as this row of the input picture contains at least two occurrences of the symbol 1. If the symbol 0 is followed by $1'$, then it is rewritten into $0'_1$ with the meaning that there is already one occurrence of the symbol 1 to the right of this tape cell in the original picture. Similarly, if the symbol 0 is followed by $0'_1$, it is also rewritten into $0'_1$. Clearly, if the first symbol in a row is not from Σ and it is neither $1'$ nor $0'_1$, then the automaton rejects.

The automaton M_{perm} can perform the above process row by row in the bottom-up order. When the symbol at position $(1,1)$ is already $0'_1$ or $1'$, the automaton can check that there is exactly one occurrence of the symbol $1'$ in each column. This time it will rewrite the columns in the right-to-left order in a similar way as the rows in the first phase. For rewriting the columns it will use the symbols $0''$, $1''$, and $0''_1$. At last, M_{perm} accepts when the field $(1,1)$ contains either $0''_1$ or $1''$. In that case, the given input P is a *square* that belongs to the language L_{perm}. Hence, we see that $L(M_{\text{perm}}) = L_{\text{perm}}$.

Given an input picture P over Σ of size $m \times n$, a det-2D-x2W-ORWW-automaton $M = (Q, \Sigma, \Gamma, \mathcal{S}, q_0, \delta, >)$ can execute at most $m \cdot n \cdot (|\Gamma| - 1)$ many cycles, as in each cycle it rewrites one of the $m \cdot n$ many symbols of the current picture by a symbol that is strictly smaller. In each cycle M can either execute up to n move-right steps, $n \cdot (m - 1)$ move-down steps, and $(n - 1)$ extended move-down steps, or m move-down steps, $m \cdot (n - 1)$ move-right steps, and $(m - 1)$ extended move-right steps. Thus, each cycle takes at most $m \cdot n$ many steps, and hence, M executes at most $m^2 \cdot n^2 \cdot (|\Gamma| - 1)$ many steps. Thus, a two-dimensional Turing machine can simulate M in time $O(m^2 \cdot n^2)$. A multi-tape Turing machine T that stores P column by column needs m steps to simulate a single move-right step of M, and it needs $m \cdot n$ steps to simulate an extended move-right step. Thus, we see from the considerations above that T may need up to $O(m^3 \cdot n^2)$ many steps to simulate M on an input picture of size $m \times n$. Hence, we obtain the following upper bound for the time complexity, where $\mathcal{L}(\mathcal{M})$ is used to denote the class of picture languages that are accepted by the automata of class $\mathcal{M}$.

Theorem 5. $\mathcal{L}(\text{det-2D-x2W-ORWW}) \subseteq \text{DTIME}((\text{size}(P))^3)$.

3   On the Language Class $\mathcal{L}(\text{det-2D-x2W-ORWW})$

First we compare our automaton to the Sgraffito automaton of [11]. A two-dimensional *Sgraffito automaton* (SA) is given by a 7-tuple $\mathcal{A} = (Q, \Sigma, \Gamma, \delta, q_0, Q_F, \mu)$, where Σ is an input alphabet and Γ is a working alphabet such that $\Sigma \subseteq \Gamma$, Q is a set of states containing the initial state q_0 and the set of final states Q_F, $\mu : \Gamma \to \mathbb{N}$ is a weight function, and $\delta : (Q \smallsetminus Q_F) \times (\Gamma \cup \mathcal{S}) \to 2^{Q \times (\Gamma \cup \mathcal{S}) \times \mathcal{H}}$ is a transition relation, where $\mathcal{H} = \{R, L, D, U, Z\}$ is the set of possible *head movements* (the first four elements denote directions (right, left, down, up) and Z represents no movement), such that the following two properties are satisfied:

1. $\mathcal{A}$ is *bounded*, that is, whenever it scans a symbol from $\mathcal{S}$, then it immediately moves to the nearest field of P without changing this symbol,
2. $\mathcal{A}$ is *weight-reducing*, that is, for all $q, q' \in Q$, $d \in \mathcal{H}$, and $a, a' \in \Gamma$, if $(q', a', d) \in \delta(q, a)$, then $\mu(a') < \mu(a)$.

Finally, $\mathcal{A}$ is *deterministic* (a 2DSA), if $|\delta(q, a)| \leq 1$ for all $q \in Q$ and $a \in \Gamma \cup \mathcal{S}$.

The notions of configuration and computation are defined as usual. In the initial configuration on input P, the tape contains $\widehat{P}$, $\mathcal{A}$ is in state q_0, and its head scans the top-left corner of P. The automaton $\mathcal{A}$ accepts P iff there is a computation of $\mathcal{A}$ on input P that finishes in a state from Q_F.

In [9] it is shown that each deterministic Sgraffito automaton can be simulated by a det-2D-3W-ORWW-automaton. By using essentially the same proof idea also the following result can be derived.

Theorem 6. $\mathcal{L}(\text{2DSA}) \subseteq \mathcal{L}(\text{stl-det-2D-x2W-ORWW}).$

We compare the det-2D-x2W-ORWW-automaton to the det-2D-3W-ORWW-automaton of [9]. For this we first establish two closure properties for the classes $\mathcal{L}(\text{det-2D-x2W-ORWW})$ and $\mathcal{L}(\text{stl-det-2D-x2W-ORWW})$.

Proposition 7. *The classes of picture languages* $\mathcal{L}(\text{det-2D-x2W-ORWW})$ *and* $\mathcal{L}(\text{stl-det-2D-x2W-ORWW})$ *are closed under transposition and complementation.*

Proof. (a) The *transpose* P^t of a picture P is obtained from P by interchanging the columns and rows. If M is a (stateless) det-2D-x2W-ORWW-automaton for $L \subseteq \Sigma^{*,*}$, then by interchanging move-right steps with move-down steps, and by transposing all the 3×3 pictures used in the description of the transition function of M, we obtain a (stateless) det-2D-x2W-ORWW-automaton M^t such that $L(M^t) = \{\, P^t \mid P \in L(M)\,\}$.

(b) Let M be a (stateless) det-2D-x2W-ORWW-automaton on Σ that accepts a language $L \subseteq \Sigma^{*,*}$. From M we obtain a (stateless) det-2D-x2W-ORWW-automaton M^c by interchanging undefined transitions and Accept transitions. Then $L(M^c) = \Sigma^{*,*} \smallsetminus L = L^c$, the complement of the language $L = L(M)$. $\qquad\square$

In [9] it is shown that the language $L_{1\text{col}}$, which is defined by

$$L_{1\text{col}} = \{\, P \in \Sigma^{2n,1} \mid n \geq 1,\, P(1,1)\dots P(n,1) = (P(n+1,1)\dots P(2n,1))^R\,\},$$

is accepted by a det-2D-3W-ORWW-automaton. The transpose $L_{1\text{col}}^t$ of this language is essentially the string language $L_{\text{pal}} = \{\, w \in \{a,b\}^* \mid |w| \equiv 0 \bmod 2 \text{ and } w = w^R\,\}$, the language of palindromes of even length, which is not regular. However, we observed above that det-2D-x2W-ORWW-automata can only accept string languages that are regular. Thus, $L_{1\text{col}}^t$ is not accepted by any det-2D-x2W-ORWW-automaton, and hence, by Proposition 7, $L_{1\text{col}}$ is not accepted by any det-2D-x2W-ORWW-automaton, either. It follows that there are some det-2D-3W-ORWW-automata which cannot be simulated by det-2D-x2W-ORWW-automata. However, also the converse holds.

Example 8. Let $L_{\text{pal},2}$ be the following picture language over $\Sigma = \{a, b, \#\}$:

$$L_{\text{pal},2} = \{\, P \in \Sigma^{2,2n} \mid n \geq 1,\, P(1,1)\dots P(1,n) = (P(1,n+1)\dots P(1,2n))^R,$$
$$P(1,i) \in \{a,b\} \text{ and } P(2,i) = \# \text{ for all } 1 \leq i \leq 2n\,\},$$

that is, $L_{\text{pal},2}$ consists of all two-row pictures such that the first row contains a palindrome of even length over $\{a,b\}$, and the second row just contains #-symbols. We claim that $L_{\text{pal},2}$ is accepted by a det-2D-x2W-ORWW-automaton.

Let $M_{\mathrm{pal},2} = (Q, \Sigma, \Gamma, \mathcal{S}, q_0, \delta, >)$ be the det-2D-x2W-ORWW-automaton that is defined by taking $\Gamma = \Sigma \cup \{a_1, a_2, b_1, b_2, \uparrow\}$ with $a > b > a_1 > b_1 > a_2 > b_2 >$ $\# > \uparrow$, and defining δ in such a way that $M_{\mathrm{pal},2}$ proceeds as follows:

Similar to the ORWW-automaton M of Example 1 of [9], $M_{\mathrm{pal},2}$ marks the letters in the first row alternatingly with indices 1 and 2, alternating between marking the first unmarked letter from the left and the first unmarked letter from the right. In order to determine which of these two cases it is currently working on, the second row is used as follows: First M scans the first row completely from left to right. If during this sweep it realizes that the first unmarked letter from the right must be marked, then it simply does this and restarts. If, however, it realizes at the end of this sweep that the first unmarked letter from the left should have been marked, then it executes an extended move-right operation at the right end of the first row, and then it replaces the letter in row 2 that is below the first unmarked letter from the left in row 1 by the symbol $\uparrow$, in this way indicating that the corresponding letter in row 1 must be marked in the next cycle. It is clear that in this way the language $L_{\mathrm{pal},2}$ is accepted.

In [9] it is shown that the language L_{dub} (see below) is not accepted by any det-2D-3W-ORWW-automaton. Using the same proof technique also the following result can be shown.

Proposition 9. $L_{\mathrm{pal},2} \notin \mathcal{L}(\text{det-2D-3W-ORWW})$.

These results yield the following incomparability result.

Corollary 10. *The class of picture languages $\mathcal{L}(\text{det-2D-x2W-ORWW})$ is incomparable under inclusion to the class of picture languages $\mathcal{L}(\text{det-2D-3W-ORWW})$.*

Actually, $L_{\mathrm{pal},2}$ also separates the det-2D-x2W-ORWW-automata from their stateless variants.

Proposition 11. $L_{\mathrm{pal},2} \notin \mathcal{L}(\text{stl-det-2D-x2W-ORWW})$.

Proof. Assume that $M = (\Sigma, \Gamma, \mathcal{S}, \delta, >)$ is a stl-det-2D-x2W-ORWW-automaton over $\Sigma = \{a, b, \#\}$ such that $L(M) = L_{\mathrm{pal},2}$. For $w = a_1 \ldots a_n$, where $n \geq 1$ and $a_1, \ldots, a_n \in \{a, b\}$, let $P_w = \begin{bmatrix} a_1 \ldots a_n \, a \, a \, a_n \ldots a_1 \\ \# \ldots \# \, \# \, \# \, \# \ldots \# \end{bmatrix} \in L_{\mathrm{pal},2}$ be an input picture. Given P_w as input, M will perform an accepting computation, which consists of a finite sequence of cycles that is followed by an accepting tail computation. We now split this computation into a finite number of *phases*, where we distinguish between four types of phases:

1. A *left-only phase* consists of a sequence of cycles in which the window of M stays on the left half of the picture.
2. An *upper-right phase* consists of a sequence of cycles in which all rewrite steps (and possibly an accept step) are performed on the right half of the picture, and in addition, in the first of these cycles, M enters the right half of the picture through a move-right step in row 1.

3. A *lower-left phase* is a sequence of cycles in which all rewrite steps (and possibly an accept step) are performed in the left half of the picture, and in addition, the first of these cycles contains an extended move-right step, that is, in the first of these cycles, M scans the first row completely from left to right, executes an extended move-right step at the right end of that row, and performs a rewrite step within the left half of the second row.

4. A *lower-right phase* is a sequence of cycles in which all rewrite steps (and possibly an accept step) are performed in the right half of the picture, and in addition, in the first of these cycles, M enters the right half of the picture through a move-right step in row 2 or through an extended move-down step.

Obviously, the sequence of cycles of the computation of M on input P_w can uniquely be split into a sequence of phases if we require that each phase is of maximum length. Thus, this computation can be described in a unique way by a string α over the alphabet $\Omega = \{O, U, L, R\}$, where O denotes a left-**O**nly phase, U stands for an **U**pper-right phase, L denotes a lower-**L**eft phase, and R stands for a lower-**R**ight phase.

Concerning the possible changes from one phase to the next there are some restrictions based on the fact that M is stateless.

- While M is in a lower-right phase (R), it just moves through the left half of the current picture after each rewrite/restart step. Thus, M cannot get into another phase until it performs a rewrite step that replaces a symbol in the first column of the right half of the picture. Only then may follow a left-only phase (O) or a lower-left phase (L). However, in a fixed column, less than $2 \cdot |\Gamma|$ many rewrite steps can be performed, and so $|\alpha|_R \leq 1 + 2 \cdot |\Gamma|$.

- When M is in a lower-left phase (L), then it can next get into a lower-right phase (R) or into an upper-right phase (U). However, when M got into the lower-left phase, then it moved all the way right across the first row. Thus, it cannot get into an upper-right phase (U) before a rewrite step is performed that replaces a symbol in the last column of the left half of the picture. As there are less than $2 \cdot |\Gamma|$ many rewrite steps that can be performed on this column, we see that $|\alpha|_L \leq 1 + |\alpha|_R + 2 \cdot |\Gamma|$.

- When M is in an upper-right phase (U), then it can next get into a lower-left phase (L), a lower-right phase (R) or a left-only phase (O). However, when M got into the upper-right phase, then it moved across the left half of the first row, and so it can get into a left-only phase only after a symbol in the first column of the right half of the picture has been rewritten. It follows that $|\alpha|_U \leq 1 + |\alpha|_L + |\alpha|_R + 2 \cdot |\Gamma|$.

- A left-only phase (O) can be followed by any other phase. Thus, we obtain that $|\alpha|_O \leq 1 + |\alpha|_R + |\alpha|_L + |\alpha|_U$.

It follows that $|\alpha| \leq |\alpha|_O + |\alpha|_R + |\alpha|_L + |\alpha|_U \leq 15 + 28 \cdot |\Gamma|$, that is, each computation of M consists of at most $15 + 28 \cdot |\Gamma|$ many phases.

Finally, we associate a *generalized crossing sequence* $\text{GCS}(w)$ with the computation of M on the input picture P_w as follows:

Let $\alpha(w) \in \{O, U, L, R\}^+$ be the description of the sequence of phases of the accepting computation of M on input P_w. Now after each letter X of $\alpha(w)$, we insert a 2-by-2 picture $\begin{pmatrix} c & d \\ e & f \end{pmatrix}$ such that $\binom{c}{e}$ is the contents of the rightmost column of the left half and $\binom{d}{f}$ is the contents of the leftmost column of the right half of the picture at the end of the phase represented by the letter X. Thus, $GCS(w)$ is a string of length at most $30 + 56 \cdot |\Gamma|$ over the finite alphabet $\Omega \cup \Gamma^{2,2}$ of size $4 + |\Gamma|^4$, that is, there are only finitely many different such crossing sequences.

If n is sufficiently large, then there are two strings $w_1, w_2 \in \{a, b\}^n$, $w_1 = a_1 \ldots a_n$ and $w_2 = b_1 \ldots b_n$, such that $w_1 \neq w_2$, but $GCS(w_1) = GCS(w_2)$. As M accepts both P_{w_1} and P_{w_2}, it follows that M will also accept on input $P' = \begin{bmatrix} a_1 \cdots a_n \ a \ a \ b_n \cdots b_1 \\ \# \cdots \# \ \# \ \# \ \# \cdots \# \end{bmatrix}$, which contradicts our assumption on M, as $P' \notin L_{\mathrm{pal},2}$. This completes the proof of Proposition 9. $\square$

Together with Example 8 this yields the following separation result.

Theorem 12. $\mathcal{L}(\mathsf{stl\text{-}det\text{-}2D\text{-}x2W\text{-}ORWW}) \subsetneq \mathcal{L}(\mathsf{det\text{-}2D\text{-}x2W\text{-}ORWW})$.

Let $\Sigma = \{0, 1\}$, and let L_{dub} denote the language of *duplicates* that consists of all pictures $P \oplus P$, where P is any quadratic picture over Σ and $P \oplus P$ denotes the column concatenation of two copies of P, that is, two copies of P are put in a row. It is shown in [11] that $L_{\mathrm{dub}} \notin \mathcal{L}(\mathsf{2SA})$, and it is shown in [9] that $L_{\mathrm{dub}} \notin \mathcal{L}(\mathsf{det\text{-}2D\text{-}3W\text{-}ORWW})$. However, by using the technique from Example 8, the following can be shown.

Proposition 13. $L_{\mathrm{dub}} \in \mathcal{L}(\mathsf{det\text{-}2D\text{-}x2W\text{-}ORWW})$.

Thus, we see that $\mathcal{L}(\mathsf{det\text{-}2D\text{-}x2W\text{-}ORWW})$ is not contained in $\mathcal{L}(\mathsf{2SA})$, but it remains the question of whether $\mathcal{L}(\mathsf{2SA}) \subset \mathcal{L}(\mathsf{det\text{-}2D\text{-}x2W\text{-}ORWW})$ holds. We continue with some more closure properties.

Theorem 14. $\mathcal{L}(\mathsf{det\text{-}2D\text{-}x2W\text{-}ORWW})$ *is closed under intersection.*

Proof. Let $M_1 = (Q_1, \Sigma, \Gamma_1, \mathcal{S}, q_0^{(1)}, \delta_1, >_1)$ be a $\mathsf{det\text{-}2D\text{-}x2W\text{-}ORWW}$-automaton on $\Sigma = \{a_1, \ldots, a_k\}$ that accepts a picture language $L_1 = L(M_1) \subseteq \Sigma^{*,*}$, and let $M_2 = (Q_2, \Sigma, \Gamma_2, \mathcal{S}, q_0^{(2)}, \delta_2, >_2)$ be a $\mathsf{det\text{-}2D\text{-}x2W\text{-}ORWW}$-automaton on Σ that accepts a picture language $L_2 = L(M_2) \subseteq \Sigma^{*,*}$. We now construct a $\mathsf{det\text{-}2D\text{-}x2W\text{-}ORWW}$-automaton $M = (Q, \Sigma, \Gamma, \mathcal{S}, q_0, \delta, >)$ such that $L(M) = L_1 \cap L_2$. Essentially, M will work as follows:

1. M first simulates M_1, that is, it behaves exactly like M_1. If M_1 should get stuck on the given input, that is, M_1 does not accept, then neither does M. If, however, M_1 accepts, then instead of accepting, M marks the position (i, j) at which M_1 accepts, using a special symbol.
2. Now M should simulate M_2. However, unless the marked position (i, j) happens to be inside the initial position of the window, M does not know that it should now simulate M_2. Therefore, it still behaves just like M_1 during the tail of its accepting computation. Hence, no rewrite will be performed, but

the marked position (i, j) will be reached eventually. Now M adds a mark also to the position (i', j') that it reached prior to the position (i, j).

However, here a problem arises if position (i, j) was reached by an extended move-right (move-down) step, as then the previous position (i', j') is not inside the window when the position (i, j) is reached. In that case, M marks the first unmarked position to the right (below) position (i, j). Continuing in this manner, row j (column i) will eventually be completely marked. Hence, when M reaches position (i', j'), then it already sees that the symbol in the next row (column) is marked, and instead of performing an extended move-right (move-down) step, the symbol at position (i', j') is marked. Thus, after finitely many cycles a path from the position (i, j), at which M_1 accepted, to the initial position $(1, 1)$ is completely marked.

3. When the symbol at position $(1, 1)$ is marked, then M starts to simulate M_2. It keeps on doing that until M_2 halts. Now M accepts iff M_2 accepts.

Obviously, with the above strategy M accepts the language $L = L_1 \cap L_2$. There are a few technical difficulties with this strategy, but they can be overcome. □

As the class $\mathcal{L}$(det-2D-x2W-ORWW) is closed under complement and intersection, we also have the following closure property.

Corollary 15. $\mathcal{L}$(det-2D-x2W-ORWW) *is closed under union.*

Actually, the construction used in the proof of Theorem 14 can be extended to stateless det-2D-x2W-ORWW-automata. Thus, we also have the following result.

Corollary 16. $\mathcal{L}$(stl-det-2D-x2W-ORWW) *is closed under union and intersection.*

4   Concluding Remarks

We have introduced a class of two-dimensional restarting automata, the det-2D-x2W-ORWW-automata, and their stateless variants, and we have seen that already the stateless automata of this type are at least as expressive as the deterministic Sgraffito automata, although they still only accept regular string languages. The classes of picture languages obtained are closed under transposition and under the Boolean operations, but it is still open whether they are closed under projection, under horizontal product, or under vertical product. Also it remains open whether the stateless det-2D-x2W-ORWW-automata accept any language that cannot be accepted by a deterministic Sgraffito automaton, or whether these two types of automata have exactly the same expressive power. In fact, it is not even known whether the stateless det-2D-x2W-ORWW-automata can be simulated by det-2D-3W-ORWW-automata with states. The det-2D-x2W-ORWW-automata (with states), however, are known to be incomparable under inclusion to the det-2D-3W-ORWW-automata, but as the former are closed under transposition, they appear to be more natural than the latter. However, it still remains to study the language class $\mathcal{L}$(det-2D-x2W-ORWW) in more detail.

References

1. Anselmo, M., Giammarresi, D., Madonia, M.: From determinism to non-determinism in recognizable two-dimensional languages. In: Harju, T., Karhumäki, J., Lepistö, A. (eds.) DLT 2007. LNCS, vol. 4588, pp. 36–47. Springer, Heidelberg (2007)
2. Blum, M., Hewitt, C.: Automata on a 2-dimensional tape. In: Proc. 8th Annual Symposium on Switching and Automata Theory (SWAT 1967), pp. 155–160. IEEE Computer Society, Washington, DC (1967)
3. Borchert, B., Reinhardt, K.: Deterministically and sudoku-deterministically recognizable picture languages. In: Loos, R., Fazekas, S., Martin-Vide, C. (eds.) LATA 2007, Preproc., pp. 175–186. Report 35/07, Research Group on Mathematical Linguistics, Universitat Rovira i Virgili, Tarragona (2007)
4. Giammarresi, D., Restivo, A.: Recognizable picture languages. International J. of Pattern Recognition and Artificial Intelligence 6, 241–256 (1992)
5. Giammarresi, D., Restivo, A.: Two-dimensional languages. In: Rozenberg, G., Salomaa, A. (eds.) Handbook of Formal Languages, vol. 3, pp. 215–267. Springer, New York (1997)
6. Jančar, P., Mráz, F., Plátek, M., Vogel, J.: Restarting automata. In: Reichel, H. (ed.) FCT 1995. LNCS, vol. 965, pp. 283–292. Springer, Heidelberg (1995)
7. Kari, J., Moore, C.: New results on alternating and non-deterministic two-dimensional finite-state automata. In: Ferreira, A., Reichel, H. (eds.) STACS 2001. LNCS, vol. 2010, pp. 396–406. Springer, Heidelberg (2001)
8. Lindgren, K., Moore, C., Nordahl, M.: Complexity of two-dimensional patterns. Journal of Statistical Physics 91, 909–951 (1998)
9. Mráz, F., Otto, F.: Ordered restarting automata for picture languages. In: Geffert, V., Preneel, B., Rovan, B., Stuller, J., Tjoa, A. (eds.) SOFSEM 2014. LNCS. Springer, Heidelberg (to appear, 2014)
10. Průša, D., Mráz, F.: Restarting tiling automata. In: Moreira, N., Reis, R. (eds.) CIAA 2012. LNCS, vol. 7381, pp. 289–300. Springer, Heidelberg (2012)
11. Průša, D., Mráz, F.: Two-dimensional sgraffito automata. In: Yen, H.-C., Ibarra, O.H. (eds.) DLT 2012. LNCS, vol. 7410, pp. 251–262. Springer, Heidelberg (2012)
12. Průša, D., Mráz, F., Otto, F.: Comparing two-dimensional one-marker automata to sgraffito automata. In: Konstantinidis, S. (ed.) CIAA 2013. LNCS, vol. 7982, pp. 268–279. Springer, Heidelberg (2013)
13. Průša, D., Mráz, F., Otto, F.: New results on deterministic sgraffito automata. In: Béal, M.-P., Carton, O. (eds.) DLT 2013. LNCS, vol. 7907, pp. 409–419. Springer, Heidelberg (2013)

Weight-Reducing Hennie Machines and Their Descriptional Complexity[*]

Daniel Průša

Czech Technical University, Faculty of Electrical Engineering,
Karlovo náměstí 13, 121 35 Prague 2, Czech Republic
prusapa1@cmp.felk.cvut.cz

Abstract. We present a constructive variant of the Hennie machine. It is demonstrated how it can facilitate the design of finite-state machines. We focus on the deterministic version of the model and study its descriptional complexity. The model's succinctness is compared with common devices that include the nondeterministic finite automaton, two-way finite automaton and pebble automaton.

Keywords: Finite automata, two-way automata, Hennie machine, descriptional complexity.

1 Introduction

Regular languages are naturally defined via finite automata. Various extensions of this basic model preserve its recognition power (nondeterminism, two-way movement, use of a pebble). However, measured in the number of states or transitions, they provide a more economic mean of a language description.

Great attention has been paid to the cost of transformations among the models [15,8,13,4]. From our point of view, the related studies usually do not cover automata which, in some restricted way, can rewrite the content of the tape. A two-way finite automaton with write-once tracks presented by Durak [2] is one of the few.

A very general representative of rewriting devices we have in mind is a Hennie machine. It is a bounded, single-tape Turing machine performing constantly many transitions over each tape field, independently on the input's length. Hennie proved that the machine recognizes only regular languages and also generalized this result on any Turing machine working in linear time [6]. Hartmanis later showed that even time $\mathcal{O}(n \log n)$ still leads to recognition of regular languages [5]. Only going beyond this time complexity allows to recognize a non-regular language.

Generality of the model causes some unpleasant properties. It is undecidable whether a given Turing machine is a Hennie machine. Moreover, there is no computable function bounding the blowup in states when transforming to a finite

---

[*] The author was supported by the Grant Agency of the Czech Republic under the project P103/10/0783.

A.-H. Dediu et al. (Eds.): LATA 2014, LNCS 8370, pp. 553–564, 2014.

automaton. The aim of this paper is to define a reasonable constructive subclass of deterministic Hennie machines and study its descriptional complexity. To achieve this, a weight-reducing property is utilized – each transition is required to lower a weight of the scanned symbol.

A two-dimensional variant of the weight-reducing Hennie machine has already been introduced in [10]. Relation to other two-dimensional models with respect to the recognition power was investigated there.

The paper is structured as follows. In the next section we demonstrate that the possibility of rewriting can greatly facilitate the design of a finite automaton. Then we define the weight-reducing Hennie machine and compare it with the original one. Section 4 focuses on the descriptional complexity. The relation to other automata is studied. The paper closes with a short summary and some open problems in Section 5.

2   Motivation

Consider a system of objects aligned in a row. The leftmost one is a transmitter that sends a signal and the rightmost one is a receiver that waits for it. The signal can be in two states, it is either *normal* or *amplified*. The objects between the transmitter and receiver are of three types: *silencer*, *reflecting silencer* and *amplifier*. The silencer changes the first amplified signal it receives to normal. After that it becomes passive and, from that time, it does not influence incoming signals at all. The reflecting silencer behaves like the silencer, but in addition, it reflects the first received amplified signal back. Finally, the amplifier changes a normal signal to amplified, however, a signal (normal or amplified) can pass trough it only k times. If it passes there $k + 1$-st time, the amplifier is burned and the signal is lost.

Now, given such a system, the question is whether a normal signal sent by the transmitter will get to the receiver. Examples are shown in Figure 1.

Fig. 1. Two systems composed of amplifiers (a), silencers (s) and reflecting silencers (r). On the left, the signal passes to the receiver provided that $k \geq 3$, on the right, it returns back to the transmitter. The signal is normal in the dashed parts of the trajectory and amplified in the solid parts.

Each system is encoded by a string over $\Sigma = \{a, r, s\}$. It is clear that systems complying with the condition can be recognized by a finite automaton. However, if we try to design such an automaton, even for $k = 3$, we find that it is not entirely easy task, despite the fact there is a deterministic solution with 8 states, depicted in Figure 2 (one "dead" state is hidden there).

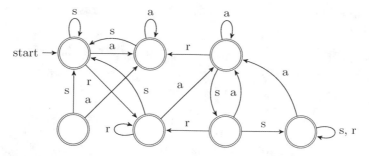

Fig. 2. A deterministic finite automaton accepting codes of systems in which the signal reaches the receiver for $k = 3$. All the displayed states are accepting. Missing transitions are heading to a hidden dead state which is rejecting.

One the other hand, it is not difficult to construct a Turing machine that tracks the signal, marks objects that become inactive and counts in amplifiers how many times the signal passed trough them. In fact, the machine will be a Hennie machine. This demonstrates that a usage of tape rewriting could simplify the design of a finite automaton, by the assumption, we have an automatic procedure that converts the rewriting device to it.

The following table lists sizes of minimal deterministic finite automata accepting our systems for greater (odd) values k. It suggests that, in this case, the dependency is bounded by an exponential function.

k	3	5	7	9	11	13	15
states	8	16	30	56	102	188	346

3 Weight-Reducing Hennie Machines

Given an input string w, a *bounded* Turing machine M operates on a tape which initially stores $\vdash w \dashv$, where $\vdash$, $\dashv$ are special end markers, not contained in the working alphabet of M. Whenever the machine reaches $\vdash$ or $\dashv$ it immediately moves the head back and does not rewrite the end marker. If w is the empty string, the head scans $\vdash$ in the initial configuration. The computation ends after performing one transition. A computation is accepting if M finishes in a final state and rejecting when it is not finite or M terminates due to non-applicability of any instruction. The language accepted by M is denoted as $L(M)$.

We say that a bounded Turing machine M a is a *Hennie machine* if there is a constant k limiting the number of transitions performed over any tape field during any computation. Let $\nu(M)$ denote the smallest such k for M.

Proposition 1 ([6]). *If M is a Hennie machine, $L(M)$ is a regular language.*

Theorem 2. *It is undecidable if a Turing machine T is a Hennie machine.*

Proof. The halting problem reduces to the stated question. Assume T has its working tape unbounded in one direction. Let w be an input and $t(w)$ the length of the computation of T for w. It is possible to construct a bounded Turing machine M fulfilling these two conditions.

- If T halts on w, M is a Hennie machine with $\nu(M) \leq t(w)$.
- If T does not halt on w, for any input v such that $|v| \geq |w|$, M either does not halt or it visits the leftmost tape field at least $|v|$ times.

The idea is to take the tape fields storing an input v as the only space available for M to perform the simulation of T on w. Thus, M memorizes w in states, writes it on the tape and erases all remaining symbols of v (if $|v| < |w|$, it halts). After that it simulates T. If the simulation exceeds the space $|v|$, M visits the leftmost tape field $|v|$ times and halts.   □

A *weight-reducing* Hennie machine is a bounded Turing machine equipped by a weight function defined on working symbols. A transition has to lower the weight of the scanned symbol. A formal definition follows.

Definition 3. A weight-reducing Hennie machine *is a tuple*

$$M = (Q, \Sigma, \Gamma, \delta, q_0, Q_F, \mu) \ where$$

- Q *is a finite set of states,*
- Γ *is a working alphabet,*
- $\Sigma \subseteq \Gamma$ *is an input alphabet,*
- $q_0 \in Q$ *is the initial state,*
- $Q_F \subseteq Q$ *is a set of final states,*
- $\delta : (Q \smallsetminus Q_F) \times (\Gamma \cup \{\vdash, \dashv\}) \to 2^{Q \times (\Gamma \cup \{\vdash, \dashv\}) \times \{\leftarrow, 0, \rightarrow\}}$ *is a transition relation, with the set of the head movements* $\{\leftarrow, 0, \rightarrow\}$,
- $\mu : \Gamma \to \mathbb{N}$ *is a weight function.*

Moreover, the following properties are fulfilled:

- $(Q, \Sigma, \Gamma, \delta, q_0, Q_F)$ *is a bounded Turing machine,*
- *the transition relation is* weight-reducing:

$$for\ all\ q, q' \in Q, d \in \{\leftarrow, 0, \rightarrow\}, a, a' \in \Gamma : \quad (q', a', d) \in \delta(q, a) \ \Rightarrow \mu(a') < \mu(a).$$

M is deterministic *(det-wr Hennie machine) iff* $|\delta(q, a)| \leq 1$ *for all* $q \in Q$ *and* $a \in \Gamma \cup \{\vdash, \dashv\}$.

Observe that the weight-reducing property of δ can be easily algorithmically verified and that $\nu(M) \leq |\Gamma|$.

Lemma 4. *Let* $M = (Q, \Sigma, \Gamma, \delta, q_0, Q_F)$ *be a Hennie machine. There is a weight-reducing Hennie machine* A *such that* $L(A) = L(M)$ *and the working alphabet of* A *has no more than* $(\nu(M) + 1)|\Gamma|$ *symbols. Moreover, if* M *is deterministic, then* A *is deterministic as well.*

Proof. Denote $k = \nu(M)$. Define $A = (Q, \Sigma, \Gamma', \delta', q_0, Q_F, \mu)$, where $\Gamma' = \Sigma \cup (\Gamma \times \{1, \ldots, k\})$ and each instruction $(q, a) \to (q', a', d)$ from δ where $a \in \Gamma$ is represented in δ' by the following instruction set:

$$(q, a) \to (q', (a', 1), d),$$
$$(q, (a, i)) \to (q', (a', i+1), d) \quad \forall i \in \{1, \ldots, k-1\}.$$

Finally, define

$$\mu(a) = k + 1 \qquad \forall a \in \Sigma,$$
$$\mu((a, i)) = k + 1 - i \quad \forall (a, i) \in \Gamma \times \{1, \ldots, k\}.$$

It is easy to see that $L(A) = L(M)$ and that every deterministic δ produces deterministic δ'. $\qquad \Box$

When designing a weight-reducing Hennie machine accepting some language L, it suffices to describe a Hennie machine M accepting L and derive $\nu(M)$. Then M can be transformed by Lemma 4. This will be applied in constructions presented in Section 4.

For a Turing machine T, it is natural to count the number of its transitions to measure the size of its description. It would not make much sense to count solely states, since each Turing machine has an equivalent with only two active states [14]. We apply this measure based on transitions also to Hennie machines.

Let $\tau(T)$ denote the number of transitions of T. Let Q be the set of states of T and Γ be its working alphabet. Note that $\tau(T) = \mathcal{O}(|Q|^2|\Gamma|)$. If T is deterministic, then $\tau(T) = \mathcal{O}(|Q||\Gamma|)$. If a Hennie machine M is transformed by Lemma 4 to a weight-reducing Hennie machine M', then $\tau(M') \leq \nu(M)\,\tau(M)$.

4   Results on Descriptional Complexity

In this section we give results on trade-offs between a det-wr Hennie machine and common models including a deterministic finite automaton (1DFA), nondeterministic finite automaton (1NFA), their two-way generalizations (2DFA, 2NFA), alternating finite automaton (1AFA) and deterministic one-pebble automaton.

Theorem 5. *There is no recursive function bounding the blowup in transitions when transforming a deterministic Hennie machine to a 1DFA.*

Proof. We utilize busy beaver function $S(n)$, defined as the maximum number of steps performed by a halting 2-state Turing machine with a binary working alphabet when started over a blank tape. It is known that $S(n)$ is noncomputable and grows asymptotically faster than any computable function [11].

For each $n > 0$, let w_n be the string over $\Sigma = \{a\}$ of the length $S(n)$. Moreover, define one-string languages $L_n = \{w_n\}$. Each L_n is accepted by a Hennie machine with $\mathcal{O}(n)$ states and $\mathcal{O}(1)$ working tape symbols. The machine works as follows. Simulate an n-state busy beaver. Whenever the beaver performs an i-th step, mark the i-th tape field and return to the original position. Accept if and only if the simulation marks all the input tape fields and does not attempt to mark the right-end marker $\dashv$.

On the other hand, a 1DFA accepting L_n has at least $|w_n| = S(n)$ states. $\quad \Box$

The following theorem is implied by results in [6]. We present a simplified proof for deterministic machines, because the procedure is essential for an automatic conversion to 1DFA.

Theorem 6. *For each n-state, m-working symbol det-wr Hennie machine, there is a $2^{2^{\mathcal{O}(m \log n)}}$-state 1DFA accepting the same language.*

Proof. Let $M = (Q, \Sigma, \Gamma, \delta, q_0, Q_F, \mu)$ be a det-wr Hennie machine such that $|Q| = n$ and $|\Gamma| = m$. Assume M can never reenter its initial state q_0. Moreover, assume M can reach a final state in Q_F only by that transition which moves the head from the right-end marker $\dashv$ to the preceding input field. Any det-wr Hennie machine can be modified to fulfill these restrictions by adding a constant number of states and working symbols.

Consider a sequence $R = ((r_1, d_1), \ldots, (r_\ell, d_\ell))$ where each $r_i \in Q$ and $d_i \in \{\leftarrow, 0, \rightarrow\}$. Such a sequence records in which states M performs transitions over some tape field, possibly including the last state in which M terminates. A pair (r_i, d_i) says that the i-th transition over the field starts in state r_i. Moreover, d_i indicates, which head movement precedes reaching r_i in the field. We can see R as a variant of the crossing sequence, however, defined over a tape field, not over the border between two neighboring fields. Let $\mathcal{R}$ be the set of all such nonempty sequences of length at most $m + 1$. Since $\nu(M) \leq m$, this covers all those sequences emerging during the computation of M.

Construct a 1NFA A with the set of states $\mathcal{R}$. Let A be in a state $R = ((r_1, d_1), \ldots, (r_\ell, d_\ell))$ and the scanned symbol be a. Define transitions by the following rules.

- R is initial iff $(r_1, d_1) = (q_0, 0)$. In such a state, A checks if R is consistent with the behavior of M over the prefix $\vdash a$.
- If R is not initial, M only checks whether it is consistent with the scanned symbol a. It is also required that $d_1 = \rightarrow$, because the field is reached first time after moving the head there from the left neighboring field.
- The next state $R' \in \mathcal{R}$ is nondeterministically guessed. It has to be consistent with transitions in R that move the A's head to the right.
- R is accepting iff $r_\ell \in Q_F$. A can enter such a state only if R is consistent with the behavior of M over the suffix $a \dashv$.

A has $|\mathcal{R}| = \sum_{i=1}^{m+1} (3n)^i = 2^{\mathcal{O}(m \log n)}$ states. If it is transformed to a minimal 1DFA, the desired automaton is obtained.   □

The next step is to prove that the trade-off between a det-wr Hennie machine and a 1DFA is really double exponential. For $n \in \mathbb{N}$, define a language B_n over $\{0, 1, \$\}$ consisting of strings $v_1 \$ v_2 \$ \ldots \$ v_j$ where $j \in \mathbb{N}$, every $v_i \in \{0, 1\}^*$, $|v_j| \leq n$ and there is $\ell < j$ such that $v_\ell = v_j$.

Informally, every string in B_n is a sequence of binary substrings which are separated by the symbol \$. Moreover, the last substring is of length at most n and it is a copy of one of the preceding substrings. For example,

$$v_1 \$ v_2 \$ v_3 \$ v_4 \$ v_5 \$ v_6 = 11\$0101110\$011\$0011\$001\$011 \in B_4$$

since $v_3 = v_6$ and $|v_6| \leq 4$.

Lemma 7. *Every B_n is accepted by a det-wr Hennie machine with $\mathcal{O}(1)$ states and $\mathcal{O}(n)$ working symbols.*

Proof. Let $\Sigma = \{0, 1, \$\}$. To accept B_n, we will construct a det-wr Hennie machine A with the working alphabet $\Gamma = \Sigma \cup \{0, 1, \$, x, f\} \times \{1, \ldots, 2n\}$. The numeric part i of each $(a, i) \in \Gamma$ is used to count the number of transitions over a tape field. In the next description, we omit technical details on it and focus rather on the role of elements in $\{0, 1, \$, x, f\}$.

Let $w \in \Sigma^*$ be an input string. Write it as $w = v_1 \$ v_2 \$ \ldots \$ v_j$ where each $v_j \in \{0, 1\}^*$. Let $s = |v_j|$ and $v_j = a_1 \ldots a_s$. A iterates trough symbols $a_s, a_{s-1}, \ldots, a_1$ and for each of them performs a traversal trough the tape to detect in which substrings v_i the symbol appears at the same position from the back. Specifically, the first iteration starts by moving the head to the right end of w. Then, a_s is memorized in the control unit and replaced on the tape by x. After that, the head moves leftwards until it scans the left-end marker. Whenever it enters a new binary substring v_j (i.e., it has passed symbol $\$$), it checks if its last symbol equals a_s. If so, it is replaced by x, otherwise it is replaced by f (indicating that the check has failed). When the left-end marker is reached, A moves the head to the right end of w and starts the next iteration by locating a_{s-1}, which is the first tape field leftwards not marked by x. The initial tape and the outcome of all iterations are illustrated by the following example.

$$11\$0101110\$011\$0011\$001\$011$$
$$1x\$010111f\$01x\$001x\$00x\$01x$$
$$xx\$01011xf\$0xx\$00xx\$0fx\$0xx$$
$$xx\$0101fxf\$xxx\$0xxx\$xfx\$xxx$$

A accepts w during the last iteration iff there is some v_ℓ ($\ell < n$) whose all symbols have been rewritten by x, including one symbol rewritten during the last iteration (this guaranteers that $|v_\ell| = |v_n|$). In the example above, A accepts since it rewrites all three symbols in v_3, each of them in one of the three iterations.

Finally, if $|v_j| > n$, A terminates and rejects, since it cannot reduce weights of symbols during the $n + 1$-st iteration. □

Lemma 8. *Every 1DFA accepting B_n has at least 2^{2^n} states.*

Proof. Encode each subset of $\{0, 1\}^n$ as a sequence of its elements separated by the symbol $\$$. There are 2^{2^n} such subsets. Let w_1 and w_2 encode two different subsets and let u be a binary substring represented in w_1 but not in w_2 (or vice versa). Then, $w_1 \$ u \in B_n$ and $w_2 \$ u \notin B_n$ (or vice versa), hence $\$ u$ is a distinguishing extension and, by Myhill-Nerode theorem, each 1DFA accepting B_n has at least 2^{2^n} states. □

We use the following proposition to show $2^{\Omega(\sqrt{n})}$ trade-off in transitions when transforming det-wr Hennie machine to 2NFA.

Proposition 9 ([7]). *Let L be a finite language over an unary alphabet accepted by a 2NFA with n states. The longest string in L has length at most $n + 2$.*

For $n \in \mathbb{N}$, define $U_n = \{a^{2^n}\}$ – a one-string language over the unary alphabet $\Sigma = \{a\}$. Every 2NFA accepting U_n has $\Omega(2^n)$ states.

Lemma 10. *There is a $\mathcal{O}(n)$-state, $\mathcal{O}(n)$-working symbol det-wr Hennie machine accepting U_n.*

Proof. For $n \in \mathbb{N}$, we first construct a deterministic Hennie machine M accepting U_n. It will have $\mathcal{O}(n)$ states and the working alphabet $\Gamma = \{a, 0, 1\}$. To process a given input $w \in \{a\}^*$, M uses a binary counter of length n. At the beginning, the counter is represented in the first n tape fields and it is initialized by value n. The least significant bit is in its leftmost field, see Figure 3(a). To perform the initialization, M memorizes in states the binary representation of n and fills the counter in each step accordingly. In the subsequent phase, it repeatedly increases the counter by one and simultaneously shifts its representation on the tape by one field to the right (the former leftmost field of the counter is rewritten to a). This guarantees that the position of the right end of the counter representation always equals the counter's value. Hence, M easily checks if the counter has been increased to value $2^n - 1$ just when there is exactly one tape field between the right end of the counter and the right-end marker $\dashv$, see Figure 3(b). The counter's increment as well as its shift is done by one traversal trough the related tape fields. This means that $\nu(M) = \mathcal{O}(n)$, hence, by Lemma 4, M can be transformed to a det-wr Hennie machine with $\mathcal{O}(n)$ states and $\mathcal{O}(n)$ working symbols. □

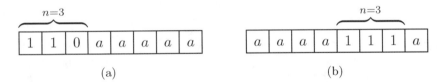

Fig. 3. A binary counter of length $n = 3$ is used by a det-wr Hennie machine to check if the input's length is $2^n = 8$. It store value 3 at the beginning (a) and value 7 at the end (b).

The acceptance of a 1NFA relates to the following problem. Given an undirected graph $G = (V, E)$ and two its vertices s, t, the undirected s-t connectivity problem (USTCON) is to determine if there is a path between s and t. USTCON is solvable by a deterministic logarithmic-space algorithm [12]. We utilize this fact in the proof of the next theorem.

Theorem 11. *Let A be an n-state 1NFA working over an input alphabet Σ of size $s = |\Sigma|$. There is a det-wr Hennie machine M accepting $L(A)$, with the number of transitions polynomial in n and s.*

Proof. Let Q be the set of states of A. Without loss of generality, A has one initial (q_I) and one accepting (q_A) state. We distinguish two cases by the length of the input string $w \in \Sigma^*$.

If $|w| \leq n$, M solves the question whether 1NFA A accepts w as an instance of USTCON problem. The related graph is depicted in Figure 4(a). Each column corresponds to one tape field and contains a vertex for each state of A. Vertices are thus pairs (q, p) where $q \in Q$ and $p \in \{1, \ldots, |w|\}$. Two vertices are connected by an edge iff they are in neighboring columns and a transition from the configuration on the left to the configuration on the right is allowed. The input is accepted iff $(q_I, 1)$ and $(q_A, |w|)$ are connected. How many states are needed for M? First, to memorize $\mathcal{O}(sn^2)$ transitions of A. Second, to provide $\mathcal{O}(\log n)$ space required to solve USTCON. This space permits polynomially many (in n) different configurations, hence polynomially many states of M are sufficient.

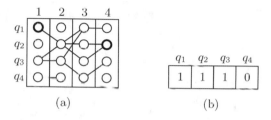

(a)   (b)

Fig. 4. (a) An undirected graph of configurations and transitions induced by a 1NFA A over an input string of length 4. State q_1 is initial, state q_2 is accepting. (b) A block representing states reachable by M when its head scans the third input's symbol.

If $|w| > n$, M has enough space to record states reachable by A on the tape. The tape is split into blocks of length n, with the exception of the last block whose length is $n + (|w| \mod n)$. The i-th field of a block stores a one-bit flag indicating the reachability of the i-th state of A, see Figure 4(b). Each block is used only when the simulated head of A is inside the block. When the head of A leaves the block, M copies the flags to the next one. This ensures that the number of transitions done by M over a tape field depends only on n (not on $|w|$). More precisely, the simulation inside a block is done in time $\mathcal{O}(n^3)$ since M carries out three nested cycles – trough fields scanned by A in the block, states of A marked as reachable and transitions of A.   □

It would be possible to apply a similar approach to the simulation of an n-state 2NFA A by a det-wr Hennie machine, however, two things will change. The question of acceptance of A reduces to the directed s-t connectivity problem (STCON). We know it is solvable by a nondeterministic algorithm in logarithmic space [9], thus, by Sawitch theorem, deterministically in space $\mathcal{O}(\log^2 n)$. However, this leads to $\mathcal{O}(n^{\log n})$ different configurations and this amount of states would be needed by a det-wr Hennie machine. The other thing is that the simulation of a two-way automaton on inputs w such that $|w| > n$ would require a different technique.

As an n-state 2NFA can be simulated by a n^2-state 1AFA [1], we also cannot expect an easy cheap simulation of 1AFA by a det-wr Hennie machine.

Nevertheless, we can show that the blowup in transitions is polynomial when simulating deterministic one-pebble automata.

Theorem 12. *Each deterministic one-pebble n-state automaton can be simulated by a det-wr Hennie machine with the number of transitions polynomial in n.*

Proof. Let A be a deterministic one-pebble automaton with the set of states Q and the input alphabet Σ. We will design a Hennie machine M simulating A. Throughout the proof, assume that A always halts, i.e., it never goes into a loop. It will be clear in the end of the construction that all looping computations of A result in a det-wr Hennie machine which terminates since it scans a symbol with the lowest weight.

Given an input $w \in \Sigma^*$. If $|w| \leq n$, M simply represents the pebble on the tape by a marker and simulates A transition by transition. There are $(|Q||w|^2)$ different configurations of A, thus M finishes in time $\mathcal{O}(n^3)$, visiting each tape field at most $n|w| = \mathcal{O}(n^2)$ times.

If $|w| > n$, a direct simulation could easily exceed the targeted polynomial number of transitions performed over a tape field. To handle this, M splits the input into blocks[1] of length n (all blocks except the last one) or $n + (|w| \mod n)$ (the last block) and computes what happens whenever the head of A moves outside a block B in a state q while the marker is left inside B. A either returns back in a state from Q or finishes, thus, two mappings $f_L, f_R : Q \to Q \times \{acc, rej\}$ are computed for each block: f_L determines the outcome when the head crosses the left border of B, while f_R relates to the right border. The constants acc and rej represent accepting and rejecting by A. An example is given in Figure 5. Values of both mappings are recorded in B. Its i-th field stores the pair $f_L(q_i)$ and $f_R(q_i)$. There are $(n + 2)^2$ different values of these pairs, required to have representatives in the working alphabet of M.

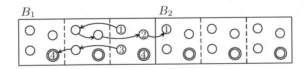

Fig. 5. The simulation of a deterministic one-pebble automaton with the set of states $Q = \{1, 2, 3, 4\}$ where 4 is a final state. All transitions done after leaving the block B_2 without the pebble are displayed for all the states. Mapping f_L for B_2 is as follows: $f_L(1) = 1$, $f_L(2) = 1$, $f_L(3) = acc$, $f_L(4) = acc$.

If all mappings are stored, M performs the simulation of A transition by transition inside blocks as well as in cases when A moves the pebble. When A leaves a block without the pebble, the corresponding mapping is used to decide what happens. This again ensures that time spent by M in a block is polynomial in n.

---

[1] Splitting points are located by counting to n in states.

The remaining task is to efficiently compute the mappings. We describe the computation of f_L. Let $B_1, B_2, \ldots, B_s$ be all the blocks from left to right. To compute f_L for B_1 is trivial. When the head of A leaves its left border, it moves to the left-end marker. Assume f_L has been already computed and represented for a block B_i. It is used to determine f_L for the next block B_{i+1}. For each state $q \in Q$, M simulates in B_i what happens when the head of A enters the rightmost field of B_i and the control unit is in state q. Whenever A is about to enter B_{i-1}, the mapping f_L of B_i is used to continue the tracking. Note that mappings f_R can be computed analogously, taking them in the reversed order.

Since time spent by M in each block is polynomial in n, it can be converted to a det-wr Hennie machine with polynomially many transitions.   □

5   Conclusion

We presented deterministic weight-reducing Hennie machines as a constructive subclass of Hennie machines. We showed that their ability to rewrite symbols can significantly facilitate design of devices accepting regular languages. When we get rid of rewriting by converting the machine to a 1DFA, the blowup in transitions is at most double exponential. This is the same order of blowup as exhibited, e.g., by alternating automata or two-way one-pebble automata [3].

The power of the model is further illustrated by the other proved trade-offs. It can simulate at a low cost a pebble used by a deterministic two-way automaton. It cannot be replaced at a low cost by a 2NFA.

Sakoda and Sipser stated two famous open problems [13]. Is it possible to simulate an n-state 1NFA or 2NFA by a 2DFA with polynomially many states? We studied these questions for a more powerful device and showed that det-wr Hennie machine can do it in the case of 1NFA. This result can be interpreted as a sort of 1NFA's determinization which keeps the size of its description small. An open question remains what is the trade-off in the case of 2NFA. We have suggested that an n-state 2NFA could be simulated by a det-wr Hennie machine with $\mathcal{O}(n^{\log n})$ transitions. Can we achieve a polynomial blowup here as for 1NFA?

References

1. Birget, J.C.: State-complexity of finite-state devices, state compressibility and incompressibility. Mathematical Systems Theory 26(3), 237–269 (1993)
2. Durak, B.: Two-way finite automata with a write-once track. J. Autom. Lang. Comb. 12(1), 97–115 (2007)
3. Globerman, N., Harel, D.: Complexity results for two-way and multi-pebble automata and their logics. Theoretical Computer Science 169, 161–184 (1996)
4. Goldstine, J., Kappes, M., Kintala, C.M.R., Leung, H., Malcher, A., Wotschke, D.: Descriptional complexity of machines with limited resources. J. UCS 8(2), 193–234 (2002)
5. Hartmanis, J.: Computational complexity of one-tape Turing machine computations. J. ACM 15(2), 325–339 (1968)

6. Hennie, F.: One-tape, off-line Turing machine computations. Information and Control 8(6), 553–578 (1965)
7. Kari, J., Moore, C.: New results on alternating and non-deterministic two-dimensional finite-state automata. In: Ferreira, A., Reichel, H. (eds.) STACS 2001. LNCS, vol. 2010, pp. 396–406. Springer, Heidelberg (2001)
8. Meyer, A.R., Fischer, M.J.: Economy of description by automata, grammars, and formal systems. In: SWAT (FOCS), pp. 188–191. IEEE Computer Society (1971)
9. Papadimitriou, C.M.: Computational complexity. Addison-Wesley, Reading (1994)
10. Průša, D., Mráz, F.: Two-dimensional sgraffito automata. In: Yen, H.-C., Ibarra, O.H. (eds.) DLT 2012. LNCS, vol. 7410, pp. 251–262. Springer, Heidelberg (2012)
11. Radó, T.: On non-computable functions. Bell System Technical Journal 41(3), 877–884 (1962)
12. Reingold, O.: Undirected connectivity in log-space. J. ACM 55(4), 1–17 (2008)
13. Sakoda, W.J., Sipser, M.: Nondeterminism and the size of two way finite automata. In: Proceedings of the Tenth Annual ACM Symposium on Theory of Computing, STOC 1978, New York, NY, USA, pp. 275–286 (1978)
14. Shannon, C.E.: A universal Turing machine with two internal states. Annals of Mathematics Studies 34, 157–165 (1956)
15. Shepherdson, J.C.: The reduction of two-way automata to one-way automata. IBM J. Res. Dev. 3(2), 198–200 (1959)

Computing with Catalan Families

Paul Tarau

Department of Computer Science and Engineering,
University of North Texas, USA
tarau@cs.unt.edu

Abstract. We study novel arithmetic algorithms on a canonical number representation based on the Catalan family of combinatorial objects.

For numbers corresponding to Catalan objects of low structural complexity our algorithms provide super-exponential gains while their average case complexity is within constant factors of their traditional counterparts.

Keywords: hereditary numbering systems, arithmetic algorithms for Combinatorial objects, structural complexity of natural numbers, run-length compressed numbers, Catalan families.

1 Introduction

Number representations have evolved over time from the unary "cave man" representation where one scratch on the wall represented a unit, to the base-n (and in particular base-2) number system, with the remarkable benefit of a logarithmic representation size. Over the last 1000 years, this base-n representation has proved to be unusually resilient, partly because all practical computations could be performed with reasonable efficiency within the notation.

While alternative *notations* like Knuth's "up-arrow" [2] or tetration are useful in describing very large numbers, they do not provide the ability to actually *compute* with them – as, for instance, addition or multiplication with a natural number results in a number that cannot be expressed with the notation anymore.

The novel contribution of this paper is a Catalan family based numbering system that *allows computations* with numbers comparable in size with Knuth's "arrow-up" notation. Moreover, these computations have a worst case complexity that is comparable with the traditional binary numbers, while their best case complexity outperforms binary numbers by an arbitrary tower of exponents factor. Simple operations like successor, multiplication by 2, exponent of 2 are constant time and a number of other operations benefit from significant complexity reductions.

For the curious reader, it is basically a *hereditary number system* [1], based on recursively applied *run-length* compression of the usual binary digit notation. To evaluate best and worst cases, a concept of structural complexity is introduced, based on the size of representations and algorithms favoring large numbers of small structural complexity are designed for arithmetic operations.

A.-H. Dediu et al. (Eds.): LATA 2014, LNCS 8370, pp. 565–575, 2014.

We have adopted a *literate programming* style, i.e. the code described in the paper forms a self-contained Haskell module (tested with ghc 7.6.3), also available as a separate file at http://logic.cse.unt.edu/tarau/research/2013/catco.hs. We hope that this will encourage the reader to experiment interactively and validate the technical correctness of our claims.

The paper is organized as follows. Section 2 introduces recursively run-length compressed natural numbers seen as a member of the Catalan family of combinatorial objects. Section 3 describes constant time successor and predecessor operations on our numbers. Section 4 describes novel algorithms for arithmetic operations taking advantage of our number representation. Section 5 defines a concept of structural complexity and studies best and worst cases. Section 6 discusses related work. and section 7 concludes the paper.

2   Recursively Run-Length Compressed Natural Numbers as Objects of the Catalan Family

The Catalan family of combinatorial objects [6,5] spans over a wide diversity of concrete representation ranging from balanced parentheses expressions and rooted plane trees to non-crossing partitions and polygon triangulations.

2.1   The "cons-list"-View of Catalan Objects

For simplicity, we will pick here as a representative of the Catalan family a language of balanced parentheses.

We fix our set of two parentheses to be {L,R} corresponding to the Haskell data type Par.

```
data Par = L | R deriving (Eq,Show,Read)
```

Definition 1. *A Dyck word on the set of parentheses* {L,R} *is a list consisting of n L's and R's such that no prefix of the list has more L's than R's.*

The set of Dyck words is a well-known member of the Catalan family of combinatorial objects [6]. Let $\mathbb{T}$ be the language obtained from the set Dyck words on {L,R} with an extra L parenthesis added at the beginning of each word and an extra R parenthesis added at the end of each word. Assuming syntactic well-formedness, we represent the language $\mathbb{T}$ in Haskell as the type T and we will call its members *terms*.

```
type T = [Par]
```

It is convenient to view $\mathbb{T}$ as the set of *rooted ordered binary trees* through the operations cons and decons, that ensure syntactic well-formedness:

```
cons :: (T,T) → T
cons (xs,L:ys) = L:xs++ys
```

```
decons :: T→(T,T)
decons (L:ps) = count_pars 0 ps where
  count_pars 1 (R:ps)  = ([R],L:ps)
  count_pars k (L:ps) = (L:hs,ts) where (hs,ts) = count_pars (k+1) ps
  count_pars k (R:ps) = (R:hs,ts) where (hs,ts) = count_pars (k-1) ps
```

The forest of subtrees corresponds to the toplevel balanced parentheses composing an element of $\mathbb{T}$ as defined by the bijections to_list and from_list.

```
to_list :: T → [T]
to_list [L,R] = []
to_list ps = hs:hss where
  (hs,ts) = decons ps
  hss = to_list ts
```

We will call *subterms* the terms extracted by to_list.

```
from_list :: [T]→T
from_list [] = [L,R]
from_list (hs:hss) = cons (hs,from_list hss)
```

We will assume for purposes of complexity analysis that an ordered rooted tree data structure is used for the language $\mathbb{T}$, *a representation ensuring that the* from_list *and* to_list *operations are constant time.*

2.2   The Catalan Encoding of Natural Numbers

We are ready for an arithmetic interpretation of the language $\mathbb{T}$, associating a unique natural number to each of its terms t:

- The term $t=$[L,R] corresponds to zero
- if xs is obtained by applying the to_list operation to t, then each x on the list xs counts the number of $b \in \{0,1\}$ digits, followed by *alternating* counts of 1-b and b digits, with the conventions that the most significant digit is 1 and the counter x represents x+1 objects.
- the same principle is applied recursively for the counters, until [L,R] is reached.

One can see this process as run-length compressed base-2 numbers, unfolded as an object of the Catalan family, after applying the encoding recursively.

By convention, as the last (and most significant) digit is 1, the last count on the list xs is for 1 digits. The following simple fact allows inferring parity from the number of subterms of a term.

Proposition 1. *If the length of* to_list x *is odd, then* x *encodes an odd number, otherwise it encodes an even number.*

Proof. Observe that as the highest order digit is always a 1, the lowest order digit is also 1 when length of the list of counters is odd, as counters for 0 and 1 digits alternate.

This ensures the correctness of the Haskell definitions of the predicates odd_ and even_, the last defined true for terms different from [L,R] only.

```
oddLen [] = False
oddLen [_] = True
oddLen (_:xs) = not (oddLen xs)

odd_ :: T→Bool
odd_ x = oddLen (to_list x)

even_ :: T→Bool
even_ x =  f (to_list x) where
  f [] =False
  f (y:ys) = oddLen ys
```

Note that while these predicates work in time proportional to the length of the list representing a term in $\mathbb{T}$, *with a (dynamic) array-based list representation that keeps track of the length or keeps track of the parity bit explicitly, one can assume that they can be made constant time with an optimal data structure choice , as* we will do in the rest of the paper, while focusing, for simplicity, on the language of balanced parentheses $\mathbb{T}$.

Definition 2. *The function* $n : \mathbb{T} \to \mathbb{N}$ *shown in equation* (1) *defines the unique natural number associated to a term of type* $\mathbb{T}$.

$$n(a) = \begin{cases} 0 & \text{if } a = [\text{L},\text{R}], \\ 2^{n(x)+1}n(xs) & \text{where } (x,xs) = \text{decons } a, \text{ if } a \text{ is even\_}, \\ 2^{n(x)+1}n(xs) - 1 & \text{where } (x,xs) = \text{decons } a, \text{ if } a \text{ is odd\_.} \end{cases} \quad (1)$$

For instance, the computation of [L,L,R,L,L,R,L,R,R,R] $= 14$ expands to $(2^{0+1}(2^{(2^{0+1}(2^{0+1}-1))+1} - 1))$. The Haskell equivalent is:

```
type N = Integer

n :: T→N
n ([L,R]) = 0
n a | even_ a = 2^(n x + 1)*(n xs) where (x,xs) = decons a
n a | odd_ a = 2^(n x + 1)*(n xs+1)-1 where (x,xs) = decons a
```

The following example illustrates the values associated with the first few natural numbers.

```
0: [L,R]
1: [L,L,R,R]
2: [L,L,R,L,R,R]
3: [L,L,L,R,R,R]
4: [L,L,L,R,R,L,R,R]
5: [L,L,R,L,R,L,R,R]
```

Definition 3. *The function* $t : \mathbb{N} \to \mathbb{T}$ *defines the unique term of type* $\mathbb{T}$ *associated to a natural number as follows:*

```
t :: N→T
t 0 = [L,R]
t k | k>0 = zs where
  (x,y) = if even k then split0 k else split1 k
  ys = t y
  zs = if x==0 then ys else cons (t (x-1),ys)
```

It uses the helper functions split0 and split1 that extract a block of contiguous 0 digits and, respectively, 1 digits from the lower end of a binary number.

```
split0 :: N→(N,N)
split0 z | z> 0 && even z = (1+x,y) where
  (x,y) = split0  (z 'div' 2)
split0 z = (0,z)
```

```
split1 :: N→(N,N)
split1 z | z>0 && odd z = (1+x,y) where
  (x,y) = split1  ((z-1) 'div' 2)
split1 z = (0,z)
```

They return a pair (x,y) consisting of a count x of the number of digits in the block, and the natural number y representing the digits left over after extracting the block. Note that div, occurring in both functions, is integer division.

The following holds:

Proposition 2. *Let* id *denote* $\lambda x.x$ *and* $\circ$ *function composition. Then, on their respective domains*

$$t \circ n = id, \quad n \circ t = id \tag{2}$$

Proof. By induction, using the arithmetic formulas defining the two functions.

Figure 1 shows the DAG obtained by folding together identical subterms at each level for the term corresponding to the natural number 12345, where we have mapped of L symbols to strings built of '(' and R symbols mapped to ')' characters, for readability. Note that integer labels mark the order of the edges outgoing from a vertex.

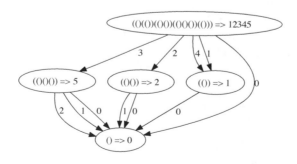

Fig. 1. The DAG illustrating the term associated to 12345

The constants e and u correspond to the natural numbers 0 and 1. The predicates e_ and u_ are used to recognize them.

```
e = [L,R]
u = [L,L,R,R]


e_ [L,R] = True
e_ _ = False


u_ [L,L,R,R] = True
u_ _ = False
```

3   Successor (s) and Predecessor (s')

We will now specify successor and predecessor on data type $\mathbb{T}$ through two mutually recursive functions, s and s'.

```
s x | e_ x = u -- 1
s x | even_ x = from_list (sEven (to_list x)) -- 7
s x | odd_ x = from_list (sOdd (to_list x)) -- 8


sEven (a:x:xs) |e_ a = s x:xs -- 3
sEven (x:xs) = e:s' x:xs -- 4


sOdd [x]= [x,e] -- 2
sOdd (x:a:y:xs) | e_ a = x:s y:xs -- 5
sOdd (x:y:xs) = x:e:(s' y):xs -- 6


s' x | u_ x = e -- 1
s' x | even_ x = from_list (sEven' (to_list x)) -- 8
s' x | odd_ x = from_list (sOdd' (to_list x)) -- 7


sEven' [x,y] |e_ y = [x] -- 2
sEven' (x:b:y:xs) | e_ b = x:s y:xs -- 6
sEven' (x:y:xs) = x:e:s' y:xs -- 5


sOdd' (b:x:xs) | e_ b = s x:xs -- 4
sOdd' (x:xs) = e:s' x:xs -- 3
```

Note that the two functions work *on a block of* 0 *or* 1 *digits at a time*. They are based on simple arithmetic observations about the behavior of these blocks when incrementing or decrementing a binary number by 1. The following holds:

Proposition 3. *Denote* $\mathbb{T}^+ = \mathbb{T} - \{e\}$. *The functions* $s : \mathbb{T} \to \mathbb{T}^+$ *and* $s' : \mathbb{T}^+ \to \mathbb{T}$ *are inverses.*

Proof. It follows by structural induction after observing that patterns for rules marked with the number -- k in s correspond one by one to patterns marked by -- k in s' and vice versa.

More generally, it can be shown that Peano's axioms hold and as a result $<\mathbb{T}, e, s>$ is a *Peano algebra*.

Note also that if parity information is kept explicitly, the calls to odd_ and even_ are constant time, as we will assume in the rest of the paper.

Proposition 4. s and s' are constant time, on the average.

Proof. Observe that the average size of a contiguous block of 0s or 1s in a number of bitsize n has the upper bound 2 as $\sum_{k=0}^{n} \frac{1}{2^k} = 2 - \frac{1}{2^n} < 2$. As on 2-bit numbers we have an average of 0.25 more calls, we can conclude that the total average number of calls is constant, with upper bound $2 + 0.25 = 2.25$.

A quick empirical evaluation confirms this. When computing the successor on the first $2^{30} = 1073741824$ natural numbers, there are in total 2381889348 calls to s and s', averaging to 2.2183 per computation. The same average for 100 successor computations on 5000 bit random numbers oscillates around 2.22.

4   A Few Other Constant Time Operations

We will now describe algorithms for constant on the average basic operations that take advantage of our number representation and can be used as "building blocks" for various arithmetic operations.

Doubling a number db and reversing the db operation (hf) are quite simple. For instance, db proceeds by adding a new counter for odd numbers and incrementing the first counter for even ones.

```
db x | e_ x = e
db xs | odd_ xs = cons (e,xs)
db xxs | even_ xxs = cons (s x,xs) where (x,xs) = decons xxs


hf x |e_ x = e
hf xxs = if e_ x then xs else cons (s' x,xs) where  (x,xs) = decons xxs
```

Note that such efficient implementations follow directly from simple number theoretic observations.

For instance, exp2, computing an exponent of 2 , has the following definition in terms of s'.

```
exp2 x | e_ x = u
exp2 x = from_list [s' x,e]
```

as it can be derived, for $k = 0$, from the identity

$$(\lambda x.2x + 1)^n(k) = 2^n(k + 1) - 1 \qquad (3)$$

Proposition 5. *The operations* db,hf *and* exp2 *are constant time, on the average.*

Proof. As s,s' are constant time, on the average, the proposition follows by observing that at most one call to s,s' is made in each definition.

Due to space constraints we will just mention that algorithms favoring numbers with large contiguous blocks of 0s and 1s in their binary representations can be devised for various arithmetic operations along the lines of [8]. For instance, addition of odd numbers, will be based on identity 4.

$$(\lambda x.2x + 1)^k(x) + (\lambda x.2x + 1)^k(y) = (\lambda x.2x + 2)^k(x + y) \tag{4}$$

5   Structural Complexity

Arguments similar to those about the average behavior of s and s' can be carried out to prove that *the average complexity of other arithmetic operations matches their traditional counterparts*, using the fact, shown in the proof of Prop. 3, that the average size of a block of contiguous 0 or 1 bits is at most 2.

To evaluate the best and worst case space requirements of our number representation, after defining the bitsize of a term as

```
bitsize x = sum (map (n.s) (to_list x))
```

we introduce here a measure of *structural complexity*, defined by the function tsize that counts the nodes of a term of type $\mathbb{T}$ (except the root).

```
tsize x =foldr add1 0 (map tsize xs) where
  xs = to_list x
  add1 x y = x + y +1
```

It corresponds to the function $c : \mathbb{T} \to \mathbb{N}$ defined as follows:

$$c(t) = \begin{cases} 0 & \text{if } t = \text{ e,} \\ \sum_{x \in \text{xs}} (1 + c(x)) & \text{if xs = to\_list t.} \end{cases} \tag{5}$$

The following holds:

Proposition 6. *For all terms $t \in \mathbb{T}$,* tsize t $\leq$ bitsize t.

Proof. By induction on the structure of t, observing that the two functions have similar definitions and corresponding calls to tsize return terms inductively assumed smaller than those of bitsize.

The following example illustrates their use:

```
*CatCo> map (tsize.t) [0,100,1000,10000]
[0,7,9,13]
*CatCo> map (tsize.t) [2^16,2^32,2^64,2^256]
[5,6,6,6]
*CatCo> map (bitsize.t) [2^16,2^32,2^64,2^256]
[17,33,65,257]
```

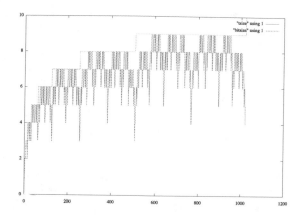

Fig. 2. Structural complexity (lower line) bounded by bitsize (upper line)

Figure 2 shows the reductions in structural complexity compared with bitsize for an initial interval of $\mathbb{N}$, from 0 to $2^{10} - 1$.

Next we define the higher order function `iterated` that applies k times the function f.

```
iterated f a x |e_ a = x
iterated f k x = f (iterated f (s' k) x)
```

We can exhibit, for a given bitsize, a best case

```
bestCase k = iterated wterm k e where wterm x = cons (x,e)
```

and a worst case

```
worstCase k = iterated (s.db.db) k e
```

The following examples illustrate these functions:

```
*CatCo> bestCase (t 4)
[L,L,L,L,L,R,R,R,R,R]
*CatCo> n it
65535
*CatCo> bitsize (bestCase (t 4))
16
*CatCo> tsize (bestCase (t 4))
4
*CatCo> worstCase (t 4)
[L,L,R,L,R,L,R,L,R,L,R,L,R,L,R,R]
*CatCo> n it
85
*CatCo> bitsize (worstCase (t 4))
7
*CatCo> tsize (worstCase (t 4))
7
```

The function `bestCase` computes the iterated exponent of 2 (tetration) and then applies the predecessor to it. For $k = 4$ it corresponds to

$$(2^{(2^{(2^{(2^{(2^{0+1}-1)+1}-1)+1}-1)+1}}-1) = 2^{2^{2^2}} - 1 = 65535.$$

The average space-complexity of the representation is related to the average length of the *integer compositions* of the bitsize of a number. Intuitively, the shorter the composition in alternative blocks of 0 and 1 digits, the more significant the compression is.

6   Related Work

Several notations for very large numbers have been invented in the past. Examples include Knuth's *arrow-up* notation [2], covering operations like *tetration* (a notation for towers of exponents). In contrast to our approach, such notations are not closed under arithmetic operations, they cannot be used as a replacement for ordinary binary or decimal numbers.

The first instance of a hereditary number system, at our best knowledge, occurs in the proof of Goodstein's theorem [1]. Another hereditary number system is Knuth's TCALC program [3] that decomposes $n = 2^a + b$ with $0 \le b < 2^a$ and then recurses on a and b with the same decomposition. Given the constraint on a and b, while hereditary, the TCALC system is not based on a bijection between $\mathbb{N}$ and $\mathbb{N} \times \mathbb{N}$ and therefore the representation is not canonical. In [10] a similar (non-canonical) exponential-based notation called "integer decision diagrams" is introduced, providing a compressed representation for sparse integers, sets and various other data types.

This paper is an adaptation of our online draft at the Cornell `arxiv` repository [8], which describes a more complex hereditary number system (based on run-length encoded "bijective base 2" numbers, first introduced in [4] pp. 90-92 as "m-adic" numbers). In contrast to [8], we are using here the familiar binary number system, and we represent our numbers as lists of balanced parentheses rather than the more complex data structure used in [8].

Arithmetic computations based on a member of the Catalan family (ordered rooted binary trees) are described in [9]. In [7] a type class mechanism is used to express computations on hereditarily finite sets and hereditarily finite functions. However likewise [9] and [7], and by contrast to those proposed in this paper, they only compress "sparse" numbers, consisting of relatively few 1 bits in their binary representation.

7   Conclusion

We have provided in the form of a literate Haskell program a specification of a number system based on a member of the Catalan family of combinatorial objects.

We have shown that *computations* that favor giant numbers with *low structural complexity*, are performed in constant time. We have also studied the best and worst case structural complexity of our representations and shown that, as structural complexity is bounded by bitsize, computations and data representations are within constant factors of conventional arithmetic even in the worst case.

References

1. Goodstein, R.: On the restricted ordinal theorem. Journal of Symbolic Logic 9(2), 33–41 (1944)
2. Knuth, D.E.: Mathematics and Computer Science: Coping with Finiteness. Science 194(4271), 1235–1242 (1976)
3. Knuth, D.E.: TCALC program (December 1994), http://www-cs-faculty.stanford.edu/~uno/programs/tcalc.w.gz
4. Salomaa, A.: Formal Languages. Academic Press, New York (1973)
5. Sloane, N.J.A.: A000108, The On-Line Encyclopedia of Integer Sequences (2013), Published electronically at http://oeis.org/A000108
6. Stanley, R.P.: Enumerative Combinatorics. Wadsworth Publ. Co., Belmont (1986)
7. Tarau, P.: Declarative modeling of finite mathematics. In: PPDP 2010: Proceedings of the 12th International ACM SIGPLAN Symposium on Principles and Practice of Declarative Programming, pp. 131–142. ACM, New York (2010)
8. Tarau, P.: Arithmetic Algorithms for Hereditarily Binary Natural Numbers (June 2013), http://arxiv.org/abs/1306.1128
9. Tarau, P., Haraburda, D.: On Computing with Types. In: Proceedings of SAC 2012, ACM Symposium on Applied Computing, PL track, pp. 1889–1896. Riva del Garda (Trento), Italy (March 2012)
10. Vuillemin, J.: Efficient Data Structure and Algorithms for Sparse Integers, Sets and Predicates. In: 19th IEEE Symposium on Computer Arithmetic, ARITH 2009, pp. 7–14 (June 2009)

Complexity of a Problem Concerning Reset Words for Eulerian Binary Automata

Vojtěch Vorel

Charles University in Prague, Czech Republic
vorel@ktiml.mff.cuni.cz

Abstract. A word is called a reset word for a deterministic finite automaton if it maps all states of the automaton to one state. Deciding about the existence of a reset word of given length for a given automaton is known to be a NP-complete problem. We prove that it remains NP-complete even if restricted on Eulerian automata over the binary alphabet, as it has been conjectured by Martyugin (2011).

1 Introduction

A *deterministic finite automaton* is a triple $A = (Q, X, \delta)$, where Q and X are finite sets and δ is an arbitrary mapping $Q \times X \to Q$. Elements of Q are called *states*, X is the *alphabet*. The *transition function* δ can be naturally extended to $Q \times X^\star \to Q$, still denoted by δ. We extend it also by defining $\delta(S, w) = \{\delta(s, w) \mid s \in S, w \in X^\star\}$ for each $S \subseteq Q$.

For a given automaton $A = (Q, X, \delta)$, we call $w \in X^\star$ a *reset word* if $|\delta(Q, w)| = 1$. If such a word exists, we call the automaton *synchronizing*. Note that each word having a reset word as a factor is also a reset word.

The *Černý conjecture*, a longstanding open problem, claims that each synchronizing automaton has a reset word of length $(|Q| - 1)^2$. However, there are many weaker results in this field, see e.g. [6,7] for recent ones.

Various computational problems arises from study of synchronization:

- *Given an automaton, decide if it is synchronizing.* Relatively simple algorithm which could be traced back to [1] works in polynomial time.
- *Given a synchronizing automaton and a number d, decide if d is the length of shortest reset words.* This has been shown to be both NP-hard [2] and coNP-hard. More precisely, it is DP-complete [5].
- *Given a synchronizing automaton and a number d, decide if there exists a reset word of length d.* This problem is of our interest. Lying in NP, it is not so computationally hard as the previous problem. However, it is proven to be NP-complete [2]. Following the notation of [4], we call it SYN. Assuming that $\mathcal{M}$ is a class of automata and membership in $\mathcal{M}$ is polynomially decidable, we define a restricted problem:

SYN($\mathcal{M}$)
Input: synchronizing automaton $A = ([n], X, \delta) \in \mathcal{M}$, $d \in \mathbb{N}$
Output: does A have a reset word of length d?

A.-H. Dediu et al. (Eds.): LATA 2014, LNCS 8370, pp. 576–587, 2014.

An automaton $A = (Q, X, \delta)$ is *Eulerian* if

$$\sum_{x \in X} |\{r \in Q \mid \delta(r, x) = q\}| = |X|$$

for each $q \in Q$. Informally, there must be exactly $|X|$ transitions incoming to each state. An automaton is *binary* if $|X| = 2$. The classes of Eulerian and binary automata are denoted by $\mathcal{EU}$ and $\mathcal{AL}_2$ respectively.

Previous results about various restrictions of SYN can be found in [2,3,4]. Some of these problems turned out to be polynomially solvable, others are NP-complete. In [4] Martyugin conjectured that $\text{SYN}(\mathcal{EU} \cap \mathcal{AL}_2)$ is NP-complete. This conjecture is confirmed in the rest of the present paper.

2 Main Result

Proof Outline. We prove the NP-completeness of $\text{SYN}(\mathcal{EU} \cap \mathcal{AL}_2)$ by polynomial reduction from 3 SAT. So, for arbitrary propositional formula ϕ in 3-CNF we construct an Eulerian binary automaton A and a number d such that

$$\phi \text{ is satisfiable} \Leftrightarrow A \text{ has a reset word of length } d. \tag{1}$$

For the rest of the paper we fix a formula $\phi = \bigwedge_{i=1}^{m} \bigvee_{\lambda \in C_i} \lambda$ on n variables where each C_i is a three-element set of literals, i.e. subset of

$$L_\phi = \{x_1, \ldots, x_n, \neg x_1, \ldots, \neg x_n\}.$$

We index the literals by the mapping κ defined by

$$\kappa : x_1 \mapsto 0, \ldots, x_n \mapsto n - 1, \neg x_1 \mapsto n, \ldots, \neg x_n \mapsto 2n - 1.$$

Let $A = (Q, X, \delta)$, $X = \{a, b\}$. Because the structure of the automaton A will be very heterogenous, we use an unusual method of description. The basic principles of the method are:

- We describe the automaton A via labeled directed multigraph G, representing the automaton in a standard way: edges of G are labeled by single letters a and b and carry the structure of the function δ. Paths in G are *labeled* by words from $\{a, b\}^*$.
- There is a collection of labeled directed multigraphs called *templates*. The graph G is one of them. Another template is SINGLE, which consists of one vertex and no edges.
- Each template T$\neq$SINGLE is a disjoint union through a set PARTS$_T$ of its proper subgraphs (the *parts* of T), extended by a set of additional edges (the *links* of T). Each $H \in$ PARTS$_T$ is isomorphic to some template U. We say that H *is of type* U.

– Let q be a vertex of a template T, lying in subgraph $H \in \text{PARTS}_T$ which is
of type U via vertex mapping $\rho : H \to U$. The *local adress* $\text{adr}_T(q)$ is a finite
string of identifiers separated by "|". It is defined inductively by

$$\text{adr}_T(q) = \begin{cases} H \mid \text{adr}_U \, \rho(q) & \text{if } U \neq \text{SINGLE} \\ H & \text{if } U = \text{SINGLE}. \end{cases}$$

The string $\text{adr}_G(q)$ is used as regular vertex identifier.

Having a word $w \in X^*$, we denote a t-th letter of w by w_t and define the set
$S_t = \delta(Q, w_1 \ldots w_t)$ of *active states at time* t. Whenever we depict a graph, a
solid arrow stands for the label a and a dotted arrow stands for the label b.

Description of the Graph G

Let us define all the templates and informally comment on their purpose.
Figure 1 defines the template ABS, which does not depend on the formula ϕ.

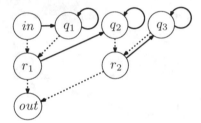

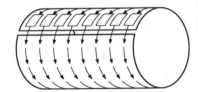

Fig. 1. Template ABS   **Fig. 2.** A barrier of ABS parts

The state *out* of a part of type ABS is always inactive after application of a
word of length at least 2 which does not contain b^2 as a factor. This allows us
to ensure the existence of a relatively short reset word. Actually, large areas of
the graph (namely the CLAUSE(...) parts) have roughly the shape depicted in
Figure 2, a cylindrical structure with a horizontal barrier of ABS parts. If we use
a sufficiently long word with no occurence of b^2, the edges outgoing from the
ABS parts are never used and almost all states become inactive.

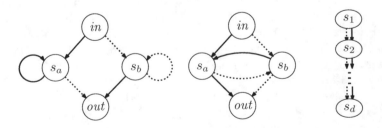

Fig. 3. Templates CCA, CCI and PIPE(d)

Figure 3 defines simple templates CCA, CCI and PIPE(d) for each $d \geq 1$. If we secure constant activity of the *in* state, the activity of the *out* state depends exactly on the last two letters applied. In the case of CCA it gets inactive if and only if the two letters were equal. In the case of CCI it works oppositely, equal letters correspond to active *out* state. One of the key ideas of the entire construction is the following. Let there be a subgraph of the form

$$\text{part of type PIPE}(d)$$
$$\downarrow a,b$$
$$\text{part of type CCA or CCI} \qquad (2)$$
$$\downarrow a,b$$
$$\text{part of type PIPE}(d).$$

Before the synchronization process starts, all the states are active. As soon as the second letter of an input word is applied, the activity of the *out* state starts to depend on the last two letters and the pipe below keeps a record of its previous activity. We say that a part H of type PIPE(d) *records* a sequence $B_1 \ldots B_d \in \{0,1\}^d$ *at time* t, if it holds that

$$B_k = 1 \Leftrightarrow H|s_k \notin S_t.$$

In order to continue with defining templates, let us define a set M_ϕ containing all literals from L_ϕ and some auxiliary symbols:

$$M_\phi = L_\phi \cup \{y_1, \ldots, y_n\} \cup \{z_1, \ldots, z_n\} \cup \{q, q', r, r'\}.$$

We index the $4n + 4$ members of M_ϕ by the following mapping μ:

$\nu \in M_\phi$	q	r	y_1	x_1	y_2	x_2	$\ldots$	y_n	x_n
$\mu(\nu)$	1	2	3	4	5	6	$\ldots$	$2n+1$	$2n+2$

$\nu \in M_\phi$	q'	r'	z_1	$\neg x_1$	z_2	$\neg x_2$	$\ldots$	z_n	$\neg x_n$
$\mu(\nu)$	$2n+3$	$2n+4$	$2n+5$	$2n+6$	$2n+7$	$2n+8$	$\ldots$	$4n+3$	$4n+4$

The inverse mapping is denoted by μ'. For each $\lambda \in L_\phi$ we define templates INC(λ) and NOTINC(λ), both consisting of $12n + 12$ SINGLE parts identified by elements of $\{1,2,3\} \times M_\phi$. As depicted by Figure 4, the links of INC(λ) are:

$$(1,\nu) \xrightarrow{a} \begin{cases} (2,r) & \text{if } \nu = \lambda \text{ or } \nu = r \\ (2,\nu) & \text{otherwise} \end{cases} \qquad (1,\nu) \xrightarrow{b} \begin{cases} (2,\lambda) & \text{if } \nu = \lambda \text{ or } \nu = r \\ (2,\nu) & \text{otherwise} \end{cases}$$

$$(2,\nu) \xrightarrow{a} \begin{cases} (3,q) & \text{if } \nu = \lambda \text{ or } \nu = q \\ (3,\lambda) & \text{if } \nu = r \\ (3,\nu) & \text{otherwise} \end{cases} \qquad (2,\nu) \xrightarrow{b} \begin{cases} (3,r) & \text{if } \nu = \lambda \text{ or } \nu = q \\ (3,\lambda) & \text{if } \nu = r \\ (3,\nu) & \text{otherwise} \end{cases}$$

Note that we use the same identifier for an one-vertex subgraph and for its vertex. The structure of NOTINC(λ) is clear from Figure 5.

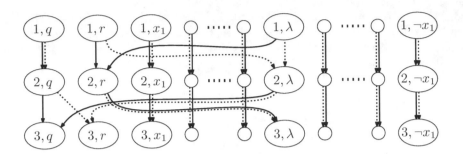

Fig. 4. Template $\text{INC}(\lambda)$

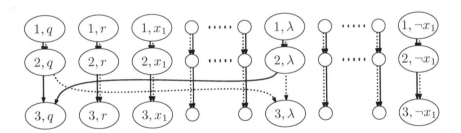

Fig. 5. Template $\text{NOTINC}(\lambda)$

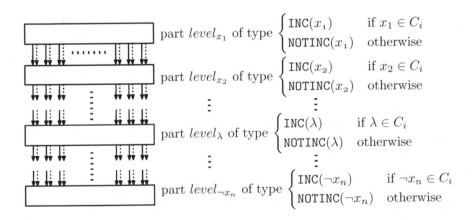

part $level_{x_1}$ of type $\begin{cases} \text{INC}(x_1) & \text{if } x_1 \in C_i \\ \text{NOTINC}(x_1) & \text{otherwise} \end{cases}$

part $level_{x_2}$ of type $\begin{cases} \text{INC}(x_2) & \text{if } x_2 \in C_i \\ \text{NOTINC}(x_2) & \text{otherwise} \end{cases}$

part $level_{\lambda}$ of type $\begin{cases} \text{INC}(\lambda) & \text{if } \lambda \in C_i \\ \text{NOTINC}(\lambda) & \text{otherwise} \end{cases}$

part $level_{\neg x_n}$ of type $\begin{cases} \text{INC}(\neg x_n) & \text{if } \neg x_n \in C_i \\ \text{NOTINC}(\neg x_n) & \text{otherwise} \end{cases}$

Fig. 6. Template TESTER

The key property of such templates comes to light when we need to apply some two-letter word in order to make the state $(3, \lambda)$ inactive assuming $(1, r)$ inactive. If also $(1, \lambda)$ is initially inactive, we can use the word a^2 in both templates. If it is active (which corresponds to the idea of unsatisfied literal λ), we discover the difference between the two templates: The word a^2 works if the type is NOTINC(λ), but fails in the case of INC(λ). Such failure corresponds to the idea of unsatisfied literal λ occuring in certain clause of ϕ.

For each clause (each $i \in \{1, \ldots, m\}$) we define a template TESTER(i). It consists of $2n$ serially linked parts, namely $level_\lambda$ for each $\lambda \in L_\phi$, each of type INC(λ) or NOTINC(λ). The particular type of each $level_\lambda$ depends on the clause C_i as seen in Figure 6, so exactly three of them are always of type INC($\ldots$). If the corresponding clause is unsatisfied, each of its three literals is unsatisfied, which causes three *failures* within the levels. Three failures imply at least three occurences of b, which turns up to be too much for a reset word of certain length to exist. Clearly we still need some additional mechanisms to realize this vague vision.

Figure 7 defines templates FORCER and LIMITER. The idea of template FORCER is simple. Imagine a situation when $q_{1,0}$ or $r_{1,0}$ is active and we need to deactivate the entire forcer by a word of length at most $2n + 3$. Any use of b would cause an unbearable delay, so if such a word exists, it starts by a^{2n+2}.

The idea of LIMITER is similar, but we tolerate some occurences of b here, namely two of them. This works if we assume $s_{1,0}$ active and it is neccesary to deactivate the entire limiter by a word of length at most $6n + 1$.

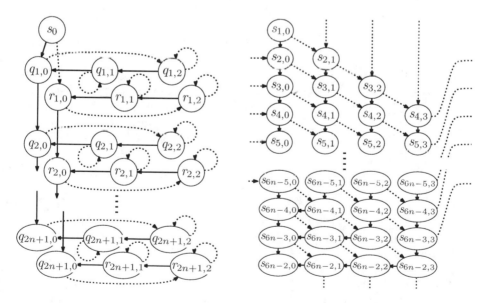

Fig. 7. Templates FORCER and LIMITER respectively

We also need a template PIPES(d, k) for each $d, k \geq 1$. It consists just of k parallel pipes of length d. Namely there is a SINGLE part $s_{d',k'}$ for each $d' \leq d$, $k' \leq k$ and all the edges are of the form $s_{d',k'} \longrightarrow s_{d'+1,k'}$.

The most complex templates are CLAUSE(i) for each $i \in \{1, \ldots, m\}$. Denote

$$\alpha_i = (i-1)(12n-2),$$
$$\beta_i = (m-i)(12n-2).$$

As shown in Figure 8, CLAUSE(i) consists of the following parts:

- Parts $sp_1, \ldots, sp_{4n+6}$ of type SINGLE.
- Parts $abs_1, \ldots, abs_{4n+6}$ of type ABS. All the template have a shape similar to Figure 2, including the barrier of ABS parts.
- Parts $pipe_2$, $pipe_3$, $pipe_4$ of types PIPE$(2n-1)$ and $pipe_6$, $pipe_7$ of types PIPE$(2n+2)$.
- Parts cca and cci of types CCA and CCI respectively. Together with the pipes above they realize the idea described in (2). As they form two constellations which work simultaneously, the parts $pipe_6$ and $pipe_7$ typically record mutually inverse sequences. We interpret them as an assignment of the variables $x_1, \ldots, x_n$. Such assignment is then processed by the tester.
- A part ν of type SINGLE for each $\nu \in M_\phi$.
- The part $tester$ of type TESTER(i).
- A part $\overline{\lambda}$ of type SINGLE for each $\lambda \in L_\phi$. While describing the templates INC(λ) and NOTINC(λ) we claimed that in certain case there arises a need to make the state $(3, \lambda)$ inactive. This happens when the border of inactive area moves down through the tester levels. The point is that any word of length $6n$ deactivates the entire tester, but we need to ensure that some tester columns, namely the $\kappa(\lambda)$-th for each $\lambda \in L_\phi$, are deactivated one step earlier. If some of them is still active just before the deactivation of tester finishes, the state $\overline{\lambda}$ becomes active, which slows down the sychronizing process.
- Parts $pipes_1$, $pipes_2$ and $pipes_3$ of types PIPES$(\alpha_i, 4n+4)$, PIPES$(6n-2, 4n+4)$ and PIPES$(\beta_i, 4n+4)$ respectively. There are multiple clauses in ϕ, but multiple testers cannot work in parallel. That is why each of them is padded by a passive PIPES$(\ldots)$ part of size depending on particular i. If $\alpha_i = 0$ or $\beta_i = 0$, the corresponding PIPES part is not present in cl_i.
- Parts $pipe_1$, $pipe_5$, $pipe_8$, $pipe_9$ of types PIPE$(12mn+4n-2m+6)$, PIPE(4), PIPE$(\alpha_i + 6n - 1)$, PIPE(β_i) respectively.
- The part $forcer$ of type FORCER. This part guarantees that only the letter a is used in certain segment of the word w. This is nessesary for the data produced by cca and cci to safely leave the parts $pipe_3$, $pipe_4$ and line up in the states of the form ν for $\nu \in M_\phi$, from where they shift to the tester.
- The part $limiter$ of type LIMITER. This part guarantees that the letter b occurs at most twice when the border of inactive area passes through the tester. Because each usatisfied literal from the clause requests an occurence of b, only a satisfied clause meets all the conditions for a reset word of certain length to exist.

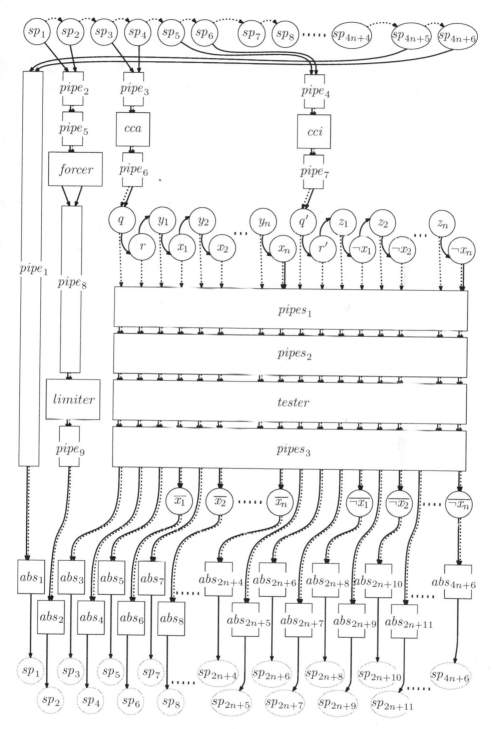

Fig. 8. Template CLAUSE(i)

Links of CLAUSE(i), which are not clear from the Figure 8 are

$$\nu \xrightarrow{a} \begin{cases} pipes_1|s_{1,\mu(\nu)} & \text{if } \nu = \neg x_n \\ \mu'\left(\mu\left(\nu\right)+1\right) & \text{otherwise} \end{cases} \qquad \nu \xrightarrow{b} pipes_1|s_{1,\mu(\nu)}$$

for each $\nu \in M_\phi \setminus \{\neg x_n\}$ and

$$pipes_3|s_{\beta_i,k} \xrightarrow{a,b} \begin{cases} \overline{\mu'(k)} & \text{if } \mu'(k) \in L_\phi \\ abs_{k+2}|in & \text{otherwise} \end{cases} \qquad \overline{\lambda} \xrightarrow{a,b} abs_{\mu(\lambda)+2}|in$$

for each $k \in \{1, \ldots, 4n+4\}$, $\lambda \in L_\phi$.

We are ready to form the whole graph G, see Figure 9. For each $i, k \in \{1, \ldots m\}$ there are parts cl_k, abs_k of types CLAUSE(i) and ABS respectively and q_k, r_k, r'_k, s_1, s_2 of type SINGLE. The edge incoming to a cl_i part ends in $cl_i|sp_1$, the outgoing one starts in $cl_i|sp_{4n+6}$. When no states outside ABS parts are active within each CLAUSE($\ldots$) part and no out, r_1 nor r_2 state is active in any ABS part, the word b^2ab^{4n+m+7} takes all active states to s_2 and completes the sychronization. Graph G does not fully represent the automaton A yet, because there are

- $8mn + 4m$ vertices with only one outgoing edge, namely $cl_i|abs_k|out$ and sp_l for each $i \in \{1, \ldots, m\}$, $k \in \{1, \ldots, 4n+6\}$, $l \in \{7, \ldots, 4n+4\}$,
- $8mn + 4m$ vertices with only one incoming edge: $cl_i|\nu$ and $cl_i|pipes_1|(1, \nu')$ for each $i \in \{1, \ldots, m\}$, $\nu \in M_\phi \setminus \{q, q'\}$, $\nu' \in M_\phi \setminus \{x_n, \neg x_n\}$.

But we do not need to specify the missing edges exactly, let us just say that they somehow connect the relevant states and the automaton A is complete. Let us set $d = 12mn + 8n - m + 18$ and prove that the equivalence (1) holds.

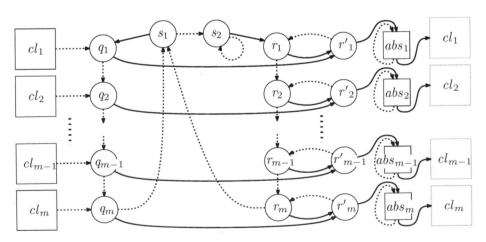

Fig. 9. The graph G

From an Assignment to a Word. At first let us suppose that there is an assignment $\xi_1, \ldots, \xi_n \in \{0, 1\}$ of the variables $x_1, \ldots, x_n$ (respectively) satisfying the formula ϕ and prove that the automaton A has a reset word w of length d.

For each $j \in \{1, \ldots, n\}$ we denote

$$
\sigma_j = \begin{cases} a & \text{if } \xi_j = 1 \\ b & \text{if } \xi_j = 0 \end{cases}
$$

and for each $i \in \{1, \ldots, m\}$ we choose a satisfied literal $\overline{\lambda}_i$ from C_i. We set

$$
w = a^2 (\sigma_n a)(\sigma_{n-1}a) \ldots (\sigma_1 a) \, aba^{2n+3} b \left(a^{6n-2} v_1\right) \ldots \left(a^{6n-2} v_m\right) b^2 ab^{4n+m+7},
$$

where for each $i \in \{1, \ldots, m\}$ we use the word

$$
v_i = u_{i,x_1} \ldots u_{i,x_n} u_{i,\neg x_1} \ldots u_{i,\neg x_n},
$$

denoting

$$
u_{i,\lambda} = \begin{cases} a^3 & \text{if } \lambda = \overline{\lambda}_i \text{ or } \lambda \notin C_i \\ ba^2 & \text{if } \lambda \neq \overline{\lambda}_i \text{ and } \lambda \in C_i \end{cases}
$$

for each $\lambda \in L_\phi$. We see that $|v_i| = 6n$ and therefore

$$
|w| = 4n + 8 + m(12n - 2) + 4n + m + 10 = 12mn + 8n - m + 18 = d.
$$

Let us denote

$$
\gamma = 12mn + 4n - 2m + 9.
$$

Because the first occurence of b^2 in w starts by γ-th letter, we have:

Lemma 1. *No state of a form* $cl\ldots|abs\ldots|out$ *or* $abs\ldots|out$ *lies in any of the sets* $S_2, \ldots, S_\gamma$.

Let us fix an arbitrary $i \in \{1, \ldots, m\}$ and describe a growing area of inactive states within cl_i. The following claims follows directly from the definition of w. Note that the claim 7 relies on the fact that b occurs only twice in v_i.

Lemma 2

1. *No state of the form* $sp\ldots$ *lies in any of the sets* $S_2, \ldots, S_\gamma$.
2. *No state from* $pipe_2$ *or* $pipe_3$ *or* $pipe_4$ *lies in any of the sets* $S_{2n+1}, \ldots, S_\gamma$.
3. *No state from* cca *or* cci *or* $pipe_5$ *lies in any of the sets* $S_{2n+5}, \ldots, S_\gamma$.
4. *No state from* $pipe_6$ *or* $pipe_7$ *or* $forcer$ *lies in any of the sets* $S_{4n+7}, \ldots, S_\gamma$.
5. *No state* ν *for* $\nu \in M_\phi$ *lies in any of the sets* $S_{4n+8}, \ldots, S_\gamma$.
6. *No state from* $pipes_1$ *or* $pipes_2$ *or* $pipe_8$ *lies in any of the sets* $S_{10n+\alpha_i+6}, \ldots$ $\ldots, S_\gamma$.
7. *No state from* $limiter$ *or* $tester$ *lies in any of the sets* $S_{16n+\alpha_i+6}, \ldots, S_\gamma$
8. *No state from* $pipe_1$ *or* $pipe_9$ *or* $pipes_3$ *lies in any of the sets* $S_{\gamma-1}, S_\gamma$.

For each $\lambda \in L_\phi$ we ensure by the word $u_{i,\lambda}$ that the $\kappa(\lambda)$-th tester column is deactivated in advance, namely at time $t = 16n + \alpha_i + 5$. The advance allows the following key claim to hold true.

Lemma 3. *No state $cl_i|\overline{\lambda}$ for $\lambda \in L_\phi$ lies in any of the sets $S_{\gamma-1}, S_\gamma$.*

We see that within cl_i only states from the ABS parts can lie in $S_{\gamma-1}$. Since $w_{\gamma-2}w_{\gamma-1} = a^2$, no state r_1, r_2 or out from any ABS part lies in $S_{\gamma-1}$. Now we easily check that all the states possibly present in $S_{\gamma-1}$ are mapped to s_2 by the word $w_\gamma \ldots w_d = b^2 ab^{4n+m+7}$.

From a Word to an Assignment. Since now we suppose that there is a reset word w of length $d = 12mn + 8n - m + 18$. The following lemma is not hard to verify.

Lemma 4

1. *Up to labeling there is unique pair of paths having length at most $d-2$, leading from $cl_1|pipe_1|s_1$ and $cl_2|pipe_1|s_1$ respectively to a common end. They are of length $d - 2$ and meet in s_2.*
2. *The word w starts by a^2.*

The second claim implies that for each $i \in \{1, \ldots, m\}$ it holds that $cl_i|pipe_1|s_1 \in S_2$, so it follows that

$$\delta(Q, w) = \{s_2\}.$$

Let us denote $\overline{d} = 12mn + 4n - 2m + 11$ and $\overline{w} = w_1 \ldots w_{\overline{d}}$. The following lemma holds, because no edges labelled by a are available for final segments of the paths described in the first claim of Lemma 4.

Lemma 5

1. *The word w can be written as $w = \overline{w}b^{4n+m+7}$ for some word $\overline{w}$.*
2. *For any $t \geq \overline{d}$, no state from any $cl_\ldots$ part lie in S_t, except for the $sp_\ldots$ states.*

The next lemma is based on properties of the parts $cl_\ldots|forcer$ but to prove that no more a follows the enforced factor a^{2n+1} we also need to observe that each $cl_\ldots|cca|out$ or each $cl_\ldots|cci|out$ lies in S_{2n+4}.

Lemma 6. *The word $\overline{w}$ starts by $\overline{u}a^{2n+1}b$ for some $\overline{u}$ of length $2n + 6$.*

Now we are able to write the word $\overline{w}$ as

$$\overline{w} = \overline{u}a^{2n+1}b \left(\overline{v}_1 v_1' c_1\right) \ldots \left(\overline{v}_m v_m' c_m\right) w_{\overline{d}-2}w_{\overline{d}-1}w_{\overline{d}},$$

where $|\overline{v}_k| = 6n - 2$, $|v_k'| = 6n - 1$ and $|c_k| = 1$ for each k and denote $d_i = 10n + \alpha_i + 6$. At time $2n + 5$ the parts $cl_\ldots|pipe_6$ and $cl_\ldots|pipe_7$ record mutually inverse sequences. Because there is the factor a^{2n+1} after $\overline{u}$, at time d_i we find the information pushed to the first rows of testers:

Lemma 7. *For each $i \in \{1, \ldots, m\}$, $j \in \{1, \ldots, n\}$ it holds that*

$$cl_i|tester|level_{x_1}|(1, x_j) \in S_{d_i} \Leftrightarrow$$
$$cl_i|tester|level_{x_1}|(1, \neg x_j) \notin S_{d_i} \Leftrightarrow w_{2n-2j+2} \neq w_{2n-2j+3}.$$

Let us define the assignment $\xi_1, \ldots, \xi_n \in \{0, 1\}$. By Proposition 7 the definition is correct and does not depend on i:

$$\xi_j = \begin{cases} 1 & \text{if } cl_i|tester|level_{x_1}|(1, x_j) \notin S_{d_i} \\ 0 & \text{if } cl_i|tester|level_{x_1}|(1, \neg x_j) \notin S_{d_i}. \end{cases}$$

The following lemma holds due to $cl_{\ldots}|limiter$ parts.

Lemma 8. *For each $i \in \{1, \ldots, m\}$ there are at most two occurences of b in the word v_i'.*

Now we choose any $i \in \{1, \ldots, m\}$ and prove that the assignment $\xi_1, \ldots, \xi_n$ satisfies the clause $\bigvee_{\lambda \in C_i} \lambda$. Let $p \in \{0, 1, 2, 3\}$ denote the number of unsatisfied literals in C_i.

As we claimed before, all tester columns corresponding to any $\lambda \in L_\phi$ have to be deactivated earlier than other columns. Namely, if $cl_i|tester|level_{x_1}|(1, \lambda)$ is active at time d_i, which happens if and only if λ is not satisfied by $\xi_1, \ldots, \xi_n$, the word $v_i'c_i$ must not map it to $cl_i|pipes_3|s_{1,\mu(\lambda)}$. If $cl_i|tester|level_\lambda$ is of type $\text{INC}(\lambda)$, the only way to ensure this is to use the letter b when the border of inactive area lies at the first row of $cl_i|tester|level_\lambda$. Thus each unsitisfied $\lambda \in C_i$ implies an occurence of b in corresponding segment of v_i':

Lemma 9. *There are at least p occurences of the letter b in the word v_i'.*

By Lemma 8 there are at most two occurences of b in v_i', so we get $p < 2$ and there is at least one satisfied literal in C_i.

References

1. Černý, J.: Poznámka k homogénnym experimentom s konečnými automatmi. Matematicko-fyzikálny časopis 14(3), 208–216 (1964)
2. Eppstein, D.: Reset sequences for monotonic automata. SIAM J. Comput. 19(3), 500–510 (1990)
3. Martyugin, P.: Complexity of problems concerning reset words for some partial cases of automata. Acta Cybern. 19(2), 517–536 (2009)
4. Martyugin, P.: Complexity of problems concerning reset words for cyclic and eulerian automata. In: Bouchou-Markhoff, B., Caron, P., Champarnaud, J.-M., Maurel, D. (eds.) CIAA 2011. LNCS, vol. 6807, pp. 238–249. Springer, Heidelberg (2011)
5. Olschewski, J., Ummels, M.: The complexity of finding reset words in finite automata. In: Hliněný, P., Kučera, A. (eds.) MFCS 2010. LNCS, vol. 6281, pp. 568–579. Springer, Heidelberg (2010)
6. Steinberg, B.: The Černý conjecture for one-cluster automata with prime length cycle. Theoret. Comput. Sci. 412(39), 5487–5491 (2011)
7. Trahtman, A.N.: Modifying the upper bound on the length of minimal synchronizing word. In: Owe, O., Steffen, M., Telle, J.A. (eds.) FCT 2011. LNCS, vol. 6914, pp. 173–180. Springer, Heidelberg (2011)

Probabilistic ω-Regular Expressions

Thomas Weidner[*]

Institut für Informatik, Universität Leipzig, 04009 Leipzig, Germany
weidner@informatik.uni-leipzig.de

Abstract. We introduce probabilistic ω-regular expressions which are an extension to classical regular expressions with semantics taking probabilities into account. The main result states that probabilistic ω-regular expressions are expressively equivalent to probabilistic Muller-automata. To obtain better decidability properties we introduce a subclass of our expressions with decidable emptiness and approximation problem.

1 Introduction

Regular expressions are used in nearly every field of theoretical computer science. Kleene's famous theorem states that regular languages and finite automata define the same class of languages [14]. This result has been transferred to various other settings. Two notable examples are the weighted setting by Schützenberger [21] and the setting of infinite words by Büchi [4]. Kleene-type results with weights on infinite words have been obtained in [11,16].

Probabilistic automata are another classical concept in theoretical computer science. Introduced by Rabin [19], they add probabilistic branching to deterministic finite automata. This model has proven very successful and has nowadays a broad range of applications including speech recognition [20], prediction of climate parameters [18], or randomized distributed systems [10]. In 2005, probabilistic automata have been extended to infinite words by Baier and Grösser [1]. This concept led to manifold further research [2,6,7,8,9,22]. We consider the quantitative behavior of a probabilistic ω-automaton, i.e., a function mapping infinite words to probability values.

In 2012, Kleene's result has been transferred to probabilistic pebble automata by Bollig, Gastin, Monmege and Zeitoun [3]. These automata are more powerful than classical probabilistic automata, but still only work on finite words. As an additional result they obtain probabilistic regular expressions which are equivalent to probabilistic automata when input words are equipped with an end marker.

In this work, we extend the ideas of Bollig, Gastin, Monmege and Zeitoun to define probabilistic ω-regular expressions (pωRE) for infinite words. These expressions arise from the addition of an ω-operator. Intuitively, E^ω denotes the limit of the probability that a word starts with E^n, for $n \to \infty$. We prove the expressive equivalence of pωRE and probabilistic Muller-automata by giving effective constructions.

[*] Partially supported by DFG Graduiertenkolleg 1763 (QuantLA).

A.-H. Dediu et al. (Eds.): LATA 2014, LNCS 8370, pp. 588–600, 2014.

Due to this equivalence, undecidability results can be gained from the probabilistic ω-automata setting. In particular, the approximation problem, which asks given an automaton and a small number $\varepsilon > 0$ for an approximation of the image of the automaton by finitely many points, is undecidable; even for probabilistic automata on finite words [17].

Furthermore, we introduce almost limit-deterministic expressions, which are a subclass of pωRE, where star iteration may only be applied to deterministic expressions or expressions whose semantics are bounded away from 1. Also ω-iteration may only be applied to deterministic expressions. We show that these expressions can be translated into almost limit-deterministic probabilistic Muller-automaton and solve the approximation problem for this class of automata.

2   Preliminaries

For the rest of the work let Σ be an arbitrary, finite alphabet. The set Σ^* contains all finite words over Σ and Σ^ω comprises all infinite words. For convenience we write Σ^∞ for $\Sigma^* \cup \Sigma^\omega$. For a word $w \in \Sigma^\infty$ let $|w| \in \mathbb{N} \cup \{\infty\}$ be the word's length and $\text{pre}(w) \subseteq \Sigma^*$ the set of all prefixes of w. For a language $L \subseteq \Sigma^\infty$ let $\mathbb{1}_L \colon \Sigma^\infty \to \{0, 1\}$ be its characteristic function.

Given a finite set X, we denote the set of all distributions on X by $\Delta(X)$, i.e. all functions $d \colon X \to [0, 1]$ such that $\sum_{x \in X} d(x) = 1$. For $Y \subseteq X$ we define $d(Y) = \sum_{y \in Y} d(y)$.

Let Ω be a set. A system of subsets $\mathcal{A} \subseteq 2^\Omega$ is a *σ-algebra*, if it contains the empty set and is closed under complement and countable union. A set $M \in \mathcal{A}$ is called *measurable*. A mapping $\mathbb{P} \colon \mathcal{A} \to [0, 1]$ is a *probability measure* if $\mathbb{P}(\Omega) = 1$ and $\mathbb{P}(\bigcup_{k \geq 1} M_k) = \sum_{k \geq 1} \mathbb{P}(M_k)$ for every pairwise disjoint family $(M_k)_{k \geq 1}$ of measurable sets. For $\Omega = \Sigma^\omega$ we consider the smallest σ-algebra containing the cone sets $\rho \Sigma^\omega$ for $\rho \in \Sigma^*$. This σ-algebra is sometimes called the Borel-σ-algebra on Σ^ω. A measure, which is given on all such cone sets, is uniquely determined.

Probabilistic automata generalize deterministic automata by employing a probability distribution to choose the next state, instead of using a fixed one.

Definition 1. *A probabilistic automaton is a tuple $A = (Q, \delta, \mu, F)$ with*
- *Q a non-empty, finite set of states*
- *$\delta \colon Q \times \Sigma \to \Delta(Q)$ the transition probability function*
- *$\mu \in \Delta(Q)$ the initial probabilities*
- *$F \subseteq Q$ a set of final states*

Instead of $\delta(p, a)(q)$ we simply write $\delta(p, a, q)$. For a word $w = w_1 \cdots w_n \in \Sigma^*$, the behavior $\|A\| \colon \Sigma^* \to [0, 1]$ of A is defined by

$$\|A\|(w) = \sum_{q_0, \ldots, q_{n-1} \in Q,\, q_n \in F} \mu(q_0) \prod_{i=1}^{n} \delta(q_{i-1}, w_i, q_i).$$

Probabilistic ω-automata have been introduced in [1]. The authors consider Büchi- and Rabin-acceptance in their work. Muller-acceptance is better suited for our approach, though it is as expressive as Rabin-acceptance.

Definition 2. *A probabilistic Muller-automaton is a tuple* $A = (Q, \delta, \mu, \mathrm{Acc})$ *where* Q, δ, *and* μ *are defined as in Theorem 1 and* $\mathrm{Acc} \subseteq 2^Q$ *is a Muller-acceptance condition. For a word* $w = w_1 w_2 \cdots \in \Sigma^\omega$, *the behavior of* A *is defined using the unique measure* $\mathbb{P}^w_A$ *on* Q^ω *given by*

$$\mathbb{P}^w_A(q_0 \cdots q_n Q^\omega) = \mu(q_0) \prod_{i=1}^n \delta(q_{i-1}, w_i, q_i).$$

The behavior $\|A\|\colon \Sigma^\omega \to [0,1]$ *of* A *is then* $\|A\|(w) = \mathbb{P}^w_A(\rho \in Q^\omega \; ; \; \inf(\rho) \in \mathrm{Acc})$, *where* $\inf(\rho)$ *designates the set of states occurring infinitely often in* ρ.

The existence and uniqueness of the measure $\mathbb{P}^w_A$ follows from the Ionescu-Tulcea extension theorem, cf. [15]. We extend the definition of $\mathbb{P}^w_A$ to finite words $w \in \Sigma^*$ by letting $\mathbb{P}^w_A(q_0 \cdots q_n) = \mu(q_0) \prod_{i=1}^n \delta(q_{i-1}, w_i, q_i)$ if $n = |w|$ and $\mathbb{P}^w_A(q_0 \cdots q_n) = 0$ if $n \neq |w|$. Then $\mathbb{P}^w_A$ is a measure on Q^*.

Let $M \subseteq Q^\omega$ or $M \subseteq Q^*$ a fixed set. We define the function $\mathbb{P}_A(M)$ on Σ^ω, resp. Σ^*, by $\mathbb{P}_A(M)(w) = \mathbb{P}^w_A(M)$.

3 Probabilistic ω-Regular Expressions

In this section we introduce the syntax and semantics of probabilistic ω-regular expressions, establish some basic results, and give an example of use.

Definition 3. *The set* $p\omega RE$ *of all probabilistic* ω-regular expressions *is the smallest set* $\mathcal{R}$ *satisfying*

1. $\Sigma^\omega \in \mathcal{R}$
2. *If* $\emptyset \neq A \subseteq \Sigma$ *and* $(E_a)_{a \in A} \in \mathcal{R}^A$, *then* $\sum_{a \in A} a E_a \in \mathcal{R}$
3. *If* $p \in [0,1]$ *and* $E, F \in \mathcal{R}$, *then* $pE + (1-p)F \in \mathcal{R}$ *and* $pE \in \mathcal{R}$
4. *If* $E\Sigma^\omega \in \mathcal{R}$ *and* $F \in \mathcal{R}$, *then* $EF \in \mathcal{R}$
5. *If* $E\Sigma^\omega + F \in \mathcal{R}$, *then* $E^* F + E^\omega \in \mathcal{R}$, $E^\omega \in \mathcal{R}$ *and* $E^* F \in \mathcal{R}$
6. *The set* $\mathcal{R}$ *is closed under distributivity of* $\cdot$ *over* $+$, *associativity, and commutativity of* $+$ *and multiplication by probability values.*

Before we define the semantics of these expressions, we give the corresponding operations on functions in general. Let $f, g \colon \Sigma^\infty \to [0,1]$. We define for $w \in \Sigma^\infty$:

$$(fg)(w) = \sum_{uv=w} f(u)g(v), \quad f^*(w) = \sum_{n \geq 0} f^n(w), \quad f^\omega(w) = \lim_{n \to \infty} (f^n \mathbb{1}_{\Sigma^\infty})(w),$$

where f^n is defined using the above product operation and f^0 as $\mathbb{1}_{\{\varepsilon\}}$. Note that in general the infinite sums may not converge or are not bounded by 1.

Definition 4. *Let* $w \in \Sigma^\infty$. *The semantics of probabilistic* ω-regular expressions *are inductively defined by* $\|\Sigma^\omega\|(w) = \mathbb{1}_{\Sigma^\omega}(w)$ *and*

$$\|p\|(w) = \begin{cases} p & \text{if } w = \varepsilon \\ 0 & \text{otherwise} \end{cases} \qquad \|a\|(w) = \begin{cases} 1 & \text{if } w = a \\ 0 & \text{otherwise} \end{cases}$$

$$\|E + F\|(w) = \|E\|(w) + \|F\|(w) \qquad \|EF\|(w) = (\|E\| \|F\|)(w)$$

$$\|E^*\|(w) = \|E\|^*(w) \qquad \|E^\omega\|(w) = \|E\|^\omega(w),$$

where $a \in \Sigma$ *and* $p \in [0,1]$.

One can show that the set of all expressions E with $\|E\|$ well-defined and $\|E\|(w) \leq$ 1 for all $w \in \Sigma^\omega$ satisfies all conditions in Theorem 3. As pωRE is the smallest set of expressions that satisfies all these conditions, this set is already the whole set pωRE. Similarly, one shows that the semantics of a probabilistic ω-regular expressions without any probability values coincides with the characteristic function of the language defined by the expression in the classical sense.

There is a connection to the probabilistic regular expressions defined in [3]: Let pRE be the set of all probabilistic regular expressions as in [3]. We then have

$$pRE = \{E \text{ subexpression} \mid E\Sigma^\omega \in p\omega RE\}.$$

Furthermore, the semantics defined on pRE in this work and in [3] coincide. Thus we can consider pRE a fragment of pωRE.

The syntax definition given in Theorem 3 is aimed to be minimal, but some more general rules to derive new expressions can be shown.

Lemma 5. *The following statements hold:*
1. *If $E + F \in p\omega RE$, then $E \in p\omega RE$.*
2. *If $E\Sigma^\omega + F \in p\omega RE$ and $G \in p\omega RE$, then $EG + F \in p\omega RE$.*

Example 6. We consider a communication device for sending messages. At every point of time either a new input message becomes available or the device is waiting for a new message. When a new message is available the device tries to send this message. Sending a message may fail with probability $\frac{1}{3}$. In this case the message is stored in an internal buffer. The next time the device is waiting for a message, sending the stored message is retried. Intuitively, as sending a buffered message has already failed once, it seems to be harder to send this message. So sending a buffered message is only successful with probability $\frac{1}{2}$. The buffer can hold one message.

To build an expression for this model, we consider the two letter alphabet $\Sigma = \{w, i\}$ for the events "wait" and "input message". The expression is built bottom up. We first construct the expression for the case that the buffer is already full. The notation $E \rightsquigarrow F$ is used to symbolize, that F can be constructed from E by application of Theorems 3 and 5. Underlined expressions have zero semantics and are left out in subsequent steps. We derive

$$\Sigma^\omega \rightsquigarrow w\Sigma^\omega + i\Sigma^\omega \rightsquigarrow \tfrac{1}{2}w\Sigma^\omega + \tfrac{1}{2}w\Sigma^\omega + \tfrac{2}{3}i\Sigma^\omega$$
$$\rightsquigarrow (\tfrac{1}{2}w)^*\tfrac{1}{2}w\Sigma^\omega + (\tfrac{1}{2}w)^*\tfrac{2}{3}i\Sigma^\omega + \underline{(\tfrac{1}{2}w)^\omega}$$
$$\rightsquigarrow ((\tfrac{1}{2}w)^*\tfrac{2}{3}i)^*(\tfrac{1}{2}w)^*\tfrac{1}{2}w\Sigma^\omega + \underline{((\tfrac{1}{2}w)^*\tfrac{2}{3}i)^\omega}.$$

Let the expression $((\tfrac{1}{2}w)^*\tfrac{2}{3}i)^*(\tfrac{1}{2}w)^*\tfrac{1}{2}w$ be denoted by B. Intuitively, this expression describes the probability of returning to an empty buffer without overflowing the buffer until then. For the full expression we continue

$$\Sigma^\omega \rightsquigarrow w\Sigma^\omega + i\Sigma^\omega \rightsquigarrow w\Sigma^\omega + \tfrac{2}{3}i\Sigma^\omega + \tfrac{1}{3}i\Sigma^\omega \rightsquigarrow w^*\tfrac{2}{3}i\Sigma^\omega + w^*\tfrac{1}{3}i\Sigma^\omega + w^\omega$$
$$\rightsquigarrow (w^*\tfrac{2}{3}i)^*w^*\tfrac{1}{3}i\Sigma^\omega + (w^*\tfrac{2}{3}i)^*w^\omega + \underline{(w^*\tfrac{2}{3}i)^\omega}$$

$$\rightsquigarrow (w^*\tfrac{2}{3}i)^*w^*\tfrac{1}{3}iB\Sigma^\omega + (w^*\tfrac{2}{3}i)^*w^\omega$$

$$\rightsquigarrow \left((w^*\tfrac{2}{3}i)^*w^*\tfrac{1}{3}iB\right)^*(w^*\tfrac{2}{3}i)^*w^\omega + \left((w^*\tfrac{2}{3}i)^*w^*\tfrac{1}{3}iB\right)^\omega$$

In the expression $(w^*\tfrac{2}{3}i)^*w^*\tfrac{1}{3}iB$ the term $(w^*\tfrac{2}{3}i)^*$ describes the iteration of waiting an arbitrary time, then sending an incoming message successfully with probability $\tfrac{2}{3}$. The term $w^*\tfrac{1}{3}i$ describes arbitrary waiting and then failing to send an incoming message with probability $\tfrac{1}{3}$. Hence the message is stored in the buffer and B describes the probability to empty the buffer again without overflowing the buffer.

4   The Main Theorem

Theorem 7. *Let $f\colon \Sigma^\omega \to [0,1]$. The following statements are equivalent*
1. $f = \|A\|$ *for some probabilistic Muller-automaton A.*
2. $f = \|E\|$ *for some probabilistic ω-regular expression E.*

In the next subsection, we give some results for prefix-free languages which are required for our proof. The following two subsections give a sketch of the two directions of the proof of Theorem 7.

4.1   Prefix-Free Languages

Definition 8. *Let Q be an alphabet. A language $L \subseteq Q^*$ is called* prefix-free *if* $\mathrm{pre}(w) \cap L = \{w\}$ *for all $w \in L$.*

Prefix-free languages, also called prefix languages, are a known concept in language- and coding theory. For example, Prefix-free languages have been used to define determinism for generalized automata [13]. Prefix-freeness allows for unique concatenation, i.e. if L is prefix-free, K any language, and $w \in LK$, then there is exactly one pair $u \in L$, $v \in K$ with $w = uv$. More generally, given a prefix-free language $L \subseteq Q^*$ and a word $w \in Q^\infty$ there is either a decomposition $w = w_1 w_2 \cdots$ for unique $w_i \in L$ ($i \geq 1$) or $w = w_1 \cdots w_n v$ for unique $w_i \in L$ and $\mathrm{pre}(v) \cap L = \emptyset$. The class of prefix-free languages is closed under concatenation.

Let $q_1, \ldots, q_n \in Q$. For the proof of Theorem 7 we consider words, that contain all letters $q_1, \ldots, q_n$ in this order and are minimal with that property, w.r.t. the prefix ordering. Such words will be used to denote minimal runs of an automaton through states $q_1, \ldots, q_n$, and, eventually, to construct expressions for loops in the automaton. Formally, for a sequence $q_1 \cdots q_n \in Q^*$ we define the prefix-free language $M_{q_1, \ldots, q_n}$ by

$$M_{q_1, \ldots, q_n} = \prod_{i=1}^n (A \setminus \{q_i\})^* q_i \qquad \text{and} \qquad M_\varepsilon = \{\varepsilon\}. \tag{1}$$

We consider prefix-free sets of paths in an automaton. These sets allow interchanging rational operations and the automaton's probability measure. In the following, for an automaton $A = (Q, \delta, \mu, \mathrm{Acc})$ and $q \in Q$, let A_q denote the automaton $(Q, \delta, \mathbb{1}_{\{q\}}, \mathrm{Acc})$.

Lemma 9. *Let A be a probabilistic Muller-automaton, Q its state space, $L \subseteq Q^*$ and $q \in Q$ such that Lq is prefix-free. Let further $w \in \Sigma^\omega$ and $K \subseteq Q^\omega$ measurable. Then the following statements hold:*

1. $\sum_{uv=w} \mathbb{P}_A^u(Lq) \leq 1$.
2. $\mathbb{P}_A(LqK) = \mathbb{P}_A(Lq) \cdot \mathbb{P}_{A_q}(qK)$.
3. $\mathbb{P}_{A_q}\big(q(Lq)^*\big) = \big(\mathbb{P}_{A_q}(qLq)\big)^*$.
4. $\mathbb{P}_{A_q}\big(q(Lq)^\omega\big) = \big(\mathbb{P}_{A_q}(qLq)\big)^\omega$.

4.2 From Automata to Expressions

For the rest of this subsection let $A = (Q, \delta, \mu, \mathrm{Acc})$ be a fixed probabilistic Muller-automaton. We may assume that $\mu = \mathbb{1}_{\{\iota\}}$ for a state $\iota \in Q$. Let $p, q \in Q$ and $R, X \subseteq Q$. We inductively construct expressions $E_{p,q}^X$, $E_{p,\inf=R}^X$, and E_p^X with either $X = \emptyset$ or $p \in X$ such that the following equalities hold:

$$\big\|E_{p,q}^X\big\| = \mathbb{P}_{A_p}(pX^*q)$$

$$\big\|E_{p,\inf=R}^X\big\| = \mathbb{P}_{A_p}(\rho \in pX^\omega \ ; \ \inf(\rho) = R)$$

$$E_p^X = \sum_{r \notin X} E_{p,r}^X \Sigma^\omega + \sum_{R \subseteq X} E_{p,\inf=R}^X \cdot$$

If $X = \emptyset$, such expressions can be constructed directly by definition.

Assume $X = \{x_1, \ldots, x_n\}$ with $n \geq 1$. For convenience let $x_{n+1} = x_1$. By induction hypothesis, there exist expressions $E_{x_i}^{X \setminus \{x_{i+1}\}} \in \mathrm{p}\omega\mathrm{RE}$ with

$$E_{x_i}^{X \setminus \{x_{i+1}\}} = \sum_{r \notin X} E_{x_i,r}^{X \setminus \{x_{i+1}\}} \Sigma^\omega + E_{x_i,x_{i+1}}^{X \setminus \{x_{i+1}\}} \Sigma^\omega + \sum_{R \subseteq X \setminus \{x_{i+1}\}} E_{x_i,\inf=R}^{X \setminus \{x_{i+1}\}} \cdot \quad (2)$$

To ease notation we make the following definitions:

$$C_k = E_{x_1,x_2}^{X \setminus \{x_2\}} E_{x_2,x_3}^{X \setminus \{x_3\}} \cdots E_{x_{k-1},x_k}^{X \setminus \{x_k\}} \qquad C = C_n E_{x_n,x_1}^{X \setminus \{x_1\}}.$$

The expression C denotes a full iteration through all states in X in order starting in state x_1, and C_k denotes a partial iteration only until state x_k. For $k = 1$ the expression C_k is just the empty expression. By Theorem 9, the behavior of these expressions can be written using the $M...$ languages from (1) in Theorem 8:

$$\|C\| = \mathbb{P}_{A_{x_1}}(x_1 M_{x_2,\ldots,x_n,x_1}) \quad \text{and} \quad \|C_k\| = \mathbb{P}_{A_{x_1}}(x_1 M_{x_2,\ldots,x_k}).$$

By repeated application of Theorem 5 on (2) for $i = 1, \ldots, n$, we obtain that the following expression $\tilde{E}_{x_1}^X$ is a $\mathrm{p}\omega\mathrm{RE}$:

$$\tilde{E}_{x_1}^X = \sum_{r \notin X} \sum_{k=1}^n C_k E_{x_k,r}^{X \setminus \{x_{k+1}\}} \Sigma^\omega + C\Sigma^\omega + \sum_{\substack{R \subsetneq X}} \sum_{\substack{k \in \{1,\ldots,n\} \\ x_{k+1} \notin R}} C_k E_{x_k,\inf=R}^{X \setminus \{x_{k+1}\}}$$

Next, we apply Item 5 in Theorem 3 to $\tilde{E}_{x_1}^X$ and obtain

$$E_{x_1}^X = \sum_{r \notin X} \sum_{k=1}^n C^* C_k E_{x_k}^{X \setminus \{x_{k+1}\}} \Sigma^\omega + \sum_{R \subsetneq X} \sum_{x_{k+1} \notin R} C^* C_k E_{x_k,\inf=R}^{X \setminus \{x_{k+1}\}} + C^\omega.$$

Using Theorem 9 one can show that the following equalities hold:

$$\left\| \sum_{k=1}^n C^* C_k E_{x_k,r}^{X \setminus \{x_{k+1}\}} \right\| = \mathbb{P}_{A_{x_1}}(x_1 X^* r)$$

$$\left\|\sum_{x_{k+1}\notin R} C^* C_k E_{x_k,\inf=R}^{X\setminus\{x_{k+1}\}}\right\| = \mathbb{P}_{A_{x_1}}(\rho \in x_1 X^\omega \; ; \; \inf(\rho) = R)$$

$$\|C^\omega\| = \mathbb{P}_{A_{x_1}}(\rho \in x_1 X^\omega \; ; \; \inf(\rho) = X).$$

For $X = Q$ the expression E_ι^Q has the form $E_\iota^Q = \sum_{R\subseteq Q} E_{\iota,\inf=R}^Q$. By Theorem 5 we can restrict the sum in E_ι^Q to summands $R \in \mathrm{Acc}$ and obtain the desired expression E with $\|E\| = \|A\|$.

Example 10. Consider the automaton $A = (\{q_1,q_2\}, \delta, \mathbb{1}_{\{q_1\}}, \{\{q_1,q_2\}\})$ over $\Sigma = \{a,b\}$ where $\delta(q_1,a,q_1) = 1/2 = \delta(q_1,a,q_2)$, and $\delta(q_2,a,q_2) = \delta(q_2,b,q_1) = 1$. This is a partial automaton, which can be made into a probabilistic Muller-automaton by adding a sink state. To keep the size of the immediate expressions small, we continue to work with the partial automaton. The following expressions occur in the construction of E_A:

$$E_{q_1}^\emptyset = \tfrac{1}{2}a\Sigma^\omega + \tfrac{1}{2}a\Sigma^\omega + b\Sigma^\omega \qquad E_{q_2}^\emptyset = a\Sigma^\omega + b\Sigma^\omega$$

$$E_{q_1}^{\{q_1\}} = (\tfrac{1}{2}a)^* \tfrac{1}{2}a\Sigma^\omega + (\tfrac{1}{2}a)^\omega \qquad E_{q_2}^{\{q_2\}} = a^*b\Sigma^\omega + a^\omega$$

$$E_{q_1}^{\{q_1,q_2\}} = \left((\tfrac{1}{2}a)^*\tfrac{1}{2}aa^*b\right)^\omega + \left((\tfrac{1}{2}a)^*\tfrac{1}{2}aa^*b\right)^*(\tfrac{1}{2}a)^*\tfrac{1}{2}aa^\omega + \left((\tfrac{1}{2}a)^*\tfrac{1}{2}aa^*b\right)^*(\tfrac{1}{2}a)^\omega$$

The first summand is the wanted expression E with $\|E\| = \|A\|$.

4.3   From Expressions to Automata

In order to handle distributivity and commutativity we adopt the idea of using terms from [3]. The terms of an expression are the summands, when the expression is maximally expanded by using distributivity. As summands may occur more than once, we use multisets instead of sets. The notation $\{\!\{x_1,\ldots,x_n\}\!\}$ denotes the multiset containing $x_1,\ldots,x_n$. Formally

Definition 11. *The multiset of* terms *of a (subexpression of a) probabilistic ω-regular expression is inductively defined by* $\mathcal{T}(\Sigma^\omega) = \{\!\{\Sigma^\omega\}\!\}$ *and*

$$\mathcal{T}(p) = \{\!\{p\}\!\} \qquad \mathcal{T}(E^*) = \{\!\{E^*\}\!\} \qquad \mathcal{T}(E+F) = \mathcal{T}(E) \cup \mathcal{T}(F)$$

$$\mathcal{T}(a) = \{\!\{a\}\!\} \qquad \mathcal{T}(E^\omega) = \{\!\{E^\omega\}\!\} \qquad \mathcal{T}(EF) = \mathcal{T}(E)\mathcal{T}(F)$$

The set of head terms $\mathcal{HT}(E)$ *and* tail terms $\mathcal{TT}(E)$ *is given by*

$$\mathcal{HT}(E) = \{\!\{E \text{ subexpression} \mid E\Sigma^\omega \in \mathcal{T}(E)\}\!\} \qquad \mathcal{TT}(E) = \mathcal{T}(E) \setminus \mathcal{HT}(E)\Sigma^\omega.$$

For a multiset $M \subseteq \mathcal{T}(E)$ let $\mathcal{E}(M) = \sum_{E\in M} E$.

We use Muller-automata with final states as we need to handle finite and infinite words, i.e., automata $A = (Q,\delta,\mu,\mathrm{Acc},F)$ such that $(Q,\delta,\mu,\mathrm{Acc})$ is a probabilistic Muller-automaton and $F \subseteq Q$ are final states. The behavior $\|A\|$ of A is then defined on infinite and finite words by either using Acc or F as acceptance condition. For a set $X \subseteq Q$ let $A[X]$ denote the probabilistic automaton (Q,δ',μ,X) where δ' equals zero on $X \times \Sigma$ and agrees with δ everywhere else.

For an expression $E \in \mathrm{p\omega RE}$ with $\mathcal{HT}(E) = \{\!\{E_1,\ldots,E_n\}\!\}$, we construct an automaton $A_E = (Q,\delta,\mu,\mathrm{Acc},F)$ such that

1. Every set in Acc is disjoint from F.
2. All states in F are sinks, i.e. all outgoing transitions are self-loops.
3. $\|A\|(w) = \|\mathcal{E}(\mathcal{TT}(E))\|(w)$ for all $w \in \Sigma^\omega$.
4. There is a partition $X_1 \dot\cup \cdots \dot\cup X_n = F$ such that $\|A[X_i]\|(w) = \|E_i\|(w)$ for all $w \in \Sigma^*$ and $1 \le i \le n$.

Note that such an automaton can be easily transformed into an automaton A'_E with $\|A'_E\| = \|E\|$ by letting $\text{Acc}' = \text{Acc} \cup \{\{q\} \mid q \in F\}$.

The proof now proceeds by showing that the set of all expressions such that an automaton A_E, as defined above, can be constructed equals the set of all probabilistic ω-regular expressions. The constructions are based on the constructions given in [3] and are therefore left out here.

5   Decidability Results

There are many decidability results known for probabilistic ω-automata. Unfortunately most of them state that the problem in question is undecidable. The next proposition states that the emptiness problem under probable and almost sure semantics, and the approximation problem is undecidable.

Proposition 12 (see [2,5,17]). *The following problems are undecidable:*
1. *Given $E \in p\omega RE$, is $\|E\|(w) > 0$ for some $w \in \Sigma^\omega$.*
2. *Given $E \in p\omega RE$, is $\|E\|(w) = 1$ for some $w \in \Sigma^\omega$.*
3. *Given E and $\varepsilon > 0$, such that either $\|E\|(w) < \varepsilon$ for all $w \in \Sigma^\omega$ or $\|E\|(w) \ge 1 - \varepsilon$ for a $w \in \Sigma^\omega$, decide which is the case.*

6   Almost ω-Deterministic Expressions

The poor decidability properties of probabilistic ω-automata motivate the search for expressive subclasses of automata with better decidability properties. Considerable work has already be done in this direction, c.f. [8,9,12].

We introduce almost ω-deterministic expressions, where ω-iteration may only be applied to deterministic expressions and every star-iteration has to be deterministic or a probability less than 1 to repeat the iteration.

Definition 13. *Given a subexpression E of a probabilistic ω-regular expression, we call E*
1. *deterministic if it does not contain any probability values other than 0 or 1,*
2. *permeable if for every subexpression F^* of E we have $F = F_1 p F_2$ such that $p < 1$ and $F_1 F_2 \Sigma^\omega$ is a probabilistic regular expression.*
We say an expression E is almost ω-deterministic if every subexpression F^* of E is either deterministic or permeable, and every subexpression F^ω is deterministic.

In order to find an analogous concept for probabilistic Muller-automata, we consider automata where every path will almost surely be eventually deterministic, i.e the set of all paths ρ such that $\inf(\rho)$ is deterministic has probability 1. A similar concept, limit-deterministic probabilistic Büchi automata, has been considered in [23].

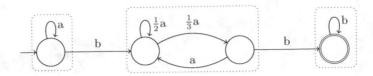

Fig. 1. Automaton for Fig. 1, SCCs are enclosed by dashed boxes

Definition 14. *Given a probabilistic Muller-automaton $A = (Q, \delta, \mu, \mathrm{Acc})$ a set of states $M \subseteq Q$ is called*
1. *deterministic if $\delta(p, a, q) \in \{0, 1\}$ for all $p, q \in M$ and $a \in \Sigma$,*
2. *permeable if $\mathbb{P}^w_{A_q}(M^\omega) = 0$ for all $w \in \Sigma^\omega$ and $q \in M$.*

We say A is an almost limit-deterministic probabilistic Muller automaton, *for short: ALDPMA, if every strongly connected component (SCC) is either deterministic or permeable, where we consider the underlying transition graph, i.e. there is an edge from state p to state q if $\delta(p, a, q) > 0$ for some $a \in \Sigma$.*

In an ALDPMA every bottom SCC, i.e. SCCs with no outgoing transition, has to be deterministic, but there may be other deterministic SCCs.

Example 15. Consider the almost ω-deterministic expression

$$\mathrm{a}^* \mathrm{b} \left(\tfrac{1}{2}\mathrm{a}\right)^* \tfrac{1}{2}\tfrac{2}{3}\mathrm{a} \left(\mathrm{a} \left(\tfrac{1}{2}\mathrm{a}\right)^* \tfrac{1}{2}\tfrac{2}{3}\mathrm{a}\right)^* \mathrm{bb}^\omega.$$

An equivalent almost limit-deterministic Muller-automaton is depicted in Fig. 1.

Theorem 16. *Let E be an almost ω-deterministic expression. There is an effectively constructible ALDPMA A with $\|E\| = \|A\|$.*

Proof. The proof works by extending the proof of Theorem 7 carefully.

We use Muller-automata with final states $A = (Q, \delta, \mu, \mathrm{Acc}, F)$ that are almost limit-deterministic with the following properties in addition to the properties in Section 4.3:
5. For every $E' \in \mathcal{HT}(E)$ with corresponding acceptance set $X \subseteq F$ let $R = \{q \in Q \mid X \text{ is reachable from } q\}$. If E' is deterministic, resp. permeable, then R is deterministic, resp. permeable.
6. $\mu(Q) \subseteq \{0, 1\}$ if there is at least one deterministic expression in $\mathcal{HT}(E)$.

Using a similar inductive construction as in the general case, we obtain an ALDPMA A which is equivalent to E. $\qquad\square$

The next theorem states that the image of an ALDPMA can be approximated arbitrarily close by a finite, computable set. This does not hold in general even for probabilistic automata on finite words [17]. In the following for $Y, Z \subseteq \mathbb{R}^{n \times m}$ let $\mathrm{B}_\varepsilon(Y)$ be the ε-neighborhood with respect to the row-sum-norm $\|\cdot\|_\infty$, resp., the absolute value if $n = m = 1$. The Hausdorff distance $d_\mathrm{H}(Y, Z)$ is the infimum of all $\varepsilon > 0$ with $Y \subseteq \mathrm{B}_\varepsilon(Z)$ and $Z \subseteq \mathrm{B}_\varepsilon(Y)$.

Theorem 17. *Given an ALDPMA A and an $\varepsilon > 0$, there is an effectively computable finite set $V \subseteq [0, 1]$ such that $d_\mathrm{H}(\|A\|(\Sigma^\omega), V) \leq \varepsilon$.*

Proof. The proof uses induction on the number of SCCs in A and removes a top SCC C in the induction step. To ensure $V \subseteq B_\varepsilon(\|A\|(\Sigma^\omega))$ we need not only to approximate $\|A\|$ in the induction, but vectors of acceptance probabilities. In case that C is permeable the probability of all paths which stay long enough in C is arbitrarily small. In the other case, C is deterministic, further difficulties arise: Intuitively the deterministic SCC may delay runs starting at one state q_1 before they leave C, whereas other runs, starting in a state $q_2 \neq q_1$ may leave C immediately. To solve this problem we also approximate all transition probabilities when reading finite words.

Formally, define for any automaton A with state space Q two functions: $\nu_A \colon \Sigma^\omega \to [0,1]^Q$ and $\Theta_A \colon \Sigma^* \to [0,1]^{Q \times Q}$ which are given by $\nu_A(w)_q = \|A_q\|(w)$ and $\Theta_A(w)_{p,q} = \mathbb{P}^w_{A_p}(Q^*q)$, where A_q is the automaton A but with initial state q. We show by induction on $n \geq 1$ that for every n holds: Given an ALDPMA A with n SCCs, an $\varepsilon > 0$ and regular languages $L_1 \subseteq \Sigma^\omega$ and $L_2 \subseteq \Sigma^{'*}$ one can effectively compute finite sets $Y \subseteq [0,1]^Q$ and $Z \subseteq [0,1]^{Q \times Q}$ such that $d_H(Y, \nu_A(L_1)) \leq \varepsilon$ and $d_H(Z, \Theta_A(L_2)) \leq \varepsilon$.

In the induction base case, the whole automaton A is deterministic and $\nu_A(L_1)$ and $\Theta_A(L_2)$ can be directly computed. Assume $n > 1$. We consider a top SCC C in Q, i.e. $\delta(q,a,p) = 0$ for all $q \in Q \setminus C$, $a \in \Sigma$, and $p \in C$. Let $Q' = Q \setminus C$. We can decompose $\Theta_A(w)$ and $\nu_A(w)$ for any $w_1 \in \Sigma^*$ and $w_2 \in \Sigma^\omega$ into components:

$$\Theta_A(w_1) = \begin{pmatrix} M_1(w_1) & M_2(w_1) \\ 0 & M_3(w_1) \end{pmatrix}, \qquad \nu_A(w_2) = \begin{pmatrix} \nu_1(w_2) \\ \nu'(w_2) \end{pmatrix},$$

where $M_1(w) \in [0,1]^{C \times C}$, $M_2(w) \in [0,1]^{C \times Q'}$, $M_3(w) \in [0,1]^{Q' \times Q'}$, $\nu_1(w) \in [0,1]^C$, and $\nu'(w) \in [0,1]^{Q'}$. First, assume that C is permeable. We then have $\|M_1(w)\|_\infty \to 0$ as $|w| \to \infty$. Choose N large enough, that $\|M_1(w)\|_\infty \leq \frac{\varepsilon}{2}$ for all w with $|w| \geq N$. Furthermore, by induction hypothesis there are finite, computable sets $Y'(u^{-1}L_1)$ and $Z'(u^{-1}L_2)$ with $d_H(Y'(u^{-1}L), M_3(u^{-1}L)) \leq \frac{\varepsilon}{2}$ and $d_H(Z'(u^{-1}L), \nu'(u^{-1}L)) \leq \frac{\varepsilon}{2}$. Thus we can choose Y and Z as

$$Z = \bigcup_{u \in \Sigma^N} \begin{pmatrix} M_2(u) \\ M_3(u) \end{pmatrix} Z'\left(u^{-1}L_2\right) \cup \Theta_A\left(L_2 \cap \Sigma^{<N}\right), \quad Y = \bigcup_{u \in \Sigma^N} \begin{pmatrix} M_2(u) \\ M_3(u) \end{pmatrix} Y'\left(u^{-1}L_1\right).$$

Now, we consider the case that C is deterministic. We first construct the set Z for an arbitrary L_2 and then move to Y. For $ua \in \Sigma^+$ ($a \in \Sigma$) let $\widetilde{M_2}(ua) = M_1(u)M_2(a)$ and $\widetilde{M_2}(\varepsilon) = 0$. Note that $\widetilde{M_2}(K)$ is finite and computable for every regular language K. One shows for every $w \in \Sigma^*$ that there is a decomposition $w = u_1 \cdots u_n v$ with $n \leq |C|$ such that $u_1 \cdots u_i$ ($1 \leq i \leq n$) are the only prefixes w' of w with $\widetilde{M_2}(w') \neq 0$. Hence $M_2(w) = \sum_{i=1}^n \widetilde{M_2}(u_1 \cdots u_i) M_3(u_{i+1} \cdots u_n v)$. Let $\equiv_{L_2}$ be the syntactic congruence relation of L_2. By induction hypothesis there are sets $Z'(L')$ for every recognizable language L' such that $d_H(Z'(L'), M_3(L')) \leq \varepsilon$. We choose Z by

$$Z = \bigcup_{n=1}^{|C|} \; \bigcup_{\substack{D_1, P_1, \ldots, D_n, P_n, D' \\ [u_1], \ldots, [u_n] \in \Sigma^* / \equiv_{L_2}}} \; \prod_{i=1}^n \begin{pmatrix} D_i & P_i \\ 0 & Z'\left(K^{D_1 \cdots D_{i-1}}_{D_i, P_i, u_i}\right) \end{pmatrix} \cdot \begin{pmatrix} D' & 0 \\ 0 & Z'\left(T^{D_1 \cdots D_n}_{D'}\right) \end{pmatrix}$$

where the D_i range over $M_1(\Sigma^*)$ and the P_i over $\widetilde{M}_2(\Sigma^*)$. The recognizable languages K and T are given by

$$K^{D_0}_{D,P,u} = \left\{ w \in \Sigma^* \;\middle|\; w \equiv_{L_2} u,\; M_1(w) = D,\; \widetilde{M}_2(w) = P, \right.$$
$$\left. D_0\widetilde{M}_2(w') = 0 \text{ for all strict prefixes } w' \text{ of } w \right\},$$

$$T^{D_0}_D = \left\{ w \in \Sigma^* \;\middle|\; M_1(w) = D,\; D_0\widetilde{M}_2(w') = 0 \text{ for all prefixes } w' \text{ of } w \right\}.$$

We are now ready to construct the set Y. For the acceptance probabilities we also need to consider paths which never leave C. Hence let B be the restriction of A to state space C. For a vector $\tau \in \{0,1\}^C$ define the regular language $L_\tau = \{w \in \Sigma^\omega \mid \nu_B(w) = \tau\}$. We construct Y by

$$Y = \bigcup_{\tau,[u]_\tau} \binom{\tau}{0} + Z([u]_\tau) \left(\begin{matrix} 0 \\ Z'(u^{-1}(L_1 \cap L_\tau)) \end{matrix} \right),$$

where τ ranges over $\{0,1\}^C$ and $[u]_\tau = \{w \in \Sigma^* \mid w^{-1}(L_1 \cap L_\tau) = u^{-1}(L_1 \cap L_\tau)\}$ ranges over all $u \in \Sigma^*$ (but there are only finitely many $[u]_\tau$). This completes the induction. Thus, we have constructed a finite computable set $Y \subseteq [0,1]^Q$ such that $d_H(Y, \nu_A(\Sigma^\omega)) \leq \varepsilon$. Hence $d_H(\mu Y, \mu\nu_A(\Sigma^\omega)) \leq \varepsilon$ and $\mu\nu_A(\Sigma^\omega) = \|A\|(\Sigma^\omega)$. $\square$

Corollary 18. *All problems stated in Theorem 12 are decidable for almost ω-deterministic expressions.*

Proof. For the approximation problem we use Theorem 17. For the two emptiness problems note that we only need to consider paths with stay eventually in a deterministic component. Hence we only need to consider countable many paths and can apply standard graph algorithms.

7   Future Work

Our work does not provide the equivalence between almost ω-deterministic expressions and almost limit-deterministic automata. We conjecture that these two concepts are indeed expressively equivalent.

In [3], a Kleene theorem was not only shown for probabilistic automata but for probabilistic pebble automata. A natural question is, whether our result can be generalized to include pebbles.

The natural star-free fragment of pωRE consists only of finite, convex combinations of star-free languages. Is there a more expressive fragment of pωRE that exhibits similar properties as classical star-free languages? For example, is there some notion of counter-freeness for probabilistic ω-automata, or a suitable probabilistic linear time logic? Is there a connection to a first-order-fragment of probabilistic MSO, as introduced in [24]?

Considerable research has constructed subclasses of probabilistic ω-automata, like hierarchical automata [5] or structurally simple automata [9], with better decidability properties. One may ask if such a subclass corresponds to a natural fragment pωRE, or if new subclasses of automata can be derived by choosing appropriate fragments of pωRE.

References

1. Baier, C., Grösser, M.: Recognizing ω-regular languages with probabilistic automata. In: Proc. LICS, pp. 137–146. IEEE (2005)
2. Baier, C., Bertrand, N., Größer, M.: On decision problems for probabilistic büchi automata. In: Amadio, R.M. (ed.) FOSSACS 2008. LNCS, vol. 4962, pp. 287–301. Springer, Heidelberg (2008)
3. Bollig, B., Gastin, P., Monmege, B., Zeitoun, M.: A probabilistic kleene theorem. In: Chakraborty, S., Mukund, M. (eds.) ATVA 2012. LNCS, vol. 7561, pp. 400–415. Springer, Heidelberg (2012)
4. Büchi, J.R.: On a decision method in restricted second order arithmetic. In: Int. Congress for Logic, Methodology and Philosophy of Science, pp. 1–11 (1962)
5. Chadha, R., Sistla, A.P., Viswanathan, M.: Power of randomization in automata on infinite strings. In: Bravetti, M., Zavattaro, G. (eds.) CONCUR 2009. LNCS, vol. 5710, pp. 229–243. Springer, Heidelberg (2009)
6. Chadha, R., Sistla, A.P., Viswanathan, M.: Probabilistic büchi automata with non-extremal acceptance thresholds. In: Jhala, R., Schmidt, D. (eds.) VMCAI 2011. LNCS, vol. 6538, pp. 103–117. Springer, Heidelberg (2011)
7. Chatterjee, K., Doyen, L., Henzinger, T.A.: Probabilistic weighted automata. In: Bravetti, M., Zavattaro, G. (eds.) CONCUR 2009. LNCS, vol. 5710, pp. 244–258. Springer, Heidelberg (2009)
8. Chatterjee, K., Henzinger, T.: Probabilistic automata on infinite words: Decidability and undecidability results. In: Bouajjani, A., Chin, W.-N. (eds.) ATVA 2010. LNCS, vol. 6252, pp. 1–16. Springer, Heidelberg (2010)
9. Chatterjee, K., Tracol, M.: Decidable problems for probabilistic automata on infinite words. In: LICS, pp. 185–194. IEEE Computer Society (2012)
10. Cheung, L., Lynch, N., Segala, R., Vaandrager, F.: Switched PIOA: Parallel composition via distributed scheduling. TCS 365(1-2), 83–108 (2006)
11. Ésik, Z., Kuich, W.: Finite automata. In: Droste, M., Kuich, W., Vogler, H. (eds.) Handbook of Weighted Automata. EATCS Monographs, pp. 69–104. Springer (2009)
12. Fijalkow, N., Gimbert, H., Oualhadj, Y.: Deciding the value 1 problem for probabilistic leaktight automata. In: LICS, pp. 295–304. IEEE Computer Society (2012)
13. Giammarresi, D., Montalbano, R.: Deterministic generalized automata. TCS 215(1-22), 191–208 (1999)
14. Kleene, S.C.: Representation of events in nerve nets and finite automata. In: Shannon, Mccarthy (eds.) Automata Studies, pp. 3–41. Princeton Univ. Press (1956)
15. Klenke, A.: Probability Theory: A Comprehensive Course. Springer (December 2007)
16. Kuich, W., Rahonis, G.: Fuzzy regular languages over finite and infinite words. Fuzzy Sets and Systems 157(11), 1532–1549 (2006)
17. Madani, O., Hanks, S., Condon, A.: On the undecidability of probabilistic planning and related stochastic optimization problems. Artif. Intell. 147(1-2), 5–34 (2003)
18. Mora-Lpez, L., Morales, R., Sidrach de Cardona, M., Triguero, F.: Probabilistic finite automata and randomness in nature: a new approach in the modelling and prediction of climatic parameters. In: Proc. International Environmental Modelling and Software Congress, pp. 78–83 (2002)

19. Rabin, M.: Probabilistic automata. Information and Control 6(3), 230–245 (1963)
20. Ron, D., Singer, Y., Tishby, N.: The power of amnesia: Learning probabilistic automata with variable memory length. Machine Learning 25, 117–149 (1996)
21. Schützenberger, M.P.: On the definition of a family of automata. Information and Control 4(2-3), 245–270 (1961)
22. Tracol, M., Baier, C., Größer, M.: Recurrence and transience for probabilistic automata. In: FSTTCS, vol. 4, pp. 395–406. Schloss Dagstuhl - Leibniz-Zentrum fuer Informatik (2009)
23. Vardi, M.Y.: Automatic verification of probabilistic concurrent finite state programs. In: Foundations of Computer Science, pp. 327–338 (1985)
24. Weidner, T.: Probabilistic automata and probabilistic logic. In: Rovan, B., Sassone, V., Widmayer, P. (eds.) MFCS 2012. LNCS, vol. 7464, pp. 813–824. Springer, Heidelberg (2012)

On the State Complexity
of Semi-quantum Finite Automata*

Shenggen Zheng[1], Jozef Gruska[1], and Daowen Qiu[2]

[1] Faculty of Informatics, Masaryk University, Brno, 602 00, Czech Republic
zhengshenggen@gmail.com, gruska@fi.muni.cz
[2] Department of Computer Science, Sun Yat-sen University,
Guangzhou 510006, China
issqdw@mail.sysu.edu.cn

Abstract. Some of the most interesting and important results concerning quantum finite automata are those showing that they can recognize certain languages with (much) less resources than corresponding classical finite automata. This paper shows three results of such a type that are stronger in some sense than other ones because (a) they deal with models of quantum finite automata with very little quantumness (so-called semi-quantum one- and two-way finite automata); (b) differences, even comparing with probabilistic classical automata, are bigger than expected; (c) a trade-off between the number of classical and quantum basis states needed is demonstrated in one case and (d) languages (or the promise problem) used to show main results are very simple and often explored ones in automata theory or in communication complexity, with seemingly little structure that could be utilized.

1 Introduction

An important way to get deeper insights into the power of various quantum resources and operations is to explore the power of various quantum variations of the basic models of classical automata. Of a special interest is to do that for various quantum variations of the classical finite automata, especially for those that use limited amounts of quantum resources: states, correlations, operations and measurements. This paper aims to contribute to such a line of research.

There are several approaches how to introduce quantum features to classical models of finite automata. Two of them will be dealt with in this paper. The first one is to consider quantum variants of the classical *one-way (deterministic) finite automata* (1FA or 1DFA) and the second one is to consider quantum variants of the classical *two-way finite automata* (2FA or 2DFA). Already the

* Work of the first and second authors was supported by the Employment of Newly Graduated Doctors of Science for Scientific Excellence project/grant (CZ.1.07./2.3.00/30.0009) of Czech Republic. Work of third author was supported by the National Natural Science Foundation of China (Nos. 61272058, 61073054).

A.-H. Dediu et al. (Eds.): LATA 2014, LNCS 8370, pp. 601–612, 2014.

very first attempts to introduce such models, by Moore and Crutchfields [18] as well as Kondacs and Watrous [14] demonstrated that in spite of the fact that in the classical case, 1FA and 2FA have the same recognition power, this is not so for their quantum variations (in case only unitary operations and projective measurements are considered as quantum operations). Moreover, already the first model of *two-way quantum finite automata* (2QFA), namely that introduced by Kondacs and Watrous, demonstrated that quantum variants of 2FA are much too powerful – they can recognize even some *non-context free languages* and are actually not really finite in a strong sense [14]. Therefore it started to be of interest to introduce and explore some "less quantum" variations of 2FA and their power [2, 3].

A "hybrid" – quantum/classical – variations of 2FA, namely, *two-way finite automata with quantum and classical states* (2QCFA), were introduced by Ambainis and Watrous [2]. For this model they showed, in an elegant way, that already an addition of a single qubit to the classical model can much increase its power. A 2QCFA is essentially a classical 2FA augmented with a quantum memory of constant size (for states of a fixed Hilbert space) that does not depend on the size of the (classical) input. In spite of such a restriction, 2QCFA have been shown to be even more powerful than *two-way probabilistic finite automata* (2PFA) [2, 26]. A one-way version of 2QCFA was studied in [25], namely *one-way finite automata with quantum and classical states* (1QCFA).

Number of states is a natural complexity measure for finite automata. In case of quantum finite automata by that we understand the number of the basis states of the quantum space – that is its dimension. In case of hybrid, that is quantum/classical, finite automata, it is natural to consider both complexity measures – number of classical and also number of quantum (basis) states – and, potentially, trade-offs between them.

State complexity is one of the important research fields of computer science and it has many applications, e.g., in natural language and speech processing, image generation and encoding, etc. Early in 1959, Rabin and Scott [21] proved that any n-state *one-way nondeterministic finite automaton* (1NFA) can be simulated by a 2^n-state *one-way deterministic finite automaton* (1DFA). Salomaa [23] began to explore state complexity of finite automata in 1960s. The number of states of finite automata used in applications were usually small at that time and therefore investigations of state complexity of finite automata was seen mainly as a purely theoretical problem. However, the numbers of states of finite automata in applications can be huge nowadays, even millions of states in some cases [12]. It becomes therefore also practically important to explore state complexity of finite automata. State complexity of several variants of finite automata, both one-way and two-way, were deeply and broadly studied in the past thirty years [1–5, 8, 9, 16, 17, 24–26].

In this paper we explore the state complexity of semi-quantum finite automata and their space-efficiency comparing to the corresponding classical counterparts. We do that by showing that even for several very simple, and often considered, languages or promise problems, a little of quantumness can much decrease the state complexity of the corresponding semi-quantum finite automata. The first of these problems will be one of the very basic problem that is explored in communication complexity. Namely, the promise version of strings equality problem [7].

We use a promise problem to model the promise version of strings equality problem. For the alphabet $\Sigma = \{0, 1, \#\}$ and $n \in \mathbf{Z}^+$, let us consider the promise problem $A_{EQ}(n) = (A_{yes}(n), A_{no}(n))$, where $A_{yes}(n) = \{x\#y \mid x = y, x, y \in \{0,1\}^n\}$ and $A_{no}(n) = \{x\#y \mid x \neq y, x, y \in \{0,1\}^n, H(x,y) = \frac{n}{2}\}$. ($H(x,y)$ is the Hamming distance between x and y, which is the number of bit positions on which they differ.)

Klauck [13] has proved that, for any language, the state complexity of exact quantum/classical finite automata, which is a general model of one-way quantum finite automata, is not less than the state complexity of 1DFA. Therefore, it is interesting and important to find out whether the result still holds for interesting cases of promise problems or not[1]. Applying the communication complexity result from [7] to finite automata, for any $n \in \mathbf{Z}^+$, we prove that there exists a promise problem $A_{EQ}(n)$ that can be solved exactly by a 1QCFA with n quantum basis states and $\mathbf{O}(n)$ classical states, whereas the sizes of the corresponding 1DFA are $2^{\Omega(n)}$.

As the next we will consider state complexity of the language $L(p) - \{a^{kp} \mid k \in \mathbf{Z}^+\}$. It is well know that, for any $p \in \mathbf{Z}^+$, each 1DFA and 1NFA accepting $L(p)$ has at least p states. Ambainis and Freivalds [3], proved, using a non-constructive method, that $L(p)$ can be recognized by a one-way measure-once quantum finite automaton (MO-1QFA) with one-sided error ε with $poly\left(\frac{1}{\varepsilon}\right) \cdot \log p$ basis states (where $poly(x)$ is some polynomial in x). This bound was improved to $\mathbf{O}(\frac{\log p}{\varepsilon^3})$ in [6] and to $4\frac{\log 2p}{\varepsilon}$ in [4]. That is the best result known for such a mode of acceptance and it is an interesting open problem whether this bound can be much improved. If p is a prime, $L(p)$ can not be recognized by any one-way probabilistic finite automaton (1PFA) with less than p states [3]. For the case that p is not a prime, Mereghetti el at. [17] showed that the number of states of a 1PFA necessary and sufficient for accepting the language $L(p)$ with isolated cut point is $p_1^{\alpha_1} + p_2^{\alpha_2} + \cdots + p_s^{\alpha_s}$, where $p_1^{\alpha_1} p_2^{\alpha_2} \cdots p_s^{\alpha_s}$ is the prime factorization of p. Mereghetti el at. [17] also proved that $L(p)$ can be recognized by a 2 basis states MO-1QFA with isolated cut point. However, this mode of acceptance often leads to quite different state complexity outcome than one-sided error and error probability acceptance modes.

Concerning two-way finite automata, for any prime p, p states are necessary and sufficient for accepting $L(p)$ on two-way deterministic finite automata (2DFA) and two-way nondeterministic finite automata (2NFA) [16]. For the case that p is not prime, the number of states necessary and sufficient for accepting $L(p)$ on 2DFA and 2NFA is $p_1^{\alpha_1} + p_2^{\alpha_2} + \cdots + p_s^{\alpha_s}$ [16], where $p_1^{\alpha_1} p_2^{\alpha_2} \cdots p_s^{\alpha_s}$ is the prime factorization of p. Yakaryilmaz and Cem Say [24] showed that there exists

---

[1] Ambainis and Yakaryilmaz showed in [5] that there is a very special case in which the superiority of quantum computation to classical one cannot be bounded.

a 7-state *one-way finite automaton with restart* (1QFAR) which accepts $L(p)$ with one-sided error ε and expected running time $\mathbf{O}(\frac{1}{\varepsilon}\sin^{-2}(\frac{\pi}{p})|w|)$, where $|w|$ is the length of input w. For any n-state 1QFAR $\mathcal{M}_1$ with expected running time $t(|w|)$, Yakaryilmaz and Cem Say [24] also proved that there exists a 2QCFA $\mathcal{M}_2$ with n quantum basis states, $\mathbf{O}(n)$ classical states, and with expected runtime $\mathbf{O}(t(|w|))$, such that $\mathcal{M}_2$ accepts every input string w with the same probability as $\mathcal{M}_1$ does. Therefore, $L(p)$ can be recognized with one-sided error ε by a 2QCFA with 7 quantum basis states and a constant number of classical states.

In this paper we prove that the language $L(p)$ can be recognized with one-sided error ε in a linear expected running time $\mathbf{O}(\frac{1}{\varepsilon}p^2|w|)$ by a 2QCFA $\mathcal{A}(p,\varepsilon)$ with 2 quantum basis states and a constant number of classical states. We also show that the number of states needed for accepting $L(p)$ on a polynomial time 2PFA is at least $\sqrt[3]{(\log p)/b}$, where b is a constant.

The problem of checking whether the length of input string is equal to a given constant $m \in \mathbf{Z}^+$, is extensively studied in literatures as well. For any $m \in \mathbf{Z}^+$ and any finite alphabet Σ, it is obvious that the number of states of a 1DFA for accepting the language $C(m) = \{w \mid w \in \Sigma^m\}$ is at least m. Freivalds [9] showed that there is an ε error probability 1PFA accepting $C(m)$ with $\mathbf{O}(\log^2 m)$ states. Ambainis and Freivalds [3] proved that $C(m)$ can be recognized by an MO-1QFA with $\mathbf{O}(\log m)$ quantum basis states. Yakaryilmaz and Cem Say [24] showed that there exists a 7-state 1QFAR $\mathcal{M}$ which accepts $C(m)$ with one-sided error ε and the expected running time $\mathbf{O}((\frac{1}{\varepsilon})^m|w|)$. Thus, for $w \in C(m)$, the expected running time of $\mathcal{M}$ is an exponential of m. The 1QFAR $\mathcal{M}$ can only work efficiently on a very small m.

In this paper we prove that the language $C(m)$ can be recognized with one-sided error ε in a polynomial expected running time $\mathbf{O}(\frac{1}{\varepsilon}m^2|w|^4)$ by a 2QCFA $\mathcal{A}(m,\varepsilon)$ with 2 quantum basis states and a constant number of classical states. The expected running time is a polynomial of m. We show also that the number of states needed for accepting $C(m)$ on a polynomial 2PFA is at least $\sqrt[3]{(\log m)/b}$, where b is a constant.

Since 1QCFA and 2QCFA have both quantum and classical states, it is interesting to ask when there is some trade-off between these two kinds of states. We prove such a trade-off property for the case a 1QCFA accepts the language $L(p)$. Namely, it holds that for any integer p with prime factorization $p = p_1^{\alpha_1}p_2^{\alpha_2}\cdots p_s^{\alpha_s}$ ($s > 1$), for any partition I_1, I_2 of $\{1, \ldots, s\}$, and for $q_1 = \prod_{i \in I_1} p_i^{\alpha_i}$ and $q_2 = \prod_{i \in I_2} p_i^{\alpha_i}$, the language $L(p)$ can be recognized with a one-sided error ε by a 1QCFA $A(q_1, q_2, \varepsilon)$ with $\mathbf{O}(\log q_1) = \mathbf{O}(\sum_{i \in I_1} \alpha_i \log p_i)$ quantum basis states and $\mathbf{O}(q_2) = \mathbf{O}(\prod_{i \in I_2} p_i^{\alpha_i})$ classical states.

The paper is structured as follows. In Section 2 Semi-quantum finite automata models involved are described in some details. State complexities for the string equality problems will be discussed in Section 3. State succinctness for two families of regular languages is explored in Section 4. A trade-off property for 1QCFA is demonstrated in Section 5.

2   Models

We introduce in this section the models of 1QCFA and 2QCFA. Concerning some basic concepts and also notations on quantum information processing for this paper we refer the reader to Section 2.1 in [27]. Concerning more on quantum information processing we refer the reader to [22], and concerning more on classical and quantum automata [10, 11, 19].

2QCFA were introduced by Ambainis and Watrous [2] and explored also by Yakaryilmaz, Qiu, Zheng and others [24–26]. Informally, a 2QCFA can be seen as a 2DFA with an access to a quantum memory for states of a fixed Hilbert space upon which at each step either a unitary operation is performed or a projective measurement and the outcomes of which then probabilistically determine the next move of the underlying 2DFA.

Definition 1. *A 2QCFA $\mathcal{A}$ is specified by a 9-tuple*

$$\mathcal{A} = (Q, S, \Sigma, \Theta, \delta, |q_0\rangle, s_0, S_{acc}, S_{rej}) \tag{1}$$

where:

1. *Q is a finite set of orthonormal quantum basis states.*
2. *S is a finite set of classical states.*
3. *Σ is a finite alphabet of input symbols and let $\Sigma' = \Sigma \cup \{\textcent, \$\}$, where $\textcent$ will be used as the left end-marker and $\$$ as the right end-marker.*
4. *$|q_0\rangle \in Q$ is the initial quantum state.*
5. *s_0 is the initial classical state.*
6. *$S_{acc} \subset S$ and $S_{rej} \subset S$, where $S_{acc} \cap S_{rej} = \emptyset$ are sets of the classical accepting and rejecting states, respectively.*
7. *Θ is a quantum transition function*

$$\Theta : S \setminus (S_{acc} \cup S_{rej}) \times \Sigma' \to U(H(Q)) \cup O(H(Q)), \tag{2}$$

   where $U(H(Q))$ and $O(H(Q))$ are sets of unitary operations and projective measurements on the Hilbert space generated by quantum states from Q.
8. *δ is a classical transition function. If the automaton $\mathcal{A}$ is in the classical state s, its tape head is scanning a symbol σ and its quantum memory is in the quantum state $|\psi\rangle$, then $\mathcal{A}$ performs quantum and classical transitions as follows.*
   (a) *If $\Theta(s, \sigma) \in U(H(Q))$, then the unitary operation $\Theta(s, \sigma)$ is applied on the current state $|\psi\rangle$ of quantum memory to produce a new quantum state. The automaton performs, in addition, the following classical transition function*

$$\delta : S \setminus (S_{acc} \cup S_{rej}) \times \Sigma' \to S \times \{-1, 0, 1\}. \tag{3}$$

   If $\delta(s, \sigma) = (s', d)$, then the new classical state of the automaton is s' and its head moves in the direction d.

(b) If $\Theta(s, \sigma) \in O(H(Q))$, then the measurement operation $\Theta(s, \sigma)$ is applied on the current state $|\psi\rangle$. Suppose the measurement $\Theta(s, \sigma)$ is specified by operators $\{P_1, \ldots, P_n\}$ and its corresponding classical outcome is from the set $N_{\Theta(s,\sigma)} = \{1, 2, \ldots, n\}$. The classical transition function δ can be then specified as follow

$$\delta : S \setminus (S_{acc} \cup S_{rej}) \times \Sigma' \times N_{\Theta(s,\sigma)} \to S \times \{-1, 0, 1\}. \qquad (4)$$

In such a case, if i is the classical outcome of the measurement, then the current quantum state $|\psi\rangle$ is changed to the state $P_i|\psi\rangle/\|P_i|\psi\rangle\|$. Moreover, if $\delta(s, \sigma)(i) = (s', d)$, then the new classical state of the automaton is s' and its head moves in the direction d.

The automaton halts and accepts (rejects) the input when it enters a classical accepting (rejecting) state (from $S_{acc}(S_{rej})$).

The computation of a 2QCFA $\mathcal{A} = (Q, S, \Sigma, \Theta, \delta, |q_0\rangle, s_0, S_{acc}, S_{rej})$ on an input $w \in \Sigma^*$ starts with the string $\mathcal{c}x\$$ on the input tape. At the start, the tape head of the automation is positioned on the left end-marker and the automaton begins the computation in the classical initial state and in the initial quantum state. After that, in each step, if its classical state is s, its tape head reads a symbol σ and its quantum state is $|\psi\rangle$, then the automaton changes its states and makes its head movement following the steps described in the definition.

The computation will end whenever the resulting classical state is in $S_{acc} \cup S_{rej}$. Therefore, similarly to the definition of accepting and rejecting probabilities for 2QFA [14], the accepting and rejecting probabilities $Pr[\mathcal{A} \text{ accepts } w]$ and $Pr[\mathcal{A} \text{ rejects } w]$ for an input w are, respectively, the sums of all accepting probabilities and all rejecting probabilities before the end of computation on the input w.

Remark 2. 1QCFA are one-way versions of 2QCFA [25]. In this paper, we only use 1QCFA in which a unitary transformation is applied in every step after scanning a symbol and an measurement is performed after scanning the right end-marker. Such model is an measure-once 1QCFA and corresponds to a variant of MO-1QFA.

Three basic modes of language acceptance to be considered here are the following ones: Let $L \subset \Sigma^*$ and $0 < \varepsilon < \frac{1}{2}$. A finite automaton $\mathcal{A}$ recognizes L with a *one-sided error* (an *error probability*) ε if, for $w \in \Sigma^*$, (1) $\forall w \in L$, $Pr[\mathcal{A} \text{ accepts } w] = 1 \ (\geq 1 - \varepsilon)$ and (2) $\forall w \notin L$, $Pr[\mathcal{A} \text{ rejects } w] \geq 1 - \varepsilon$.

Obviously, one-sided error acceptance is stricter than an error probability acceptance.

Language acceptance is a special case of so called promise problem solving. A *promise problem* is a pair $A = (A_{yes}, A_{no})$, where $A_{yes}, A_{no} \subset \Sigma^*$ are disjoint sets. Languages may be viewed as promise problems that obey the additional constraint $A_{yes} \cup A_{no} = \Sigma^*$.

A promise problem $A = (A_{yes}, A_{no})$ is solved by exactly by a 1QCFA $\mathcal{A}$ if (1) $\forall w \in A_{yes}$, $Pr[\mathcal{A} \text{ accepts } w] = 1$ and (2) $\forall w \in A_{no}$, $Pr[\mathcal{A} \text{ rejects } w] = 1$.

3   State Succinctness for Promise Problem $A_{EQ}(n)$

Theorem 3. *The promise problem $A_{EQ}(n)$ can be solved exactly by a 1QCFA $\mathcal{A}(n)$ with n quantum basis states and $\mathbf{O}(n)$ classical states, whereas the sizes of the corresponding 1DFA are $2^{\Omega(n)}$.*

A1: Description of the behavior of $\mathcal{A}(n)$ when solving the promise problem $A_{EQ}(n)$.
1. Read the left end-marker $\cent$, perform U_s on the initial quantum state $|1\rangle$, change its classical state to $\delta(s_0, \cent) = s_1$, and move the tape head one cell to the right.
2. Until the currently scanned symbol σ is not $\#$, do the following:
   2.1 Apply $\Theta(s_i, \sigma) = U_{i,\sigma}$ to the current quantum state.
   2.2 Change the classical state s_i to s_{i+1} and move the tape head one cell to the right.
3. Change the classical state s_{n+1} to s_1 and move the tape head one cell to the right.
4. While the currently scanned symbol σ is not the right end-marker $\$$, do the following:
   2.1 Apply $\Theta(s_i, \sigma) = U_{i,\sigma}$ to the current quantum state.
   2.2 Change the classical state s_i to s_{i+1} and move the tape head one cell to the right.
5. When the right end-marker is reached, perform U_f on the current quantum state, measure the current quantum state with $M = \{P_i = |i\rangle\langle i|\}_{i=1}^n$. If the outcome is $|1\rangle$, accept the input; otherwise reject the input.

Proof. Let $x = x_1 \cdots x_n$ and $y = y_1 \cdots y_n$ with $x, y \in \{0,1\}^n$. Let us consider an MO-1QCFA $\mathcal{A}(n)$ with n quantum basis states $\{|i\rangle : i = 1, 2, \ldots, n\}$. $\mathcal{A}(n)$ will start in the quantum state $|1\rangle = (1, 0, \ldots, 0)^T$. We use classical states $s_i \in S$ $(1 \leq i \leq n+1)$ to point out the positions of the tape head that will provide some information for quantum transformations. If the classical state of $\mathcal{A}(n)$ will be s_i $(1 \leq i \leq n)$ that will mean that the next scanned symbol of the tape head is the i-th symbol of $x(y)$ and s_{n+1} means that the next scanned symbol of the tape head is $\#(\$)$. The automaton proceeds as Algorithm **A1**, where

$$U_s|1\rangle = \frac{1}{\sqrt{n}} \sum_{i=1}^{n} |i\rangle; \tag{5}$$

$$U_{i,\sigma}|i\rangle = (-1)^{\sigma}|i\rangle \quad \text{and} \quad U_{i,\sigma}|j\rangle = |j\rangle \text{ for } j \neq i; \tag{6}$$

$$U_f(\sum_{i=1}^{n} \alpha_i|i\rangle) = (\frac{1}{\sqrt{n}} \sum_{i=1}^{n} \alpha_i)|1\rangle + \cdots. \tag{7}$$

Transformations U_s and U_f are unitary, where the first column of U_s is $\frac{1}{\sqrt{n}}(1, \ldots, 1)^T$ and the first row of U_f is $\frac{1}{\sqrt{n}}(1, \ldots, 1)$.

The quantum state after scanning the left end-marker is $|\psi_1\rangle = U_s|1\rangle = \sum_{i=1}^{n} \frac{1}{\sqrt{n}}|i\rangle$, the quantum state after Step 2 is $|\psi_2\rangle = \sum_{i=1}^{n} \frac{1}{\sqrt{n}}(-1)^{x_i}|i\rangle$, and the quantum state after Step 4 is $|\psi_3\rangle = \sum_{i=1}^{n} \frac{1}{\sqrt{n}}(-1)^{x_i+y_i}|i\rangle$. The quantum state after scanning the right end-marker is therefore

$$|\psi_4\rangle = U_f \left(\sum_{i=1}^{n} \frac{1}{\sqrt{n}}(-1)^{x_i+y_i}|i\rangle \right) = U_f \frac{1}{\sqrt{n}} \begin{pmatrix} (-1)^{x_1+y_1} \\ (-1)^{x_2+y_2} \\ \vdots \\ (-1)^{x_n+y_n} \end{pmatrix} \tag{8}$$

$$= \begin{pmatrix} \frac{1}{n}\sum_{i=1}^{n}(-1)^{x_i+y_i} \\ \vdots \\ \\ \end{pmatrix}. \tag{9}$$

If the input string $w \in A_{yes}(n)$, then $x_i = y_i$ for any i and $|\frac{1}{n}\sum_{i=1}^{n}(-1)^{x_i+y_i}|^2 = 1$. The amplitude of $|1\rangle$ is 1, and that means $|\psi_4\rangle = |1\rangle$. Therefore the input will be accepted with probability 1 at the measurement in Step 5.

If the input string $w \in A_{no}(n)$, then $H(x,y) = \frac{n}{2}$. Therefore the probability of getting outcome $|1\rangle$ in the measurement in Step 5 is $|\frac{1}{n}\sum_{i=1}^{n}(-1)^{x_i+y_i}|^2 = 0$.

The deterministic communication complexity for the promise version of equality problem is at least $0.007n$ [7]. Therefore, the sizes of the corresponding 1DFA are $2^{\Omega(n)}$ [15].

4   State Succinctness for 2QCFA

State succinctness for 2QCFA was explored by Yakaryilmaz, Zheng and others [24, 26]. In [26], Zheng et al. showed the state succinctness for polynomial time 2QCFA for families of promise problems and for exponential time 2QCFA for a family of languages. In this section, we show the state succinctness for linear time 2QCFA and polynomial time 2QCFA for two families of languages.

4.1   State Succinctness for the Language $L(p)$

Theorem 4. *For any $p \in \mathbf{Z}^+$ and $0 < \varepsilon \leq \frac{1}{2}$, the language $L(p)$ can be recognized with one-sided error ε by a 2QCFA $A(p, \varepsilon)$ with 2 quantum basis states and a constant number of classical states (not depending on p) in a linear expected running time $\mathbf{O}(\frac{1}{\varepsilon}p^2 n)$, where n is the length of input.*

Proof. The main idea of the proof is as follows: Consider a 2QCFA $A(p, \varepsilon)$ with 2 orthogonal quantum basis states $|q_0\rangle$ and $|q_1\rangle$. $A(p, \varepsilon)$ starts computation in the initial quantum state $|q_0\rangle$ and with the tape head on the left end-marker. Every time when $A(p, \varepsilon)$ reads a symbol 'a', the current quantum state is rotated by the angle $\frac{\pi}{p}$. When the right end-marker \$ is reached, $A(p, \varepsilon)$ measures the

A2: Description of the behavior of $\mathcal{A}(p,\varepsilon)$ when recognizing the language $L(p)$. Repeat the following ad infinity:

1. Move the tape head to the right of the left end-marker.
2. Until the scanned symbol is the right end-marker, apply U_p to the current quantum state and move the head one cell to the right.
3 Measure the current quantum state in the basis $\{|q_0\rangle, |q_1\rangle\}$.
   3.1 If quantum outcome is $|q_1\rangle$, reject the input.
   3.2 Otherwise apply $U_{p,\varepsilon}$ to the current quantum state $|q_0\rangle$.
4 Measure the quantum state in the basis $\{|q_0\rangle, |q_1\rangle\}$. If the result is $|q_0\rangle$, accept the input; otherwise apply a unitary operation to change the quantum state from $|q_1\rangle$ to $|q_0\rangle$ and start a new iteration.

current quantum state. If the resulting quantum state is $|q_1\rangle$, the input string is rejected, otherwise the automaton proceeds as Algorithm **A2**, where

$$U_p = \begin{pmatrix} \cos\frac{\pi}{p} & \sin\frac{\pi}{p} \\ \sin\frac{\pi}{p} & \cos\frac{\pi}{p} \end{pmatrix} \quad and \quad U_{p,\varepsilon} = \begin{pmatrix} \dfrac{1}{\sqrt{p^2/4\varepsilon}} & -\dfrac{\sqrt{p^2/4\varepsilon-1}}{\sqrt{p^2/4\varepsilon}} \\ \dfrac{\sqrt{p^2/4\varepsilon-1}}{\sqrt{p^2/4\varepsilon}} & \dfrac{1}{\sqrt{p^2/4\varepsilon}} \end{pmatrix}. \qquad (10)$$

See Section 4.1 in [27] for a detail proof. (Similar unitary matrixes of U_p and proof methods can be found in[3, 4, 17, 24].)

Theorem 5. *For any integer p, any polynomial expected running time 2PFA recognizing $L(p)$ with error probability $\varepsilon < \frac{1}{2}$ has at least $\sqrt[3]{(\log p)/b}$ states, where b is a constant.*

In order to prove this theorem, we need

Lemma 6 ([8]). *For every $\varepsilon < 1/2$, $a > 0$ and $d > 0$, there exists a constant $b > 0$ such that, for any integer c, if a language L is recognized with an error probability ε by a c-state 2PFA within time an^d, where $n = |w|$ is the length of input, then L is recognized by some DFA with at most c^{bc^2} states.*

Proof. Assume that a c-state 2PFA $\mathcal{A}(p)$ recognizes $L(p)$ with an error probability $\varepsilon < 1/2$ and also within a polynomial expected running time. According to Lemma 6, there exits a 1DFA that recognizes $L(p)$ with c^{bc^2} states, where $b > 0$ is a constant. As we know, any DFA recognizing $L(p)$ has at least p states. Therefore,

$$c^{bc^2} \geq p \Rightarrow bc^2 \log c \geq \log p \Rightarrow c^3 > (\log p)/b \Rightarrow c > \sqrt[3]{(\log p)/b}. \qquad (11)$$

4.2   State Succinctness for the Language $C(m)$

Theorem 7. *For any $m \in \mathbf{Z}^+$ and $0 < \varepsilon \leq \frac{1}{2}$, the language $C(m)$ can be recognized with one-sided error ε by a 2QCFA $\mathcal{A}(m,\varepsilon)$ with 2 quantum basis states and a constant number of classical states (not depending on m) in a polynomial expected running time $\mathbf{O}(\frac{1}{\varepsilon}m^2n^4)$, where n is the length of input.*

A3: Description of the behavior of $\mathcal{A}(m,\varepsilon)$ when recognizing the language $C(m)$. Repeat the following ad infinity:

1. Move the tape head to the left end-marker, read the end-marker ¢, apply $U_{¢}$ on $|q_0\rangle$, and move the tape head one cell to the right.
2. Until the scanned symbol is the right end-marker, apply U_α to the current quantum state and move the tape head one cell to the right.
3.0 When the right end-marker is reached, measure the quantum state in the basis $\{|q_0\rangle, |q_1\rangle\}$.
    3.1 If quantum outcome is $|q_1\rangle$, reject the input.
    3.2 Otherwise repeat the following subroutine two times:
        3.2.1 Move the tape head to the first symbol right to the left end-marker.
        3.2.2 Until the currently read symbol is one of the end-markers simulate a coin-flip and move the head right (left) if the outcome of the coin-flip is "head" ("tail").
4. If the above process ends both times at the right end-marker, apply $U_{m,\varepsilon}$ to the current quantum state and measure the quantum state in the basis $\{|q_0\rangle, |q_1\rangle\}$. If the result is $|q_0\rangle$, accept the input; otherwise apply a unitary operation to change the quantum state from $|q_1\rangle$ to $|q_0\rangle$ and start a new iteration.

Proof. The main idea of the proof is as follows: Consider a 2QCFA $\mathcal{A}(m,\varepsilon)$ with 2 orthogonal quantum basis states $|q_0\rangle$ and $|q_1\rangle$. $\mathcal{A}(m,\varepsilon)$ starts computation with the initial quantum state $|q_0\rangle$. When $\mathcal{A}(m,\varepsilon)$ reads the left end-marker ¢, the current quantum state will be rotated by the angle $-\sqrt{2}m\pi$ and every time when $\mathcal{A}(m,\varepsilon)$ reads a new symbol $\sigma \in \Sigma$, the state is rotated by the angle $\alpha = \sqrt{2}\pi$ (notice that $\sqrt{2}m\pi = m\alpha$). When the right end-marker \$ is reached, $\mathcal{A}(m,\varepsilon)$ measures the current quantum state with projectors $\{|q_0\rangle\langle q_0|, |q_1\rangle\langle q_1|\}$. If the resulting quantum state is $|q_1\rangle$, the input string w is rejected, otherwise, the automaton proceeds as Algorithm **A3**, where

$$U_{¢} = \begin{pmatrix} \cos m\sqrt{2}\pi & \sin m\sqrt{2}\pi \\ -\sin m\sqrt{2}\pi & \cos m\sqrt{2}\pi \end{pmatrix}, \quad U_\alpha = \begin{pmatrix} \cos\sqrt{2}\pi & -\sin\sqrt{2}\pi \\ \sin\sqrt{2}\pi & \cos\sqrt{2}\pi \end{pmatrix}, \quad (12)$$

$$U_{m,\varepsilon} = \begin{pmatrix} \frac{1}{\sqrt{2m^2/\varepsilon}} & -\frac{\sqrt{2m^2/\varepsilon-1}}{\sqrt{2m^2/\varepsilon}} \\ \frac{\sqrt{2m^2/\varepsilon-1}}{\sqrt{2m^2/\varepsilon}} & \frac{1}{\sqrt{2m^2/\varepsilon}} \end{pmatrix}. \quad (13)$$

See Section 4.2 in [27] for a detail proof. (A similar proof method can be found in [2].)

Remark 8. Using the above theorem and the intersection property of languages recognized by 2QCFA [20], it is easy to improve the result from [26] related to the promise problem[2] $A^{eq}(m)$ to a language $L^{eq}(m) = \{a^m b^m\} = L^{eq} \cap C(2m)$, where the language $L^{eq} = \{a^n b^n \mid n \in \mathbf{N}\}$. Therefore, the open problem from [26] is solved.

---

[2] See page 102 in [26].

It is obvious that the number of states of a 1DFA to accept the language $C(m)$ is at least m. Using a similar proof as of Theorem 5, we get:

Theorem 9. *For any integer m, any polynomial expected running time 2PFA recognizing $C(m)$ with error probability $\varepsilon < \frac{1}{2}$ has at least $\sqrt[3]{(\log m)/b}$ states, where b is a constant.*

The sizes of 1PFA and 1QFA recognizing languages $L(p)$ or $C(m)$ with an error ε depend on the error ε in most of the papers. For example, in [4], the size of MO-1QFA accepting $L(p)$ with one-sided error ε is $4\frac{\log 2p}{\varepsilon}$. If $\varepsilon < \frac{4}{p}$, the state complexity advantage of MO-1QFA disappears. However, in our model, the sizes of 2QCFA do not depend on the error ε, which means that 2QCFA have state advantage for any $\varepsilon > 0$.

5 A Trade-Off Property of 1QCFA

Quantum resources are expensive and hard to deal with. One can expect to have only very limited number of qubits in current quantum system. In some cases, one cannot expect to have enough qubits to solve a given problem (or to recognize a given language). It is therefore interesting to find out whether there are some trade-off between needed quantum and classical resources. We prove in the following that it is so in some cases. Namely, we prove that there exist trade-offs in case 1QCFA are used to accept the language $L(p)$.

Theorem 10. *For any integer $p > 0$ with prime factorization $p = p_1^{\alpha_1} p_2^{\alpha_2} \cdots p_s^{\alpha_s}$ ($s > 1$), for any partition I_1, I_2 of $\{1, \ldots, s\}$, and for $q_1 = \prod_{i \in I_1} p_i^{\alpha_i}$ and $q_2 = \prod_{i \in I_2} p_i^{\alpha_i}$, the language $L(p)$ can be recognized with one-sided error ε by a 1QCFA $A(q_1, q_2, \varepsilon)$ with $\mathbf{O}(\log q_1) = \mathbf{O}(\sum_{i \in I_1} \alpha_i \log p_i)$ quantum basis states and $\mathbf{O}(q_2) = \mathbf{O}(\prod_{i \in I_2} p_i^{\alpha_i})$ classical states.*

Proof. See Section 5 in [27] for a detail proof.

References

1. Ambainis, A., Nayak, A., Ta-Shma, A., Vazirani, U.: Dense quantum coding and quantum automata. Journal of the ACM 49(4), 496–511 (2002)
2. Ambainis, A., Watrous, J.: Two-way finite automata with quantum and classical states. Theoretical Computer Science 287, 299–311 (2002)
3. Ambainis, A., Freivalds, R.: One-way quantum finite automata: strengths, weaknesses and generalizations. In: Proceedings of the 39th FOCS, pp. 332–341 (1998)
4. Ambainis, A., Nahimovs, N.: Improved constructions of quantum automata. Theoretical Computer Science 410, 1916–1922 (2009)
5. Ambainis, A., Yakaryilmaz, A.: Superiority of exact quantum automata for promise problems. Information Processing Letters 112(7), 289–291 (2012)
6. Bertoni, A., Mereghetti, C., Palano, B.: Some formal tools for analyzing quantum automata. Theoretical Computer Science 356, 14–25 (2006)
7. Buhrman, H., Cleve, R., Massar, S., de Wolf, R.: Nonlocality and Communication Complexity. Rev. Mod. Phys. 82, 665–698 (2010)

8. Dwork, C., Stockmeyer, L.: A time-complexity gap for two-way probabilistic finite state automata. SIAM J. Comput. 19, 1011–1023 (1990)
9. Freivalds, R.: On the growth of the number of states in result of determinization of probabilistic finite automata. Automatic Control and Computer Sciences 3, 39–42 (1982)
10. Gruska, J.: Descriptional complexity issues in quantum computing. J. Automata, Languages Combin. 5(3), 191–218 (2000)
11. Hopcroft, J.E., Ullman, J.D.: Introduction to Automata Theory, Languages, and Computation. Addision-Wesley, New York (1979)
12. Kiraz, G.A.: Compressed Storage of Sparse Finite-State Transducers. In: Boldt, O., Jürgensen, H. (eds.) WIA 1999. LNCS, vol. 2214, pp. 109–121. Springer, Heidelberg (2001)
13. Klauck, H.: On quantum and probabilistic communication: Las Vegas and one-way protocols. In: Proceedings of the 32th STOC, pp. 644–651 (2000)
14. Kondacs, A., Watrous, J.: On the power of quantum finite state automata. In: Proceedings of the 38th FOCS, pp. 66–75 (1997)
15. Kushilevitz, E.: Communication Complexity. Advances in Computers 44, 331–360 (1997)
16. Mereghetti, C., Pighizzini, G.: Two-way automata simulations and unary languages. J. Autom. Lang. Comb. 5, 287–300 (2000)
17. Mereghetti, C., Palano, B., Pighizzini, G.: Note on the Succinctness of Deterministic, Nondeterministic, Probabilistic and Quantum Finite Automata. RAIRO-Inf. Theor. Appl. 35, 477–490 (2001)
18. Moore, C., Crutchfield, J.P.: Quantum automata and quantum grammars. Theoretical Computer Science 237, 275–306 (2000)
19. Paz, A.: Introduction to Probabilistic Automata. Academic Press, New York (1971)
20. Qiu, D.: Some Observations on Two-Way Finite Automata with Quantum and Classical States. In: Huang, D.-S., Wunsch II, D.C., Levine, D.S., Jo, K.-H. (eds.) ICIC 2008. LNCS, vol. 5226, pp. 1–8. Springer, Heidelberg (2008)
21. Rabin, M.O., Scott, D.: Finite automata and their decision problems. IBM J. Research and Development 3(2), 115–125 (1959)
22. Nielsen, M.A., Chuang, I.L.: Quantum Computation and Quantum Information. Cambridge University Press, Cambridge (2000)
23. Salomaa, A.: On the reducibility of events represented in automata. In: Annales Academiae Scientiarum Fennicae, volume Series A of I. Mathematica 353 (1964)
24. Yakaryilmaz, A., Cem Say, A.C.: Succinctness of two-way probabilistic and quantum finite automata. Discrete Mathematics and Theoretical Computer Science 12(4), 19–40 (2010)
25. Zheng, S.G., Qiu, D.W., Li, L.Z., Gruska, J.: One-way finite automata with quantum and classical states. In: Bordihn, H., Kutrib, M., Truthe, B. (eds.) Languages Alive. LNCS, vol. 7300, pp. 273–290. Springer, Heidelberg (2012)
26. Zheng, S.G., Qiu, D.W., Gruska, J., Li, L.Z., Mateus, P.: State succinctness of two-way finite automata with quantum and classical states. Theoretical Computer Science 499, 98–112 (2013)
27. Zheng, S.G., Gruska, J., Qiu, D.W.: On the state complexity of semi-quantum finite automata. arXiv:1307.2499 (2013)

Author Index